Springer Geophysics

The Springer Geophysics series seeks to publish a broad portfolio of scientific books, aiming at researchers, students, and everyone interested in geophysics. The series includes peer-reviewed monographs, edited volumes, textbooks, and conference proceedings. It covers the entire research area including, but not limited to, applied geophysics, computational geophysics, electrical and electro-magnetic geophysics, geodesy, geodynamics, geomagnetism, gravity, lithosphere research, paleomagnetism, planetology, tectonophysics, thermal geophysics, and seismology.

More information about this series at http://www.springer.com/series/10173

Salwa F. Elbeih · Abdelazim M. Negm ·
Andrey Kostianoy
Editors

Environmental Remote Sensing in Egypt

Editors
Salwa F. Elbeih
Engineering Applications Department
National Authority for Remote Sensing
and Space Sciences
Cairo, Egypt

Abdelazim M. Negm
Faculty of Engineering
Zagazig University
Zagazig, Egypt

Andrey Kostianoy
P.P. Shirshov Institute of Oceanology
Russian Academy of Sciences
Moscow, Russia

ISSN 2364-9119 ISSN 2364-9127 (electronic)
Springer Geophysics
ISBN 978-3-030-39595-7 ISBN 978-3-030-39593-3 (eBook)
https://doi.org/10.1007/978-3-030-39593-3

This Springer imprint is published by the registered company Springer Nature Switzerland AG
The registered company address is: Gewerbestrasse 11, 6330 Cham, Switzerland

Preface

This volume came into conception to highlight the use of remote sensing (RS) and its applications in research and monitoring of the environment in Egypt. This unique volume is authored by experts in the topic from Egypt, Canada, Russia, and USA, to present the results and findings of their research and the state-of-the-art knowledge related to the book title. The volume is divided into six parts and contains 19 chapters including introductory and conclusion chapters written by more than 31 authors.

Part I is an *Introduction* to the environmental remote sensing in Egypt, where the editors and authors present a general overview to the topic and theme of the book. It consists of three chapters. Chapter 1 is titled "Introduction to 'Environmental Remote Sensing in Egypt'." It presents an overview of the book to highlight the main technical elements of each chapter, while Chap. 2 is titled "Overview for Recent Applications of Remote Sensing in Egypt." It summarizes recent applications of remote sensing in Egypt regarding the number, types, and objectives of the published documents related to the remote sensing field. In addition, academic, industrial, governmental, and non-governmental sponsors and institutions, laboratories, and research centers involved in the topic are highlighted. On the other hand, Chap. 3 is about "Egypt's Environment from Satellite." It presents 32 images and pictures that take the reader on a tour of Egypt. They are showing the Nile River, lakes, depressions, Lake Nasser Khores, deserts, and oasis from space. These images reflect various sites in Egypt from space like you have never seen before.

Part II of the volume entitled *Environmental Applications* contains four chapters. Chapter 4 which is titled "Environmental Applications of Remote Sensing in Egypt: A Review and an Outlook" includes a review of environmental applications of optical/infrared remote sensing data in Egypt and presents an outlook at the future development of the applications to catch up with the technology development and modern uses. The specific objectives were (1) to expand the domain of remote sensing environmental applications in Egypt and (2) to persuade scientists to deliver new research novelty related to the environmental parameters and processes, while Chap. 5 is titled "Radar Remote Sensing Applications in Egypt." It introduces key concepts of radar remote sensing and provides a review of present applications in Egypt as well as an outlook at future development of the applications. On the other hand, Chap. 6 is about the "Application of Remote Sensing for Monitoring Changes in Natural Ecosystems: Case Studies from Egypt." It presents recent remote sensing and GIS techniques applications for monitoring environmental and ecological changes in Egypt. The last chapter in Part II is titled "Hyperspectral Based Assessment of Mosquito Breeding Water in Suez Canal Zone, Egypt." It presents an assessment of the physico-chemical and spectral characteristics contributing to mosquito breeding in Suez Canal Zone. This assessment will help in determining which parameter is more effective for mosquito abundance and to determine preventive measures to eliminate future mosquito proliferation and disease transmission.

Part III entitled *Water Quality and Pollution* contains six chapters from Chaps. 8 –13. Chapter 8 with the title "Satellite Image Data Integration for Groundwater Exploration in Egypt" shows that radar microwave, thermal infrared data, and SRTM DEMs are ideal for use in the detection and mapping of water-bearing structures. In this chapter, space data integration is proven to be a promising approach that can help in groundwater exploration efforts in the desert land of Egypt, where freshwater access is essential for the development of the region, while Chap. 9 is titled "Monitoring and Protection of Egyptian Northern Lakes Using Remote Sensing Technology." It reviews the role of satellite remote sensing imagery to assess and protect the northern lakes of Egypt through: (1) monitoring the present status of the northern lakes of Egypt, (2) addressing the developed techniques and procedures for using satellite imagery in studying the coastal lakes on a regional scale, and (3) protecting the conservations of lakes' resources in addition to management activities. On the other hand, Chap. 10 is titled "An Overview of the Environmental and Anthropogenic Hazards Threatening Lake Nasser, Egypt National Water Reservoir." It analyzes the environmental and anthropological hazards that threaten Lake Nasser as Egypt's National Water Reservoir with the aid of remote sensing. Moreover, Chap. 11 is titled "Oil Pollution in the Mediterranean Waters of Egypt" to provide satellite SAR imagery acquired in 2017–2019 to demonstrate detection of oil spills in the Mediterranean waters of Egypt related to shipping activities and offshore gas exploration. While Chap. 12 is about "Oil Pollution in the Northern Red Sea: A Threat to the Marine Environment and Tourism Development." It describes development of the coastal tourism in Egypt, shipping activities connected to the passage using the Suez Canal, crude oil delivery to oil terminals, and development of oil fields in the northern part of the Red Sea. A series of SAR images demonstrated that oil pollution is a serious threat to the marine environment and tourism development in Egypt. The last chapter in this part is titled "Satellite Monitoring of the Nile River and the Surrounding Vegetation."

It shows the capabilities of high-resolution satellite optical imagery to monitor seasonal and interannual variability of natural vegetation and agricultural fields around the River Nile.

Part IV entitled *Environment and Climate Change* contains two chapters. In Chap. 14 under the title "Remote Sensing and Modeling of Climate Changes in Egypt," the authors emphasize the vulnerability of Egypt to climate changes and the need for large-scale, global systems for monitoring, modeling, assessment, and follow-up of mitigation and adaptation measures using remote sensing and GIS techniques. While in Chap. 15 under the title "Qualitative and Quantitative Assessment of Land Degradation and Desertification in Egypt, Based on Satellite Remote Sensing: Urbanization, Salinization and Wind Erosion," the author presents and discusses the problems of desertification and land degradation, highlighting the difference between the two terms and aspects to be assessed.

Part V entitled *Hydrology and Geomorphology* contains three chapters from Chaps. 16–18. Chapter 16 is titled "Landscapes of Egypt." It shows peculiarities of two types of landscapes typical for Egypt: (1) physical landscapes and (2) anthropogenic landscapes, while Chap. 17 under the title "Integrated Watershed Management of Grand Ethiopian Renaissance Dam via Watershed Modeling System and Remote Sensing" demonstrates the capabilities of remote sensing and modeling for integrated watershed management related to the GERD construction. On the other hand, Chap. 18 is titled "Quantitative Analysis of Shoreline Dynamics Along the Mediterranean Coastal Strip of Egypt. Case Study: Marina El-Alamein Resort." It focuses on monitoring, analyzing, and quantifying shoreline dynamics along the Egyptian Mediterranean coastal strip from Sidi Abd El-Rahman to El-Arish over three decades using RS and GIS technologies with special reference to Marina El-Alamein resort's shoreline.

The last part in this book (Part VI) is the "Update, Conclusions, and Recommendations of 'Environmental Remote Sensing in Egypt'." The chapter highlights the main conclusions and recommendations of the chapters presented in the book. Also, some findings from recently published research works related to the environmental remote sensing in Egypt are presented.

Volume Editors would like to thank all those scientists, authors, and researchers who contributed to the book *Environmental Remote Sensing in Egypt* and made it a great source of information and knowledge for all those who are interested in satellite remote sensing. Much appreciation and great thanks are also owed to the book series editors and the Springer team who worked hard during the production of this volume. Volume Editors would be grateful to receive any comments to improve future editions of this book. Comments, feedback, suggestions for improvement, or new chapters for the next editions are much welcomed and can be sent directly to Volume Editors.

Last but not least, Abdelazim M. Negm acknowledges the partial financial support from the Academy of Scientific Research and Technology (ASRT) of Egypt via the bilateral collaboration Italian (CNR)–Egyptian (ASRT) project titled

"Experimentation of the new Sentinel missions for the observation of inland water bodies on the course of the Nile River," while Andrey Kostianoy was partially supported in the framework of the P.P. Shirshov Institute of Oceanology RAS budgetary financing (Project N 149-2019-0004).

Cairo, Egypt — Salwa F. Elbeih
Zagazig, Egypt — Abdelazim M. Negm
Moscow, Russia — Andrey Kostianoy
November 2019

Contents

Part VI Conclusions

Part I
Introduction

Chapter 1
Introduction to "Environmental Remote Sensing in Egypt"

Abdelazim M. Negm, Andrey Kostianoy and Salwa F. Elbeih

Abstract This chapter introduces the book "Environmental Remote Sensing in Egypt". The main technical elements covered in the chapters are presented. The information in this chapter cover topics which include environmental applications of remote sensing, applications of radar remote sensing, monitoring changes in natural ecosystems, groundwater exploration, monitoring and protection of Egyptian Northern Lakes, environmental hazards threatening Lake Nasser, oil pollution in the Mediterranean and Red Seas, Nile River and Nile Delta monitoring using remote sensing, monitoring and modeling of climate changes, land degradation and desertification, landscapes of Egypt, shoreline dynamics and monitoring of the Nile River.

Keywords Egypt · Environment · Remote sensing · Monitoring · Water resources · Climate change · Oil pollution · Ecosystem · Water quality · Desertification · Red Sea · Nile River · Mediterranean · Lakes · Radar · Landscape

1.1 Background

Egypt is a unique country in the world as it has many features that are not found anywhere in the world. Remote sensing is used, in this book, to present most of

---

A. M. Negm (✉)
Water and Water Structures Engineering Department, Faculty of Engineering, Zagazig University, Zagazig 44519, Egypt
e-mail: amnegm85@yahoo.com; amnegm@zu.edu.eg

A. Kostianoy
P. P. Shirshov Institute of Oceanology, Russian Academy of Sciences, 36, Nakhimovsky Pr., Moscow 117997, Russia
e-mail: kostianoy@gmail.com

S.Yu. Witte Moscow University, 12, Build. 1, 2nd Kozhukhovsky Pr., Moscow 115432, Russia

S. F. Elbeih
Engineering Applications Department, National Authority for Remote Sensing and Space Sciences (NARSS), Cairo, 1564 Alf Maskan, Egypt
e-mail: saelbeih@gmail.com; saelbeih@narss.sci.eg

S. F. Elbeih et al. (eds.), *Environmental Remote Sensing in Egypt*,
Springer Geophysics, https://doi.org/10.1007/978-3-030-39593-3_1

the unique features of Egypt's environment and discuss them including the related state-of-the-art review. Otherwise, it would be a difficult task without the use of the remote sensing approach. Scientists from Egypt and abroad put their great efforts for about two years to produce this amazing book for all who are interested in Egypt and its unique features. The book contains 17 different chapters in addition to the introduction (this chapter) and the conclusions/recommendations chapter which closes the book. The conclusion chapter includes findings from recent publications including Mohamed et al. (2018), Selim and El Raey (2018), Basheer et al. (2019), El-Alfy et al. (2019), Sayed et al. (2019), and Sowilem et al. (2019) among others.

In this introduction chapter, a summary is presented to indicate briefly what will be presented in each following chapter without mentioning many details. For sure, the interested readers will consult the chapters for further details.

1.2 Themes of the Book

The book intends to address in more detail the following main themes:

- Environmental Applications
- Water Quality and Pollution
- Environment and Climate Change
- Hydrology and Geomorphology.

1.3 Chapters' Summaries

In the following subsection, brief descriptions of the 19 chapters of the book are presented.

1.3.1 Introduction to the Book and Its Themes

This chapter is to introduce the book by presenting the basic technical elements of the book without going to the conclusions and recommendation presented in each chapter. Chapter 2 is devoted to an "**Overview for Recent Applications of Remote Sensing in Egypt**". Recently, the Egyptian government has developed advanced strategies to improve and expand the application of satellite remote sensing in several engineering and environmental fields. The National Authority for Remote Sensing and Space Sciences (NARSS) is the pioneering Egyptian institution in the area of "Satellite Remote Sensing". NARSS employs the state-of-the-art space technologies and applications to monitor and investigate water, land, and natural resources of Egypt. During the previous decade, NARSS has been funded by 341 research and development projects from several funding agencies. According to the Scopus database, about 150 authors from NARSS have published 487 documents from 1996 to 2019. The document types could be classified into articles (67.6%), conference papers (24.0%), book chapters (2.9%), reviews (1.8%), and others (3.7%). The search (in July 2019) in the Scopus database using the keywords "Remote", "Sensing", and "Egypt" shows 517 documents that have recently been published during this decade (i.e., 2010–2019). This number is higher than that reported during 2000–2009 (181 documents) and 1990–1999 (67 documents). The academic disciplines and fields of study covered essential areas in geology, agriculture, engineering, environment, water, marine sciences, mineral resources, and space archaeology. The outputs of this chapter are beneficial to undergraduate and graduate students, researchers, decision-makers, stakeholders, and several public and private sectors. Some of the useful sources used to prepare this chapter include Hassan et al. (2019), NARSS (2019), Ramzi and El-Bedawi (2019), Sayed et al. (2019) and Scopus (2019) among others listed in the chapter.

Chapter 3 presents "**Egypt's Environment from Satellite**". The authors presented thirty-one figures (images and photos) that were identified to cover Egypt's environment. They can be grouped into three groups. The first group of figures portrays Egypt, Nile, topography, agricultural land, and desertification, in particular. The second group deals with the largest Nile islands, main wadis, and Egypt's top canyon. The third group, however, shed light on Egypt's canals, depressions, oases, and protected areas. The 31 figures (images and photos) take the reader on a tour of Egypt such as lakes, depressions, Khores, oasis… etc. These images reflect the various sites within Egypt from above as you have never seen before. These sites and features include agricultural lands of Egypt and its estimation on the 17th of January 2019, sand dune encroachment areas which include west of the Nile Delta (255 km^2), El-Faiyum and Wadi El-Rayyan Depression (480 km^2), Southwest of El-Minya governorate (350 km^2), El Kharga Oasis (400 km^2) and High Dam Lake (800 km^2), islands of the Nile from geomorphological features, regions with highest rainfall in Egypt and suitable for sustainable development using groundwater like North Sinai, regions in the desert suitable for horizontal expansion like El-Farafra Oasis, and areas of natural reserves in Egypt.

1.3.2 Environmental Applications

This part is presented in four chapters from 4 to 7. Chapter 4 is entitled "**Environmental Applications of Remote Sensing in Egypt: A Review and an Outlook**". It includes a review of the environmental applications of optical/infrared remote sensing data in Egypt. It also presents an outlook on the future development of the applications to catch up with technology development and new uses. This chapter highlights a range of environmental remote sensing applications in Egypt. It covers environmental geology, land use/land cover, water resources, agriculture and biomass, arid environment, ocean and coastal areas, and atmospheric pollution.

Customary applications along with modern applications are presented. A wide range of publications is summarized and quoted in order to point out key groups that work on remote sensing in Egypt. Many Applications are linked to the current national development programs. A future outlook of remote sensing applications in Egypt is also suggested. The topic "**Radar Remote Sensing Applications in Egypt**" is covered in Chap. 5. The chapter introduces key concepts of radar remote sensing with its three facets of synthetic aperture radar, scatterometer and altimeter. A review of present applications in Egypt has been presented as well as an outlook on the future development of the applications. It also introduces basic information about the three satellite radar remote sensing systems; namely synthetic aperture radar (SAR), scatterometer and radar altimeter. It aims at reinforcing the background about this relatively new technology within the Egyptian remote sensing community. It then moves on to summarize current applications of SAR in Egypt which encompass geological/morphological, hydrological, archaeological, and agricultural applications. The two emerging SAR technologies; polarimetry and interferometry are introduced with a brief theoretical background. Though their applications are still limited, the chapter introduces samples of these applications. An outlook at future applications of SAR and scatterometer data is presented. It entails the fusion of these radar sensors with optical data and more focus on arid region applications.

On the other hand, Chaps. 6 and 7 present case studies related to environmental applications. Chapter 6 is entitled "**Application of Remote Sensing for Monitoring Changes in Natural Ecosystems: Case Studies from Egypt**". It presents an overview of the use of satellite-based remote sensing for the detection and description of environmental pollutants. The potential and limitations of the application of remote sensing technologies in arid ecosystems of the Egyptian desert are discussed. The use of traditional field surveys entails considerable efforts to offer accurate spatial and temporal coverage of the natural ecosystems. However, large quantities of global data have recently become readily accessible due to the development of remote-sensing methods. Recently, in Egypt, urbanization, anthropogenic activities, and exponential population growth have resulted in severe reductions in water bodies and agricultural lands. Hence, the subject of remote sensing should be comprehensively investigated to provide a complete evaluation of ecological and environmental conditions in various regions of Egypt. Moreover, this chapter considers the state-of-the-art for remote sensing for mapping the environmental variables at different locations in Egypt, including deserts, oases, sand dunes, salt marshes, fish farms, reed vegetation, and agricultural lands. To sum up, Chap. 6 attempts to overcome the problems of the complexity and high cost of data acquisition, unavailability of specialized search engines, and the absence of a satisfactory geographic database. It also provides useful information to the public and private sectors, policymakers, and stakeholders.

Chapter 7 is entitled "**Physico-chemical and Spectral Assessment of Mosquito Breeding Waters in Suez Canal Zone: Approach of Remote Sensing**". The main aim of this chapter are to assess physicochemical and spectral characteristics contributing to mosquito breeding in the Suez Canal Zone, which will help in determining the preventive measures to eliminate future mosquito proliferation and disease transmission. Fifty-two different sites were sampled, in February and April 2016, for mosquito larvae. This is because mosquitoes have greater importance in terms of major public health problems. Some highly adaptable insects continue to co-exist with the man and transmit many diseases such as Malaria, Filariasis, Rift Valley Fever, West Nile Virus, Dengue Fever. The favourable environmental unique features characterizing the Suez Canal area assist in the flourishing of different mosquito species and their spatial wide-spread in and around the area. Mosquitoes prefer to breed in all sorts of running or stagnant water bodies.

1.3.3 Water Quality and Pollution

This theme is covered into 6 chapters from 8 to 13. Chapter 8 discusses "**Satellite Image Data Integration for Groundwater Exploration in Egypt**". The authors show the importance of using Synthetic Aperture Radar (SAR), thermal infrared data and digital elevation models in detection and mapping of groundwater-bearing basins. These tools reveal the buried paleodrainge courses, river deltas, and lake basins that may have acted as preferential flow paths and sites for subsurface water in arid and hyper-arid sandy deserts. The delineation of these ancient fluvial features arc signals for potential locations for groundwater accumulation areas. In addition, they might aid in future exploration for oil and gas reservoirs across the Egyptian desert.

Chapter 9 is entitled "**Monitoring and Protection of Egyptian Northern Lakes Using Remote Sensing Technology**". It integrates the existing scientific knowledge in order to obtain an ecological understanding of the lake. It also implements a systematic decision support procedure in such a way that final judgments can be as objective as possible. This is mainly to find practical solutions for the environmental problems of the lakes via utilizing remotely sensed data. This data can be used as a tool to detect, monitor and evaluate changes in ecosystems to develop management strategies for ecosystem resources, satellite, and airborne systems. It offer a plausible monitoring system for large scale, earth surface viewing and provide a usable database for change detection studies. Remote sensing data can be used to span temporal and spatial scales ranging from local systems to aggregated global systems.

Also, Chap. 10, which is entitled "**An Overview of the Environmental and Anthropogenic Hazards Threatening Lake Nasser, Egypt National Water Reservoir,**" focuses on Lake Nasser Environment. Lake Nasser is acting as Egypt's National Water Reservoir. Since its creation, Lake Nasser is undergoing several changes and threats. With filling the Lake basin with water, though changes from year to year, the area of the Lake expanded, the water level increased, new forms developed such as local delta, islands, and khors (embayment). Deposition of Nile silt and aeolian sand results in a gradual decrease in the capacity of the Lake basin. In the meantime, Lake Nasser is exposed to several environmental and anthropogenic hazards. Since about 80% of water needs comes from Lake Tana (Ethiopia), changes in the volume of water reaching Egypt from Ethiopia is one of the most dangerous environmental hazards that threaten water storage in the Lake basin. Of the anthropogenic hazards, comes the construction of new dams on the upper reaches of the Nile as the most dangerous hazard.

Chapters 11 and 12 discussed the oil pollutions in large water bodies, including the Mediterranean Sea and the Red Sea waters. Chapter 11 is entitled "**Oil Pollution in the Mediterranean Waters of Egypt**". It shows that oil pollution is a serious environmental problem for the Mediterranean waters of Egypt because the number of detected oil spills and their size is very large. The authors present a state of the art review of the oil pollution in the Mediterranean Sea, shipping activities in the coastal zone of the Mediterranean, offshore oil/gas fields and infrastructure which is out of control of the local authorities, oil pollution events in tabular form with detailed information including date (from 1 September 1982 to 21 May 2017), coordinates of each event, Ship/Flag to indicate the country of origin, accident type, and oil spill volume whenever it is known. Satellite monitoring of oil pollution can help the authorities to take the needed action to minimize or prevent the oil pollution in the waters of the Mediterranean. The presented text in the chapter is supported by remarkable pictures and satellite images.

Chapter 12 with the title "**Oil pollution in the Northern Red Sea: A Threat to the Marine Environment and Tourism Development**" comes to describe the development of the coastal tourism in Egypt, shipping activities connected to the passage using the Suez Canal, crude oil delivery to oil terminals, and development of oil fields in the northern part of the Red Sea. The main shipping routes going along the Egyptian coastline and tourist resort areas in the Red Sea with intense traffic, as well as drilling at offshore oil fields, represent a threat to the unique marine environment. Based on the vast experiences of the authors and the available evidence, they presented a review of the Red Sea and tourist resorts, the Suez Canal and shipping in the Red Sea, offshore oil fields and infrastructure, oil pollution events, the use of satellite monitoring and its usefulness in detecting the oil pollution information. The use of high-resolution satellite images enables the authors to show the mesoscale water dynamics in the Red Sea, and the cases of oil pollution detected in the Suez Canal area and the resort areas as well. All texts are supported with clear pictures of high-resolution satellite images, which increase the usefulness of the chapter and the presented information.

The last chapter in this section is Chap. 13 with the title "**Satellite Monitoring of the Nile River and the Surrounding Vegetation**". The authors demonstrated modern capabilities of satellite remote sensing technologies in environmental monitoring of the Nile River Valley, the Nile Delta, and important agricultural zones. High-resolution satellite imagery is very effective in monitoring the whole Nile River area because today, one can use, for example, freely available OLI Landsat-8 optical imagery to trace seasonal and inter-annual variability of the vegetation or desertification processes in different parts of the river with a spatial resolution of 30 m. The course of the Nile was divided into 19 scenes of OLI Landsat-8 images acquired in July–September 2018 to show spatial and vegetation peculiarities of every sub-region. Besides, the NASA Giovanni online data system v.4.30, developed and maintained by the NASA Goddard Earth Sciences Data and Information Services Center (GES DISC), was used to show seasonal and inter-annual (2000–2018) variability of NDVI as a measure of vegetation health. The authors did the analysis for the whole Egypt, the area around the Nile Valley, the Nile Delta, and two important agricultural areas of Egypt located along the Nile in the southern and northern parts of the country.

1.3.4 Environment and Climate Change

This theme is covered into two chapters that are Chaps. 14 and 15. Chapter 14 presents an overview of "**Remote Sensing and Modelling of Climate changes in Egypt**". It describes some of the efforts carried out in Egypt as one of the most vulnerable countries. Authors started by the efforts to identify and reduce greenhouse gases in the atmosphere, and with some preliminaries of the satellite technology for receiving and analyzing data on the atmosphere. Remote sensing and GIS were also useful in identifying what mitigation measures, early warning systems and adaptation projects could be useful. A demonstration of success stories of energy conservation and coastal risk assessment is also presented. The role of RS and GIS for vulnerability assessment, mitigation, and adaptation was then emphasized. The vulnerability of Egypt resources (water, agricultural, touristic, urban, and socioeconomic), have been described by remote sensing and GIS techniques. Recommendations for urgent needs, for each resource for adaptation measures, were then discussed.

On the other hand, Chap. 15 is entitled "**Qualitative and quantitative assessment of land degradation and desertification in Egypt based on Satellite remote sensing: Urbanization, salinization and wind erosion**". It highlights three active land degradation processes (i.e., urban encroachment, salinization, and wind erosion), commonly observed in the Egyptian territories. Both the descriptive and quantitative approaches are followed and merged, showing advantages of combining both approaches in assessment, sizing and combating preparedness. Multiscale and multispectral space satellite sensors, in addition to thematic maps, supply valuable information concerning landscape, vegetation type, and quality and land use/cover, as inputs to FAO-UNEP provisional methodology to assess aspects of each desertification processes. The same space data serve the EU-MEDLUS methodology assessing environmental sensitivity. Space data are used in computing the Soil Quality Index (SQI), Vegetation Quality Index (VQI), and Management Quality Index (MQI). Climate Quality Index (CQI) may be computed using meteorological satellite data. It is approved that the Egyptian territory is susceptible to very high-to-high desertification sensitivity. However, the Nile Valley is moderately sensitive due to cultivated vegetation cover. Combating desertification measures are essential for sustainable agriculture. Special concerns have to be taken at the desert oases, wadis and interference zone due to their role in decreasing food gaps and accelerating agriculture expansion.

1.3.5 Hydrology and Geomorphology

The last theme of the book is covered into three chapters, 16, 17 and 18. Chapter 16 is about the "**Landscapes of Egypt,**" which are recognized on the landmass of Egypt as two types: (1) physical landscapes and (2) anthropogenic landscapes. Most of the first type landscapes are inherited from past environments, mainly from past wetter climates or from tectonics. Therefore, the physical ones can be divided into present-day and inherited landscapes. Only the aquatic and dunes/sand types belong to present-day landscapes, whereas all other physical types (Mountain Landscape, Structural Domes Landscape, Fluvial Landscape, Karst Landscape, Fluvio-Marine Landscape, and Ridge-Depression Landscape) are inherited ones. There are only two types of anthropogenic (urban and rural) landscapes. The urban-type represents some areas of the landmass of Egypt, where people changed the physical face of the landscape by obliterating the physical features and constructing new features for population agglomerations. The new features are mainly houses, roads, and other ones to make life viable for people who will be living in. The best example for urban landscape is Cairo Conurbation and Alexandria City, where a house is built beside another house, separated only by roads, railway lines, and other public utilities. There are two sub-types of rural landscape: the Nile Valley sub-type and the desert oases subtype. Although both subtypes have a green surface, because both are irrigated and cultivated by field crops, they use two different types of water: the Nile water and groundwater. In addition to the green fields, the rural landscape is characterized by certain other features (canals, villages, small towns, and roads) that are necessary for the economy of this landscape.

Moving from landscape to the hydrology, Chap. 17 with the title "**Integrated Watershed Management of Grand Ethiopian Renaissance Dam via Watershed Modeling System and Remote Sensing**" are made using remote sensing and GIS with the help of the Watershed Modeling System (WMS) Program and Global Mapper software. The watershed and its sub-basins delineation process were carried out. In the outlet of the catchment area sub-basins as well as at the catchment outlet, flood hydrographs were developed. HEC-RAS program researched the method of delineation of the floodplain. The HEC-1 model is used to create a hydrological modeling rain-flow form on the Grand Ethiopian Renaissance Dam (GERD) watershed. This research is an attempt to help to address and estimate the upper Blue Nile (UBN) basin's water budget using a fresh hydrological modeling structure and remote sensing information. This model is then used to forecast the basin's hydrological reaction to climate change and land-use situations. The present Renaissance Dam site (506 m amsl) allows the creation of a 100 m deep reservoir with a total 17.5 km^3 storage; overflow will occur at the lake's level of (606 m masl) from the northwestern portion of the advanced lake downstream into the Renaissance. Building the spillway dam to control the overflow area can create a 180 m profound lake that will store up to 173 km^3 in a 3130 km^2 lake. Moreover, the GERD is built in a mountainous region where data on important environmental variables has been gathered and where on-site data condition, dangers, and environmental effects are missing. Remote sensing has been used to map and predict the flooded region in Sudan and Egypt in case of GERD failure.

The last chapter in this section is Chap. 18 with the title "**Quantitative Analysis of Shoreline Dynamics Along the Mediterranean Coastal Strip of Egypt. Case Study: Marina El-Alamein Resort**". This chapter provides a comprehensive historical shoreline analysis for the Egyptian Mediterranean coastal strip from Sidi Abd El-Rahman to El-Arish, with special reference to Marina El-Alamein resort's shoreline. Besides developing predictive shoreline scenarios along Marina El-Alamein shoreline (for years 2023, 2057, and 2107) to aid decision-makers in addressing coastal issues affecting the area more efficiently in a relatively short time. Results generated shoreline change-maps for the entire strip for 30 years. Quantitatively, the entire coastal strip from Sidi Abd El-Rahman to El-Arish had dynamically changed over time. Results indicated that the greatest displacement in shoreline position was about 1717.66 m, with an average of 102.17–175.78 m in 30 years. About 14% of transects displayed stability in shoreline positions, whereas 76% of transects experienced shoreline displacement up to 250 m. Moreover, the coastal strip suffered medium erosion with an average of −0.484 to 0.28 m/year. About 40% of the coastal strip understudy hosts five lagoons. During the study period, 1987–2017, the overall trend of shoreline change rate (EPR) indicated very high accretion in El-Alamein and Edku Lagoons, very high erosion in Burullus and Manzala Lagoons and high erosion in Bardawil Lagoon. Constructing land structures along Marina El-Alamein shoreline during the last 3 decades interrupted the shoreline stability greatly where seaward shifting was predominant. Shoreline evolution model predicted that the western inlet might be closed in the future. Consequently, careful planning, coupled with long-term considerations, are required when developing coastal engineering structures to ensure the sustainability of available natural resources.

The book ends with the conclusions and recommendations chapter numbered 19.

This book may be regarded as a follower of the book "Environmental Remote Sensing and GIS in Iraq" edited by Al-Quraishi and Negm (2020) and published in the "Springer Water" book series of Springer International Publishing. It will be followed by the next book entitled "Environmental Remote Sensing and GIS in Tunisia" edited by Allouche and Negm (2020) and published in the same book series in 2020.

Acknowledgements The editors who wrote this chapter would like to acknowledge the authors of the chapters for their efforts during the different phases of the book, including their inputs in this chapter.

Abdelazim M. Negm would like to acknowledge the partial financial support from the Academy of Scientific Research and Technology (ASRT) of Egypt via the bilateral collaboration Italian (CNR)–Egyptian (ASRT) project titled "Experimentation of the new Sentinel missions for the observation of inland water bodies on the course of the Nile River".

Andrey Kostianoy was partially supported in the framework of the P.P. Shirshov Institute of Oceanology RAS budgetary financing (Project N 149-2019-0004).

References

Allouche FK, Negm AM (2020) Environmental remote sensing and GIS in Tunisia. Springer International Publishing (under production processes)

Al-Quraishi A, Negm AM (2020) Environmental remote sensing and GIS in Iraq. Springer International Publishing, p 529

Basheer MA, El Kafrawy SB, Mekawy AA (2019) Identification of mangrove plant using hyperspectral remote sensing data along the red sea, Egypt. Egypt J Aquat Biol Fish 23(1):27–36

El-Alfy MA., Hasballah AF, El-Hamid HTA, El-Zeiny AM (2019) Toxicity assessment of heavy metals and organochlorine pesticides in freshwater and marine environments, Rosetta area, Egypt using multiple approaches. Sustain Environ Res 29:19. https://doi.org/10.1186/s42834-019-0020-9

Hassan A, Belal A, Hassan M, Farag F, Mohamed E (2019) Potential of thermal remote sensing techniques in monitoring waterlogged area based on surface soil moisture retrieval. J Afr Earth Sci 155:64–74

Mohamed ES, Saleh AM, Belal AB, Gad A (2018) Application of near-infrared reflectance for quantitative assessment of soil properties. Egypt J Remote Sens Space Sci 21(1):1–14

NARSS. Available https://www.narss.sci.eg/. Accessed 20 July 2019

Ramzi A, El-Bedawi M (2019) Towards integration of remote sensing and GIS to manage primary health care centers. Appl Comput Inform 15(2):109–113

Sayed E, Riad P, Elbeih S, Hagras M, Hassan A (2019) Multi criteria analysis for groundwater management using solar energy in Moghra Oasis, Egypt. Egypt J Remote Sens Space Sci 22:227–235

Scopus. Available https://www.scopus.com. Accessed 20 July 2019

Selim N, El Raey M (2018) EIA for lake maryut sustainable development alternatives using GIS, RS, and RIAM software. Assiut Univ Bull Environ Res 21(1)

Sowilem MM, El-Zeiny AM, Mohamed ES (2019) Mosquito larval species and geographical information system (GIS) mapping of environmental vulnerable areas, Dakhla Oasis, Egypt. Int J Environ Clim Change 9(1):17–28

Chapter 2
Overview for Recent Applications of Remote Sensing in Egypt

Mahmoud Nasr, Salwa F. Elbeih, Abdelazim M. Negm and Andrey Kostianoy

Abstract Remote sensing is the science and art used to collect, organize, and analyze reasonable information about a specific phenomenon, pattern, or region without direct or physical contact with the study area. In Egypt, the development of remote sensing has recently been supported, maintained, and widened via various research projects, which have been sponsored and funded by national and international organizations. These collaboration projects have been applied in various fields such as agriculture, coastal management, hydrology, ecology, meteorology, geoarchaeology, and geology. The research areas also included geography, environmental monitoring, and flood prediction. Accordingly, this chapter summarizes the recent applications of remote sensing in Egypt regarding (a) the number, types, and objectives of published documents, (b) the academic, industrial, governmental, and non-governmental sponsors, and (c) the institutions, laboratories, and research centers. The outputs of this work could be beneficial for the decision-makers and stakeholders to improve the social, ecological, financial, and environmental conditions in Egypt.

M. Nasr (✉)
Sanitary Engineering Department, Faculty of Engineering, Alexandria University, Alexandria 21544, Egypt
e-mail: mahmoud-nasr@alexu.edu.eg; mahmmoudsaid@gmail.com

S. F. Elbeih
Engineering Applications Department, National Authority for Remote Sensing and Space Sciences (NARSS), Cairo, 1564 Alf Maskan, Egypt
e-mail: saelbeih@gmail.com; saelbeih@narss.sci.eg

A. M. Negm
Water and Water Structures Engineering Department, Faculty of Engineering, Zagazig University, Zagazig 44519, Egypt
e-mail: amnegm85@yahoo.com; amnegm@zu.edu.eg

A. Kostianoy
P.P. Shirshov Institute of Oceanology, Russian Academy of Sciences, 36, Nakhimovsky Pr., Moscow 117997, Russia
e-mail: kostianoy@gmail.com

S.Yu. Witte Moscow University, 12, Build. 1, 2nd Kozhukhovsky Pr., Moscow 115432, Russia

S. F. Elbeih et al. (eds.), *Environmental Remote Sensing in Egypt*,
Springer Geophysics, https://doi.org/10.1007/978-3-030-39593-3_2

Keywords Egypt · Funding institutions · Grant projects · Published documents · Remote sensing

2.1 Introduction

Remote sensing is an efficient approach used to acquire, gather, and handle a large amount of temporal and spatial land-surface observations using satellite sensors, images, and measurements (Hassan 2013; Mohamed and Elmahdy 2017). The remotely-sensed imagery can be acquired regarding acoustic wave technology, variation in electromagnetic energy, force-sensing mechanisms, and other data detection, collection, and processing methods (Ghassemian 2016; Zhu et al. 2017). Recently, the Egyptian government has developed advanced strategies to improve and expand the application of satellite remote sensing in several engineering and environmental fields (Abdelkareem and El-Baz 2018). Accordingly, this chapter represents the number, types, and aims/scopes of documents about the field of "Remote Sensing" that have recently been published by the Egyptian researchers. The governmental and non-governmental sponsors that have funded the grant projects covering the application of remote sensing in various environmental, engineering, agricultural, industrial, and ecological sectors are demonstrated. Moreover, the institutions, laboratories, and research centers that are collaborating in these projects are listed.

2.2 National Authority for Remote Sensing and Space Sciences (NARSS) in Egypt

2.2.1 *Establishment of NARSS*

The National Authority for Remote Sensing and Space Sciences (NARSS) is the pioneering Egyptian institution in the remote sensing platforms. In 1971, NARSS started in Egypt through an American-Egyptian joint project, which was affiliated to the Egyptian Academy of Scientific Research and Technology (ASRT) (NARSS 2019). In 1991, NARSS worked under the umbrella of the Ministry of State for Scientific Research as a national authority for remote sensing. In 1994, NARSS was reorganized, following the presidential decree No. 261 for 1992 under the authority of the Ministry of Scientific Research, to cover two major sectors, i.e., Remote Sensing and Space Sciences. Nowadays, NARSS employs the state-of-the-art in space technologies to monitor and assist the water, land, and natural resources of Egypt. It also generates maps and spatial data to record the environmental changes and to identify, control, and avoid unexpected hazards.

NARSS includes eight scientific divisions, which can be summarized as follows (NARSS 2019):

(a) Geological Applications and Mineral Resources. The division activities include the generation of geological and geomorphological information databases, evaluation of the non-renewable natural resources (e.g., minerals, and groundwater), and prediction of hazard events (e.g., earthquakes, landslides, and flash floods). The division also employs Geographic Information Systems (GISs) to generate basic layers such as roads, streams, railway lines, and important places that can support the planning process and the development of new communities.
(b) Aviation and Aerial Photography. This sector provides mapping and surveying services to NARSS projects. It is responsible for flight and Light Detection and Ranging (LiDAR) data processing, aerial photography sheet film production, and photogrammetry and image analysis.
(c) Environmental Studies and Land Use. The strategic goal of this division is to develop the capabilities of national scientists and specialists for the applications of remote sensing and related technologies. The division also attempts to generate high-quality technical reports and scientific articles for addressing sustainable development objectives.
(d) Scientific Training and Continuous Studies. This sector offers advanced and technical courses to NARSS staff, external scientists and researchers, and regional organizations from the Arab and African countries. It prepares highly qualified graduates that can provide technical and informational supports to decision-makers. For instance, in 2018, a professional diploma was launched via the collaboration between NARSS and Ain Shams University to prepare qualified candidates that can cope with the development of modern technologies in remote sensing and GIS.
(e) Engineering Applications and Water Resources. This division establishes the use of remote sensing in the fields of groundwater monitoring and assessment, water pollution detection, irrigation and hydraulics works, traffic evaluation and management, archaeological survey and exploration, geotechnical studies, and drinking water supply and sewage networks.
(f) Space Sciences and Strategic Studies. The main aim of this sector is to establish a scientific and research base, enabling Egypt to join the space age and manufacture its satellites. This objective is considered through acquiring technological knowledge and capabilities, as well as regarding the gradual construction of small-sized and infrastructure-related satellites.
(g) Data Reception, Analysis, and Receiving Station Affairs. This division aims at generating, processing, and archiving various satellite digital data, photomaps, mosaics, and atlases.
(h) Agriculture Applications, Soils, and Marine. This division conducts several types of research and consultancies on the agricultural sectors, soil science, irrigation management, food security, and marine fields for various governmental and non-governmental agencies.

2.2.2 Contribution of NARSS to the Field of Remote Sensing

During the previous decade, NARSS has been funded by about 340 research and development projects. The grant projects designate the application of remote sensing and related technologies in geology, agriculture, engineering, environment, water, marine sciences, mineral resources, and space archaeology. According to the Scopus database (Scopus 2019), about 150 authors from NARSS have published 487 documents during 1996–2019. The document types are classified into articles (67.6%), conference papers (24.0%), book chapters (2.9%), reviews (1.8%), and others (3.7%). As shown in Fig. 2.1, the number of publications from NARSS has significantly ($p < 0.05$) increased from 107 documents in 2009 to 487 documents in 2019. Moreover, the slope of the linear regression line during 2009–2019 (y = 37.16x – 74,561; R^2: 0.996) is 6-fold greater than that during 1996–2009 (y = 6.02x – 12,041; R^2: 0.676). This pattern suggests that the studies on "Remote Sensing" are becoming an important and essential topic of research in Egypt.

Based on the Scopus database (Scopus 2019), several countries have contributed to the published articles of NARSS from 1996 to 2019. Amongst them, the leading country is Egypt, which participated in about 485 documents. Further, United States, Saudi Arabia, Germany, and the United Kingdom were involved in 47, 33, 21, and 17 published papers, respectively. The published works that contained the affiliation "National Authority for Remote Sensing and Space Sciences" in Scopus during 1996–2019 were funded by more than 15 sponsors and research organizations. Moreover, several institutions and funding sources, such as Western Michigan University, Deutsche Forschungsgemeinschaft, National Natural Science Foundation of China, National Science Foundation, Science and Technology Development Fund, Consiglio Nazionale Delle Ricerche, Hanyang University, Ministry of Education of

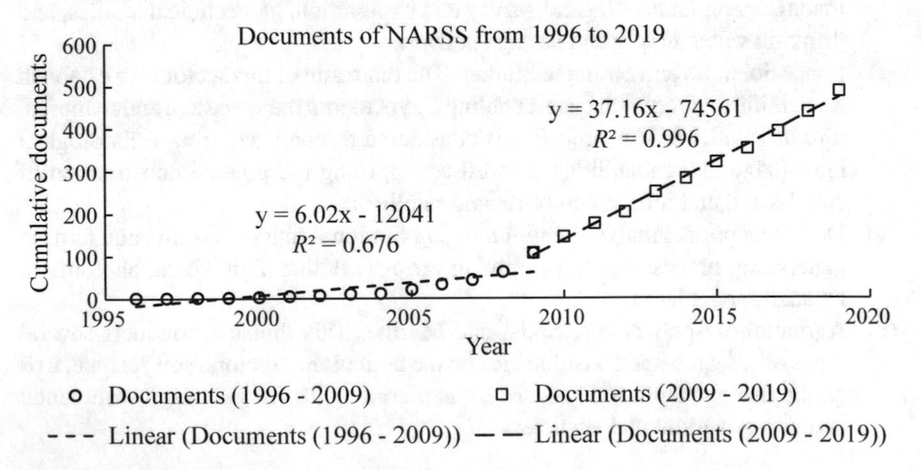

Fig. 2.1 Cumulative number of documents in Scopus database containing "National Authority for Remote Sensing and Space Sciences" from 1996 to 2019 (Scopus 2019)

the People's Republic of China, and National Aeronautics and Space Administration, were acknowledged in the funding statements of the published documents. The most significant number of NARSS documents was published in the Egyptian Journal of Remote Sensing and Space Sciences (74 papers), followed by the Arabian Journal of Geosciences (34 documents), and the Journal of African Earth Sciences (23 documents). These journals cover all aspects of remote sensing, GIS, space sciences, sedimentology, oceanography, natural hazards, hydrogeology, tectonics, geology, environmental sciences, and sustainable development.

Recently, NARSS has collaborated with several organizations and local entities, such as Urban Planning Authority, Building Cooperatives Authority, Academy of Scientific Research and Technology, National Defense Council, General Authority for Drinking Water and Sewage, General Organization for Physical Planning, New Urban Cities Authority, General Authority for Tourism Development, National Telecommunication Regulatory Authority, General Authority for Construction and Housing Cooperatives, General Authority for Fish Resources Development, National Authority for the Development of Sinai, and General Petroleum Company. Moreover, the list of laboratories involved in the remote sensing research of NARSS comprises Satellite Images Receiving Station at Aswan; Meteorological Receiving Station, NARSS building, Cairo; Ground Control Station, New Cairo City, Cairo; Image Processing Laboratory; Modeling and Assimilation Laboratory; Survey Instrumentation Facilities; GIS Laboratory; Space Payload Laboratory.

2.3 Survey for Recent Publications of Remote Sensing in Egypt

The search (on July 2019) in the Scopus database (Scopus 2019) using the keywords "Remote", "Sensing", and "Egypt" shows 517 documents that have recently been published during this decade (i.e., 2010–2019). This number is higher than that reported during 2000–2009 (181 documents) and 1990–1999 (67 documents). The pattern in Fig. 2.2 implies that the Egyptian government is currently exerting considerable efforts for the widespread application of remote sensing in various research and industrial sectors. The publications about "Remote sensing in Egypt" during 2010–2019 include 386 articles, 73 conference papers, 26 book chapters, and 13 reviews. These documents aim at addressing, transferring, and providing the most advanced technologies in the fields of remote sensing and space sciences.

The funding sponsors about "Remote sensing in Egypt" during 2010–2019 include National Science Foundation (10 documents), Science and Technology Development Fund (STDF) (10 papers), Ministry of Higher Education, Egypt (9 papers), Ministry of Higher Education and Scientific Research (7 documents), and National Aeronautics and Space Administration (5 documents). About 84 documents were published in the Egyptian Journal of Remote Sensing and Space Sciences, followed by 46, 27, and 21 in the Arabian Journal of Geosciences, Journal of African Earth

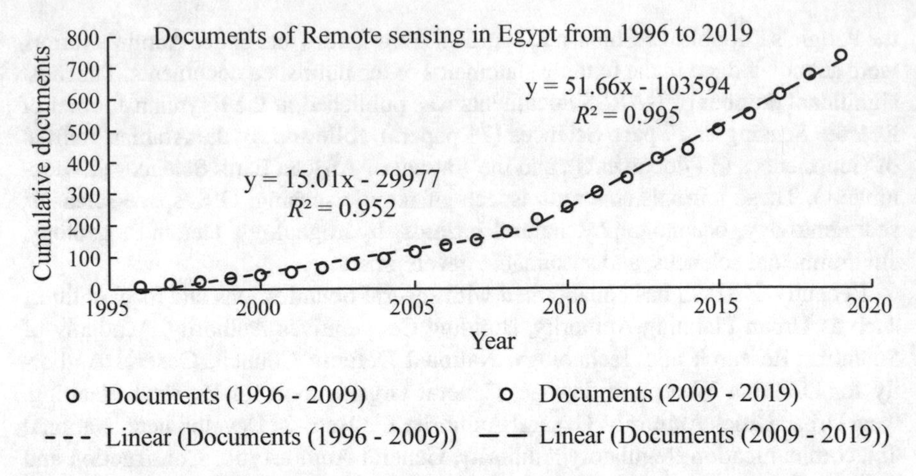

Fig. 2.2 Cumulative number of documents in Scopus database containing the keywords "Remote", "Sensing", and "Egypt" from 1996 to 2019 (Scopus 2019)

Sciences, and Environmental Earth Sciences, respectively. Furthermore, Egypt is a member/partner in the European Union (EU) Framework Programme, and the U.S. Agency for International Development (USAID) programs.

Based on the observations mentioned above, some examples of the recent publications during 2010–2019 can be summarized as follows:

In 2010, Hereher (2010) used remote sensing to determine the areas located at the Nile Delta of Egypt, which were negatively influenced by the sea-level rise in 2006. The study depicted that a 1 m sea-level rise would cause flooding to more than 25% of the Nile Delta region. Moreover, the shoreline would be shifted by 60–80 km southward due to the rise of the sea level by 2 m.

In 2011, Belal and Moghanm (2011) applied remote sensing to generate land use and land cover maps of several areas in Al-Gharbiya Governorate located at Middle of Nile Delta, Egypt. The study also monitored the changes in agricultural lands and water bodies, and it was demonstrated that urbanization during 1972–2005 caused the reduction of productive farm areas by 5.84–7.17%.

In 2012, Yones et al. (2012) used the remote sensing approach to limit cotton leafworm, namely *Spodoptera littoralis*, in Ezbet Shalaqan, Al-Qalyubiya Governorate, Egypt during 2006. The study demonstrated that 174.85–197.59 degree-days were the optimum timing for pest control, avoiding dense sprays of pesticides.

In 2013, Hassan (2013) used the remote sensing technique to estimate the evaporation rate and water balance of Lake Nasser located in Upper Egypt. The study depicted that the daily evaporation rate of Lake Nasser ranged between 6.0 and 6.7 mm during October 1998–October 2000.

In 2014, Hereher (2014) employed remote sensing to assess the sand drift potential along the Nile Valley and Delta, Egypt. The study demonstrated that the longitudinal and transverse dunes were primarily influenced by the wind environment and topography (i.e., duration, energy, and direction of the wind).

In 2015, Gabr et al. (2015) used the remote sensing approach to map the gold and gold-related alteration zones at El-Hoteib, South Eastern Desert, Egypt. The study demonstrated that the monitored area contained smithsonite and hemimorphite and that the gold content varied from 0.3 to 6.4 g/t in the altered rock and declined to 0.3–0.9 g/t in the related quartz veins.

Elbeih and Soliman (2015) mapped various swelling clay minerals near Sohag-Safaga highway, Eastern Desert of Egypt by integrating field samples with Advanced Spaceborne Thermal Emission and Reflection Radiometer (ASTER) satellite imageries. It was depicted that the soil of the study area contained smectite clay minerals (montmorillonite), reflecting a swelling property. Accordingly, special considerations should be given to the existing and the future planned projects within the area.

In 2016, Abdalla et al. (2016) employed remote sensing to illustrate the influences of human activities on the archaeological sites in Southern Egypt. The study also described the characterization and evolution of social activities, causing groundwater level increase, loss of agricultural/arable lands, and recharging and pollution of groundwater.

In 2017, El-Zeiny et al. (2017) investigated the environmental variables associated with the spatial distribution of mosquito species at Suez Canal Zone using field surveys along with remote sensing techniques. The developed GIS model predicted that the most abundant mosquito species were *Culex pipiens* and *Ochlerotatus detritus*. Moreover, Ismailia Governorate could comprise the most areas suffering from environmental risks of mosquito-borne diseases.

Mohamed and Elmahdy (2017) used remote sensing to obtain data about site condition, geology, and tectonics of the Grand Ethiopian Renaissance Dam (GERD), as well as the impact of the dam construction on the Sudanese and Egyptian environments. The study succeeded to collect relevant information about the factors affecting dam stability and failure. Moreover, the study outputs could predict the flooded parts in Sudan and seawater disturbance in the Nile Delta of Egypt due to the constructed dam. The dam also could influence the amount of sediments, Nile water evaporation rate, water quality, and the bird and mammal life in the study area.

In 2018, Abdelkareem and El-Baz (2018) applied remote sensing to identify the changes of hydrothermal zones in the middle part of the Eastern Desert of Egypt. The study succeeded to define the area of high alteration zone, and it depicted that the hydrothermal alteration profile was associated with gold and massive sulphide mineralization.

Elbeih and El-Zeiny (2018) assessed the temporal and spatial variations in the groundwater quality parameters at the west of Sohag Governorate, Egypt. The status of groundwater was obtained using two multispectral Landsat images and a set of spectral derived land use indices. It was reported that the increased sources of groundwater pollution could be linked to the high coverage of cultivated and urban lands (>75%). Moreover, groundwater quality was negatively influenced by urbanization, agricultural expansion, land degradation, usage of phosphate fertilizers, and other natural and anthropogenic activities.

In 2019, Hassan et al. (2019) used the remote sensing technique to monitor waterlogged zones (water bodies and wet sabkhas) in Ismailia Governorate. The study demonstrated that water table levels, land use/land cover, and soil properties, elevation, and texture influenced the seasonal and spatial variations of soil moisture content. Moreover, several sabkhas (salt-encrusted mudflats) were converted into fish farms and water bodies over 1998–2015 due to excessive surface irrigation and water table rise; i.e., a pattern that increased the surface soil moisture.

Moreover, in 2019, Ramzi and El-Bedawi (2019) used the remote sensing method to investigate, locate, monitor, distribute, and manage the primary health care centers (PHCCs) at El-Salam medical region, Cairo, Egypt. The work indicated that the existence of several public and private health services coped well with the shortage of PHCCs in the study area.

In Sayed et al. (2019), a multi-criteria analysis model supported by GIS was employed to select the optimum locations for extracting and pumping groundwater using solar energy. The GIS-based model was used to generate an informative map for the solar energy distribution in Moghra Oasis, Western Desert of Egypt. The model input parameters included solar radiation, area accessibility, depths to groundwater, topography, salinity, and land-use. The output of the model revealed that the optimum site for the installation of the Photovoltaic panels would be near the Nile Delta and outside the oil fields, Qattara Depression, and Moghra Lake.

2.4 Conclusions and Recommendations

This chapter represents an essential survey about the recent applications of remote sensing in Egypt. According to the Scopus database, the number of publications about "Remote sensing in Egypt" has increased from 181 documents during 2000–2009 to 517 documents during 2010–2019. The publications during 2010–2019 comprise 386 articles, 73 conference papers, 26 book chapters, and 13 reviews. The published documents attempt to address, transfer, and sustain the most advanced technologies in the fields of remote sensing and space sciences in Egypt. The academic disciplines and fields of study cover essential areas in geology, agriculture, engineering, environment, water, marine sciences, mineral resources, and space archaeology. The outputs of this chapter are beneficial to undergraduate and graduate students, researchers, decision-makers, stakeholders, and several public and private sectors.

Based on the chapter's outputs, it's recommended that:

- Additional governmental and non-governmental sponsors should fund the grant projects to cover the application of remote sensing in various sectors.
- The Egyptian government should exert additional efforts to ensure the widespread application of remote sensing in various research and industrial fields.
- More institutions, laboratories, and research centers should be developed to incorporate scientists and specialists that can apply remote sensing and related technologies.

– More attention needs to be paid to expand the application of satellite remote sensing that can monitor and assess the Mediterranean and Red seas surrounding Egypt.
– There is a need to promote international collaborations in the field of satellite remote sensing with different countries and international organizations.

Acknowledgements The first author would like to acknowledge Nasr Academy for Sustainable Environment (NASE). The third author would like to acknowledge the partial financial support from the Academy of Scientific Research and Technology (ASRT) of Egypt via the bilateral collaboration Italian (CNR)–Egyptian (ASRT) project titled "Experimentation of the new Sentinel missions for the observation of inland water bodies on the course of the Nile River".

References

Abdalla F, Moubark K, Abdelkareem M (2016) Impacts of human activities on archeological sites in southern Egypt using remote sensing and field data. J Environ Sci Manag 19(2):15–26
Abdelkareem M, El-Baz F (2018) Characterizing hydrothermal alteration zones in Hamama area in the central Eastern Desert of Egypt by remotely sensed data. Geocarto Int 33(12):1307–1325
Belal A, Moghanm F (2011) Detecting urban growth using remote sensing and GIS techniques in Al Gharbiya governorate, Egypt. Egypt J Remote Sens Space Sci 14(2):73–79
Elbeih S, El-Zeiny A (2018) Qualitative assessment of groundwater quality based on land use spectral retrieved indices: case study Sohag governorate, Egypt. Remote Sens Appl Soc Environ 10:82–92
Elbeih S, Soliman N (2015) An approach to locate and map swelling soils around Sohag—Safaga road, Eastern Desert, Egypt using remote sensing techniques for urban development. Egypt J Remote Sens Space Sci 18:S31–S41
El-Zeiny A, El-Hefni A, Sowilem M (2017) Geospatial techniques for environmental modeling of mosquito breeding habitats at Suez Canal Zone, Egypt. Egypt J Remote Sens Space Sci 20:283–293
Gabr S, Hassan S, Sadek M (2015) Prospecting for new gold-bearing alteration zones at El-Hoteib area, South Eastern Desert, Egypt, using remote sensing data analysis. Ore Geol Rev 71:1–13
Ghassemian H (2016) A review of remote sensing image fusion methods. Inf Fusion 32(Part A):75–89
Hassan M (2013) Evaporation estimation for Lake Nasser based on remote sensing technology. Ain Shams Eng J 4(4):593–604
Hassan A, Belal A, Hassan M, Farag F, Mohamed E (2019) Potential of thermal remote sensing techniques in monitoring waterlogged area based on surface soil moisture retrieval. J Afr Earth Sci 155:64–74
Hereher M (2010) Vulnerability of the Nile Delta to sea level rise: an assessment using remote sensing. Geomatics Nat Hazards Risk 1(4):315–321
Hereher M (2014) Assessment of sand drift potential along the Nile Valley and Delta using climatic and satellite data. Appl Geogr 55:39–47
Mohamed M, Elmahdy S (2017) Remote sensing of the Grand Ethiopian Renaissance Dam: a hazard and environmental impacts assessment. Geomatics Nat Hazards Risk 8(2):1225–1240
NARSS (2019) Available: https://www.narss.sci.eg/. Accessed 20 July 2019 [Online]
Ramzi A, El-Bedawi M (2019) Towards integration of remote sensing and GIS to manage primary health care centers. Appl Comput Inf 15(2):109–113
Sayed E, Riad P, Elbeih S, Hagras M, Hassan A (2019) Multi criteria analysis for groundwater management using solar energy in Moghra Oasis, Egypt. Egypt J Remote Sens Space Sci 22(3):227–235

Scopus (2019) Available: https://www.scopus.com. Accessed 20 July 2019 [Online]
Yones M, Arafat S, Abou Hadid A, Abd Elrahman H, Dahi H (2012) Determination of the best timing for control application against cotton leaf worm using remote sensing and geographical information techniques. Egypt J Remote Sens Space Sci 15(2):151–160
Zhu X, Tuia D, Mou L, Xia D-S, Zhang L, Xu F, Fraundorfer F (2017) Deep learning in remote sensing: a comprehensive review and list of resources. IEEE Geosc Remote Sens Mag 5(4):8–36

Chapter 3
Egypt's Environment from Satellite

El-Sayed E. Omran and Abdelazim M. Negm

Abstract One way of enjoying Egypt's heritage and ancient history is to visit tourist attractions, national museums and local cities. However, there is always another way of looking at things. The shape of Egypt is different from near to its view from a distance. Usually, distinctive pictures reflect the place, so this chapter present 32 figures (images and pictures) that take you on a tour of Egypt such as lakes, depressions, khores, oasis… etc. These images reflect the various sites within Egypt from above like you have never seen before. The overall Egypt's image can be divided based on satellite images from above to three groups. First group of images describe general view of Egypt, Nile, topography, agricultural land, and desertification. The second group deals with the most important Nile islands, major wadis, and top canyon in Egypt. However, the third group shed the light on Canals, depressions, Oases, and protected areas in Egypt.

Keywords Egypt · Satellite · Nile islands · Wadis · Canyons · Depressions · Oases

3.1 Overall Egypt's Satellite Image

Egypt is a Mediterranean country, which has a unique geographical position in North-east Africa, at the crossroads of Europe and Asia, on the Mediterranean and Red Sea, and its connection to Sub Saharan Africa through the Nile Valley. It occupies an area of about one million Km^2, between Lat 22° and 32° N and Long 25° and 35° E. It is bordered on the North by the Mediterranean Sea, on the South by the Republic of Sudan, on the West by Libya, and on the East by Palestine, Gulf of Aqaba and

E.-S. E. Omran (✉)
Soil and Water Department, Faculty of Agriculture, Suez Canal University, Ismailia 41522, Egypt
e-mail: ee.omran@gmail.com

Department of Natural Resources, Institute of African Research and Studies, Aswan University, Aswan, Egypt

A. M. Negm
Water and Water Structures Engineering Department, Faculty of Engineering, Zagazig University, Zagazig 44519, Egypt
e-mail: amnegm85@yahoo.com; amnegm@zu.edu.egs

S. F. Elbeih et al. (eds.), *Environmental Remote Sensing in Egypt*,
Springer Geophysics, https://doi.org/10.1007/978-3-030-39593-3_3

the Red Sea. Geographically, Egypt can be distinguished into four main geographic regions; Nile Valley & Delta, Sinai Peninsula, Eastern Desert and Western Desert. In terms of Aridity Index, 86% of the total area of Egypt is classified as hyper-arid and the remaining as arid (Hussein 2006).

One way of enjoying Egypt's heritage and ancient history is to visit tourist attractions, national museums and local cities. However, there is always another way of looking at things that is why Egyptian Streets brings you a set of photos revealing Egypt's most fascinating landscapes from an eagle eye perspective. Photographic images of the Earth from above have been recorded since the late 19th century, initially by cameras attached to balloons, kites, or pigeons. This remotely sensed information has been used for economic analysis since at least the 1930s. Satellite data is widely useful in variety aspect of life to determine several morphological appearances such as land use, land cover, bedding, fault, formation boundary of an area to trace the presence of natural resources, especially in remote area. The first advantage is simply that remote sensing technologies can collect panel data at low marginal cost, repeatedly, and at large scale on proxies for a wide range of hard-to-measure characteristics.

The shape of Egypt is different from near to its view from a distance. Satellite images or aerial photographs reflect the stunning beauty and amazing precision. You can check this high accuracy both from cities and from places around Egypt, as a wonderful view of the places in this beautiful world! "A picture is worth a thousand words" is an English language-idiom. It refers to the notion that a complex idea can be conveyed with just a single picture. This picture conveys its meaning or essence more effectively than a description does. A picture can convey what might take ten pages in a book. A satellite image tells a story just as well as, if not better than, many written words (Fig. 3.1). Usually, distinctive pictures reflect the place, so the goal of this chapter is to shed light on the Egyptian environment by presenting 32 satellite images and pictures. These pictures represent the different Egyptian locations, showing Egypt from above as you have never seen before. It take you on a tour of Egypt such as lakes, depressions, khores, oasis… These images reflect the various sites within Egypt. Figure 3.1 shows Egypt from above like you've never seen before.

3.1.1 Egypt Gift of the Nile and Egyptians

Herodotus has said, "Egypt is the Gift of the Nile". Since about 6 million years ago, the Nile in Egypt was drilled as a result of unusual circumstances caused the Mediterranean basin drying due to higher Gibraltar (which linking the Mediterranean with the Atlantic Ocean). Thus the arrival of water to the Mediterranean Sea was stopped. It transforms into a lake, which its water was evaporated until dried (Said 1993). It was connected as a result of earth movements and volcanic eruptions. This associated with the formation of the great African Rift Valley, which led to the formation of Lakes Tana and Victoria. In addition, the grace of God and a gift

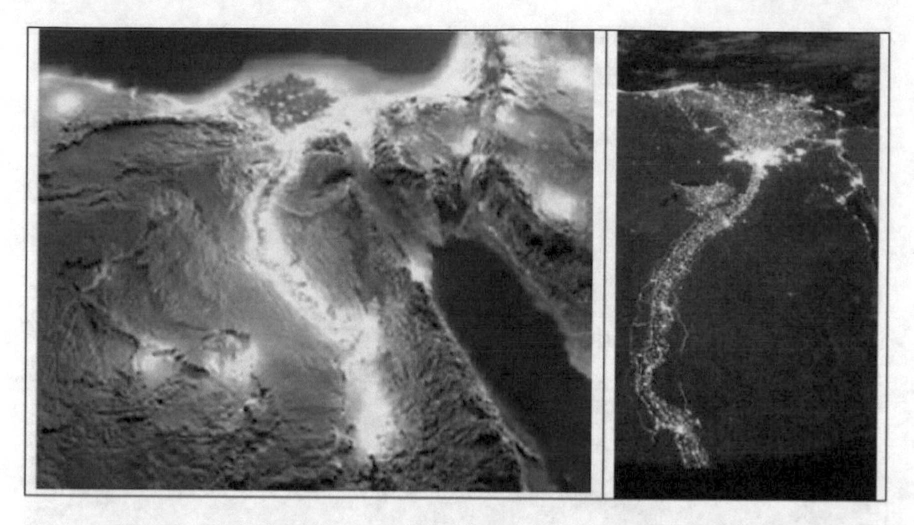

Fig. 3.1 From space, the Nile looks like a cosmic King Cobra made of light (NASA satellite image https://www.pinterest.com/pin/325103666827948341/)

from God to Egypt, caused these ground movements in the turning the water from the Ethiopian highlands to the Nile instead of from the Red Sea as it was in the past. With this shift and connected with the river in Africa, the water has been arrived from the sources. Otherwise, the Nile River has dried up and become like other desert valleys such as the Wadi Qena and Assiouti and the Egyptian man was not able to settle down in the valley and not to establish a stable agricultural civilization (Hozaien 1991).

So the author, believe that "Egypt is the Gift of the Nile and Egyptians together". The river passes through several countries starting from the upstream and downstream so. There was no giant civilization influenced in human like ancient civilization. Nile water cannot make a civilization, but it is the Egyptian man, who controls Nile runoff and digging canals and drains. It has monitored the Egyptian priests' stars to calculate the flood, and thus created astronomy. Is there found the people other than the Egyptian have a calendar for four thousand years BC?

The River Nile has two sources of water, which is; plateau tropical lakes and Ethiopian highlands with summer rain. Plateau lakes have five large lakes connected to the Nile. Some of these lakes are great rift (groove) lakes in the center of the west (Edward, George, and Albert), and others are large shallow (downward) lakes such as Victoria and Kyoga (Awad 1998). Lake Victoria (Fig. 3.2) is the largest lake with an area of about 69,000 km^2 (more than double the Delta area and the valley of the River Nile). Lakes Victoria, Kyoga and Albert have its present form because of earth movements that made up the Great African Rift. This is one of the most prominent geographical phenomena on the Earth's surface, where a length of nearly three thousand kilometers, and its depth varies from place to place (Said 1993).

The rivers of the core of the Nile, which originates from the Ethiopian highlands, include Blue, Atbara and Sobat (Fig. 3.2), which stems from the volcanic area. Igneous rocks (basalt) are covered the bulk of the plateau in the Ethiopian highlands.

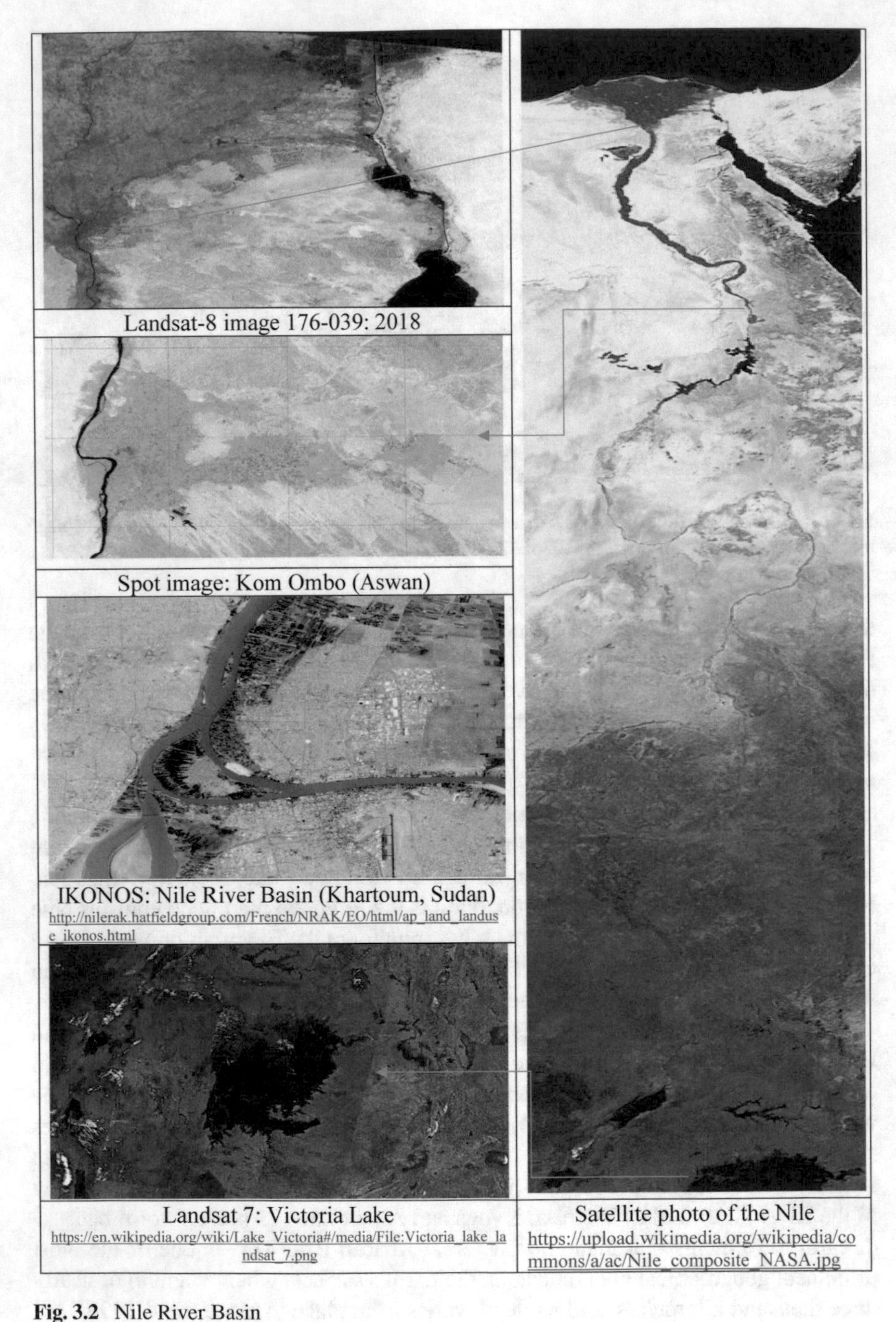

Fig. 3.2 Nile River Basin

The average rainfall is between 1000 and 1400 mm a year. The Ethiopian highlands are considered an important source of most of the Nile water, while the amount of water that reaches the very few plateau lakes, even though it low, however, it has particularly important because they make the river flow sustained throughout the year. The lack of water arriving from tropical plateau was due to the loss of river water in the marshes and the various bodies in the region, where the biggest losses of the water were to the sea. Nile River enters the Egyptian territory near Adindan on the Egyptian-Sudanese border and holds the river for a distance of about 1536 km until emptying into the Mediterranean Sea. The total length of the river from Aswan to Ras Delta at Cairo is about 981 km.

3.1.2 *The Topography of Egypt*

Egypt comprises four main geographical units (Said 1990): The Nile Valley and the Delta; the Western Desert; the Eastern Desert; and the Sinai Peninsula.

The altitude of Egypt ranges from 133 m below sea level in the Libyan Desert to 2629 m above in the Sinai Peninsula. The Nile Delta is a broad, alluvial land, sloping to the sea for some 160 km, with a 250-km maritime front between Alexandria and Port Said. South of Cairo, most of the country (known as Upper Egypt) is a tableland rising to some 460 m. The narrow valley of the Nile is enclosed by cliffs as high as 550 m as the river flows about 981 km from Aswan to Cairo. A series of cascades and rapids at Aswan, known as the First Cataract (the other cataracts are in the Sudan), forms a barrier to movement upstream.

The country's bulk is covered by the Sahara, normally called the Libyan Desert north of Aswan. The Arab Desert extends east of the Nile to the Red Sea. The Western Desert is made up of low-lying dunes of sand and many depressions.

The landscape is dotted by Kharga, Siwa, Farafra, Bahariya and other large oases; another lowland, the Qattara Depression, is an inhospitable region of highly saline lakes and soils covering some 23,000 km^2. The Nile River, on which human existence depends, is the exceptional topographic feature because its annual floods provide the water needed for agriculture. The floods, which generally lasted from August to December, caused the river level to rise about 5 m before the completion of the Aswan High Dam in 1970. However, floodwaters can now be collected, allowing for year-round irrigation and reclaiming approximately 1 million feddans (approximately 1.04 million acres) of land. Damming the Nile resulted in Lake Nasser being created, a reservoir 292 km long and 9–18 km wide (WorldFish 2018).

The Nile is about 1530 km lies within Egypt. On entering Egypt from the Sudan a little north of Wadi Haifa (Fig. 3.3), the Nile flows more than 300 km in a narrow valley, with cliffs of sandstone and granite on both its eastern and western sides before reaching the First Cataract, about 7 km upstream of Aswan. North of Aswan the Nile Valley broadens and the flat strips of cultivated land, extending between the river and the cliffs bounding the valley on both sides, gradually increase in width

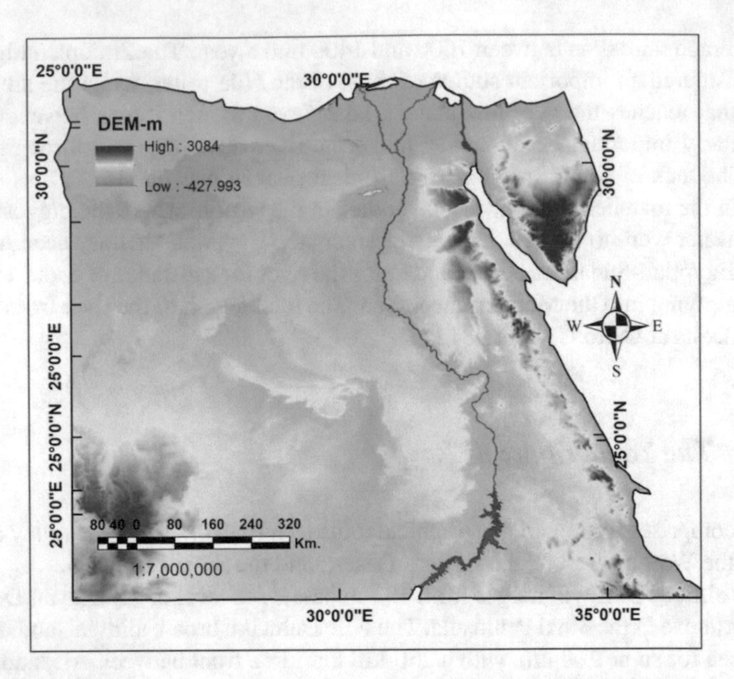

Fig. 3.3 Digital elevation model of Egypt showing major flats, uplands, and depressions in the country

northwards. The average width of the flat alluvial floor of the Nile Valley between Aswan and Cairo is about 10 km and that of the river itself about 0.75 km.

Closely connected with the River Nile is the El-Faiyum Depression (c. 1700 km^2) which lies a little to the west of the Nile Valley and to which it is connected by a narrow channel through the distant hills. The lower part of the depression (25 m below sea level, 200 km^2) is occupied by a shallow saline lake called Lake (Birket) Qarun. The depression floor slopes downward to the lake in a northwesterly direction from about 23 m above sea level. It is a rich alluvial land irrigated by the Bahr Yusuf Canal that enters it from the Nile. After passing Cairo, the Nile takes a north-westerly direction for some 20 km, then divides into two branches, each of which meanders separately through the delta to the sea: the western branch (c. 239 km long) reaches the Mediterranean Sea at Rosetta and the eastern one (c. 245 km long) at Damietta.

Eastern Desert (c. 223,000 km^2) extends eastwards from the Nile Valley to the Red Sea. It consists essentially of a great backbone of high mountains more or less parallel to the Red Sea. It is dissected by deeply incised valleys (wadis), some of which drain westward to the Nile and others eastward to the Red Sea. The desert region east of the Nile's topographic characteristics vary from those to the west of the Nile. The Eastern Desert is fairly mountainous. The elevation increases sharply from the Nile, and a downward-sloping sandy plateau gives way to arid, defoliated, rocky hills running north and south between the Sudan border and the Delta within 100 km. The hills are more than 1900 m high. The most important feature of the region is the rugged mountain eastern chain, the Red Sea Hills, which expands east from the Nile Valley to the Suez Gulf and the Red Sea. This extremely high region has a natural drainage pattern that, due to inadequate rainfall, rarely works. It also has an irregular, sharply cut wadis complex that stretches westward toward the Nile. The desert environment stretches to the Red Sea coast as far as possible. Sinai Peninsula (c. 61,000 km^2) is separated from the Eastern Desert by the Gulf of Suez. It is a complex of high mountains intensely dissected by deep canyon-like wadis draining to the Gulf of Suez, the Gulf of Aqaba and to the Mediterranean Sea.

Western Desert stretches westward from the Nile Valley to the border of Libya with an area (exclusive of El-Faiyum) of some 681,000 km^2, more than two thirds of that of the whole of Egypt. Its surface is for the most part composed of bare rocky plateau and high-lying stony and sandy plains with few distant drainage lines. True mountains are to be seen only in the extreme southwestern part where the highest peak of Gebel El-Uweinat is 1907 m. In northern and central parts of the Western Desert, the plateau surface is broken at intervals by great depressions and oases.

The topography of Egypt as showed in Fig. 3.3 reveals that agricultural expansion, after securing irrigation water, could be oriented mainly toward the Western Desert, which has several depressions such as Qattara Depression along the northern third of this desert.

3.1.3 The Egypt's Agricultural Lands

The satellite image of Egypt (Fig. 3.4) shows two characteristics of the country: the proportion of the country that is comprised of the Desert and the location of the Nile River. The land of Egypt was covered by the sea from the north since about 60 million years ago (before the arrive Nile). It covered a large part of the north of Sudan. The sea began to retreat until its beach became the line running between El-Faiyum and Siwa, and thereafter seaside has become close to the current position. Egypt with its lands is endowed with varied agro-ecological zones, which represents the greater majority of cultivated lands of the Nile Valley, as well as, most of the reclaimed desert lands, mainly, on the western and eastern fringes of the Delta in addition to relatively limited areas at on fringes of the Valley in Upper Egypt (total areas over 7.5 million feddans).

Actually, there is no periodic census for agricultural land in Egypt. The last formal agricultural census was carried out in early 1960s; however, inventory of cultivated area is carried out by traditional methods of surveying. In addition, urban encroachment upon cultivated land makes it difficult to make an accurate inventory by traditional surveying. The area of cultivated land has been controlled by the amount of water allowed to Egypt to discharge from the Lake Nasser through the High Dam (55.5×10^9 m^3) after the 1959 Egyptian-Sudan Treaty of sharing Nile water. At that time population in Egypt was about 26 million and the total area of cultivated land counted about 6 million feddan (Abu Zeid 1993) (1 feddan equals 1.038 acre). The current population of Egypt is 100,349,475 as of January 17, 2019, based on the

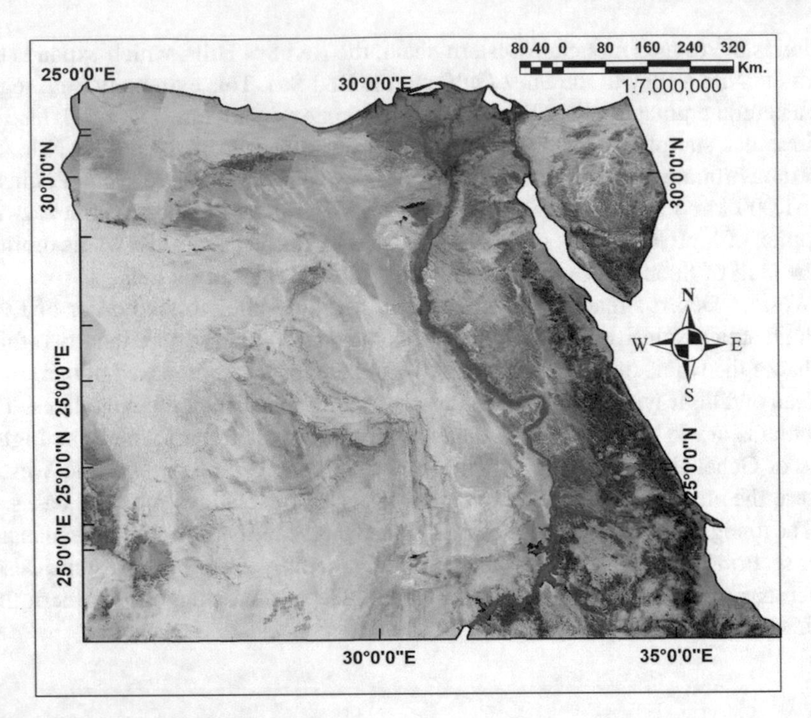

Fig. 3.4 Agricultural area (green color) of Egypt at January 2019 compiled from 4 MODIS Aqua Enhanced Vegetation Index (EVI)

latest United Nations estimates (https://www.worldometers.info/world-population/egypt-population/). The water allowance to Egypt has been the same since 1959. This country equals one million square km and has its population living in less than 5% of the total area.

One of the most successful applications of remote sensing is in the agronomic field (Omran 2016). The availability of at least four decades of digital data in multiple wavebands of the spectrum (visible, near infrared, and thermal bands) and their large ground coverage makes remote sensing superior to field-based studies. The premise for using digital data in monitoring agricultural land is based upon the unique interaction of vegetation biomass with solar electromagnetic radiation, which differs from other land cover components, e.g. water and bare deserts. The chlorophyll of the green leaf strongly absorbs the red radiation (630–690 nm), whereas the leaf's cellular structure strongly reflects the near infrared radiation (760–900 nm).

Vegetation indices are quantitative surrogates of the vegetation vigor (Omran 2018). They are mathematical models enhancing the vegetation signal for a given pixel. Most vegetation indices were calculated from the reflection in two bands of the spectrum: the visible red and near infrared reflection (Aronoff 2005). These indices were utilized effectively in long term monitoring of vegetation in semi-arid.

MODIS products have been widely used quantitatively for time series analysis of vegetation dynamics (Beck 2006). Satellite images with moderate spatial resolution and wide geographic coverage (MODIS Aqua data) helped in the estimation of Egypt's entire agricultural land. The country, which is situated entirely in the arid region exclusively, relies on irrigated agriculture. MODIS is a recent generation of satellites, which are a part of NASA's Earth Observing System that offers considerable potential for rapid, repetitive and a near-daily global coverage of Earth's resources. Most recent land cover mapping studies has operated the enhanced vegetation index (EVI) data from MODIS instead of the most traditional NDVI data because of atmospheric and canopy background corrections incorporated into EVI's calculation (Hereher 2013; Wardlow and Egbert 2010).

This chapter was utilized MODIS Aqua EVI instead of NDVI data for the inventory and mapping the current agricultural land in Egypt. EVI was the algorithm operated to estimate the vegetated area of the country. Time series analysis of satellite data revealed the vigor pattern of cultivated land. Figure 3.4 shows a mosaic image of Egypt on 17th January 2019 compiled from 4 MODIS tiles. Agricultural area (green color) of Egypt on (Fig. 3.4) 17th January 2019 is estimated at 10.15 million Acres (4.09% of Egypt area).

The ancient Egyptian civilization was placed on the banks of the Nile River. Most of agricultural areas are concentrated in the Delta and Valley (Fig. 3.5). The majority of the planting areas were located west of the Nile. While the cultivated areas east of the Nile were much lower than in the west Bank of the Nile (Fig. 3.5). The river between Aswan and Cairo always has a tendency to adhere to the right side of the valley. It does not rotate to the left side of the river until it reverts to the right side. This phenomenon is not clear in Qena, where the river runs away from east to west. However, it is apparent from a distance in particular between Manfalut (Asyut) and north of Cairo. In Minya and Beni Suef, for example, the sedimentary plain is situated on the left side. However, on the right side, the river is hardly separated from the desert. This implies that the river through its sediment on the left side and dissolves in the carving of a small part of the right side so that the plain on the left. On the right side, there is a cliff and a desert plateau.

This phenomenon is also not clear in Kom Ombo, which is extended on the east side of the Nile. Kom Ombo is an agricultural town (Fig. 3.5), producing mostly irrigated sugar cane and corn, and unremarkable but for the unusual double temple of Ptolemaic date situated picturesquely high on its banks above the river Nile. The town has ancient origins, of which virtually nothing beyond the temple is to be seen today (awaiting excavation!).

3.1.4 Landsat Surface Temperature of Egypt

Land Surface Temperature (LST) is the temperature of the earth's ground that is in direct contact with the measuring instrument (generally measured in Kelvin). LST is the surface temperature of the Earth's crust that absorbs, reflects and refracts the

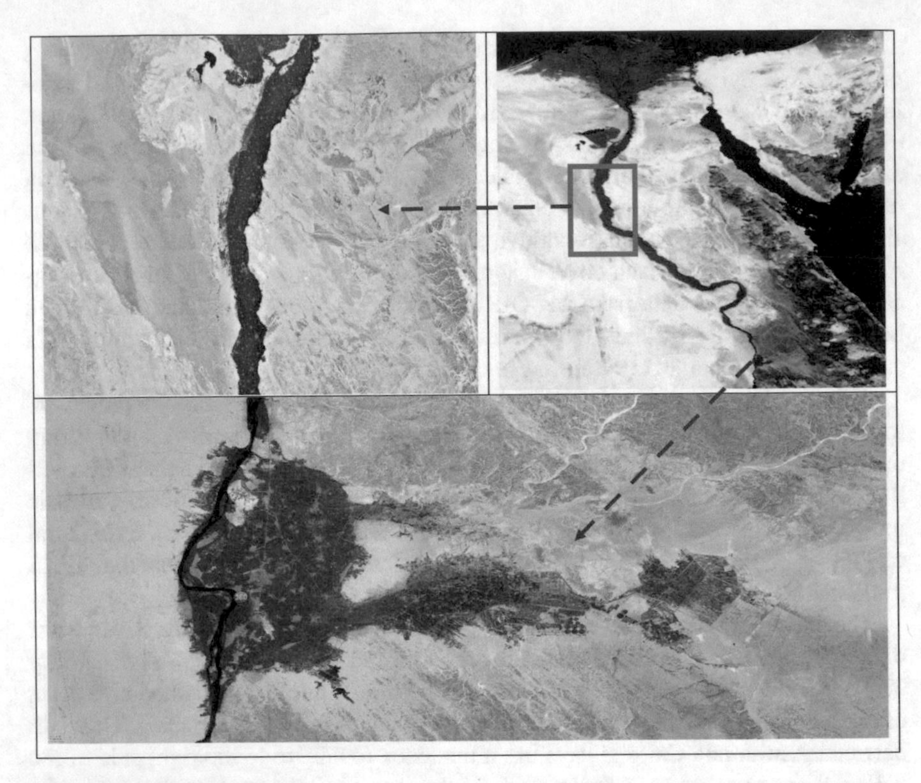

Fig. 3.5 Satellite image (Sentinel-3A satellite and google map) shows the majority of the planting areas were located west of the Nile. The river between Aswan and Cairo always tends to adhere to the right side of the valley

heat and radiation from the sun. LST changes with climate change and other human activities where predicting exactly becomes difficult. Global urbanization has risen considerably in greenhouse gases and reshaped the landscape, which has significant climatic consequences across all scales as a result of the concurrent conversion of natural land cover and the introduction of urban products, i.e. anthropogenic surfaces.

Land surface temperature (LST) is an significant variable in many fields such as climate change, urban land use/land cover, studies of heat equilibrium and a main input for climate models. Provisional surface temperature for energy equilibrium and hydrological modeling is an significant parameter. LST information are also helpful for tracking crop and vegetation health, as well as extreme heat occurrences such as natural disasters (e.g. volcanic eruptions, wild fires) and surveys of urban heat islands. The Landsat Provisional Surface Temperature is processed to 30-m spatial resolution in projection using the World Geodetic System 1984 (WGS84) datum (Fig. 3.6).

In the Eastern Desert, LST is generally higher than in the Western Desert. This can be attributed to the lithology of the bedrocks in these regions, where crystalline low-albedo basement rocks make up the majority of the Eastern Desert and absorb

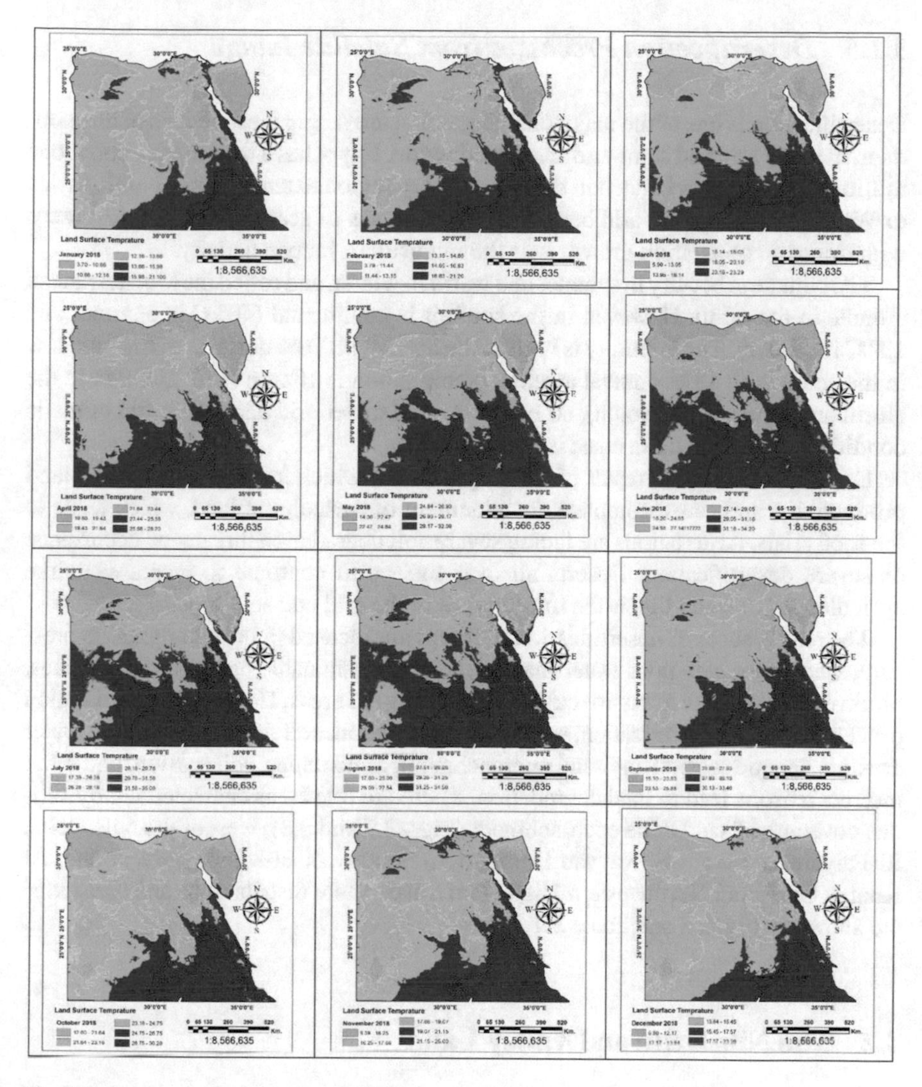

Fig. 3.6 Landsat surface temperature of Egypt from January to December 2018

a great deal of solar energy, while the Western Desert, covered by sedimentary successions of sandstones and calcareous rocks of higher albedo, suggests lower LST. The depressions within the Western Desert are also noted to have greater LST than the adjacent lands. This is evident in the depressions of Dakhla, Kharga, Toshka and Qattara. The maximum LST in Egypt coincides with the latitudinal position from north to south (Fig. 3.6).

3.1.5 *Desertification Processes from Satellite Image*

Desertification is one of the major challenges that are facing the activities of development in many arid and semi-arid areas worldwide. Egypt has a total area of about one million km^2. Egypt is located in the severely dry region extended from North Africa to West Asia, where the arid belt, under the stresses of scarcity of water resources with dramatic population growth thus the intensity of human activity.

The climate is hot-dry in summer and warm in winter, and called the Mediterranean climate in the north. However, in the south it is continental (40° C in summer and 13° C in winter). The humidity is high in the north with 70% during summer and low in the south with 13%. Annual average precipitation is 10 mm (150–200 mm in the North and 2 mm in the South), so Egypt located under arid and hyper arid climatic conditions. The Nile delta coast is about 300 km.

Desertification is the result of soil degradation, which leads to a decline in land productivity, and cause complete abandonment of agricultural land, which leads to the food crisis. Arid regions are facing severe soil degradation threatened occurrence of severe desertification. Deserts all over the world continue to increase, unlike agricultural land, and this is the most severe in the arid and semi-arid.

The main causes of desertification in Egypt are focused in the demographic pressure, water shortage, poor water management, unsustainable agricultural practices, biodiversity loss, and intensive cultivation in rain-fed areas. However, desertification processes include urbanization, salinization, pollution, soil fertility depletion, water erosion, genetic erosion and sand encroachment. Sand dunes, soil erosion, and other indirect reasons lead to land degradation. There are five areas characterized by serious coverage of sand dune encroachment (Figs. 3.7 and 3.8); west of the Nile Delta, El-Faiyum and Wadi El-Rayyan Depression, Southwest El-Minya governorate, Al Kharga Oasis and Northwestern High Dam Lake. There distributions and areas km^2 are shown in Fig. 3.8 and Table 3.1.

3.2 The Nile Delta and Valley Land

The Nile Valley system extends from the Mediterranean shores of the Nile Delta to the north till Aswan in the south over an area extending from 22° to 32° latitude north under arid and hyperarid conditions.

Egyptian Nile River basin includes of Delta and Valley, El-Faiyum, and Lake Nasser. Fluvial sediment and soil that cover the Delta and Valley and some parts of El-Faiyum are the result of the annual precipitation for the silt from African headwaters of the Nile before the construction of the High Dam. Fluvial sedimentary soils of El-Faiyum was considered of the Egyptian Nile River Basin, due to contact El-Faiyum with Nile River by Joseph Sea. As for the area of Lake Nasser, the situation is different. It has been put the Lake Nasser area within the Egyptian Nile River basin as a geographic and not because the soils are sedimentary, because most soils

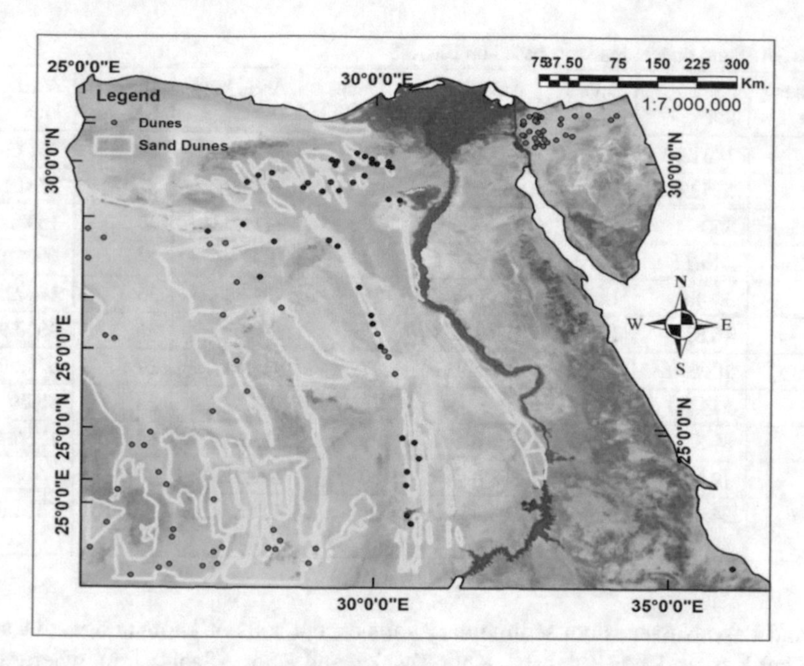

Fig. 3.7 A map of sand dunes sites in the Western Desert

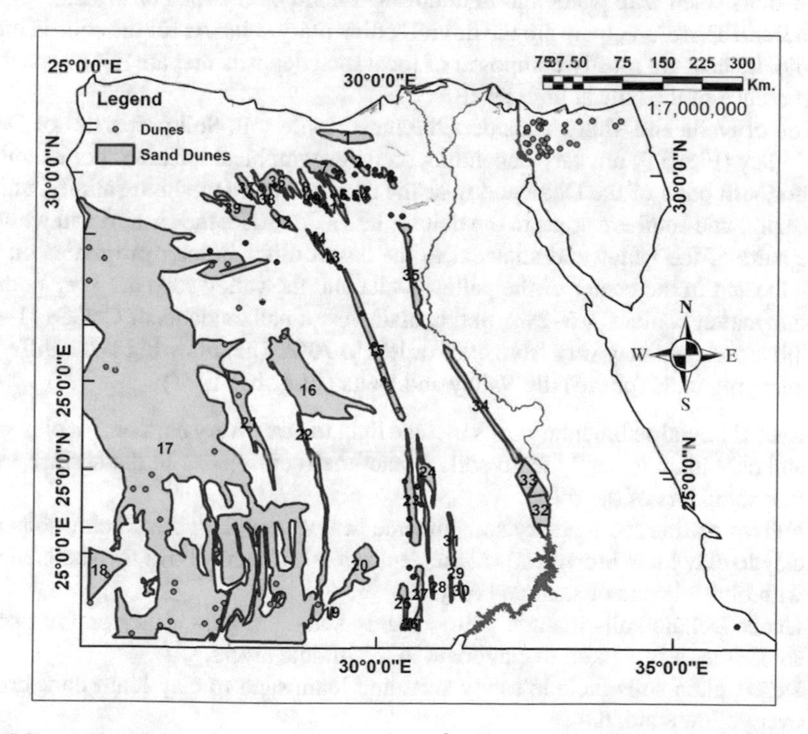

Fig. 3.8 A map of sand dunes locations and area km^2

Table 3.1 Sand dunes area km^2 based on Fig. 3.8

Dunes NO	Area, km^2	Dunes NO	Area, km^2	Dunes NO	Area, km^2	Dunes NO	Area, km^2
1	1431.50	11	145.84	21	3203.62	31	561.09
2	535.84	12	562.85	22	308.09	32	1731.61
3	350.21	13	55.20	23	2111.97	33	1892.22
4	246.74	14	326.46	24	1999.54	34	1992.19
5	34.40	15	2935.67	25	704.39	35	1447.00
6	46.89	16	10,013.54	26	83.20	36	2036.99
7	2096.46	17	162,538.10	27	921.44	37	679.50
8	574.37	18	3085.93	28	198.37	38	85.69
9	89.30	19	60.16	29	173.63	39	1766.40
10	79.44	20	2587.76	30	189.92		
Min	34.40 km^2	Max	162,538.096 km^2	Sum	209,883.52 km^2	Mean	5381.63

around Lake Nasser is not sedimentary soils except soils of khors and wadis, such as Khor Karkar, Khor Kalabsha, Khor Toshka, and Khor Allaqi … etc. where were these khors filled with water and mud during a flood before the construction of the High Dam. Therefore, there are the fluvial sedimentary soils. As for the soils in areas around the lake are mostly composed of local rock deposits that are not affected by the presence of the Nile at high levels.

Soil of Delta and Valley considers the most fertile soil. Soil properties of Delta and Valley (Fig. 3.9) are vary depending on its geographical location. For example, the northern parts of the Delta and near the northern lakes are high salinity, unlike the central and southern areas in the delta is less salty. Also, the eastern and western edge parts of the valley and adjacent to the desert differ in their properties on the parts located in the center of the valley. Delta and the valley soils are very poor in organic matter content (0.5–2%), and contain very small amounts of $CaCO_3$ (1–4% or a bit more), and clay vary from 40% or less to 70%. The following is the different soil mapping units for the Nile Valley and Delta (El-Nahal 1977):

1. Recent fluvial sedimentary soils include light to heavy very dark brown clay soils and clay loam to sandy loam soils, sometimes-coarse sand in the surface layer (the shoulders of the river).
2. Fulivio marine sedimentary soils include heavy, very dark brown clay soils and clay to clay loam brown soils at the depth of 50–120 cm above the shells mixed with bluish layers of sand and clay.
3. Under Deltaic soils include yellow sandy soils and gray sandy or faint soils, sometimes sandy loam to clay loam in the middle layers.
4. Desert plain soils include sandy soils and loam sand to clay loam dark brown over yellow sand, flat.
5. River terraces soils include faint sandy limestone soils mixed with pebble and gravel of various sizes and undulating surface.

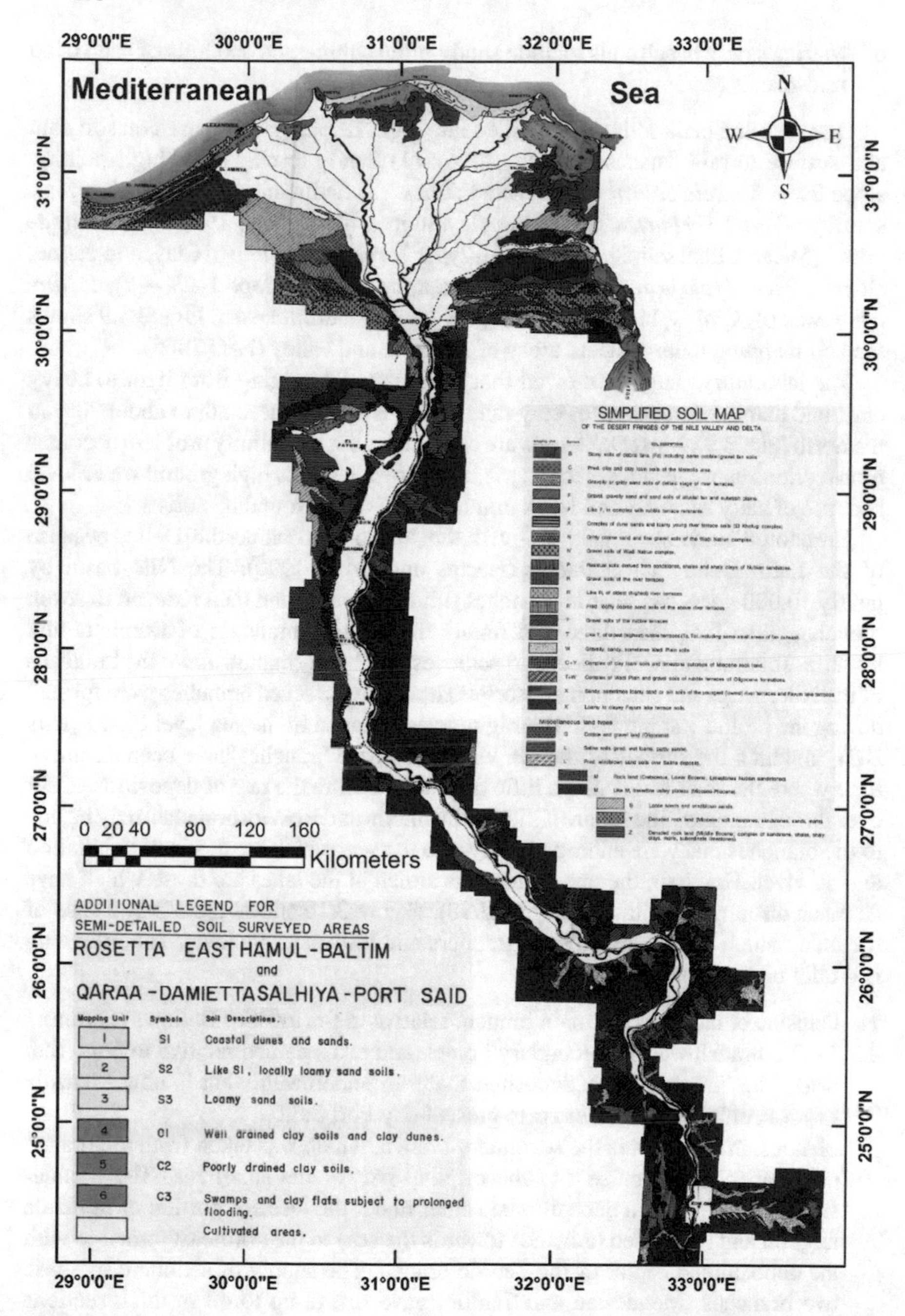

Fig. 3.9 Simplified soil mapping units adjacent areas of the delta and valley (FAO 1965)

6. Marine sandy beach soils include sandy often submerged and hills of sand deep mid-rise.

However, the delta soils were divided into eight mapping units based on soil data and surface terrain database (Rahim 2006): (1) Sandy to clay soils, high salinity, slope 2%—*Aquolic Salorthids*. (2) Sand dunes. (3) Sedimentary marine, clay, high salinity—*Vertic Ustifluvents*. (4) Fine clay, high salinity, slope 1%—*Typic Ustifluvents*. (5) Sand, high salinity, slope 2%—*Typic Torripsamments*. (6) Clay, non-saline, slope 1–2%—*Typic Torrerts*. (7) Sandy loam, non-saline, slope 1–2%—*Typic Torrifluvents*. (8) Clayey, high salinity, slope 2%—*Typic Salitorrents*. Figure 3.9 shows the soil mapping units adjacent areas of the delta and valley (FAO 1965).

The laboratory analyzes showed that the Delta soil ranging from light to heavy clay, and that the percentage of clay ranging from 40% in the south to about 70% in the north (Fig. 3.9). Most Delta soils are non-saline, but the salinity problem becomes more serious as we get closer to the sea, lakes, beach due to high ground water level because of salty seawater and lakes nominated in the surrounding soils.

Herodotus since about 500 BC was the first one to name the Delta, because of the Latin Delta (▼) character (Harms and Wray 1990). The Nile basin by nearly 10,000 years has several branches (nine branches) and then reduced to seven branches, then five, then three and finally the existing branches of Damietta and Rosetta. Toussoun (1922) collect and achieves much information about the branches of the Nile. Maps and ancient manuscripts show that the seven branches were formed during the period leading up to the big increase happen in the sea level (5000 years BC), in which the surface of the sea was low. These branches have been silting in time where the river was acting a little bit, and therefore the rate of deposition of silt over these branches. The failure silting of Damietta and Rosetta branches may be due to the branches that were ending directly into the sea, which are destined to inflation and survival. However, the one which was aimed at the lakes are those which have as much atrophy and silting (Shahin 1978). Figure 3.10 shows the old branches of the Nile, which were drawn by geographers and historians ancients. The following is a brief on these branches:

1. Pelusiac branch was the main branch, relative to the town of Bellows (Farma).
2. Tanitic branch was the secondary branch and takes named relative to Sais (Sun Sea). The Tanitic branch deposited material underneath what is now Manzala Lagoon, and prograded across to present day Port Said.
3. Mendesian branch was the secondary branch, which was taken from the branch of Alsabenity its course and the original part of the small sea. The Mendesian branch formed a depositional center under the western portion of Manzala Lagoon and prograded (advance towards the sea) to the northeast to merge with the depositional center of the Tanitic branch. The supply of sediment by these two branches (Mendesian and Tanitic) gave rise to up to 40 m thick sections of Holocene deposits, currently underlying Manzala Lagoon. Around 1500 to 2500 years ago. These two branches silted up allowing the Damietta branch to become the largest contributor of Nile sediment flux (Stanley and Warne 1993).

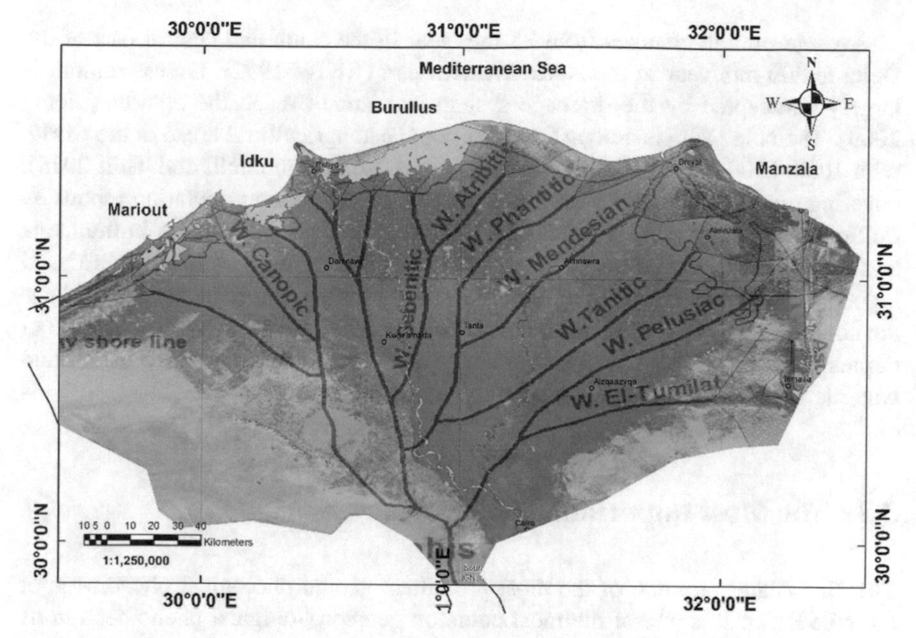

Fig. 3.10 Old branches and historic cities of the Delta (Said 1993)

4. Phatnitic branch was the secondary branch of Sebennytic, which was taking its course in line with the lower part of the Damietta branch.
5. Sebennytic branch was the main branch in the center of the Delta, and starts at the top Delta and ends at Borollos tower and its name relative to Spenatios (Samanoud).
6. Bolbitine branch was the secondary branch, which was subdivided from canopic near Damanhur being in the lower part of the Rosetta branch.
7. Canopic branch was the third main western branch and was ended at Canopus (Abu Qir).

Nile Delta starts in branching north of the City of Cairo (Fig. 3.10). The river is divided to Rashid (239 km) and Damietta (245 km) branches. The average width of the Rosetta branch 500 m, while the Damietta branch of about 270 m. Damietta branch and its drained water much lower than the discharge of the Rosetta branch, which discharges water up to one and half the Damietta branch. Rosetta is erosional branch, while the Damietta is deposional branch (Hamdan 1984). Environmental conditions have helped on the speed of the Nile Delta composition. These factors are leveling the surface of the deposition area, clay aggregates, which facilitates deposition by salt in seawater, and finally the lack influenced the Mediterranean Sea coast marine currents because it is one of the semi-enclosed seas. The length of the Delta from the south to the north (about 170 km), while the base up to about 220 km, an area of about 22,000 km^2 (Fig. 3.10), twice the area of the valley.

Average rainfall changes from 25 mm/year in the south and central part of the Delta to 200 mm/year in the North Western part (RIGW 1992). Excess rainfall is largely intercepted by the drainage system and cannot reach the aquifer (Morsy 2009). The Nile Delta is among the most populated agricultural areas in the world, with 1080–1500 inhabitants per km^2 (Aquastat 2013; Tamburelli and Thill 2013), covering nine governments with about 43% of the Egyptian population (about 39 million inhabitants) (CAPMAS 2016). The 'old' cultivated lands are in the Delta and 'new' commonly reclaimed desert in the east and the west of the Delta. Coastal lagoons and lake (Manzala, Borollus, Edku and Mariut), salt marshes and sabkhas are comprised the northern part of the Nile Delta. They are usually a low surface of depression composed by the removal of dry and loose material down to the ground water level or the capillary water zone (Al-Agha et al. 2015).

3.3 The Most Important Nile Islands

The Nile islands are one of the most prominent geomorphological phenomena of the Nile River. It is one of the most common geomorphological phenomena in its dimensions and movement. This is due to the natural factors of water flow and associated changes in the load of carving, transport and sedimentation, as well as the human factors, which are the establishment of dams, arches (Al Qanater) and stone heads on the river.

Islands of the Nile from Geomorphological features are considered distincting more in the river Nile in Egypt. It amounts 507 islands (Table 3.2, Figs. 3.11 and 3.12), which occupies areas 144.05 km^2.

The area from High Dam to Kom Ombo includes a group of islands and barriers which numbered 145 during the decline of the river at its lowest level in March, when excluding the islands less than 10 m^2, ranging from rocky islands such as

Table 3.2 The most important Nile Islands

Profile	Length, km	Islands number	% from total Islands
From High Dam to Esna Dam	160	49	9.7
From Esna Dam to Naga Hammadi Dam	190	78	15.4
From Naga Hammadi Dam to Assiut Dam	180	109	21.5
From Assiut Dam to Mina	140	84	16.6
From Mina to Beni Suef	120	103	20.3
From Beni Suef to Delta	141	53	10.4
Rosetta Branch	215	15	3
Damietta Branch	213	16	3.1
Total	**1359**	**507**	**100**

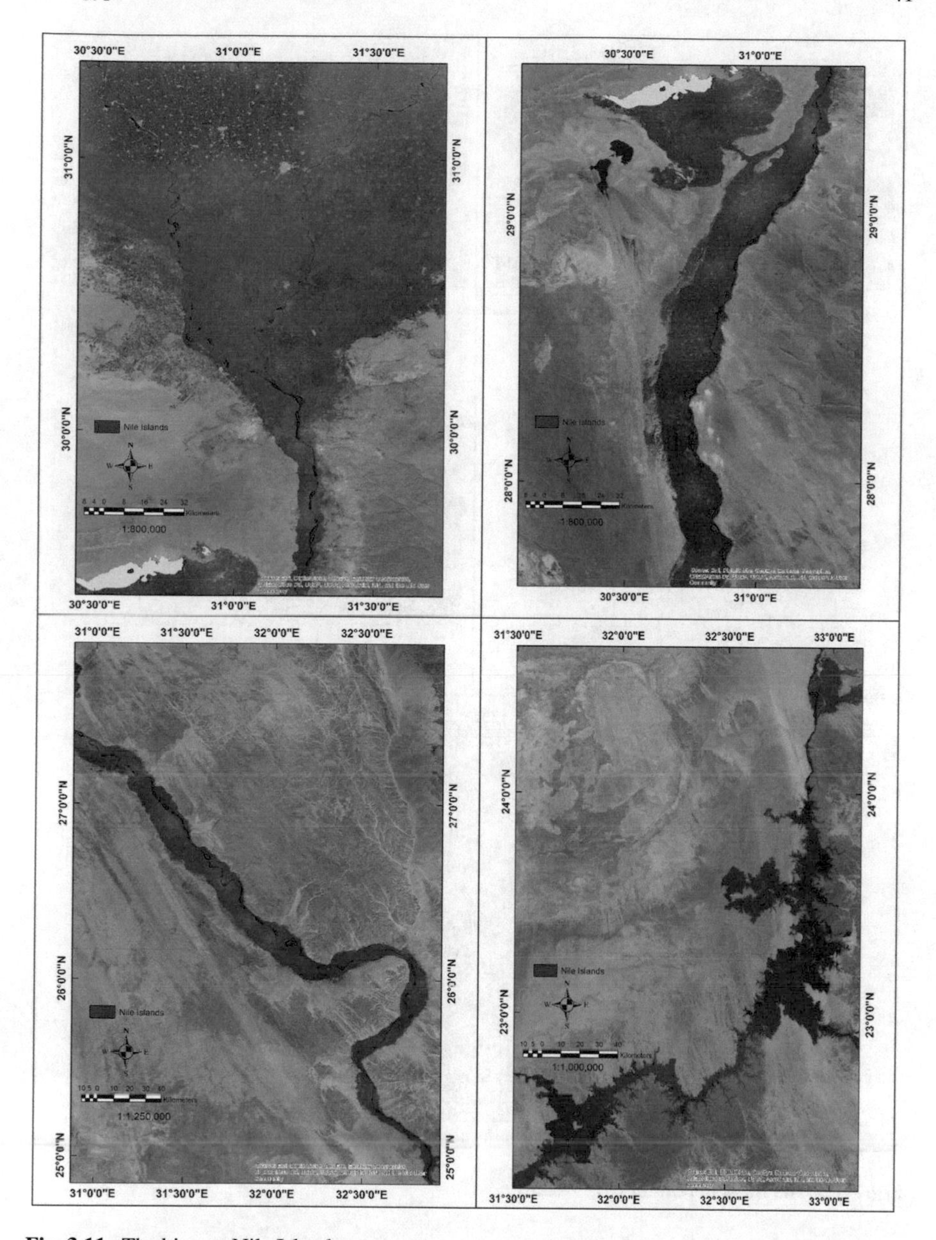

Fig. 3.11 The biggest Nile Islands

Fig. 3.12 The most important Nile Islands

the islands of Lake Elephants Al-Heisa—Philae—Agilika—Awad and islands north reservoir such as International Garden—El hips—Sohail and others. Sedimentary islands such as Al tiwasa—Mansuriya and other islands. Rocky-sedimentary islands, such as the island of Aswan, Kitchener. The populated islands reached only 8 Islands, and gaining the region the importance of a major tourist attraction which includes many of the monuments and nature reserves and charming landscape. In addition to natural hazards resulting from human activities ranging from projects set, the river to the character built on these islands, many of the environmental problems, which limits the development and utilization of optimal exploitation.

The island of Mansouriya is one of the largest Nile islands in Egypt in the area and islands rock/sedimentary such as Aswan Island and the Bharif. The region is gaining significant tourist importance, which includes many of the Monuments and nature reserves such as the Salouga Reserve, Ghazal and the Plant Island, which are predominant agriculture and related activities. There are only 5 islands, namely the Beja, Hissa and Awwad Islands, located in Lake Philae. Sohail Islands, Aswan and Amun from the rocky islands, and the Mansouriya and Harbiya Islands from the sedimentary islands. Human activities vary from island to island depending on the environmental potential of nature, and natural obstacles limit the human exploitation of other islands.

The island's soil varies in texture and classified to three types namely: Sandy loamy; Sandy clay; Sandy. The similarity in the soil texture of the sedimentary islands is due to the similarities in the conditions of their origin, where they formed from the silt transferred with the floodwater over the years. In the rocky islands, a new factor is also introduced: mechanical weathering of the granite rocks.

3.3.1 Sedimentary Islands

This sedimentary Island group (Figs. 3.11 and 3.12) includes the island of Mansuriya and the southern islands group up to the island of Kubaniya. It is clear from the topographic maps and satellite images of the islands that the importance of curvature of the river and the turns and dry valleys on the banks in the emergence of many islands, where the river is free of big islands distance of 13.9 km from South of the warp until the north of the Kubaniya, which may also return in its inception to the dumping of sediments at the mouth of the valley of the Kubaniya in the west, and then become the river free of the islands to the island of Bahreif for a distance of 6 km.

After Bahreif Island, river appears empty until reaches Rocky islands for a distance of 7 km. The factors influencing in the Islands genesis are that slow stream that led to increase the area of the Islands. The slower the current led to the increase in the area of the islands, and the more uneven the width of the channel, the higher the speed of the current. In a reverse relationship, the number of islands depends on the area of the aqueduct in an inverse relationship and the numerical abundance of the islands depends on the coefficient of the shape of the channel (depth/width) (Taha 1997). Chorley (1973) shows the slope of the water surface decreases towards the downstream and thus increases the shearing.

If there is one or some of the deep barriers, especially at the sharp turns, the high barriers towards the estuary appear after the turn and divide the river into two. The rivers, which are divided is less efficiency than the river, which is replaced by the smaller rivers. The slope becomes deeper and the barriers show higher and dry, allowing it to grow plants and by repeating the process of fragmentation of the river, the winding river becomes braided. So the braided stream becomes more inclined (although it is equal in water flow) (Chorley 1973).

Thus, the turning point is at the turn of Mansouriya, which is the factor responsible for the formation and expansion of islands followed by side valleys and then the decline and density of portable materials (Taha 1997). All these factors led to calm water flow and lack of efficiency and cannot forget the impact of irrigation projects. The river become unable to transfer the load of the valleys coming from the banks and the load of the river itself. It took the sedimentary component of these islands.

The island of Mansouriya is the largest and most important sedimentary islands in the region. This island is one of the ancient and permanent islands of geomorphology. It has been proven over the past two centuries that there is little variation that may be attributed to any variation in water levels or accuracy of measurement (Taha 1997). Contour map of the island shows that it was inflated and increased due to the fusion of at least two islands together, along with other factors such as the turn of the river where the island was distributed and the speed of the current stream was affected by its horizontal texture.

The sculpted works on the concave side led to the increase of locally carved materials. The carved materials are from the right side to build the southern part of the island, while the carved materials are from the left side to build the northern part of the island. Some of the secondary factors may be responsible for the island's inflation. To the south-east of the island lies the valley of Wadi Kom Ombo, which increases the density of the materials carried by the river stream locally, leading to any increase (the rates of island sedimentation and the amount of the bottom of the island) (Taha 1997).

3.3.2 Sedimentary-Rocky Islands North of the Aswan Reservoir

Rocky islands (Figs. 3.11 and 3.12) are divided into the southern sector of the river, which has 9 major islands such as the island of Aswan and plants island where covered by the Nile sediment so that it could be cultivated. These islands at the beginning of the establishment were rocky, but with the continuation of the river sediment covered with thick soil of silt and worked to inflate and increase the area to enable the people to reclaim. The river still dumps its load, especially after the reservoir and High Dam, which may make the rocky layer completely covered with silt. Braided rivers have more depth and breadth than the curved or other types of river streams.

The gradient velocity from the surface to the bottom is sharp and increases pressure and dredging (**erosion**). The pattern is combined with the river load to sharpen the sharp edges of the barriers at the lower levels of the river. The bottom materials are made from the sand grains or the roughness along with the weakness of the cohesion (the banks increase the breadth and the shallowness) and thus the incarnation or braided (Chorley 1973).

Where braided rivers are usually bristling with sandy banks or easy-to-break materials, and plants play an important role in transport (in the stabilization of sediments on the islands (Leopold and Millerck 1964).

3.3.3 *Rocky Island*

The most famous of which is the island of Hissa, Beja, Awwad, Elephantine, which was the palace of Anas presence (Figs. 3.11 and 3.12). In the north of the dam, Suhail island and Salouga, consist of thick and fine granite and in some layers Mica and Hornbland then the igneous rock known as Syanite. These rocks are also found on both sides of the river (Awad 2005). They are called Gendellic islands where there is a Cataracts in the Nile River from the Sblogha gorge in northern Sudan to the first waterfall south of Aswan. This is due to the modern watercourse. The River managed to carve soft formations to crystalline rock, which was to be a long time so that the river can be removed it.

When rock formations intersect rivers, whether inclined slightly toward the lower parts of the riverbed, or in vertical or semi-vertical positions, a series of cataracts or rapids along the course of the river, as the water of the river descends above or around these formations (Safi al-Din 2001). The rocky islands representing the last cataracts in the Nile River intercept at the end of its journey from the upstream to the estuary, though it is called the first gondola (cataracts). Therefore, it is possible to agree on the name (Aswan waterfall). Its formation is not due to the interception of crystalline and volcanic rocks, despite their existence. It is due to the presence of meandering in the rocks. Its general direction is from north to south. It formed narrow valleys, and between them, there are islands that divide the course into two or more sections in the river, which runs for 12 km.

3.3.4 *Immersion of Islands*

After the city of Aswan, the river is running 1200 km without being intercepted by the jungle, gorge or other obstacles. After Aswan begins to form the fertile sedimentary plain and it seems that this narrow and then suddenly widens in Kom Ombo. This phenomenon called belly cow, where the flood plane more than 35-km^2. This is due to an old dry lake, where the valley of Shu'ayt and Wadi Al-Khayrat were located. But it narrows again and expands again at Edfu and continues to expand even to Qena (Awad 2005).

The flood plain is flat and wide areas on both sides of the river aqueduct, where active sedimentation occurs on both sides of the river during aging and during the flood. When this happens in the soils known as Natural Levees, the coarse material is deposited in the nearest area. From the river, the river is often flooded so that its water can be applied to these bridges and the flood plain drowns and leaves its deposits on its surface (Mahsoub 1996).

The islands are considered one of the most sensitive places to change the level of water. By its location, it is the first inhabited islands that receive floodwater coming from the south after the construction of the High Dam and the displacement of the villages Nubia. The flood comes in the summer to sink them all and destroy them. This is in addition to the modern islands that emerged after the construction of the dams and are still growing and immersion in spite of its large area, due to the deposition of silt on the aspects or border more than the vertical deposition, which raises the water level. Although some of the islands were severely damaged by the construction of the High Dam, where they were blocked by floodwaters, there are other islands that have emerged and flourished in life, such as Salouga and Ghazal, which was completely submerged before the construction of the High Dam and other islands.

Another problem facing these islands now is the rise in water level when the discharge of the surplus of the reservoir in Lake Nasser. At the start of the water year comes the flood. If the water level continues to rise, and exceeded the level of vision, the height of 178 m in Lake Nasser, the Ministry of Water Resources and Irrigation begins after considering the possibility of discharging excess water to maintain the safety of the High Dam. The Committee of the River Revenue tries to increase the discharge from the High Dam to find a balance between providing the needs and washing the Nile River, while maintaining the safe levels in front of the High Dam. A drainage program adapted to the need of safe water, and with the efficiency of the watercourse, including the possibility of sinking some of the low areas that are planted by the violation, in addition to the existence of more than 150 small islands within the Nile River. Low levels subject to water covering annually during the season of maximum need. While 144 other large islands that are not covering up to 162 billion cubic meters in the High Dam Lake.

3.4 The Major Wadis of Egypt

Even though the country has many major wadis (Fig. 3.13), the Nile is the only continuous river in Egypt. Some dry tributaries or wadis intercept the Nile as it crosses the deserts of the East. The wadis drain their water to the coast of the Red Sea. A wadi is an Arabic term that describes a watercourse in North Africa and Arabia which has a dry bed except when it rains and often forms an oasis. The watercourse may be a channel, a stream, a valley, or simply a course followed by water throughout rainfall periods. Typically, Wadis are situated on the slightly sloping and almost flat desert areas. They start in the fans' distal portions and extend to the Sabkhas or playas inland. They're a trend along a fan terminus' basin axes. Due to lack of continuous water flow, there are no continuous channels. Wadis tend to show a braided-stream pattern due to water deficiency and sediment abundance. The wadis have water surfaces that are intermittent or ephemeral. Generally sudden and rare

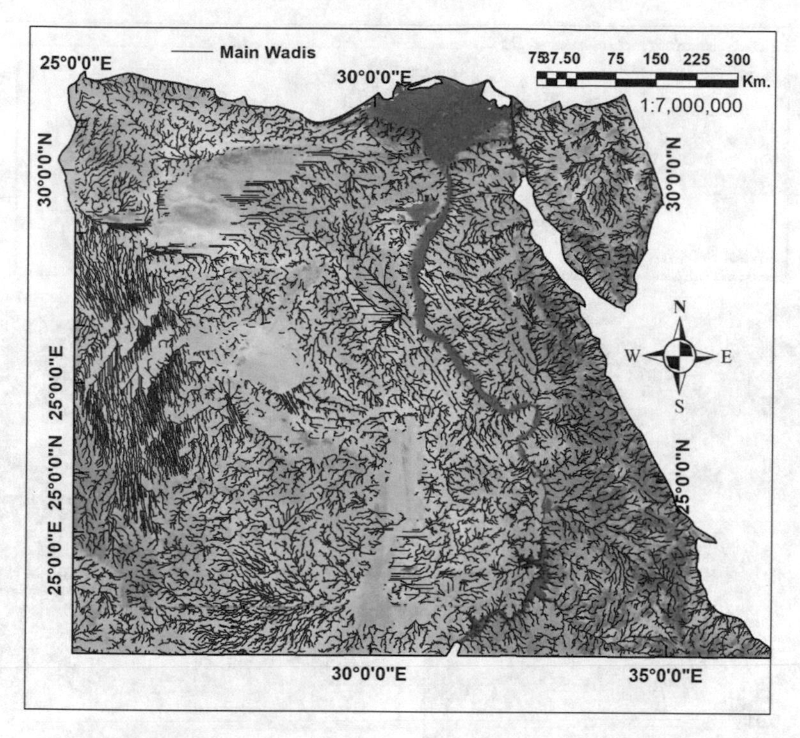

Fig. 3.13 The Major Wadis of Egypt

heavy rainfall results in flash floods in the deserts. This water percolates deep into the streamed, leading to energy loss and massive deposits. Furthermore, the sudden loss of flow velocity and rainwater flow into the porous sediment contributes to a rapid drying up of the watercourse. The channels will flow to the Red Sea until the rain stops if the rains are simultaneous. During the dry season, the watercourses dry up, and only the oasis remains.

From an economic point of view, Wadis and their oases supply people and animal populations with habitats. Settlements emerge around a wadi that has an oasis. Agriculture begins as people settle down. For example, the Feiran Wadi (Fig. 3.14) attracts thousands of visitors because of the huge dates harvests which grow on its course. Other crops are wheat, barley, vineyards and palms. The Wadis are also valuable Egyptians' trade and transportation routes. Throughout their migration, caravans and nomadic people follow Wadis to recharge supplies of water and food once they attain a wadi oasis. Wadis are a unique and distinctive environment system defined by the diversity, variety, and wealth of natural attractions from an ecological significance. The wadis have some of the greatest wetlands with mammals, birds, and amphibians harboring their oasis. The Wadis of Egypt have a very rich cultural importance for the Egyptians, based on a cultural point of view. For example, The Wadi Feiran is an major cultural background for the movement of the Israelites from

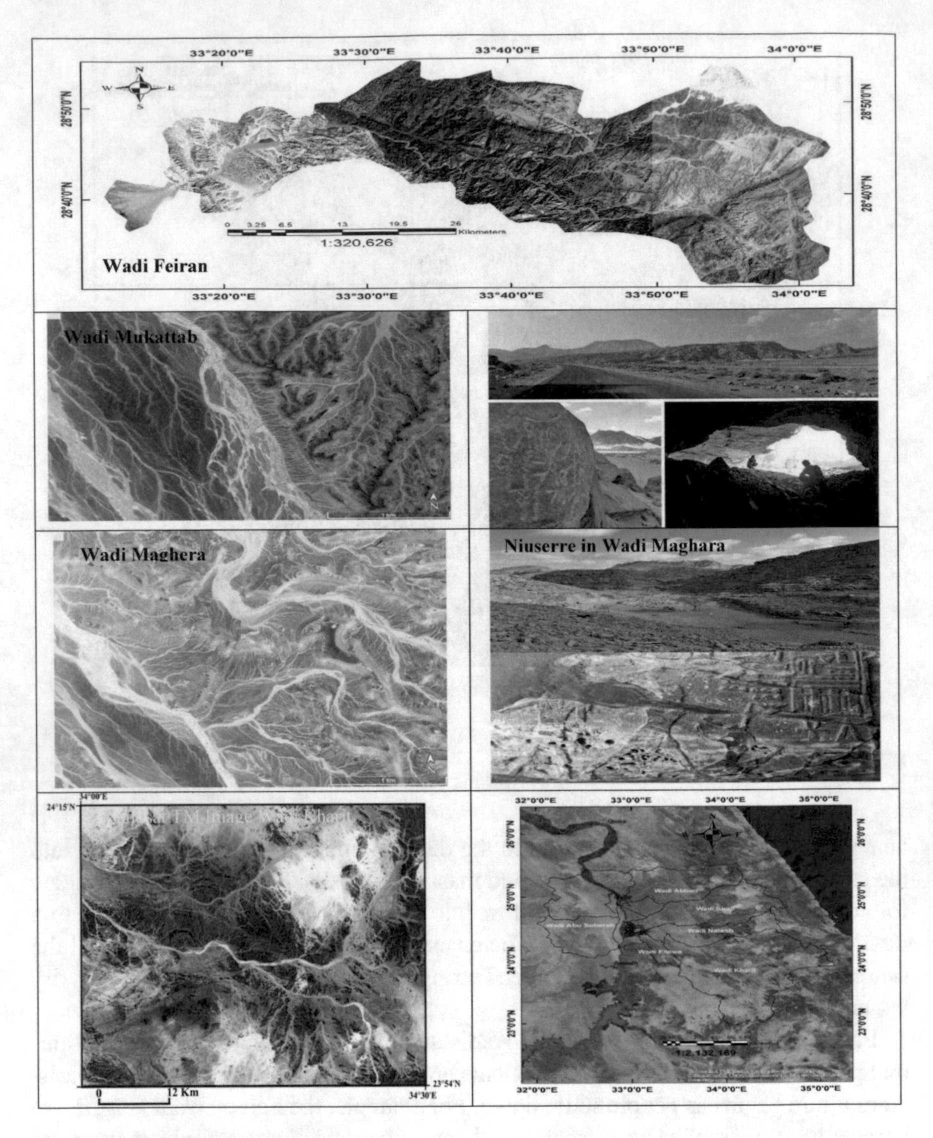

Fig. 3.14 Major Wadis of Egypt

Egypt to their promised and Holy land. The Inscription Valley has types of writing from an ancient people culture that existed 2–3 centuries BC. The inscriptions usually tell of the history of civilization and humanity as a whole.

3.4.1 Wadi Feiran

The Wadi Feiran (Fig. 3.14), which called the Wadi Paran in the Sinai Peninsula runs for 130 km. It discharges into the Gulf of Suez in the Red Sea some 29 km southeast of Abu Rodeis. The Oasis of Wadi Feiran, which known as the Pearl of Sinai is the biggest in Sinai. The Wadi oasis covers four kilometers bounded with palm trees and vines. Corn, barley, wheat, tamarind and tobacco are growing, but the main crop is the dates. The Wadi is the biblical Rephidim followed by the Israelites when they left Egypt. Amalek's battle was also fought here.

3.4.2 Wadi Mukattab

Wadi Mukattab (Fig. 3.14) called the Valley of Inscription in the Sinai Peninsula that is located on the main road between Wadi Maghera and Wadi Feiran. The steep rocky walls nearby to the Wadi have carvings. Most of the writings somewhere in the 2nd or 3rd centuries AD dating back to the Nabataea. Some of the different scripts are the vertical signs that indicate distinctive single sounds from Semitic and Egyptian scripts.

3.4.3 Wadi Maghera

Wadi Maghera (Fig. 3.14) is also another wadi in the Sinai Peninsula, an old turquoise mining area under the Pharaoh regime. On its rocky slopes, it also has ancient inscriptions. Some of the earliest quotes engraved in the valley are those left by copper miners in the Sinai on the development of copper and turquoise mines, dating back to over two millennia BC.

3.4.4 Wadi El-Kharit

El-Kharit (Fig. 3.14), one of Egypt's utmost desert wadis, has its main source at Gebel Ras on the Nile-Red Sea's main river watershed. It requires a course northwest and collects on its way the drainage from the many wadis. The minute it debuts from the plains of Kom Ombo, it heads to the Nile and joins the Valley of the Nile in the same geographical location as the Great Wadi Shait. Tributaries such as Natash, Antar, Khashab, Garara, Hamamid, and Abu are 260 km away.

3.4.5 Wadi Shait

Wadi Shait forms one of the westward drainage of the Red Sea Hills of the other El-Kharit in Eastern desert. The Shait Wadi joins the great course of River Nile close to Ridisiya village. Shait sources from the Red Sea Mountains. After about 200 km, it discharges into the Nile River Valley a small distance north of Kom Ombo. The main trunk of Shait has some wells like the Bir Helwat and Bir Salam.

3.4.6 Wadi Abbad

Wadi Abbad (Fig. 3.14) transverses the dry plateau of the Eastern Desert. The water that runs through the wadi valley originates from the Red Sea Mountains. Runoffs collect in the upstream tributaries. They are voluminous enough; they may reach the main trunks of the Wadi to flow downstream, which is a very rare occurrence. The dominant vegetation along the course is ephemerals.

3.5 Top 10 Canyons in South Sinai

A canyon is categorized as a deep valley between mountains or hills, usually with jagged rocky walls and a channel that runs through it. Depending on this, you might claim that in South Sinai there are hundreds of canyons as you find narrow valleys throughout the country. Water, indeed, runs only after rain for a longer or shorter time in most of them seasonally. There may be several months of running underground waterways, resurfacing at bottlenecks, filling pools. Actuality, you find a few locations with continuous water pools or streams, and there used to be many more in the past, according to locals and maps. Unfortunately, the Sinai suffered a decades-long drought, though the last few years were better than average—let us hope the trend continues.

3.5.1 White Canyon

The White Canyon (Fig. 3.15) is situated beside Ein Khudra Oasis and is part of the well-known landmarks within Sinai. From the St. Katherine-Dahab-Nuweiba road it can be easily reached on foot. The canyon is sculpted in a plain's soft rock, beginning with a sudden deep drop as a crack. The canyon begins between tremendously narrow walls, after which opens up a little before reaching an other deep drop. That little scrambling is engaged in a crack over a hill and down, and you attain the lower part of the canyon connecting to the oasis.

Fig. 3.15 Top Canyons in South Sinai

3.5.2 *Arada Canyon (Double Canyon)*

An other desert canyon near St. Katherine-Dahab-Nuweiba Road, Arada (Double) Canyon can still be expanded from Wadi Arada's roadside settlement. It is carved in the base of the Guna Plateau, involving, as this name suggests, of two canyons that can be linked by creating a circuit. It includes some scrambling. The hike is easily doable, and a lot of fun.

3.5.3 *Coloured Canyon*

Most likely the best known site in the Sinai after the Monastery of St. Katherine and Mt. Sinai, the Colored Canyon is relatively near the town of Nuweiba. The canyon is under an open plain starting from several wadis. It runs between very tall and very narrow rock walls, and a little scrambling is engaged at one point. A big rock used to be here, as saw on the photos, and people had to climb through a whole under it. However, the rock is no longer there, strong floods of 2013 washed it away like pebble.

3.5.4 *Kharaza*

Situated in the High Mountains near St Katherine's town, Kharaza consists of a few interlinked water pools in a short but beautiful canyon of granite. There's usually a little water here, and it's a challenge to get through after rain—never mind staying dry. The surfaces of smooth and steep granite are tricky. The canyon is just off a large open wadi at the base of the Mt. Katharina range, known as Wadi Mathar. You find additional little canyon nearby, more of a gap cutting through a granite hill, connecting the area to Wadi Umm Serdi.

3.5.5 *Wadi Sagar*

Wadi Sagar (Fig. 3.15) is a short wadi that literally cuts through a granite mountain range. It links the higher Abu Tuweita area to Wadi Tinya, which is at the base of Jebel Abbas Basha. You find a small fountain in the small canyon running between vertical walls that receives water from a crack. There is also a fig tree, grafted on a local tree. It's one of the mountain's most attractive sights.

3.5.6 *Ubugiya*

The small canyon known as Ubugiya can be seen next to the Bedouin settlement of Abu Seila and draining the area to Wadi Itlah, which starts at the town of St. Katherine. Water often flows through the canyon, starting at a cliff and continuing in a narrow, remarkably straight stretch. There is one tricky point where the road leads a few meters above the canyon floor over a smooth and steep rock surface, but some rocks were cemented on the rock as steps to make it safer. Also in the narrow part there is a smaller tangible water tank that gathers water, and a small garden at a wider sandy spot and a few trees.

3.5.7 *Wadi Isla Gorge*

Wadi Isla (Fig. 3.15) is part of the traditional caravan way that links the town of St. Katherine to the coastal city of El-Tur. The Bedouin used to take their products along this route—almond, apricot, other fruits—and bring back products and supplies until quite recently. It was also a location from which they could get bamboo, used for buildings, because in the narrow part of the wadi there is a stream that supports dense vegetation. The long meandering wadi is pretty narrow all the way, but it gets very narrow towards the end, leading between very high vertical granite walls. Wadi Isla is located in a remote part of the wider mountain ranges. It is a trek that is longer and more demanding.

3.5.8 *Closed Canyon*

It is a small cul-de-sac situated between the oasis of Ein Kudra and Wadi Ghazala in a branch of the sandy Wadi Khudra. It begins in a small crack, running between steep vertical walls in a sandstone range. You can just squeeze through the canyon at points that are extremely narrow. The road leads to the end under big rocks, before reaching a small open area surrounded by an unreachable wall. From here, you have to go back the same way you came in. Closed Canyon can be visited en-route from Ein Kudra oasis to Jebel Mileihis.

3.5.9 *Wishwashi Canyon*

It is a little wadi off Wadi Milha, along the walking route to the Colored Canyon. The wadi gets very narrow just a little up from the mouth, with big boulders blocking the way. There is a piece of rope and ladder at tricky parts, and water might be present in pools. The canyon is unreachable further up; you have to return the same way you came in. It is a pretty place, but after rain when a pool forms, it is simply amazing. You can visit the canyon on the Red Sea coast from the Ras Shaitan area.

3.5.10 *Salama Canyon*

Occasionally named the "Little Colored Canyon," this is a nice place that begins as a crack in a rocky area of Wadi Ghazala, which is otherwise sandy. It gets a little deeper, and a little scrambling is included in the hike. The lower part of the short canyon is wider, connecting to the larger Wadi Disco in which a tiny Bedouin settlement is located. If you visit the area of Wadi Ghazala, both the mountain and the canyon are worth seeing. Ras Ghazala, as well recognized as El Taor, is on St. Katherine-Dahab-Nuweiba road, easily identifiable on a hill with a smiley face from the rusty truck.

3.6 Canals in Egypt

3.6.1 *El-Salam Canal from Above*

El-Salam (peace) Canal project in northern Sinai (Fig. 3.16) aimed to be fed from the Damietta branch with drainage water from lower El-Serw and Bahr Hadous drain near its outfall (Mostafa et al. 2005). El-Salam Canal project irrigates 220,000 acres west of Suez Canal and 400,000 acres in Sinai. El-Salam Canal is the giant project in the agricultural development and national projects, in addition to security and strategic dimensions in connection with the reconstruction of the Sinai. El-Salam Canal transmits 4.45 billion cubic meters a year, which is a mixture of water from river and agricultural drainage (2.11 billion m^3/y of fresh Nile water mixed with 1.905 billion m^3/y of water from Bahr Hadous and 0.435 billion m^3/y of El-Serw drain). The ratio of Nile water to drainage water is about 1:1 (Gadallah et al. 2014), which is determined to reach total dissolved solids in the range of 1000–1200 mgl^{-1} to be appropriate for cultivated crops. The wastewater salinity is not constant throughout the year. In Bahr Hadous, salinity varies from 1540 to 2983 ppm approximately, while the range of salinity at El-Serw drain between 908 and 1696 ppm (Dri et al. 1993). The total length of the El-Salam Canal 242 km, 87 km west of the Suez Canal and 155 km in the Sinai. Canal path passes in the provinces of Dakahlia, Sharqia, Port Said, Ismailia and the Sinai. It crosses the canal through a culvert under the Suez Canal to Sinai at kilometer 27.8 southern city of Port Said, and siphon consists of four tunnels.

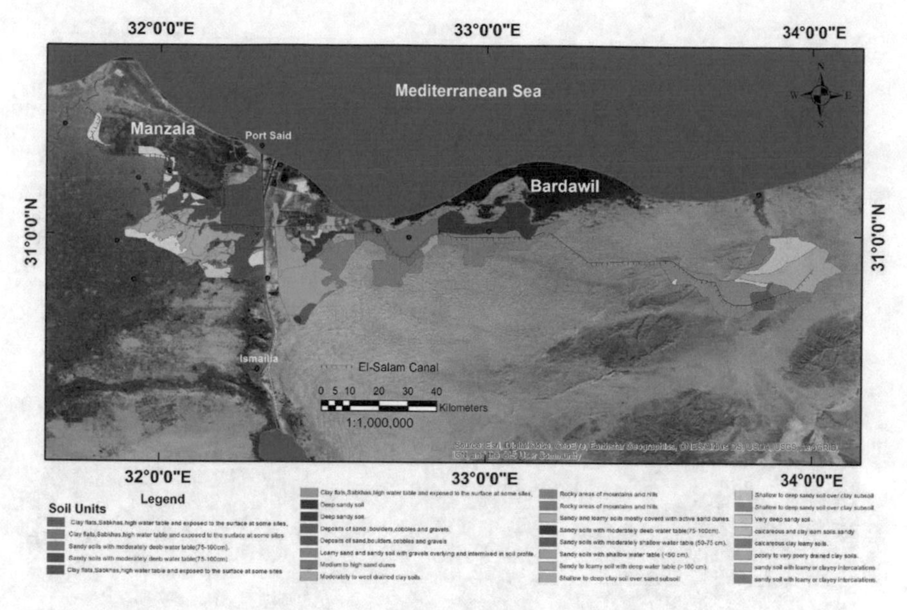

Fig. 3.16 Soil of El-Salam Canal project in northern Sinai

3.6.2 *New Suez Canal: View from Above*

The Suez Canal and its encompassing region (Fig. 3.17) especially Sinai are considered as ones of the most critical unexploited economic resources. It stayed ignored for a long time and was not involved within the economic development system despite its extraordinary financial significance to Egypt (Molouk 2015; Hidayat 2012). The New Suez Canal of Egypt, which was propelled on 5 August 2014 are projected to be 72 km long (News 2014; Post 2014) whereas the whole canal is 164 km long. Six new channels for cars and trains are additionally intended to end the isolation of the Sinai Peninsula, linking it better to the Egyptian heartland (Oxford Business Group 2015). It was likely that the creation of a new canal would maximize the value of the current canal and its by-passes and double the longest possible parts of the waterway to enable traffic in both directions and reduce the waiting time for transiting ships. However, the development of New Canal has 'double trouble' range of impacts, at local and regional scales, on both the environmental and ecosystem of the area. One of these impacts is a waterlogged area. Water-logged area, which is an environmental problem is found throughout the world (Quan 2010; Qureshi 2008; Minar et al. 2013; Chowdary 2008) and Egypt (Omran 2012a; b). The New Suez Canal area has been facing drastic waterlogging problems, which the out-of-date procedures for delineation it are time consuming.

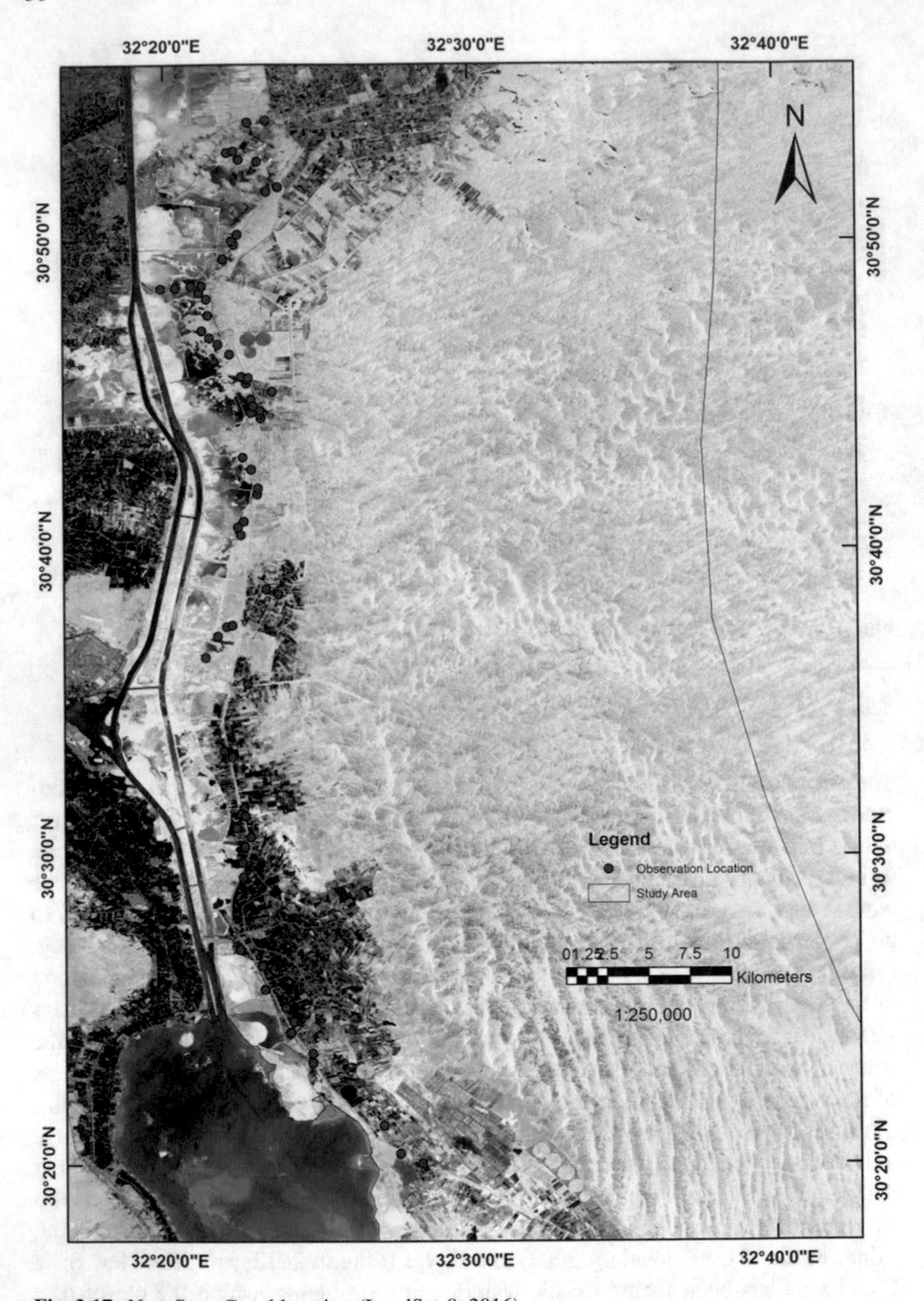

Fig. 3.17 New Suez Canal location (LandSat-8, 2016)

3.7 Lakes in Egypt

3.7.1 Northwestern and Eastern Coastal Lakes and Lagoons

Five natural lakes lie adjacent to the Mediterranean Sea (Fig. 3.18); Lake Mariut, Edku, Borollus and Manzala (deltaic section) (Wahaab and Badawy 2004) and Bardawil (North Sinai). These lakes are kept separate from the sea by narrow splits and are not more than 2 m deep. They provided fish and recreation. Part of Lake Mariut was once used as a landing place for seaplanes. These Lakes are the most productive lakes in Egypt, which have fresh, brackish and saline or hyper-saline water. The depth of these lakes is ranging from 50 to 180 cm. In addition, they are internationally important sites for wintering of the migrating birds, providing valuable habitat for them and an important natural resource for fish production in Egypt (Shaltout and Khalil 2005).

The current pattern of these lakes is changing rapidly, due to natural developments and, commonly, to man's activities (e.g., fishing and agricultural practices). Unfortunately, most of these lakes have deteriorated sharply over the last twenty years due to the wastewater being discharged into them. Three types of wastewater contributed to the problem (Wahaab and Badawy 2004): domestic sewage, untreated industrial effluents and agricultural drainage water. The first used to be discharged directly into the sea. The second has increased dramatically due to the growth of new industries. The third has also increased after the building of the High Dam because the agricultural land has been switched to producing more than one crop a year. The construction of the Aswan High Dam in 1964 is the driving force for a continuous evolution of the Delta Lakes.

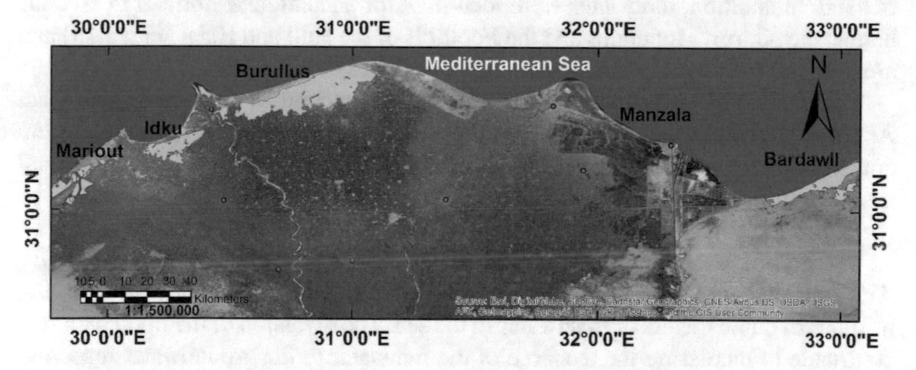

Fig. 3.18 Map showing the five Egyptian Northwestern and Eastern Coastal Lakes

3.7.2 Ancient Rivers and Lakes

3.7.2.1 El-Arish Wadi

The Sinai Peninsula is considered the northeastern frontier of Egypt, yet it has not received its development, which was limited to coastal cities in the south, while the north suffers from lack of resources and poor economic conditions. However, climate information confirms that North Sinai is one of the best areas in the country that can be developed sustainably using renewable groundwater sources, because the region receives the highest rainfall in all of Egypt. There is no alternative to the use of renewable groundwater as one of the most important natural sources of life in the Sinai. Moreover, northern Sinai is an ideal place for agricultural and population development. It is characterized by its moderate climate and is relatively close to densely populated cities such as El-Arish and the Suez Canal cities, in addition to its Bedouin population.

The northern Sinai desert and the ancient rivers and lakes in the valley of El-Arish were filled with water in the rain ages, which ended about 5000 years ago. The question, which arises now, is "how to benefit from the seasonal floods in the development of North Sinai." The potential of using satellite images and radar information to detect ancient river paths indicate the concentration of groundwater beneath the surface as new sources for future generations (Fig. 3.19).

The integration of spatial data and topographical data with field emphasis on geophysics equipment and field studies, as well as the treatment and interpretation of a wide range of radar data is showing that the ancient course of Wadi El-Arish was passing through the El-Sirr and El-Guarir area at a length of 110 km with 500–3000 m width. The sediment of this ancient river were discovered under a thin cover of sand. In addition, three sites were identified for ancient lakes formed in lagoons behind the Khirim Mountains and the Foothills of the Hull and Halal (Fig. 3.19) and are believed to contain groundwater (AbuBakr 2013).

The creation of a two-kilometer-long channel and a depth of six meters at the Wadi Abu Suweir, southwest of Tala Al-Bodin, is essential in order to reach the current course of Wadi El-Arish in its old course. This channel will turn the course of seasonal flood into a flat area west of Mount Halal, where nearly half a million acres can be reclaimed to provide a stable life and investment in agricultural production.

The proposed canal will reduce the risk of devastating floods in the city of El-Arish and increase the community's benefit from monsoon rains instead of losing millions of cubic meters of freshwater in the sea. The diversion of the flood path will contribute to increasing the recharge of the rainwater to the groundwater reservoir, which qualifies the rise of the groundwater level and helps to replenish the reservoir. Therefore, the Egyptian government can plan for a new crossing whose purpose is to develop the northern Sinai Peninsula and raise its people.

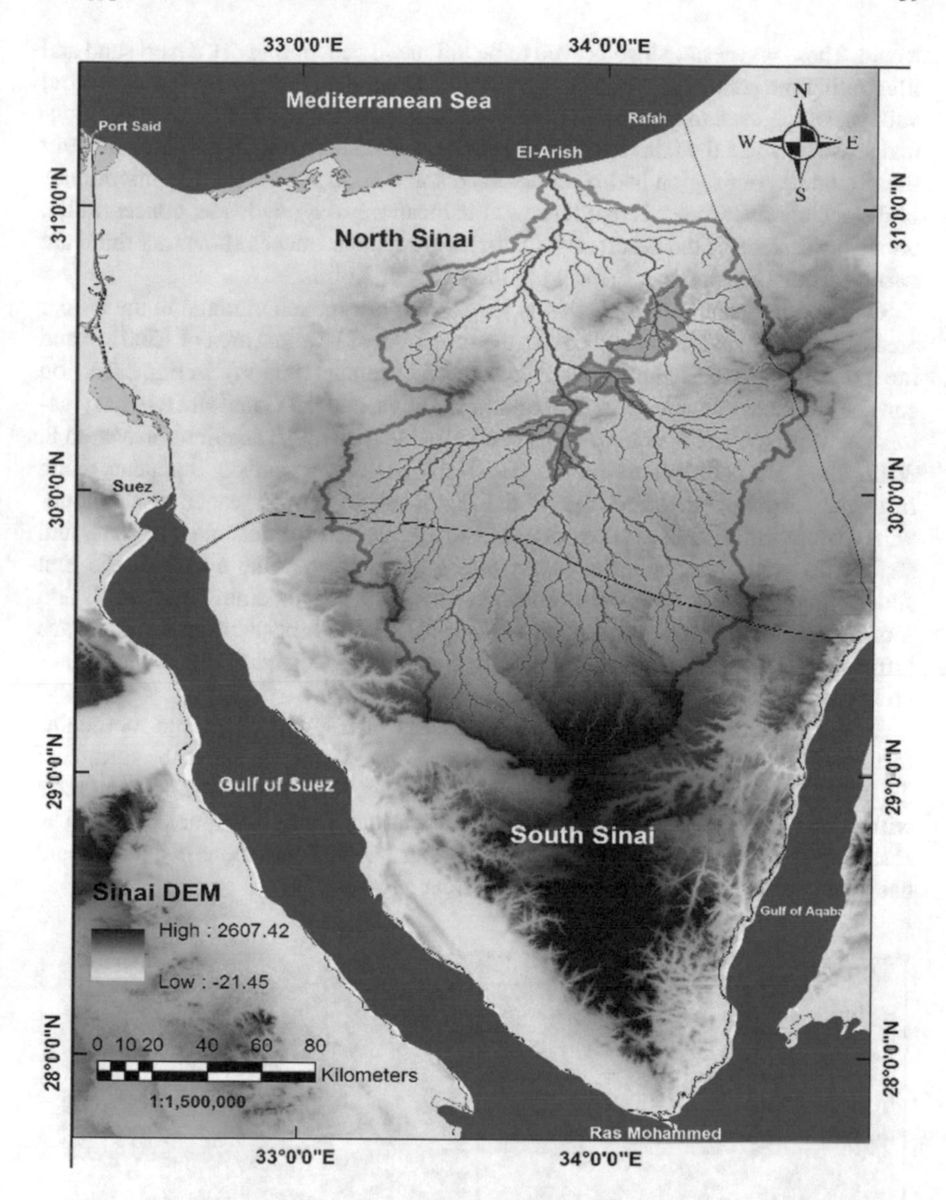

Fig. 3.19 Ancient river paths of Wadi El-Arish in the Sinai Peninsula

3.7.2.2 East Uweinat

On the one hand, another ancient lake in southwestern Egypt, which extracted abundant groundwater. There are currently more than 1000 wells in the area of East Uweinat. These wells are used in agriculture to produce wheat, chickpeas, beans and others. Radar images have shown us geological formations in many parts of the world. These waves have also proved to be unique in penetrating dry desert sand and illustrating the submerged terrain of the sand. The radar shows the paths of the old valleys, which used to be rivers where water had been abundant in previous geological seasons, when the rains had fallen heavily, and disappeared under the sand after the drought in our region had been resolved some 5000 years ago. The importance of the ancient river routes is that it shows the locations of groundwater concentration under the surface of the desert, as in Libya where radar images show that there are two wadis that reach under the sand to the Oasis of Kufra.

On the other hand, Mount Uweinat (Fig. 3.20) is a mountain range in the border area of Libya Egypt Sudan, and covers the mountain of Uweinat area of 1500 km and most of its area is located in Libya High Mount Uweinat (1934 m). In Fig. 3.20, you can see the Jabal Arkanu (*top*), a mountain in Libya, and the Gabal El Uweinat (*bottom*) in Multispectral Instrument (*left*) and NIR/SWIR (*right*) acquired on March 8, 2017. They are both located on the Gilf Kebir plateau, a sandstone plateau rising from the Libyan Desert and one of the driest places on the planet. Both mountains feature different kinds of rock. They are made up of sandstone, siltstone, and shale with granite intrusionsin their western parts. If you take a look at the first slider, you can see a natural color image, made up of data from bands 4, 3, and 2 of Sentinel-2A's Multispectral Instrument (*MSI*) and a near-infrared/short wave infrared (*NIR/SWIR*) image, using data from bands 12, 8, and 2. Sandstone shows up in reddish hues, granite in blue hues.

In the western part of the mountain consists of a group of granite mountains formed from the igneous rocks formed by the fusion of magma under the surface of the earth, the rank of the body of a collar up to a diameter of some 25 km and ends with three valleys to the west are Karkur "Karkour Idris" and Karkour Ibrahim. The eastern part of the sandstone ends with "Karkur Moore," and there is a permanent oasis known as the eye of the Prince "or" Bear Moore."

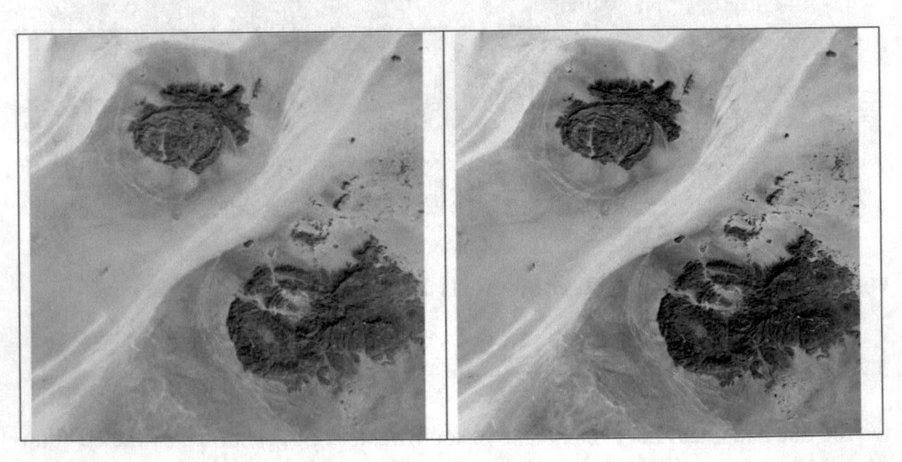

Fig. 3.20 Mount Uweinat image of Sentinel-2A's Multispectral Instrument (left) and NIR/SWIR (right) acquired on March 8, 2017

The region is known for its collection of ancient primitive drawings on the mountain rocks. The earliest discoveries in the region were made by Ahmed Pasha Hassanein, who drew his first maps in 1932. The region is mostly dry, not the graphics embodied in its infancy, which tell the forms of life from a time ago, including the sea of ostrich, giraffe, deer, and acacia trees. Despite this, the number of plant species present in the region is not less than 55 species of plants.

3.7.3 Southern Egypt's Lakes

3.7.3.1 Toshka Lowland Development

The Egyptian government prepares in its current strategy to reclaim 1.6 million ha (4 million feddan) from the desert in 4 years through four stages starting from June 2014. The first stage was to reclaim 1 million feddan, raised to be 1.5 million feddan in August 2015. About 90% of the 1.5 million feddan depend on nonrenewable and deep groundwater through 5000 water wells. The new reclamation project will be carried out in three sub-phases in nine areas of the Western Desert, including Toshka, Farafra Oasis, Dakhla Oasis, Bahariya Oasis, Qattara Depression (Moghra), West Minya, and East Owainat. In addition to the 540,000 feddan in Toshka, about 100,000 feddan depending on surface water and 32,000 feddan depending on groundwater (through 102 wells) are included in the "1.5 Mio. Feddan Project."

Further to the south, a mega reclamation project; the South Valley Development Project (popularly known as Toshka project) was implemented to convey 5.5×10^9 m^3 from Lake Nasser behind the High Dam at southern Egypt to cultivate 0.5 million feddan by 2017 and resettle 7 million people in the Western Desert (Lonergan and Wolf 2001). According to the Landsat image of September 24, 2017, the total cultivated area was 24,000 ha (60,000 feddan), representing only 11.1% of the target.

Toshka (Fig. 3.21) is Nubian word with two syllables. The first is "Touche," which is a plant that grows in depression, "Key" meaning home or place. Toshka means plant or flowers places (Saleh 1998). The Toshka area is the first sites to gather the population in the desert, by prehistoric man, and the region of archaeological value, including the effects of and information about the origin of human civilization. Many dry valleys cut surface of Lower Nubia, which discharged, into the lake, and the valleys are Toshka and Wadi Aldka, where the headwaters are located in the plains of Nubia. The region is surrounded by hills in some parts of the Nubian Sandstone, where there hilly landscape form a hierarchical area shape or continuous sandstone ridges.

Toshka Lakes were made in the 1980s and 1990s to convey water from Lake Nasser to the Sahara Desert through a synthetic canal. Between 1998 and 2002, a series of unusually high-volume floods of the Nile increased the water level of Lake Nasser/Nubia to beyond the capacity of the High Dam Lake. This problem prompted construction of a draw-down channel linking Lake Nasser with the Toshka Depression adjacent to Lake Nasser. The New Valley project, which started

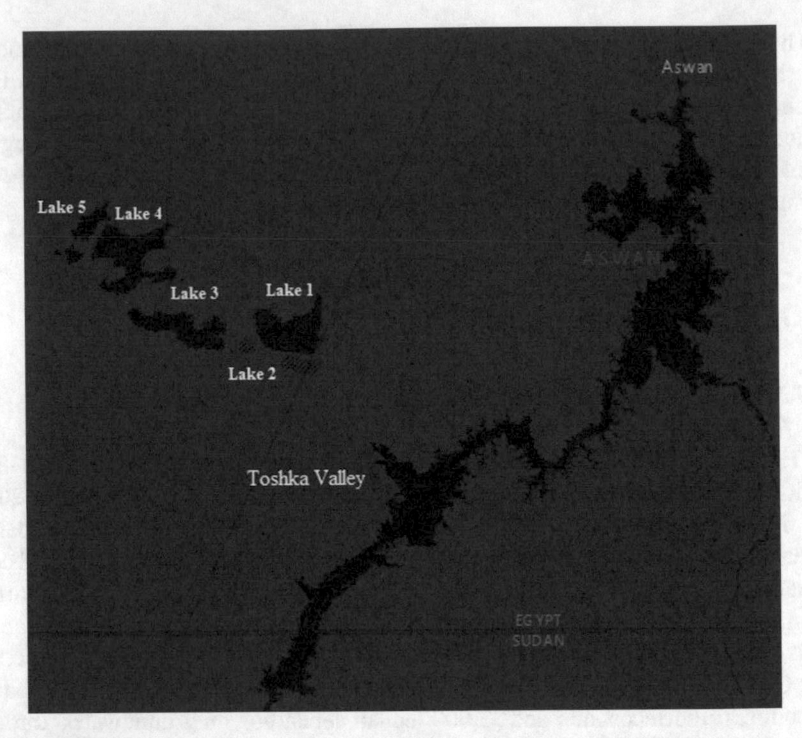

Fig. 3.21 Toshka Lakes were created by the diversion of water from Lake Nasser through a man-made canal into the Desert

in 2008 and ended in 2013, failed to reach its objectives due to poor soils, sand dune encroachment, and excessive evaporative losses.

Toshka Depression is a massive 6000 km^2 of relatively low land. This domain comprises 5 unnamed lakes (Fig. 3.21) and several shallow spots at elevation below water level in Lake Nasser. The area of those low-level spots exceeds 150,000 acres and can be developed, with limited risk, for agricultural production; seasonal crops and livestock.

Flooding of the Toshka depression made five fundamental lakes with a most extreme surface area of around 1450 km^2—around 25.26 billion cubic meters of water. By 2006, the amount of water stored was decreased by 50%. In June 2012, water filled just the lowest parts of the main western and eastern basins—representing a surface area of 307 km^2, or approximately 80% smaller than in 2002. Water is totally absent from the central basin. Here in a "new valley," in the hope of helping a home for millions of Egyptians, agriculture and industrial communities could be evolved. Five major lakes (Fig. 3.21) evolved in the process of this development from the diversion of water from Lake Nasser through the canals. The Toshka Project was to be finished by 2020. Apparently, things have not gone as planned in Egy pt in recent years, and ambitious, costly undertakings like this have probably come to a halt.

Figure 3.21 indicate that this project had come to a halt, with water in the associated Toshka Lakes evaporating in the hot desert extends.

Managing the proposed fresh water fisheries (1256 km^2) requires maintaining water salinity stable. Therefore, these lakes cannot be used as endorheic lakes and continuous flow of fresh water has to be maintained, implying some of the water has to leave the lakes, forming return streams. These streams will be used to irrigate all the lowland that has been already developed for agricultural use, as per phase I. Estimated average rate of evaporation from the lakes is 100 m^3/ s, or a total of 3 km^3/yr. Evaptranspiration of the cultivated 1190 km^2 (293,000 acres) of low land is about the same rate for a total of 2.6 km^3/yr. This implies that the agrarian development requires about 6 km^3/yr or 11% from Egypt's share of in the Nile water.

3.7.3.2 High Dam Reservoir

The Aswan High Dam (AHD) was constructed in 1968 at a distance of 7 km south of Aswan City. It is a rock-filled dam made of granite rocks and sands and provided with a vertical cutoff wall consisting of very impermeable clay. Lake Nasser (22° 31′–23° 45′ N and 31°30′–33°15′ E) reached its operating level of 175 m (asl) in 1975, with a total amount of 121 $\times$ 10^9 m^3 of stored water. In southern Egypt and northern Sudan, Lake Nasser is a vast reservoir. Nasser Lake only refers to the much larger portion of the lake in Egyptian territory (83% of the total). At present, Lake Nasser is the world's largest man-made lake. The region around the reservoir is a potential area for land reclamation and new settlements, urgently needed by Egypt's expanding population (Abu Zeid 1987). On the one hand, the Aswan High Dam contributed greatly to the economic development of Egypt. It supplies 15% more irrigation water and 2100 MW electric power and protecting the lower reaches of the Nile from seasonal floods. On the other hand, its environmental impacts are serious. From a demographic point of view, building this dame necessitated permanent relocation of more than 100,000 Nubians people, 600 km from their homes. After more than 40 years, many of these people are still suffering the anguish of their displacement. May the history overlook the inconvenience of some of the people for the benefit of the multitude, as the high dam commissioned to do? History will condemn us, if the rising sea forces millions of the Nile Delta inhabitants to flee their homes and land because the solution to prevent it is costly.

Lake Nasser (Fig. 3.22) has a number of known as khors. There are 100 important khors in Lakes Nasser and Nubia combined. Their total length when the lake is full is nearly 3000 km and their total surface area is 4900 km^2 (79% of total lake surface). In volume, they contain 86.4 km^3 water (55% of total lake volume).

3.7.3.3 High Dam Lake Khors

Even though the Lake Nasser reservoir reaches its maximum storage capacity, the depressions west of Lake Nasser in Egypt's southwestern desert are used as a natural

Fig. 3.22 Aswan High Dam Reservoir—Lake Nasser of Egypt and Lake Nubia of Sudan

flood diversion basin to reduce potential downstream damage to the Nile Valley caused by exceptional flooding (Kim and Sultan 2002). The major strategic units and the ground water table level in Lake Nasser are shown in the hydrogeologic map (Fig. 3.23). Before the High Dam was built, 93% of the total annual suspended load of 124 million tons of sediment flowed directly into the Mediterranean Sea every year. AHD came in aid of this problem and retained 98% of the load within the reservoir, with only 2.5 million tons making it to the Sea yearly.

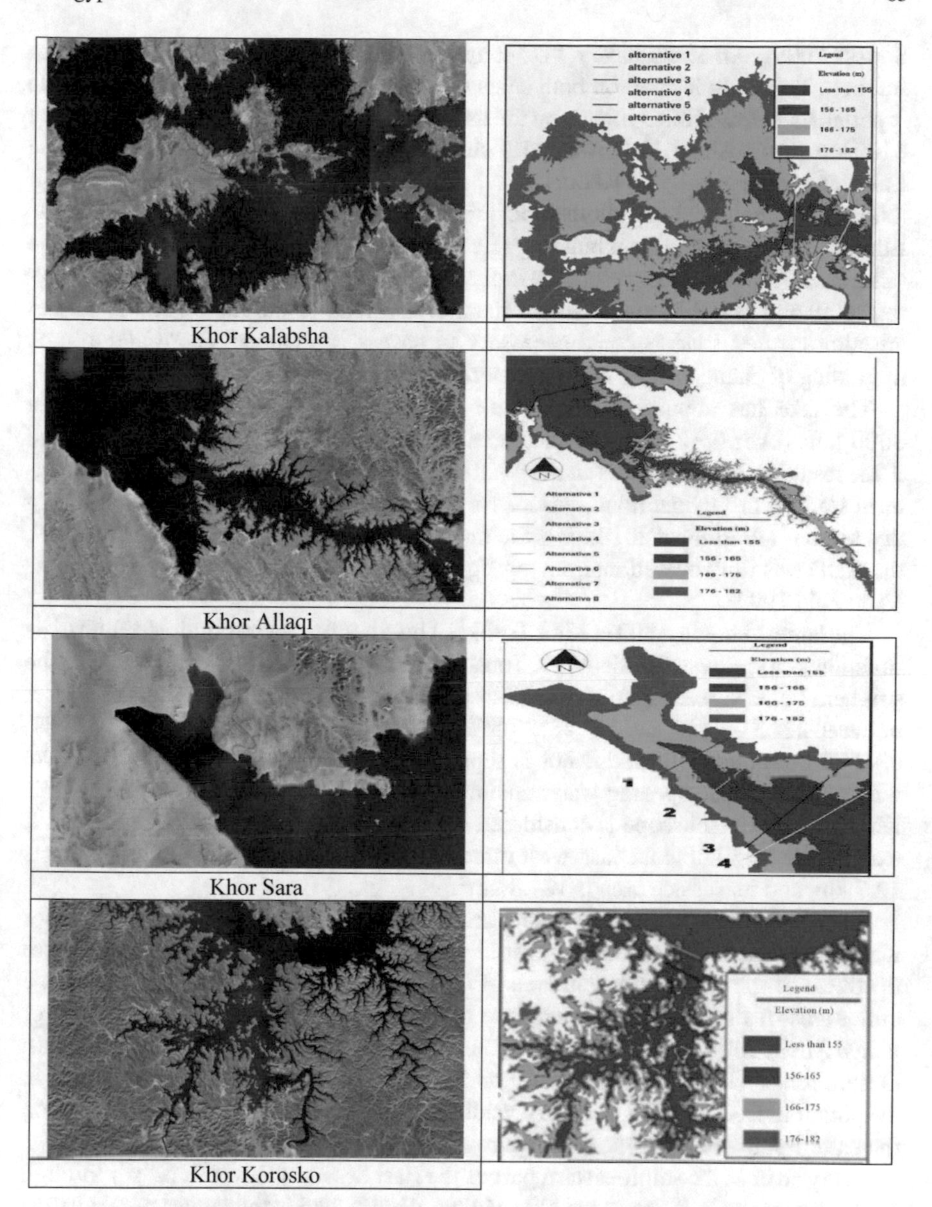

Fig. 3.23 Lake Nasser and its main Khors in east and west shoreline (Elba et al. 2014)

Lake Nasser that covers an area of approximately 6000 km^2, has a major impact on the area's geomorphology in terms of filling the surrounding Khors (side extensions) and the nearby depressions. The Lake is featured by the presence of numerous dendritic inlets, or side extensions of the reservoir, known as khors (Mostafa and

Soussa 2006). All khors have a "U" shape in cross section, with a flat sandy central belt. These are located on both sides of the reservoir. van Zwieten et al. (2011) reported that, these Khors impact on the length of the shoreline, which is estimated at 8700 km. Khors Allaqi, Kalabsha and Toshka are the largest (El-Shabrawy 2009). The Lake, and consequently its Khors (Figs. 3.22 and 3.23), may have fluctuated in their total surface areas and morphometerical features; due to the population increase, the Ethiopian Renaissance Dam building (Bedawy 2014), in addition to climate change (Elsaeed 2012). As water resources in Egypt are becoming scarce (Abdin and Gaafar 2009) and as public awareness of the morphometry of the Lake's Khors studies are almost rare, it has become necessary to have a continuous record and a good upgrading of changes in its morphometerical features.

The lake has about 100 khors with 12,000 km of shoreline covering about 3000 km^2 (over 45% of the surface area of the lake), as shown in Figs. 3.22 and 3.23, resulting in high losses in evaporation (Said 1990). This lake uses as a long-term storage throughout flood seasons for the Nile water to be used in the following dry seasons (WorldFish 2018). It is the main source of Egypt's fresh water. In 1969, the AHD was fully operational, saving Egypt from major floods and severe droughts (Abu Zeid 1993).

The largest khor in AHD is Khor Toshka. This khor cannot be eliminated due to its function as a spillway for the AHD. Toushka, also called Toushka west, is one of the southern Lake Nasser's Khors. The Khor is located to the south-west of Lake Nasser between 22° 31′ 41.3″ and 22° 43′ 27.5″ N Latitudes, and 31° 32′ 19.7″ and 31° 47′ 08.7″ E Longitude. Toushka Khor is separated from the main channel of the Lake with a natural wide opening water channel of 11.25 km width, with no outlet. This Khor with its funnel shape is considered one of the largest Khors in Lake Nasser. It extends for 34.5 km in an east–west direction. The maximum width of this Khor is 12.7 km, and its surface area is 184.8 km^2.

Khor Kalabsha and Khor El-Allaqi are the largest khors (Elba et al. 2014). Khor Kalabsha is the lake's second biggest khor. It's in the Western Desert about 30 km upstream of the AHD. It has an area of 600 km^2, approximately 10 percent of the entire area of the lake. Every year, due to evaporation, it loses about 2700 mm of water. About 100 km upstream of the AHD, Khor El-Allaqi expands into the Eastern Desert. It has a large area of 500 km^2 and experiences annual losses of 2500 mm in evaporation. El-Allaqi Khor is a dry wadi that lies about 180 km south of Aswan on the eastern shore of Lake Nasser in the road leading to Sudan. This wadi is basically a large dry river in the south-eastern part of the East desert of Egypt. It passes through the hills near the Red Sea to the Nile Valley. Wadi Allaqi is the largest desert valley in the east and extends to more than 150 km in the north-west of the south East of the tributaries of the Red Sea towards the Nile Valley about 180 km. The length of about 150 km and is within the administrative boundaries of Aswan and extends to the borders of the northern Sudan, ie the valley is located within the borders of the northern Sudan. The valley is located within the borders within the country of Nubia and the valley is one of the most important sources of gold where it is very rich.

Furthermore, Khor Sara and Khor Korosko have high evaporation rates with their small surface areas (Elba et al. 2014). Khor Korosko is located approximately 180 km upstream of the AHD in the Eastern Desert and covers approximately 103 km^2 at 182 m water level. With a rocky bottom, it's steep sided and fairly narrow. Every year, it loses about 3000 mm by evaporation. Khor Sara is located in the western desert about 325 km upstream of the AHD. It is one of the AHD miniature khors and has only 50 km^2 of surface area. It suffers from high losses of around 3100 mm of evaporation per year.

3.8 Depressions and Oases

Deserts are one of the most notable characteristics of Western Sahara. It is rare to find a similar area where such a number of major depressions occur. These depressions (Fig. 3.24) are scattered on the surface of the plateau in the far north near the sea and the far south and west on the border. Some of these depressions are uninhabited and of an immortal type such as a Qattara Depression. Depressions in Western Sahara can be divided by geographical location into three groups: (A) group of northern depressions, including Wadi Al-Natroun, Qattara Depression, Siwa Oasis. (B) The

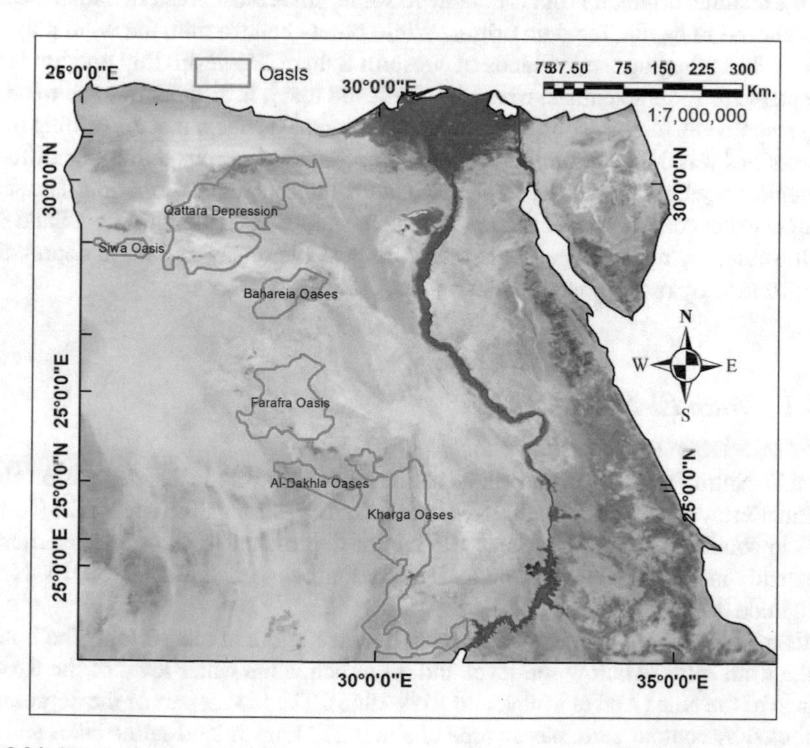

Fig. 3.24 Depressions and Oases

group of central depressions includes the Bahareia Oases and the Farafra Oasis. (C) The group of the southern depressions, which include Kharga, Darb al-Arba'een, Al-Dakhla and Abu Munqar.

The remaining six depressions contain restricted agricultural production and permanent settlements, all of which have fresh water supplied by the Nile or local groundwater. Siwa Oasis is isolated from the rest of Egypt, near the Libyan border and west of Qattara, but has lived since ancient times. For more than 1000 years, Siwa's cliff-hung Amun Temple has been renowned for its oracles. Among the many illustrious people that visited the temple in the pre-Christian era were Herodotus and Alexander the Great. The other major oases form a topographic chain of basins extending from the El-Faiyum Oasis (sometimes called the El-Faiyum Depression) which lies 60 km southwest of Cairo, south to the Bahariya, El-Farafra, and Dakhla oases before reaching the country's largest oasis, Kharga. For centuries, sweet water artesian wells in the El-Faiyum Oasis have permitted extensive cultivation in an irrigated area that extends over 1800 km^2.

Most of the northern depressions are located below sea level, while the central and southern depressions is located above sea level. The order of depressions is descending by area as follows (Fig. 3.24): Qatara, Farafra, Kharga, Dakhla, Bahariea, El-Faiyum, Siwa, Rayyan and Natroun. As for the origin and composition of the depressions, the views differed between the tectonic origin and the Aeolian origin. For the tectonic origin, it is unacceptable to some, since most Western Sahara deserts are believed to be not Tectonic origin. While others believe that, the wind played a major role in digging the lowlands of Western Sahara. However, this opinion is not acceptable to some scientists, where it is believed that it is difficult to wind to be the only responsible for the drilling of these depressions. It is likely that the drilling of the depressions was done by running water during the rainy periods during the different geological ages, and wind may have a role in the final stages of formation. The act of running water confirms the existence of ancient waterways under the Great Sand Sea, as illustrated by radar images. It is therefore possible to say that these depressions were formed by running water and wind together.

3.8.1 Wadi El-Natrun Lakes

Wadi El-Natrun and its alkaline inland saline lakes are an elongated depression about 90 km northwest of Cairo. Wadi El-Natrun Lake is a salt-alkaline water lake (Natron salt) in Wadi El-Natron area near El-Qattara depression in Matruh Governorate. It extends in a northwest by southeast direction between Latitudes 30°15′ N and Longitude 30°30′ E (Fig. 3.25).

Its average length is about 60 km and average width about 10 km. The bottom of the wadi is 23 m below sea-level and 38 m below the water-level of the Rosetta branch of the Nile (Abd el Malek and Rizk 1963). The lowest part of the depression, encircled by contour zero, has an area of about 272 km^2. Inland saline lakes and salt crusts occupy the area surrounded by contour zero (Abu Zeid 1984).

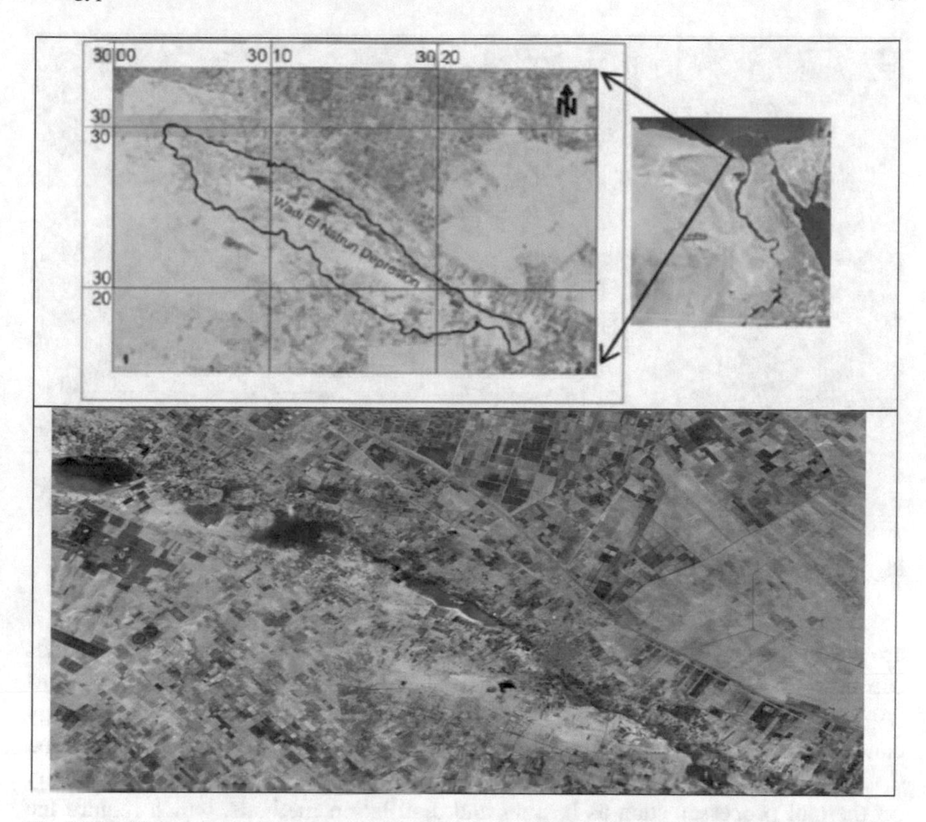

Fig. 3.25 ETM+ Landsat-7 image of the Wadi El-Natrun Depression, with its alkaline inland saline lakes and its extended northward on both sides of the Cairo-Alexandria desert road

3.8.2 Qattara Depression from the Satellite

Qattara (Fig. 3.26) is a huge depression located in Egypt in the Western Desert in Matruh. The lowest drop is 134 m below sea level. Extending from east to west, the eastern tip of the Mediterranean Sea approaches the El-Alamein area, about 298 km long and 80 km wide at its widest point. Transforming the mid-Saharan Qattara Depression region (30,000 km^2) into a completely self-sufficient and habitable community for 3–5 million people, generating well paid jobs for a multitude of citizens, working in advanced industries and modern agricultural activities. This project relies primarily on technology for hypersaline osmotic power generation of 3–4 GW of power, which will be used in constructing the new Qattara city; desalinate billions of cubic meters of brackish and agricultural drainage water to cultivate one million acres.

Solar Ponds as a technique to Desalinate of Qattara Depression Seawater is presented. The techniques, which currently used in seawater desalination, such as reverse osmosis, distillation and electrolysis, need to be managed with large quantities of energy, obtained through fossil fuel or electric power. The use of solar energy faces the problem of storage and use during the night and on cloudy days, but solar ponds

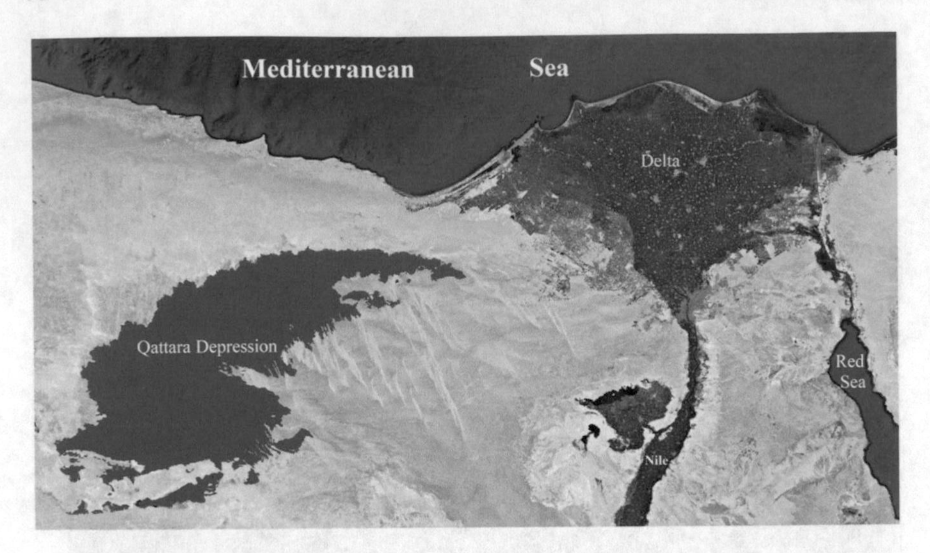

Fig. 3.26 Qattara Depression

are one of the new innovative systems for storing solar thermal energy. Solar Ponds can be defined as solar pools of saline ponds with variable salinity concentrations (Agha 1994). Solar ponds are one of the technologies through which thermal energy can be obtained for many applications, foremost of which is seawater desalination.

The idea is based on the existence of an energy source, because it relies mainly on thermal processes such as heating and distillation methods, which require the heat is usually in the form of steam generated inside large boilers. Solar ponds, as a renewable and cheap source of energy, can contribute to providing the required heat to complete seawater desalination, as an alternative to fossil fuels (diesel) or currently used high-cost, renewable energy, as well as the contribution of some to the pollution of the environment.

The economic return from the use of solar ponds as a source of thermal energy required to de-saline seawater is a useful and proven return. In addition, the cost of producing a cubic meter of water desalinated with solar ponds is virtually non-existent because it uses a natural source of energy (solar radiation), unlike conventional stations where the cost of production is high because of its dependence on fossil fuel sources or electricity. There is no doubt that solar ponds are economically and environmentally superior to other sources of energy currently used in seawater desalination.

The solar pond consists of a flat surface of water and relatively large salt that can be divided into the following layers (Fig. 3.27) (Deba 2011).

1. Upper Convective Zone. A low salinity layer exposed to wind and air currents, and is therefore affected by convection.

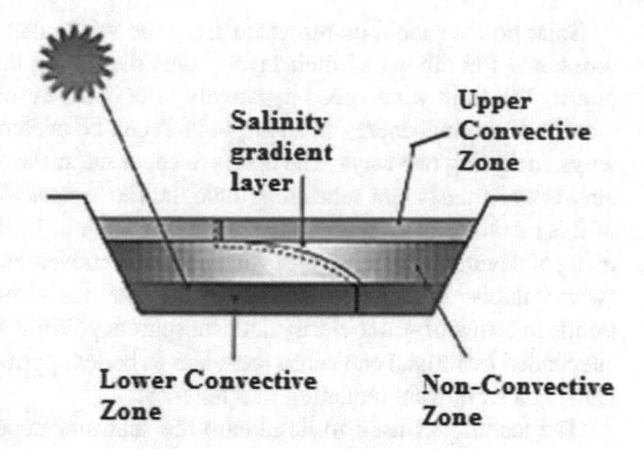

Fig. 3.27 Solar pond layers

2. Non-Convective Zone. Salinity in this layer is low salinity at the top surface adjacent to the surface layer of salinity to high salinity at the end, and because of the salinity gradient in this layer, they tend to generate pregnancy currents from the gradient. The temperature tends to increase in the downward direction, and this layer is called a lapping layer or salinity gradient layer.
3. Lower Convective Zone. It begins at the end of the layer of bore, which ends with the gradient of salinity. This layer is exposed to load, where the collected solar energy is stored, called the lower load layer or the energy storage layer.

The upper layers (the surface, the middle) allow sunlight to reach the bottom layer, heating them up. While the top two layers are not activated for thermal radiation, which makes them work like the glass cover of the solar collector. The saline intermediate layer remains stable and is not connected to heat despite the warming of the bottom of the pond. The increase in salinity is accompanied by a gradual increase in saline density, which prevents the rise of high-temperature water.

There are many factors affecting the localization of solar ponds, the most important of which are salt water, solar radiation, and wind. Saline water is one of the most important requirements for constructing solar ponds. The vertical difference in the salinity of these ponds and the subsequent temperature difference are used to form a layer of salty water at the bottom of the pond, which is higher than the upper layers. Generating temperature differences between the bottom of the pond and its surface can be utilized by heat exchangers, whether in heating for seawater desalination and heating of swimming pools, or in the management of thermal cycle to generate electricity.

Solar radiation plays an important role in determining the most suitable locations for solar ponds, which act as flat assemblies of solar radiation, allowing the bottom layer to be heated to about 90 °C often. These ponds also act as a large thermal reservoir; they allow storage of heat each of the daytime for use at night and summer heat for winter use. The greater the depth, the more the solar pool becomes more effective because of the presence of a large water mass and large thermal capacity accompanying (Mujahid 2002).

Solar ponds should be protected from the winds that cause their saline water to cross, and the mixing of their layers, thus disrupting the basic properties of these ponds. The high wind speed negatively affects the efficiency of the collection and storage of thermal energy in solar pools. It can be overcome this problem in several ways, including two ways: The first is to cover the surface of the pond with a floating network of transparent tubes to reduce the movement of surface water. Experience of this method was at Ohio State University, USA in 1976 (Fig. 3.28). The second is using of devices to increase the resistance of air movement, and was tried in Australia (www.fakieh-rdc.org). It also requires constant monitoring and maintenance of solar ponds in terms of water clarity and transparency. Wind loaded with dust (and other suspended materials) can cause the water to be less permeable to the solar radiation, causing a significant reduction in efficiency.

The techniques used to desalinate the seawater depending on the solar ponds, although the technology of evaporation flash is the best; because of the advantages, the most prominent of which are the following:

1. It is a long-standing technique.
2. High coefficient of its performance, and the simplicity of its creation (Mujahid 2002).
3. It consumes heat energy at relatively low temperatures, making it suitable for mixing with several solar thermal systems. Solar ponds are at the forefront of these systems. They have the potential to provide heat energy at a temperature of 60–80 °C throughout the year.

Fig. 3.28 A model of a desalination plant operating in a multi-stage flash evaporator through a solar pool (El Paso station) (University of Texas at El Paso 2002)

4. The design of the desalination unit is compatible with the thermal source of the solar ponds, which have a daily temperature change. They contain a self-regulating regulator to evaporate, enabling them to operate under a wide range of source temperatures to ensure efficient control (Hawas and Kamel 2004).

The idea of using solar ponds to desalinate seawater by means of evaporative evaporation is because water boils at low temperatures as it continues to be subjected to low pressure. The process of desalination in this technique through several stages can be summarized as follows:

1. Initially, the seawater is heated by a heat exchanger, feeding the thermal energy derived from the bottom layer of the solar pond.
2. Then enter the seawater into the pressure chamber until it happens to a direct boiling, or what is called blinking, and part of it becomes vapor. The evaporation process reduces the temperature of the remaining amount of salt water. The remaining amount is pushed into a second chamber with a pressure lower than the initial. Thus, the process is repeated. Any amount of water flashes to vapor, while the remaining water temperature decreases again pushed into, which third and fourth room .... etc., until the end of the last stage.
3. The steam generated by the flashing process is intensified to obtain fresh water by touching the heat exchanger through which the salt water passes before entering the heating chamber. Thus, part of the energy used is recovered by the heat extracted from the steam when condensed and converted into fresh water. This heat is transferred through the heat exchanger to the seawater, thus gaining part of the thermal energy required for boiling. Figure 3.28 illustrates the direction of the flow of water as a model for a desalination plant operating in the form of solar evaporation through a solar pool set up by University of Texas.

3.8.3 *Siwa Oasis*

Oasis of Siwa is one of the lowlands of Western Sahara (Fig. 3.29). The total area of the Siwa is 1088 km^2. It is about 18 m below sea level. Siwa is located between 29°12′11.52″ N and 25°31′10.74″ E. Siwa is located 65 km from the Libyan border, 300 km west of Marsa Matruh city and 600 West of the Nile Valley. The oasis has more than 220 water springs, and eye water flows about 25,005,000 cubic meters per day. The total discharge of the eyes is about 185,000 cubic meters per day. Siwa is dominated by the desert continental climate. It is very hot in summer, but its winter is very hot during the day, and the most dangerous thing that is exposed to Siwa is the floods despite its scarcity, but it causes serious damage.

Eyes and wells such as Ain Tzazrt, Ain Dakrur, Ain Qureisht, Ain Hammam, Ain Tamosa, Ain Khamisa, Ain Aljerba, Ain Shifa, and Ain Mashandt.

- Ein Kilopatra is one of the most famous tourist attractions of Siwa, also known as Ain Juba or Ain Shams, a stone bath filled with natural hot spring water.

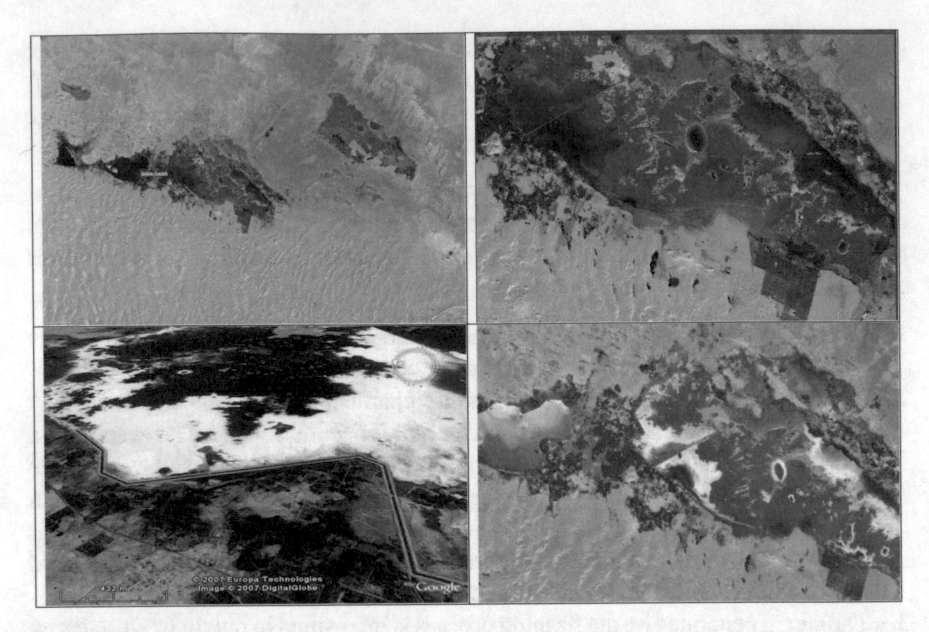

Fig. 3.29 Siwa Oasis, which is a centre of salt production as visible in Sentinel2 image (upper) and shows salinisation and huge water channels (lower) facing east (google image at co-ordinates: 25°32′3.60″ E, 29°13′21.74″ N)

- Ein Fattanais is about 6 km west of Siwa and is located on the island of Patanas overlooking the salt lake and surrounded by palm trees and desert landscapes.
- One eye is also called the "Great Sand Sea Well", a hot spring fountain 10 km from the oasis near the Libyan border in the heart of the Great Sand Sea.
- Ein Kigar is the most famous of them with a temperature of 67 °C and contains several mineral and sulfuric elements. The eyes of the water used for therapeutic purpose of rheumatic psoriasis are widespread.

Siwa has four salty lakes, which are the modern (Contemporary) Lake northeast of the oasis, the olive lake east of the oasis, the Siwa Lake west of the city of Shali and Lake Al- Maraki east. It has an area of 960 feddans, Siwa Lake west of Shali city with an area of 3600 feddans, and Al-Maraki Lake west of the 700-acre Bahi al-Din oasis. The oasis includes a number of other lakes, including Taghagin Lake, Middle Lake, and Shayta Lake. Lake Watanas is a tourist attraction located 5 km west of Siwa, with Watanas island surrounded by a three-way lake. Some of these lakes have recently been turned into saline land after the land became impregnated due to agricultural drainage. It is not the salt of ordinary salt, but one of the best types of salt used by European countries abroad because it is used in many industries and used in airports to melt the ice.

3.8.4 El-Faiyum and Wadi El-Rayyan Lakes

Wadi El-Rayyan Protected Area (Fig. 3.30) is located in the western part of the El-Faiyum Governorate, about 200 km southwest of Cairo. The protected area was established in 1989, and today is 1759 km^2 and home to Wadi El-Hitan Valley of the Whales World Heritage Site, designated in 2005. Wadi El-Rayyan was the first national park in Egypt to have a management plan.

Wadi El-Rayyan is a unique natural protectorate in El-Faiyum Governorate, Egypt. The valley of Wadi El-Rayyan is an area of 1759, 113 km^2 of which are the dominating

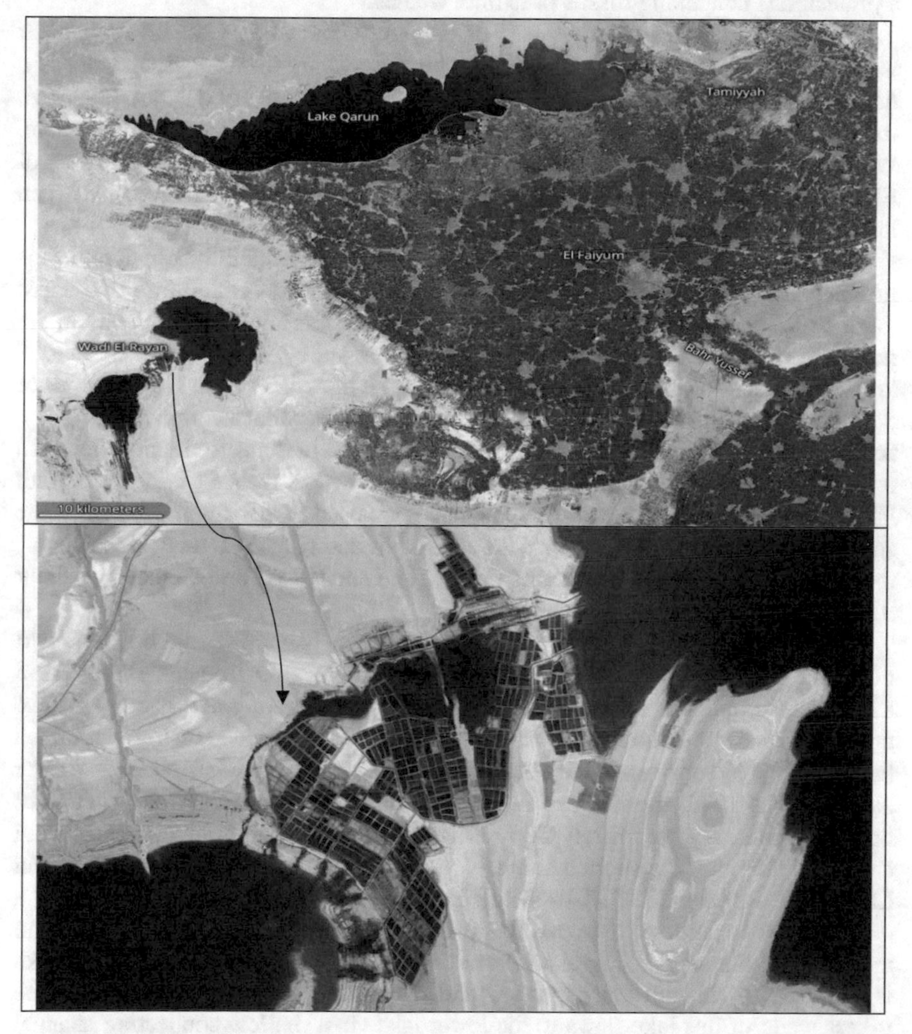

Fig. 3.30 Wadi El-Rayyan upper and lower lakes (upper image) as visible in Sentinel2 acquired on 31 Jul 2017 and fish farm (lower image)

water body of the Wadi El Rayan lakes. It is located about 65 km southwest of El-Faiyum city and 80 km west of the Nile River. The Wadi has been used for man-made lakes from agricultural drainage which has made a reserve of the two separate Wadi El-Rayyan Lakes. The reserve is composed of a 50.90 km^2 upper lake and a 62.00 km^2 lower lake, with waterfalls between the two. Among the springs, there are three sulphur springs at the southern side of the lower lake, with extensive mobile sand dunes. Wadi El-Rayyan Waterfalls are considered to be the largest waterfalls in Egypt.

At the south and southeast of the springs is Gabal Manqueer El-Rayyan, where marine fossils and archeological remains are found. Gabal Madwera, near the lower lake, is known for its extensive dune formations. In the northwest is Wadi El-Hitan, a protectorate containing fossils of extinct whales.

A low-lying depression southwest of Faiyum is known as Wadi El-Rayyan. It is separated from Faiyum by a thick barrier. The valley of Wadi El-Rayyan is a unique nature reserve located in southwest Faiyum. It boasts not one but rather two freshwater lakes - an upper and a lower one with a waterfall running in between them. Wadi El-Rayyan depression is situated in the Western Desert, 40 km southwest of El-Faiyum Province, and has an expected area of 703 km^2. It is situated between latitudes 28°45′ and 29°20′N and longitudes, 30°15′ and 30°35′E.

It is believed that Wadi El-Rayyan has never been connected to El-Faiyum. This means that it is free of Nile sediments. The Depression is divided into two parts: Wadi El-Rayyan (Al-Kabeer) in the south and Wadi El-Rayyan (Al-Saghir) in the north. The first lake (upper lake) is in the Masakhat Valley, with an area of about 65 km^2 and a depth of 22 m. The second lake has an area of 110 km^2 and a depth of 34 m (Al-Thamami 2008). The area called the whales Valleys compared to whale fossils. The capacity of the Wadi El-Rayyan reservoir is up to 18 m below sea level at about 2 billion cubic meters. The channel was cut from El-Faiyum to Wadi El-Rayyan to serve as a bank. The channel is exposed to 9.5 km at the end of the southwestern tip of the El-Faiyum to the edge of the desert, and tunnel tunneled below the barrier rock limestone separating the low, and about 8 km and diameter of about 3 m (Drc 2004).

The Wadi El-Rayyan project does not receive all the Faiyum sewage, but only part of it. The drainage of the lands north of El-Faiyum governorate goes to the Qarun Lake, and the drainage of the southern lands is transferred to Wadi El-Rayyan. The drainage of water in Wadi El-Rayyan changes the mountains around it to an industrial lake where it is considered the second industrial (Egyptian man) lake after Lake Nasser. Since 1973, the depression has been used as a reservoir for agricultural drainage water. Approximately 200 million cubic meters of drainage water from cultivated lands are transported annually via El Wadi Drain to the Wadi El-Rayyan Lakes (El-Shabrawy 2001). Two synthetic lakes (i.e., upper and lower) joined by a channel were constructed at two different altitudes (Fig. 3.30). The upper lake has an area of approximately 53 km^2 at an elevation of 10 m below sea level. The upper lake is completely filled with water and surrounded by dense vegetation (Saleh 1985). The excess water of this lake flows to the lower lake via a shallow connecting channel (Sayed and Abdel-Satar 2009). The lower lake is greater than the upper lake and has an estimated area of approximately 110 km^2 at an elevation of 18 m below sea level

(Mansour and Sidky 2003). The recorded maximum water depth in the lower lake is 33 m (Abd-Ellah 1999). The inflow of water to the lower lake varied from 17.68 $\times 10^6$ m^3 (March 1996) to 3.66 $\times 10^6$ m^3 (July 1996), with a total annual inflow of 127.2 $\times 10^6$ m^3/year (Abd-Ellah 1999). The area between these two lakes is used for fish farming.

3.8.5 El-Farafra Oasis Development: From Lake to Sand

Egypt is paid a great attention for the establishment of the new settlements and land reclamation projects, to overcome the over population crisis. Farafra Oasis is one of the prime targets for horizontal expansion plans in the desert (ASRT 1989). The depression of Farafra Oasis is located in the middle part of the Western Desert nearly 650 km to the southwest of Cairo (Fig. 3.31). El-Farafra depression is the second largest depression in the Western Desert of Egypt. It lies essentially between longitudinal 27° 20′ and 29° 00′ E, and latitude 26° 28′ and 27° 40′ N, it has a triangular shape with a total area inside the rims of the depression about 10,000 km^2.

The Farafra depression is a semi-closed basin with an irregular shape, where it is bounded by escarpments from the north, east and west. The depression floor gradually rises southward and merges into the plateau that forms the northern escarpment for Dakhla and Abu Minqar Oases (Fig. 3.31). The depression is mainly underlain by Palaeocene–Eocene sedimentary rocks, where the chalk of Upper Cretaceous (Khoman Formation) and Palaeocene (Tarawan Formation) covers large areas in the north, and the shale (Dakhla Formation) is mainly exposed in the south. Lower Eocene formations (El-Naqb and Farafra limestone) cap prominent escarpments. The eastern part of the Farafra depression is covered by sand dunes. These dunes generally extend southeast for about 200 km from the northeast corner of the depression to the northern escarpment of Dakhla depression. Interdune areas become wider southwards and eastwards, and therefore the decrease in sand supply dismantles the widely separated dune ridges into separate barchans before reaching the scarps of Dakhla depression (Embabi 2004). Sand sheets also cover large surface areas in different places and have consequently obscured most of the fluvial channels developed at wetter climatic phases.

Geomorphologically, it is characterized by certain conspicuous geomorphic units, namely; plateaux, badland and depression floor, where the latter is divided into the following sub-geomorphic units (Fig. 3.31); chalky plain, out- wash plain, Aeolian accumulations (sand sheets and sand dunes), playas, El-Hattias, sabkhas, alluvial fans, dry Wadis, mud flats, foot-hill slopes and undifferentiated complexes. El-Shazly (1976) and UNDP/UNWSCO (2001) concluded that the geomorphologic features displayed in the studied area includes: hills, ridges, lacustrine deposits, local marshes, crescentic sand dunes, and sand sheets.

Many of the formations of the White Desert-Farafra are sculpted by the harsh desert winds into weird shapes, which constantly change over time.

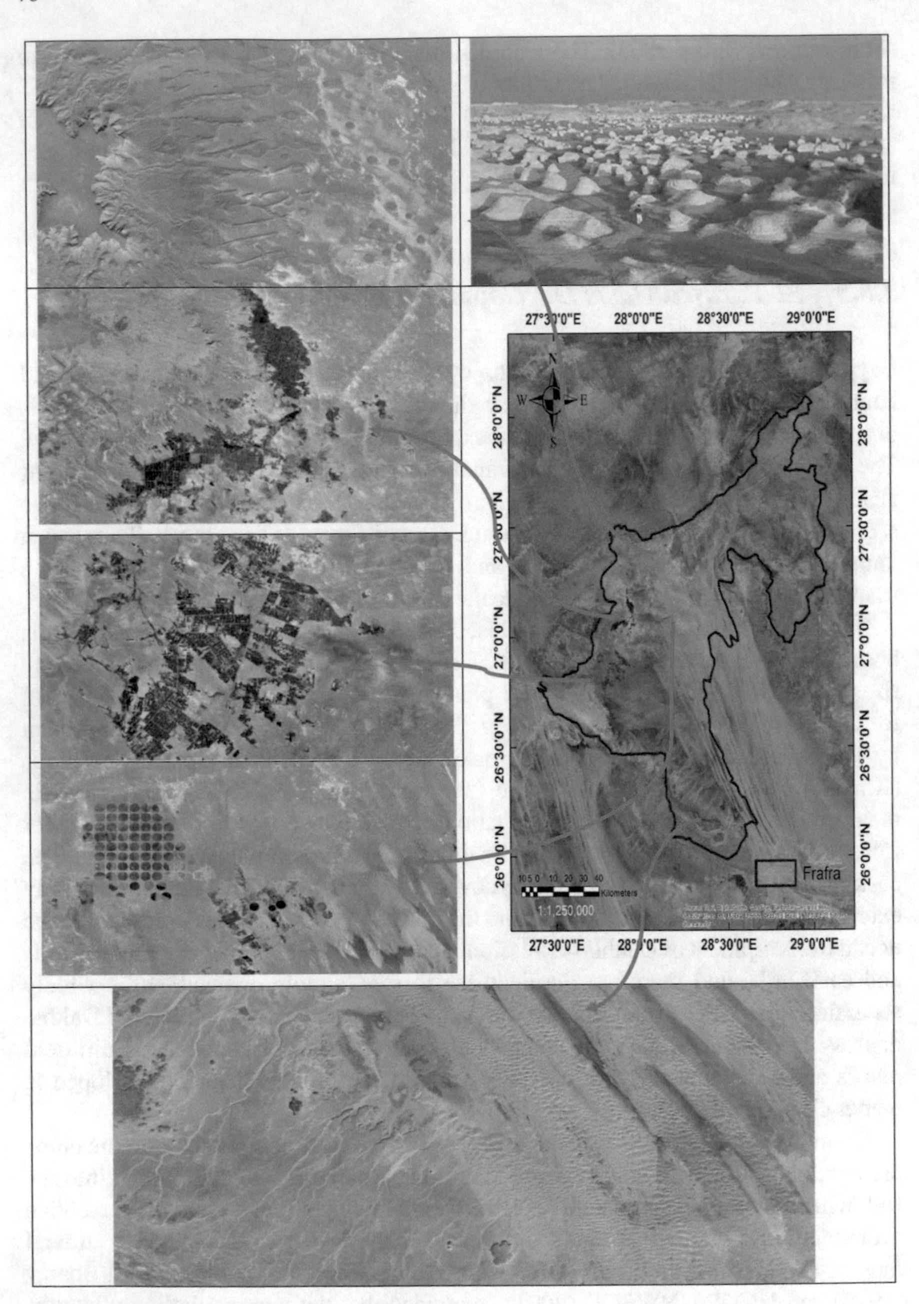

Fig. 3.31 El-Farafra Depression development

3.9 Protected Areas in Egypt

There are 30 nature reserves in Egypt (Fig. 3.32), which cover 12% of Egyptian land. Those nature reserves were built according to the laws no. 102/1983 and 4/1994 for protection of the Egyptian nature reserve. Egypt announced a plan from to build 40 nature reserves from 1997 to 2017, to help protect the natural resources and the culture and history of those areas. The largest nature reserve in Egypt is Gebel Elba (35,600 km^2) in the southeast, on the Red Sea coast.

3.9.1 Ras Mohammed

Ras Mohammed (Fig. 3.32) is the headland on the Sinai Peninsula's southernmost tip overlooking the Suez Gulf and the Aqaba Gulf. Tiran and Sanafir Islands. Littoral habitats include a community of mangroves, salt marshes, intertidal flats, a diversity of shoreline configurations and coral reef ecosystems, which are recognized internationally as some of the best in the world. A variety of desert habitats including mountains and wadis, gravel plains and sand dunes are also available. The boundaries of this National Park extend to the southern boundary of the Nabq Protected Area on the Suez Gulf from a point opposite the Qad Ibn Haddan lighthouse. The area (480 km^2) includes the island of Tiran and all shorelines fronting the Sharm el Sheikh tourism development area.

3.9.2 Zaranik Protected Area

Zaranik Protected Area (250 km^2, 68% water surface and 32% sand dunes) is located at the eastern end of Lake Bardawil on the Mediterranean coast of Sinai. The Protected Area (Fig. 3.32) is bordered from the north by the Mediterranean, from the south by the main Qantara-El-Arish road, from the east by tourist development areas, and from the west by Lake Bardawil.

3.9.3 Al Ahrash Protected Area

Al Ahrash Reserve is located in Egypt's northeastern corner, bordering the vast expanded sand dune area that reaches 60 m above sea level. The reserve is densely covered with numerous acacia trees, various camphor trees, bushes and pastoral plants, which help in the protection sector to fix sand dunes. They also help maintain an important characteristic of Mediterranean coast ecology that has been exposed to processes of development that change its original natural formations. This high

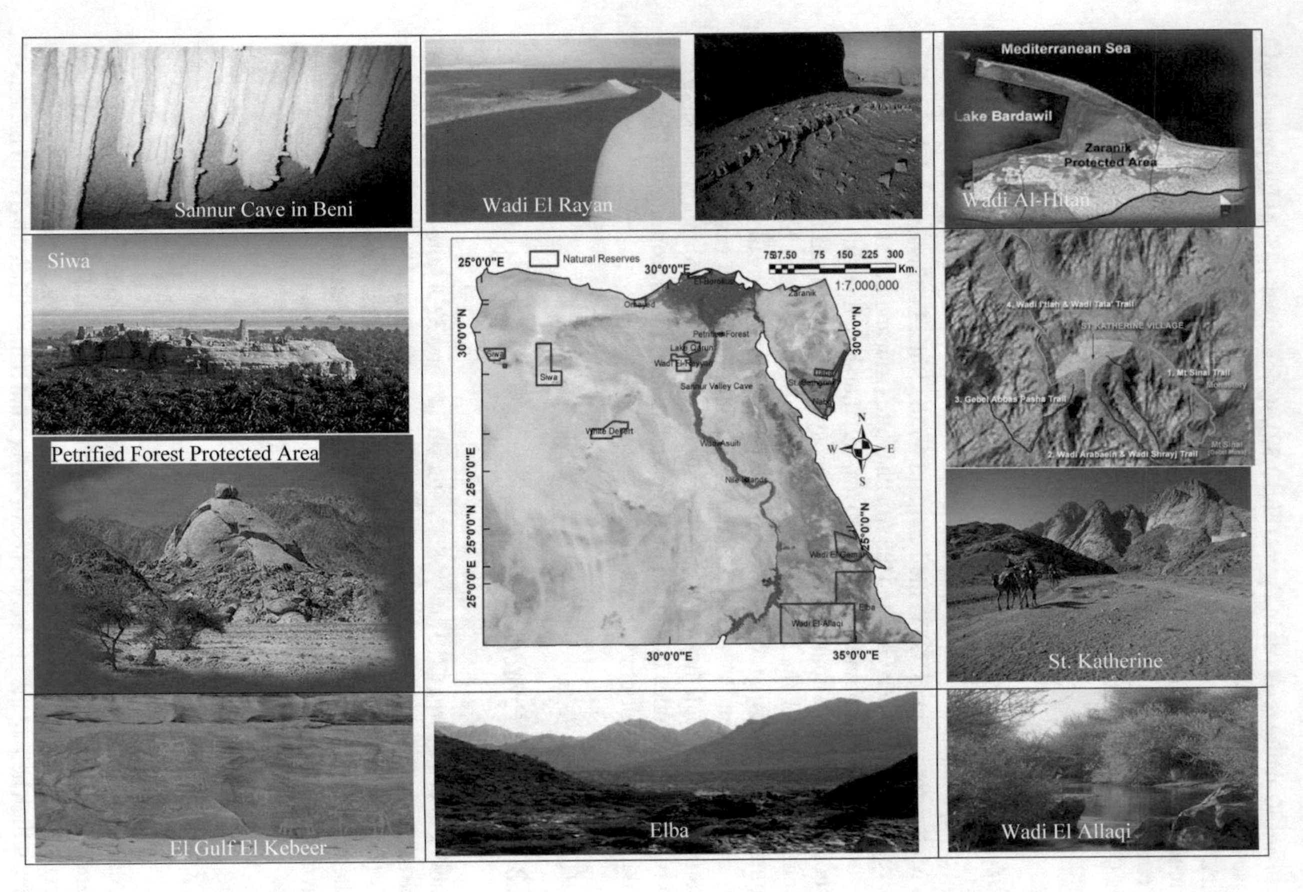

Fig. 3.32 Natural reserves (protected) areas

plant cover density also helps to increase the soil's water content to preserve the soil water and its salinity. In addition, due to its natural resources it causes the attraction of tourist activity. As one of the areas expected to have positive effects on soil protection, dense plant cover, and water resources, Al Ahrash Reserve is one of the sources that the state is interested in working for its conservation and protection.

3.9.4 The Elba Protected Area

The Elba Protected area (35,600 km^2) is an extensive and complex area comprising a number of ecosystems: The mangroves of the Red Sea, the Red Sea 22 islands, coral reefs, coastal sand dunes, coastal salt marshes, coastal desert plains and a cluster of coastal mountains (Jabal Elba, Jabal Ebruq and Al Daeeb). Jabal Elba is the single igneous mountain rising up to 1437 m. Its summit is a "mist oasis" where a considerable part of precipitation is contributed in the form of dew or mist and clouds, creating unique and rare ecosystem not found anywhere else in Egypt. Having an unusually high rainfall, this area is unique and can get up to 400 mm of precipitation per year. The Gabal Elba Nature Reserve contains mangroves and coastal forests in coastal strips.

3.9.5 Salouga and Ghazal

The two islands of Salouga and Ghazal in the River Nile are north of Aswan dam. Islands of Asbournati and Amoun and the plantation garden at Aswan are north of the reserve vicinity. As for Sohail island, it lies in its southern area. The reserve area is characterized by predominantly expanding kinds of bushes that survived after the establishment of the old Aswan dam and the High Dam. The huge greenery in the area comprises 94 kinds of various plants. The favorable natural conditions in the area provided wildlife chances for resident as well as migrating birds.

3.9.6 St. Catherine National Park

Conservation of biodiversity is the justification of the rationale of the National Park. Over geological time, this occurrence has increased, the biodiversity of the world is now richer than ever in its evolutionary history. At the same time, there is a loss of global biological diversity many times faster than ever before, largely due to human activities. St. Catherine (Fig. 3.32) National Park (5750 km^2) comprises much of the central part of South Sinai, a mountainous region of Precambrian igneous and metamorphic rock that involves the highest peaks of Egypt (St. Catherine, Moussa, Serbian, Umm Shomer and Tarbush). Mountain St. Catherine is Egypt's highest peak

2624 m above sea level. The Sinai massif includes some of the oldest rocks in the world. Approximately 80% of the rocks are 600 million years old.

3.9.7 Lake Qarun Nature Reserve

The lake is located in the province of El-Faiyum, 40 km long, 5.7 km wide and 34–43 m below sea level with an average depth of 4.2 m. Groundwater on the bottom of the lake seems to flow continuously from a number of sub-surface springs. A slightly sloping sandy plain stretches north and upward from the lakeshore to attain sea level 7 km north of the shoreline. Due to the presence of marine, fluvial and continental environments all in one area, the lake is an valuable archeological site with a unique collection of fossil fauna and flora dating back to about 40 million years. The reserve (230 km^2) comprises many monuments such as As-Sagha (goldsmiths) palace, which dates back to the Pharaonic Middle Kingdom and is located on the northern part. It looks like a small temple with seven compartments made of rosy stone. There is Abu Lifa Monastery, built in the monastic era on an elevated spot in the bosom of a mountain, three kilometers away from As-Sagha palace to keep monks safe from Roman oppression. Greek monuments involve relics from the old town of Skitnopius, once the point of departure for commercial caravans bound to the south of the desert. Fossils, many of which are kept at the Museum of Agriculture in Cairo, include those of an odd animal found to have lived on Mount Qatrani. Only in Lake Qarun was a carcass of the extinct mammal that lived in this area about 35 million years ago discovered. Ape Egpotothyx, the world's oldest ape dating back 32 million years ago, is the second major fossil in the reserve. This animal is the link between the apes of ancient and modern times.

3.9.8 Wadi El-Rayyan

Wadi El-Rayyan (1759 km^2) depression is located in the western desert of Egypt, about 65 km southwest of the town of El-Faiyum and 80 km west of the Nile River. The reserve is composed of:

- Upper lake (50.90 km^2).
- Lower lake (62.00 km^2).
- Falls between the two lakes.
- Springs, of which three are sulphur springs at the southern side of the lower lake, with extensive mobile sand dunes.
- Jabal Manqueer El-Rayyan at the south and southeast of the springs where marine fossils and archeological remains are found.
- Jabal Madwera near the lower lake, which is known for its extensive dune formations.

3.9.9 Wadi El-Allaqi

Wadi El-Allaqi (30,000 km^2) is a valley formed by the drying up of a large river, 275 km in length with an average width of 1 km. The reserve is divided into three zones: the core zone, the buffer zone and the transition zone.

3.9.10 Wadi Asuiti

The Asuiti Valley (50,000 feddans) begins as tributaries, the most essential of which is the area, which usually begins from the Qena Valley in the south and goes north. Among them is the Habib valley, which ends on both sides in the plateaus. The valley of Asuiti then expands west to meet the valley of the Nile in the form of a delta whose base is parallel to the Nile.

3.9.11 Petrified Forest Protected Area

The forest reserve is located at the border gate in the northeastern corner of Egypt, with dunes expanding up to 60 m above sea level. It is widely covered by high density of acacia trees, bushes and pastoral plants, which all work to fix the dunes to preserve the Mediterranean coast's important environmental feature that has been subjected to development projects that have transfigured its natural components. The high density of acacia helps to increase the soil water, which maintains the soil water and its natural qualities. Because of its natural resources, it also forms a kind of tourist attraction in the region. The site is one of the sources in the wildlife, animal and plant domains that the state aims to preserve and protect for its value and wealth.

3.9.12 Sannur Valley Cave

This cave is located ten kilometers southeast of Beni-Suef's big city. There are several quarries in the area, some of which have been discovered and used in ancient times by pharaohs, and others are modern and presently exploited.54 large cavities leading to a caves in the bottom of the earth due to the ongoing alabaster drilling operations. They incorporate ups and downs known geological formations. The cave spreads over an area of 700 m, 15 m wide and 15 m deep. The cave's most important characteristic is the value of its natural formations, which are the world's rarest. They are also important for researchers to carry out detailed comparative studies on variations in ancient environmental conditions.

3.9.13 Nabq

The Nabq Managed Resource Protected Area (600 km^2), located 35 km north of Sharm El-Sheikh, is an outstanding natural area containing varied ecosystems and habitat types. The largest Coastal Protected area on the Gulf of Aqaba, it contains a variety of ecosystems in the Sinai Peninsula.

3.9.14 Abu Galum

It is one of Egypt's most beautiful nature reserves (500 km^2). Its dramatic mountains of granite end abruptly on a narrow coastal plain, facing rich coral reefs. The region often includes narrow sinuous valleys, springs of fresh water, dunes of coastal sand, alluvial fans of gravel, elevated fossil reefs and low-lying semi sabkha. Deserts and wadis of high altitude: The reserves of South Sinai in general and Abu Galum in particular have varied types of ecosystems and habitats. Management of these areas is based on the premise that physical and biological processes link all contained habitats. Deserts of high altitude and connecting wadi systems form watersheds of catchment, providing habitats at lower elevations with fresh water. The highlands include a multitude of flora and fauna supporting micro-habitats, which are well suited to this environment. Flash floods will be washed via wadis during rare winter rains, carrying seeds and organic materials to lowland areas. There is also transportation of gravel and sands. These will often be deposited in downstream areas setting up new plant growth areas. Small, shaded hillside indentations will retain water for long periods of time and serve as water reservoirs for local fauna. These animals move to lowland areas when water and forage are scarce.

3.9.15 Taba

Taba Reserve is located in the southwest of Taba, with an area of 3590 km^2. Taba reserve involves mountain passages and valley network geological formation caves. The area includes some natural springs surrounded by Bedouins visited plantation gardens. It is a habitat of 25 mammalian species, 50 rare resident bird species, and 24 reptile species. There are 480 types of extinct species as for plants. It is also defined by ancient times that date back to 5000 years ago, rare wildlife and traditional Bedouin heritage.

3.9.16 El-Borollus Lake

It is located northeast of the Nile River Rosetta branch. It expands 10 km in length and 6–17 km in width It is about 460 km square and is regarded the second largest natural lake in Egypt. El-Borollus Lake is dominated by a number of environments, with salt swamps and sand plains being the most important of them. High dunes cover the coasts of the lake. Thus, 135 types of land and water plants are considered a natural location. In addition, receiving migrating wild birds is convenient. It intends to preserve El-Manzala Lake's biodiversity, rehabilitate species that have been extinguished as a result of human activities monitoring the lake's environmental change, and protect the humid areas. The reserve also aims to promote the tourism of the environment and conduct scientific and applied research. It also aims to preserve natural resources, particularly economic revenue resources.

3.9.17 Degla Valley

Degla Valley is considered part of the Northern Plateau, known in Egypt as a main distinguishing geographical environment. The Degla valley is situated between 56°, 29′ N and 24°, 31′ E latitudes. It is 30 km long, stretching from east to west. Degla valley begins in the form of small tributaries where rainfall water flows into the valley's surrounding hills. The area of the reserve is about 60 km^2. The Degla Valley's remarkable resources are its general scene, rich in plant and animal life. The valley is covered by a permanent protective plantation layer with 64 plant types. In this area, traces of the accessibility of deers have been newly recorded as well as 20 types of reptiles including Egyptian tortoises that are threatened with extinction. In addition to migrating and visiting birds in winter as well as the resident and visiting birds in summer, there are also 12 types of eastern desert birds.

3.10 Conclusions: Seeing the Whole Egypt from Space

Thirty-two figures (images and photos) were identified in the current chapter, which can be grouped to three groups.

First group of pictures portray Egypt, Nile, topography, agricultural land, and desertification in particular. The second group deals with the largest Nile islands, main wadis, and Egypt's top canyon. The third group, however, shed light on Egypt's canals, depressions, oases, and protected areas. Conclusions adapted to these groups can be summarized as follow:

1. While the author, believe that "Egypt is the Gift of the Nile and Egyptians together". The river passes through several countries starting from the upstream and downstream so. There was no giant civilization influenced in human like

ancient civilization. Nile water cannot make a civilization, but it is the Egyptian man, who controls Nile runoff and digging canals and drains. Is there found the people other than the Egyptian have a calendar for four thousand years BC?

2. The topography of Egypt reveals that agricultural expansion, after securing irrigation water, could be oriented mainly toward the Western Desert, which has several depressions such as Qattara Depression along the northern third of this desert.
3. This chapter was utilized MODIS Aqua EVI instead of NDVI data for the inventory and mapping the current agricultural land in Egypt. Time series analysis of satellite data revealed the vigor pattern of cultivated land. Agricultural area of Egypt on 17th January 2019 is estimated at 10.15 million feddan (4.09% of Egypt area).
4. The majority of the planting areas were located west of the Nile. While the cultivated areas east of the Nile were much lower than in the west Bank of the Nile. The river between Aswan and Cairo always has a tendency to adhere to the right side of the valley. It does not rotate to the left side of the river until it reverts to the right side.
5. The main causes of desertification in Egypt are focused in the demographic pressure, water shortage, poor water management, unsustainable agricultural practices, biodiversity loss, and intensive cultivation in rain-fed areas. There are five areas characterized by serious coverage of sand dune encroachment; west of the Nile Delta (255 km^2), El-Faiyum and Wadi El-Rayyan Depression (480 km^2), Southwest El-Minya governorate (350 km^2), Al Kharga Oasis (400 km^2) and Northwestern High Dam Lake (800 km^2.).
6. Herodotus since about 500 BC was the first one to name the Delta, because of the Latin Delta (▼) character. The Nile basin by nearly 10,000 years has several branches (nine branches) and then reduced to seven branches, then five, then three and finally the existing branches of Damietta and Rosetta. These branches have been silting in time where the river was acting a little bit, and therefore the rate of deposition of silt over these branches. The failure silting of Damietta and Rosetta branches may be due to the branches that were ending directly into the sea, which are destined to inflation and survival.
7. The Nile islands are one of the most prominent geomorphological phenomena of the Nile River. It is one of the most common geomorphological phenomena in its dimensions and movement. This is due to the natural factors of water flow and associated changes in the load of carving, transport and sedimentation, as well as the human factors, which are the establishment of dams, arches)Al Qanater) and stone heads on the river.
8. Islands of the Nile from Geomorphological features are considered distincting more in the river Nile in Egypt. It amounts 507 islands, which occupies areas 144.05 km^2.
9. From an economic point of view, wadis and their oases provide people and animal populations with habitats. Settlements are emerging around an oasis-laden wadi. Agriculture begins as people settle down. For instance, Feiran's Wadi attracts thousands of visitors because of the large dates harvests that grow along

its course. Wadis create a unique and distinctive environment system defined by the diversity, variety, and wealth of natural attractions from an ecological significance. The wadis have some of the best mammals, birds and amphibians in their oasis. Based on a cultural point of view, the Wadis of Egypt have a very rich cultural significance to the Egyptians.

10. There are hundreds of canyons in South Sinai, as you find narrow valleys all over the region. Water, however, only flows seasonally in most of them after rain, for a longer or shorter time.
11. Five natural lakes lie adjacent to the Mediterranean Sea; Lake Mariut, Edku, Borollus and Manzala (deltaic section) and Bardawil (North Sinai). These lakes are kept separate from the sea by narrow splits and are not more than 2 m deep. They provided fish and recreation.
12. The current pattern of these lakes is changing rapidly, due to natural developments and, commonly, to man's activities (e.g., fishing and agricultural practices). Unfortunately, most of these lakes have deteriorated sharply over the last twenty years due to the wastewater being discharged into them.
13. Climate information confirms that North Sinai is one of the best areas in the country that can be developed sustainably using renewable groundwater sources, because the region receives the highest rainfall in all of Egypt. There is no alternative to the use of renewable groundwater as one of the most important natural sources of life in the Sinai. Moreover, northern Sinai is an ideal place for agricultural and population development. It is characterized by its moderate climate and is relatively close to densely populated cities such as El-Arish and the Suez Canal cities, in addition to its Bedouin population.
14. Lake Nasser has a number of known as khors. There are 100 important khors in Lakes Nasser and Nubia combined. Their total length when the lake is full is nearly 3000 km and their total surface area is 4900 km^2 (79% of total lake surface). In volume, they contain 86.4 km^3 water (55% of total lake volume). Lake Nasser and its Khors are key players in the environmental planning, restoration and management of water resources in Egypt. The knowledge of the Lake's Khors is based on observation of the dimensions of their water basins.
15. Deserts are one of the most notable characteristics of Western Sahara. It is rare to find a similar area where such a number of major depressions occur. Some of these depressions are uninhabited and of an immortal type such as a Qattara Depression. Most of the northern depressions are located below sea level, while the central and southern depressions is located above sea level. The order of depressions is descending by area as follows: Qatara, Farafra, Kharga, Dakhla, Bahariea, El-Faiyum, Siwa, Rayyan and Natroun. As for the origin and composition of the depressions, the views differed between the tectonic origin and the Aeolian origin.
16. Transforming the mid-Saharan Qattara Depression region (30,000 km^2) into a completely self-sufficient and habitable community for 3–5 million people, generating well paid jobs for a multitude of citizens, working in advanced industries and modern agricultural activities. This project relies primarily on technology for hypersaline osmotic power generation of 3–4 Gigawatts of power, which

will be used in constructing the new Qattara city; desalinate billions of cubic meters of brackish and agricultural drainage water to cultivate one million acres.

17. The economic return from the use of solar ponds as a source of thermal energy required to de-saline seawater is a useful and proven return. In addition, the cost of producing a cubic meter of water desalinated with solar ponds is virtually non-existent because it uses a natural source of energy (solar radiation), unlike conventional stations where the cost of production is high because of its dependence on fossil fuel sources or electricity. There is no doubt that solar ponds are economically and environmentally superior to other sources of energy currently used in seawater desalination.
18. Egypt is paid a great attention for the establishment of the new settlements and land reclamation projects, to overcome the over population crisis. Farafra Oasis is one of the prime targets for horizontal expansion plans in the desert. The depression of Farafra Oasis is located in the middle part of the Western Desert nearly 650 km to the southwest of Cairo. El-Farafra depression is the second largest depression in the Western Desert of Egypt.
19. There are 30 nature reserves in Egypt, which cover 12% of Egyptian land. Those nature reserves were built according to the laws no. 102/1983 and 4/1994 for protection of the Egyptian nature reserve. Egypt announced a plan from to build 40 nature reserves from 1997 to 2017, to help protect the natural resources and the culture and history of those areas. The largest nature reserve in Egypt is Gebel Elba (35,600 km^2) in the southeast, on the Red Sea coast.

Acknowledgements Abdelazim M. Negm acknowledges the partial financial support from the Academy of Scientific Research and Technology (ASRT) of Egypt via the bilateral collaboration Italian (CNR)–Egyptian (ASRT) project titled; Experimentation of the new Sentinel missions for the observation of inland water bodies on the course of the Nile River.

References

Abd el Malek Y, Rizk SG (1963) Bacterial sulphate reduction and the development of alkalinity. III. Experiments under natural conditions. J Appl Bacteriol 26:20–26

Abd-Ellah R (1999) Physical limnology of El-Fayoum depression and their budget. Ph.D. thesis. Faculty of Science, South Valley University, 140 pp

Abdin A, Gaafar I (2009) Rational water use in Egypt. Options Mediterranean. An 88, 2009—Technological Perspectives for Rational Use of Water Resources in the Mediterranean Region, pp 11–27

Abu Zeid KA (1984) Contribution to the geology of Wadi El Natrun area and its surroundings. M.Sc. thesis. Faculty of Science, Cairo University

Abu Zeid M (1987) Environmental impact assessment for Aswan High Dam. In: Biswas AK, Geping Q (eds) Environmental impact assessment for developing countries, London, pp 168–190

Abu Zeid M (1993) Egypt's water resources management and policies. In: Faris M, Khan M (eds) Sustainable agriculture of Egypt. Lynne Rienner Publishers, Inc., Boulders, Colorado, pp 71–79

AbuBakr M et al (2013) Use of radar data to unveil the paleolakes and the ancestral course of Wadi El-Arish, Sinai Peninsula, Egypt. Geomorphology 194:34–45

Agha K (1994) Solar ponds and their applications, center for solar energy studies. J Energy Life Tripoli (3)

Al-Agha DE, Closas A, Molle F (2015) Survey of groundwater use in the central part of the Nile Delta. Water and salt management in the Nile Delta: Report No. 6, 2015

Al-Thamami A (2008) The features and civilization of Fayoum Coptic and Islamic. Family Library

Aquastat (2013) FAO. https://www.FAO.org/nr/water/aquastat/irrigationmap/egy/index

Aronoff S (2005) Remote sensing for GIS managers. ESRI Press, Redlands, California

ASRT (1989) Encyclopedia of the Western Desert of Egypt. Academy of Scientific Research and Technology 4th part, Desert Research Institute

Awad M (1998) The Nile river. Egyptian General Book Authority

Awad MA (2005) River Nile. Egyptian General Book Organization (Library of the family), Cairo

Beck P et al (2006) Improved monitoring of vegetation dynamics at very high latitudes: a new method using MODIS NDVI. Remote Sens Environ 100(3):321–334

CAPMAS (2016) Central agency for public mobilization and statistics. https://www.capmas.gov.eg/

Chorley RJ (1973) Introduction to fluvial processes. Methuen & CO LTD, Great Britain

Chowdary VM et al (2008) Assessment of surface and sub-surface waterlogged areas in irrigation command areas of Bihar state using remote sensing and GIS. Agric Water Manag 95:754–766

Deba GS et al (2011) Optimal design of solar pools. Tishreen Univ J Res Sci Stud (Engineering Sciences Series), Tripoli 33(6)

DRC (2004) D.R.C., Project of the use of low quality water and its impact on the characteristics of low quality water and its impact on the characteristics of land and vegetation in Wadi Al Rayyan (**First Progress Report**)

DRI (1993) D.R.I., drainage water, Volume III. Drainage water reuse project

El Bedawy R (2014) Water resources management: alarming crisis for Egypt. J Manag Sustain 4(3):108–124

Elba E, Dalia F, Brigitte U (2014) Modeling high Aswan Dam reservoir morphology using remote sensing to reduce evaporation. Int J Geosci 5:156–169

El-Nahal MA et al (1977) Soil studios on the Nile Delta. Egypt J Soil Sci 17(1):55–65

Elsaeed G (2012) Effects of climate change on Egypt's water supply. In: Fernando HJS et al (ed) National security and human health implications of climate change. NATO Science for Peace and Security Series C: Environmental Security, pp 337–347

El-Shabrawy G (2001) Ecological studies on macrobenthos of Lake Qarun, El-Fayum Egypt. J Egypt Acad Soc Environ Dev 2:29–49

El-Shabrawy MG (2009) Lake Nasser–Nubia. In: Dumont HJ (ed) The Nile: origin, environments, limnology and human use. © Springer Science + Business Media B.V.

El-Shazly E (1976) Geology and ground water potential of Kharga and Dakhla Oases area, Western Desert, Egypt from NASA landsat-1 satellite images. Academy of Scientific Research, Cairo

Embabi NS (2004) The geomorphology of Egypt. Landform and evolution: The Nile valley and Western Desert, vol 1. The Egyptian geographical society, special pub., Cairo, Egypt

FAO (1965) Simplified map of the Desert fringes of the Nile valley and Delta. High Dam Soil Survey Project. Ministry of Agricultuie

Gadallah H et al (2014) Application of Forward/Reverse Osmosis Hybrid System for Brackish Water Desalination using El-Salam Canal Water, Sinai, Egypt, Part (1): FO Performance. In: 2014 4th international conference on environment science and engineering IPCBEE, vol 68 © (2014). IACSIT Press, Singapore, p 2

Hamdan G (1984) The personality of Egypt. Dar Al Hilal

Harms JC, Wray JL (1990) Nile Delta. In: Said R (ed) Chapter 17. Geology of Egypt

Hawas A H, Al-Mansouri KA (2004) Operation of desalination plant by a dual system of solar pool and wind turbine. Center for Solar Energy Studies, Tripoli

Hereher T (2013) The status of Egypt's agricultural lands using MODIS Aqua data. Egypt J Remote Sens Space Sci 16:83–89

Hidayat S (2012) Sinai channel, a new vision for the world navigation parallel to the Suez Canal. Al-Ahram

Hozaien S (1991) Egyptian civilization. Dar Al Shorouk. ISBN: 9770900252
Hussein IAG (2006) Desertification challenge in Egypt. In: Environment, health and sustainable development (IAPS 19 Conference Proceedings on CD-Rom). Alexandria, Egypt
Kim J, Sultan M (2002) Assessment of the long-term hydrologic impacts of Lake Nasser and related irrigation projects in southwestern Egypt. J Hydrol Hydromechanics 262:68–83
Leopold LB, Wolman MG, Miller JP (1964) Fluvial processes in geomorphology. W.H. Freeman and Company, London
Lonergan S, Wolf A (2001) Moving water to move people. Water Int 26:589–596
Mahsoub SM (1996) Natural geography foundations and modern concepts. Arab Thought House
Mansour S, Sidky M (2003) Ecotoxicological studies. 6. The first comparative study between Lake Qarun and Wadi El-Rayan wetland (Egypt), with respect to contamination of their major components. Food Chem 82:181–189
Minar MH, Hossain B, Shamsuddin MD (2013) Climate change and coastal zone of Bangladesh: vulnerability, resilience and adaptability. Middle-East J Sci Res 13(1):114–120
Molouk KE (2015) The expected economic effects of the new Suez Canal project in Egypt. Eur J Acad Essays 1(12):13–22
Morsy WS (2009) Environmental management to groundwater resources for Nile Delta region. Ph.D. thesis, Faculty of Engineering, Cairo University, Egypt
Mostafa M, Soussa H (2006) Monitoring of Lake Nasser using remote sensing and GIS techniques. ISPRS, Netherlands, 5p
Mostafa H, El Gamal F, Shalby A (2005) Reuse of low quality water in Egypt. In: Hamdy A, El Gamal F, Lamaddalena N, Bogliotti C, Guelloubi R (eds) Non-conventional water use: WASAMED project (Options Méditerranéennes: Série B. Etudes et Recherches; n. 53). CIHEAM/EU DG Research, Bari, pp 93–103
Mujahid MM (2002) Energy resources in Egypt and prospects for development. The Academic Library, Cairo
News C (2015) New Suez Canal project proposed by Egypt to boost trade. Net. 5 Aug 2014
Omran EE (2012a) A neural network model for mapping and predicting unconventional soils at a regional level. Appl Remote Sens J 2(2):35–44
Omran EE (2012b) Detection of land-use and surface temperature change at different resolutions. J Geogr Inf Syst 4(3):189–203
Omran ESE (2016) Early sensing of peanut leaf spot using spectroscopy and thermal imaging. Arch Agron Soil Sci 1–14
Omran ESE (2018) Remote estimation of vegetation parameters using narrowband sensor for precision agriculture in arid environment. Egypt J Soil Sci 58(1):73–92
Oxford Business Group (2015) Digging for victory: citizens raise around $9bn for Suez project in eight working days
Post H (2014) Egypt plans to dig New Suez Canal in effort to boost trade
Quan RS et al (2010) Waterlog-ging risk assessment based on land use/cover change: a case study in Pudong New Area, Shanghai. Environ Earth Sci 61, 1113–1121
Qureshi AS et al (2008) Managing salinity and waterlogging in the Indus basin of Pakistan. Agric Water Manag 96(1):1–10
Rahim SI (2006) Compilation of a soil and terrain data base of the Nile Delta at Scale 1:100000. J Appl Sci Res 2(44):226–231
RIGW (1992) Groundwater resources and projection of groundwater development. Water security project, (WSP), Cairo.
Safi al-Din M (2001) The crust of the earth. Dar Ghraib for printing, Cairo.
Said R (1990) The geology of Egypt. Balkema, Rotterdam, 734p
Said R (1993) The River Nile; geeology, hydrology and utilization. Pergamon Press, Oxford
Saleh M (1985) Ecological investigation of inorganic pollutants in El-Faiyum and El-Raiyan aquatic environment. Supreme Council of Universities, FRCU Report, pp 1–54
Saleh I (1998) Project Toshka. Human Investment Development, Family Library

Sayed M, Abdel-Satar A (2009) Chemical assessment of Wadi El-Rayan lakes Egypt. Am-Eurasian J Agric Environ Sci 5(1):53–62
Shahin AAW (1978) Some of the geological phenomena in the Nile Delta. Arab Geographical Mag (11):9–26
Shaltout KH, Khalil MT (2005) Lake Burullus (Burullus protected area) (No. 13). Publication of national biodiversity unit
Stanley DJ, Warne AG (1993) Nile Delta: recent geological evolution and human impact. Sci Technol 260:628–634
Taha MM (1997) Geomorphology of sedimentary islands in Egypt. Geogr J 29(1)
Tamburelli P, Thill O (2013) The Nile metropolitan area. Berlage-Institute, TU Delft
Toussoum O (1922) Memire sur les annciennes branches du Nil. Imprimeric d'Instit. Francais. Epoque ancienne. T.IV.D'archeologie Orientale. Cairo
UNDP/UNESCO (2001) Joint project for the capacity building of the Egyptian geology survey and mining authority and the national authority for remote sensing space science for the sustainable development of the south valley and Sinai
University of Texas at El Paso (2002) Thermal desalination using MEMS and salinity-gradient solar pond technology. El Paso, Texas
van Zwieten PAM et al (2011) Review of tropical reservoirs and their fisheries—the cases of Lake Nasser, Lake Volta and Indo-Gangetic Basin reservoirs. FAO Fisheries and Aquaculture Technical Paper (No. 557). Rome, FAO, p 148
Wahaab R, Badawy M (2004) Water quality assessment of the River Nile system: an overview. Biomed Environ Sci 17:87–100
Wardlow BD, Egbert SL (2010) A comparison of MODIS 250-m EVI and NDVI data for crop mapping: a case study for southwest Kansas. Int J Remote Sens 31:805–830
WorldFish (2018) Management plan for the Lake Nasser fishery: stock assessment study. WorldFish, Penang, Malaysia (**Program Report: 2018–20**) (2018)

Part II
Environmental Applications

Chapter 4
Environmental Applications of Remote Sensing in Egypt: A Review and an Outlook

Mohammed E. Shokr

Abstract Remote sensing has been used to cover a wide range of environmental applications in Egypt. The customary description of the applications has been introduced in several publications. This chapter is written to highlight key applications that have been traditionally used as well as modern applications that link local environmental issues to their regional and global sources. The latter is particularly important, given the pressing issues of climate change and the rapid growth in population and human activities that already have their remarkable impacts on the environment, especially around mega cities and coastal areas. Traditional applications involve using optical remote sensing for mapping natural resources (e.g. minerals, crops, groundwater, and aquatic ecosystem), as well as monitoring key environmental phenomena (e.g. desertification, coastal erosion, sand dune movement and atmospheric aerosols). Newly potential applications signify ventures in the fields of air quality, renewable energy, regional resources and climate-related geophysical processes. The material in this chapter, though not comprehensive, aims at both reviewing and expanding the domain of remote sensing environmental applications in Egypt, with an intention to persuade scientists to deliver new research novelty related to the environmental parameters and processes.

Keywords Remote sensing · Earth observation satellites · EO applications in Egypt · Optical and thermal infrared sensors

4.1 Introduction

Egypt is one of a few countries in the world that has a profound historical origin. The ancient Egyptian civilization is known for its advancement in agriculture, architecture, astronomy, and advanced use of natural (particularly mineral) resource. These achievements had never been detachable from the influence of the environment.

M. E. Shokr (✉)
Environment and Climate Change Canada, 4905 Dufferin St., Toronto, Ontario M35T4, Canada
e-mail: mo.shokr.ms@gmail.com; mohammed.shokr@canada.ca

National Authority for Remote Sensing and Space Sciences (NARSS), 23 Joseph Tito Street, El-Nozha El-Gedida, P.O. Box: 1564 Alf Maskan, Cairo, Egypt

S. F. Elbeih et al. (eds.), *Environmental Remote Sensing in Egypt*,
Springer Geophysics, https://doi.org/10.1007/978-3-030-39593-3_4

While the famous say of the Greek historian Herodotus "Egypt is the gift of the Nile" is commonly circulated, it is also passable to say that Egypt is the gift of its environment. The Nile River provides 95% of water available to the country, but the right environment around its passages has also been an indispensable factor in flourishing the great civilization of Egypt. For thousands of years, Egyptians lived in complete harmony with their natural environment. Yet, there has been an upsurge of the population in the past few decades (estimated to be 94.8 Million according to the formal census in 2017 by the Central Agency for Public Mobilization and Statistics) at a rate of increase near 2.5%. This, along with the rapid growth of human activities, have caused substantial changes of the environment, sometimes in damaging ways. This is where remote sensing has emerged as a prime tool to monitor the environment and assess the impacts of its change on many aspects of life.

Geographically speaking, Egypt is located at the eastern margin of the Sahara, Africa's largest desert. About 95% of Egypt is desert with the dry environment. However, the Nile River, which originates mainly from rainfall on Ethiopian Hills and equatorial lakes in Tanzania, Kenya, and Uganda, ends in a vast delta in Egypt as it flows into the Mediterranean Sea after traveling some 6800 km. Egyptians thrived along the river for more than 3000 years. The black soil carried by the river enables Egyptians to develop a prosperous agriculture-based economy. As a mid-latitude country, Egypt is practically covered daily by wide-swath satellite sensors such as NOAA's Advanced Very High-Resolution Radiometer (AVHRR), the Moderate Resolution Imaging Spectroradiometer (MODIS), and the Advanced Microwave Scanning Radiometer (AMSR-E followed by AMSR2). Less frequent coverage is achieved by mid-resolution sensors such as Landsat and Sentinel-2. A map of Egypt showing places that will be referred to in this chapter is shown in Fig. 4.1.

A map of population distribution in Egypt is presented in Fig. 4.2. The majority of the population is concentrated in the Nile Delta (ND) region, including the two major cities of Cairo and Alexandria (Fig. 4.1), as well as the Nile Valley (NV); a narrow strip of the Nile's fertile valley extending from the south to Cairo of Aswan, also called Wadi. Less population density is found along the north coast of the Mediterranean Sea and the east coast of the Red Sea. Small communities are clustered in the Western desert around a few oases. This distribution defines priorities of remote sensing mainly as the government continues to encourage migration to newly reclaimed lands in the desert. Areas of dense populations should be monitored for land use changes and impacts of anthropogenic activities on the environment. On the other hand, remote areas of light or no population should be monitored for mapping natural resources and their potential for future land reclamation. Generally speaking, populated areas need fine-resolution data while coarse-resolution data might be sufficient for rural areas.

From the climatic point of view, Egypt has long dry summers and relatively short winters (December–February). Summer is almost cloud-free, and winter is characterized by frequent passages of upper westerly troughs associated with surface depressions at a rate of 3–5 per month. These depressions generate thick layers of low and medium clouds covering the northern part of Egypt, and it may extend to Cairo and southern areas (El-Fandy 1948). This limits the precipitation season to winter

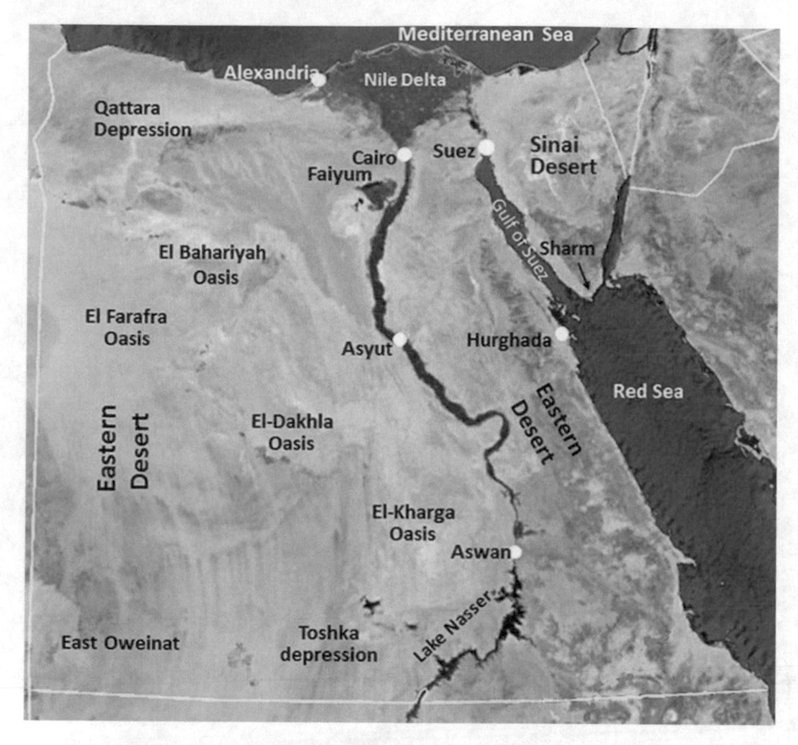

Fig. 4.1 Map of Egypt with a few sites used in the material covered in this chapter

when rain varies between 200 mm annually in the wettest areas along the north coast to nearly 0 mm in the central and southern parts of the country. The average over the entire country is around 20 mm annually.

Under almost cloud-free and precipitation-free skies with a sunshine duration that varies between 3300 h along the northernmost part and 4000 h farther in the interior, optical remote sensing applications become most suitable. A well-known sand/dust storm phenomenon, called *Khamsin*, is triggered by strong, dry wind (speed up to 120 km/h) that sweeps unobstructed across vast desert areas of the south or southwest, picking up sand and dust particles. This causes very poor visibility and air quality accompanied with rising in temperature by 10 °C or more within 2 h. Applications of remote sensing to monitor this phenomenon and air quality, in general, have been addressed in several studies using optical remote sensing (Solomos et al. 2018; El-Ameen et al. 2011; Marey et al. 2011).

The use of remote sensing data for monitoring and modelling environmental processes has remarkably grown worldwide in the past three decades. The wide range of spectral, radiometric and spatiotemporal resolution has expanded the environmental applications at several fronts including crop and watershed mapping, coastal erosion, land cover change, evaporation and evapotranspiration, water and air quality,

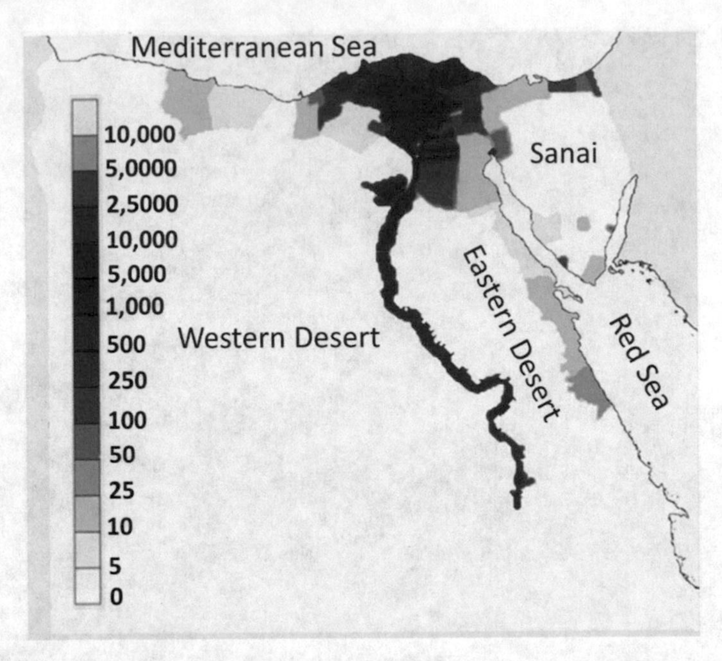

Fig. 4.2 Population density of Egypt in 2013. The scale in the color bar is the population/km^2 (*source* https://sedac.ciesin.columbia.edu/gpw/)

and energy flux estimation. Remote sensing data have also been used as input for environment and climatic processes modeling.

The rest of this chapter reviews vital applications that use remote sensing for environmental monitoring and modeling in Egypt and provides a few hints to potential applications that may hopefully receive priorities in the future. This includes broadening the range of the applications and using more comprehensive retrieval approaches such as physical and radiative transfer modelling to improve the retrieval accuracies (may be considered as vertical expansion). The spatial domain of remote sensing applications in Egypt may also be reconsidered by stepping up from local to regional. The availability of the vast data volume, free of charge, from many sensors, including radar, should be a motivation to pursue these new directions. Discussions in this chapter are limited to the use of satellite (not airborne) remote sensing.

4.2 Impacts of Global Environmental Aspects on Local Environment

"Think regionally and think globally"; this is a theme that the author of this chapter has frequently communicated to different research and application groups of remote sensing in Egypt, including undergraduate and graduate students. It merely entails studying any local environmental phenomenon in terms of its regional/or global origin.

In many cases, applications of remote sensing in Egypt are limited to mapping resources and monitoring processes within the borders of Egypt. Connection to global or regional environmental phenomena is usually pursued through exploring their impacts on local environmental aspects, without addressing the global view of the phenomena; namely their origin, causes, mechanisms, and extent. This limits the contribution of the remote sensing studies in Egypt to the wider international community. Expanding the range of research and applications of remote sensing to encompass regional and global aspects will not only furnish a tangible contribution to international efforts but will also add more depth to the local studies. The following paragraphs address a few connections between local environmental phenomena to their regional and global origins briefly.

At a regional scale, the completion of the Grand Ethiopian Renaissance Dam (GERD) will have impacts on the environment and water resources in Egypt. Aside from the fact that this has been a thorny political issue, remote sensing can be used to assess and quantify the impacts of this major project. It has been documented that the flow of Nile water to Egypt will be reduced by as much as 25% during the period of filling the reservoir upstream of the dam (Stanley and Clemente 2017), which is estimated to be within 3–4 years. Extending the remote sensing studies in Egypt to cover the GERD site is undoubtedly warranted. An example of this effort is presented in Mohamed and Elmahdy (2017), who used the Shuttle Radar Topographic Mission (SRTM) and the Advanced Radar Observation Satellite Phased Array L-band Synthetic Aperture Radar (ALOS-PALSAR) to assess the geological and meteorological factors that may lead to the GERD failure.

Expanding the geographic domain of remote sensing studies is also manifested in utilizing the data in regional climate models to explore, for example, possible future changes of the rainfall over the lakes that feed the River Nile at its origin in response to the recent global warming. Some modeling endeavors have been undertaken by researchers in the Egyptian Meteorological Authority (EMA) and the Nile Forecast Center (NFC) of the Ministry of Irrigation and Water Resources (https://www.emwis-eg.org/).

The Nile Delta Plain is only about 1 m above present sea level. The strata underlying the plain is being compacted while the supply of new sediment to re-nourish the delta is lacking since the 1970s as a result of building the High Dam in Aswan. This, combined with the sea level rise (SLR), has raised the alarm of possible erosion or even sinking of the northern margin of the Mediterranean coastal area. This possibility has been studied using radar remote sensing, and it has been estimated that northern third of the delta is subsiding at a rate of about 4–8 mm per year (Geological Society of America 2017).

Using satellite altimetry data to address the SLR in the Mediterranean and explore its impacts on the coastal environment in Egypt, Shaltout et al. (2015) found that it rose from 1993 to 2011 by approximately 3.0 cm decade^{-1}. Other studies on the impacts of SLR on flooding in coastal areas and the rise of soil salinity in the ND region are presented in Said et al. (2012). El-Raey (1997) predicted that the city of Rosetta, located on the Mediterranean coast, would sink in a matter of decades unless climate change is addressed on a global scale. The prime cause of the SLR is the retreat of massive glaciers in the polar regions. The Egyptian remote sensing community will be in a better position if some effort is directed to study melting events in the polar regions in relation to SLR, not just the impacts of SLR on Egypt's territories. This is precisely where the recommendation presented in the opening of this section should come into force.

The decline of sea ice in the Arctic, whether viewed as a cause or a consequence of climate change, will increase the heat flux from the ocean to atmosphere in autumn and early winter, accompanied with a parallel increase in air temperature, moisture and cloud cover in the lower troposphere. This will likely increase precipitation in Europe, the Mediterranean, and East Asia (Vihma 2014). Would that affect the frequency of flash floods in coastal areas, Sinai and Upper Egypt? Once again, the answer requires participation in studies of the global and regional water cycle and the earth's energy budget.

Another global-scale study field that deserves the attention of the remote sensing community in Egypt is the recent accelerated changes in the delicate balance of the atmosphere, especially when instigated by anthropogenic causes. The changes have altered the solar radiation processes (e.g. absorption, scattering, and re-emission) and consequently the earth's energy budget. Changes are instigated mainly by the rise in the emission of greenhouse gases (GHG) emission and the release of several kinds of anthropogenic aerosols from industries and mega cities. It should be noted that data of GHG emissions in Egypt have been published and updated continuously (Egyptian Environmental Affairs Agency 2016). Significant air pollution events worldwide leave their local impacts in many regions. For example, any disturbance in the world's largest tropical rain forest, the Amazon basin, may lead to tangible changes in the global carbon budget. This has affected the diversity and complexity of the ecosystem worldwide. Remote sensing is undoubtedly a prime tool for detecting and monitoring air quality.

The above discussions point to the need for expanding the spatial domain of remote sensing studies beyond the geographic borders of Egypt and increasing the diversity of the applications to address regional and global phenomena that impact the local environment.

4.3 Common Environmental Applications of Satellite Remote Sensing in Egypt

A few major environmental applications of remote sensing are summarized in this section. This includes Environmental geology, land use/land cover (LULC) classification, water resources management, agriculture and biomass, arid environment monitoring, sea water and coastal zone applications, and atmospheric remote sensing application. The material aims to be an entry point to applications covered elsewhere in this book. Only optical and thermal infrared data are addressed in this chapter. For this category, the following sensors/satellites are often used: Landsat, SPOT, MODIS, and NOAA-AVHRR. Fine resolution sensors such as RapidEye have been lately used particularly for agriculture applications. Landsat is the most extensively used satellite because of its fine resolution (from 15 to 30 m), the availability of the data free of charge, and its long record that is dated back to 1972. Reviews of environmental applications of remote sensing are presented in Melesse et al. (2007). Locations of study areas in most of the research publications quoted in the following sections are indicated in the map of Egypt (Fig. 4.1).

4.3.1 Environmental Geology

Geology is concerned with the composition and the processes that govern the evolution of the solid earth. Environmental geology is an applied science concerned with human interest in surface and subsurface layers of the earth as well as the interaction between natural habitat and the surface components, particularly the lithosphere and hydrosphere. It is well known that the land surface is an expression of subsurface geology. This theme is used in locating potential sites of groundwater; an essential subject of environmental geology, especially in arid regions. Natural hazards pertinent to geology (as opposed to hazards pertaining to meteorology) are another subject. These include floods, landslides, subsidence, and earthquakes. In general, there is always a need for geological data to support studies in the fields of hydrosphere, biosphere and cryosphere, lithosphere and anthrosphere (the part of the environment made or modified by humans).

Geological and environmental geology applications are the earliest category of remote sensing applications in Egypt, dated back to 1970s. Numerous studies have been published on geological applications. Though this is not an application addressed in this chapter, it is worth mentioning quoting a few review books and papers on remote sensing applications in in geo-exploration (El-Baz 1984; Prost 2001; Gupta 2003; Freek et al. 2012).

Remote sensing studies of environmental geology in Egypt revolve around a few practical-driven applications including geo-environmental hazard assessment, underground and surface water, flash floods and seismic activities. The last two are geo-environmental hazards. Both have been studied in the north western area of the Gulf of Suez (Arnous et al. 2011) using Landsat data along with ancillary topographical and geological data. Using morphometric analysis, the study confirmed the high probability of flash flooding in the area. A highly demanded information by urban developers is to identify, monitor and remediate unstable land localities. Youssef et al. (2009) addressed this concern by demonstrating the use of high-resolution QuickBird satellite imagery (0.61 m spatial resolution) for mapping the recent developments and the slope instability hazard zones in Sharm El-Sheikh/Ras-Nasrani area (a most attractive tourist resorts in Egypt). A notable number of geo-environmental studies of key development areas at the Red Sea coast, particularly the Gulf of Suez, and south Sinai regions have been performed (Omran 2006; Yehia et al. 2002). Many of these studies originate in academic institutions in the Suez Canal region.

The subjects of surface water and ground water are common between geologists and hydrologists. Geologists study the surface and subsurface structure (geomorphology) to determine the water paths and possible locations of its accumulation (natural water harvesting). Hydrologists, on the other hand, remain interested in these aspects in addition to the geophysical processes of evaporation, rainfall and water cycle. In arid regions, surface water is essential for human and animal consumption while ground water is used for both purposes in addition to irrigation of vegetation areas.

Integration of geomorphology, remote sensing and geographic information system (GIS) to study surface water development in key development areas has recently been addressed in a few studies. One such geographic area is the north coast, which has been used by many Egyptians as a summer resort. Lately, the government decided to expand few small towns to full-scale cities (e.g. the under-construction city of New Alamin) to alleviate the population load from the crowded cities (Cairo and Alexandria have a population of 20 million and 4.7 million, respectively, while about 39 million live in the ND region). Obviously, water resources have become an issue for the new cities, and this has attracted the remote sensing community to provide data and recommend solutions. Yousif and Bubenzer (2011) studied surface water development in Ras El-Hekma area of the northwestern coast (45 km east of Marsa Matruh) to exploit it. They recommended establishing water reservoirs in the vicinity of watersheds downstream of the drainage basin to store the water before draining to the sea. Other relevant water harvesting studies are presented in Ali et al. (2007) and El-Asmar and Wood (2000) using satellite remote sensing data.

Groundwater is more of an issue in geology than hydrology. That is because the aquifer locations are defined not only by the subsurface structures (e.g. folds, faults, and foliation) but also the type of deposits in the arid region. Geologists have been active in identifying paleo-drainage networks in the Egyptian desert and use them as proxy indicators of potential locations of groundwater. For example, Ramadan et al. (2006) suggested the use of Radarsat-1 in combination with optical data from Landsat satellite to map lithological and structural features, including fluvial beneath the desert sand cover, in the East Oweinat district in the South Western Desert (Refer to the chapter titled Radar Remote Sensing Applications in Egypt in this book). The study showed that the merged image brought up the buried drainage network. Another important point about using remote sensing data in any geological application should be noted. While the data surely carry information, the interpretation is usually supported by ancillary data. A striking example was illustrated during the above-mentioned study by Ramadan et al. (2006). The picture in Fig. 4.3 (courtesy of late Dr. T. Ramadan) shows ancient drawings on the rocks of a plateau in the study area. The inset shows human bodies that appear to be swimming. This was used as an indicator of the presence of river water in the area (now dry channels) that was barely visible in optical data yet better identified in radar data.

Fig. 4.3 Ancient drawings on a rock in the study area of Ramadan et al. (2006) showing swimmers. This is used as evidence of the presence of paleo rivers in the area, now have become dry water beds buried under the sand (photo courtesy of Late Dr. T Ramadan, NARSS)

4.3.2 Land Use and Land Cover Classification

Though LULC was not the first application of remote sensing data in Egypt, yet it has become one of the most common. The blooming of this application has been driven by the accumulation of longer records of satellite data and the growing demand for monitoring changes in LULC in relation to:

(1) the rapid population growth,
(2) the expansion of land reclamation projects, and
(3) the possible impacts of climate change, especially on coastal areas.

These motivations are met with the increasing availability of image classification and change detection techniques in the commercial remote sensing processing software. Land cover classes are estimated from spectral signature of land surfaces, mainly in the optical (visible and near-infrared bands) and thermal infrared spectra.

The ND extends 160 km from south to north and covers 240 km of the Mediterranean coastline. It is one of the largest deltas of the world and the main agriculture land in Egypt. Despite this economic importance, geospatial data at fine resolution level are still needed to update the land cover and study the effects of sea level rise on soil characteristics. While the government has been pushing to "green the desert' in Egypt, the already green area of the ND has been seriously threatened by the encroachment of urban settlements on agricultural land. High-density settlements across the delta are expanding at an unprecedented rate since January 2011.

Remote sensing has been used to monitor this phenomenon. A notable effort at this front was initiated by NARSS under the auspices of the Ministry of Agriculture and Land Reclamation to generate maps of LULC of the ND and the NV regions.

Maps are generated using a set of RapidEye satellite images acquired during winter (usually between January and March). RapidEye is a constellation of 5 satellites, carrying visible/near-infrared sensor, with a spatial resolution of 5 m. An example of a map generated from January 2014 data, along with a sample of RapidEye image, are shown in Fig. 4.4. According to the map, the cultivated and urbanized area is 2.261 and 0.371 million hectares, respectively (i.e. as of January 2014, 14% of the ND was covered by settlements). Further work is underway at NARSS to generate seasonal LULC maps on a regular basis using the constellation of Dove CubeSat (3U, 5 kg, 40 mm aperture optical payload with 3 m resolution) that make up Plant Lab's constellation satellites.

Detection of land cover continues to be a popular application of remote sensing data in Egypt. Figure 4.5 shows changes in land cover around Cairo revealed by two scenes of the same area: from Spot-2 satellite acquired in September 1995 and EgyptSat-1 in acquired September 2007. Both satellites have same spectral bands (in μm wavelength): B1-Green (0.50–0.59), B2-Red (0.61–0.68) and B3-Near Infrared (0.79–0.89). The remarkable expansion of the city eastward is visible. EgyptSat-1 was the first Egyptian earth observation satellite, launched in April 2007 but communication with it was lost in July 2010. The spatial resolution of the cameras was 7.8 m. Almost all uses of the data were demonstrated by local researchers with applications limited to the Egyptian landscape and environment. Most of the processing revolved around image analysis (quantitatively and qualitatively) using non-calibrated data.

Most of the LULC studies use optical remote sensing combined with GIS tools. El-Gammal et al. (2010) applied remote sensing change detection techniques and ARC-GIS software to Landsat images acquired between 1972 and 2008 to capture changes in the shoreline of Lake Nasser (Fig. 4.1) and relate changes to the surrounding geomorphic, geological, and climatic factors. El-Asmar et al. (2013) applied remote sensing indices to quantify changes in the surface area of Burullus lagoon during the period 1973–2011. Emara et al. (2016) present temporal and spatial patterns of

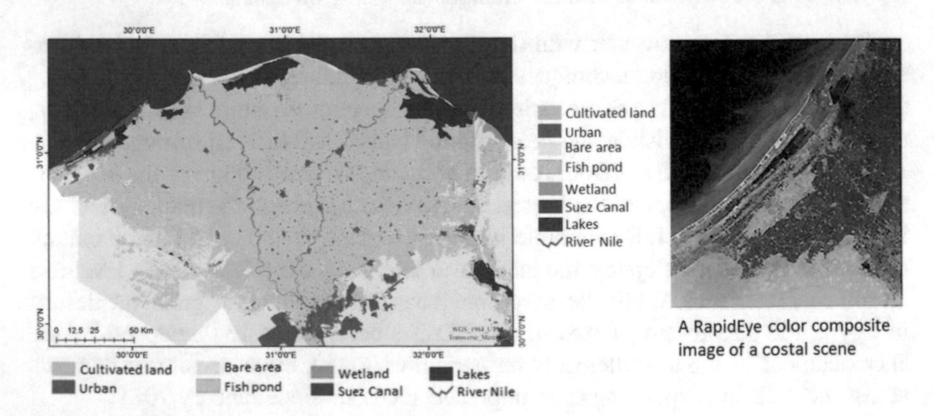

Fig. 4.4 Land cover classification map of the ND region in January 2014 at a scale of 1:50,000 generated using RapidEye satellite images (an example of the scene acquired on January 10, 2014, is shown in the right panel) (courtesy of S. Arafat, NARSS)

Fig. 4.5 Change detection of land cover in an area surrounding Cairo, revealed by comparing the SPOT-2 satellite image acquired in September 1995 (left panel) against EgyptSat-1 image acquired in September 2007. Note the expansion of urban areas east of Cairo (blue-greenish color) and the new roads. The RGB color channels are assigned to B3, B2 and B1 bands (see text). Image courtesy of the Image Analysis Laboratory of NARSS

LULC to identify the process of settlement encroachment in the central area of the ND between 1990 and 2016 using Landsat images followed by GIS-based techniques. They found that urban area increased by 8.8% of the total area during the study period. Several other studies of LULC in different areas in Egypt are achieved (e.g. Shalaby and Tateishi 2007; Hegazy and Kaloop 2015; Farrag et al. 2016).

4.3.3 Water Resources and Management

Water resources and management are significant issues in Egypt. They involve a few stakeholders from agriculture, environmental, industrial, municipal, and urban planning authorities. Nile river and groundwater are the only sources amid the extensive arid landscape of Egypt. Remote sensing has been used in practical applications to meet the challenges and explore opportunities for sustainable water management.

Abu Salem et al. (2017) integrated remote sensing data from Landsat-5, -7 and -8 with chemical analysis of water to assess changes and study water quality in two lakes formed mainly from agriculture wastewater in a depression west of Cairo called Wadi El Raiyan. This and similar studies aimed at assessing the deteriorating effect of contaminated water on agriculture. Other studies were performed to determine the eutrophication profile of Lake Burullus (Farag and El-Gamal 2012), the water level in Tushka Lake, resulting from spill-over of excess water from Lake Nasser in Aswan region (Bastawesy et al. 2008), water quality in Lake Timsah in Suez Canal area (Saad El-Din et al. 2013) and Lake Burullus (El-Kafrawy et al. 2015), and trends of environmental changes of water and aquatic vegetation in Mariut Lake (ND north)

(El-Kafrawy 2017). A comprehensive study to estimate water discharge/balance from Lake Nasser using Landsat TM/ETM+ and altimetry data is presented in Muala et al. (2014).

Water management in newly reclaimed areas has received particular attention lately. Groundwater (not far beneath the ground) is the prime water source in these areas. A review on applications of remote sensing and GIS for groundwater mapping in Egypt is presented in Elbeih (2015). It highlights indicators derived from remote sensing data such as vegetation surface water, geological structure, and water runoff. Generally speaking, geological features such as faults, folds, and fractures in hard-rock appear in remote sensing images as lineaments and can, therefore, be used as proxy indicators of groundwater storage bodies. Elbeih (2015) stated that "A key to the RS of groundwater is the recognition that shallow groundwater flow is usually driven by surface forcing and parameterized by geologic properties that can be inferred from surface data". Ground water is mainly detected by radar remote sensing, which is covered in Chap. 5.

Waterlogging is another hydrological phenomenon that has frequently been studied in Egypt. It is worth pointing to the difference between surface water and waterlogging. The former originates from natural water sources and is determined by surface structure while the latter originates from wastewater (agriculture or sanitary drainage) in local low lying areas by seepage of water. The water usually originates from excessive irrigation which also adds to groundwater. Figure 4.6 shows two photos of surface water located in Jordan and waterlogging located in Abu Sweir, Egypt, about 25 km east of Ismailia (about 100 km north of Suez). Discrimination between these two entities in satellite images can be achieved based on their spectral signatures in the optical data along with other ancillary information. This approach has been adopted in Kaiser et al. (2013) using Landsat and GIS data. The latter is used to identify the high morphology land, water table, and plant stress. The study found a positive correlation between agriculture activities and water logging, with a rate of groundwater rising between 3 and 8 cm per year. Other waterlogging studies include El Bastawesy et al. (2013), who assessed waterlogging in agriculture mega-projects

Fig. 4.6 Photographs of surface water in a desert area in Jordan (left) (courtesy of V. Singhroy, Canada Center for Remote Sensing) and waterlogging area in Abu-Sweir, Egypt (right) (from Kaiser et al. 2013)

in the Western desert of Egypt and El Abd and El Osta (2014) who assessed it in the newly reclaimed area northeast of El Fayoum.

4.3.4 Agriculture and Biomass

Crop classification is a traditional application of remote sensing in Egypt. Data from optical sensors are usually used for this purpose since many crops can be identified by their unique spectral signature or some derived ratios. The uncertainty of the classification is caused by the overlapping of the signature from different crops. Recently, more applications on crop classification and monitoring have evolved in Egypt recently in response to the national demands for the expansion of land reclamation and food security.

A notable Crop classification map of the ND and NV regions were generated for the 2 winter seasons of 2014 and 2015 by the Agricultural Application Division of NARSS using a large set of high-resolution images (5 m) from RapidEye satellite. The approach entails using atmospheric correction scheme followed by the ISODAT unsupervised classification in ERDAS-Imagine software to generate 100 classes in a given image, then using ground observations (truth data) to re-group them into four classes of wheat, clover, sugar cane and sugar beet. An example of a map showing the distribution of the crops in the Middle sector of the NV region is shown in Fig. 4.7. Note the narrow band of agriculture area (about 20 km in width) around the Nile and the numerous small areas of homogeneous crop cover (typically a few hundreds of square meters). These small and fragmented areas require very high resolution data to map. Work is underway at NARSS to use the higher resolution images (3 m) available from the CubeSat constellation series developed by Planet Labs to generate monthly crop classification maps. This approach will also be used to assess the utility of Sentinel-2 data with its higher spectral data (12 channels) and coarser resolution (10 m or 20 m) for crop classification (I. Farag, NARSS, Personal communication).

The first database of spectral reflectance from agriculture crops in the visible and near-infrared was compiled by the Agriculture Application Group at NARSS using a ground-based spectroradiometer system. The system covers the spectral range from 350 to 2500 nm at 1 nm resolution and includes metadata that describes the site environment and measurement processes (Arafat et al. 2013a, b). The database has been incorporated in a web-based system with a friendly user interface (https://www.spectraldb.narss.sci.eg/spectral). The available data covers a wide variety of agriculture land in Egypt including field crops and different soil types. Figure 4.8 includes a set of photographs of the spectroradiometer during data collection at different configuration (hang-held or mounted on a mobile platform). An example of the spectral signature of the wheat crop at different growth stages is also shown. The spikes around 1400 nm and 1850–1950 nm represent noise from the H_2O absorbing atmospheric windows.

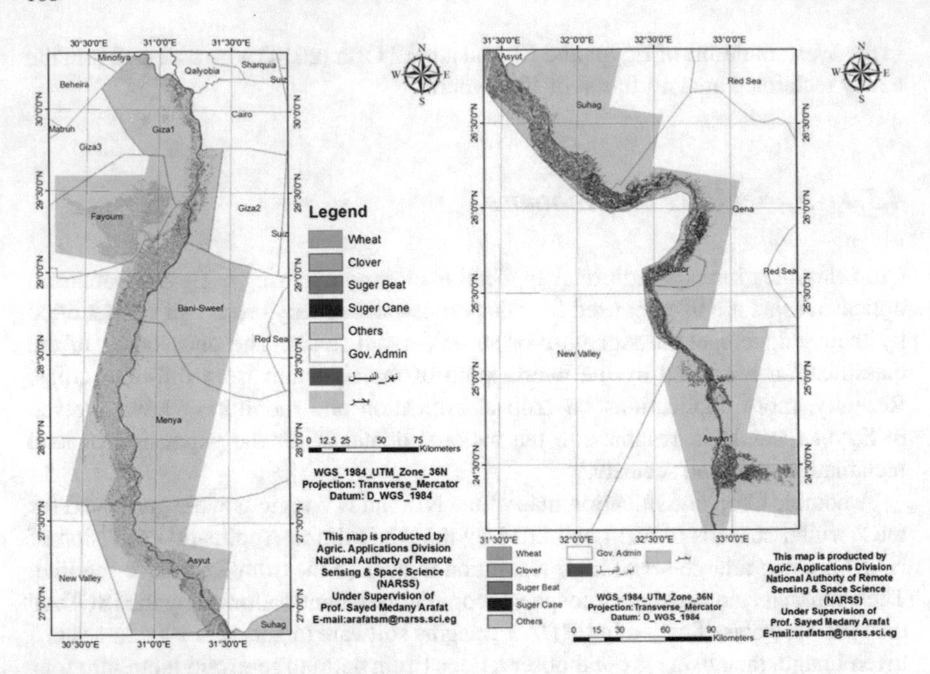

Fig. 4.7 Crop classification map of the agriculture area around the Nile Valley in the Middle and Upper Egypt (left and right, respectively), generated using a series of RapidEye satellite images (courtesy of S. Arafat, NARSS)

Fig. 4.8 A set of photographs showing different arrangements for the spectral database collection using spectroradiometer (left) and an example of the spectral signature of the wheat crop at different growth stages (right) (*source* Arafat et al. 2013a, b)

So far, the spectral database has been used to identify the optimal spectral bands and spectral patterns that can discriminate between different crops. Discrimination becomes possible if a unique reflectance pattern is established for different crops. Arafat et al. (2013a, b) identified an optimal spectral signature to discriminate between two winter crops (wheat and clover) and two summer crops (maize and rice). A similar study was presented by Aboelghar et al. (2013) to discriminate between four potato varieties. These studies furnish the first step towards using a future spaceborne hyperspectral system to map crop types. So far, no airborne or spaceborne hyperspectral sensor is available to use against the established database for crop discrimination.

Soil quality is an important component of the agriculture domain though it also intersects with the geology domain. As reported in El Nahry et al. (2015), soil degradation in Egypt is triggered by four human-induced factors: salinity, alkalinity, compaction and waterlogging in addition to erosion resulting from sea water level rise in the Mediterranean. The same reference used Landsat TM and ETM+ data to study the effect of each factor on soil quality in the northern part of the ND. Increasing human activities in the agricultural area may lead to salinization of soil. The study of Arnous and Green (2015) used multi-temporal Landsat data and GIS to monitor and map the waterlogged and salt-affected areas in the eastern ND. The study pursued an approach that can be considered representative of many remote sensing studies in Egypt. It involves using Landsat images (the most commonly used data in Egypt) followed by digital image processing techniques to enhance the images then application of supervised and non-supervised image classification methods and finally defining spectral ranges or indices to identify the land cover classes. In their work, Arnous and Green (2015) used the well-known Normalized Difference Vegetation Index (NDVI) and Normalized Difference Water Index (NDWI) as well as the less-known Normalized Difference Salinity Index (NDSI).

The increasing demand for water in Egypt, driven by the high rate of population increase and the need for more agriculture products, requires careful management of water resources including assessment of water fluxes. The agricultural sector consumes more than 80% of water resources under surface irrigation (El-Shirbeny et al. 2014). To improve management of irrigation water, accurate estimates of evapotranspiration (a term denotes evaporation and transpiration from plants) has become a set goal of a few hydrological and agronomical research. It is often difficult to quantify this parameter experimentally because it requires expensive instrumentation. Therefore, hydrological modeling with input from remote sensing is usually used.

One of the well-known models is based on surface energy balance equations (Bastiaanssen et al. 1998; Psilovikos and Elhag 2013). The satellite data input includes surface temperature, albedo, and emissivity of longwave radiation. Psilovikos and Elhag (2013) used this model with remote sensing data downloaded from the European Space Agency (ESA) to forecast daily evapotranspiration over ND. Farg et al. (2012) estimated crop evapotranspiration using SPOT-4 satellite data integrated with meteorological data. Simonneaux et al. (2010) combined MODIS images with climatic data to estimate annual evapotranspiration of crops in the ND region. Crop water requirements and the potential crop evapotranspiration have also been estimated using Landsat data in El-Shirbeny et al. (2014). It is worth noting that many studies of evapotranspiration in the ND and Nile Basin, in general, have been initiated by researches in Europe and USA.

The concept of precision farming has emerged worldwide since the early 1990s. It is based on relating spatial and temporal variability of crop production to plant and soil conditions. Examples include land preparation, grading, fertilizers, the degree of pest infestation, disease infection, weed competition and pesticides and herbicides (Tran and Nguyen 2006). Non-uniform application of fertilizers, for example, may lead to a spatial variation of crop production. Remote sensing is a fundamental tool for detecting spatial and temporal variability of crop and soil conditions such as crop water and moisture stress, soil salinity, leaf density, crop diseases and nutrient deficiencies, among other parameters. Data from high spectral and spatial resolution sensors are usually used in precision farming. Only limited research has been conducted on this subject in Egypt. A notable academic thesis that addresses remote sensing for precision farming in the Nile Valley of Egypt is presented by Elmetwalli (2008).

4.3.5 Monitoring Arid Environment

Remote sensing was primarily developed for applications in remote areas; namely the largest deserts and the polar regions of the earth. There are three vast deserts in Egypt; Western Desert (WD), Eastern Desert (ED) and the Sinai Desert. The WD extends over 680,000 km^2, which is roughly two-third of the area of the country. About 74% of this area is covered by sand dunes (Besler 2008). It contains mineral resources, hydrocarbons, underground water and artifacts buried under the sand. Some areas are suitable for land reclamation. The ED is mostly rocky and mountainous. The Sinai Peninsula is the land bridge between Asia and Africa and has an area of 60,000 km^2. It features a few world-class tourist resorts, mineral resources (e.g. silica sand, limestone, manganese, uranium, and coal), igneous rocks (granite and basalt) in addition to petroleum and marine resources. With the recent accelerated development of land reclamation and new desert settlements, remote sensing has been intensively used to support the sustainable development of mega projects in the WD and Sinai areas.

Perhaps the most impending issue for any arid region is the water resources. A drought index was developed and applied to monitor the drought in El-Kharga and El-Dakhla Oases (Fig. 4.1) for the period 2003–2014 (Ebaid 2015). It combines land surface temperature and soil moisture parameters, both derived from blue, red and near-infrared bands of Landsat and TM 5 (2003) and Landsat 8 (2014) data. An integrated remote sensing approach to improve estimation of renewable water resources is presented in (Milewski et al. 2009). It involves extraction of spatial and temporal data from a wide range of satellite optical, thermal IR and microwave observation to estimate precipitation, soil moisture, reservoir volume and flow in large river channels. The approach was used to estimate the rainfall-runoff and groundwater recharge in Saini and the eastern deserts.

Sand dune movements pose a potential hazard in desert areas where population exists or is expected. They migrate due to saltation, a phenomenon responsible for sand transport using surface rolling wind. This causes the crest to become unstable and sand avalanches to occur down the slip face of the dune (Abou El-Magd et al.

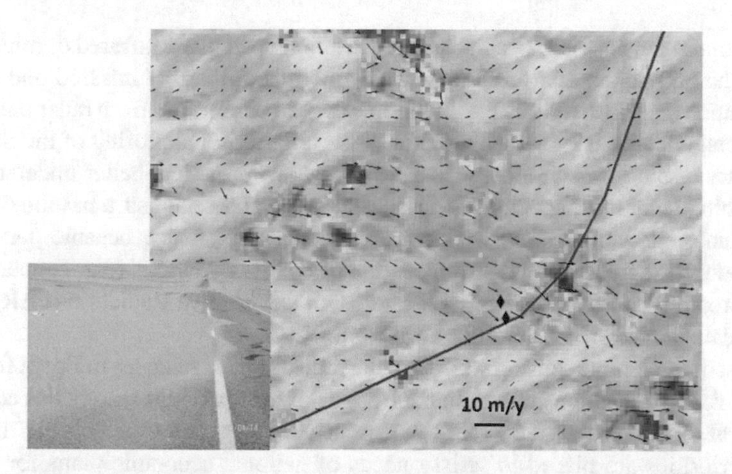

Fig. 4.9 Magnitude and direction of a sand dune displacement (represented by the arrows) at a location in north Sinai. The displacement scale is shown. The background is an ASTER image used for the computation of the displacement and the colors have correlation with the displacement. The black diamonds represent a GPS location for the filed photos. The inset is a photo showing part of the road (marked by the red line) covered by the sand (courtesy of I. Abu-Elmagd, NARSS)

2013). A few studies on using remote sensing to detect and measure sand dune migration in Egypt's deserts have been published. Hermas et al. (2012) used two SPOT 4 panchromatic images, 9 months apart, with a sub-pixel correlation technique that employs Fast Fourier analysis to determine the mean sand dune movement at a location in north Sinai (Fig. 4.9). The study detected that the lateral movements along the crest lines of linear dunes ranged from 4.0 to 20.1 m/year.

Abou El-Magd et al. (2013) studied also the sand dune morphology and movement in Toshka depression in the south Western Desert (see Fig. 4.1). The authors used two sets of optical satellite data; Landsat TM in 2000 and Spot 4 in 2006, to trace the geographic location of each sand dune in a pair of images. A notable finding was the impact of the sand dune movement on the spillway canals of Toshka and Sheikh Zayied that feed the agriculture planned areas in Toshka depression. This was a national megaproject that started in the 1980s and continued through1990s. Other sand dune studies using remote sensing have been published (Gifford et al. 1979; El-Banna 2004; Hosny and Abdelmoaty 2009).

4.3.6 Sea Water and Coastal Zone Applications

Oceanographers, climatologists, coastal managers, fishers and marine operators need information about ocean and coastal areas. Depending on the need of each group, the information may include surface wind, wave height and spectrum, oil slicks, ocean current, sea surface temperature (SST) and coastal water turbidity (due to both organic and inorganic material). The first three items are usually obtained using radar remote sensing. The ocean current can be mapped using infrared or microwave data. The SST and water turbidity are obtained from thermal infrared and optical observations, respectively. The estimation of ocean parameters from radar data is not a conventional application in Egypt nor is the long-term monitoring of the SST and the water turbidity. Such monitoring, however, can provide a better understanding of the biological marine activities in coastal areas and establish a baseline against which unusual events can be identified and measured. A few oceanic parameters are produced regularly and distributed free of charge by NASA (e.g. concentration of chlorophyll and another pigment, Photosynthetic Active Radiation (PAR), etc.). These data, however, are not fully utilized in Egypt.

Most of the remote sensing studies in the field of ocean sciences in Egypt focus on coastal zone applications rather than the status of the open seas (Egypt has coastline length of approximately 2500 km, half of it stretches along the coast of the Red Sea). Priorities are placed to satisfy needs of national economic plans for coastal area development. In a study that did not use remote sensing data, Abdel-Latif et al. (2012) linked the tourism economy to the environmental sensitivity of the coastal resources; e.g. shoreline erosion, water pollution, irrational land use and deterioration of natural resources and habitats. Protecting the sensitive coral systems in the Red Sea is studied in Masria et al. (2016), while mitigating the degraded Mediterranean ecosystem caused by land-based pollution is addressed in El-Kafrawy (2017). The agency responsible for coastal zone management is the Egyptian Environmental Affairs Agency.

4.3.7 Atmospheric Remote Sensing Applications

When one thinks of remote sensing of the atmosphere, the first notion that comes to mind would be about the weather forecast and air quality. Indeed, remote sensing has greatly advanced the operational programs in both fields in the past four decades. Moreover, the first civilian satellite, NASA's TIROS-1, was a weather satellite launched on April 1, 1960. It paved the way for many dedicated meteorological satellites to measure temperature and humidity profile, precipitation, rainfall, cloud systems, air pollution, among other parameters to monitor the weather systems. Atmospheric remote sensing is a major subject in meteorology and is covered in several books (e.g. Stephens 2003). In Egypt, it is mainly demonstrated in studies of atmospheric aerosol with limited investigations in the fields of atmospheric gases, cloud physics, air chemistry or meteorological parameter profiles.

Remote sensing of aerosols has received a notable share in the publications of the Egyptian remote sensing community (El-Askary and Kafatos 2008; El-Metwally et al. 2008, 2010; Marey et al. 2011; Shokr et al. 2017). Egypt has the ninth highest mean annual concentration of aerosol particulate matter (PM_{10}) in the world. Sources of aerosols over Egypt comprise blown sand and dust from the Sahara, emission of aerosols from industrial parks and heavy-populated urban areas, burning of biomass residues, and sea salt carried along with pollutants originating in Europe across the

Mediterranean by northerly winds. Anthropogenic sources have become increasingly severe over Egypt's mega cities (Cairo's population is approaching 20 million during the day). Aerosol optical depth (AOD) from NASA's MODIS satellite was used in Shokr et al. (2017) to examine its seasonal and spatial distribution over Egypt during a 12-year period (2003–2014) and assess the air quality level over selected cities and region. Figure 4.10 (from the same reference) presents the seasonal average of AOD over the entire country. It reveals a seasonal cycle with a peak in the spring and early summer (with values between 0.4 and 0.5), and a minimum in early winter with values <0.35.

Many aerosol parameters are readily available from a few spaceborne sensors including MODIS, the Multi-angle Imaging SpectroRadiometer (MISR), the Sea-Viewing Wide Field-of-view Sensors (SeaWiFS), the POLarization and Directionality of the Earth's Reflectance (POLDER; 1996–2003), and the Total Ozone Mapping Spectrometer (TOMS). A full review on using remote sensing for aerosol monitoring can be found in Lenoble et al. (2013). It is recommended to expand the research in this field in Egypt using such extensive data sets, which is available for free download.

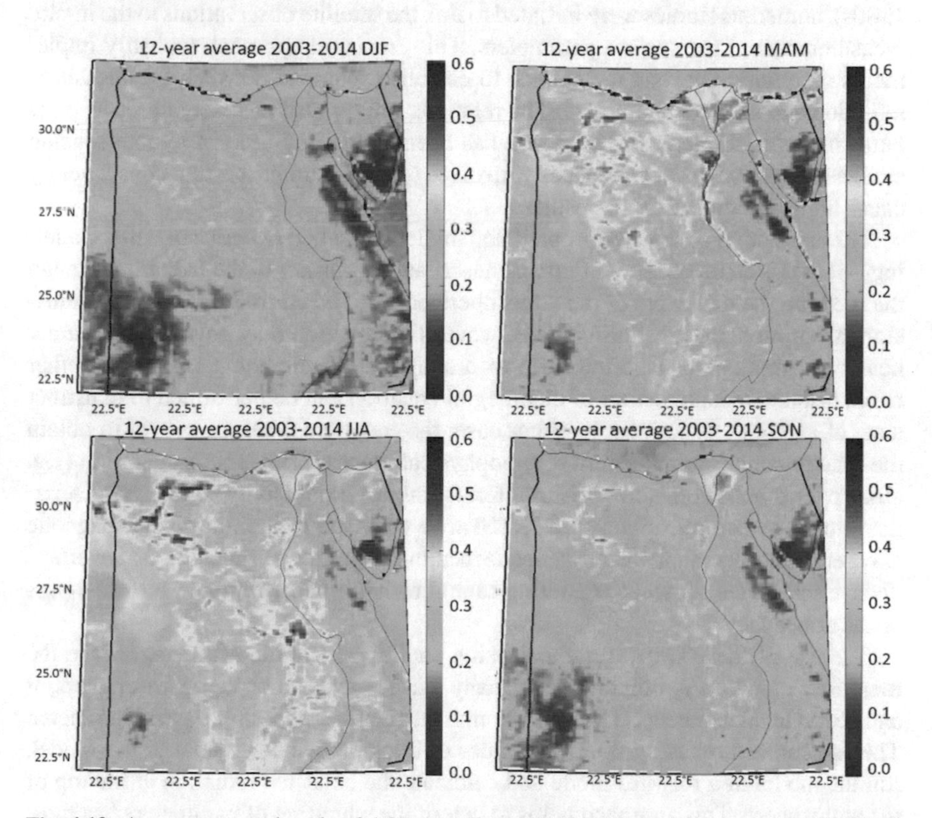

Fig. 4.10 Average seasonal cycle of AOD over Egypt obtained from daily coverage of MODIS data (2003–2014). Starting from top right in clockwise circle the panels represent maps of winter, spring, summer and autumn (*source* Shokr et al. 2017)

4.4 An Outlook of Remote Sensing Applications in Egypt

This section presents a few themes to support the future progress of remote sensing research and applications in Egypt. The material is discussed from the perspective of someone who has more than three decades of experience and exposure to international applications of remote sensing and whose scientific involvement has also spanned almost 20 years of experience with the remote sensing community in Egypt.

4.4.1 Modeling Approach in Remote Sensing

There are three types of modeling that are usually connected to remote sensing research and applications: physical models, electromagnetic wave propagation models, and empirical models. The last one is the simplest and most commonly used. In the early stage of exploring the potential of satellite remote sensing (the 1970s and 1980s), numerous studies were initiated to link the satellite observations to the in-situ measurements of the surface parameters. This empirical approach typically implements regression analysis techniques to establish relations between the measured radiation and surface parameters. The relations can be used inversely to retrieve the latter from the former. This approach has been commonly used in remote sensing studies in Egypt. However, it usually involves severe assumptions that consequently cause high uncertainty of the results.

The electromagnetic wave propagation modeling (called radiative transfer modeling—RTM) is an integral equation that describes the transfer of the radiation through the medium; be the layers of the atmosphere or the solid earth. It involves the emission, absorption, and scattering of the wave. The equation may be simplified under heavy assumptions to take the form of a simple algebraic equation. On the other hand, discrete reconstruction of the integral equation can be performed to construct a set of algebraic equations. In either case, the equation(s) can be solved to obtain the state parameters that govern the geophysical processes under consideration (e.g. atmospheric pollution, soil moisture, terrestrial net productivity, snow characteristics, etc.). The complications of the RTM arise when the path of the electromagnetic wave encounters a highly heterogeneous medium (e.g. heavy vegetation that overlays diverse soil layers) or when scattering cannot be neglected within the surface layers or the atmosphere.

The use of the RTM in remote sensing studies entails the following. Since the measured radiation is influenced by many surface and atmospheric parameters, it can be modeled using an RTM to determine the relative influence of each parameter. This is achieved by using different values of the parameters as input to the model, run the model in a forward mode and calculate the expected radiation at the top of the atmosphere. This approach helps to determine which set of parameters has most influence on the satellite measurements and therefore most amenable to retrieval. Moreover, a database of the expected radiation can be generated used later with an input of satellite measurement (observation) to search for the best set of parameters that generate this particular observation. This is an inverse problem that requires a search mechanism through a huge database. This approach has been used in retrieval of many parameters including the aerosol optical properties from MODIS data. A review of RTM for remote sensing applications can be found in Kuznetsov et al. (2012). So far, this approach has not been pursued in applications in Egypt.

Physical modeling of the earth's radiation and its impacts are presented mostly in the form of weather and climate models. A model is a comprehensive set of differential equations that simulate the mechanical, thermal, and radiative interactions of the important drivers of climate, including atmosphere, oceans, land surface and ice. They can be used to predict the future climate or simulate existing conditions (in retroactive mode). Remote sensing serves the modeling task when used as initial and boundary conditions or is assimilated in the solution to produce more accurate information.

The most famous application of this approach is in the remote sensing data assimilation in weather forecast models to generate information on the wind speed and direction, cloud liquid water contents, the chance of severe weather events and precipitation type/amount within several hour periods. Numerical Weather Prediction (NWP) is a major study field, which manipulated vast data sets and performs complex calculations using most powerful supercomputers. The Egyptian Meteorological Authority (EMA) use NWP in their forecast program. They feed their model with atmospheric parameters derived from many sources, including remote sensing, obtained or interpolated at the model's grid.

At the academic research front, efforts of using remote sensing data in weather and climate modeling have been undertaken in Cairo University since 2006. In an early study, Kandil et al. (2006) used the Mesoscale Atmospheric Model (MM5) to simulate an air pollution episode over the ND and Cairo regions, known as "black cloud". It frequently occurs in the fall and is associated with the thermal inversion phenomenon (featuring an increase of air temperature with height instead of the expected decrease). Results indicate that an inversion system was present over Cairo at a layer extending between 300 and 800 m above the ground. Abdel-Kader (2007) used the same model to study the factors that contribute to thermal inversion using the available remote sensing data from NARSS. A highpoint conclusion was the extent of this phenomenon across some 40,000 km^2 in ND region.

Scenarios of climate change induced by the land cover change were the subject of another study by Elsayed (2007). The author fed basic satellite-derived parameters (vegetation index plus land and sea surface temperature) into the MM5 model to capture the expected weather in case of the fictitious scenario of fully urbanizing the ND region. The study noted that the percentage of the urban area in the ND region increased from 5.7% in 1990 to 10.55% in 2003. Given the scenario of full urbanization, the air temperature was found to be increased by more than 6 °C in most of the governorates of the ND. Sultan (2011) continued to use MM5 but with using nudging-based data assimilation of remote sensing observations to study its effect on the output of the numerical weather forecast. The study concluded that the ensemble mean is better and more correlated to observations than a single control

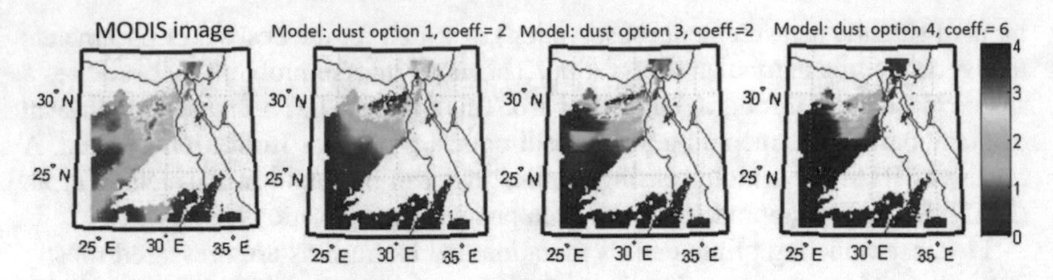

Fig. 4.11 Data showing a severe sand and dust storm, originated in the WD and transported to cover Cairo and the ND. MODIS image of the event (far left) is shown along with three simulations using WRF-Chem model with different emission schemes and different coefficients (see text). Image is courtesy of M. Eltahan, Cairo University

forecast and the accuracy of the results is proportional to the number and quality of the ensemble members.

In a recent study, El-Tahan et al. (2018) used the WRF-chem model to reproduce the AOD of a severe dust/sand storm that blew from the Western Desert and covered the ND and part of Upper Egypt. The study attempted to improve the bias of the model (which always underestimates the AOD with respect to the satellite observations) by examining three dust emission schemes and tuning the coefficients in each scheme. Two storm events were studies, occurred on 22 January 2004 and 31 March 2013. Although the tuning led to significant improvement of the results, the values were different between the two events. Apparently, it depends on the microphysical characteristics of the sand/dust particles. Figure 4.11 shows the AOD from MODIS image of January 2004 event (far left panel) and the output AOD from three runs of the model using three dust schemes (indicated by dust option in the figure) with a different coefficient for each scheme. Apparently, the model with dust option 1 and coefficient 2.0 produces the closet result to the map from MODIS data.

Strengthening the research on using climate modelling with input from remote sensing data (for initial/boundary conditions or data assimilation) is expected to advance the applications of remote sensing data in Egypt. It can indeed be used in studying the impacts of changes of land use on future climate scenarios, and developing the necessary expertise for more sophisticated applications of NWP in the region.

4.4.2 Horizontal and Vertical Expansion of Remote Sensing Applications

Satellite EO data are usually acquired over large areas that extend beyond political borders. As the Canadian astronaut Marc Garneau puts it "Looking down from space we see only the global context – no borders – just a continuum of atmosphere, ocean, and land". This is particularly true for satellite data obtained from medium and coarse resolution sensors. Large-Scale geophysical phenomena that traverse political borders include aerosol dispersion, sand storm paths, ocean currents, weather systems, and regional climatic impacts. For those reasons, promotion of regional and global applications of remote sensing data will be favorable.

Applications of satellite earth observations in Egypt are expected to expand horizontally and vertically. Horizontal expansion entails broadening the range of applications; spatial-wise and parameter-wise. Vertical expansion entails improving the retrieval of the parameters using more comprehensive approaches, including modeling and data assimilation. Each aspect is addressed briefly in the following.

Spatial coverage of the applications is expected to encompass the regions of North Africa, Middle East, and the Nile Basin countries. An example is illustrated in the mosaic of the spectral albedo from MODIS data, which covers North Africa and the Arabian Peninsula region (Fig. 4.12). The importance of this regional view is demonstrated by the variability of the albedo. The dark areas mark rocks that are absorptive of solar radiation while bright areas mark sand. Regular update of this map is important to account for the surface ablation and deposit of new particles (Shokr 2010).

The importance of regional studies is also demonstrated by the example of the water cycle. The Middle East and North Africa (MENA) region is one of the most water-scarce regions in the world. MENA countries should adapt to meet the challenge of growth population and the adverse effects of global warming precipitation and evaporation (Droogers et al. 2012). A regional study of water resources is the only approach to understand and mitigate the expected decline of water in the MENA region as the temperature is expected to increase by 3–5 °C by 2050 (Parry et al. 2007). This requires a combined use of the hydrological model and remote sensing data obtained simultaneously over a grid covering the entire region. Such project was initiated by NASA in collaboration with the World Bank and USAID to addressing water resource issues and adapting to climate change impacts for improved decision making and societal benefit (Habib et al. 2012). It involves running the Land Data Assimilation System (LDAS) with input from NASA's satellites.

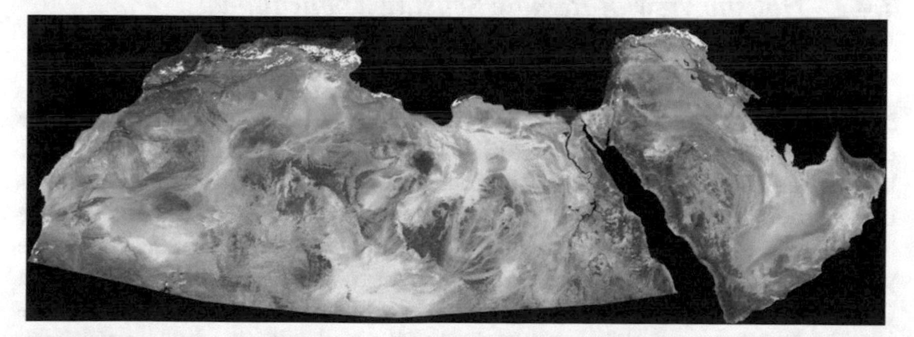

Fig. 4.12 Map of broadband albedo over the North Africa and Arabian Peninsula region from MODIS passes over a one-day period (22 April 2002). The color scale is not given, but dark colors indicate low albedo and vice versa (image is courtesy of Elena Tsvetsinskaya, Boston University)

Another example that requires using the regional perspective of remote sensing data is the climate change impacts on the water in the Nile Basin. Will the change result in more or less rain? Would the surface topography lead to flooding somewhere? Will an increase of biomass emission along the Nile Basin affect the atmospheric heating and cooling in the region? These and similar questions can be addressed through collaborative regional applications that foster the exchange of remote sensing data on hydrological, meteorological, and environmental at regional scale between national centers to the benefit of all countries.

Broadening the range of the retrieved parameters is another aspect of the suggested "horizontal expansion" of remote sensing endeavors. It entails retrieving more parameters and addressing more geophysical processes to develop a comprehensive analysis of the given phenomenon (e.g. hydrological cycle). New parameters that are recommended to be addressed by the Egyptian remote sensing community include ocean salinity, ocean currents, cloud and aerosol parameters, photosynthetic active radiation, atmospheric parameter profiles, and more biophysical parameters. Studying global phenomena such as the earth's radiation budget will be a worthwhile additional effort that contributes to the current international studies on using remote sensing to understand the earth system.

As mentioned before, vertical expansion of remote sensing applications in Egypt entails improving the retrieval of the parameters using more involving approaches and employing the results in more sophisticated applications. While empirical models are commonly used in Egypt to relate surface parameters to remote sensing radiative measurements, the use of more sophisticated techniques to improve the accuracy of the parameter retrieval is expected to grow. Results should be employed in physical modelling to enhance the forecast of key environmental parameters. Assimilation of remote sensing data in weather or pollution transfer models will improve the accuracy of the model at short term forecast (a few days). It will also further our understanding of the earth system processes.

4.4.3 Holistic View of Remote Sensing Applications: Studying the Earth System

Modern use of remote sensing applications has been demonstrated in studying the earth as a system. The earth system has diversified components including atmosphere, hydrosphere, biosphere, geosphere, lithosphere, cryosphere, and anthrosphere (populated areas). These components interact with each other. Studies of Earth Science System (ESS) started in the early 1980s with the creation of the ESS committee within NASA, and that led to the development of NASA's Earth Observing System (EOS) program. The program comprised a series of satellites designed for simultaneous, long-term and global observations of the land surface, ocean, atmosphere, biosphere, and clouds.

This holistic approach of using remote sensing data reveals links not only between parameters at regional and microclimate scales but also between global and regional phenomena. Connection between the two scales will help understanding processes occurring within the ecosystem. This includes, for example, anthropogenic gas and aerosol emissions due to recent accelerating growth of human activities. Subsequently, this should facilitate the prediction of their global environmental impacts. Moreover, integrating coincident data from different sensors supports a few cross-cutting themes such as hydrologic cycle, carbon cycle, and global warming. It gives a complete answer to many questions that would otherwise not be possible from using a single source at a local scale.

Satellite programs have been developed in the past three decades to endorse the above theme. Multi-channels sensors such as MODIS were designed to carry a synergistic instrument payload that measures water in its gaseous, liquid, and solid forms in addition to atmospheric and surface temperatures, land and ocean vegetation, and other aspects of the global climate system. Later, the satellite "constellation" concept emerged where a series of satellites carrying different sensors to measure parameters representing different Earth's components are put in the same orbit, spaced apart by a few minutes. An example is NASA's A-Train satellite constellation which has six satellites with different but complementary missions (L'Ecuyer and Jiang 2010). The synergistic and coordinated measurements provide comprehensive information about land cover, atmospheric constituents, air quality, ozone, carbon dioxide, aerosol radiative properties, clouds, and oceanic state.

Applications of EO data in Egypt have yet to make use of the satellite constellation data; namely the simultaneous retrieval of a range of geophysical parameters to construct a holistic view of the earth system processes. Not only will this contribute to the accelerating international effort at this front but it will also help to better understand the mechanisms that govern the impact of global phenomena on local processes as explained in Sect. 4.2 of this chapter.

To further illustrate this point, it suffices to note that geological and climatological information, both retrieved from remote sensing, are integrated to estimate the erosion of surface material that subsequently causes a change of the surface albedo. This information, along with other remote-sensing-driven information about aerosols and air quality, will feed into studying the radiation budget of the earth system; a crucial subject in the context of the current climate change. Hydrology parameters may also be combined with agriculture parameters to assess the evapotranspiration and soil moisture. Likewise, agriculture information can be integrated with marine information to explain the delivery and diffusion of nutrients into the coastal water through storms (hence, developing blooms of phytoplankton). These are topics that support the theme of the holistic view of remote sensing applications.

4.5 Conclusions

Remote Sensing applications in Egypt started in the mid-1970s on a purely experimental basis. The activities began by establishing a remote sensing center under a joint project between Egypt and the United States of America. Applications were mainly oriented to support needs of national projects in urbanization and agriculture and to generate/update geological and morphological maps of the country. Nevertheless, in the past three decades remote sensing has become a source of remarkably detailed environmental management information in Egypt. The users' community has expanded to include, in addition to the government, industries keen on environmental sustainability as well as academic institutions keen on developing information from remote sensing data. This also paved the way for the expansion of this technology in the MENA region.

This Chapter presents a quick review of remote sensing applications in Egypt as well as a future outlook of the applications. Historically, the applications started in the field of geology and agriculture. Today, the realm of applications has expanded greatly to cover various aspects of natural resources exploration, water resources, urban planning, land cover changes, crop monitoring and forecasting, coastal environment management, air pollution monitoring, arid region reconnaissance among other applications.

Examples of recent undertakings of remote sensing research and applications are highlighted in this chapter. Geo-environmental hazard in key areas for urbanization has been assessed. Integration of geomorphology, remote sensing and GIS has been used for exploration of surface and ground water. Land cover classes have been estimated from the spectral signature of the surface. Trends of environmental changes of water and aquatic vegetation in major lakes and lagoons along the northern coast have been identified. Sand dune movement in the Western desert and Sinai regions has been monitored. The first database of spectral reflectance from agriculture crops and soil was compiled using ground-based spectroradiometer and later used for crop classification. Areal monitoring of selected crops in the Nile Delta and the Wadi regions has been implemented. Tourist economy in coastal areas has been linked to environmental sensitivity using indices obtained from remote sensing data. Remote sensing of aerosols has received a notable share in the publications of the Egyptian remote sensing community.

At the core of the future outlook of remote sensing applications in Egypt are the two issues of:

(1) The expansion of the domain of applications and the quality of the retrieved parameters.
(2) The holistic view of the earth system with its governing processes.

The expansion requires enlarging the spatial domain of applications of remote sensing data (at least to a regional scale covering MENA or tropical Africa). In addition, the number of geophysical parameters sought to be retrieved should include climate-related parameters affecting the earth's radiation budget. An example would be the reflected radiation in the short wave (optical) spectrum and the emitted radiation in long wave (thermal infrared) spectrum. Finally, this expansion entails improving the accuracy of the retrieved parameters using more comprehensive modeling and data assimilation approaches.

The holistic view involves using remote sensing data to study the earth as a system. This should incorporate earth's physics and advanced physical models and electromagnetic wave propagation models in the analysis and utilization of remote sensing data.

So far, optical remote sensing is mostly used in Egypt, particularly from Landsat. Yet, radar remote sensing applications are currently growing particularly in the fields of hydrology and water resources exploration (see Chapter Radar Remote Sensing Applications by M. Shokr in this book). Fine- and medium-resolution data are commonly used (e.g. around 30 m from Landsat and 250–500 m from MODIS). No much use of coarse-resolution data from AVHRR and similar sensors (few kilometers) has emerged, and no use at all of passive microwave data (resolution of few kilometers or tens of kilometers). Fine and medium resolution match the relatively small area of Egypt. Coarser resolution data will be more on demand when the domain of application is expanded to cover regional or global scale.

The increasing accessibility of remote sensing data with improvement of their spatial resolution, along with the availability of more data processing software will further accelerate the data applications in Egypt. Improving the availability and quality of data from microsatellite constellation (CubeSat) missions in the future will open new ventures of applications.

4.6 Recommendations

It is recommended to reinforce the collaboration between the Egyptian remote sensing community and their counterparts in the MENA region to study region-wide phenomena such as sand storms, pollutant transport, water resources management and impacts of regional climate change. Partnership with international organizations is crucial for capacity building but the collective and coordinated work of the regional institutions will be more rewarding in terms of adjusting the scheme of the research and the output to the common needs of the partners. With the spatial expansion of the applications, it is necessary to explore the utilization of the coarse-resolution data from passive microwave and scatterometer systems. None of these systems have been used in Egypt so far probably because of their coarse resolution (a few km at best) though the data are available daily and free of charge. Advanced applications of remote sensing for environment have been developed recently to address a few issues related to national mega-projects. However, it is recommended to explore applications in earth systems sciences in order to contribute to the intensive international effort to understand interactions between the earth's components and track/model impacts of the current global warming. Remote sensing data should also be used to support short-term forecast models such as weather, pollutant transport and land surface processes (e.g. energy balance, carbon emission, and nutrient fluxes) on a local to regional scale. This is the essence of the remote sensing data assimilation, which is another recommended research front.

Acknowledgements The author would like to thank Dr. S. Arafat (NARSS) for his reviewing the first draft of this chapter. Thanks are also extended to Drs. I. Abu-Elmagd and E. Farag for the data and the graphs they provided. The author acknowledges also the fruitful discussions with scientists from NARSS and other Egyptian institutions involved in using satellite remote sensing over many years that led to the compilation of the information presented in this chapter.

References

Abdel-Kader M (2007) A numerical study of the thermal inversion phenomenon over Cairo and the Nile Delta. M.Sc. Thesis, Aerospace Engineering Department, Cairo University, Giza, Egypt

Abdel-Latif T, Ramadan ST, Galal AM (2012) Egyptian coastal development through economic diversity for its coastal cities. HBRC J 8(3):252–262

Aboelghar M, Arafat S, Farag E (2013) Hyperspectral measurements as a method for potato crop characterization. Int J Adv Remote Sens GIS 2(1):122–129

Abou El-Magd I, Hassan O, Arafat S (2013) Quantification of sand dune movement in the south western part of Egypt using remotely sensed data and GIS. J Geogr Inf Syst 5:48–508

Abu Salem HS, Abu Khatita A, Abdeen M, Mohamed EA, El Kammar AM (2017) Geo-environmental evaluation of Wadi El Raiyan lakes, Egypt, using remote sensing and trace element techniques. Arab J Geosci 10:224. Accessible through https://doi.org/10.1007/s12517-017-2991-3

Ali AOT, Rashid M, El Naggar S, Abdul-Al A (2007) Water harvesting options in the drylands at different spatial scales. Land Use Water Resour Res 7:1–13

Arafat SM, Farag E, Shokr M, Al-Kzaz G (2013a) Internet-based spectral database for different land covers in Egypt. Adv Remote Sens 2:85–92. https://doi.org/10.4236/ars.2013.22012

Arafat SM, Aboelghar MA, Ahmed EF (2013b) Crop discrimination using field hyperspectral remotely sensed data. Adv Remote Sens 2:6–70

Arnous M, Abouleda H, Green DR (2011) Geo-environmental hazards assessment of the north western Gulf of Suez, Egypt. J Coast Conserv 15:37–50. https://link.springer.com/article/10.1007/s11852-010-0118-z

Arnous MO, Green DR (2015) Monitoring and assessing waterlogged and salt-affected areas in the Eastern Nile Delta region, Egypt, using remotely sensed multi-temporal data and GIS. J Coast Conserv 19(3):369–391

Bastawesy M, Khalaf FI, Arafat SM (2008) The use of remote sensing and GIS for the estimation of water loss from Tusha lakes, southwest desert, Egypt. J Afr Earth Sci 52(3):73–80

Bastiaanssen WGM, Menenti M, Feddes RA, Holtslag AAM (1998) Remote sensing surface energy balance algorithm for land (SEBAL): 1. Formulation. J Hydrol 212–213(1–4):198–212

Besler H (2008) The Great Sand Sea in Egypt: formation, dynamics and environmental change: a sediment-analytical approach. Elsevier, Amsterdam, pp 1–30. ISBN 978-0-444-52941-1

Droogers P, Immerzeel WW, Terink W, Hoogeveen J, Bierkens MFP, van Beek LPH, Debel B (2012) Water resources trend in the Middle East and North Africa towards 2050. Hydrol Earth Syst Sci 16:3101–3114

Egyptian Environmental Affairs Agency (2016) Egypt third national communication. Report submitted to the United Nations Framework Convention on Climate Change (UNFCCC). https://unfccc.int/files/national_reports/non-annex_i_parties/biennial_update_reports/application/pdf/tnc_report.pdf

El Abd EA, El Osta MM (2014) Waterlogging in the new reclaimed areas Northeast El Fayoum, Western Desert, Egypt, reasons and solutions. J Water Resour Prot 6:1631–1645
El-Askary H, Kafatos M (2008) Dust storm and black cloud influence on aerosol optical properties over Cairo and the Greater Delta Region Egypt. Int J Remote Sens 29(24):7199–7211
El-Asmar M, Wood P (2000) Quaternary shoreline development: the northwestern coast of Egypt. Quat Sci Rev 19:1137–1149
El-Asmar HM, Hereher ME, El Kafrawy SB (2013) Surface area change detection of the Burullus Lagoon, North of the Nile delta, Egypt, using water indices: a remote sensing approach. Egypt J Remote Sens Space Sci 16:119–123
Ebaid HM (2015) Drought monitoring using remote sensing approach, Western Desert, Egypt. Int J Geomatics Geosci 6(3):1638–1652
El-Ameen M, Farragallah A, Essa MA (2011) Mineralogical composition of Khamsin wind dust at Assiut, Egypt. Assiut Univ Bull Environ Res 14:95–107
Elbeih SF (2015) An overview of integrated remote sensing and GIS for groundwater mapping in Egypt. Ain Shams Eng J 6(1):1–15
El-Banna MS (2004) Geological studies emphasizing the morphology and dynamics of sand dunes and their environmental impacts on the reclamation and developmental areas in northwest Sinai, Egypt. Ph.D. Dissertation, Cairo University, Egypt
El Bastawesy M, Ali RR, Faid A, El Osta M (2013) Assessment of water logging in agricultural megaprojects in the closed drainage basins of the Western Desert of Egypt. J Hydrol Earth Syst Sci 17:1493–1501
El-Baz F (ed) (1984) Desert and arid lands: remote sensing of earth resources and environment. Martinus Nijhoff Publishers, The Hauge, Boston and Lancaster
El-Fandy MG (1948) Baroclinic low of Sybrus. Q J R Meteorol Soc 72:291–306
El Gammal EA, Salem SM, El Gammal AEA (2010) Change detection studies on the world's biggest artificial lake (Lake Nasser, Egypt). Egypt J Remote Sens Space Sci 13:89–99
El-Kafrawy S (2017) Monitoring the environmental changes of Mariout Lake during the last four decades using remote sensing and GIS techniques. MOJ Ecol Environ Sci 2(5):00037. https://doi.org/10.15406/mojes.2017.02.00037
El-Kafrawy SB, Khalafallah A, Omar M, Khalil M, Yehia A, Allam M (2015) An integrated field and remote sensing approach for water quality mapping in Lake Burullus, Egypt. Int J Environ Sci Eng 6(15):15–20
Elmetwalli AMH (2008) Remote sensing as a precision farming tool in the Nile Valley, Egypt. Ph.D. Dissertation, School of Biological and Environmental Sciences, University of Stirling, Stirling, UK, 330 p
El-Metwally M, Alfaro SC, Abdel Wahab M, Chatenet B (2008) Aerosol characteristics over urban Cairo: seasonal variations as retrieved from Sun photometer measurements. J Geophys Res 113:D14219
El-Metwally M, Alfaro SC, Abdel Wahab MM, Zakey AS, Chatenet B (2010) Seasonal and inter-annual variability of the aerosol content in Cairo (Egypt) as deduced from the comparison of MODIS aerosol retrievals with direct AERONET measurements. Atmos Res 97:14–25
El Nahry AH, Ibrahim MM, El Baroudy AA (2015) Assessment of soil degradation in North part of Nile Delta using remote sensing and GIS techniques. In: ISPRS—International Archives of the Photogrammetry, Remote Sensing and Spatial Information Sciences, Volume XL-7/W3, 3rd international symposium on remote sensing of environment, Berlin, Germany, pp 1461–1467
El-Raey M (1997) Vulnerability assessment of the coastal zone of the Nile Delta of Egypt to the impacts of sea level rise. Ocean Coast Manag 37(1):29–40
Elsayed AA (2007) Modeling land cover induced climate changes using satellite data. M.Sc. Thesis, Aerospace Engineering Department, Cairo University
El-Shirbeny MA, Ali AA, Saleh NH (2014) Crop water requirements in Egypt using remote sensing techniques. J Agric Chem Environ 3:57–65

EL-Tahan M, Shok M, Sherif AO (2018) Simulation of severe dust events over Egypt using tuned dust schemes in Weather Research Forecast (WRF-Chem). Atmosphere 9:246. https://doi.org/10.3390/atmos9070246

Emara S, Khadr M, Zeidan B (2016) Assessment of land use/cover change using remote sensing and GIS in the Nile Delta, Egypt. In: Proceedings of third international environmental forum, environmental pollution: problem & solution, Tanta University, Egypt, 12–14 July 2016, pp 10–15

Farag H, El-Gamal A (2012) Assessment of the eutrophic status of Lake Burullus (Egypt) using remote sensing. Int J Environ Sci Eng 2:1–7

Farg E, Arafat SM, Abdel El-Wahed MS, El-Gindy AM (2012) Estimation of evapotranspiration ET_c and crop coefficient K_c of wheat, in south Nile Delta of Egypt using integrated FAO-56 approach and remote sensing data. Egypt J Remote Sens Space Sci 15(1):83–89

Farrag AA, El-Sayed EA, Megahed HA (2016) Land use/land cover change detection and classification using remote sensing and GIS techniques: a case study at Siwa Oasis, Northwestern Desert of Egypt. Int J Adv Remote Sens GIS 5(3):1649–1661

Freek D et al (2012) Multi- and hyperspectral geologic remote sensing: a review. Int J Appl Earth Obs Geoinf 14(1):112–128

Geological Society of America (2017) Looming crisis of the much decreased fresh-water supply to Egypt's Nile Delta. ScienceDaily, March 13. https://www.sciencedaily.com/releases/2017/03/170313135006.htm

Gifford AW, Warner DM, El-Baz F (1979) Orbital observations of sand distribution in the Western Desert of Egypt. In: El-Baz F, Warner DM (eds) Apollo-Soyuz Test Project, summary science report, vol 2, Earth observations and photography, NASA Sp-412, Washington DC, pp 219–236

Gupta RP (2003) Remote sensing geology. Springer-Verlag, Berlin and Heidelberg, 32 p

Habib S, Kfour C, Peters M (2012) Water information system platforms addressing critical societal needs in the MENA region. In: Proceedings of international geoscience and remote sensing symposium (IGARSS), pp 2767–2770. https://doi.org/10.1109/IGARSS.2012.6350859

Hegazy IR, Kaloop MR (2015) Monitoring urban growth and land use change detection with GIS and remote sensing techniques in Daqahlia governorate Egypt. Int J Sustain Built Environ 4:117–124

Hermas E, Leprince S, Abou E-M (2012) Retrieving sand dune movements using sub-pixel correlation of multi-temporal optical remote sensing imagery, northwest Sinai Peninsula Egypt. Remote Sens Environ 121:51–60

Hosny MM, Abdelmoaty MS (2009) Assessment the hazard of sand dune movements on the irrigation canals, Toshka Project. In: Proceedings of the 13th international water technology conference, Hurghada, pp 311–321

Kaiser MF, El Rayes A, Ghodief K, Geriesh B (2013) GIS data integration to manage waterlogging problem on the eastern Nile Delta of Egypt. Int J Geosci 4:680–687

Kandil H, Abdelkader M, Abdul Moaty A, Elhadid B, Sherif A (2006) Simulation of atmospheric temperature inversions over Greater Cairo using the MM5 mesoscale atmospheric model. Egypt J Remote Sens Space Sci 9:15–30

Kuznetsov A, Melnikova I, Pozdnyakov D, Seroukhova O, Vasilyev A (2012) Remote sensing of environment and radiation transfer: an introductory survey. Springer, Berlin, Heidelberg, p 203

L'Ecuyer TS, Jiang J (2010) Touring the atmosphere aboard the A-train. Phys Today 63:36–41

Lenoble J, Remer L, Tanré D (eds) (2013) Aerosol remote sensing. Springer, Heidelberg, Germany; New York, NY, USA; London, UK, 328 p

Marey HS, Gille JC, El-Askary HM, Shalaby EA, El-Raey ME (2011) Aerosol climatology over Nile Delta based on MODIS, MISR and OMI satellite data. Atmos Chem Phys 11:10637–10648

Masria A, Negm AM, Iskander M (2016) Assessment of Nile Delta coastal zone using remote sensing. In: Negm A (ed) The Nile Delta: the handbook of environmental chemistry, vol 5. Springer, Cham

Melesse AM, Weng Q, Thenkabail PS, Senay GB (2007) Remote sensing sensors and applications in environmental resources mapping and modelling. Sensors 7(12):3209–3241

Milewski A, Sultan M, Yan E, Becker R, Abdeldayem A, Soliman F, Gelil KA (2009) A remote sensing solution for estimating runoff and recharge in arid environments. J Hydrol 373(1–2):1–14

Mohamed MM, Elmahdy SI (2017) Remote sensing of the Grand Ethiopian Renaissance Dam: a hazard and environmental impacts assessment. Geomatics Nat Hazard Risk. https://dx.doi.org/10.1080/19475705.2017.1309463

Muala E, Mohamed YA, Duan Z, van der Aag P (2014) Estimation of reservoir discharge from Lake Nasser and Roseires reservoir in the Nile Basin using satellite altimetry and imagery data. Remote Sens 6(8):7522–7545. https://doi.org/10.3390/rs6087522

Omran A (2006) Geo-environmental studies of North Western Gulf of Suez region, Egypt. M.Sc. Thesis, Suez Canal University, Faculty of Science, Geology Department, Ismailia, Egypt, 247 p

Parry M, Canziani O, Palutikof J, van der Linden P, Hanson C (eds) (2007) Climate change 2007, impacts adaptation and vulnerability. Cambridge University Press, Cambridge, New York, Melbourne, Madrid, Delhi, 9987 p

Prost G (2001) Remote sensing for geologist: a guide to image interpretation. Taylor and Francis Publishers, Sussex, UK, 440 p

Psilovikos A, Elhag M (2013) Forecasting of remotely sensed daily evapotranspiration data over Nile Delta region, Egypt. Water Resour Manag 27(12):4115–4130

Ramadan TM, Nasr AH, Mahmood A (2006) Integration of Radarsat-1 and Landsat TM images for mineral exploration in East Oweinat District, South Western Desert, Egypt. In: ISPRS Commission VII mid-term symposium "Remote sensing: from pixels to processes", Enschede, The Netherlands, 8–11 May, pp 244–249. https://www.isprs.org/proceedings/XXXVI/part7/PDF/236.pdf

Saad El-Din M, Gaber A, Koch M, Ahmed RS, Bahgat I (2013) Remote sensing applications for water assessment in Lake Timsah, Suez Canal, Egypt. J Remote Sens Technol 1(3):61–74

Said MA, Moursy ZA, Radwan AA (2012) Climate change and sea level oscillations off Alexandria, Egypt. In: Proceedings of the international conference on marine and coastal ecosystem, MarCoastEcs2012, Tirana, Albania, pp 25–28, 353–359

Shalaby A, Tateishi R (2007) Remote sensing and GIS for mapping and monitoring land-cover and land-use changes in the Northwestern coastal zone of Egypt. Appl Geogr 27:28–41

Shaltout M, Tonbol K, Omstedl A (2015) Sea-level change and projected future flooding along the Egyptian Mediterranean coast. Oceanologia 57(4):293–307

Shokr M (2010) Potential directions for applications of satellite earth observations data in Egypt. Egypt J Remote Sens Space Sci 14:1–13

Shokr M, El-Tahan M, Ibrahim A, Steiner A, Gad N (2017) Long-term, high-resolution survey of atmospheric aerosols over Egypt with NASA's MODIS data. Remote Sens 9(10):1027. https://doi.org/10.3390/rs9101027

Simonneaux V, Abdrabbo MAA, Saleh SM, Hassanein MK, Abou-Hadid AF, Chehbouni A (2010) MODIS estimates of annual evapotranspiration of irrigated crops in the Nile delta based on the FAO method: application to the Nile river budget. In: Proceedings of the SPIE, vol 7824, Remote sensing for agriculture, ecosystems, and hydrology XII, 78241S

Soloms S et al (2018) From tropospheric folding to Khamsin and Foehn winds: how atmospheric dynamics advanced a record-breaking dust episode in Crete. Atmosphere 9(7):240. https://doi.org/10.3390/atmos9070240

Stanley J, Clemente PI (2017) Increased land subsidence and sea-level rise are submerging Egypt's Nile Delta coastal margin. GSA Today 25(5). https://doi.org/10.1130/GSATG312A.1

Stephens GL (2003) Remote sensing of lower atmosphere: an introduction. Oxford University Press, 544 p. ISBN-13: 978-0195081886

Sultan H (2011) Ensemble forecasting and data assimilation in numerical weather modeling for Egypt. M.Sc. Thesis, Aerospace Engineering Department, Cairo University

Tran DV, Nguyen NV (2006) The concept and implementation of precision farming and rice integrated crop management systems for sustainable production in the twenty-first century. Int Rice Comm Newsl (FAO) 55:91–102

Vihma T (2014) Effects of Arctic Sea ice decline on weather and climate: a review. Surv Geophys 35:1175–1214. https://doi.org/10.1007/s10712-014-9284-0

Yehia M, Hamdan A, Hassan O, El-Etr H (2002) A regional study of the drainage basins of the Gulf of Suez and an assessment of their flash flood hazard. Egypt J Remote Sens Space Sci 5:77–98

Yousif M, Bubenzer O (2011) Integrated remote sensing and GIS for surface water development. Case study of Ras El Hekma area, northwester coast of Egypt. Arab J Geosci 6(4):1295–1306. https://link.springer.com/article/10.1007%2Fs12517-011-0433-1

Youssef AM, Maerz NH, Hassan AM (2009) Remote sensing applications to geological problems in Egypt: case study, slope instability investigation, Sharm El-Sheikh/Ras-Nasrani Area, Southern Sinai. Landslides 6:353. https://doi.org/10.1007/s10346-009-0158-3

Chapter 5
Radar Remote Sensing Applications in Egypt

Mohammed E. Shokr

Abstract Radar remote sensing is a rapidly developing topic with a fast-growing facets of technology and range of applications. Exploratory research studies started in Egypt in the late 1990s on opportunity basis when data were available free of charge. Radar sensors transmit radiation in the microwave bands and use the measured backscatter to infer properties of the earth's surface. The surface parameters that influence the backscatter are different from those influencing the reflection and emission from optical and thermal infrared bands, respectively. Therefore, radar data usually offer distinctive perspectives on the surface and subsurface properties. This chapter provides an overview of the three spaceborne radar sensors with focus on the most versatile one; the synthetic aperture radar (SAR). The other two are scatterometers and radar altimeters. Due to the increasing availability of space-borne SAR systems and the accessibility of free data, SAR users' community is expanding in Egypt. A brief theoretical background of SAR is presented, followed by applications of single and dual channel systems, then the newly-developed polarimetric and interferometric systems. The most frequent SAR applications in Egypt are in the fields of geology, hydrology and archeology. No operational applications have been developed yet. Brief information on relevant applications of radar altimetry data are also presented along with an outlook on the future use of radar remote sensing in Egypt.

Keywords Radar remote sensing · Earth observation satellites · SAR applications in Egypt · Ground water · Geological and hydrological applications in Egypt

---

M. E. Shokr (✉)
Environment and Climate Change Canada, 4905 Dufferin Street, Toronto, ON M3H-5T4, Canada
e-mail: mo.shokr.ma@gmail.com; mohammed.shokr@canada.ca

National Authority for Remote Sensing and Space Sciences (NARSS), 23 Joseph Tito Street, El-Nozha El-Gedida, P.O. Box: 1564 Alf Maskan, Cairo, Egypt

S. F. Elbeih et al. (eds.), *Environmental Remote Sensing in Egypt*,
Springer Geophysics, https://doi.org/10.1007/978-3-030-39593-3_5

5.1 Introduction

Radar remote sensing is the active version of the microwave remote sensing (sub-millimeter to several tens of centimeter wavelength). It is "active" because the sensor transmits radar pulses and use the measured return to infer properties of the earth's surface. The return, called backscatter, is mainly influenced by the structural features of the ground area (including surface roughness) and its complex dielectric constant, which is mainly affected by the amount of moisture, salinity, and voids in the sub-surface layer. These influences are different from those that influence the reflection in optical or emission in thermal infrared frequencies (namely the optical and thermal properties of the subsurface). Hence radar offers distinctive perspectives of the earth's cover, yet it complements information retrieved from optical and thermal sensors.

Passive microwave (PM) sensors, on the other hand, measure the radiation emitted from the surface and sub-surface layer (within what is called penetration depth) of the electromagnetic wave, modulated by the atmosphere. The information content of the measurements is mainly a function of the emissivity of the ground cover and its physical temperature; both within the radiating layer. This, in turn, depends on the physical composition of the material, the soil moisture, salinity and degree of heterogeneity. The information can be used to identify the surface cover and its physical temperature (if the emissivity is known). Due to their coarse spatial resolution (a few kilometers or tens of kilometers), PM sensors are not used in Egypt though the data are available for free. The data may be useful only if regional applications are pursued.

Table 5.1 includes the frequency and wavelength of all bands used in microwave remote sensing (active and passive). Radar sensing covers the frequency range from P to Ku bands while passive microwave sensing starts from the L band (for the Soil Moisture and Ocean Salinity satellite (SMOS) sensor) up to the 183 GHz channel of the Special Sensor Microwave Imager/Sounder (SSMIS) satellite. As a rule of thumb,

Table 5.1 Frequency and wavelength of operational microwave sensors (active and passive)

Band symbol	Frequency (GHz)	Wavelength (cm)
P	0.3–1.0	100–30
L	1.0–2.0	30–15
S	2.0–4.0	15–7.5
C	4.0–8.0	7.5–3.75
X	8.0–12.5	3.75–2.4
Ku	12.5–18.0	2.4–1.67
K	18.0–27.0	1.67–1.10
Ka	27.0–40.0	1.10–0.75
V	40.0–75.0	0.75–0.40
W	75.0–110	0.40–0.27
mm	110–300	0.27–0.10

the higher the wavelength, the deeper the penetration of the signal into the surface. However, the depth depends also on the properties of the material (moisture and salinity) and the various modes of its interaction with the electromagnetic radiation. Depending on the ground cover and the moisture contents the penetration depth varies between a few millimeters to tens of centimeters (more penetration in dry medium).

In general, the atmosphere is transparent to microwave signal and more so for higher wavelengths. All radar sensors are not affected by the atmospheric contents except for the X band sensors which are affected by the atmosphere in presence of heavy rain. Similarly, the effect of the atmosphere on the PM measurements become tangible only for the V band and higher frequency signals.

Radar remote sensing is grouped into two categories; imaging and non-imaging (also called profile sensing). An imaging sensor provides a two-dimensional array of pixels from which an image of the ground scene is produced. The most popular sensor in this category is the Synthetic Aperture Radar (SAR). When the first L-band SAR flew onboard the short-lived NASA's Seasat satellite in 1978 (operated for only 106 days before a major failure in the electric system ended the mission) it produced astounding images, not only of the ocean surface for which it was designed but also of many types of earth's cover. Images of arid areas show, for the first time, sub-surface features. That was particularly interesting for archeological discoveries in Egypt.

Following the legacy of Seasat, plans were proposed in the 1980s to develop the next satellite-based SAR systems (by the European Space Agency—ESA, the Canadian Space Agency—CSA and the National Space Development Agency of Japan—NASDA, which is now called Japan Aerospace Exploration Agency—JAXA). At that time the remote sensing community in Egypt debated the potential advantages of using SAR for environmental applications in light of the growing and successful use of the optical and infrared sensors (mainly from Landsat and NOAA-AVHRR series then). The consensus was not in favor of using SAR because its "selling" features, namely its ability to capture the scene through cloud cover and during night did not appear as such. In Egypt, the sky is clear 72% of the time in most areas except for the coastal regions. Moreover, no long nights are experienced in winter (unlike the polar and sub-polar regions). At that time only the difficulty of SAR image interpretation was recognized.

Nevertheless, SAR was later recognized to be an indispensable tool for geological applications when its advantages for revealing subsurface information in the Egyptian arid landscape was proven. This emerged in the late 1990s using the European C-band SAR onboard ERS-1/2 satellites, the Japanese Phased Array L-band Synthetic Aperture Radar onboard the Advanced Land Observing Satellite (PALSAR-ALOS) and the Canadian C-band Radarsat.

Unlike electromagnetic wave in the optical and thermal infrared spectral regions, radar signal can penetrate a dry medium up to few tens of centimeters, depending on the wavelength of the signal and the dryness of the medium (the longer the wavelength and the drier the medium the more penetration). That is how it can capture subsurface information, especially in arid and semi-arid regions. While the interest in SAR started to grow in Egypt in the 1990s, the challenge of image interpretation remained. This is caused by the peculiar geometric distortions resulting from the side-looking viewing geometry of SAR and the fact that it is a distance-based rather than angular-based data acquisition (which is the case with the optical systems, including human vision). This leads to forms of geometric distortions not prone to interpretation by human vision system. Examples include foreshortening and layover distortions associated with rich topographic areas (https://www.nrcan.gc.ca/node/9325). Special training on SAR image interpretation and processing had to be offered to Egyptian users since 1999.

The spaceborne radar remote sensing includes three systems; SAR, scatterometery, and radar altimetry (RA). Table 5.2 includes a list of major satellite platforms that carry radar sensors. SAR may have fine modes operating at 1 m resolution (e.g. TerraSAR-X system), but the typical resolution is around 10 m (Radarsat-2 and Sentinel-1) or coarser. Scatterometers have a much coarser resolution, measured in tens of kilometers. However, the 4.5 km resampled resolution shown in Table 5.2 refers to a reconstructed resolution produced by the Microwave Earth Remote Sensing Laboratory of Brigham Young University (BYU) in Utah, U.S.A. The reconstructed data sets are available from a few scatterometer and PM systems from ftp://ftp.scp.byu.edu/pub/data/.

In addition to being part satellite payload, SAR also operated onboard two missions of the space shuttle Endeavour; the Spaceborne Imaging Radar-C/X-Band Synthetic Aperture Radar (SIR-C/X-SAR) in 1994 and the Shuttle Radar Topography Mission (SRTM) in 2000. The latter generated digital elevation model of the earth that has become a standard elevation data for many applications. SAR data from ERS-1/2, Radarsat-1/2, Envisat, PALSAR-ALOS, and Sentinel-1 satellites have been used in Egypt. However, it is fair to say that the data have been underused so far.

The rest of this chapter includes material on SAR applications in Egypt with information on the use of the two recent technological advances: radar polarimetry and interferometry. Scatterometer data have no applications so far, and altimeter data have limited use. Yet, a brief description of each system is included for the sake of completeness of the presentation.

5.2 Spaceborne Radar Systems

This section includes brief information on the basics of the three radar systems: SAR, scatterometer, and RA. More focus is placed on SAR. Detailed information about his system can be found in several textbooks (e.g. Elachi 1988; Oliver and Quegan 2004; Richards 2009; Barnes 2015). A comprehensive review of its applications in arid regions is presented in Elsherbini (2011). Detailed information about scatterometer is included in Long (2014). Description of radar altimetry with data processing and examples of applications are included in Rosmorduc et al. (2018). It is worth mentioning that two chapters on scatterometery and radar altimetry are presented with excellent illustrations in Ulaby and Long (2014).

Table 5.2 Major space-borne radar sensors (as of 2018) and their basic information

Satellite/Sensor	Country/Agency	Sensor type	Band	Pol.	Lifespan	Res. (m)	Swath (km)
Seasat	U.S.	SAR	L	HH	1978	25	100
Almaz-1	Russia	SAR	S	HH	1991–1992	10–100	350
ERS-1/ERS-2	ESA	SAR	C	VV	1991–2011	30	100
ERS-1 and -2	ESA	Scatterometer	Ku	VV	1991–2011	25/50	500
JERS-1	Japan	SAR	L	HH	1992–1998	18 × 24	75
Radarsat-1	Canada	SAR	C	HH	1995–2013	10–100	50–500
QuikSCAT/SeaWinds	U.S.	Scatterometer	Ku	HH, VV	1999–2009	4.5 km Resamp.	1400 (i), 1836 (o)
Jason-1/ -2	NASA/CNES	Altimeter	Ku		2001–13/2008–	11.2 km (along) ×5.1 km (across)	Profile
Envisat	Europe	SAR	C	S, D	2002–2012	30–500	5–406
PALSAR-ALOS	Japan	SAR	L	S, D, Q	2006–	7–100	20–350
ASCAT	ESA	Scatterometer	C	VV	2006–	12.5 km	1100
Radarsat-2	Canada	SAR	C	S, D, Q	2007–	3–100	10–500
TerraSAR-X and TanDEM-X	Germany	SAR	X	S, D, Q	2007–	1–18	100 × 150
COSMO-SkyMed series	Italy	SAR	X	S, D	2007–	15	30–40
OceanSat-2/OSCAT	India	Scatterometer	Ku	HH, VV	2009	4.5 km Resamp.	1400 (i), 1836 (o)
Cryosat-2	ESA	Altimeter	Ku	–	2010	300 × 1500 for first pixel, footprints spaced by 350 m	Profile

(continued)

Table 5.2 (continued)

Satellite/Sensor	Country/Agency	Sensor type	Band	Pol.	Lifespan	Res. (m)	Swath (km)
RISAT	India	SAR	C	S, D, Q, C	2012–	1–50	10–225
Sentinel-1	ESA	SAR	C	S, D	2014	5 × 20	250
SCATSAT-1/OSCAT-2	India	Scatterometer	Ku	HH, VV	2016–	4.5 km Resamp.	1400
Jason-3	NASA, CNES, NOAA, EUMETSAT	Altimeter	Ku		2016–	2–10 km depending on sea wave height	Profile
Radarsat Constellation Mission (RCM)	Canada	SAR	C	S, D, Q	2018	Variety	Variety

The systems are listed in chronological order

Note For the polarization: S, D, Q, and C refer to single, dual, quad and circular polarization; H and V refer to horizontal and vertical polarization. The symbols (i) and (o) that appear in the scatterometer swath denote the inner and outer beams of the scatterometer according to Fig. 5.5. Profile sensors collect data from a footprint at nadir only (no swath)

5.2.1 *Synthetic Aperture Radar (SAR)*

Monostatic SAR systems use a single antenna that functions as transmitter of radar pulses and receiver of the return signal after interaction with the ground cover. All spaceborne SAR systems are monostatic except for the SRTM that used two antennae (bistatic). Figure 5.1 shows the geometry of transmitted and received radar signals and a few parameters involved in the radar equations that follow. For the convenience of figure interpretation, the transmitted and received points are shown at different locations on the graph (but in reality, they use the same antenna).

5.2.1.1 Radar Equations

For "point target", i.e. a strongly scattering target located against a non-scattering background, the radar equation for the received power usually takes the forms

$$P_r = \left[\frac{P_t}{4\pi R_t^2} G_t\right] \sigma \left[\frac{1}{4\pi R_r^2} A_r\right] \tag{5.1}$$

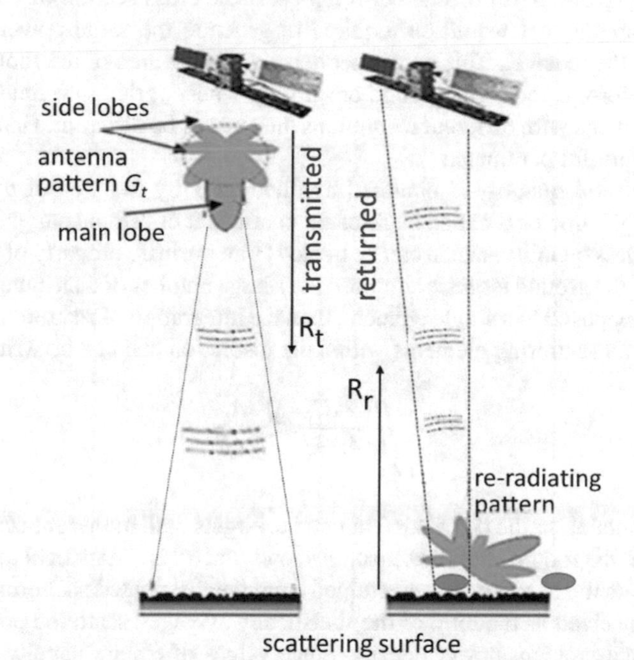

Fig. 5.1 An illustration of a radar transmitted pulse, its interaction with the surface and the backscatter that travels back to the receiving antenna. Parameters used in the "radar equations" (included in the text) are shown. For the ease of interpretation, the transmitting and receiving antennae are made separate in the figure and so is the receiving and the scattering footprints. Constructed by M. Shokr

The RHS of the equation is composed of three terms. The first represents the power distribution P_t of the main lobe of the transmitted signal (i.e. excluding the side lobes) at distance R_t from the antenna. The power intensity is inversely proportional to the square of the distance and is determined also by the antenna G_t within the antenna pattern. The third term represents the power of the received backscattering. A_r is the effective aperture of the antenna and R_r is the distance from the ground target to the antenna. The middle term, σ, represents the effect of the ground target on the balance of power between the transmitted and received. This term has units of area, hence called radar cross section. The incident energy is intercepted by the "effective receiving area" of the ground target A_{rs}. When this happens, the target absorbs a fraction of it, denoted p_a. This excites electric currents on the scattering elements within the target, which then becomes an antenna re-radiating with its own antenna pattern G_{rs}. The re-radiating power is resented σ and it takes the form

$$\sigma = A_{rs}(1 - p_a)G_{rs} \tag{5.2}$$

In this equation presents σ characterizes the target in terms of its ability to scatter the incident radar signal. The equation implies that the scattering from the target is isotropic, therefore, σ is considered to represent the cross section of an equivalent isotropic scatterer that would be required to generate the same power density as observed by the receiver. This parameter depends on the area of the footprint of the senor. Therefore, if the same ground cover (e.g. wheat agriculture field) is viewed with SAR sensors with different resolutions then σ will be different. Hence, σ is not a practically useful parameter.

A more useful quantity is obtained by dividing σ by the area of the observed terrain. This parameter is called backscatter coefficient or "sigma naught" (σ^o). It is a dimensionless quantity, which can be treated as an intrinsic property of the surface. Moreover, if the ground target is composed of a large number of distributed scattering elements as opposed to a single element, then the integration of scattered differential power from all scattering elements within the resolution cell can be written as:

$$p_r = \iint\limits_{A_o} \frac{P_t G_t}{4\pi R_r^2} \sigma^o \frac{A_r}{4\pi R_r^2} ds \tag{5.3}$$

This is a model for the distributed target; i.e. targets with numerous scattering elements within the footprint of the sensor. It shows that σ^o is a statistical measurement that represents the average power returned from the distributed scattering elements. It can be conceived as the ratio of the statistically averaged scattered power density to the average incident power density. Since values of σ^o are usually very small, they are presented in decibels (dB) rather than linear scale. When averaged over a number of pixels (sampling area), the σ^o must be calculated from its values in the linear power scale then converted to dB using the equation

$$\sigma^0 = 10 \log_{10}\left(\sigma^0_{linear}\right) \tag{5.4}$$

The standard deviation of σ^o over the sampling area is expressed in dB using the equation

$$SD\sigma^0_{db} = 10\log_{10}\left[\frac{Mean(\sigma^0_{linear}) - SD(\sigma^0_{linear})}{Mean(\sigma^0_{linear})}\right] \quad (5.5)$$

5.2.1.2 Factors Affecting Measured Backscatter

The measured backscatter is affected by two sets of factors: one pertaining to the sensor and the other to the ground cover. The sensor's parameters include the frequency, polarization and incidence angle of the transmitted signal. Lower frequencies (i.e. higher wavelengths) have more penetration into the medium and therefore are affected by its internal structure (see Table 5.1 for the radar frequencies and their designations). Moreover, higher wavelengths are not affected by small-size elements within the volume (i.e. elements with characteristic size less than one order of magnitude of the radar wavelength). In this case SAR backscatter will have a smaller component of volume scattering. Figure 5.2 illustrates what each microwave frequency "see" of a pine tree. The high-frequency X-band sees only the top canopy of the forest (or just a thin depth of the land). Backscatter from medium frequencies such as C- and L-bands is triggered by the surface and volume compositions while the lower frequencies (P and VHF) have higher penetration with no sensitivity to the heterogeneous composition of the branches and leaves of the tree. Once again, the effective scattering elements within the volume are those having dimensions of the same order as of the wavelength.

The effect of the incidence angle of the radar beam is manifested in higher backscatter at steep angles (near range) than shallow angles (far range). This is true for any surface. An example of three Radarsat-2 images of Lake Bardawil, north coast of Sinai is presented in Fig. 5.3. The images were acquired from the same orbit

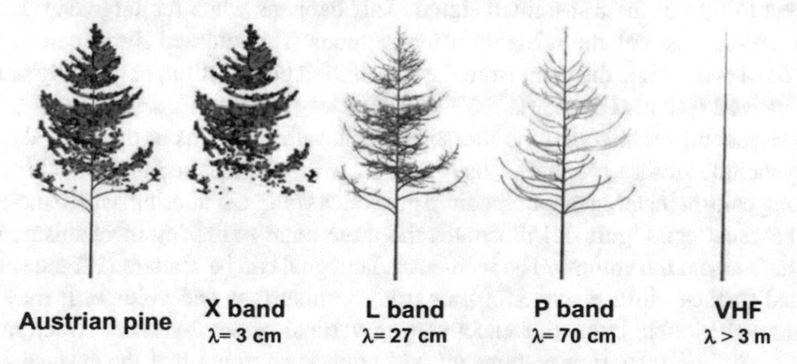

Fig. 5.2 Effect of the frequency of the radar signal on the captured information from a pine tree. *Source* https://whrc.org/wp-content/uploads/2016/02/Walker_SAR_Veg_Mapping.pdf

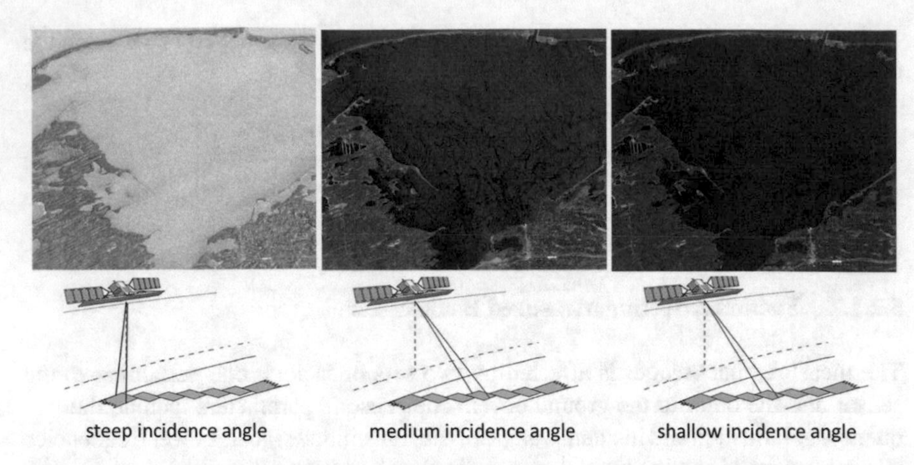

Fig. 5.3 Three scenes of Lake Bardawil, north Saini, from Radarsat-2 HH polarization, acquired from the same orbit at three different dates in early 2009, showing the effect of the incidence angle on the measured backscatter. Images courtesy of the Canadian Space Agency

in different dates using different standard beams (i.e. different incidence angles) as shown. The figure reveals the extreme high and low backscatter from the lake when captured by the steepest and shallowest angle radar beams, respectively. The medium angle range reveals the greatest level of details. This lake features shallow water with wet salt marshes, saline sand flats, sabkha, and sand dunes (sabkha is flat salty soil that can be sandy or muddy). That engenders the texture visible in the middle image in Fig. 5.3. One conclusion from the shown example is that if the incidence angle varies across the swath (when the swath becomes wide as in the case of ScanSAR wide mode of 400–500 km) then it has to be accounted for before proceeding with the image analysis.

Backscatter is also affected by the polarization of the transmitted radar signal (definition of polarization follows in the next section). Some point targets (also called pure targets) or surfaces features change the polarization of the scattered signal with respect to that of the transmitted signal. This happens when the target/surface has structures, or the volume is highly heterogeneous. The returned signal can be fully polarized even if it is different from the transmitted polarization, or partially or even unpolarized (see next section).

The second set affecting the measured backscatter pertains to the ground cover. They include surface roughness, dielectric properties of the medium and its internal decomposition (heterogeneous medium generates volume scattering, which increases the backscatter). Figure 5.4 illustrates the three radar scattering mechanisms from the surface and the volume. The incident radar signal can be scattered off the surface (called surface, diffuse scattering), or from both surface and volume. It may also go through double bounce if encountering vertical or semi-vertical structures. In nature, surface scattering features an odd bounce provided that the surface is not very rough. Double bounce is generated from upright or semi-upright structures

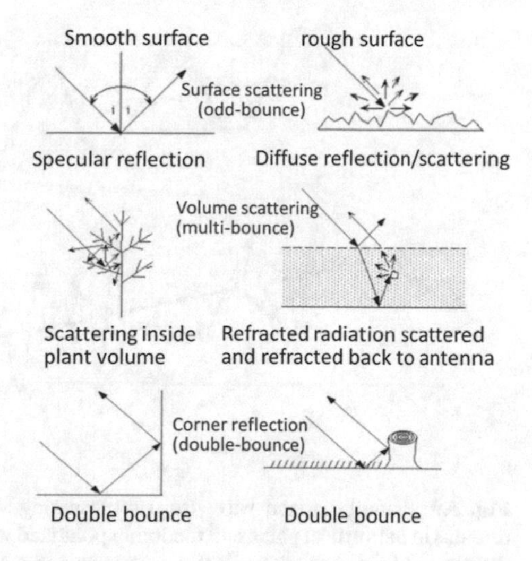

Fig. 5.4 The three scattering mechanisms of radar signal when interacts with ground cover. Compiled by M. Shokr

(natural or man-made) while multi-bounce is triggered by very rough surface or heterogeneous volume composition such as plants' leaves, branches, twigs, trunks in addition to soil in vegetation areas.

5.2.1.3 Radar Polarimetry

Two important developments of SAR technology have been achieved during the past three decades: polarimetry and interferometry. A brief theoretical background of radar polarimetry is presented in this section and of radar interferometry in the next section.

The essence of radar polarimetry is the polarized signal of the transmitted radar pulses. A polarized EM wave has a predictable alignment of its electric field vector as it propagates. It can be confined to a fixed plane (in this case the wave is called plane or linearly polarized) or oscillates in a predictable time-varying plane (e.g. elliptic polarization). Otherwise, the wave will be unpolarized (i.e. polarization varies randomly with time as the wave propagates) as in the case of sunlight. Illustrations of vertically and randomly polarized signals are show in Fig. 5.5. A common polarization configuration of SAR system is horizontal (H) or vertical (V).

In classical literature, the polarization state of a plane wave is expressed by a vector, known as the Stokes vector. Its elements are a function of the orientation ψ and ellipticity χ of the polarized signal (definitions are presented in many sources on polarimetric SAR systems such as Canada Centre for Remote Sensing 2003). The vector can be written as,

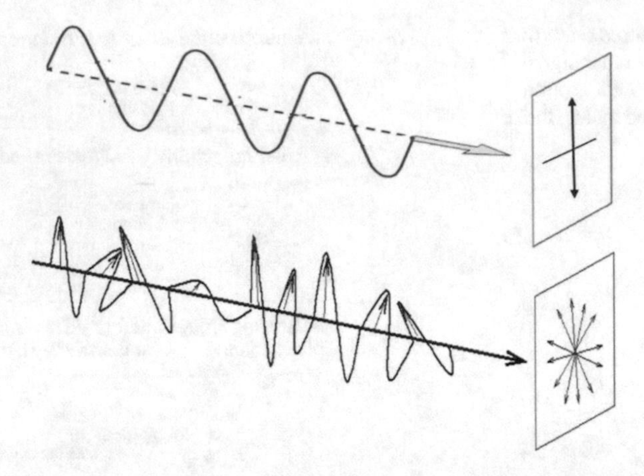

Fig. 5.5 Plane polarized wave (top) representing vertical polarization where the electric filed remains in the vertical plane and randomly polarized wave (bottom) (also called unpolarized) where the electric field changes randomly as the wave propagates. Compiled by M. Shokr

$$\begin{bmatrix} I_o \\ Q \\ U \\ V \end{bmatrix} = \begin{bmatrix} |E_v|^2 + |E_h|^2 \\ |E_v|^2 - |E_h|^2 \\ 2Re\{E_v E_h^*\} \\ 2Im\{E_v E_h^*\} \end{bmatrix} = \begin{bmatrix} I_0 \\ I_0 \cos 2\psi \cos 2\chi \\ I_0 \sin 2\psi \cos 2\chi \\ I_0 \sin 2\chi \end{bmatrix} \tag{5.6}$$

where I_0, Q, U and V are known as Stokes parameters and the superscript * refers to the complex conjugate of the vector. For fully polarized wave, the Stokes parameters are related by the equation:

$$I_o^2 = Q^2 + U^2 + V^2 \tag{5.7}$$

Therefore, only three parameters are independent. The parameter I_o^2 represents the total power in a completely polarized signal. For a partially polarized wave, the total power is greater than the summation of the squares of the other three terms. Therefore, the degree of polarization d_p of a signal is then given by the ratio

$$d_p = \frac{\sqrt{Q^2 + U^2 + V^2}}{I_0^2} \tag{5.8}$$

Upon scattering off a natural surface, the polarized radar signal often exhibits a change of polarization. The received signal may be partially polarized or completely unpolarized. The degree of polarization depends on the surface structure/roughness and can be derived using Eq. (5.8) (Touzi et al. 2004). The basic physical process responsible for the change of polarization (i.e. depolarization) of the scattered signal is the number of bounces (Fig. 5.4). Single bounce is produced by smooth surface and that maintains the same polarization as of the incident radar signal. Double bounce is produced by upright structures and that changes the polarization orthogonally, i.e. the incident H polarized signal will be reflected as V polarized signal yet both are still polarized. Multiple random bounces are produced either by rough surfaces or presence of inclusions in otherwise homogeneous composition. That leads to random polarization (partially polarized or unpolarized signal). Forest or any thick plant is a good example of this mechanism.

In dual-polarization SAR, the system transmits pulses with a selected polarization (e.g. horizontal H) and receives the component of the backscatter in the same polarization (e.g. HH) or in the orthogonal polarization (e.g. HV). In the first case the return is called co-polarized and in the second cross-polarized. Often the antenna can be switched, upon user's request, between the two transmitting modes of H and V. Therefore, the user may get one of two pairs of SAR data: HH and HV or VV and VH. The reciprocity theorem dictates that the backscatter in HV and VH are equal. Such system has been available from a few spaceborne SAR systems (e.g. Envisat, Radarsat-2 and Sentinel-1).

In fully-polarimetric SAR, the system alternates between transmitting horizontally- and vertically-polarized signals and receive the return from the two components for each transmitted signal. That is how it ends up receiving four scattering components (S_{hh}, S_{hh}, S_{vh}, and S_{vv}) from each pixel. Moreover, each component is recorded in terms of magnitude and phase; i.e. represented in the form of a complex number. That leads to each pixel represented by 2×2 matrix, called scattering or Sinclair matrix, whose elements are complex numbers. The matrix describes the transformation of the electric field of the incident wave to the scattered wave,

$$\begin{bmatrix} E_h^s \\ E_v^s \end{bmatrix} = [S] \begin{bmatrix} E_h^i \\ E_v^i \end{bmatrix} = \begin{bmatrix} S_{hh} & S_{hv} \\ S_{vh} & S_{vv} \end{bmatrix} \begin{bmatrix} E_h^i \\ E_v^i \end{bmatrix} \tag{5.9}$$

where E is the electric field intensity, superscripts i and s denote the incident signal (transmitted) and the scattered signal (i.e. received); respectively. Subscripts h and v denote the vertical and horizontal polarizations, respectively. A few useful parameters can be derived directly from combinations of elements of the scattering matrix. Examples include the total power of the backscatter (called SPAN), the co- and cross-polarization ratios, the cross-polarization correlation function and co-polarization phase difference. Definitions and explanations can be found in Zebker and van Zyl (1991).

More importantly, a few decompositions of the scattering matrix have been developed to describe the scattering mechanism from a "pure target"; that is a single scattering element within a background of a non-scattering surface cover (Touzi 1992). For these targets, also called coherent targets, both the incident and scattered waves are completely polarized. They are not common in nature. Examples of the decompositions suitable for coherent target characterization include Pauli, Krogager and Cameron. The reader can find more details about the formulation and uses of each decomposition in ESA Earth Online document https://earth.esa.int/documents/653194/656796/Polarimetric_Decompositions.pdf.

Pauli decomposition is based on a commonly-used scattering vector expression, derived from the scattering matrix, known as Pauli target vector.

$$K_P = \frac{1}{\sqrt{2}}[(S_{hh} + S_{vv}), (S_{hh} - S_{vv}), 2S_{hv}]^T \tag{5.10}$$

It takes the form

$$[S] = \begin{bmatrix} S_{hh} & S_{hv} \\ S_{vh} & S_{vv} \end{bmatrix} = \alpha[S]_a + \beta[S]_b + \gamma[S]_c \tag{5.11}$$

where α, β, and γ are the weighting factor of each term in the RHS; $[S]_a$, $[S]_b$ and $[S]_c$.

$$[S]_a = \frac{1}{\sqrt{2}}\begin{bmatrix} 1 & 0 \\ 0 & 1 \end{bmatrix} \tag{5.12}$$

$$[S]_b = \frac{1}{\sqrt{2}}\begin{bmatrix} 1 & 0 \\ 0 & -1 \end{bmatrix} \tag{5.13}$$

$$[S]_c = \frac{1}{\sqrt{2}}\begin{bmatrix} 0 & 1 \\ 1 & 0 \end{bmatrix} \tag{5.14}$$

$[S]_a$ correspond to the scattering from a sphere (referred to single- or odd-bounce scattering element). $[S]_b$ correspond to the scattering from a dihedral oriented at 0 degrees (referred to double or even-bounce scattering element). $[S]_c$ correspond to the scattering from a diplane oriented at 45°. This matrix is expressed in the linear orthogonal polarization (*hv* and *h*), i.e. the target returns wave polarization in orthogonal direction with respect to the incident wave. This mechanism usually results from volume scattering or scattering from a highly rough surface. Based on these definitions, the square of each term in the RHS of Eq. (5.11) (α, β, and γ) represents contributions from the odd-bounce, double-bounce and multi-bounce scattering mechanisms (Fig. 5.4).

It is worth reiterating that Pauli, Krogager and Cameron decompositions, which are all based on the scattering matrix, are most useful in the case of coherent targets; i.e. when the incident and the scattered waves are completely polarized. Nevertheless, they have been used to infer qualitative physical information from non-coherent targets. These are called "distributed targets" where numerous scattering elements coexist within the footprint. As a rule of thumb, man-made targets such as power towers and buildings are coherent whereas natural targets are distributed.

The above discussions indicate that when working with SAR polarimetric data it is important to know whether the backscatter is generated by a coherent target or distributed targets. For the latter, target can only be characterized statistically. Second-order polarimetric descriptors are usually used for this purpose. This is because distributed targets are associated with speckle noise. A commonly-used form of the descriptor is expressed in the form 3 × 3 Hermitian average covariance or coherency matrix (the two representations are equivalent).

The coherency matrix $[T]$ is derived from the Pauli vector (Eq. 5.10) as follows

$$[T] = K_P K_P^{*T} \tag{5.15}$$

where the superscript K_P^{*T} is the conjugate transpose of $\left[K_p\right]$. In a statistical form, the matrix takes the form

$$\langle [T] \rangle = \frac{1}{2} \begin{bmatrix} \left\langle |S_{hh}|^2 + 2Re(S_{hh}S_{vv}^*) + |S_{vv}|^2 \right\rangle & \left\langle |S_{hh}|^2 - 2Im(S_{hh}S_{vv}^*) - |S_{vv}|^2 \right\rangle & \left\langle 2S_{hh}S_{hv}^* + 2S_{vv}S_{hv}^* \right\rangle \\ \left\langle |S_{hh}|^2 + 2Im(S_{hh}S_{vv}^*) - |S_{vv}|^2 \right\rangle & \left\langle |S_{hh}|^2 - 2Re(S_{hh}S_{vv}^*) + |S_{vv}|^2 \right\rangle & \left\langle 2S_{hh}S_{hv}^* - 2S_{vv}S_{hv}^* \right\rangle \\ \left\langle 2S_{hv}S_{hh}^* + 2S_{hv}S_{vv}^* \right\rangle & \left\langle 2S_{hv}S_{hh}^* + 2S_{hv}S_{vv}^* \right\rangle & \left\langle 4|S_{hv}|^2 \right\rangle \end{bmatrix} \tag{5.16}$$

The symbol ⟨■⟩ denotes the average over a selected window size (e.g. 5 × 5) surrounding the given pixel. The diagonal elements are the square of the elements in the Pauli vector. Therefore they represent the three scattering mechanisms mentioned above. A few polarimetric decomposition parameters are derived from the coherency matrix. As an example, two parameters called entropy (denoted by H) and anisotropy (denoted by A) are derived from its eigenvalues, and a third one called alpha-angle (α) is derived from its eigen vector (Lee and Pottier 2009). The three parameters furnish a commonly-used set, called H/A/α or Cloude-Pottier decomposition parameters. They provide information about the dominance or the degree of coexistence of the three scattering mechanisms. This information identifies the geometric structure of the distributed target.

Other decompositions which are commonly used in processing polarimetric data from distributed targets include Freeman-Durden, Huynen, Yamaguchi, and Touzi. Details on these decompositions and their derived parameters are covered in several papers and books (Cloude and Pottier 1997; Freeman and Durden 1998; Touzi et al. 2004; Lee and Pottier 2009; Cloude, 2010; van Zyl 2011). They are implemented in commercial software such as PolSARpro (a multi-polarization SAR data processing software, developed under contract for ESA). This particular software is commonly used in SAR polarimetric research in Egypt. Cloude-Pottier decomposition (Cloude and Pottier 1997) is the most frequently used.

5.2.1.4 Radar Interferometry

SAR Interferometry (InSAR) is a technology to estimate the digital elevation of the surface (land or glaciers). Differential SAR interferometry (DInSAR) is an extension to estimate the subsidence (surface displacement) resulting after oil and gas extraction, soil compression occurs in delta plains or coastal areas, slow motion of glaciers in the polar regions, or land deformation after earthquakes.

The essence of this technology is the coherency of the transmitted radar pulses from the SAR antenna; meaning that pulses have the same phase. This feature allows measuring the change of phase of the backscatter with respect to the "reference" phase of the transmitted signal. The "phase shift" can be converted to a measure of the digital elevation model (DEM) of the observed surface, its subsidence or rate of displacement (deformation) in case of deformation. As an example of estimating the elevation suppose that the satellite flies at an altitude of 800 km. If the distance to the ground exceeds 800 km by 1 cm, the wave will have to cover an additional 2 cm in round trip. This constitutes a 40% of the C-band wavelength (5 cm). Then the phase of the reflected wave will be off by 40% of a cycle when it reaches the satellite. Accordingly, the measurement of phase provides a way to measure the distance to a target with cm precision. A review of this technology is presented in Hein (2004) and Ketelaar (2009). This technology has been used in limited but rather important applications in Egypt. The applications, along with more theoretical background, is presented in Sect. 5.3.3.

For mapping terrain elevation, the backscatter from each pixel is recorded in terms of the amplitude and phase in a SAR mode called Single Look Complex (SLC). If a single antenna is used (which is the case in all operational SAR systems so far), records from two passes of the same orbit (after a complete repeat cycle) should be obtained. However, the two passes must be displaced by a distance of a few hundreds of meters. This is called baseline. The phase difference between the two passes can be converted to the height of the pixel (digital elevation). A review of this technique with details of the calculations is presented in Bamler and Hartl (1998). For the purpose of this descriptive introduction, it suffices to mention the factors that affect the change of phase. If the two satellite passes P1 and P2 fly over a piece of land that does not undergo any deformation (the left panel of Fig. 5.6), then the change of phase coming from the same pixel between the two passes can be written as

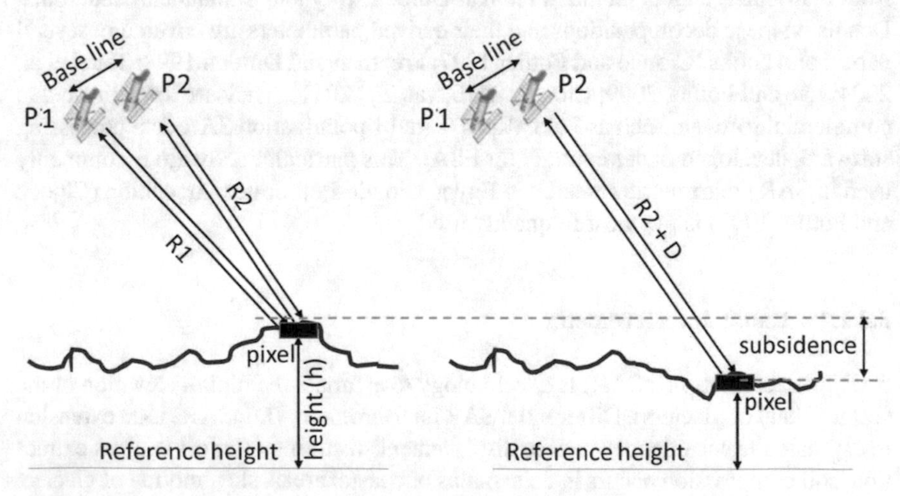

Fig. 5.6 Schematic diagram showing geometries for InSAR with the two passes P1 and P2 (left) and DInSAR with P1 same as in the left panel (before land subsidence) and P2 is acquired after land subsidence (right). Constructed by M. Shokr

$$\Delta\varphi = \left(\frac{4\pi.R2}{\lambda} + \varphi_{g,p2}\right) - \left(\frac{4\pi.R1}{\lambda} + \varphi_{g,p1}\right) \quad (5.17)$$

The first term in each bracket represents the change of phase due to the distance from the pixel to the antenna (called topography effect) and the second term φ_g represents the phase shift due to scattering that takes place at the pixel from each pass. The scattering is caused by surface geometry and subsurface dielectric properties. With simple geometrical manipulation, the height (with respect to a reference datum) can be calculated from the measured $\Delta\varphi$. It can also be seen that $\Delta\varphi$ depends on the distances $R1$ and $R2$, which accordingly depends on the baseline (Fig. 5.6). Large baseline may imply significant change in the incidence angle between the two passes, which adds undesirable change to the phase due to change of incidence angle. This phase is not introduced in the above equation simply because it cannot be accounted for during calculations.

Calculations of height using Eq. 5.17 requires that $\varphi_{g,p1}$ and $\varphi_{g,p2}$ to be equal (so they can be cancelled out). This means that the pixel must not undergo any change in physical conditions, i.e. no growing plants, change in soil moisture, modifications in urban structures, etc. Coherency is used to measure how close these two terms are. Rapid and random change in the position/or composition of the pixel is detected as low coherency. Examples include vegetation areas, ocean surface, snow cover, soil erosion, rapid urban development between the two satellite passes, etc. In such cases SAR interferometry will not produce accurate results. On the other hand, pixels of high coherence such as those from established urban sites or desert landscape (excluding mobile sand dunes) are suitable for both InSAR and DInSAR. Coherence is mainly used as an interferometric quality check but it has been also used as a terrain classification parameter (ftp://ftp.ccrs.nrcan.gc.ca/ad/MAS/globesar/eng/adv_int_e.pdf).

In the DInSAR technology, the second satellite pass P2 must take place after some sort of land displacement/subsidence D has been experienced (as shown in the right panel of Fig. 5.6). This displacement will be manifested as an additional component of phase change. It can still be calculated using Eq. 5.17 but after replacing the tern $R2$ by $(R2+D)$. The calculation becomes possible if the digital elevation of the pixel, which implies $R2$, is known. This can be obtained either from a previous InSAR calculations of elevation or through an existing DEM (e.g. from STRM data). DInSAR uses the following equation for the change of phase, which is more comprehensive than Eq. 5.17,

$$\Delta\varphi = \left(\frac{4\pi.R2}{\lambda} - \frac{4\pi.R1}{\lambda}\right) + \frac{4\pi.D}{\lambda} + \varphi_{g,p2} - \varphi_{g,p1} + \varphi_{atm,p2} - \varphi_{atm,p1} + \varphi_{orb,p2} - \varphi_{orb,p1} + \varphi_{noise} + 2k\pi \quad (5.18)$$

The first and the second terms represent the phase change due to topography and displacement, respectively. The second term is the unknown. The phase change due to the atmosphere is represented by φ_{atm} (the wave speed through the atmosphere is lower than in vacuum, depending on the air temperature, pressure and the water vapor column). This term can be neglected if the atmosphere is clear. φ_{orb} is the phase component due to errors in orbital calculations and φ_{noise} is the phase noise. The last term exists because the phase is bounded within the range $(-\pi, \pi)$, where k is an integer value (called phase ambiguity). All the terms in Eq. 5.18 can be estimated, at least approximately and the two terms of φ_g have to be equal to cancel out as mentioned before.

This last condition led to an important development in SAR interferometry in the late 1990s by stacking images for the same ground area and searching for pixels that return consistent and stable radar backscatter (i.e. exhibiting sufficiently high coherence). This theme led to the development of a few algorithms that are grouped under the title of persistent scatterer (PS) (Ferretti et al. 2001; Colesanti et al. 2003; Hooper et al. 2004). A review of this approach is offered in Crosetto et al. (2016).

5.2.2 Satellite Scatterometery

Scatterometers constitute a class of active microwave (i.e. radar) sensors. A scatterometer is a non-scanning or scanning radar sensor that observes the same part of the surface from different viewing angles (two or three). It acquires measurements at very coarse resolution footprints (tens of kilometers). It is originally developed to estimate winds over ocean surface (Pan et al. 2003) though many land applications have been developed later. The scatterometer may operate in the C-band, but most systems operate in the Ku band (Table 5.1). With its coarse resolution and subsequently wide swath, a scatterometer provides more frequent coverage, which is particularly useful in polar regions. Currently, the operating scatterometer systems include the Advanced Scatterometer (ASCAT) onboard European EUMETSAT satellite and the Indian Oceansat-2 scatterometer (OSCAT-2) onboard the Indian Oceansat-2 satellite (see Table 5.1). QuikSCAT satellite (operated from 1999 to 2009) carried the Seawinds scatterometer. This sensor scanned the surface in circular patterns (Fig. 5.7) and acquired data in two co-polarized modes (HH and VV) at incidence angles 46° and 54°, respectively. The image is constructed from the sequential backscatter return from the two radar beams. Circular viewing geometry has the advantage of data acquisition at constant incidence angle, so no correction is needed to account for the variation of the angle across the swath of the image. The data from QuikSCAT and OSCAT are reconstructed into finer resolution of 4.5 km at BYU (see the website in Sect. 5.1). Scatterometer data have not been used in Egypt, yet this brief introduction is included to complete the coverage of satellite radar sensors.

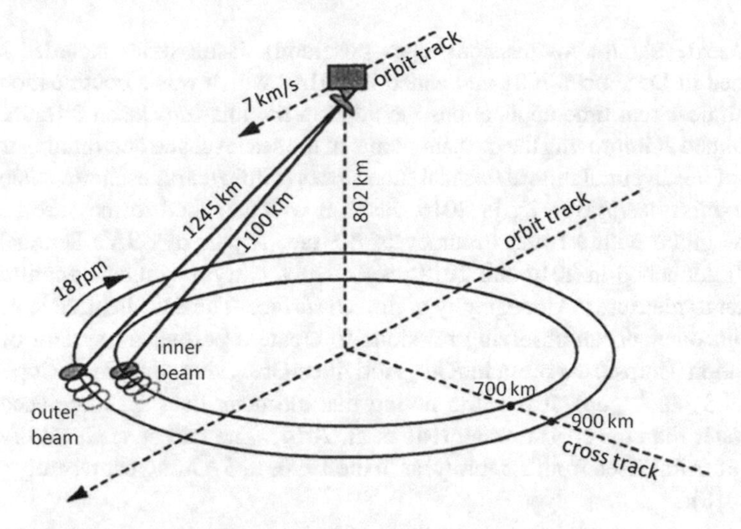

Fig. 5.7 Viewing geometry of the SeaWinds scatterometer with its rotating antennae that acquire data in HH and VV from the inner and outer bean, respectively. *Source* Weissman et al. (2003)

5.2.3 Radar Altimeters (RA)

"Radar altimetry measures the time taken by a radar pulse to travel from the satellite antenna to the surface and back to the satellite receiver" (https://www.altimetry.info/radar-altimetry-tutorial/). Combined with precise satellite location data, altimetry measurements yield sea-surface heights. The accuracy of the measurements is discussed in details in Shum et al. (1995). Satellite radar altimeters go back to 1991 and continues since then. With this long enough record, altimeter data can be used to conclude something about sea level change over nearly three decades. Satellite RAs started when ESA included the first instrument on the ERS-1 satellite in 1991, then in 1995 on ERS-2. An advanced version was later flown on ESA's ENVISAT satellite, which was decommissioned in 2012.

A current operational RA system is called SAR Interferometric Radar Altimeter (SIRAL) onboard ESA's Cryosat-2 satellite. This sensor is dedicated to measuring the sea level and monitor changes in glacier and ice shelf heights in the polar regions. The satellite was launched in April 2010, almost five years after Cryosat-1 was lost in a launch failure. SIRAL is a Ku-band (13.4 GHz) senor with footprint dimensions 300 m in the along-track and about 1.5 km in the cross-track direction (Wingham et al. 2006). It should be noted that the repeat cycle of Cryosat-2 is 369 days with track spacing of over 4 km. Therefore, mapping sea level over a large area requires numerous passes acquired over several weeks.

Other satellite programs that carry RA systems dedicated to measuring sea level include TOPEX/Poseidon, Jason and Sentinel-3. TOPEX/Poseidon (TP), a joint venture of NASA and the French space agency, CNES, was launch in 1992 and ended its mission in 2006. It measured ocean surface topography to an accuracy of 4.2 cm (https://sealevel.nasa.gov/missions/topex-poseidon). Jason series included Jason-1 (launched in December 2001 and ended in 2013), which was a continuation of TP but with near-real time applications included in the mission. Jason-2 (launched in 2008) aimed at improving the measurements of the sea level and determining the variability of ocean circulation at decadal time scales (https://earth.esa.int/web/eoportal/satellite-missions/j/jason-2). In 2016, Jason-3 was launched to measure sea level over the globe with a better accuracy of 3.3 cm. A pair of ESA's Sentinel-3 (-A and -B), launched in 2016 and 2018 respectively, carry a synthetic aperture radar altimeter to measure the topography of the sea surface. The data are used in conjunction with other ocean-observing missions to create a permanent system of ocean observation (https://www.esa.int/Our_Activities/Observing_the_Earth/Copernicus/Sentinel-3_stacks_up). It is worth noting that altimeter data are more accurate in open water than the coastal water (Rio et al. 2014). The rate of sea level rise in the satellite era has risen from 2.5 mm/year in the 1990s to 3.4 mm/year recently (Nerem et al. 2018).

5.3 Applications of Radar Remote Sensing in Egypt

The value of SAR data for subsurface information was realized in 1982 when images of part of the Western Desert in Egypt was acquired from the L-band SAR onboard the Space Shuttle Columbia (SIR-A). Figure 5.8 shows an image from this mission overlaid on a wider Landsat MSS image (SIR-A image had swath width is 50 km). The figure clearly demonstrates the powerful capability of radar images in revealing the subsurface information. Here, dry watercourses from a wetter paleo-period is clearly visible. This is referred to as "radar river". Archaeologists provided evidence of occupation of his area when discovered a significant number of stone tools near the site. Since the discovery of this sand buried system, radar images have been exploited worldwide to map subsurface features beneath a sandy cover of extremely low humidity and dielectric loss in arid regions.

The use of SAR data in Egypt started with geological applications in the late 1990s then progressed to include applications in the fields of hydrology and archeology. In general, SAR data have been underused in Egypt but it is expected to be on the rise since Sentinel-1 data became available free of charge and many good SAR processing tools also became available through open access (e.g. Sentinel-1 Toolbox, MapReady, and PolSARpro in addition to a few SAR processing modules in major satellite image processing software packages).

5.3.1 Single- and Dual-Polarization SAR

The most frequently-used SAR data in Egypt include the European ERS-1/2, Envisat, the Canadian Radarsat-1/2 and the Japanese ALOS-PALSAR. Many groups have

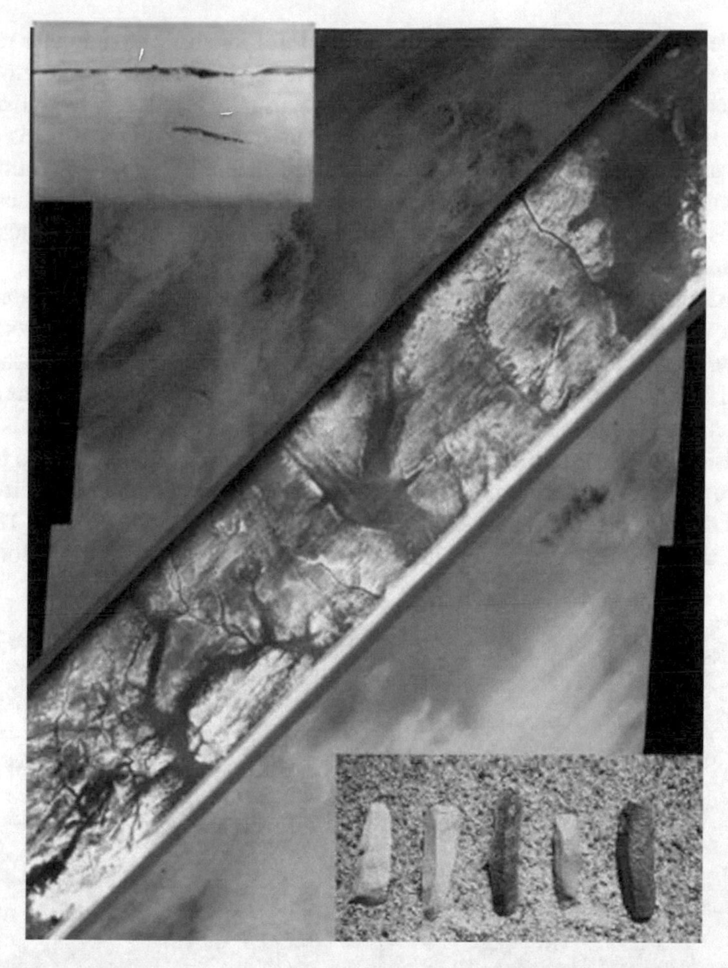

Fig. 5.8 SIR-A radar image of the Western Desert of Egypt laid over a Landsat MSS image. The radar image shows a fluvial landscape under the sand from wetter paleo-climatic episodes. Upper left inset shows the ground seen on the site. The lower right inset shows typical artifacts collected. *Source* Elachi et al. (1984)

recently started to download ESA's Sentinel-1 images, but no studies have yet been completed. Except for the L-band ALOS data, all other data are obtained from C-band systems. Most of SAR studies are conducted on exploration basis (as opposed to operational use) to examine the utility of the data in a few applications that include land cover classification and geological features identification. The L-band data are more suitable for the arid landscape of Egypt although they are widely available. Combining C- and L-band data may improve the classification and information retrieval, but this approach has not been pursued for environmental applications in Egypt yet.

Single-channel SAR data may not be very useful for surface parameter retrieval because the single backscatter observation is usually influenced by many parameters. Therefore, studies used single-channel SAR usually fail to achieve satisfactory classification accuracies. Three approaches are used to increase the dimensionality of SAR data. The first one employs multi-temporal data. This approach can support, for example, operational crop mapping and in-season crop acreage estimates. Crop growth is identified by its phenological stages, which are usually manifested in the radar backscatter.

An example of a multi-temporal composite color image of Cairo and its surrounding area was obtained from three ERS-2 images (Fig. 5.9). It reveals a composite image where the RGB combines three dates: 26 September 2002 (R), 2 May 2002 (G), and 4 January 2004 (B). Cairo is shown in white color (bottom-center) as a result of the persistent high backscatter from the three dates. The Nile Delta area is shown in reddish color because in September (the red channel) the summer crops become fully grown and therefore return the highest backscatter. Part of the eastern desert appears dark, indicating persistent low backscatter from the three dates. The Nile River is also visible in dark color. The circular patterns in the upper right corner are

Fig. 5.9 A composite color of 100 km swath (25 m resolution) from ERS-2 SAR. Cairo is shown in the bright area at the bottom. The RGB is generated from the three acquisition dates (see text). *Source* ESA Earth Watching—20 years of SAR website

irrigated cultivated fields. Different colors are generated because the soil moisture changed between the acquisition dates. A large volume of multi-temporal SAR data is readily available from repeat-cycle orbits from Sentinel-1 satellite but still to be put into use to identify changes in the land cover in Egypt.

The second approach to increase the dimensionality of SAR data is through multi-frequency channels. This approach was made available from a few airborne systems such as NASA-JPL DC-8 SAR system (Lou et al. 1996) and space shuttle missions (in April 1994 a multi-frequency SAR sensor was flown onboard the Space Shuttle Endeavour). It has not been yet made available from an operational satellite. Perhaps this should be a direction for future development of spaceborne SAR systems. The third approach comprises multi-polarization systems; namely dual and fully-polarimetric SAR (also called Quad-pol SAR) as described in Sect. 5.2.1.3. A limited volume of dual-polarization data (co- and cross-polarization) became available in Egypt from Envisat, Radarsat-2 and recently Sentinel-1 satellites, yet they have not been used in notable studies. Fully polarimetric SAR data have been used in a handful of studies (Sect. 5.3.2).

5.3.1.1 Geological and Geomorphological Applications

SAR images of geological sites usually feature numerous lineaments of different length. They represent a range of structural attributes (e.g. faults and folds), buried channels, or just effects of terrain relief (radar shadow). It is up to the interpreter to determine the source of the lineament whether it be geomorphologic or structural. Generally speaking, there is over-representation of short lineaments across the look direction of SAR due to enhanced micro-relief and radar shadow effects and under-representation of lineaments along the look direction. These factors complicate the interpretation of geological scenes in SAR images.

The knowledge of the study area and other ancillary data such as seismic, magnetic, and gravity enables the drawing of accurate conclusions from the analysis of SAR images. For that reason, the interpretation of SAR images in geological applications may be regarded as an art; i.e. it relies largely on the subjective skills of the interpreter who can successfully fuse all information as opposed to using objective digital image analysis. However, image enhancement techniques are usually applied to produce images more appealing to visual interpretation. This includes methodologies to improve contrast, remove speckle, accentuate edges, connect fragmented lines, and account for undesirable atmospheric effects.

Most of the geological application of imaging radar in Egypt involves data fusion with another remote sensing data. Paillou et al. (2003) combined the use of SIR-A L-band SAR, ground penetration radar (GPR) at 900 MHz and a priori knowledge of the geological context to retrieve information about subsurface structures down to a couple of meters deep in the Bir Safsaf region in south-central Egypt. The authors claim that comparison of the total power from GPR against the scattering coefficient from the L-band SAR opens the way to the new interpretation of SAR images.

Ramadan et al. (2006) fused Landsat TM and Radarsat-1 SAR data for mineral exploration in the East Oweinat district in south western Egypt. Landsat TM images proved to be useful in the surface mapping of lithological structures while Radarsat-1 images reveal fluvial features beneath a surface cover of the desert sand. The principal component analysis was used for merging the two data sets to enhance the interpretation of geological features. Gaber et al. (2010) combined Landsat TM+ with Radarsat-1 then with PALSAR images of Wadi Feiran basin, Sanai Peninsula, to classify wadi deposits based on grain size distribution and predominant rock composition. The study found that the C-band Radarsat-1 enables better detection of textural variations in the area while the L-band PALSAR was better in defining the boundaries of each deposit class. This and similar studies aimed at defining the deposits of the wadis from which favorable areas of groundwater recharge can be identified.

5.3.1.2 Hydrological Applications

The prime subject of hydrological applications of SAR data in Egypt is groundwater. This is related to key geomorphology information in the area. Several authors have utilized this relationship to identify potential drilling sites in SAR images for groundwater abstraction. SAR images usually detect buried channels, paleo-drainage networks, underfit valleys, natural ridges, and banks, alluvial deposits and fans; all have a bearing on the permeability of the rocks and therefore recharge conditions. The following discussions are intended to offer hints on a few studies of SAR hydrological applications.

Early studies on SAR applications for groundwater in Egypt were undertaken by researchers outside Egypt. One of the earliest study was presented in Robinson (2002). It used SIR-C and Radarsat SAR data to map structural and fluvial features in south-western Egypt in order to identify new groundwater resources in fracture rock aquifer settings. Five types of channel morphologies have been identified along with flood features that follow regionally extensive faults (up to 50 km in length). A follow-up study used C-band Radarsat-1 SAR and topographic data to map groundwater distributions in the Nubian aquifer, which underlies East Oweinat and Tushka regions (Robinson 2008). The study confirmed that SAR could penetrate the desert sand and reveal subsurface groundwater-related features such as ancient rivers, streams, faults, and fractures. It estimated the penetration depth of the signal to be on the order of half a meter.

A major national project, called Development Corridor, was proposed to develop a north-south strip in Western Desert in Egypt for sustainable development (El-Baz 2007). The project may enable a gradual expansion of urban and agricultural areas away from the Nile Valley into the Western Desert. A U.S.-Egypt project was initiated in 2010 to assess natural resources in a sector within the corridor near Aswan and Kom Ombo. The exploration and assessment of untapped groundwater resources in this sector are presented in Koch et al. (2012). The authors used Radarsat-1 to reveal alluvial fans and some of the fracture systems that seem to be promising in bringing groundwater from the Nubian aquifer (and possibly excess water from the River Nile or Lake Nasser) into the sector under consideration. The L-band ALOS-PALSAR images were also used to reveal buried extension of paleo-channels, which have implications in terms of potential water resources in the area.

More recently, Sultan et al. (2013) assess groundwater in El Qaa Plain, south Sinai by using SAR indirectly through interferometric-generated DEM from the SRTM to identify subsurface structures such as faults and paleo-drainages that control the geometry of groundwater aquifers. Additionally, the study used Landsat ETM+ to map lithological structures. Koch et al. (2013) combined multispectral data from the Along Track Scanning Radiometer (ASTR) onboard the European ERS satellite and the L-band PALSAR in addition to limited ground surveys with GPR to locate potential areas for accumulating groundwater in El-Gallaba Plain, west of Aswan City.

5.3.1.3 Agricultural Applications

To the best knowledge of the author of this chapter no agricultural applications of single- or dual-channel SAR have been undertaken so far. However, one application was completed using polarimetric SAR (see next section). At this point, it is worth noting the advantage of using the regularly available Sentinel-1 images in monitoring the phenological stages of crops in the Nile Delta region. Data are available approximately once per week if downloaded from the same repeat orbit from both Sentinel-1A and Sentinel-1B, or with higher frequency if different orbits covering the same area are used.

Figure 5.10 shows an example of Seninel-1A image of the Nile Delta with an enlargement of an area in the Kafr El-Sheikh governorate in the Nile Delta region.

Fig. 5.10 Sentinel image of the Nile Delta acquired on July 2017 (left) with enlargement of the red box that marks different agriculture fields in the Sakha Agriculture Research Station (right). The inset shows the rice field from which the time series of the backscatter (Fig. 5.11) was obtained. Image courtesy of ESA and the map is prepared by A. El-leithy of Zewail City of Science and Technology

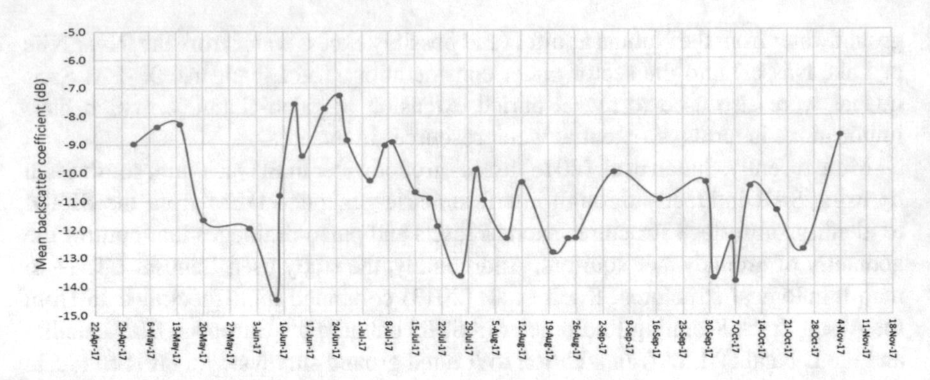

Fig. 5.11 Time series of backscatter coefficient (dB) from the rice field marked in the inset in Fig. 5.10. Generated by A. El-leithy of Zewail City of Science and Technology

The area encompasses the Sakha Agriculture Research Station. This is the largest research station in the Middle East, established at the beginning of 1960 to develop research and technologies to improve agricultural productivity. The right panel in the figure shows the spatial coverage of the experimental farm with the fields of experimental crops marked. The area in the inset marks a rice field in the summer of 2017 (planted on 5 May and harvested on 5 October 2017). The time series of the backscatter coefficient (VV polarization) from this field is presented in Fig. 5.11. This information should be linked to the plant and soil parameters (including the growth stages and irrigation condition) in order to parameterize the backscatter and later use it to infer plant parameters. Figure 5.11 shows that the backscatter is lowest on the days when the field is flooded with irrigation water. The periodicity of the backscatter is probably related to the irrigation water cycle. The backscatter in the early stage of the plant is higher than that when the plant becomes fully grown. These qualitative observations have triggered research questions that are yet to be answered if and when in-situ data and observations are obtained coincidently with the SAR data acquisition.

5.3.1.4 Archeological Applications

SAR data have been found to be useful for archeological applications, especially in arid areas. Egypt has it both; the arid land and the numerous ancient treasures buried under the sand. According to Chapman and Blom (2013), successful analysis of SAR data for this application is subject to a few surface conditions. The cover material must be very dry and of homogeneous and fine-grained relative to the radar wavelength. It should also be less than 2 or 3 m thick. Moreover, beneath this cover must be a radar reflective surface. Otherwise the radar image will not reveal subsurface features. Chapman and Blom (2013) also offered a comprehensive review of SAR applications in archeology. Field studies carried out in Egypt confirmed that the L-band SAR of SIR-A mission could penetrate in the Selima drift sand sheet at the southern border of Egypt to depths of at least 1 m and in sand dunes to 2 or more meters (McCauley et al. 1982).

Almost all of SAR archeological applications in Egypt use the Japanese L-band JERS-1 SAR or ALOS-PALSAR. Most of the studies originated outside Egypt; namely in Europe and Japan. The Japan Aerospace Exploration Agency (JAXA) became interested in verifying what believed to be archeological information discovered using JERS-1 and SIR-A SAR images of Egyptian deserts. This was observed at a location between Abu Rawash and Dahshur on the west bank of the Nile River. Ground truth in Egypt was carried out in collaboration with the National Authority of Remote Sensing and Space Sciences (NARSS). The study by Masahiro et al. (2008) summarizes the potential and difficulties in using JERS-1 and PALSAR systems to identify the archeological site. Soil moisture and viewing geometry of SAR may contribute to the difficulties. Nevertheless, the project yielded promising results.

Preliminary results from a project to conduct archaeological surveys in areas of historical interest in north Saini, Egypt was initiated by the Italian Ministry of Education, Universities, and Research. PALSAR images were used to identifying roads and partially buried tracks under the sand cover (Maura et al. 2015). These features usually return higher backscatter and higher interferometric coherence than the surrounding sand.

A few important observations regarding SAR applicability for archeological exploration are made in Stewart et al. (2016). The authors emphasize the importance of regular and systematic survey to enable better distinction of archeological features. They also underscore the importance of ground-based information to support the interpretation of the images. Because of SAR artifacts (including range and azimuth ambiguities), the use of other remote sensing data may be vital to aid the interpretation. An informative coverage of SAR applications in archeology in Egypt is found in the Ph.D. thesis of Stewart (2017).

5.3.1.5 Dual-Polarization SAR

As mentioned before, dual-polarization SAR provides an extra dimension to reduce the unambiguity in the surface cover classification and the retrieval of surface properties from radar data. Most of the recent operational SAR offer dual polarization (namely, Envisat, Radarsat-2, ALOS-PALSAR, and Sentinel-1).

Gaber et al. (2017a) presented what seems to be the only published study on using dual-polarization SAR in Egypt. It uses data from ALOS-PALSAR (HH and HV polarizations). Instead of using traditional polarization ratios, the authors calculated the 2×2 covariance matrix at each pixel and derived the H/A/α polarimetric decomposition parameters (see Sect. 5.2.1.3). They fed the three parameters along with the element from the covariance matrix and generated a supervised classification map using predefined training sites derived from the geological map, field work, and Google Earth. Results include classification of surface sediments in terms of grain size as well as the dominant and the secondary radar scattering mechanisms.

In honor of the late Dr. Talaat Ramadan, a leading geologist in Egypt who undertook his work with passion, it is worth noting a project he started but never ended. A set of Dual-polarization images of North Sinai were obtained through the Earth Observation Application Development Program (EOADP) of the Canadian Space Agency (CSA) to explore the utility of the data for identifying mineral resources.

Figure 5.12 shows a scene acquired on 9 February 2009 of north Sinai just west of Suez Canal (the straight narrow strip at the left side of the image) in HH and HV polarizations. Enlargements of areas labels a, b, c, d, and e are also shown. In the area (a) a few ships appear, and the contrast against the sea water (clutter) is better in the cross-polarization image (HV). In (b) an agriculture area appears where both HH and HV hold different information. Apparently, the integration of the two parameters (e.g. through a ratio) may hold better potential for crop classification. The area in (c) is part of the desert with sand dunes visible in the HH but not HV image. The

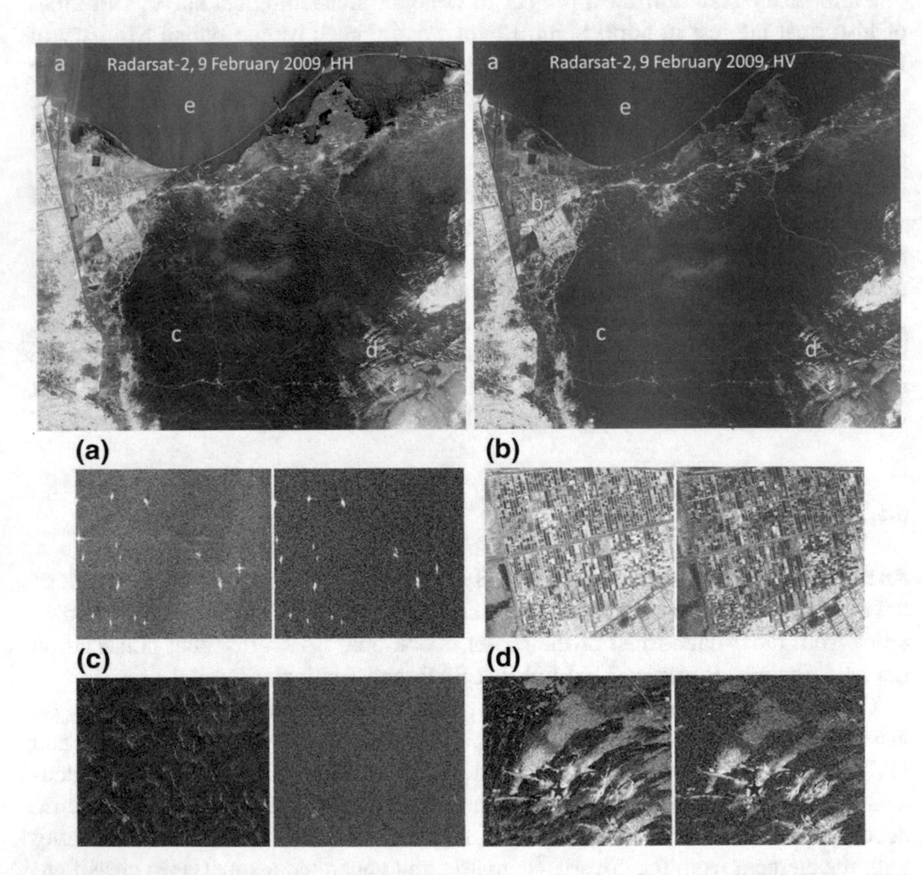

Fig. 5.12 Radarsat-2 images of a scene in north Sinai acquired on February 2009 in HH and HV polarizations (top two panels) and enlargements of the areas labeled (**a**), (**b**), (**c**) and (**d**) appear in the bottom panels. For each pair, the left image is from HH polarization and the right from VV polarization. The high backscatter area marked by the red asterisk is a paleo lake. Images credit MDA, Canada

area in (d) features high backscatter batches. It was presumed to originate from a rough layer along the path of the radar signal while penetrating through the sand. After some ground verification work, it was confirmed that at least the area marked by the asterisk is a paleo lake. In area (e), which features open sea, it is obvious that HH is more sensitive to the wind-triggered surface roughness than HV. That is why the former is usually used to retrieve surface wind over the ocean, which has been a standard product from several SAR systems.

5.3.2 Polarimetric SAR Applications

Polarimetric SAR applications in Egypt started fairly recently using data from the world first fully polarimetric L-band SAR system, PALSAR. Later a limited data set from the C-band Radarsat-2 was provided under the SOAR-Africa program of the CSA. So far, all studies aimed at either classifying ground cover types using the best polarimetric parameters or delineating surface and subsurface structures using false color images from polarimetric decompositions. Linking surface properties to the backscatter set (HH, HV and VV) and polarimetric decomposition parameters (e.g., H, A and Alpha) has not been pursued though it is supposed to be the first step to demonstrate the use of SAR polarimetric parameters. This step has not been pursued perhaps because it relies on a comprehensive field data gathering program preferably coincidently with the satellite overpass if the scene changes on daily or weekly basis (as in the case of agriculture).

Gaber et al. (2011) used fully polarimetric ALOS/PALSAR data to detect and delineate subsurface structures in a desert plain west of the city of Aswan to reveal the potential for groundwater. The authors constructed linear, elliptical and circular polarization images from the four basic orthogonal polarizations (HH, HV, VH, and VV). They found that circular polarization to be the best for revealing buried faults. This has been the first study to experiment with circular polarization in Egypt. Gaber et al. (2015) combined the use of data from the two polarimetric SAR from ALOS-PALSAR and Radarsat-2 with four optical/infrared data sets from different satellites in supervised and unsupervised classification methods to compare their accuracy for mapping surface sediment in the non-vegetated El-Gallaba Plain of the Western Desert of Egypt, in terms of grain size and surface roughness. Results concluded that the area has a high potential for agricultural production using its groundwater resources, which are also confirmed by the same data.

Zeyada et al. (2016) examined Radarsat-2 quad-pol data to evaluate the accuracy of three supervised classification methods in discriminating between four crops: rice, maize, grape and cotton in an agricultural area in the middle of the Nile Delta. Results indicate that combining the most commonly used polarimetric parameters from Pauli, Cloude–Pottier, and Freeman–Durden decompositions along with the three fundamental backscatter coefficients using the Support Vector Machine classification approach strikes an appropriate balance that minimizes the effect of the training error on the classification accuracy without going to over-fitting classification. Figure 5.13 from the same study shows that some decompositions such as Freeman–Durden are not visually useful while Pauli decomposition is best for visual interpretation.

Recently, Abdeen et al. (2018) used quad-pol Radarsat-2 data combined with Landsat-8 in unsupervised and supervised classification methods to map different geological units in the central part of Suez Canal. Polarimetric target decomposition and Wishart unsupervised classification with five input classes were performed. The study was also the first (in Egypt) to experiment with the polarization signatures and found a high correlation with the examined classes in terms of their surface roughness and mineral composition.

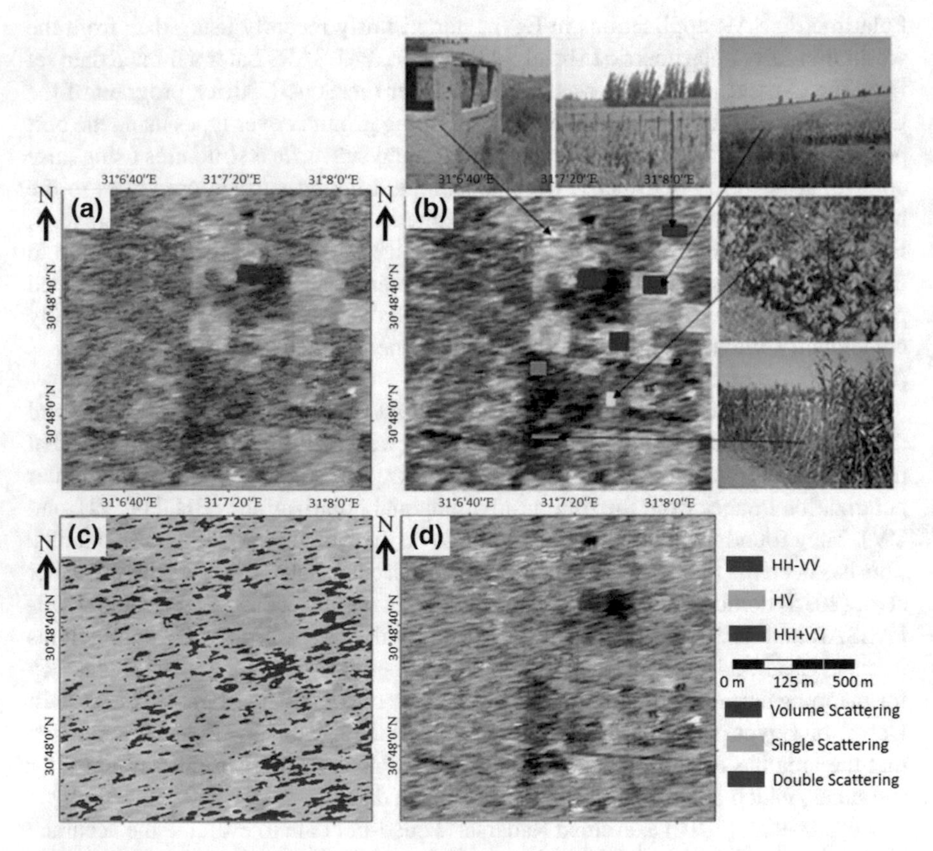

Fig. 5.13 Images of an agricultural area in Al-Jimmeza village, Gharbia Governorate, in the Nile Delta region derived from Quad-pol Radarsat-2 acquired on 8 August 2013. **a** H/A/α decomposition, **b** Pauli decomposition, **c** Freeman–Durden decomposition, and **d** composite color image of backscatter from the three basic polarization HH, HV and VV. The set color clues above the scale apply to the three decomposition images (**a**, **b** and **c**) and the set below applies to the backscatter image in (**d**). Photographs of different sites are linked to their physical locations as shown in (**b**). Adapted from Zeyada et al. (2016)

5.3.3 Interferometric SAR Applications

Applications of InSAR is gaining ground in Egypt mainly due to the eminent threat of the subsidence of the Nile Delta region. As mentioned in Sect. 5.2.1.4, the essence of radar interferometry is the acquisition of a pair of satellite images of the same area from the same orbit after one or multiple repeat cycles or from two antennae on the same platform that look at the scene with slightly different angles. Examples from the former arrangement include images from the European satellites ERS, ENVISAT (Monti-Guarnieri and Tebaldini 2008), the Canadian satellites Radarsat-1 and Radarsat-2, the Japanese satellite series ALSO-PALSAR and the German satellite TerraSAR-X. The example of the interferometric image pairs from two antennae on the same platform is the SRTM mission (see Sect. 5.1), which generate the most complete high-resolution digital topographic database of the earth from 56° S to 60° N (Farr et al. 2007).

The earliest study using InSAR data in Egypt was presented in Masoud et al. (2003) to generate DEM of Safaga area on the Red Sea using interferometric SAR pair from the Japanese satellite JERS-1 acquired on 12 January 1994 and 10 June 1996. In another study, a series of InSAR data from the European satellite ENVISAT was used to pinpoint peculiar uplift spots in an area called Abu Rawash, north-west of Giza pyramids plain (Parcharidis et al. 2012). The same study used DEM generated from SRTM to simulate the topographic phase and therefore retrieve the surface deformation. A color-coded map of displacement/year of the area was generated for the investigated period (2003–2009), and the nominal range was found to be within ±5 mm/year. To support the geological study and palaeo-environmental interpretation, Salvini et al. (2015) produced DEM from COSMO-SkyMed imagery to delineate paleo-drainage in a wide area surrounding the Siwa and Al-Jaghbub oases of the Western Sahara Desert.

In a pioneering study, Gaber et al. (2014) used eight L-band ALOS/PALSAR scenes of the city of Port Said acquired from 12 November 2007 to 4 April 2010 to monitor the building and ground deformation using the DInSAR approach. A persistent scattering technique was applied to select points with no backscatter change. The authors found the total displacement to be 17 mm during the investigated period. A more comprehensive study was later undertaken, using the same technique, to estimate land subsidence in the same area and link it to sea-level rise (Gaber et al. 2017b). The authors used a stereo pair of ALOS/PRISM to generate an accurate DEM that minimizes the residual topographic effect. Moreover, they used 347 distributed ground control points to calibrate the DEM and co-register the ALOS/PALSAR images. The study estimated the subsidence to be about 28 mm on the average during the investigated period (2007–2010). It predicted that this rate might induce serious environmental impacts, given the observed sea level rise especially in the northern part of the city. The estimated rate of displacement for pixels that have coherence >0.3 are overlaid on an image obtained from WorldView-2 satellite (Fig. 5.14). More details on using radar interferometry for studying subsidence in Port Said area is included in Darwish (2017).

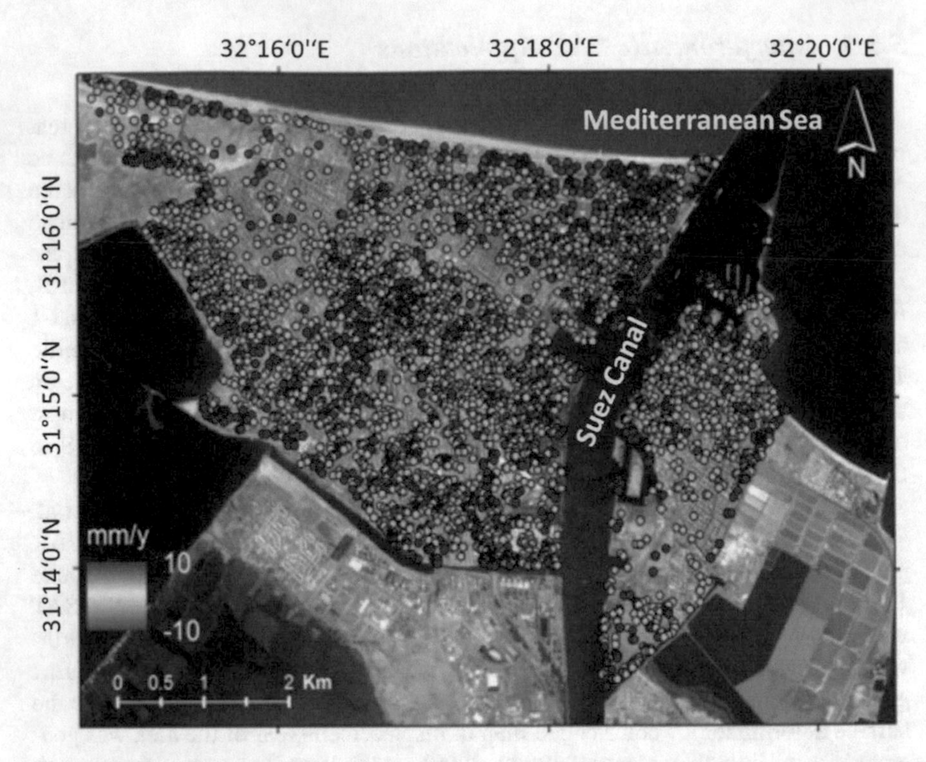

Fig. 5.14 The rate of displacement (mm/year) for pixels of high coherence (>0.3) in the Port-Said city are overlaid on the WorldView-2 image. *Source* Gaber et al. (2017b)

In a study on land subsidence in Alexandria with links to sea level rise, Wöppelmann et al. (2013) used 49 SAR descending scenes acquired by the Envisat satellite between 2003 and 2010 with the PS technique and estimated that the coastline has been subject to land subsidence of 0.4 mm year^{-1} on average and up to 2 mm year^{-1} locally. This average is less than values reported in El-Raey (2010) (2 mm annually at Alexandria to about 2.5 mm annually at Port Said). Based on their findings, Wöppelmann et al. (2013) reasoned that land subsidence in the area of Alexandria to be dominated by tectonic setting or gravitational collapse episodes on timescales of multi-century or millennia. On decadal to century timescales, they suggested that subsidence rates would be likely moderate and affected by natural compaction and dewatering.

So far, InSAR and DInSAR techniques have limited use in Egypt though interest in this technology is growing. The limited availability of interferometric SAR pairs is certainly a constraint. Studies that have been conducted so far were undertaken on opportunity basis when data became available.

5.3.4 *Radar Altimetry Applications*

The interest in RA in Egypt has emerged mainly because of the concern about the sea level rise (SLR) in the Mediterranean Sea. El-Raey (2010) estimated that an SLR of 0.5 m would cause 30% of the cities of Alexandria and Port Said as well as nearly all the Nile Delta beaches to be eroded. Sea level is rising mainly from melting glaciers and ice sheets in the polar regions and partly because of expansion of ocean water in response to global warming. Global mean sea level rose by a rate of 1.8 mm year^{-1} from 1961 to 2003 as measured from a network of tide gauges (IPCC Fourth Assessment Report available from https://www.ipcc.ch/publications_and_data/ar4/wg1/en/ch5s5-5-2.html). A higher rate of 3.1 mm year^{-1} was estimated for the period 1993–2003 as measured from a combination of tide gauges and satellite altimetry. However, the International Panel on Climate Change (IPCC) was unable to determine whether this rate was due to decadal variability of the oceans or an acceleration in sea-level rise (National Research Council 2012).

Radar altimeters have been a prime tool for mapping variations in sea level (also called ocean surface topography). Gridded data of this parameter are compiled from available satellite altimeters and distributed by the Archiving, Validation, and Interpretation of Satellite Oceanographic (AVISO) from the French Space Agency (CNES) https://www.aviso.altimetry.fr/en/data.html.

A comprehensive study is presented in Shaltout et al. (2015) to estimate and predict the sea level change along Egypt's Mediterranean coast using a 21-year record of ocean surface topography (1993–2013) from AVISO. The study found that the sea level in the investigated region rose by approximately 3.1 cm decade^{-1} and exhibited an annual variation between −17 and 8 cm. It is affected by variation in ocean surface level west of the Gibraltar Strait as well as sea-surface temperature. The same study used the Geophysical Fluid Dynamics Laboratory model from which the authors predicted that Egypt's Mediterranean coast would experience substantial SLR of up to 22 cm by the end of this century (Shaltout et al. 2015).

Finally, it is worth noting that RA data can provide a view of the Earth's interior and its gravity field over the marine regions. That is because the marine surface can be considered as a geoid. Deviations of geoid height, measured by altimeters, are related to the topography of marine floor, and that reflect tectonic settings and subsurface structure as indicated in Zahran and Saleh (2006). This study compared geoid maps of the northern part of the Red Sea region against geological and tectonic settings and found that some high-resolution information fairly well correlated with known structural and tectonic features.

5.4 An Outlook on Future Radar Applications in Egypt

Satellite radar applications are expected to grow in Egypt for a number of reasons. Most importantly is the fact that radar is the most suitable sensor to reveal subsurface information in an arid landscape. About 94% of Egypt's land is covered by desert or low elevation mountains. Numerous ancient treasures and groundwater reservoirs and waterways are buried under the sand. Another reason is the increasing availability of the free SAR data and processing software. Additionally, more people have become trained on using the data after a few training workshops were locally offered (between 1999 and 2017) and a number of academic dissertations have been completed in the past ten years or so.

From the brief coverage of SAR studies and applications addressed in Sects. 5.3.1 and 5.3.2 it can be concluded that in most cases the objectives are set to either map/classify ground cover or identify surface features that can later be used as proxy indicators of the subsurface contents. These objective dictate a research methodology based on image classification and pattern recognition techniques, perhaps with the support of visual or field observations in the case of geological features. This is an acceptable exploratory research route, but it would be more constructive to start the research endeavors by linking the observed backscatter from SAR to the surface and subsurface properties that engender it. This approach would facilitate image interpretation and later classification. In addition, it would facilitate the construction of a backscatter database to develop a statistical model that can be used inversely to retrieve surface parameters. This approach should be stressed especially with the current efforts to explore the potential of dual-polarization and full polarization SAR data.

Fusion of radar and optical remote sensing data offers unique spectral and structural characteristics of the ground cover. Most of the studies that have adopted this approach confirmed that it provided results with higher accuracy than using either one of the two datasets (Joshi et al. 2016). While this approach has been used in a limited number and scope in Egypt so far, it is expected to grow given the availability of the free Sentinel-1 and Sentinel-2 data.

A significant advancement in the field of SAR architecture has been achieved by developing a hybrid polarity system known as compact polarimetry (CP) (Raney 2007). This system transmits a single polarization (mostly circularly) and receives two linear orthogonal signals (for example RH and RV, where R refers to Right circular polarization). While it does not provide the full suite of polarization information as the fully polarimetric (FP) SAR systems, the CP provides a much wider swath coverage (possibly 400 km as opposed to the few tens of kilometer from the FP SAR). The CP mode will be available in the near future from the Canadian Radarsat Constellation Mission (RCM), to be launched by the end of 2018. It is too early to persuade the Egyptian remote sensing community to consider this new development, but it is expected to be widely used in the future.

Scatterometer data (Table 5.2) are available free of charge through a few sources (e.g. from the Brigham Young University website ftp://ftp.scp.byu.edu/pub/data/). The data are usually calibrated into backscatter coefficient and offered based on a 3-day mosaic that covers several regions of the world, including the relevant region of North Africa. An example is presented in Fig. 5.15. These data have not been used in Egypt because of their coarse resolution (4.5 km at best) and wide coverage. Almost all studies of remote sensing in Egypt address phenomena/resources within

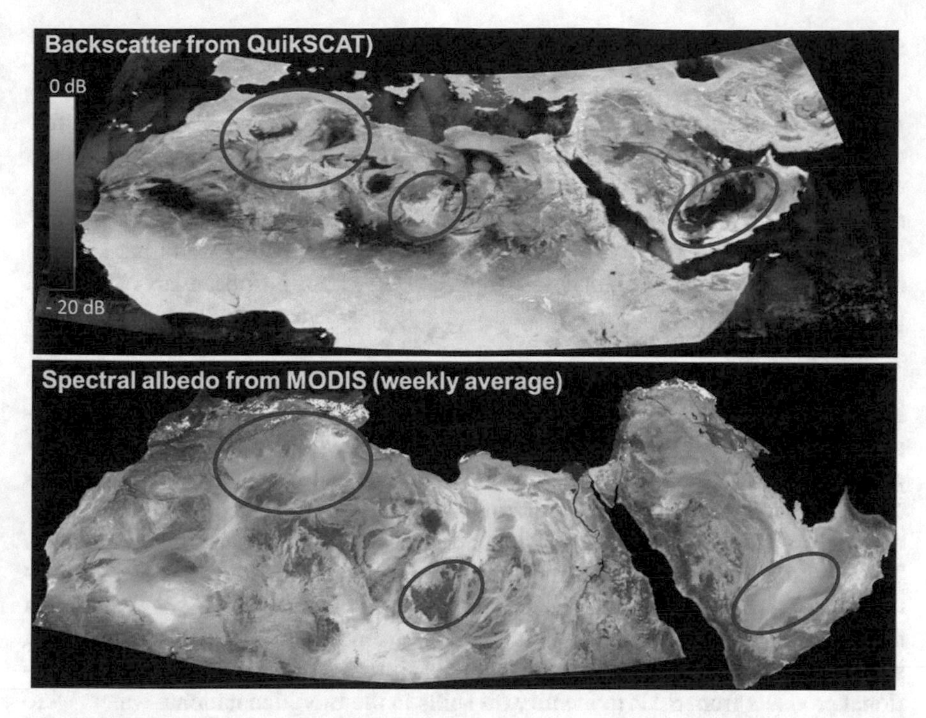

Fig. 5.15 Mosaic of backscatter from a 3-day coverage of QuikSCAT scatterometer (top) and the spectral albedo from MODIS onboard NASA's Terra and Aqua satellites, acquired over a 16-day period (April 7–22, 2002). Note the visual correspondence between the two parameters in the circled areas. *Source* Scatterometer image is from Microwave Earth Remote Sensing Laboratory of Brigham Young University and albedo image is courtesy of Elena Tsvetsinskaya, Boston University

by the geographic boundaries of the country. The value of scatterometer data will be appreciated more if regional studies are pursued. It is recommended to consider this spatial scale in the future as it serves to address regional and global phenomena that trigger the local phenomena of interest.

The backscattering distribution in Fig. 5.15 reveals very low values (dark areas) representing flat smooth desert surface and very high values (bright areas) representing the dense vegetation areas of tropical Africa south of Sahara (south of latitude 10° N). Between these two extremes, surfaces of different roughness/or dielectric properties exist. The figure also includes a nearly coincident map of surface albedo albeit with slightly different areal coverage) obtained from MODIS optical sensor onboard NASA's satellites Terra and Aqua. The combination of the two images can be used to better identify the surface cover. This is evident from the contrasting signatures in marked areas (low albedo in areas that feature high backscatter and vice versa). The availability of the long record of scatterometer data (starting in the mid-1990s) is useful in detecting changes over short and long timescales. Data fusion between scatterometer and optical observations will allow estimation of more surface parameters at better accuracies.

5.5 Conclusions and Recommendations

Applications of radar remote sensing in Egypt have focused on SAR data. Only few attempts have been undertaken in using radar altimeter data for sea level estimates while no applications of scatterometer data have commenced. SAR applications started in the late 1990s when data from the European satellites ERS-1/2 became available yet had to cope with challenge of image interpretation and lack of free data. Later, data from the Japanese ALOS-PALSAR, the European ENVISAT and the Canadian Radarsat systems became available and the exploration of SAR data in a few applications boosted.

For the most part, SAR data have been used from single- and dual-channel sensors in exploratory studies to demonstrate their capability in retrieving surface and sub-surface parameters. The predominant applications are in the fields of geology, morphology, hydrology and archaeology. This is expected, as Egypt's landscape features are mainly arid and semi-arid conditions. Qualitative interpretation of SAR images reveals subsurface features that do not appear in optical and thermal infrared images. The subsurface penetration of radar signal has served greatly in locating potential sites of groundwater. Sand dune identification and their dynamic behavior have also been studied using SAR. The L-band data available from the Japanese ALOS-PALSAR system have been used to verify archaeological information in selected sites. Operational products from SAR to identify oil spills in the Egyptian national waters were used briefly through a collaborative program with the ESA. Other well-established SAR-based marine products such as surface wind and ocean wave parameters have not been used yet.

Explorations and applications of SAR interferometric data are gaining ground in Egypt mainly due to the eminent threat of the subsidence of the Nile Delta region. Data have been used to monitor subsidence along the northern coast of the highly populated ND region. This is in addition to the traditional use of the data in generating topographic mapping of mountainous areas. Polarimetric SAR data, on the other hand, have started its applications to classify agricultural crops in the ND and map different geological units in the central part of Suez Canal.

In general, SAR has not realized its full potential use in Egypt so far. More applications are expected to emerge as open access data, processing software and image interpretation capabilities are becoming more available. To advance the applications, it is important to conduct field measurements of surface parameters coincidely with the SAR overpasses or shortly after the overpasses if the surface conditions remain unchanged. The link between surface characteristics and the engendered backscatter is at the core of understanding the information that can possibly be retrieved from SAR data.

It is recommended to use the coarse resolution data from scatterometer and passive microwave data to retrieve information at regional scale. Chapter "Environmental Applications of Remote Sensing in Egypt: A Review and an Outlook" by M. Shokr in this book includes more justifications on pursuing remote sensing studies at regional and global scales.

Acknowledgements The author would like to thank all the staff at the National Authority of Remote Sensing and Space Sciences (NARSS) in Egypt who have shared their experience of using SAR data over nearly 20 years. This information is reflected in many passages in this chapter. A special note of appreciation goes to the Late Prof. T. Ramadan who was a pioneer in using SAR data for geological applications.

References

Abdeen MM, Gaber A, Shokr M, El-Saadawy OA (2018) Minimizing the labelling ambiguity during classification process of the geological units covering the central part of the Suez Canal Corridor, Egypt, using their radar scattering response. Egypt J Remote Sens Space Sci 21(1):S55–S66

Bamler R, Hartl P (1998) Synthetic aperture radar interferometry. Inverse Prob 14:R1–R54

Barnes CF (2015) Synthetic aperture radar: wave theory foundations, analysis and algorithms, 1st edn. Barnes, 624 p. ISBN-13: 978-0692313732

Canada Centre for Remote Sensing (2003) Advanced radar polarimetry tutorial. Available online https://pdfs.semanticscholar.org/2298/476a4ef796c142bf764b579ffb0211bf1b69.pdf

Chapman B, Blom RG (2013) Synthetic aperture radar, technology, past and future applications to archaeology. In: Comer DC, Harrower MJ (eds) Mapping archeological landscapes from space. Springer, New York, Heidelberg, London, pp 113–131

Cloude SR (2010) Polarization applications in remote sensing. Oxford University Press, Oxford, New York, 466 p

Cloude SR, Pottier E (1997) An entropy based classification scheme for land applications of polarimetric SAR. IEEE Trans Geosci Remote Sens 35(1):68–78

Colesanti C, Ferretti A, Novali F, Prati C, Rocca F (2003) SAR monitoring of progressive and seasonal ground deformation using the permanent scatterers technique. IEEE Trans Geosci Remote Sens 41:1685–1701

Crosetto M, Monserral O, Cuevas-González M, Devanthéry N, Crippa B (2016) Persistent scatterer interferometry: a review. ISPRS J Photogrammetry Remote Sens 115:78–89

Darwish N (2017) Evaluating land subsidence in port said city using radar interferometry. M.Sc. thesis, Geology Department, Faculty of Sciences, Port Said University, Egypt

Elachi C (1988) Spaceborne radar remote sensing: applications and techniques. IEEE Press, NY, USA

Elachi C, Roth LE, Schaber GG (1984) Spaceborne radar subsurface imaging in hyper-arid regions. IEEE Trans Geosci Remote Sens 22:383–388

El-Baz F (2007) Use of a desert strip west of the Nile Valley for sustainable development in Egypt. Bull Tethys Geol Soc 2:1–10

El-Raey M (2010) Impacts and implications of climate change for the coastal zones of Egypt. In: Michel D, Pandya A (eds) Coastal zones and climate change. Henry L. Stimson Center, Washington, DC, pp 31–50

Elsherbini AA (2011) Radar remote sensing of arid regions. Ph.D. dissertation, Electrical Engineering Department, The University of Michigan, Ann Arbor, Michigan, USA

Farr TG, Rosen PA, Caro E, Crippen R, Duren R, Hensley S et al (2007) The shuttle radar topography mission. Rev Geophys 45(2). https://doi.org/10.1029/2005RG000183

Ferretti A, Prati C, Rocca F (2001) Permanent scatterers in SAR interferometry. IEEE Trans Geosci Remote Sens 39(1):8–20

Freeman T, Durden SL (1998) A three-component scattering model for polarimetric SAR data. IEEE Trans Geosci Remote Sens 36(3):963–973. https://doi.org/10.1109/36.673687

Gaber A, Koch M, El-Baz F (2010) Textural and compositional characterization of Wadi Feiran deposits, Sinai Peninsula, Egypt, using Radarsat-1, PALSAR, SRTM and ETM+ data. Remote Sens 2(1):52–75

Gaber A, Koch M, Geriesh MH, Sato M (2011) SAR remote sensing of buried faults: implications for groundwater exploration in the Western Desert of Egypt. Sens Imaging Int J 12(3–4):133–151

Gaber A, Darwish N, Sultan Y, Arafat S, Koch M (2014) Monitoring the building stability in Port-Said city, Egypt using differential SAR interferometry. Int J Environ Sustain 3(1):14–22

Gaber A, Soliman F, Koch M, El-Baz F (2015) Using full-polarimetric SAR data to characterize the surface sediments in desert areas: a case study in El-Gallaba Plain, Egypt. Remote Sens Environ 162:11–28

Gaber A, Amarah BA, Abdelfattah M, Ali S (2017a) Investigating the use of dual-polarized and large incidence angle of SAR data for mapping the fluvial and Aeolian deposits. NRIAG J Astron Geophys 6(2):349–360

Gaber A, Darwish N, Koch M (2017b) Minimizing the residual topography effect on interferograms to improve DInSAR results: estimating land subsidence in Port-Said city, Egypt. Remote Sens 9:752. https://doi.org/10.3390/rs9070752

Hein A (2004) Processing of SAR data: fundamentals, signal processing, and interferometry. Springer, Berlin, Heidelberg, 291 p

Hooper A, Zebker H, Segall P, Kampes B (2004) A new method for measuring deformation on volcanoes and other natural terrains using InSAR persistent scatterers. Geophys Res Lett 31:L23611. https://doi.org/10.1029/2004GL021737

Joshi N, Baumann M, Ehammer A, Fensholt R, Grogan K, Hostert P et al (2016) A review of the application of optical and radar remote sensing data fusion to land use mapping and monitoring. Remote Sens 8(70). https://doi.org/10.3390/rs8010070

Ketelaar VBH (2009) Satellite radar interferometry. Springer, The Netherlands, 255 p. ISBN 9781402094286

Koch M, Gaber A, Burkholder B, Geriesh MH (2012) Development of new water resources in Egypt with Earth observation data: opportunities and challenges. Int J Environ Sustain 1(3):1–12

Koch M, Gaber A, Gereish MH, Zaghloul E, Arafat SM, AbuBakr M (2013) Multisensor characterization of subsurface structures in a desert plain area in Egypt with implications for groundwater exploration. In: Proceedings of SPIE remote sensing conference, vol 8887, pp 23–26, Dresden, Germany

Lee JS, Pottier E (2009) Polarimetric radar imaging: from basics to applications. CRC Press Taylor and Francis Group, Florida, USA, 422 p

Long D (2014) Radar scatterometers. In: Njoku EG (ed) Encyclopedia of remote sensing. Encyclopedia of earth sciences series. Springer, New York, NY. https://doi.org/10.1007/978-0-387-36699-9

Lou Y, Kim Y, van Zyl J (1996) The NASA/JPL airborne synthetic aperture radar system. The sixth annual JPL airborne earth science workshop, vol 2, pp 51–56, 4–8 Mar 1996, NASA-CR-203428

Masahiro E, Nakano R, Shimoda H, Sakata Y, Zaghloul EA, Shimada M (2008) A study for archeological exploration using spaceborne SAR. Japan Aerospace Exploration Agency (JAXA) report. Available online https://repository.exst.jaxa.jp/dspace/bitstream/a-is/15522/1/65135075.pdf

Masoud A, Raghavan V, Shinji M, Kiyoji S (2003) JERS-1 interferometric SAR DEM generation and validation in Safaga area, Red Sea coast of Egypt. J Geosci Osaka City Univ 46(13):207–216

Maura A, Stewart C, Lemmens K (2015) Satellite radar in support to archaeological research in Egypt: tracing ancient tracks between Egypt and Southern Levant across North Sinai. In: Capriotti Vittozzi G (ed) Egyptian curses 2. A research on ancient catastrophes. Archaeological heritage & multidisciplinary egyptological studies, vol 2, pp 197–221, Rome

McCauley JF, Schaber GG, Breed CS, Grolier MJ, Haynes CV, Issawi B et al (1982) Subsurface valleys and geoarcheology of the eastern Sahara revealed by shuttle radar. Science 218(4576):1004–1020

Monti-Guarnieri A, Tebaldini S (2008) On the exploitation of target statistics for SAR interferometry applications. IEEE Trans Geosci Remote Sens 46(11):3436–3443

National Research Council (2012) Measured global sea-level rise, chapter 2. In: Sea level rise of California, Oregon and Washington: past, present and future. The National Academic Press, Washington, DC

Nerem RS, Beckley BD, Fasullo JT, Hamlington BD, Masters D, Mitchum GT (2018) Climate change driven accelerated sea level rise detected in the altimeter era. PNAS 115(9):2022–2025
Oliver C, Quegan S (2004) Unerstanding synthetic aperture radar images. SciTech Publishing, Herdon, Virginia, USA, 512 p
Paillou P, Grandjean G, Baghdadi N, Heggy E, August-Bernex T, Achache J (2003) Subsurface imaging in South-Central Egypt using low-frequency radar: Bir Safsaf revisited. IEEE Trans Geosci Remote Sens 41(7):1672–1684
Pan J, Yan X, Zheng Q, Liu WT, Klemas VV (2003) Interpretation of scatterometer ocean wind vector EOFs over the Northwestern Pacific. Remote Sens Environ 84(1):53–68
Parcharidis I, Poscolier M, Seleem TA (2012) SAR interferometry monitoring over the folded area of Abu Rawash, Egypt. 4th EARSeL workshop on remote sensing and geology, Mykonos, Greece, 24–25 May
Ramadan TM, Nasr AH, Mahmood A (2006) Integration of Radarsat-1 and Landsat TM images for mineral exploration in East Oweinat District, South Western Desert, Egypt. In: ISPRS commission VII mid-term symposium "Remote sensing: from pixels to processes", pp 244–249, Enschede, The Netherlands, 8–11 May.
Raney RK (2007) Hybrid-polarity SAR architecture. IEEE Trans Geosci Remote Sens 45(11):3397–3404
Richards A (2009) Remote sensing with imaging radar. Springer, Heidelberg, Dordrecht, London, New York, 376 p
Rio MH, Pascual A, Poulain PM, Menna M, Barceló B, Tintoré J (2014) Computation of a new mean dynamic topography for the Mediterranean Sea from model outputs, altimeter measurements and oceanographic in situ data. Ocean Sci 10:731–744
Robinson CA (2002) Application of satellite radar data suggest that the Kharga Depression in south-western Egypt is a fracture rock aquifer. Int J Remote Sens 23(19):4101–4113
Robinson CA (2008) Understanding the distribution of groundwater resources using synthetic aperture radar data over Southwest Egypt. In: Proceedings of international geoscience and remote sensing symposium (IGRASS'08), vol 3, pp 7–11, Boston, USA
Rosmorduc V, Benveniste J, Bronner E, Dinardo S, Lauret O, Maheu C et al (eds) (2018) Radar altimetry tutorial. ESA online publication. https://www.altimetry.info/filestorage/Radar_Altimetry_Tutorial.pdf
Salvini R, Carmignani L, Francioni M, Casazzza P (2015) Elevation modelling and palaeo-environmental interpretation in the Siwa area (Egypt): application of SAR interferometry and radargrammetry to COSMO-SkyMed imagery. CATENA 12:46–62
Shaltout M, Tonbol K, Omstedt A (2015) Sea-level change and projected future flooding along the Egyptian Mediterranean coast. Oceanologia 57(4):293–307
Shum CK, Ries JC, Tapley BD (1995) The accuracy and applications of satellite altimetry. Geophys J Int 121:321–336
Stewart C (2017) Archeological prospection using spaceborne synthetic aperture radar. Ph.D. thesis, Department of Civil Engineering and Computer Science, The University of Rome Tor Vergata. https://www.disp.uniroma2.it/geoinformation/students/geoinformation-dissertations/CStewart_thesis_150dpi.pdf
Stewart C, Montanaro R, Sala M, Riccardi P (2016) Feature extraction in the North Saini desert using spaceborne synthetic aperture radar: potential for archeological applications. Remote Sens 8(825). https://doi.org/10.3390/rs8100825
Sultan A, Abdel Rahman N, Ramadan TM, Salem SM (2013) The use of geophysical and remote sensing data analysis in the groundwater assessment of El Qaa Plain, South Sinai, Egypt. Aust J Basic Appl Sci 7(1):394–400
Touzi R (1992) Extraction of point target response characteristics from complex SAR data. IEEE Trans Geosci Remote Sens 30(6):1158–1161
Touzi R, Boerner WM, Lee JS, Lueneburg E (2004) A review of polarimetry in the context of synthetic aperture radar: concept and information extraction. Can J Remote Sens 30(3):380–407

Ulaby FT, Long DG (2014) Microwave radar and radiometric remote sensing. University of Michigan Press, Michigan, 984 p. ISBN 978-0-472-11953-6
van Zyl JJ (2011) Synthetic aperture radar polarimetry. Wiley, Hoboken, New Jersey, 312 p
Weissman DE, Bourassa MA, O'Brien JJ, Tongue JS (2003) Calibrating the QuiKscat/sea winds radar for measuring rainrate over the oceans. IEEE Trans Geosci Remote Sens 41(12):2814–2820
Wingham D, Francis C, Baker S, Bouzinac C, Brockley D, Cullen R et al (2006) CryoSat: a mission to determine the fluctuations in Earth's land and marine ice fields. Adv Space Res 37:841–871
Wöppelmann G, Le Cozannet G, de Michele M, Raucoules D, Cazenave A, Garcin M et al (2013) Is land subsidence increasing the exposure to sea level rise in Alexandria, Egypt? Geophys Res Lett 40:2953–2957
Zahran KH, Saleh S (2006) Contribution of satellite altimetry data in the geophysical investigation of the Red Sea region, Egypt. Acta Geophys 54(3):303–318
Zebker FA, van Zyl JJ (1991) Imaging radar polarimetry: a review. Proc IEEE 79(11):1583–1606
Zeyada H, Ezz MM, Nasr AH, Shokr M, Harb HM (2016) Evaluation of the discrimination capability of full polarimetric SAR data for crop classification. Int J Remote Sens 37(11):2585–2603

Chapter 6
Application of Remote Sensing for Monitoring Changes in Natural Ecosystems: Case Studies from Egypt

Marwa Waseem A. Halmy, Manal Fawzy and Mahmoud Nasr

Abstract In Egypt, monitoring and assessing the changes in natural biodiversity by the traditional site-based methods involve considerable effort and costs. Alternatively, remote sensing can be used as a promising technique to provide complete coverage of habitat, and vegetation species at a specific study area over a given period. Hence, this chapter considers the state-of-the-art of remote sensing for mapping the environmental variables at different locations in Egypt, including deserts, oases, sand dunes, saltmarshes, fish farms, reed vegetation, and agricultural lands. Moreover, the detections of land use/land cover (LULC), soil properties, spatial rainfall distribution, and surface runoff, as well as the management of flood and water resources, by geographical information systems (GIS) were demonstrated. The chapter undertakes useful information and knowledge about the Egyptian environment, giving multiple benefits to researchers, policy planners, and stakeholders. The study objectives are illustrated regarding previous articles reported in the literature.

Keywords Ecological monitoring · Egyptian environment · Habitat detection · Landsat satellite · Remote sensing

6.1 Introduction

Recently, in Egypt, urbanization, anthropogenic activities, and exponential population growth have resulted in severe reductions in water bodies and agricultural lands (Halmy 2019). Remotely sensed data, including satellite images, and aerial photos

---

M. W. A. Halmy · M. Fawzy
Environmental Sciences Department, Faculty of Science, Alexandria University, Alexandria 21511, Egypt
e-mail: marwa.w.halmy@alexu.edu.eg

M. Fawzy
e-mail: dm_fawzy@yahoo.com

M. Nasr (✉)
Sanitary Engineering Department, Faculty of Engineering, Alexandria University, Alexandria 21544, Egypt
e-mail: mahmoud-nasr@alexu.edu.eg; mahmmoudsaid@gmail.com

S. F. Elbeih et al. (eds.), *Environmental Remote Sensing in Egypt*,
Springer Geophysics, https://doi.org/10.1007/978-3-030-39593-3_6

can be used to monitor the surface features and human activities at various regions in Egypt (Shalaby and Tateishi 2007). Remote sensing also plays a significant role in environmental and conservational applications, especially for monitoring the changes in biodiversity and natural ecosystems (El-Asmar et al. 2013). Remote sensing technologies offer periodic repeat coverage of satellite-based maps, which can be used to understand and asses the management of natural resources (Hegazy and Kaloop 2015). Hence, the subject of remote sensing should be comprehensively investigated to provide a complete evaluation of ecological and environmental conditions in various regions.

The use of traditional field surveys entails considerable efforts to offer accurate spatial and temporal coverages of natural ecosystems (Halmy et al. 2015). The integration of in situ observations with remotely sensed data is useful to obtain inclusive outcomes, such as site allocation, terrestrial biodiversity, and habitat classification (Maxwell et al. 2018). This trend could be due to the inaccessibility to some regions having steep slopes, dense mangrove forests, polluted area, and water body obstacle (El-Asmar et al. 2013; Belal and Moghanm 2011). Moreover, a standardized sampling protocol should be provided to evaluate the spatial habitat heterogeneity, species-habitat relationships, and other environmental properties (Abdel-Kader 2018). For instance, Wang et al. (2012) investigated the integration of in-situ sampling with remote sensing in the University of Wyoming King Air (UWKA) to study the cloud microphysical properties and dynamical processes. Moreover, the sampling protocol can be linked to species distribution models to forecast the patterns of species across the place, time, and attributes using environmental and geographic data (Faid and Abdulaziz 2012). Furthermore, physical variables such as elevation, slope, texture, orientation, and temperature have also proven to be appropriate inputs to prepare high-resolution images (Halmy et al. 2019).

Recently, large quantities of global data have become readily accessible due to the development of remote-sensing methods (Ghassemian 2016). Moreover, earth-surveying techniques have been integrated with remote-sensing technologies to attain sufficient information about the biodiversity profile over time in a given region (Araujo Barbosa et al. 2015). However, still, the application of remote sensing to monitor the environmental and biodiversity changes in Egypt requires broad studies (Afifi and Semary 2018). Hence, this chapter provides an overview of the recent applications of remotely sensed data for monitoring the biodiversity and ecosystem changes in some regions of Egypt. Moreover, this chapter represents the use of satellite-based remote sensing for the detection and description of environmental pollutants. The potential and limitations of the application of remote sensing technologies in the arid ecosystems of the Egyptian desert are discussed. This work attempts to overcome the problems of the complexity and high cost of data acquisition, the unavailability of specialized search engines, and the absence of satisfactory geographic database.

6.2 Recent Applications of Remote Sensing: Case Studies in Egypt

This work attempts to obtain a broad and realistic view of the recently published articles regarding the application of remote sensing in Egypt during the last few decades. The search in Scopus database using the terms "remote", "sensing", and "Egypt" resulted in 60 documents for 1980–1990, 68 documents for 1991–2000, 211 papers for 2001–2010, and 470 documents for 2011–2019. The publications included about 70% articles, 15% conference papers, 5% book chapters, 2% review, and 8% other document types. This trend implies that the Egyptian government is exerting high efforts for the widespread application of remote sensing at various regions. In this context, this chapter represents an essential survey that can be helpful to the public and private sectors in Egypt.

Shalaby and Tateishi (2007) applied the remote sensing, and geographic information system (GIS) approaches to monitor the LULC change in the Northwestern coast of Egypt between 1987 and 2001. The study demonstrated that agricultural practices and tourist development plans caused a severe change in the land cover in the study area. This pattern was linked to land degradation, loss of vegetation cover, water shortage, and wind erosion.

Belal and Moghanm (2011) monitored the LULC change in the Middle of Nile Delta, Egypt, using remote sensing and GIS techniques during 1972–2005. The study demonstrated that the urban areas expanded by 5.8–7.2%, leading to the loss of productive cultivated lands. This pattern was mainly due to the exponential population growth. The findings of the article could be beneficial for the government and decision-makers to detect the illegal application of agricultural areas in Delta and Nile Valley.

Abdelkareem et al. (2012) described the evolution of paleo-rivers in the Nile basin of the eastern Sahara using Shuttle Imaging Radar (SIR-A) data. The location of the African Plate regarding the Earth's equator and the significant shifts in paleoclimate have converted various rivers into dry channels obscured by sand deposits. The study also displayed the evolution stages of the Nile River and the change in its deposits from pure and silica-rich sandstone to kaolinite-rich sediments.

Faid and Abdulaziz (2012) investigated the LULC change in the desert of Kom Ombo area, South Egypt regarding urban expansion and agricultural development. Their study demonstrated that the agricultural sector improved by 39.2% through the years 1988–2008 with an average rate of 8.7 km^2 per year. Moreover, the study observed a total expansion in the urbanization of about 28.0 km^2 and a 70% increase in the canal length over the same period. The observations of this work could be useful in establishing policies and strategies for sustainable natural resource management.

AbuBakr et al. (2013) used remote sensing and GIS to define the past shape and flow direction of Wadi El-Arish, i.e., the most extensive ephemeral drainage system in the Sinai Peninsula, Egypt. Moreover, the study attempted to describe the reasons that caused the deviation of the study area from its original course. It was revealed that Wadi El-Arish shifted from northwest to northeast due to the recent uplifting of the Syrian Arc System that blocked the water flow across the main drainage course. It was suggested that a canal connecting the present drainage course with the previous one should be constructed to redirect the occasional runoff for sustainable agriculture development.

El-Asmar et al. (2013) applied the remote sensing approach to detect the changes in the surface area of the Burullus Lagoon, North of the Nile Delta, Egypt during 1973–2011. It was demonstrated that about 43% of the lagoon's surface area was reduced due to the impacts of reclamation activities for aquaculture, agricultural wastes rich in fertilizers and nutrients, and the movement of sand dunes from the coastal line. In addition, anthropogenic activities, population stresses, and soil pollution have led to the environmental degradation of Burullus Lake.

Elhag et al. (2013) applied the remote sensing method to monitor and understand the LULC changes in the Nile Delta region of Egypt from 1984 through 2005. The land-use information was described regarding the agricultural area, urban zone, desert region, fish farm, and surface water. The study found that urban and farming lands increased by almost 6.0% and 6.5%, respectively. A large portion of the desert area was changed to agricultural land due to the reclamation processes and human intervention. Moreover, the patterns of land cover were influenced by cropland degradation, desertification, and urban encroachment. Their work also proposed that remote sensing was a cost-effective technique that could acquire enough information about land development patterns and processes.

Gabr and El Bastawesy (2015) used field investigations and multiple sets of remote sensing data to estimate the hydrological parameters of flash flood events that affected Ras Sudr, Sinai, Egypt, during 2010. Results obtained from the Shuttle Radar Topography Mission (SRTM) and GIS depicted that the peak flow rate was 70 m^3/s, with a total discharge of 5.7×10^6 m^3. The study suggested that the extreme flash flood could be mitigated via (a) using an alluvial fan to adjust the natural flow dispersion, (b) building small dams at the fingertip channels, and (c) transferring the resulting discharge into a single channel.

Halmy et al. (2015) investigated the LULC distribution in the north-western desert of Egypt using the Cellular Automata (CA)-Markov chain technique during 1988–2011. The study demonstrated that built-up, resorts, cropland, and quarrying areas expanded by about 150%, 250%, 200%, and 120%, respectively. This pattern was influenced by agriculture intensification, land degradation, and deforestation. The proposed model predicted expansion in quarries, urbanization of the landscape, and growth in residential areas for 2023.

Hegazy and Kaloop (2015) studied the LULC change in Daqahlia governorate, Egypt using remote sensing data and GIS during 1985–2010. The work indicated that the rate of urbanization resulted in the loss of water bodies and agricultural areas. For instance, the urban land expanded by about 32% (i.e., from 28 to 255 km^2) along with a decrease in the agricultural sector by 33%. This pattern was influenced by the exponential population growth, unorganized land expansion, and increased immigration. The article would provide beneficial strategic plans for the economy and energy use to similar areas in Egypt.

Abdel-Kader (2018) investigated the seasonal and spatial variations of LULC in the northwest coast of Egypt during 2001–2016. The study depicted that the driest (89 mm/year) and the wettest (690 mm/year) years were 2010 and 2016, respectively. Moreover, the vegetation cover at Barrani increased to 38%, which could be linked to the impact of climate and human interactions.

Afifi and El Semary (2018) employed remote sensing and GIS to determine the impact of exhaustive and continuous cropping on soil degradation at the northern part of the Nile Delta during 1961–2016. The study demonstrated that the intensive application of soil for rice cultivation and poor land management caused a significant reduction in land capability and quality.

Yousif et al. (2018) used remote sensing and geological data to detect the occurrence of groundwater at the Western Desert of Egypt. Field studies, geophysical data, microfacies analysis, and GIS application were also demonstrated. The groundwater of Nubian sandstone was found under the confined condition, whereas that of Middle Eocene limestone and Oligocene sandstone was described as an unconfined aquifer. Moreover, the study clarified the infiltration and recharge reasons of groundwater during the pluvial time.

Bakr and Afifi (2019) investigated the LULC change in the Northern Nile Delta, Egypt, using maximum likelihood classifier for the years of 1972, 1984, 2003, and 2016. Accurate thematic maps were obtained using the post-classification tools of sieve classes, majority analysis, and clump classes. The study depicted that the agricultural area increased by approximately 10% between 1972 and 2003 due to the reclamation contribution. Concurrently, the urban land expanded from 5 to 9% during the same period. In addition, the fish farms stretched from 4% in 1983 to 11% by 2016, whereas the area of the Burullus Lake reduced from 480 to 222 km^2 during 1972–2016. These changes affected the rice cultivation and productivity in the monitored area.

Ghoraba et al. (2019) applied the Red List of Ecosystems (RLE) protocol to describe the disruption of the biotic processes in the Burullus Lake located at the north of the Nile Delta of Egypt. The LULC of salt marshes, fish farms, reed vegetation, agricultural lands, sand plain, and bare soil at 1973, 1978, 1999, 2003, 2014, 2015, and 2016 were represented using supervised classification of Landsat images. The study demonstrated that the remotely sensed approach succeeded to cope with the data insufficiency. The lake was influenced by nutrient-rich multisource discharges, which degraded the integrity and natural quality of the ecosystem. The threats impacting the Burullus wetland included biological resource use, pollution from domestic and urban wastewater, human intrusion and disturbances, transportation and service corridors, residential and commercial development, and agriculture and aquaculture.

Halmy (2019) used the Floristic quality (FQ) indices to investigate the effect of anthropogenic practices due to the LULC changes on the environmental quality of the northwestern coast of Egypt. The anthropogenic disturbance indices (ADIs) during 2011–2015 was reported for sand dunes, non-saline depression, salt marshes, and coastal dunes. However, in the past, the Mediterranean coastal desert ecosystem was influenced by the grazing and rain-fed agriculture practices. These factors provided negative impacts on the existing natural habitats, environmental integrity, vegetation quality, and species structure.

Halmy et al. (2019) used remotely sensed data to monitor the composition and distribution of alien plant species at the northwestern coastal desert of Egypt from 2011 to 2014. The collected environmental variables included soil type, vegetation index, topographic roughness index, and distances to roads, coast, resorts, and irrigation canals. The study depicted that the involvement of alien species into new regions due to human activities caused severe issues and threats to the biodiversity system. The data obtained via the species distribution modeling (SDM) approach predicted that at least one alien species could infest over 40% of the study region.

6.3 Changes in Coastal Saltmarsh Distribution as an Indicator of Climate Change

6.3.1 Problem Statement

Recently, satellite measurements have revealed a rise in both absolute sea level and sea level relative to land along the Egyptian coastline. Sea level rise due to climate change is an essential factor that affects the distribution of saltmarshes in many coastal regions (Halmy et al. 2014). Saltmarshes are also influenced by multiple factors such as invasive species, environmental pollution, and LULC change (Saintilan et al. 2018). Saltmarshes link the land and sea, and they deliver various advantages to the coastal communities, such as shoreline and marine biodiversity protection, sustainable fishery support, water quality enhancement, carbon sequestration, and wildlife habitat recovery. A study by Hansen and Reiss (2015) represented the impacts of saltmarshes on ecosystem services, and it highlighted various steps that could be used to mitigate and restore saltmarshes. Deviation in the distribution of the coastal salt marshes is an essential indicator of environmental change, and it has been recently proposed as an indicator of global warming. Hence, the current study aims at representing the distribution of saltmarshes at the northwestern coast of Egypt during 1984–2014.

6.3.2 Methodology

The study area is part of the Western Desert of Egypt located at the west side of Alexandria city. It stretches about 100 km from the Mediterranean coast to the southward of the Qattara Depression and extends westward to El-Alamein city (see Fig. 6.1). Grazing and rainfed agriculture are the primary land use activities in the region.

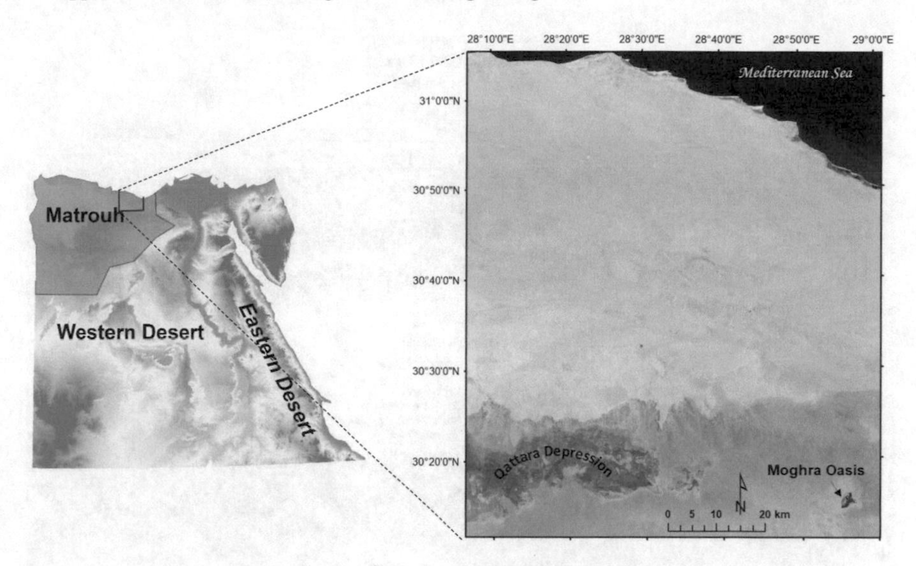

Fig. 6.1 Map of study area located at the Western Desert of Egypt (from 30° 20′ 0″ N to 31° 0′ 0″ N, and 28° 10′ 10″ E to 29° 0′ 0″ E) (*source* Halmy 2014a)

Figure 6.2 displays the flowchart of the methodology used for mapping the changes in the distribution of the coastal saltmarshes at the study area. The procedures have been reported in a previous study by Halmy et al. (2019). The major step was to obtain a time series of low-cost, high quality, and cloud-free imagery that could provide adequate information about the land surface features. Landsat 5 Thematic Mapper (TM) and Landsat 8 operational land imager (OLI) were used to acquire Landsat data during 1984–2003 and 2014, respectively. Radiometric corrections were carried out following Chander et al. (2009) for Landsat 5 TM data and Landsat 8 Data Users Handbook (Survey 2015). Topographic parameters derived from the SRTM data and vegetation indices obtained from Landsat data were included as ancillary observations to map the distribution of the coastal saltmarshes.

The Random Forest (RF) Algorithm was used as a machine learning technique for the classification of the multi-date subsets (Breiman 2001). This technique has been successfully employed for the classification of remotely sensed data, and it was reported to yield highly accurate groupings compared to the conventional classification methods (Belgiu and Drăguţ 2016). For example, a review article by Belgiu and Drăguţ (2016) represented the recent applications of the RF method to handle large data in remote sensing. Random forest models of size 500 decision trees and two variables at each split node were generated after successive trials of different combinations. The randomForest R package within the open-source R 3.0.3 (Core Team 2018) was used for conducting the classification process. The produced LULC maps were evaluated using the overall accuracy and Kappa coefficient (Ghoraba et al. 2019). The post-classification comparison approach was used to detect the LULC changes (Jensen 2005), which were used to assess the spatial distribution of saltmarshes over the study period.

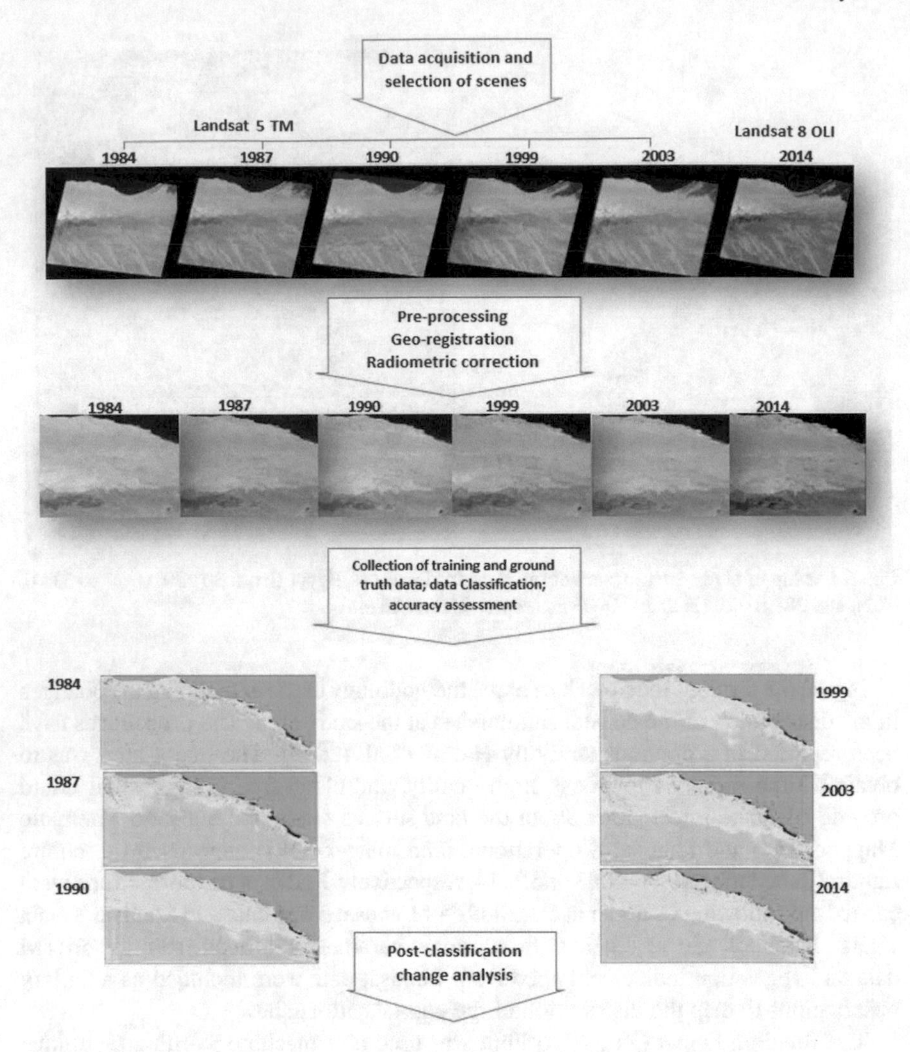

Fig. 6.2 Flowchart of the methodology used for mapping coastal saltmarshes along the northwestern coast of Egypt using Landsat 5 TM and Landsat 8 OLI (*source* Halmy 2014a)

6.3.3 Results and Discussion

Classification using the RF method resulted in highly accurate LULC maps with an overall accuracy higher than 85% and kappa coefficients over 0.81 (see Table 6.1). These results indicated that the RF technique successfully classified the current and past distribution of coastal saltmarshes in the study area.

Figure 6.3 shows that the saltmarsh areas expanded during 1984–1990, and then declined between 1990 and 2014. The loss in coastal salt marshes and wetland areas at specific periods could be attributed to the intensive human modifications as well

Table 6.1 Overall accuracy and kappa statistic for LULC classification of the years 1984, 1987, 1990, 1999, 2003 and 2014 (*source* Halmy 2014a)

Year	Overall accuracy (%)	Kappa
1984	90.53	0.87
1987	88.77	0.85
1990	86.30	0.84
1999	91.20	0.88
2003	85.33	0.82
2014	88.11	0.86

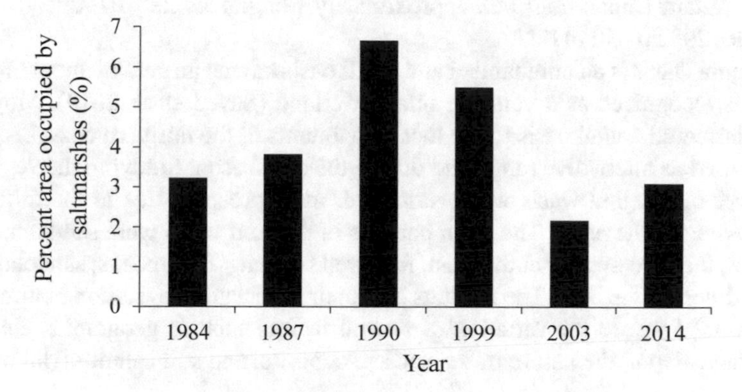

Fig. 6.3 Proportion of area occupied by coastal salt marsh as a percentage of the total area monitored during 1984, 1987, 1990, 1999, 2003, and 2014 (*source* Halmy 2014a)

as tourism and farming practices. Other factors, such as groundwater levels, flooding, and wind waves, could also influence the distribution of saltmarshes. Similarly, Halmy (2019) reported that the northwestern coast of Egypt (including salt marshes) had experienced dynamic changes due to various anthropogenic activities and the creation of artificial lands such as quarries, roads, resorts, and croplands. Their study reported that saltmarshes were dominated by the species of *Arthrocnemum macrostachyum*, *Sarcocorinia fruticose*, and *Atriplex halimus* (Halmy 2019). Consequently, the salt marsh-related habitats and natural vegetation cover were seriously influenced in the area.

The period between 1990 and 1999 has noticed the highest percentage of coastal saltmarshes (see Fig. 6.3). This observation could be linked to seawater intrusion in the coastal region due to the impacts of global warming and sea-level rise (Fagherazzi et al. 2019). A review article by Fagherazzi et al. (2019) has illustrated the conversion of agricultural fields into salt marshes due to the influence of sea-level rise. Their work also demonstrated that sea-level rise and storms profoundly influenced farming fields compared to woodlands and grasslands (Fagherazzi et al. 2019). Further investigations are required to illustrate the phenomenon of marshes migration into agricultural fields, uplands, and suburban lawns.

6.4 Modeling the Distribution of Plant Communities in Moghra Oasis

6.4.1 Problem Statement

Halmy (2014) used the potentialities offered by Landsat satellite images to detect and explore the distribution of plant communities in Moghra Oasis located at Egypt's Western Desert. As shown in Fig. 6.4, Moghra is situated at the northeastern zone of the Qattara Depression; i.e., approximately Longitudes 28° 10′–29° 10′ E and Latitudes 29° 50′–30° 41′ N.

Moghra Oasis is an uninhabited and small oasis having an area of about 630 km^2, and it is recognized as a valuable inland wetland (Sayed et al. 2019). Moghra is also considered a vital oasis to the local inhabitants of the northern coast because it can be used as alternative rangeland during the dry season. Studying the vegetation resources of Moghra Oasis was overlooked, most probably due to the difficulties in accessing to the oasis. The main habitats in the study area were sand dunes, salt marshes, the reed-swamp at the west, followed by sandy hummocks, sandplains and gravel desert at the east. The habitats and their associated vegetation communities were formed due to the variation of several factors such as geographic elevation, water table depth, the nature of the surface deposits, and soil salinity (Halmy et al. 2015).

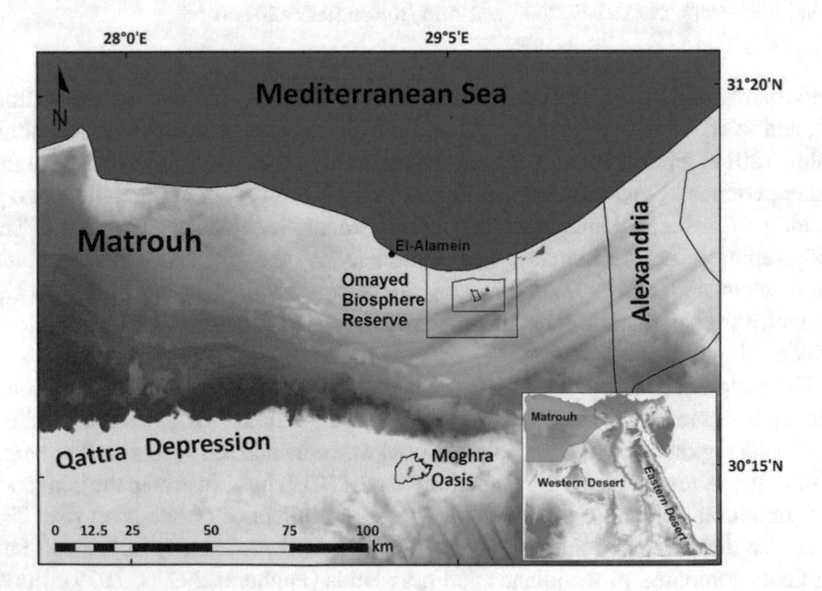

Fig. 6.4 Map of the north-western coastal desert showing the location of Moghra Oasis at the south of Omayed Biosphere Reserve and the northeast side of the Qattara Depression in the Western Desert of Egypt (*source* Halmy 2014b)

6.4.2 Methodology

Field surveys were conducted to monitor the natural plant resources of Moghra Oasis. More than 150 plots, each with an area of 400 m^2, were randomly selected to account for the significant physiographic variation, habitat types, and plant species in the study area. The Latin names of species were reported according to Boulos (1999, 2000, 2002, 2005), whereas nomenclature and identification were updated following Täckholm (1974). Non-metric multidimensional scaling (NMDS) based on Bray-Curtis dissimilarity matrices (Woods et al. 2018) was employed to analyze the vegetation data. Hierarchical clustering by Ward's method, known as a dendrogram, was applied to find groups within the data based on feature similarity and dissimilarity (Shirkhorshidi et al. 2015).

6.4.3 Results and Discussion

The clustering process depicted that the vegetation data could be classified into six major groups. The vegetation clusters were identified regarding the maximum relative frequency and the highest indicator value in each group. Several species characterized each group (see Table 6.2), and the groups could be defined as follows:

Group I was dominated by *Artemisia monosperma*, and *Minuartia geniculate*.
Group II was dominated by *Nitraria retusa*, *Tetraena alba*, and *Sporobolus spicatus*.
Group III was dominated by *Phragmites australis*, and *Juncus rigidus*.
Group IV was dominated by *Zygophyllum album*, and *Alhagi graecorum*.
Group V was dominated by *Tamarix nilotica*, and *Calligonum polygonoides*.
Group VI was dominated by *Arthrocnemum macrostachyum*, and *Halocnemum strobilaceum*.

Figure 6.5 shows the distribution of vegetation groups within Moghra Oasis using the RF model. The high overall accuracy (>90%) and Cohen's kappa coefficient higher than 0.8 indicated that the model was adequate for predicting the distribution of the identified plant communities. The relationship between species abundance and groups of sites revealed that 71% of all species were significantly associated with previously established vegetation types ($p < 0.05$). This result depicted that the spatial distribution of the major plant communities in Moghra Oasis was undergoing dynamic changes between 1984 and 2011. Moreover, the community group "Art-Min" occupied the most area of the oasis with values of 2644 and 2376 ha during 1984 and 2011, respectively. Similarly, Khelifi Touhami et al. (2019) used processing optical remote sensing data to define the spatial-morphological mapping of cultivated lands in the northern-eastern oasis, Algeria. Their work (2019) demonstrated that the satellite images were classified into several subsets (e.g., palm trees, sand, agricultural land, and buildings), and the total salt occupation rate was 24% of the study area.

Table 6.2 List of vegetation species recorded in Moghra Oasis (*source* Halmy 2014b)

Species	Group
Alhagi graecorum Boiss	I + III + IV
Artemisia monosperma Delile	I + II
Arthrocnemum macrostachyum (Moric.) K. Koch	III + VI
Calligonum polygonoides L.	I + II + V
Convolvulus lanatus Vahl	I
Cynodon dactylon (L.) Pers	IV
Ephedra alata Decne	I
Halocnemum strobilaceum (Pall.) M.Bieb	VI
Imperata cylindrica (L.) Raeusch	III + IV
Limbarda crithmoides (L.) Dumort	III
Juncus rigidus Desf	III
Minuartia geniculata (Poir.) Thell	I
Nitraria retusa (Forssk.) Asch	II
Panicum turgidum Forssk	I
Phoenix dactylifera L.	III
Phragmites australis (Cav.) Trin ex. Steud	III
Sarcocornia fruticosa (L.) A.J.Scott	III + VI
Sporobolus spicatus (Vahl) Kunth	II
Stipagrostis obtusa (Delile) Nees	I
Tamarix nilotica (Ehrenb.) Bunge	V
Zygophyllum album L.f.	I + II + IV

6.5 Conclusions

This work represents the recent applications of remote sensing and GIS techniques for monitoring the environmental and ecological changes in various regions of Egypt. The methods and techniques used for data analyses reported in the literature are illustrated. To the best of our knowledge, the study provided useful information to the public and private sectors, policymakers, and stakeholders. The main conclusions of the study are:

– Remote sensing offers appropriate spatial and temporal coverage to compensate for the limitations of ecological site-based data.
– The Egyptian government is exerting high efforts to widen the application of remote sensing at various regions.
– Multiple factors, such as grazing, agricultural activities, and environmental pollution, have caused considerable LULC changes.
– Land degradation, loss of vegetation cover, and water shortage and salinity have been reported in Delta and Nile Valley during the past few decades.

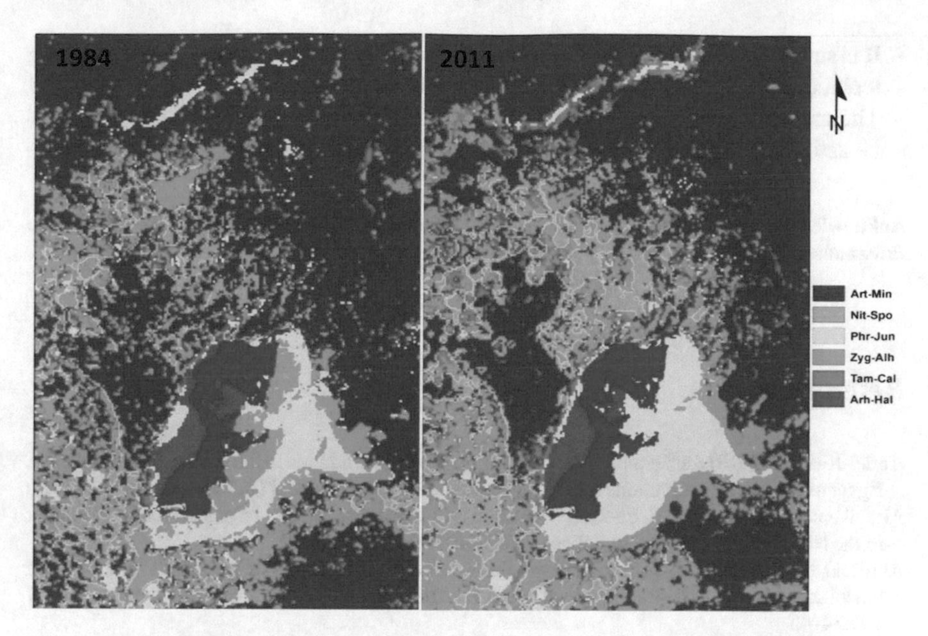

Fig. 6.5 Distribution of the identified plant communities in Moghra Oasis in 1984 and 2011 using Random forest model (*source* Halmy 2014b)

- Reclamation activities for aquaculture, agricultural wastes rich in fertilizers, and the shift of sand dunes have negatively impacted the water bodies located at the northern part of Egypt.
- The salt marsh areas along the northwestern coast of Egypt expanded during 1984–1990 and then declined between 1990 and 2014, which could be due to various anthropogenic activities and climate change.
- The spatial distribution of the major plant communities in Moghra Oasis was undergoing dynamic changes between 1984 and 2011.

6.6 Recommendations

This chapter aims at providing a comprehensive summary for the recent works performed to monitor the Egyptian ecosystem using remote sensing. Some recommendations can be addressed from the study:

- Integrating remote sensing data with field observations should be considered to map the species distribution appropriately.
- Remote sensing should be employed to detect the illegal application of agricultural areas in Delta and Nile Valley.
- Necessary policies and regulations should be established for sustainable natural resource management.

- It is suggested that land and water managers and policymakers use remote sensing techniques to perform soil erosion risk assessment.
- The carbon storage capacity of salt marshes in the coastal lands of Egypt should be assessed.

Acknowledgement The third author would like to acknowledge Nasr Academy for Sustainable Environment (NASE).

References

Abdel-Kader F (2018) Assessment and monitoring of land degradation in the northwest coast region, Egypt using Earth observations data. Egypt J Remote Sens Space Sci (In Press, Corrected Proof)

Abdelkareem M, Ghoneim E, El-Baz F, Askalany M (2012) New insight on paleoriver development in the Nile basin of the eastern Sahara. J Afr Earth Sci 62(1):35–40

AbuBakr M, Ghoneim E, El-Baz F, Zeneldin M, Zeid S (2013) Use of radar data to unveil the paleolakes and the ancestral course of Wadi El-Arish, Sinai Peninsula, Egypt. Geomorphology 194:34–45

Afifi A, El Semary M (2018) The impact of long term cropping and land use change on the degradation of heavy clay soils in the Nile Delta, Egypt. Model Earth Syst Environ 4(2):805–814

Bakr N, Afifi A (2019) Quantifying land use/land cover change and its potential impact on rice production in the Northern Nile Delta, Egypt. Remote Sens Appl Soc Environ 13:348–360

Belal A, Moghanm F (2011) Detecting urban growth using remote sensing and GIS techniques in Al Gharbiya governorate, Egypt. Egypt J Remote Sens Space Sci 14(2):73–79

Belgiu M, Drăguţ L (2016) Random forest in remote sensing: a review of applications and future directions. ISPRS J Photogrammetry Remote Sens 114:24–31

Boulos L (1999) Flora of Egypt. Azollaceae–Oxalidace, vol 1. Al Hadara Publishing, Cairo

Boulos L (2000) Flora of Egypt. Geraniaceae–Boraginaceae, vol 2. Al Hadara Publishing, Cairo

Boulos L (2002) Verbenaceae-Compositae, vol 3. Al Hadara Publishing, Cairo

Boulos L (2005) Flora of Egypt. Monocotyledons: Alismataceae–Orchidaceae, vol 4. Al Hadara Publishing, Cairo

Breiman L (2001) Mach Learn 45:5. https://doi.org/10.1023/A:1010933404324

Chander G, Markham B, Helder D (2009) Summary of current radiometric calibration coefficients for Landsat MSS, TM, ETM+, and EO-1 ALI sensors. Remote Sens Environ 113(5):893–903

de Araujo Barbosa C, Atkinson P, Dearing J (2015) Remote sensing of ecosystem services: a systematic review. Ecol Ind 52:430–443

El-Asmar H, Hereher M, El Kafrawy S (2013) Surface area change detection of the Burullus Lagoon, North of the Nile Delta, Egypt, using water indices: a remote sensing approach. Egypt J Remote Sens Space Sci 16(1):119–123

Elhag M, Psilovikos A, Sakellariou-Makrantonaki M (2013) Land use changes and its impacts on water resources in Nile Delta region using remote sensing techniques. Environ Dev Sustain 15(5):1189–1204

Fagherazzi S, Anisfeld SC, Blum LK, Long EV, Feagin RA, Fernandes A, Kearney WS, Williams K (2019) Sea level rise and the dynamics of the marsh-upland boundary. Front Environ Sci 7(FEB):25

Faid A, Abdulaziz A (2012) Monitoring land-use change-associated land development using multitemporal Landsat data and geoinformatics in Kom Ombo area, South Egypt. Int J Remote Sens 33(22):7024–7046

Gabr S, El Bastawesy M (2015) Estimating the flash flood quantitative parameters affecting the oil-fields infrastructures in Ras Sudr, Sinai, Egypt, during the January 2010 event. Egypt J Remote Sens Space Sci 18(2):137–149

Ghassemian H (2016) A review of remote sensing image fusion methods. Inf Fusion 32:75–89

Ghoraba S, Halmy M, Salem B, Badr N (2019) Assessing risk of collapse of Lake Burullus Ramsar site in Egypt using IUCN Red List of Ecosystems. Ecol Ind 104:172–183

Halmy M (2014a) Changes in the coastal salt marsh distribution as an indicator of climate change. In: The 2nd Arab-American frontiers of science, engineering, and medicine symposium. The Research Council (TRC) of Oman, Muscat, Oman, 13–15 Dec 2014

Halmy M (2014b) Modeling the distribution of plant communities of Moghra Oasis. In: The international biogeography society conference. Early career conference. Australian National University, Canberra, 7–11 Jan 2014

Halmy M (2019) Assessing the impact of anthropogenic activities on the ecological quality of arid Mediterranean ecosystems (case study from the northwestern coast of Egypt). Ecol Ind 101:992–1003

Halmy M, Gessler P, Hicke J, Salem B (2015) Land use/land cover change detection and prediction in the north-western coastal desert of Egypt using Markov-CA. Appl Geogr 63:101–112

Halmy MWA, Fawzy M, Ahmed DA, Saeed NM, Awad MA (2019) Monitoring and predicting the potential distribution of alien plant species in arid ecosystem using remotely-sensed data. Remote Sens Appl Soc Environ 13:69–84

Hansen V, Reiss K (2015) Threats to marsh resources and mitigation. In: Shroder J, Ellis J, Sherman D (eds) Coastal and marine hazards, risks, and disasters. Elsevier, Boston, pp 467–494

Hegazy I, Kaloop M (2015) Monitoring urban growth and land use change detection with GIS and remote sensing techniques in Daqahlia governorate Egypt. Int J Sustain Built Environ 4(1):117–124

Jensen JR (2005) Introductory digital image processing: a remote sensing perspective, 3rd edn. Prentice Hall, Upper Saddle River, New Jersey

Khelifi Touhami M, Bouraoui S, Berguig M-C (2019) Contribution of satellite imagery to study salinization effect of agricultural areas at Northern Eastern Oasis Algerian Region. In: El-Askary H, Lee S, Heggy E, Pradhan B (eds) Advances in remote sensing and geo informatics applications. Advances in science, technology & innovation (IEREK Interdisciplinary Series for Sustainable Development). Springer, Cham

Maxwell A, Warner T, Fang F (2018) Implementation of machine-learning classification in remote sensing: an applied review. Int J Remote Sens 39(9):2784–2817

R Core Team (2018) R: A language and environment for statistical computing. R Foundation for Statistical Computing, Vienna, Austria. Available online at https://www.R-project.org/. Accessed on Jan 2019.

Saintilan N, Rogers K, Kelleway JJ, Ens E, Sloane DR (2018) Climate change impacts on the coastal wetlands of Australia. Wetlands, 1–10

Sayed E, Riad P, Elbeih S, Hagras M, Hassan AA (2019) Multi criteria analysis for groundwater management using solar energy in Moghra Oasis, Egypt. Egypt J Remote Sens Space Sci (In Press, Corrected Proof)

Shalaby A, Tateishi R (2007) Remote sensing and GIS for mapping and monitoring land cover and land-use changes in the Northwestern coastal zone of Egypt. Appl Geogr 27(1):28–41

Shirkhorshidi A, Aghabozorgi S, Ying Wah T (2015) A comparison study on similarity and dissimilarity measures in clustering continuous data. PLoS ONE 10:e0144059. https://doi.org/10.1371/journal.pone.0144059

Survey UG (2015) U.S. Geological Survey, 2015. Landsat 8 (L8) data users handbook. Version 1.0, p 97

Täckholm V (1974) Students Flora of Egypt, 2nd edn. Cairo University, Cooperative Printing Company, Beirut

Wang Z, French J, Vali G, Wechsler P, Haimov S, Rodi A et al (2012) Single aircraft integration of remote sensing and in situ sampling for the study of cloud microphysics and dynamics. Bull Am Meteorol Soc 93(5):653–668

Woods C, Robertson S, Sinclair W, Collier N (2018) Non-metric multidimensional performance indicator scaling reveals seasonal and team dissimilarity within the National Rugby League. J Sci Med Sport 21(4):410–415

Yousif M, Sabet H, Ghoubachi S, Aziz A (2018) Utilizing the geological data and remote sensing applications for investigation of groundwater occurrences, West El Minia, Western Desert of Egypt. NRIAG J Astron Geophys 7(2):318–333

Chapter 7
Hyperspectral Based Assessment of Mosquito Breeding Water in Suez Canal Zone, Egypt

Asmaa El-Hefni, Ahmed M. El-Zeiny, Mohamed Sowilem, Manal Elshaier and Wedad Atwa

Abstract Suez Canal Zone has an old history of diseases transmitted by mosquito such as Malaria, Lymphatic filariasis, West Nile virus and Rift Valley Fever virus. Water quality of mosquito breeding habitat represents an essential determinant of whether female mosquitoes will deposit their eggs and whether the resulting stages will complete their developmental process or not. Therefore, the objective of this study is to assess physical, chemical and spectral properties of mosquito breeding habitats in the Suez Canal Zone using hyperspectral data and spectral analyses. Fifty-two different sites were sampled, during February and April 2016, for mosquito larvae and were characterized based on water temperature, pH, Oxidation-Reduction Potential (ORP), Electrical Conductivity (EC), Turbidity, Chlorophyll, Dissolved oxygen (HDO), Crude Oil (CO), Salinity, and Organic Matter (OM). Data were statistically assessed by one way ANOVA. Hyperion image and ASD Field Spectroradiometer were processed to generate an innovative spectral library for the investigated mosquito breeding habitats. Mosquito larvae were identified as *Culex* (3 spp.), *Anopheles* (1 sp.), *Ochlerotatus* (2 spp.) and *Culiseta* (1 sp.) at seven different habitats. Analyses showed that the spectral reflectance patterns were specific for each mosquito breeding habitat corresponding to the variability of water quality. Most of habitats reported high levels of total dissolved and suspended solids such as Turbidity, EC, and OM (i.e. >535.93 NTU, >16,642.05 μS/cm, and >38.31 mg/l,

A. El-Hefni (✉) · A. M. El-Zeiny · M. Sowilem
Department of Environmental Studies, National Authority for Remote Sensing and Space Sciences (NARSS), 23 Joseph Tito Street, El-Nozha El-Gedida, Alf Maskan, P.O. Box: 1564, Cairo, Egypt
e-mail: m.asmaa53@yahoo.com; asmaa.elhefni@narss.sci.eg

A. M. El-Zeiny
e-mail: narss.ahmed@gmail.com

M. Sowilem
e-mail: sowilem60@gmail.com

M. Elshaier · W. Atwa
Department of Zoology, Faculty of Science (Girls Branch), Al-Azhar University, Al Nasr Road, Opposite to Cairo International Conference Center, P.O. Box: 11751, Nasr City, Cairo, Egypt
e-mail: manalgalhoum@yahoo.com

W. Atwa
e-mail: Ahmadgalhoum@hotmail.com

S. F. Elbeih et al. (eds.), *Environmental Remote Sensing in Egypt*,
Springer Geophysics, https://doi.org/10.1007/978-3-030-39593-3_7

respectively). It can be concluded that hyperspectral data analyses help to give more feasible assessment of mosquito breeding habitats which should widely be utilized.

Keywords Water quality · Hyperspectral analyses · Mosquito breeding water · Suez Canal Zone · Egypt

7.1 Introduction

Mosquitoes (Diptera: Culicidae) have great importance in terms of severe public health problems and risks. These insects co-exist with humans and transmit many pathogens. More than half of the world's population lives under constant risk of pathogens spread by mosquitoes as estimated by the World Health Organization (WHO) (Guruprasad et al. 2014).

Suez Canal area has an old history of diseases transmitted by mosquito (Harbach et al. 1988; Morsy et al. 1990); Rift Valley Fever and malaria transmitted by *Culex pipiens* and *Anopheles pharoensis* respectively (Meegan et al. 1980; Ghoneim and Woods 1983; Kenawy 1988, 2015). The favorable environmental distinct features describing the Suez Canal Zone assist in the spatial wide-spread of various species of mosquito in and around the Zone. The recent environmental modification caused by urbanization and agricultural activities contribute to spreading the breeding sites of various mosquitos' species (Amusan et al. 2005).

Mosquitoes prefer to breed in various types of running or stagnant aquatic habitats; these habitats can be natural or artificial, sunny or partially shaded, permanent or temporary and with different sizes (Bashar et al. 2005; Imbahale et al. 2011; Liu et al. 2019). The most effective impact of urbanization is the production of many habitats that accelerates larval development of some mosquito species (Leisnham and Slaney 2009). In rural areas, the potential breeding sites are usually associated with sewage water, water logging/seepage; drainage canals, cesspits, and cesspools. More specifically, edges of shallow small water bodies, surrounded with vegetation and debris, are considered as larval micro-habitat (Rogers et al. 2002).

Physicochemical properties of mosquito breeding water greatly affect the survival and proliferation of mosquito (Garba and Olayemi 2015). A good knowledge of the physicochemical and environmental parameter affecting mosquito richness can help in planning control methods of mosquitoes (Overgaard et al. 2001; Tabbabi et al. 2015). Physicochemical parameters such as temperature, Electrical Conductivity (EC), salinity, Total Dissolved Solid (TDS), Dissolved Oxygen (DO) and pH have a significant effect on the mosquito larval abundance (Dejenie et al. 2011; Kwasi et al. 2012; Imam and Deeni 2014; Bashar et al. 2016).

Larval developmental processes and the emergence of adult stage in the aquatic habitats, play the main role in the determination of distribution pattern and abundance of mosquito species (Ali et al. 2013). In Egypt, few studies focused on the properties (chemical and physical) of mosquito breeding habitats were carried out; Kenawy and

El-Said (1990), Kenawy et al. (1996, 2013), Abdel-Hamid et al. (2011a, b), Ammar et al. (2013), Bahgat (2013), El-Naggar et al. (2013), El-Zeiny and Sowilem (2016).

Even now, conducting field survey, near the Suez Canal area, often requires special permission. Remote sensing techniques offers the potential for identifying and characterizing larval habitats on a large scale area that is more difficult or impossible using traditional ground survey methods (Hayes et al. 1985; Washino and Wood 1994; Dale et al. 1998; Hay et al. 1998; Zou et al. 2006). Hyperspectral images have found a wide range of applications in water resources management, agriculture and environmental monitoring, water quality assessment studies (Goetz et al. 1985; Smith 2001; Thenkabail et al. 2004a; Adam et al. 2010; Bioucas-Dias et al. 2013; Abou El-Magd and El-Zeiny 2014; Thenkabail et al. 2014; El-Zeiny and El-Kafrawy 2017; Omran 2018).

The diversity of mosquito species in the Suez Canal area requires a wide mosquito control program based on detailed biological and environmental information. Hence, the present study was planned to assess physicochemical and spectral characteristics contributing to mosquito breeding in the study area using spectral data investigations.

7.2 Materials and Methods

7.2.1 Study Area

Suez Canal is anthropogenic sea level waterway in Egypt, connecting the Mediterranean Sea with the Red Sea. The northern boundary of Suez Canal is Port Said; the southern terminus is Port Tawfiq at Suez city and Ismailia on its west bank, about 3 km from the half-way point. The canal passes through three lakes; Manzala Lake at the north which is protected from the canal with bedding on its western side, Timsah Lake at the middle, and the Great Bitter Lakes further south (Suez Canal guide 2010). The study area includes parts of three Governorates Fig. 7.1; Port Said, Ismailia and Suez, occupying an area of (7523.008 km^2). It extends between Lat. 29° 30′ N to 31° 30′ N and Long. 32° 10′ E to 32° 40′ E, bordered from the north by the Mediterranean Sea, from the northern east by a part Sinai Peninsula, and west and south by eastern desert.

The study area is characterized by different Land Use/Land Cover (LULC) patterns (Sowilem et al. 2017a) including; (1) Desert represented mainly by eastern and western desert, (2) Agricultural area which are distributed along the eastern and western banks of the canal, (3) Fish Farms mostly located at the southern parts of Port Said Governorate; (4) Manzala Lagoon at the Northern parts of Port Said, (5) Sabkha lands accumulated in northern parts of the study area, particularly in the eastern banks of Suez Canal, (6) Urbanized areas distributed in different parts of the study area.

According to the results of the 2015 census by the Central Agency for Public Mobilization and Statistics, the total population of Port Said, Ismailia, Suez is (666.66 thousand, 1.178 million, 622.86 thousand people) respectively (CAPMAS 2015).

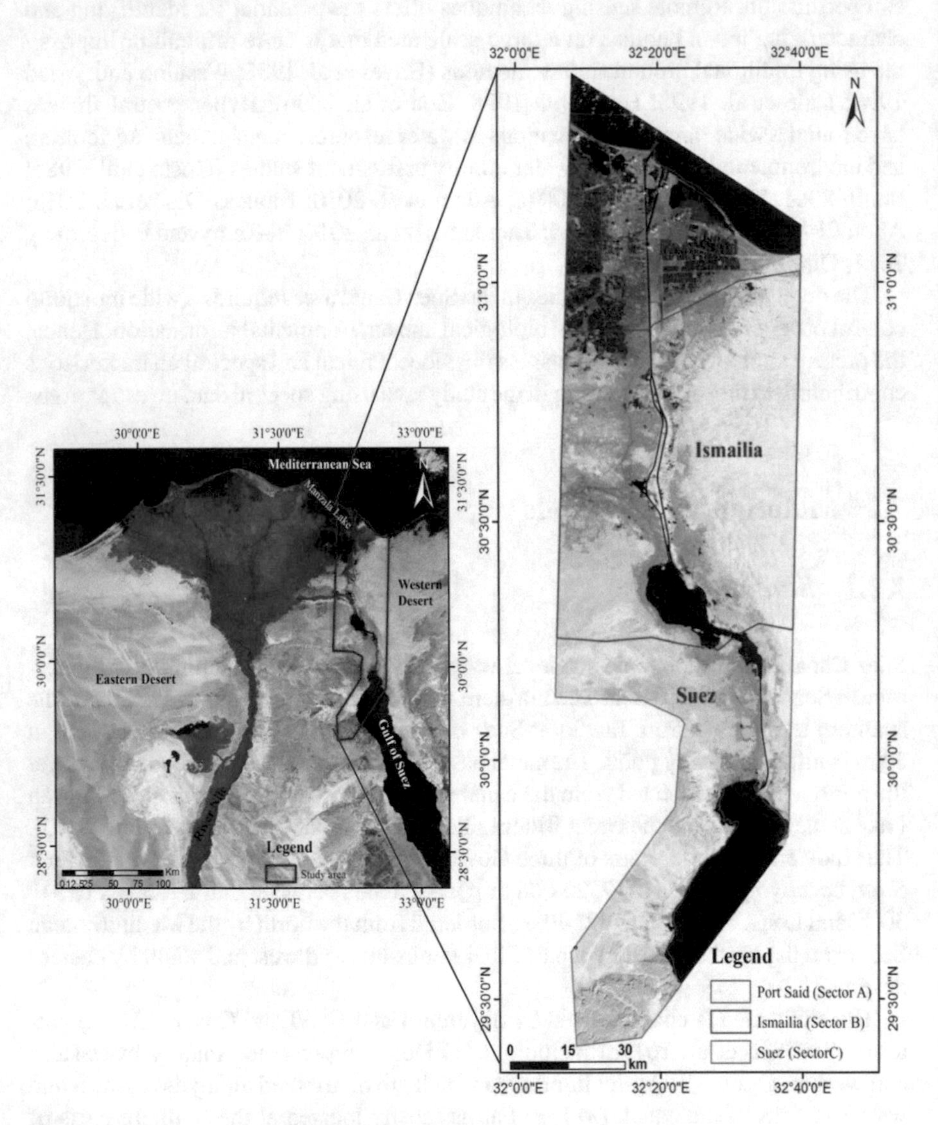

Fig. 7.1 Location map of the study area

7.2.2 *Mosquito Larval Collection*

The larval collection performed along the Suez Canal Zone for all accessible breeding locations as demonstrated in Fig. 7.2 from 52 favorable mosquito larvae locations in February and April 2016. During each survey, mosquitos' larvae were collected by a standard dipper (a small ladle 12.5-cm diameter with a 90-cm small wooden handle).

Collected samples were placed in transparent cup and preserved in 70% of Ethyl Alcohol; the larvae were identified morphologically to species level using the Keys

Fig. 7.2 Map of study area showing sampled sites. Different types of larval habitats (photos): **a** Drainage canal, **b** Seepage, **c** Sewage, **d** Agriculture drainage canal

of Harbach (1988) and Glick (1992). Procedures and precautions, regarding larval catching and transportation, were done according to WHO (1975) guidelines.

The types of breeding sites inspected for larvae presence in the Suez Canal area including; sewage water, some irrigated fields and drainage canals, cesspools, cesspits, and seepage. Physical characteristics of each breeding sites were recorded. Each visited spot was geo-referenced using GPS Magellan 320-USA (Tadesse et al. 2011).

7.2.3 *Physicochemical Characterization of Larval Habitats*

Simultaneously with larval collections, water was sampled during February and April 2016 from 52 sites along Suez Canal Zone. Physicochemical characteristics of breeding water were recorded including; water temperature, pH, Turbidity, Oxidation-Reduction Potential (ORP), Chlorophyll, Electrical Conductivity (EC), Dissolved oxygen (HDO), Crude Oil (CO), Salinity and Organic Matter (OM) content. All parameters were measured using Manta-2 instrument except OM which was recorded in the Laboratory following the standard method of water analyses (APHA 1992).

7.2.4 *Hyperspectral Data Acquisition*

Space-borne Hyperion EO-1 image was downloaded from https://glovis.usgs.gov/. The available scene was acquired on 23 April 2002. The Hyperion satellite provides a high-resolution hyperspectral imager (capable of resolving 220–242 spectral bands from 0.4 to 2.5 μm). The spatial resolution is 30 m at a swath width of 7.5 km, and provide detailed spectral mapping across all bands with high radiometric accuracy. The visible (V) and Near Infra-Red (NIR) region (from 0.4 to 1.2 μm) collects data from band 1–70, while the Short Wave Infra-Red (SWIR) region from (1.2 to 2.5 μm) detector collects data from band 71–224.

7.2.5 *Hyperspectral Data Processing*

The Hyperion image was corrected for the removal of bad bands and striping to be prepared for atmospheric correction and obtaining reflectance image (Fig. 7.3) (Kumar and Yarrakula 2017). The following steps were adopted:

(1) Removing the absorption bands and bad bands; some bands have no information and are called zero bands (1–7, 58–76, 221–242) (EO-1 User Guide 2003). The remaining 194 bands were corrected due to sensor overlap problem. Amongst the 194 bands, there are a number of water-vapor bands 120–132, 165–182, 221–224, these bands having a lot of noise so they have been ignored. After the image

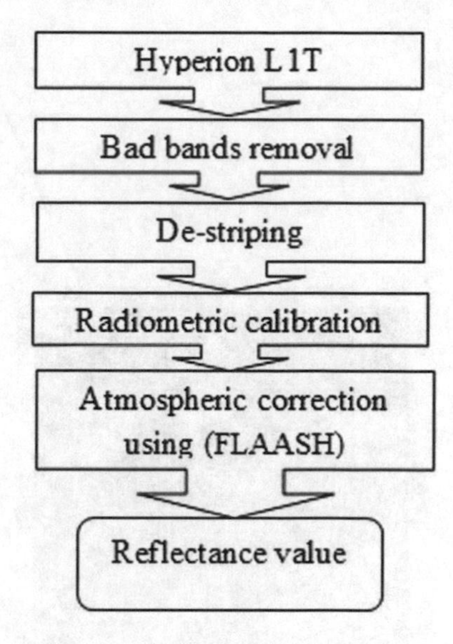

Fig. 7.3 Methodology flowchart of hyperspectral pre-processing steps

bands have been resized to 92 for atmospheric correction. (2) Removal of vertical stripe, the striping makes the image difficult to visualize, and the vertical strips are removed using local De-striping method. (3) Radiometric calibration used to convert DN value to radiance value. (4) Atmospheric correction used to convert radiance value to reflectance value using FLAASH module. The calibrated Hyperion image (Fig. 7.4) will be utilized in integration with the field identified breeding sites to generate a spectral library for different mosquito breeding habitats using ENVI 5.1 software.

On the one hand, and simultaneously with larval sampling, the in-situ spectral reflectance behavior of different larval breeding sites was recorded using hyperspectral data sets, acquired from Analytical Spectra Devices Field Spectro-radiometer (ASD FieldSpec) during February 2016.

ASD FieldSpec line of Spectro-radiometers offers hyperspectral data where the narrow intervals allow measuring the spectral signature of water body (Abou El-Magd and El-Zeiny 2014). This is a high resolution device, utilizes in the full optical spectral range (VNIR and SWIR) starting from (350 to 2500 nm). The sampling interval is 1.4 nm within the range (350–1050 nm) and 2 nm within the range (1000–2500 nm). The sampling unit is composed of three spectrometers: the VNIR spectrometer (350–1000 nm) measured by 512 bands, the $SWIR_1$ (1000–1830 nm), and the $SWIR_2$ (1830–2500 nm). Each SWIR spectrometer has 600 bands. The measured reflectance is resembled to produce a spectrum at 1 nm interval for the entire range (Danner et al. 2015).

Spectral data pre-processing was done by removal of noisy bands; because of low signal to noise ratios are found in wavelength regions (350–399 nm and 2401–2500 nm). Data smoothing took place using the Savitzky-Golay smoothing filter

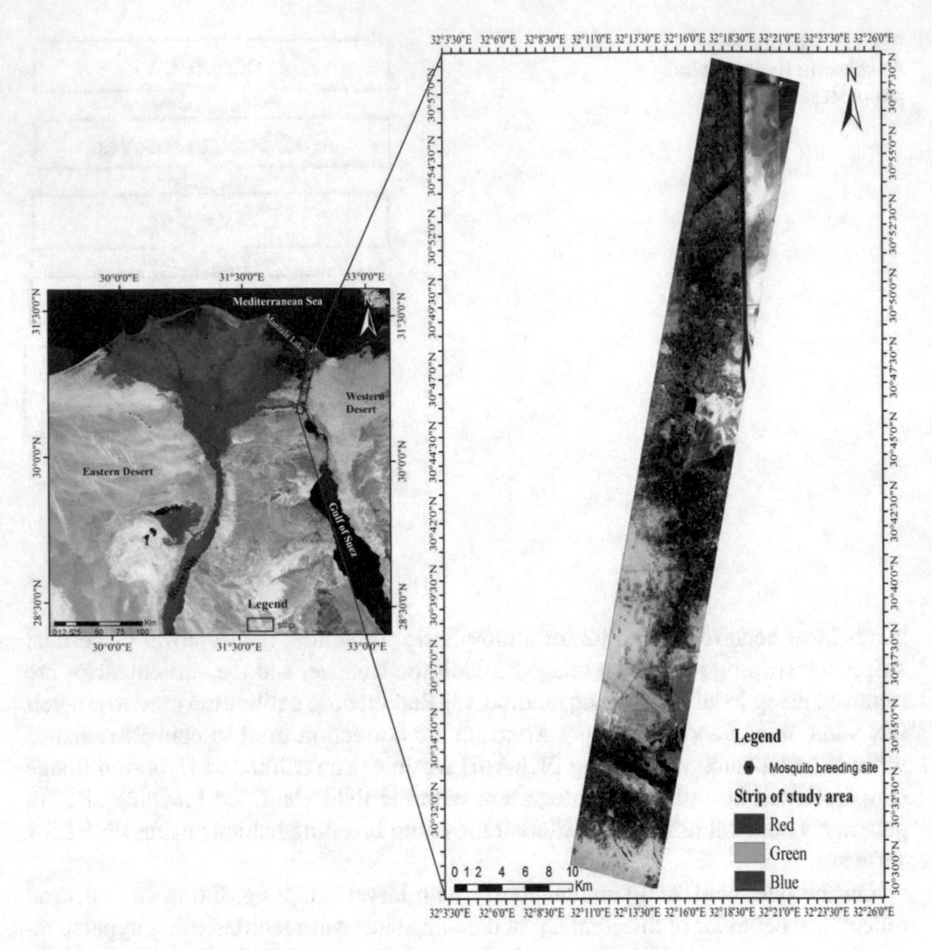

Fig. 7.4 Hyperion image showing mosquito breeding habitats

(Savitzky and Golay 1964). Then, transformation was carried out to reduce the particle size effect and enhance the spectra (Brunet et al. 2007). A total of 2001 spectral bands ranges between (400 and 2400 nm) were used for analyses.

The calibrated space-borne Hyperion image and in-situ spectroradiometer were integrated with water quality analyses to differentiate different mosquito breeding habitats and to assess each habitat individually. This will finally be utilized to generate an innovative spectral library for different mosquito breeding habitats which will be saved as ENVI library to be applicable in any similar environment.

7.2.6 *Statistical Analysis*

Mean physicochemical characteristics per breeding habitat were calculated and one–way ANOVA was used to test for significance between water quality parameters and larval species densities at ($P = 0.05$), and to obtain the best predictor parameter contributing to the richness of mosquito larval species. The statistical analysis was carried out using SPSS package (version 19). The mosquito larval density in each breeding habitat was calculated using the following formula (Banaszak and Winiewski 1999):

$$D = l/(L \times 100\%)$$

where, D: Density, l: Number of specimens of each mosquito species, L: Number of all species.

7.3 Results and Discussions

In this study, 52 different sites out of 81 total breeding sites were sampled in February and April 2016. Seven mosquito species belonging to four genera were identified; *Culex pipiens* L., *Culex perexiguus* Theobald, *Culex* (*Barraudius*) *pusillus* Macquart, *Anopheles* (*Anopheles*) *tenebrosus* Dönitiz, *Culiseta longiareolata* (Macquart), *Ochlerotatus detritus* Haliday, and *Ochlerotatus caspius* (Pallas). Species of the three genera (*Culex, Anopheles,* and *Ochlerotatus*) are vectors of several diseases.

Mosquito breed in a wide range of aquatic habitats with different physicochemical characteristics. Accessible larval breeding habitats inspected, during the study period were categorized into seven habitat types; drainage canals, sewage, seepage, irrigation canals, agriculture drainage canals, unused irrigation water basins and drainage canals of fish farms. Larval densities are varied in different habitats. *Cx. pipiens* was the most abundant species.

7.3.1 *Physicochemical Characteristics of Breeding Water*

Physicochemical characteristics of water (Temperature, pH, ORP, EC, Turbidity, Chlorophyll, HDO, CO, Salinity and OM) were measured and analyzed during February and April 2016 (Tables 7.1 and 7.2). The obtained water temperature showed the range 21.01–23.05 °C (February) and 22.31–23.94 °C (April) in all types of breeding habitats. For the collected species, the observed pH ranges among breeding habitats are 7.23–9.43 in February and 6.94–9.43 in April indicated that most mosquito breeding habitats are alkaline tendency. However, pH 6.94 restricted only to agriculture drainage canals where *Cx. pipiens* and *Cx. perexiguus* breed. Turbidity showed high variations within mosquito breeding habitats. Higher mean levels of turbidity were

Table 7.1 Physicochemical characteristics (mean, range) and larval species (density) of mosquito breeding habitats at Suez Canal Zone (February 2016)

Habitat Type	Drainage canal	Sewage	Agriculture drainage canals	Seepage	Irrigation canals	Unused irrigation water basin	Drainage Fish Farm
Temperature (°C)	22	22.10, (21.13–23.05)	21.95, (21.39 – 22.75)	22.21, (21.68–23.05)	22.37, (21.48–23.31)	22.58	22.75
pH	7.66	8.52,(7.58 – 9.43)	7.93, (7.48–8.17)	7.89, (7.23–8.98)	7.93, (7.62–8.22)	7.94	8.12
ORP (mV)	407.2	-180.15, (-244.67–78.21)	-54.28, (194.59 – 18.18)	-71.95, (-399.90–73.47)	-37.71, (-150.12 – 15.05)	-289.95	18.18
EC (uS/cm)	52941.54	12749.54, (11365.00 – 15043.6)	10723.58, (623.70 – 27720.00)	26923.43, (1047.89–89574.62)	7870.86, (501.04 – 37966.67)	2271.5	27720
Turbidity (NTU)	10.68	1157.74, (88.34–2798.44)	15.63, (6.22–29.26)	330.93, (0.38–2487.73)	129.86, (0.19 – 846.22)	6057	6.22
Chlorophyll (µg/l)	70.64	1632.53, (759.71–3016.21)	453.42, (164.05– 875.79)	2337.83, (57.85–5471.19)	1434.13, (37.43–4366.98)	439.38	875.79
HDO (%Sat)	68.39	63.05, (61.32–64.55)	73.25, (52.76–98.75)	62.47, (49.43–76.79)	62.72, (48.04–73.83)	49.31	98.75
CO (rfu)	7119.24	7173.38, (7144.88 –7224.74)	7202.58,(7159.94–7228.37)	7177.73, (7082.49–7318.20)	7184.15, (7132.08–7252.18)	7173.37	7159.94
Salinity (mg/L)	34.96	7.52, (6.47– 9.33)	6.47, (0.30–17.06)	17.73, (0.51–63.83)	4.71, (0.24–24.12)	1.16	17.06
Larval species (Density)	*Oc. detritus* (0.4)	*Cx. pipiens* (55.75), *Oc. detritus* (12.5)	*Cx. pipiens* (53.6)	*Cx. pipiens* (40.5),*Oc. detritus* (19.3), *Cx. pusillus* (0.75), *An. tenebrosus* (0.45),*Cx. perexiguus* (0.6)	*Cx. pipiens* (23), *Oc. detritus* (8.85), *Cx. pusillus* (4), *An. tenebrosus* (3.2), *Cx. perexiguus* (3)	*Cx. pipiens* (8.75)	*Oc. detritus* (5.75), *Cx. pusillus* (0.25)

Where each of the three habitats; drainage canal, unused irrigation water basin and drainage fish farm were observed in a single locality during this period

Table 7.2 Physiochemical characteristics (mean, range) and larval species (density) of mosquito breeding habitats at Suez Canal Zone (April 2016)

Habitat Type	Drainage canals	Sewage	Agriculture drainage canals	Seepage	Irrigation canals	Unused irrigation water basin	Drainage Fish Farm
Temperature (°C)	23.32, (22.43 –23.54)	23.06, (22.37–23.40)	23.07, (22.66 – 23.50)	23.25, (22.47–23.48)	22.94, (22.31–23.46)	23.39, (23.37–23.43)	23.4, (23.39 –23.41)
pH	7.50, (7.19 –7.80)	7.9, (7.43 –8.98)	7.27, (6.94–7.61)	7.81, (7.18 –8.63)	8.01,(7.71 –8.26)	8.14, (7.12 – 9.74)	8.34, (8.34 –8.35)
ORP (mV)	-34.48, (-374.03 – 547.32)	-264.75, (-375.77–176.36)	-239.4, (-154.93)–(-323.88)	9.17, (-296.02 – 238.40)	(-83.88), (-278.77 – 195.82)	(-213.15), (-312.26 – 127.44)	-154.89, (-155.18 – 158.60)
EC (uS/cm)	18952.7, (825.58 – 90858.89)	9066.38, (1064.33 –18525)	1415.45, (2457.71 – 373.20)	37621.11, (1491.73–112994.11)	17278.63, (2286.45 – 52477.5)	1094.46, (528.22 – 1804.8)	60809.16, (60800.00–60818.33)
Turbidity (NTU)	149.73,(4.09–786.02)	44.67, (4.31 – 89.10)	67.32, (3.49–131.15)	12.16, (0.19 –28.85)	40.48, (2.19 – 107.13)	4.8, (-1.59 – 15.75)	5.06, (4.44 –5.69)
Chlorophyll (µg/l)	257.8,(46.71–934.69)	631.87, (160.42–1068.0)	563.62, (141.35 – 985.89)	414.23, (136.13–1324.5)	169.85, (32.11 – 444.68)	82.99, (45.75 – 126.10)	625.29, (583.61 – 666.98)
HDO (%Sat)	58.80, (38.21–80.89)	50.57, (28.98–58.33)	52.36, (54.72–50)	68.81, (50.95–82.98)	67.95, (46.81–82.72)	55.3, (39.72–78.84)	46.54, (46 –47.08)

(continued)

Table 7.2 (continued)

Habitat Type	Drainage canals	Sewage	Agriculture drainage canals	Seepage	Irrigation canals	Unused irrigation water basin	Drainage Fish Farm
CO (rfu)	-37834.37, (-71213.1) – (-10534.3)	-47174.37, (-97830.36 – 14998.58)	-31936.75, (-57492.02) – (-6381.48)	-82734.29, (-133140.50) – (-40351.79)	-25520.65, (-33977.19) – (-15345.4)	-30982.02, (-48038.8) – (-20724.32)	-16960.5, (-170153) – (-169056)
Salinity (mg/L)	12.95, (0.40–64.89)	5.22, (0.52-10.98)	0.72, (0.18–1.26)	25.86,(0.77 –84.20)	11.03,(1.21–34.59)	0.54, (0.25 – 0.91)	40.9, (40.86 – 40.94)
Larval species (Density)	*Cx. pipiens* (796),*Oc. detritus* (1), *Cx. perexiguus* (16.6), *Cs. longiareolata* (14.9)	*Cx. pipiens* (1895), *Cx. perexiguus* (35.4), *Cs. longiareolata* (25), *An. tenebrosus* (3.6), *Cx.pusillus* (119)	*Cx. pipiens* (210), *Cx.perexiguus* (21.6)	*Cx. pipiens* (156.75), *Cx. perexiguus* (16.6), *Cx. pusillus* (1.2),*Oc. detritus* (78.9), *Oc. caspius* (29)	*Cx. pipiens* (123), *Cx. perexiguus* (12.3), *Oc. detritus* (171.6), *Oc. caspius* (0.4)	*Cx. pusillus* (0.25), *Cs. longiareolata* (13), *Cx. pipiens* (355), *Cx. perexiguus* (14.9)	*Oc. detritus* (3.6)

observed in sewage (1157.74 NTU at February) and drainage canals (149.73 NTU at April). On the one hand, ORP indicates the state of substance (oxidation or reduction). A positive reading of ORP means that a substance is an oxidizing agent, while negative ORP value indicates that a substance is a reducing agent. Most of the studied breeding habitats are reducing agent.

Maximum salinity and EC levels were very high in seepage (63.83 mg/L, 89574.62 µS/cm in February) and (84.20 mg/L, 112994.11 µS/cm in April) respectively compared to other habitats. In the study area, *Oc. detritus*, *Oc. caspius* and *Cx. pusillus* can breed in high saline water since the higher densities were recorded. Levels of HDO were extensively related to levels of organic pollutants. A mild fluctuation was observed in HDO levels recorded (48.04–98.75%Sat in February) and (28.98–82.98%Sat in April). The maximum level of HDO was reported at drainage fish farm (98.75%Sat in February) and seepage (82.98%Sat in April), while low level detected in irrigation canal (48.04%Sat in February) and sewage (28.98%Sat in April).

Crude oil (CO) is a mixture of naturally occurring hydrocarbons that is refined into heating oil, diesel, gasoline, kerosene, jet fuel, and thousands of petrochemicals. Heavier crudes yield more heat upon burning in comparison to light crudes. In the current study, CO showed high levels fluctuating from (7082.49–7318.20 rfu. in February), while recorded negative values in April. Chlorophyll levels recorded a great fluctuation, within the studied breeding habitats, ranging from (37.43–5471.19 µg/l in February) and (37.43–14506.65 µg/l in April). Seepage showed the maximal concentrations of chlorophyll (5471.19 µg/l) followed by irrigation canal (4366.98 µg/l). Organic matter content (OM), associated breeding water, fluctuated from (0–116 mg/l in February). This was clearly observed high levels of OM in sites largely exposed to human activities.

7.3.2 Hyperspectral Characteristics of Breeding Water

The ASD FieldSpec system offers hyperspectral data where the narrow intervals give the opportunity to characterize water quality. The spectral reflectance patterns of mosquito breeding sites were recorded in-situ during February 2016. This was performed in integration with conventional in-situ and laboratory investigations of breeding water to obtain a more precise assessment of water quality characteristics. It was observed that spectral patterns obtained were different to each breeding habitat, corresponding to water content (Fig. 7.5).

Most of the investigated breeding habitats reported high levels of active total dissolved and suspended solids such as Turbidity, EC, and OM (i.e., >535.93 NTU, >16,642.05 µm/cm, and >38.31 mg/l, respectively). This was explained by the high spectral reflectance in the VNIR parts of the spectrum as investigated in drainage fish farms, seepage, drainage canals, and agriculture drainage habitats. However, drainage fish farms and agriculture drainage habitats recorded a clearer reflection in the red edge region of the spectrum (700–800) than seepage due to the

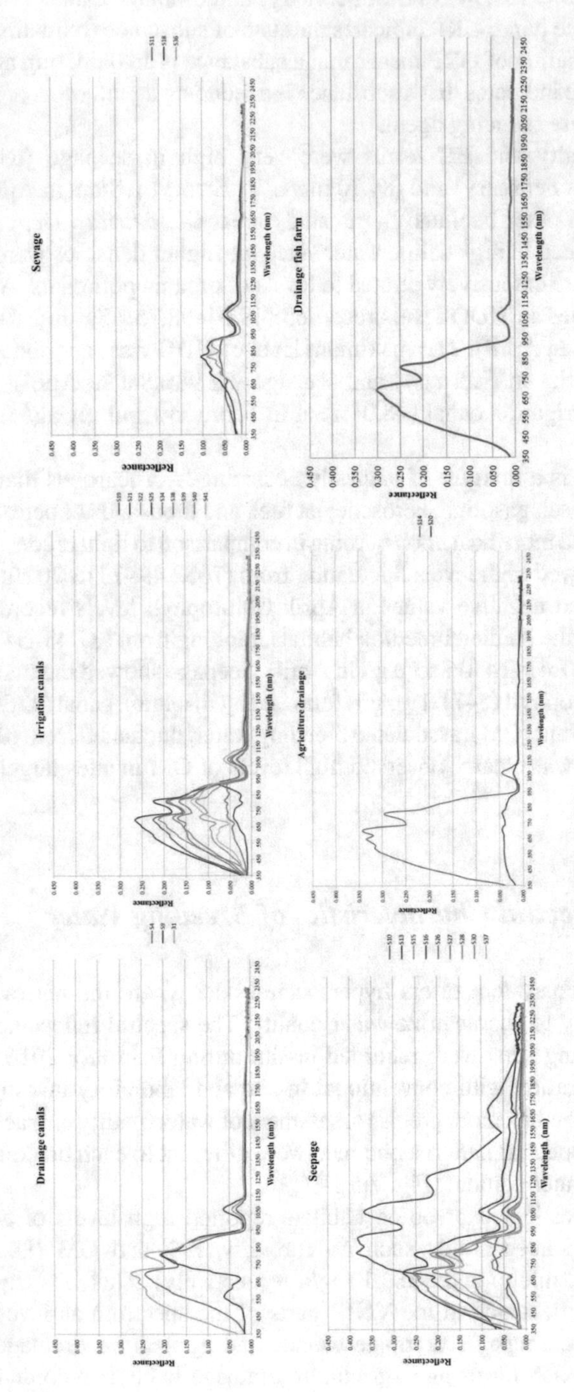

Fig. 7.5 Hyperspectral characteristics of mosquito breeding habitats during February 2016

high levels of chlorophyll; 14,503.65 μg/l. (Fig. 7.6). It is clear that the ability of water to reflect the spectrum is increased in turbid water while clear water absorbs more in the parts of VNIR. Consequently, breeding sites with saline and turbid water showed high- spectral reflection rather than clear water.

Chlorophyll is an indicator of algal biomass, which shows distinct absorption in the blue band at 440 nm and in the red band from (672–678 nm) leaving a maximum green reflectance due to scattering process. The red edge of the spectrum near (705 nm), which is characterizing the chlorophyll in the water where a clear peak reflection is observed. Variable levels of chlorophyll were observed among the investigated mosquito breeding habitats along Suez Canal Zone. Both in-situ measurements and reflectance analyses showed a great fluctuation of chlorophyll content, which could greatly affect the nutrient concentrations preferable for larval survive. Spectral library is the collections of representative spectra of different materials generated from laboratory and ground-based measurement. It is utilized as the reference against which hyperspectral imaging data are compared to identify earth surface materials composition.

A calibrated Hyperion image was processed to identify the spectral characteristics of breeding sites in three different selected habitats; seepage, drainage, and irrigation canals. Spectral library plots of the three habitats were investigated, as illustrated in Fig. 7.8. Based on this investigation, a spectral library for the three indicated habitats was generated and saved into ENVI5.1 software (Fig. 7.7). The narrow interval of spectral bands in hyperspectral image achieves high efficiency in recording the changes that occurred in the spectrum within different habitats.

7.3.3 *General Characterization of Mosquito Breeding Habitats*

Several mosquito species have been reported in Suez Canal Zone. Abdel-Hamid et al. (2011b) reported eight species at Ismailia Governorate. Ten species were collected at Suez Canal Zone (Sowilem et al. 2017b). Twelve species were previously surveyed at Ismailia Governorate; some of them have public health importance (Harbach et al. 1988; Morsy et al. 1990).

The results obtained from this study indicate that mosquito breeding habitats in the Suez Canal area were associated with shallow polluted water bodies, sewage, drainage canals, seepage surrounded by various kinds of vegetation; partially exposed to sunlight and at distance of residential areas. The relative abundance of Anophelinae and Culicinae mosquitoes may be attributed to their capacity to develop in different environments as formerly observed by Dondrop et al. (2010) and Simon Oke and Ayani (2015). As stated by Sunahara et al. (2002), the larval richness mainly depends on the quality of the habitats and each species develops in a specific environment such as fresh-water, rain-water, sewage and drainage at a specific temperature. Field

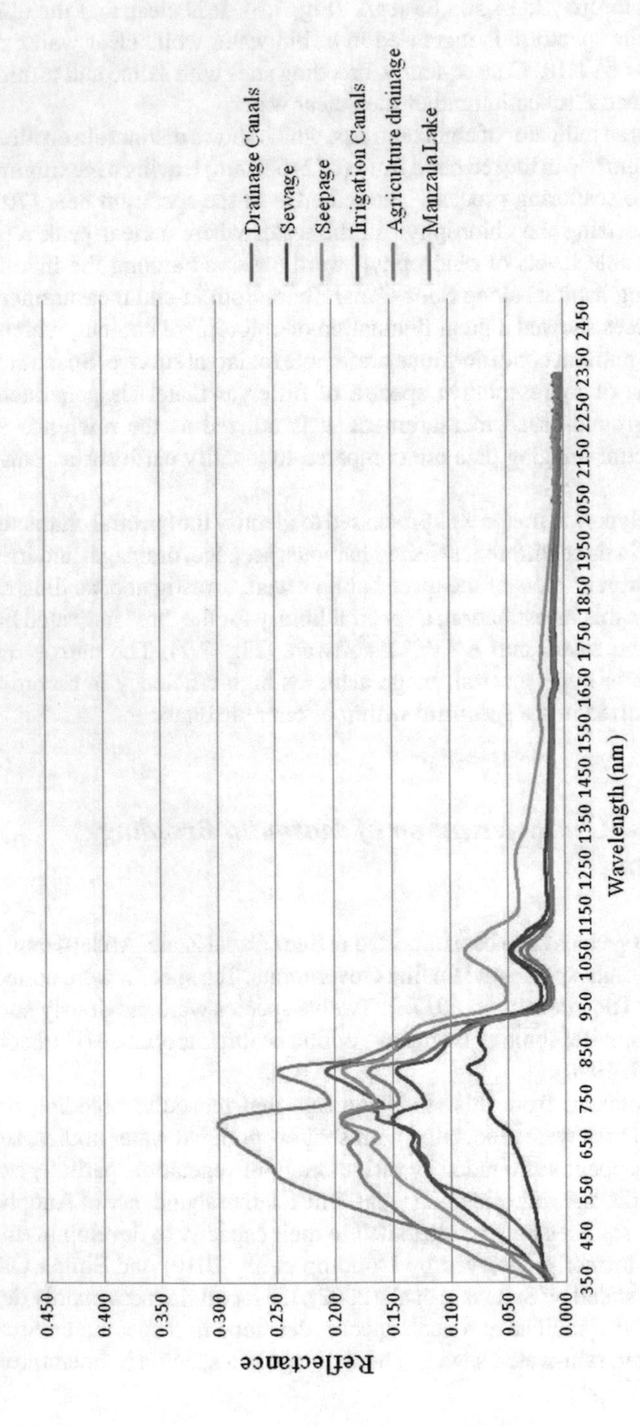

Fig. 7.6 Hyperspectral characteristics of mosquito breeding habitats based on reflectance average (February 2016)

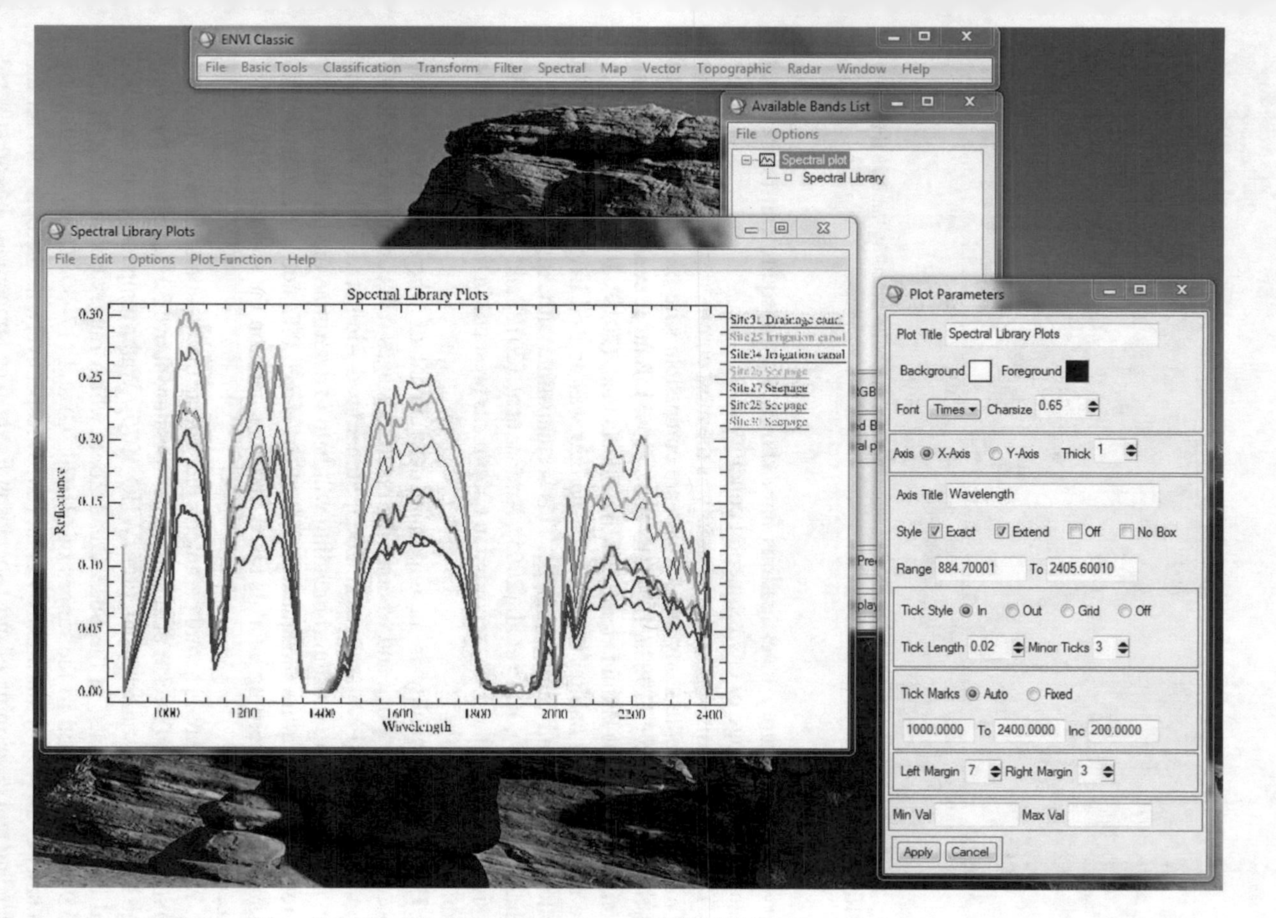

Fig. 7.7 Generated spectral library of the selected three habitat types as appeared in ENVI5.3 software

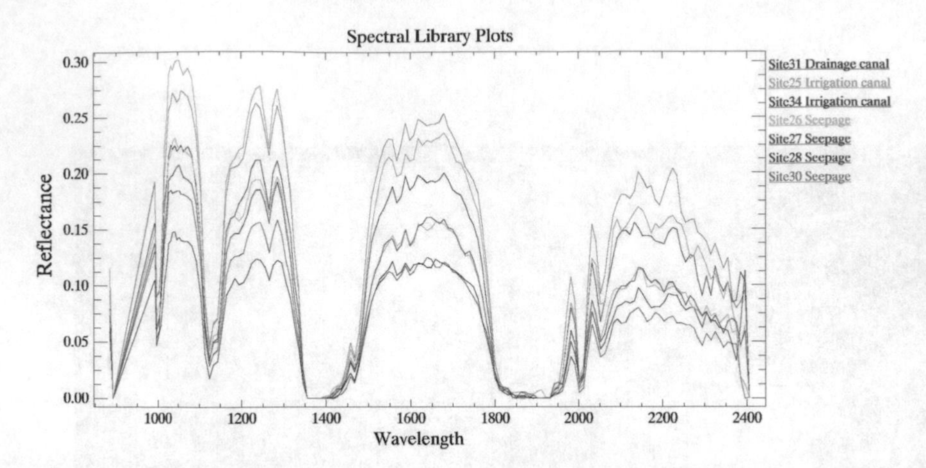

Fig. 7.8 Spectral library plots of the selected three types of mosquito breeding habitats (February 2016)

surveys and laboratory investigations have shown that mosquito can tolerate and breed in a wide range of environmental conditions.

The main filariasis vector, *Cx. pipiens* was the most common species and prevails in most habitats in the study area, which are compatible with the previous study of El-Said and Kenawy (1983a, b), Harb et al. (1993), Bahgat et al. (2004), Shaman et al. (2010), Abdel-Hamid et al. (2011b) and Liu et al. (2019). It was observed that the most preferred habitat types of *Cx. pipiens* was unused irrigation water basin followed by polluted drainage canals. This is compatible with Amr et al. (1997), Al-Khalili et al. (1999), Knio et al. (2005), Ammar et al. (2012) and Tran et al. (2013). A significant difference was recorded in *Cx. pipiens* presence in the different aquatic habitats.

Habitat water quality is an important determinant of whether or not female mosquitoes will oviposit and whether the resulting stages will complete their developmental process (Piyaratnea et al. 2005; Muhammad-Aidil et al. 2015). Assessing the factors affecting mosquito breeding will help to determine the preventive measures to mitigate future mosquito proliferation. Hence this study is undertaken to characterize the breeding habitats taken into consideration the spectral behavior of breeding habitats.

Water temperature is a vital parameter that affects the duration of the mosquito's larval development (Becker et al. 2010). Temperature above 30 °C reduces the average life of the mosquito population (WHO 1975). Water temperature of the studied breeding habitats, in the Suez Canal Zone, which ranges between (22.31 and 23.94 °C) confirmed to the ranges of (21–29 °C) and (17–30 °C) as specified by Kenawy et al. (1998) and Kenawy et al. (2013) respectively, as the optimum for the breeding and developing of most species in Egypt. The temperature has an adverse impact on *Cx. pipiens* densities (-0.85, $p > 0.05$). However, Abdel-Hamid et al.

(2011b) observed a positive correlation between larval density and temperature for total collected species including *Cx. pipiens.*

Series recording pH of mosquito aquatic habitats in several studies in Egypt indicated that all breeding habitats are alkaline with pH (7.5–9.0). Also, a correlation was found between pH and some of the mosquito species in aquatic habitats. From the present results, in the study area, indicate that most mosquito larval breeding is an alkaline tendency (>7). However, the pH (6.94) restricted only to agriculture drainage canals where *Cx. pipiens* and *Cx. perexiguus* breed, in agreement with Hamdy (1987) recorded pH (6.9–8.5) in Ismailia. On the one hand, pH has a negative effect on *Cx. perexiguus* density (−0.82, $p > 0.05$), such result agrees with that obtained by Kenawy and El Said (1990), Kenawy et al. (1998) and Abdel-Hamid et al. (2013), while, Abdel-Hamid et al. (2011b) reported that pH has a positive effect on *Cx. perexiguus* density in Ismailia Governorate.

The present resulted showed no significant correlation between turbidity and larval density. This is explained by the presence of culicine larvae in diverse habitats with turbid and slightly turbid stagnant water which is in agreement with Abd El-Meguid (1987). However, Kenawy et al. (1996) found that water turbidity significantly affect the breeding of *Cx. perexiguus* and *Cx. pipiens* ($P > 0.05$) while clear water is preferable for the species *Cs. longiareolata* in Sharkiya Governorate.

Salinity levels were very high in drainage fish farm habitats and seepage compared to other habitats. The reported species can breed in fresh and brackish water as observed and reported by Abdel-Hamid et al. (2011b) for the most prevalent species *Cx. pipiens* and *Cx. perexiguus* in Ismailia, and Kenawy et al. (2013) for five species in Cairo. Kenawy reported that *Cx. pusillus* breeds only in brackish water like the present findings, and breed under conditions very similar to those required by *Oc. caspius.* Also, Morsy et al. (1990) recorded *Cx. pusillus* in irrigation canals, irrigation water basin, and sewage which also matches with the presenting observations. Concerning elevated levels of CO might be attributed to that most of mosquito breeding are characterized by high levels of contamination resulting from domestic and agricultural drains discharged into these sites. DO is the best factor to demonstrate the effect of pollution in the aquatic habitat unless it contains toxic elements (Lester 1975). In this study there was no association significant observed between mosquito larval abundance and DO. However, some studies investigated that there was a significant association between them (Oyewole et al. 2009; Varun et al. 2013).

7.3.4 *Benefits of Using Hyperspectral Data in the Present Study*

Hyperspectral remote sensing is a comparatively new technology that is utilized in several applications (the detection and identification of minerals, vegetation, and in the field of water resources and environmental applications) (Nagy and Jung 2005). In the current study, hyperspectral imagery and data provide high-spectral resolution details, with very narrow band-widths which are very helpful for detailed terrestrial applications.

The spectral reflectance pattern is accurately corresponding with water constituents. In this study, a quantitative study of mosquito breeding sites was performed using the conventional methods of field surveys and lab analyses. However, a qualitative study was carried out depending on analyzing spectral reflectance behavior of mosquito breeding sites. The integration between conventional and advanced techniques in analyses has carried out to produce more precise results. These results are important in characterizing larval habitat and in the determination of preferable characteristics of each species. The present study represents one of the preliminary attempts in Egypt to assess the spectral behavior of various mosquito breeding sites. Spectral data can be utilized to generate a spectral library. This library has different environmental applications such as identifying and mapping the potential breeding habitat using its signature. In the present study, two different types of libraries were developed for mosquito breeding sites. The first library was developed based on the in-situ spectral measurement using a field spectroradiometer. The second library is generated from a Hyperion image based on the conducted field surveys. Both libraries were developed based on hyperspectral data sets which ensure a high-spectral accuracy within different bands.

This library has different significant applications such as;

– Discriminating spectral characteristics of various habitats.
– Mapping mosquito breeding habitats using hyperspectral data sets.
– Qualitative assessment of breeding sites in regarding of water quality.

7.4 Conclusions

Present study is an attempt to utilize the hyperspectral data sets to assess mosquito breeding water and generate a spectral library for various habitats for different purposes. Investigations showed that mosquito can breed in various aquatic environments where different species have their habitat preferences. This study has provided physicochemical analyses on mosquito's larval habitats in Suez Canal Zone. The main vector of filariasis (*Cx. pipiens)* was dominant in the study area. Results presented here demonstrate that the hyperspectral data can provide higher spectral resolution for the assessment of water quality parameters associated with breeding habitats, on a large geographic area and help in water management to mitigate the area under risk of emergence/re-emergence diseases. Generating a spectral library of different mosquito breeding habitats is very necessary for facilitating the further assessment of mosquito towards selecting the appropriate mitigation measures.

Recommendations

The present study highly recommended using remote sensing techniques for rapid and accurate assessment and monitoring towards sustainable health. Most of the developed countries are applying these techniques to mitigate mosquito problems. Results of this study have confirmed that different parts of the Suez Canal Zone are environmentally preferable for mosquito breeding and therefore the probability of diseases outbreak. It is very urgent, for decision makers, to put appropriate control plan to mitigate the area under risk of potential diseases outbreak and mosquito nuisance. To control mosquitoes effectively at a long-term, it is crucial to use many complementary management techniques including;

- Reduce breeding habitats: Removing containers of all types and human-made structures.
- Water resources management.
- Monitoring environmental factors contributing mosquito breeding habitats.
- Biological control using nematodes, fish, and bacteria *Bacillus thuringiensis* and *Bacillus sphaericus.*
- Genetic control: Modifying mosquitoes so that they can't be breed.
- Chemical insecticides against adults and/or larvae.
- Housing modification provides door and window screens to help keep insects away.
- Personal protection: Wearing protective, light-colored and soft clothing; using insect repellents and avoiding activities in areas where mosquitoes are active.

Acknowledgements The authors expresses their sincere appreciations and gratitude to the National Authority for Remote Sensing and Space Sciences (NARSS) for pasturing the study and development project "Modeling spatial/temporal distribution of mosquito breeding habitats and monitoring environmental factors associated with the proliferation in Suez Canal Zone, using remote sensing and GIS" (Period 2014–2016) through which the study data was acquired. Also, to the United States Geological Survey (https://glovis.usgs.gov/) for supporting Landsat-8 (OLI) and Hyperion images.

References

Abd El-Meguid AH (1987) The susceptibility of different mosquito types in Ismailia Governorate to some insecticides. M.Sc. thesis (Vector Control) High Institute of Public Health, Alexandria, 133 pp

Abdel-Hamid YM, Mostafa AA, Allam KM, Kenawy MA (2011a) Mosquitoes (Diptera: Culicidae) in El Gharbyia Governorate, Egypt: their spatial distribution, abundance and factors affecting their breeding related to the water, shallow depth ranging between 5–20 cm, situation of *lymphatic filariasis*. Egypt Acad J Biol Sci 3(1):9–16

Abdel-Hamid YM, Soliman MI, Kenawy MA (2011b) Mosquitoes (Diptera: Culicidae) in relation to the risk of disease transmission in El Ismailia governorate, Egypt. J Egypt Soc Parasitol 41:347–356

Abdel-Hamid YM, Soliman MI, Kenawy MA (2013) Population ecology of mosquitoes and the status of bancroftian filariasis in El Dakahlia Governorate, the Nile Delta, Egypt. J Egypt Soc Parasitol 43:103–113

Abou El-Magd I, El-Zeiny A (2014) Quantitative hyperspectral analysis for characterization of the coastal water from Damietta to Port Said, Egypt. Egypt J Remote Sens Space Sci 17:61–76

Adam E, Mutanga O, Rugege D (2010) Multispectral and hyperspectral remote sensing for identification and mapping of wetland vegetation: a review. Wetlands Ecol Manag 18:281–296

Ali N, Marjan Khan K, Kausar A (2013) Study on mosquitoes of Swat Ranizai sub division of Malakand. Pak J Zool 45(2):503–510

Al-Khalili YH, Katbeh-Bader A, Mohsen ZH (1999) Siphon index of *Culex pipiens* larvae collected from different bio-geographical provinces in Jordan. Zool Middle East 17:71–76

American Public Health Association (APHA) (1992) Standard methods for the examination of water and wastewater, 17th edn. Washington, D.C.

Ammar SE, Kenawy MA, Abd El Rahman HA, Gad AM, Hamed AF (2012) Ecology of the mosquito larvae in urban environments of Cairo Governorate, Egypt. J Egypt Soc Parasitol 42:191–202

Ammar SE, Kenawy MA, Abdel-Rahman HA, Ali AF, Abdel-Hamid YM, Gad AM (2013) Characterization of the mosquito breeding habitats in two urban localities of Cairo Governorate, Egypt. Greener J Biol Sci 3(7):268–275

Amr ZS, Al-Khalili Y, Arbaji A (1997) Larval mosquitoes collected from northern Jordan and the Jordan Valley. J Am Mosq Control Assoc 13:375–378

Amusan AAS, Mafiana CF, Idowu AB, Ola-tunde GO (2005) Sampling mosquitoes with CDC light traps in rice field and plantation communities in Ogun State, Nigeria. Tanzania Health Res Bull 7:111–116

Bahgat IM (2013) Impact of physical and chemical characteristics of breeding sites on mosquito larval abundance at Ismailia Governorate, Egypt. J Egypt Soc Parasitol 43:399–406

Bahgat IM, El Kadi GA, Sowilem MM, El Sawaf BM (2004) Host-feeding patterns of *Culex pipiens* (Diptera: Culicidae) in a village in Qalubiya Governorate and in a new settlement in Ismailia Governorate. Bull Entomol Soc Egypt 81:77–84

Banaszak J, Winiewski H (1999) Podstawyekologii. Foundation of ecology. WydawnictwoUczelniane WSP, Bydgoszcz, Poland, p 630

Bashar K, Samsuzzaman M, Ullah MS, Iqbal MJH (2005) Surveillance of Dengue vectors mosquito in some rural areas of Bangladesh. Pak J Biol Sci 8:1119–1122

Bashar K, Rahman SM, Nodi IJ, Howlader AJ (2016) Species composition and habitat characterization of mosquito (Diptera: Culicidae) larvae in semi-urban areas of Dhaka, Bangladesh. Pathog Glob Health 110(2):48–61

Becker N, Petrić D, Zgomba M, Boase C, Madon M, Dahl C, Kaiser A (2010) Mosquitoes and their control. Springer, Berlin, Heidelberg, New York, p 579

Bioucas-Dias JM, Plaza A, Camps-valls G, Scheunders P, Nasrabadi N, Chanussot J (2013) Hyperspectral remote sensing data analysis and future challenges. IEEE Geosci Remote Sens Mag 1:6–36

Brunet D, Barthès BG, Chotte JL, Feller C (2007) Determination of carbon and nitrogen contents in Alfisols, Oxisols and Ultisols from Africa and Brazil using NIRS analysis: effects of sample grinding and set heterogeneity. Geoderma 139:106–117

Central Agency for Public Mobilization and Statistics (CAMPAS) (2015) Statistical yearbook population 2015

Dale PE, Ritchie SA, Territo BM, Morris CD, Muhar A, Kay BH (1998) An overview of remote sensing and GIS for surveillance of mosquito vector habitats and risk assessment. J Vector Ecol 23(1):54–61

Danner M, Locherer M, Hank T, Richter K (2015) Spectral sampling with the ASD FieldSpec 4—theory, measurement, problems, interpretation. EnMAP field guides technical report. GFZ Data Services.

Dejenie T, Yohannes M, Assmelash T (2011) Characterization of mosquito breeding sites in and in the vicinity of tigray microdams. Ethiop J Health Sci 21:57–66

Dondrop A, Francois N, Poravuth Y, Depashish D, Aung P, Traning J, Khin M, Ariey F, Hanpithakpong W, Lee S, Ringwald P, Kamolarat S, Imwong N,Lindegardh N, Socheat D, White N (2010) Artemisinin resistance in *Plasmodium falciparum* malaria. N Engl J Med 361(5):455–467

El-Naggar A, Elbanna SM, Abo-Ghalia A (2013) The impact of some environmental factors on the abundance of mosquito's larvae in certain localities of Sharkia Governorate in Egypt. Egypt Acad J Biol Sci 6(2):49–60

El-Said S, Kenawy MA (1983a) Geographical distribution of mosquitoes in Egypt. J Egypt Public Health Assoc 58:46–76.

El-Said S, Kenawy MA (1983b) Anopheline and culicine mosquito species and their abundance in Egypt. J Egypt Public Health Assoc 58, 108–142

El-Zeiny A, El-Kafrawy S (2017) Assessment of water pollution induced by human activities in Burullus Lake using Landsat-8 operational land imager and GIS. Egypt J Remote Sens Space Sci 20(1):49–56

El-Zeiny A, Sowilem M (2016) Environmental characterization for the area under risk of mosquito transmitted diseases, Suez Canal Zone using remote sensing and field surveys. J Environ Sci 45(3–4):283–297

EO1 User Guide (2003) EO-1 User Guide v. 2.3. https://eo1.usgs.gov

Garba Y, Olayemi IK (2015) Spartial variation in physicochemical characteristics of wetland rice fields mosquito larval habitats in Minna, north Central Nigeria. In: International conference on agricultural, ecological and medical sciences, 10–11 Feb, pp 11–14

Ghoneim NH, Woods GT (1983) Rift Valley fever and its epidemiology in Egypt. J Med 14:55–79

Glick JI (1992) Illustrated key to the female Anopheles of southwestern Asia and Egypt (Diptera: Culicidae). Mosq Syst 24:125–153

Goetz AFH, Vane G, Solomon JE, Rock BN (1985) Imaging spectrometry for earth remote sensing. Science 228:1147–1153

Guruprasad NM, Jalali SK, Puttaraju HP (2014) Wolbachia-a foe for mosquitoes. Asian Pac J Trop Dis 4(1):78–81

Hamdy AM (1987) The susceptibility of different mosquito types in Ismailia Governorate to some insecticides. M.Sc. thesis in Public Health Science (Vector Control), 133 pp

Harb MR, Faris AM, Gad ON, Hafez R, Ramzy AA, Buck AA (1993) The resurgence of lymphatic filariasis in the Nile Delta. Bull World Health Organ 71(1):49–54

Harbach RE (1988) The mosquitoes of the subgenus *Culex* in the southwestern Asia and Egypt (Diptera: Culicidae). Contrib Am Entomol Inst (Ann. Arbor.) 24:240

Harbach RE, Harrison BA, Gad AM, Kenawy MA, El-Said S (1988) Records and notes on mosquitoes (Diptera: Culicidae) collected in Egypt. Mosq Syst 20:317–342

Hay SI, Snow RW, Rogers DJ (1998) Predicting malaria seasons in Kenya using multi-temporal meteorological satellite sensor data. Trans Royal Soc Trop Med Hyg 92:12–20

Hayes RO, Maxwell EL, Mitchell CJ, Woodzick TL (1985) Detection, identification and classification of mosquito larval habitats using remote sensing scanners in earth orbiting satellites. Bull World Health Org 63:361–374

Imam AA, Deeni Y (2014) Common types of Anopheles gambiae breeding habitats in north western Nigeria. J Innovative Res Eng Sci 4:496–504

Imbahale SS, Paaijmans KP, Mukabana WR, van Lammeren R, Githeko AK, Takken W (2011) A longitudinal study on Anopheles mosquito larval abundance in distinct geographical and environmental settings in western Kenya. Malaria J 10:81

Kenawy MA (1988) Anopheline mosquitoes (Diptera: Culicidae) as malaria carriers in A.R. Egypt (History and pre-sent status). J Egypt Public Health Assoc 63:67–85

Kenawy MA (2015) Review of *Anopheles* mosquitoes and malaria in ancient and modern Egypt. Int J Mosq Res 5(4):1–8

Kenawy MA, El-Said S (1990) Factors affecting breeding of culicine mosquitoes and their associations in the Canal Zone, Egypt. In: Proceedings international conference for statistics, computer science, social and demographic research, vol 1, pp 215–233

Kenawy MA, Rashed SS, Teleb SS (1996) Population ecology of mosquito larvae (Diptera: Culicidae) in Sharkiya Governorate, Egypt. J Egypt Ger Soc Zool 21E:121–142

Kenawy MA, Ammar SE, Abdel-Rahman HA (2013) Physico-chemical characteristics of the mosquito breeding water in two urban areas of Cairo Governorate, Egypt. J Entomol Acarol Res 45(3):17

Kenawy MA, Rashed SS, Teleb SS (1998) Characterization of rice field mosquito habitats in Sharkia Governorate, Egypt. J Egypt Soc Parasitol 28(2):449–459

Knio KM, Markarian N, Kassis A, Nuwayri-Salti N (2005) A two-year survey on mosquitoes of Lebanon. Parasite 12:229–235

Kumar MV, Yarrakula K (2017) Comparison of efficient techniques of hyperspectral preprocessing for mineralogy and vegetation studies. Indian J Geo-Mar Sci 46(05):1008–1021

Kwasi B, Biology A, Kumasi T (2012) Physico-chemical assessment of mosquito breeding sites from selected mining communities at the Obuasi municipality in Ghana. J Environ Earth Sci 2:123–130

Leisnham PT, Slaney DP (2009) Urbanization and the increasing risk from mosquito-borne diseases: linking human well-being with ecosystem health. In: De Smet LM (ed) Focus on urbanization trends. Nova Science Publishers, Inc., Hauppauge, NY, USA, pp 47–82

Lester WF (1975) Polluted river, river trent England. In: Whitton BA (ed) River ecology. Oxford, 725 pp

Liu X, Baimaciwang, Yue Y, Wu H, Pengcuociren, Guo Y, Cirenwangla, Ren D, Danzenggongga, Dazhen, Yang J, Zhaxisangmu, Li J, Cirendeji, Zhao N, Sun J, Li J, Wang J, Cirendunzhu, Liu O (2019) Breeding site characteristics and associated factors of *Culex pipiens* complex in Lhasa, Tibet, P. R. China. Int J Environ Res Public Health 16:1407

Meegan JM, Khalil GM, Hoogstraal H, Adham FK (1980) Experimental transmission and field isolation studies implicating *Culex pipiens* as a vector of Rift Valley fever virus in Egypt. Am J Trop Med Hyg 29:1405–1410

Morsy TA, el Okbi LM, Kamal AM, Ahmed MM, Boshara EF (1990) Mosquitoes of the genus *Culex* in the Suez Canal Governorates. J Egypt Soc Parasitol 20(1):265–268

Muhammad-Aidil R, Imelda A, Jeffery J, Ngui R, Wan Yusoff WS, Aziz S, Lim YA, Rohela M (2015) Distribution of mosquito larvae in various breeding sites in National Zoo Malaysia. Trop Med 32:183–186

Nagy Z, Jung A (2005) A case study of the anthropogenic impact on the catchment of Mogyorod-brook, Hungary. Phys Chem Earth 30:588–597

Omran ESE (2018) Remote estimation of vegetation parameters using narrowband sensor for precision agriculture in arid environment. Egypt J Soil Sci 55

Overgaard HJ, Tsuda Y, Suwonkerd W, Takagi M (2001) Characteristics of *Anopheles minimus* (Diptera: Culicidae) larval habitats in northern Thailand. Environ Entomol 10:134–141

Oyewole IO, Momoh OO, Anyasor GN, Ogunnowo AA, Ibidapo CA, Oduola OA, Awolola TS (2009) Physico-chemical characteristics of Anopheles breeding sites: impact on fecundity and progeny development. Afr J Environ Sci Technol 3(12):447–452

Piyaratnea MK, Amerasinghea FP, Amerasinghea PH, Kon-Radsena F (2005) Physico-chemical characteristics of Anopheles culicifacies and Anopheles varuna breeding water in a dry zone stream in Sri Lanka. J Vector Borne Dis 42:61–67

Rogers DJ, Randolph SE, Snow RW, Hay SI (2002) Satellite imagery in the study and forecast of malaria. Nature 415:710–715

Savitzky A, Golay MJE (1964) Smoothing and differentiation of data by simplified least squares procedures. Anal Chem 36:1627–1639

Shaman J, Day JF, Komar N (2010) Hydrologic conditions describe West Nile virus risk in Colorado. Int J Environ Res Public Health 7:494–508

Simon Oke IA, Ayani FE (2015) Relative abundance and composition of endophilic mosquitoes in Federal university of Technology Akure Hostels, Ondo state, Nigeria. Appl Sci Rep 10(3): 133–136

Smith RB (2001) Introduction to hyperspectral imaging. www.microimages.com

Suez Canal guide (2010) https://www.atlas.com.eg/scg.html. Marine Services Co. Retrieved 2 Apr

Sowilem M, El-Zeiny A, Atwa W, Elshaier M, El-Hefni A (2017a) Assessing and monitoring spatiotemporal distribution of mosquito habitats, Suez Canal Zone. Asian J Environ Ecol 4(2): 1–13

Sowilem M, Elshaier M, Atwa W, El-Zeiny A, El-Hefni A (2017b) Species composition and relative abundance of mosquito larvae in Suez Canal Zone, Egypt. Asian J Biol 3(3):1–12

Sunahara T, Ishizaka K, Mogi M (2002) Habitat size: a factor for determining the opportunity for encounters between mosquito larvae and aquatic predators. J Vector Ecol 27(1):8–20

Tabbabi A, Boussès P, Rhim A, Brengues C, Daaboub J, Bouafif N, Fontenille D, Bouratbine A, Simard F, Aoun K (2015) Larval habitats characterization and species composition of *Anopheles* mosquitoes in Tunisia, with particular attention to *Anopheles maculipennis* complex. Accepted for Publication, Published online January 5, 2015. https://doi.org/10.4269/ajtmh.14-0513

Tadesse D, Mekonnen Y, Tsehaye A (2011) Characterization of mosquito breeding sites in and in the vicinity of Tigraymicrodams. Ethiop J Health Sci 21(1):57–66

Thenkabail PS, Enclona EA, Ashton MS (2004) Accuracy assessment of hyperspectral waveband performance for vegetation analysis applications. Remote Sens Environ 91:354–376

Thenkabail PS, Gumma MK, Teluguntla P, Mohammed IA (2014) Hyperspectral remote sensing of vegetation and agricultural crops. Photogram Eng Remote Sens 80:697–709

Tran A, Ippoliti C, Balenghien T, Conte A, Gely M, Calistri P, Goffredo M, Baldet T, Chevalier V (2013) A geographical information system-based multicriteria evaluation to map areas at risk for Rift Valley fever vector-borne transmission in Italy. Transboundary Emerg Dis 60:14–23

Varun T, Sharma A, Yadav R, Agrawal OP, Sukumaran D, Vijay V (2013) Characteristics of the larval breeding sites of Anopheles culicifacies sibling species in Madhya Pradesh. Indian J Med Res 1(5):47–53

Washino RK, Wood BL (1994) Application of remote sensing to vector arthropod surveillance and control. Am J Trop Med Hyg 50(6 Suppl):134–144

World Health Organization (WHO) (1975) Manual on Practical Entomology in Malaria. PART2, Methods of techniques, 197 pp

Zou L, Miller SN, Schmidtmann ET (2006) Mosquito larval habitat mapping using remote sensing and GIS: implications of Coalbed methane development and west Nile virus. J Med Entomol 43(5):1034–1041

Part III
Water Quality and Pollution

Chapter 8
Satellite Image Data Integration for Groundwater Exploration in Egypt

Eman Ghoneim and Farouk El-Baz

Abstract The Egyptian desert, as a part of the Eastern Sahara, is among the driest places on Earth. In the past, this region underwent drastic climatic changes through the alternation of dry and wet conditions. During wet phases, when the rain was plentiful over a prolonged time period, the surface was veined by rivers and dotted by large lakes. The use of microwave and topographic satellite data has led to the identification of sandy buried ancient mega drainage systems, river deltas and lake basins. The long wavelength of radar data, in particular, enabled the penetration of the sands and revealed many of these hidden fluvial basins which represent indisputable evidence of the past pluvial conditions of the region. Many of these basins could hold large groundwater reservoirs as well as oil and natural gas at depth. This study shows that space-data integration is a promising approach that can help significantly in the groundwater exploration efforts in the desert land of Egypt, where freshwater access is essential for the developing of the region.

Keywords Synthetic aperture radar · Shuttle Radar Topographic Mission (SRTM) · Paleorivers · Nubian sandstone · Subsurface terrain

8.1 Introduction

Starting with the first space missions in the 1960s, orbiting astronauts were particularly fascinated by their view of the land of Egypt. They repeatedly relayed observations and photographs of the Nile River and its Delta; the band of verdant green contrasted sharply with the vast surroundings of bare desert surfaces. That contrast became even more pronounced by the advent of the multi-spectral images

E. Ghoneim (✉)
Department of Earth and Ocean Sciences, University of North Carolina Wilmington, 601 S. College Road, Wilmington, NC 28403, USA
e-mail: ghoneime@uncw.edu

F. El-Baz
Center for Remote Sensing, Boston University, 725 Commonwealth Ave., Boston, MA 02215-1401, USA
e-mail: farouk@bu.edu

S. F. Elbeih et al. (eds.), *Environmental Remote Sensing in Egypt*,
Springer Geophysics, https://doi.org/10.1007/978-3-030-39593-3_8

of Landsat, starting in 1972. The growing interest encouraged planning of further observations and photography on the Apollo-Soyuz mission of 1975 (El-Baz 1977; El-Baz and Warner 1979). One result of analyzing these satellite image data was the recognition of a triangular zone—to the southwest of the present Nile Delta—as a former delta of the river (Abdel-Rahman and El-Baz 1979). Field observations indicated that the whole triangular area was covered by round pebbles. This supported the premise that a changing course of the Nile in the geological past resulted in the abandonment of its ancient delta, prior to the formation of the present one.

As orbiting satellites began to complete the coverage of Egypt's 1,000,000 km^2, study of the rest of its features began in earnest. In particular, studies concentrated on the origin and evolution of sand dunes, which cover over 70% of the Western Desert of Egypt (Gifford et al. 1979). Its most extensive accumulation of linear dunes is known as the Great Sand Sea, which dominates the western half of the desert. Its dune masses, as well as all smaller accumulations of sand in this desert, consistently move southward along the prevailing wind (Bagnold 1941; Wolfe and El-Baz 1979; El-Baz 1998). That fact initiated discussion of where did all the sand in the vast dune fields and sand sheets originated. That question had no obvious answer, because from where the wind blew all exposed rocks were of limestone, which could not have been the source of the quartz sand. The only possible source of the sand grains throughout the desert was the so called Nubian Sandstone, which dominated the land to the south of all major sand accumulations (El-Baz 2000).

That geologic puzzle initiated the search for clues to uncover the reason. It was confirmed that the sand originated by the erosion of the Nubian Sandstone (El-Baz 1998). Rushing water from excessive rain in the geological past caused its breakup up to release individual grains. The latter particles would have been transported northward within the courses of the rushing rivers and streams to be deposited in low areas, north of the present-day accumulations of sand. Such low areas would have housed lakes within low topographic depressions. Most of those would have been located in the northern part of the desert, as the rivers and streams followed the tilt of North Africa. When the climate dried up, probably due to persistent northward movement of the African continent (as later explained by Abdelkareem and El-Baz 2017), the wind took over to initiate the southward transport of the sand grains, to whence they originated, which continues to this day.

In the extreme southern part of the Western Desert of Egypt there existed an exception to the aforementioned rule. An extensive flat area of surface sand dominated the border area with Sudan. It was named the Selima Sand Sheet, as it covered an area to the west of the only named features in the vicinity, the Selima Oasis in northwestern Sudan, just to the south of the border of Egypt. Here, archeological studies provided evidence of animal and human populations over an extended period of time (Wendorf and Schild 1980; Wendorf et al. 1987; Haynes 1985). Vast tracks were described to host human remains such as milling and grinding stones, hand axes and blades made of solid quartzite, along with numerous remains of ostrich eggshells and other animal-remains. These artifacts clearly pointed to the potential of extensive savanna-like environment that allowed the accumulation of enough water on the surface to form a lake that must have been sustained by wet climate periods. All such discoveries remained in the realm of interesting geologic literature.

Then, in its mission in 1981, NASA's space shuttle SIR-A mission returned images of small and large dry rivers (to the south of the region in southwestern Egypt and northwestern Sudan) that have long dried up and covered by sand (McCauley et al. 1982; Elachi and Granger 1982). During the same year, it was suggested to the Ministry of Agriculture in Egypt to drill a test well in the flat region Hence, the theory of the northward tilt of the Nubian Sandstone, and thus, the water had long flowed to the far north was still effective. Repeated requests resulted in drilling a test well in 1995, which proved the existence of massive amounts of fresh water, in the "East Oweinat" region as discussed below. In addition, they also encouraged use of the space-born data to map details of the topography of the region in the past as well as the present. This was facilitated by imaging from space utilizing a variety of instruments including those that depicted topographic variations (Ghoneim and El-Baz 2007a, b). The result represented a most-appropriate way to map sand-hidden topographic features in sand-covered desert areas. It also allowed establishing the mode of formation of the fluvial features, which resulted in the formation and transportation of the desert sand. The methodologies also allowed a better understanding of the interplay between these fluvial features and the aeolian forces that resulted in shaping, then distributing desert sands.

8.2 Near-Surface Imaging

Unlike optical sensors that receive the natural spectral radiation of the Earth's surface at a given wavelength, radar images the same surface by transmitting a microwave signal which interacts with the surface that reflects it back to the receiving antenna (Robinson et al. 2006). An important property of imaging radar is its ability to penetrate sand-covered surfaces and produce images of the shallow subsurface terrain. This theory was first suggested in 1975 by Roth and Elachi (1975). For imaging such near surface features in desert regions, however, certain conditions must be met concurrently:

I. The surface material must be fine grained (relative to the wavelength), whereas the subsurface material should be rough enough to generate backscatter. Surface material has to be physically homogeneous in order for the radar signal to penetrate without significant attenuation (Roth and Elachi 1975). For grain sizes larger than 1/5 of the wavelength, attenuation is substantial, but is small for grain sizes less than 1/10 of the wavelength (Roth and Elachi 1975).
II. The surface material needs to be extremely dry with a moisture content of less than 1% (Hoekstra and Delaney 1974). The depth of subsurface imaging will vary according to the exact moisture content of the desert sand at the time of imaging. Observation by Schaber et al. (1997) and Robinson et al. (1999) indicated that

dry desert sand has a depth of 0.3 m for radar X-band, 0.5 m for radar C-band and 2–5 m for radar L-band.

Common radar sensors that were used most recently for groundwater exploration in the Great Sahara include Radarsat-1, ALOS-Palsar and Shuttle Radar Topographic Mission (SRTM). Radarsat-1 is a Synthetic Aperture Radar (SAR) imaging system that uses a single frequency, 5.8 cm horizontally (HH) polarized C band with resolutions from 10 to 100 m (Parashar et al. 1993). The ALOS PALSAR is a SAR imaging system that has variable resolution and polarization including two fine beam modes (single polarization and dual polarization) and quad polarization. SAR data have the capability of imaging sand buried rivers and lake basins in the Sahara. Since it is difficult to determine whether the mapped fluvial features are surface or subsurface both radar and multispectral datasets should be used together to correctly identify in which surface they lie (Ghoneim et al. 2007; Robinson et al. 2017).

Unlike the SAR systems that image the surface and sub-surface roughness, the SRTM system images the topography of the land. SRTM data are distributed by the USGS EROS Data Center (EDC), as part of an international project spearheaded by the National Geospatial-Intelligence Agency (NGA) and the National Aeronautics and Space Administration (NASA). These data are available at 1 arc-second (30 m) and also sampled at 3 arc-second (90 m). Studies showed that SRTM DEM data images the paleotopography and not the upper surface of the sand. This is because the phase signal processed to retrieve height information penetrates the Sahara sand cover in exactly the same way as the magnitude signal used to produce SAR images (Ghoneim and El-Baz 2007a, b). Because of such capability, it is suggested that RADAR-derived DEMs should be used in place of optically-derived elevation data (e.g. ASTER and LiDAR) in mapping desert hydrology and paleotopography (Ghoneim et al. 2007, 2012; Robinson et al. 2017).

8.3 Case Studies

Several remote sensing studies have been conducted over the past four decades in attempts to define potential locations for near-surface water accumulation in the Egyptian deserts. In these studies different remote sensing data types and geospatial techniques were adopted and utilized. Below are examples of successful remote sensing research efforts throughout the broad Egyptian deserts.

8.3.1 Mega Paleorivers of West and Southwest Egypt

8.3.1.1 Tushka Paleoriver System

In southwestern Egypt and northwestern Sudan, lies a large flat sand covered area, some 300 km^2 (El-Baz 2008) known as the Great Selima Sand Sheet (Bagnold 1931). This area is named after the Selima Oasis on its eastern border, a prominent location along the Darb El-Arbain (the 40-day track) of camel caravans from Darfur in northwestern Sudan to the Nile Valley in Egypt (El-Baz 2008). Large buried segments of ancient river courses southwest of the Selima Sand Sheet were first captures by the first Space-borne Imaging Radar (SIR-A) (McCauley et al. 1982). El-Baz et al. (2000) and Robinson et al. (2000), based on multispectral and Radar SIR-C images, have indicated that where the Selima Sand Sheet lies, there is a closed basin that received water from the surrounding highlands. The flow direction of these ancient rivers resulted in a continuing controversy among scholars who have studied these river systems. A much recent study by Ghoneim and El-Baz (2007a, b), using the Shuttle Radar Topography Mission (SRTM) and GIS hydrological modeling routine, delineated the entire paleo-watershed of the Selima area (Fig. 8.1). This ancient fluvial system, named the Tushka mega-watershed, represents the first attempt to map one of the mega drainage basins of the Great Sahara and has provided the first complete paleo-drainage map of the region. The Tushka mega-basin, today is masked by a sand cover, has a large catchment area of about 150,000 (closer to the size of Tunisia). The basin is composed of four main tributaries that emerge from

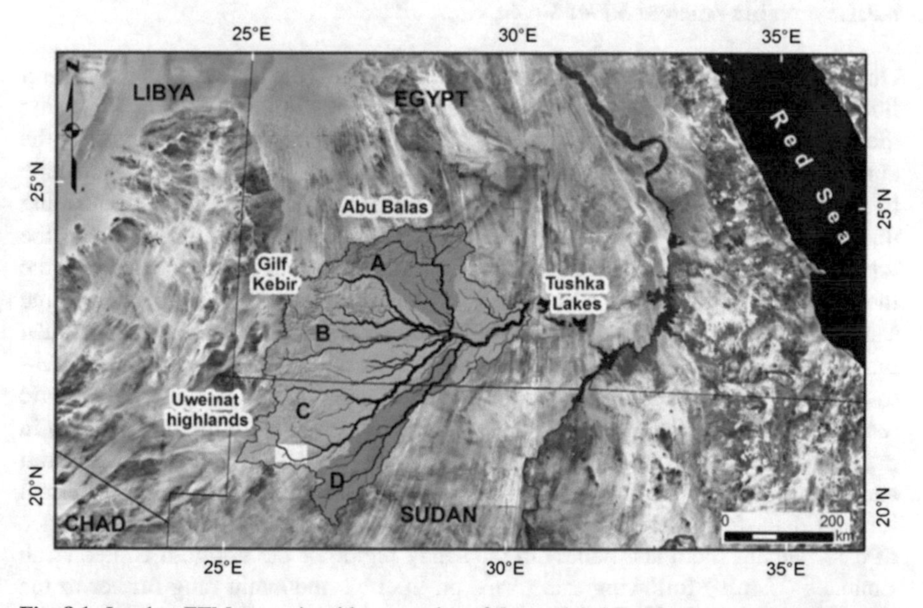

Fig. 8.1 Landsat ETM+ mosaic with an overlay of the modeled Tushka Basin and its four main sub-catchments (labeled A, B, C, and D) (After Ghoneim and El-Baz 2007a)

the highlands of Abu-Balas and Gilf Kebir plateaus, the Uweinat Mountain and the northwestern highlands of Sudan. The main channel courses of the four tributaries, which are several kilometers wide, used to collected water from a vast catchment region and drained eastward where they converged into a mainstream course. The downstream area is marked by a gentle slope surface and the presence of several exposures of late Pleistocene lake (playa) deposits, associated with Early to Mid-Paleolithic archeological sites, at Bir-Tarfawi and Bir-Sahara (Wendorf 1977). These playa deposits, which today stand 2–3 m above the surrounding surface, are among one of the largest lake deposits in Egypt (Embabi 2004). These lacustrine deposits, together with the numerous truncated spring-vents of more than 50 m in diameter (Wendorf 1977), are all testify to the presence of large lakes in this specific location of the Tushka mega-basin during wet climate conditions of the past (Ghoneim and El-Baz 2007a, b).

The downstream area of the Tushka mega-basin has been proven to contain large amounts of groundwater at depth. Today, over 500 wells were drilled by the Egyptian government within this "East Uweinat" region, to water agricultural fields using circular, spray irrigation (El-Baz 2010). The products include wheat, peanuts and other basic food crops. The wheat, in particular has proven essential for flour production in the mills of Aswan for bread that is distributed in towns of southern Egypt. The proven water resources are capable of supporting agriculture over 150,000 acres for at least 100 years (El-Baz 1988; Robinson et al. 1999, 2000; Ghoneim and El-Baz 2007a, b).

8.3.1.2 Nabta Ancient River Basin

On the southeastern side of the Tushka mega-basin, a new old river system of northern flow direction, named the Nabta basin, was mapped by Robinson et al. (2017) using the SRTM DEM, Radarsat-1 and ASTER Thermal Infrared data. In that study, the authors reported that SRTM data (although of coarser resolution at 90 m), were able to capture subsurface drainage better than the finer (15 m) ASTER-DEM confirming that SRTM maps paleotopography whereas the optical ASTER datasets map the topography of the surface of the sand. The same study has also indicated that the thermal wavelength can be used successfully in arid desert regions to detect some sand-buried channels in the same way that radar data did. The study concluded that thermal infrared spectral channels of ASTER is favored over Landsat ETM+ to take advantage of the better calibration, and increased spectral and radiometric resolutions (Robinson et al. 2017) (Fig. 8.2). In fact, the role of thermal imagery in providing information relevant to hydrogeology has been clearly documented (Pratt and Ellyett 1979; Ghoneim 2008; El-Baz 2008). Using Aster and MODIS thermal infrared data, for example, by Ghoneim (2008) revealed the frequent appearance of dark patches (cool anomalies) in the sandy region of the northern United Arab Emirates (U.A.E.) following rainstorms on the high mountain rang further to the east. These cool anomalies, which are invisible in visible wavelengths, indicate water accumulation near the surface. This is because the latent heat content of water slows

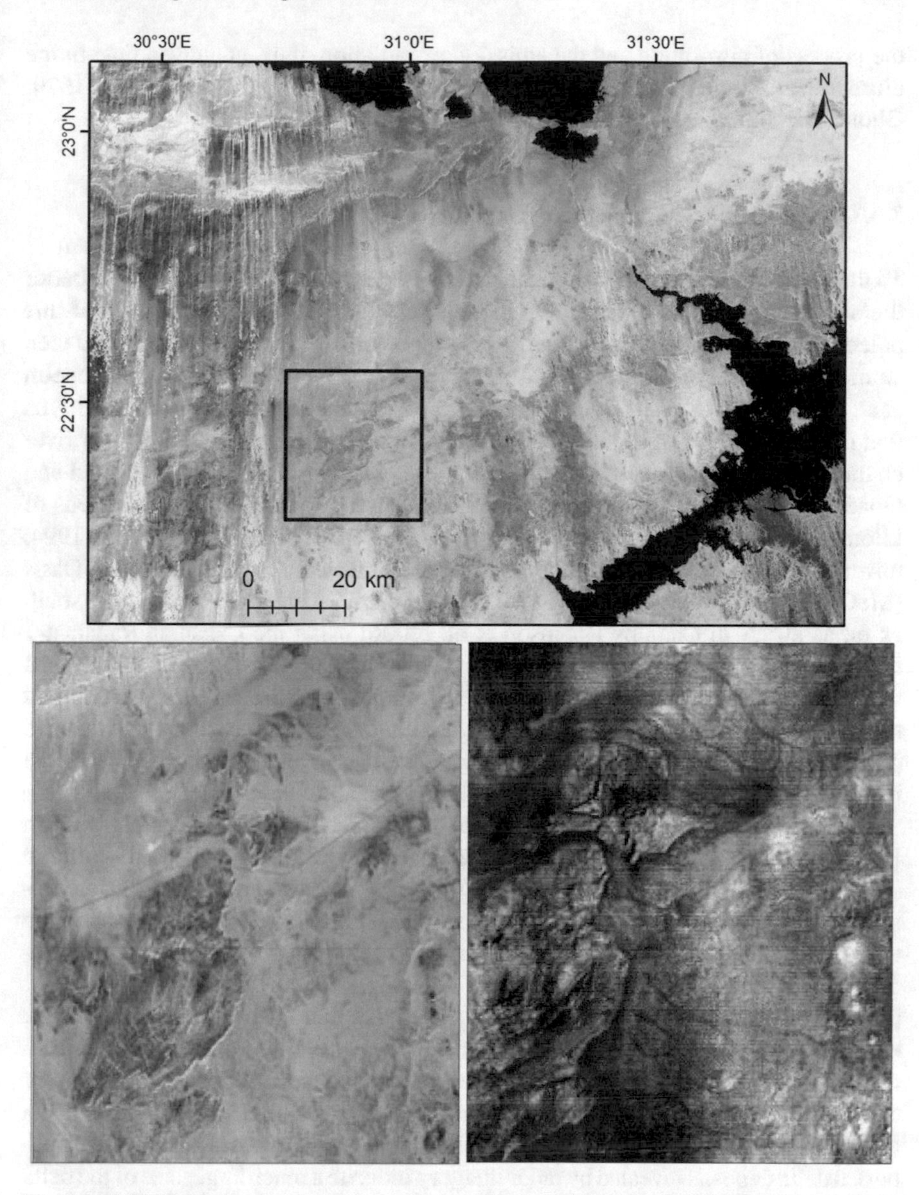

Fig. 8.2 ASTER satellite images of the study area (Top). It shows the visible ASTER image (left) and the thermal infrared image (right). Drainage paleoriver courses are obvious as dark tones in the thermal infrared scene however are not clear in the visible bands. This implies that thermal infrared wavelengths image near-surface paleochannels (After Robinson et al. 2017)

the process of absorption and the emission of radiation, thus, at a given time in the diurnal heating cycle, slowing the warming of moist soil (Pratt and Ellyett 1979; Ghoneim 2008; El-Baz 2008).

8.3.1.3 Kufrah Mega Paleoriver System

To the west of the Tushka Megabasin, a massive paleoriver system, almost one quarter the size of Egypt, was mapped by Ghoneim et al. (2012). Relict channels of this paleoriver, named Kufrah Megariver, were first identified by Pachur (1993), and then studied by several researchers (e.g., McCauley et al. 1995; El-Baz 1998; Ghoneim et al. 2007; Drake et al. 2008; Paillou et al. 2009). It was suggested that in late 1970s that the groundwater well fields in the Kufrah Oasis area are located in former river channels that are being recharged by rainfall in the Tibesti Mountains (Ahmad and Goad 1978). Pachur (1993) traced this north-trending river course for hundreds of kilometers south to 22° N. NASA SIR-C radar imagery, which was acquired in 1994, unveiled two massive tributaries of this river draining north toward the Kufrah Oasis (McCauley et al. 1995; El-Baz 1998). These tributaries were delineated farther south as far as 20° N in Chad by Robinson et al. (2006) using the Canadian Radarsat-1 imagery, and then was validated by Ghoneim et al. (2007) using the SRTM DEM hydrological delineation routine. A number of studies using the same radar and digital topography dataset have afterward confirmed that the Kufrah basin is an extensive paleoriver that drained most of the eastern flank of Libya and terminated with a large inland delta (Ghoneim and El-Baz 2008; Drake et al. 2008; Paillou et al. 2009).

Nonetheless, none of these studies have determined the full extent of the Kufrah Paleoriver and its drainage network. The later study by Ghoneim et al. (2012) offered a comprehensive picture of the entire Kufrah paleoriver system through the integration of different types of geospatial data in a GIS (Fig. 8.3). This study mapped the entire catchment of the Kufrah Paleoriver for the first time by incorporating Standard mode C-band 78 Radarsat-1 scenes, several SRTM DEM tiles, and Landsat-7 (ETM+) images. The river system flow northward and span over four countries: eastern Libya, western Egypt, northeastern Chad and northwestern Sudan. The paleoriver terminates with a large inland delta located where the Kufrah Paleoriver exited the plateau at Jabal Dalma Plateau. About two thirds of the Kufra Delta (34,000 km^2) is exposed in optical satellite imagery. However; remnant of sand-buried delta deposits revealed by radar imagery indicate a much larger size of the delta (about 47,500 km^2) (see Fig. 8.3).

This massive delta system was most likely developed during the Pliocene, when humid conditions would have promoted perennial streamflow and low energy fluvial environments (Ghoneim et al. 2012). This inland delta was probably largely similar to the modern day inland deltas of the Okavango and Niger rivers in Africa (McCarthy 1993). SRTM data reveal a deep elongated topographic basin at the distal foot of the Kufrah alluvial delta. This basin is filled by approximately 45 m of unconsolidated quartz sand with thin beds of sandy clay (Di Cesare et al. 1963). The upper part of this formation is of Upper Pliocene or Pleistocene in age and contains specific types

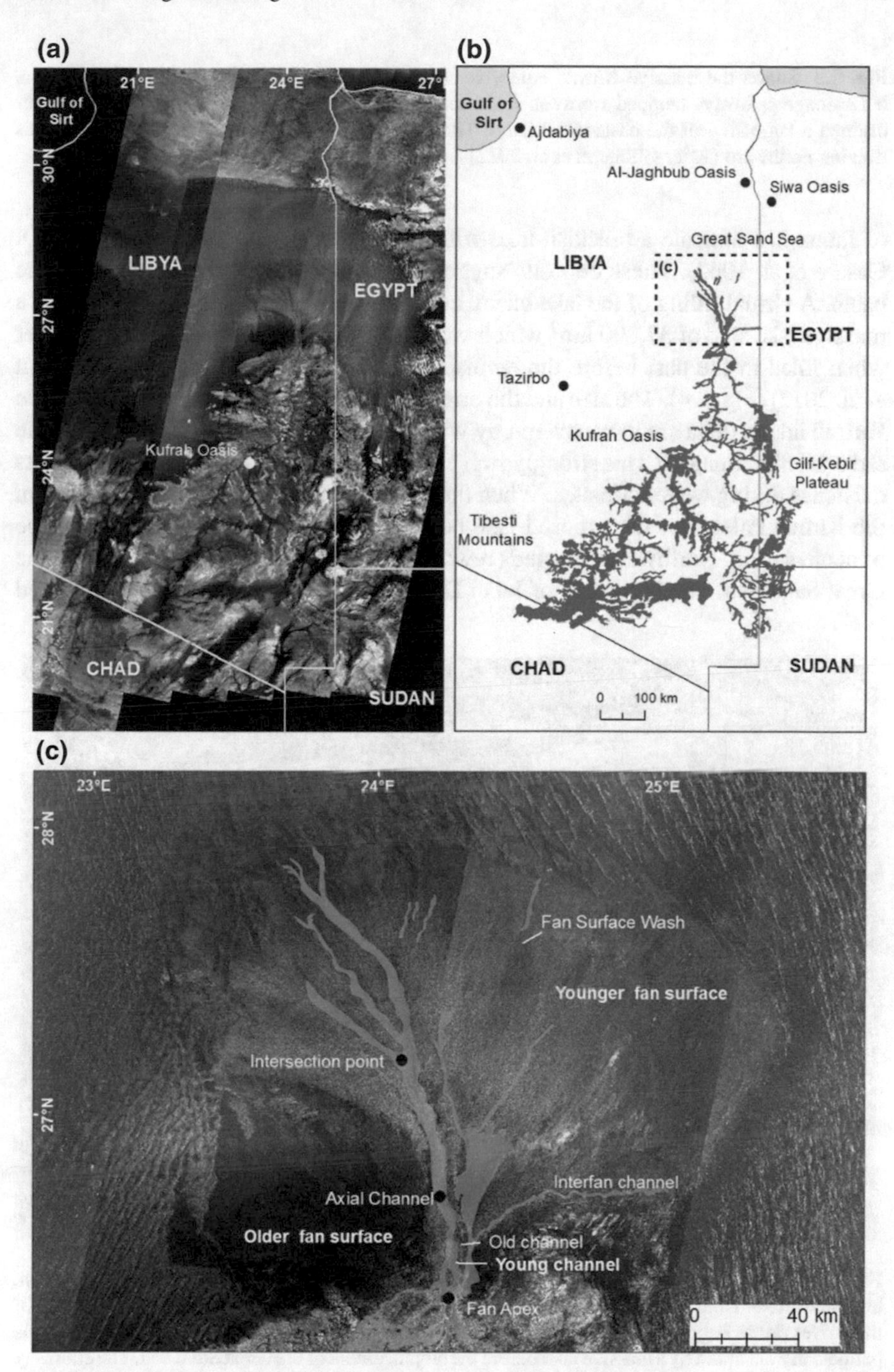
(a)
Gulf of Sirt
21°E
24°E
27°
30°N
27°N
24°N
21°N
LIBYA
EGYPT
Kufrah Oasis
CHAD
SUDAN
(b)
Gulf of Sirt
Ajdabiya
Al-Jaghbub Oasis
Siwa Oasis
Great Sand Sea
LIBYA
(c)
EGYPT
Tazirbo
Kufrah Oasis
Gilf-Kebir Plateau
Tibesti Mountains
CHAD
SUDAN
0 100 km
(c)
23°E
24°E
25°E
28°N
27°N
Fan Surface Wash
Younger fan surface
Intersection point
Interfan channel
Axial Channel
Older fan surface
Old channel
Young channel
Fan Apex
0 40 km

◂**Fig. 8.3** Shows the massive Kufrah Paleoriver system. **a** Radarsat-1 mosaic of the study area. **b** Drainage pathways mapped from the Radarsat-1 mosaic. It shows that the Kufrah Paleoriver drained a large area of the eastern Sahara. **c** The Kufrah inland delta with several distributaries flowing northward (After Ghoneim et al. 2012)

of fauna that indicate a brackish lake with limited connection to the open sea (Di Cesare et al. 1963). These deposits suggest that a vast lake formerly occupied the basin. A virtual filling of the lake basin, using SRTM topographic data, estimated a massive lake size of 37,800 km^2 which would have held around 3300 km^3 of water when filled in the past before the emplacement of the Great Sand Sea (Ghoneim et al. 2012) (Fig. 8.4). The size and the orientation of the distributary channels of the Kufrah inland delta are now covered by windblown sand, and are only visible in the radar satellite imagery This strongly suggest that they fed the massive lake during its existence during wetter climates. When the lake dried up due to reduced runoff from the Kufrah Paleoriver the exposed lake beds would have served as a major source of aeolian sand. Northwesterly winds reworked these dried lake deposits to form the Great Sand Sea. The highlands of Jabal Dalma and the Gilf-Kebir Plateaus limited

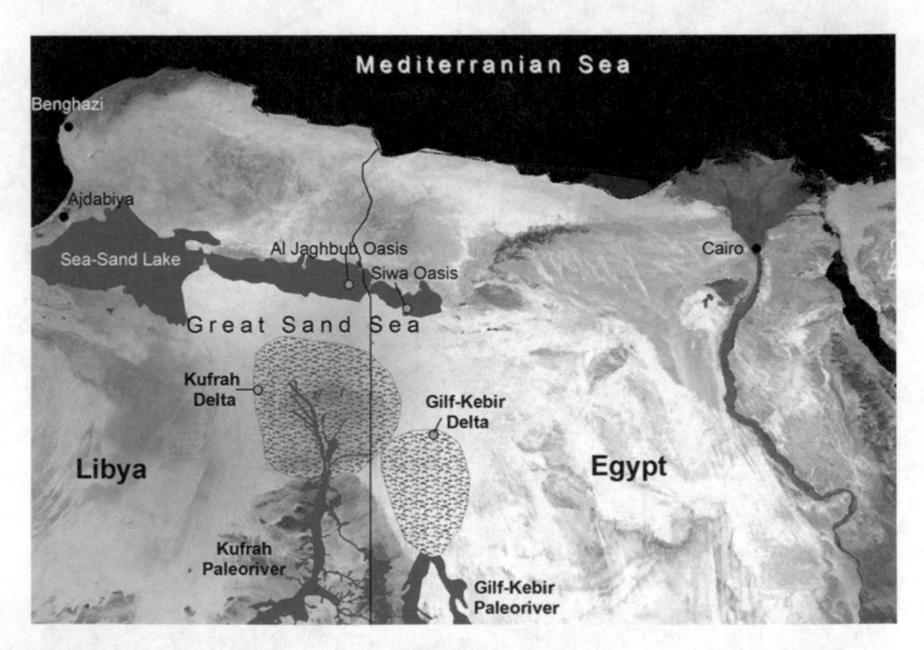

Fig. 8.4 Shows both the Kufrah and Gilf paleoriver systems. These rivers emanate from the south, trends northward and terminates with vast inland deltas. The northern trend of the distributaries of these river deltas suggest that both rivers would have drained to a topographic depression that was periodically occupied by a massive lake before the emplacement of the Great Sand Sea. The enormity of these paleofluvial systems and the lake basin provides an explanation as to why continuous extraction of groundwater in the oases of Siwa and Al-Jaghbub is possible (After Ghoneim 2011)

the southward transport of the sand dune field, enhancing the thick accumulations of sand that characterize the modern Great Sand Sea (Ghoneim et al. 2012).

The delineated paleoriver has a 950 km long main river course and a watershed area of 236,000 km^2 and, as a consequence, this is the largest paleoriver system yet identified and mapped in the Eastern Sahara of North Africa (Ghoneim et al. 2012). It is much larger than the neighboring Tushka, Northern Darfur and Howar paleoriver systems delineated by Ghoneim and El-Baz (2007a, b, 2008), respectively. Ghoneim et al. (2012) concluded that the Kufrah Paleoriver, which was the dominant drainage system in the region during the Pliocene and Pleistocene was most likely developed by progressive capture of the Sahabi River System that dominated the area in the Miocene (Griffin 2002, 2006). The same study revealed an alignment between one of the Kufrah main river branches and a paleoriver system in Chad. This could indicate that the Kufrah paleoriver may have served as an outlet from Megalake Chad to the Mediterranean Sea during wet phases of the Neogene and acted as one of the natural corridors used by human and animal populations to cross the Sahara during the Pleistocene (Fig. 8.5).

To the east of the Kufrah basin, a large paleoriver system, named Gilf Paleoriver, was mapped in Egypt by Ghoneim (2011). The two main drainage courses of this river are enormous in size and both emerge from the Gilf-Kebir Plateau. The river flow northward along the western flank of Egypt and terminates with a large inland delta south to Siwa Oasis (Ghoneim 2011) (see Fig. 8.4).

8.3.2 The Paleorivers of the Nile Valley in Egypt

8.3.2.1 The Wadi Kubbaniya Paleoriver System

To the northeast of the Tushka–Nabta Paleoriver systems, west of the Nile River bank, sand buried segment of a broad ancient river system was first discovered during a field visit by Butzer and Hansen (1968). This river segment was reported to be very rich in archeological sites mostly of middle and late Paleolithic age (Wendorf and Schild 1986). The complete paleoriver course was later mapped using various types of microwave satellite imagery including SIR-C, Radarsat-1 and ALOS-PALSAR (Hinz et al. 2003; Thurmond et al. 2004; Ghoneim et al. 2007; Gaber et al. 2015). The paleoriver, known today as Wadi Kubbaniya, is located in the Gallaba Plain area, northwest of Aswan in Upper Egypt. Wadi Kubbaniya enters the Nile Valley from the northwest through a narrow gorge, just opposite to Wadi Abu-Subeira on the eastern side of the Nile River, which is a large deep river course draining the northeastern portion of the Red Sea Hills. It is suggested that Wadi Kubbaniya was the early downstream continuation of Wadi Kharit and Wadi Abu-Subeira (Butzer and Hansen 1968), where they formed an east–west river that drained the Red Sea Hills in the east. Nowadays, Wadi Kubbaniya flows southeast into the Nile River, however, in late-Pliocene and early-Pleistocene time this drainage system may have flowed into the opposite direction. In the mid-late Pleistocene, this river was disrupted

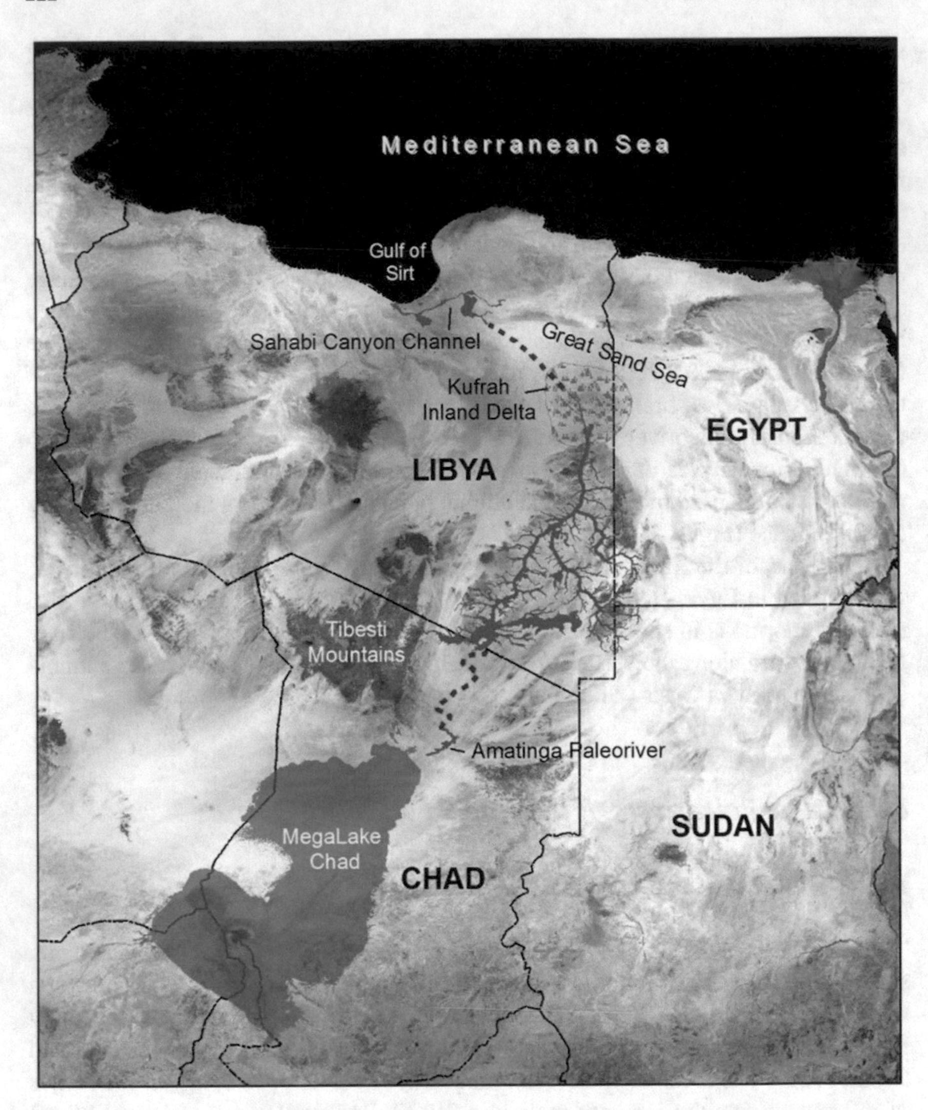

Fig. 8.5 Proposed outlet from Megalake Chad to the Mediterranean Sea through the Kufrah Paleoriver. The channel may have been active during high lake stages of the Pliocene or early Pleistocene (After Ghoneim et al. 2012)

by today's northward flowing Nile (Hinz et al. 2003) and reversed its flow direction towards the southeast to join the Nile River. This ancient river system (termed also as the Kubbaniya Nile), of Plio-Pleistocene age, was most likely responsible for the huge expanse of the Gallaba Gravel Plain deposits along the western bank of the Nile River (Issawi and Hinnawi 1980).

Field investigation near the river mouth showed that the wadi is bounded on both sides by nearly vertical scarps of Nubian sandstone, which stand from 30 to

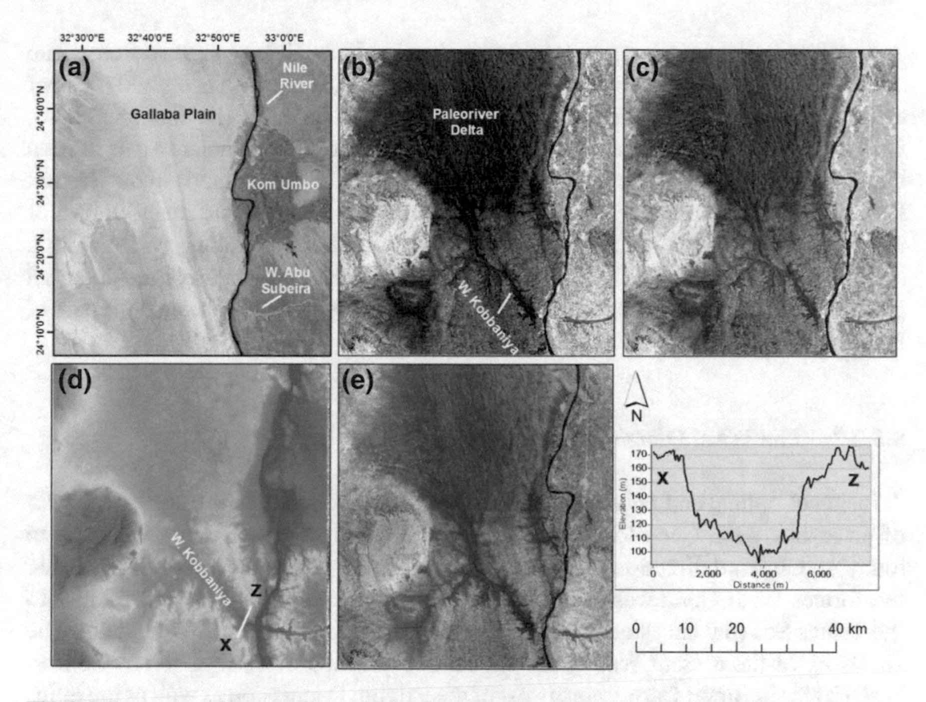

Fig. 8.6 Shows **a** multispectral Landsat-8 image, **b** Radarsat-1 microwave date, **c** Landsat-Radarsat fused image, **d** SRTM-30 m topographic data, **e** combined elevation with microwave image. Integration of space data clearly shows the paleoriver course Wadi Kubbaniya that is suggested to be a western continuation of Wadi Abu Subeira before it was disrupted by today's northward flowing of the Nile River. The 50 m deep paleoriver course (illustrated in the topographic profile) and the massive alluvial delta is barely visible in the multispectral Landsat image

50 m above the floor of the wadi (Wendorf and Schild 1986). A profile depicted with SRTM data (see Fig. 8.6) shows that this paleo-river is nearly 3 km wide and 50 m deep which provide further evidence of the near-surface imaging capabilities of the SRTM (Ghoneim et al. 2007). The northerly winds carried sand from the neighboring, barren sandstone uplands down into the ancient river course, where it was trapped and built extensive dune-fields (Wendorf and Schild 1986). Today, these sand dunes mask the surface and make it hard to identify the entire river course in multispectral satellite images. The entire paleoriver course of the Wadi Kubbaniya, however, was revealed using different types of microwave satellite imagery including SIR-C, Radarsat-1, SRTM and ALOS-PALSAR (Hinz et al. 2003; Thurmond et al. 2004; Ghoneim et al. 2007; Gaber et al. 2015) (Fig. 8.6). A northwest flow direction can be inferred from the SRTM data and radar images since several large tributaries flows northwesterly into the main Wadi Kubbaniya river course nearly at right angles (Ghoneim et al. 2007). The reversal in flow direction was caused by non-tectonic Nile River capture and the northward regional tilt. Another small southeast former palaeo-channel, which too drains towards the Nile, is also visible in the SRTM 7 km north of Wadi Kubbaniya (Ghoneim et al. 2007).

Satellite Radarsat-1 imagers revealed a massive alluvial delta at the downstream area of Wadi El-Kubanyia paleoriver system (Fig. 8.6), (Ghoneim et al. 2007; Gaber et al. 2011). This river delta is now buried under a vast flat area, (the Gallaba Plain), which is invisible in optical satellite images (Fig. 8.6). This ancient river system may have carried large quantities of water from the Red Sea highlands in the Eastern Desert west into the Gallaba gravel plain (Hinz et al. 2003). The massive size of the Wadi El-Kubanyia main river course and its alluvial delta along with the long history of fluvial processes in the region has a significant implication of the potential groundwater resources and future agricultural development in this specific part of Egypt.

8.3.2.2 The Wadi Qena Paleoriver System

A fusion of optical and radar satellite data led to the proposition of a northward flow of an earlier Wadi Qena River, following the general slope of Africa, opposite to today's southward direction (Abdelkareem et al. 2010). With the aid of SRTM data, the former Wadi Qena was suggested to flow northward (Fig. 8.7) and drain into the Tethys Sea (the ancestral Mediterranean Sea), with a massive alluvial delta. The similarity of the present Wadi Qena fluvial deposits and those reported by Seiffert et al. (2008) in Birket Qarun, northwest of the Faiyum Depression as well as the relics of a large paleochannel that drained northward at south Wadi Qena supported this hypothesis. The northward flow by the Qena River was hindered later by the tectonic uplift associated with Red Sea rifting, which forced the ancestral Qena Paleoriver to divert its pathway southward.

8.3.3 The Paleorivers of the Sinai Peninsula

8.3.3.1 Wadi El-Arish River Basin

The integration of a Radarsat-1 image with SRTM data in synergy with satellite multispectral images and field investigations resulted in the mapping of the paleoriver course of Wadi El-Arish in the northern part of Sinai Peninsula (AbuBakr et al. 2013). A buried segment of the ancestral drainage course was discovered beneath the windblown Sand far west to the present downstream area of Wadi El-Arish indicating a northwestern flow direction of the former river system. This river segment was later dammed as a result of a structural uplift? (anticlinal fold) activity in the area which blocked the northwestern pathway of the river and forced its flow to deviate to the present northeastern direction (Fig. 8.8). A DEM-based simulation was performed to remove the effect of the tectonic uplift and delineate the former drainage system of Wadi El-Arish before the diversion of its drainage course. Data analysis indicated the presence of vast plane, called the El-Sirr and El-Qawarir, north of the structural uplifted area. This plain contains the former buried river course of Wadi El-Arish,

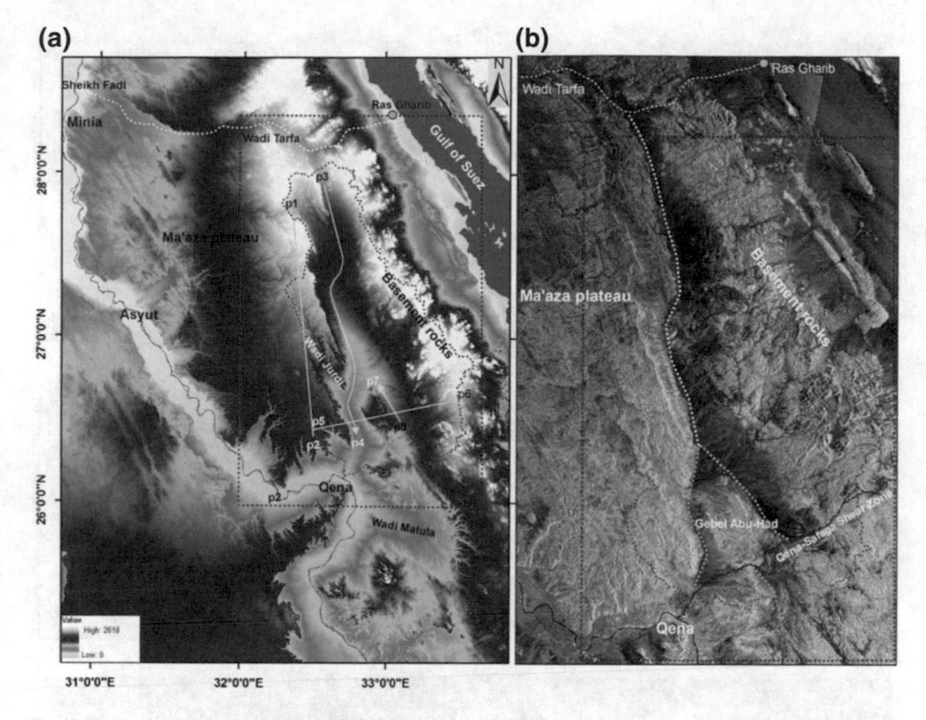

Fig. 8.7 Shows Wadi Qena basin. **a** SRTM data and **b** Radarsat-1 imagery reveal a broad headwater area of Wadi Qena, Radarsat-1 image showing the proposed Wadi Qena River trace (dashed green line), present Wadi Qena channel (dashed cyan line). The Radarsat-1 images highlights the main course with strong radar backscatter (dark color) extending beyond the northern most end of the Qena valley (After Abdelkareem and El-Baz 2015)

and is covered by fertile soil (mostly Esna shale that mainly consist of claystone) and thus most likely represents a promising area for agricultural development within the Wadi Al-Arish basin. As a result, the study proposed to dig a 2 km long and 6 m deep canal through the anticlinal fold in order to reconnect the present course of Wadi El-Arish to its former drainage course and thus redirect the surface runoff during rainy seasons to the vast plane of El-Sirr and El-Qawarir. This proposed canal would provide water for approximately 1400 km^2 of fertile land for future agricultural development (AbuBakr et al. 2013).

8.4 Discussion

In fact, the fluvial deposits of these large rivers (Fig. 8.9) and lake basins in the Egyptian deserts served as the main aeolian sources for the formation of extensive sand dune and sand sheets. The development of the sand dunes and sand sheets in Egypt, same as everywhere else in the Great Sahara, is closely linked to the

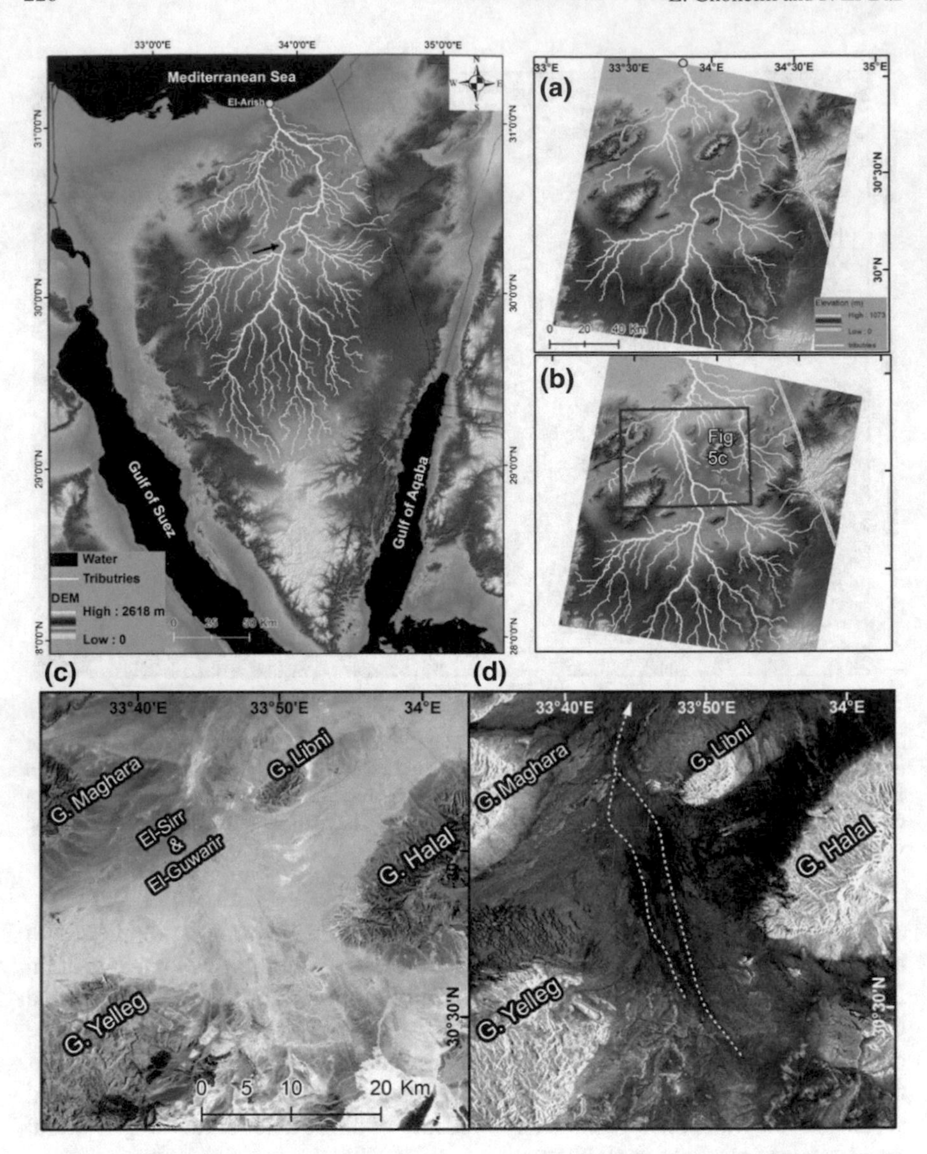

Fig. 8.8 Shows Wadi El-Arish catchment area in Sinai Peninsula. **a** Extracted drainage network derived from the original DEM showing the present course of the paleoriver. **b** Extracted drainage network derived from the modified DEM simulating the former course of Wadi El-Arish. The red box represents the area of (**c**). **c** Landsat image showing the area between the anticlinal hills of the Syrian arc system, with no surface sign of a master drainage. **d** Radarsat-1 revealing traces of two parallel tributaries trending northwest that join and flow northeast (After AbuBakr et al. 2013)

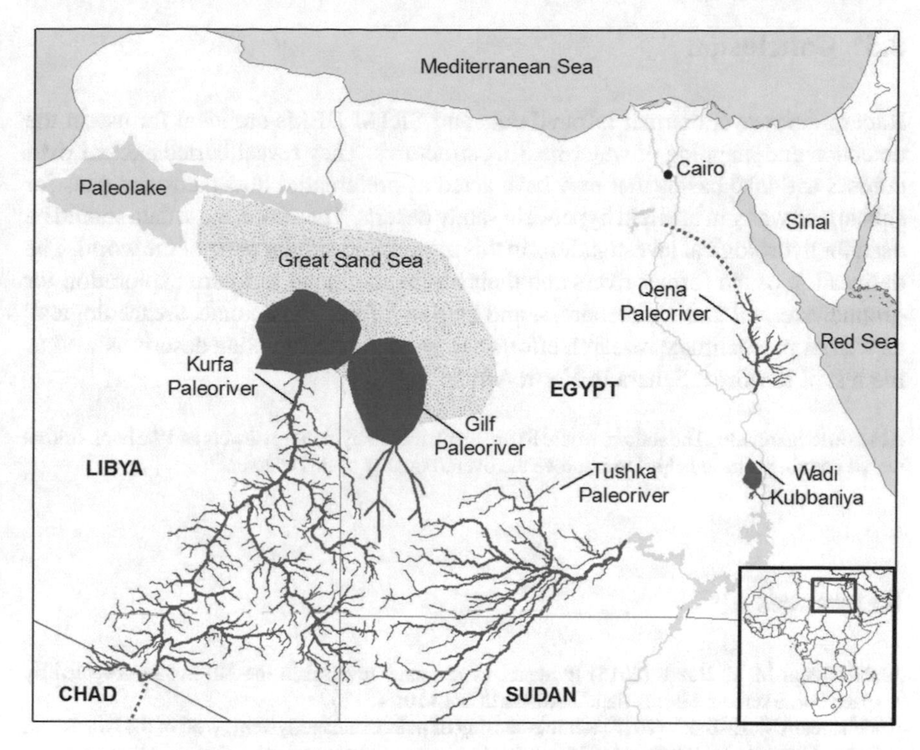

Fig. 8.9 Shows the main paleoriver systems and lake basins mapped by remote sensing and GIS in the Egyptian deserts. It is evident that the two main river systems of El-Kufra and Gilf have served as the main aeolian sources for the formation of the extensive Great Sand Sea

paleo-climatic condition of the Sahara-Mediterranean region. The climatic history of this region suggests that wet and dry phases were linked to the timing of high latitude glaciations in the Northern Hemisphere (De Menocal 2004). During the late Pliocene, when ice sheets began to form, this region became progressively drier, with wet phases associated with transitions between glacial and interglacial periods.

During humid periods, fluvial deposits from active surface runoff would have served as the key sources for aeolian deposits that later formed the vast sand cover in the region. This notion lends support to the model first proposed by Di Cesare et al. (1963) and expanded upon by El-Baz (1982). These studies emphasized that the ultimate source of aeolian sands in the Eastern Sahara is erosion of the Nubian Sandstone and Paleozoic bedrock outcrops in southern Libya and Egypt, transported northward by paleorivers that existed during humid phases of the Pliocene and Pleistocene. These deposits were later displaced southward by northerly winds that prevailed during dry phases. These areas of sand accumulation are the dominant landforms of the region today. They also have importance as they signal potential locations for groundwater accumulation (El-Baz 1998; El-Baz et al. 2000; Robinson et al. 2000; Ghoneim and El-Baz 2007a, b; Ghoneim 2008; Ghoneim et al. 2012).

8.5 Conclusion

Radar microwave, thermal infrared data and SRTM DEMs are ideal for use in the detection and mapping of water-bearing structures. They reveal buried ancient river courses and lake basins that may have acted as preferential flow paths and sites for subsurface water in arid and hyper-arid sandy deserts. Therefore, these data should be used for hydrological investigations in this region, and similar parts of the world. The delineation of the former rivers and their basins might aid to future exploration for groundwater, oil and gas resources, and help to inform neotectonic, archaeological, as well as paleoclimate research efforts across all of the Egyptian deserts as well as the rest of the Great Sahara in North Africa.

Acknowledgements The authors would like to thank the anonymous reviewers and the book editors whose comments have helped to improve the overall quality of this chapter.

References

Abdelkareem M, El-Baz F (2015) Evidence of drainage reversal in the NE Sahara revealed by space-borne remote sensing data. J Afr Earth Sci 110:245–257

Abdelkareem M, El-Baz F (2017) Remote sensing of paleodrainage systems west of the Nile River, Egypt. Geocarto Int 32(5):541–555

Abdelkareem M, Ghoneim E, El-Baz F, Askalany M (2010) Did Egypt's Wadi Qena drain northward during its earlier stages? Geol Soc Am Abstr 42(5):645

Abdel-Rahman MA, El-Baz F (1979) Detection of a probable ancestral delta of the Nile River. In: Apollo-Soyuz test project summary science report, NASA SP-412, vol II, pp 511–520

AbuBakr M, Ghoneim E, El-Baz F, Zeneldin M, Zeid S (2013) Use of radar data to unveil the paleolakes and the ancestral course of Wadi El-Arish, Sinai Peninsula, Egypt. Geomorphology 194:34–45

Ahmad M, Goad M (1978) Discovery of an ancient underground channel in the North Sarir Well Fields, Libya. In: Proceedings of symposium on investigation, exploitation, and economy of underground waters. Jagreb, Yugoslavia, pp 77–85

Bagnold RA (1931) Journeys in the Libyan Desert 1929 and 1930. Geogr J 78:13–39, 524–535

Bagnold RA (1941) The physics of blown sand and desert dunes. Methuen and Co., Ltd., London

Butzer KW, Hansen CL (1968) Desert and river in Nubia. Geomorphology and prehistoric environments at the Aswan reservoir. University of Wisconsin Press, Madison, WI

De Menocal PB (2004) African climate change and faunal evolution during the Pliocene-Pleistocene. Earth Planet Sci Lett 220:3–24

Di Cesare F, Fraanchino A, Sommaruga C (1963) The Pliocene-Quaternary of Giarabub Erg region. Revue de L'Inststut Francais du Pétrole 18:1344–1362

Drake NA, El-Hawat AS, Turner P, Armitage SJ, Salem MJ, White KH et al (2008) Palaeohydrology of the Fazzan Basin and surrounding regions: the last 7 million years. Palaeogeogr Palaeoclimatol Palaeoecol 263:131–145

Elachi C, Granger J (1982) Space-borne imaging radars probe "in depth". IEEE Spectr 19:24–29

El-Baz F (1977) Astronaut observations from the Apollo-Soyuz mission. Smithsonian Institution Press, Washington D.C.

El-Baz F (1982) Genesis of the Great Sand Sea, Western Desert of Egypt. In: 11th international conference. International Association of Sediment, Hamilton, Ontario, Canada, p 68

El-Baz F (1988) Origin and evolution of the desert. Interdisc Sci Rev 13:331–347
El-Baz F (1998) Sand accumulation and groundwater in the Eastern Sahara. Episodes 21:147–151
El-Baz F (2000) Satellite observations of the interplay between wind and water processes in the Great Sahara. Photogram Eng Remote Sens 66(6):777–782
El-Baz F (2008) Remote sensing of the earth: implications for groundwater in Darfur. The Bridge 38(3):5–13
El-Baz F (2010) Remote sensing: generating knowledge about groundwater. In: El-Ashry M, Saab N, Zeitoon B (eds) Sustainable management of a scarce resource. The Arab Forum for Environment and Development (AFED), special study, pp 201–222
El-Baz F, Warner DM (1979) Apollo-Soyuz test project summary science report, volume II: earth observations and photography. NASA SP-412
El-Baz F, Mainguet M, Robinson CA (2000) Fluvio-aeolian dynamics in the North-eastern Sahara: interrelation between fluvial and aeolian systems and implications to ground water. J Arid Environ 44:173–183
Embabi NS (2004) The geomorphology of Egypt. Landforms and evolution: The Nile Valley and the Western Desert. The Egyptian Geographical Society, 447 p
Gaber A, Koch M, Helmi M, Sato M (2011) SAR remote sensing of buried faults: implications for groundwater exploration in the Western Desert of Egypt. Sens Imaging Int J 12(3/4):133–151
Gaber A, Soliman F, Koch M, El-Baz F (2015) Using full-polarimetric SAR data to characterize the surface sediments in desert areas: a case study in El-Gallaba Plain, Egypt. Remote Sens Environ 162:11–28
Ghoneim E (2008) Optimum groundwater locations in the northern Unites Arab Emirates. Int J Remote Sens 29(20):5879–5906
Ghoneim E (2011) Ancient mega rivers, inland deltas and lake basins of the Eastern Sahara: a radar remote sensing investigation. Geol Soc Am Abs Programs 43(5):141
Ghoneim E, El-Baz F (2007a) The application of radar topographic data to mapping of a mega-paleodrainage in the Eastern Sahara. J Arid Environ 69:658–675
Ghoneim E, El-Baz F (2007b) Dem-optical radar data integration for paleo-hydrological mapping in the northern Darfur, Sudan: implication for groundwater exploration. Int J Remote Sens 28:5001–5018
Ghoneim E, El-Baz F (2008) Mapping water basins in the Eastern Sahara by SRTM data. In: IEEE international geoscience and remote sensing symposium, vol 1, 6–11 July, Boston, Massachusetts, USA, pp 1–4
Ghoneim E, Robinson C, El-Baz F (2007) Relics of ancient drainage in the Eastern Sahara revealed by radar topography data. Int J Remote Sens 28:1759–1772
Ghoneim E, Benedetti M, El-Baz F (2012) An integrated remote sensing and GIS analysis of the Kufrah Paleoriver, Eastern Sahara, Libya. Geomorphology 139:242–257
Gifford AW, Warner DM, El-Baz F (1979) Orbital observations of sand distribution in the Western Desert of Egypt. In: El-Baz F, Warner DM (eds) Apollo-Soyuz test project summary science report, volume II: earth observations and photography. NASA SP-412, pp 219–236
Griffin DL (2002) Aridity and humidity: two aspects of the late Miocene climate of North Africa. Palaeogeogr Palaeoclimatol Palaeoecol 182:65–91
Griffin DL (2006) Late Neogene Sahabi Rivers of the Sahara and their climatic and environmental implications for the Chad Basin. J Geol Soc 163:905–921
Haynes CV (1985) Quaternary studies, Western Desert, Egypt and Sudan—1979–1983 field seasons. Natl Geogr Soc Res Rep 16:269–341
Hinz E, Stern R, Thurmond A, Abdelsalam M, Abdeen M (2003). When did the Nile begin? Remote sensing analysis of paleo drainages near Kom Ombo, Upper Egypt. Eos transactions, Fall meeting supplement, Abstract, vol 84, p 46
Hoekstra P, Delaney A (1974) Dielectric properties of soils at UHF and microwave frequencies. J Geophys Res 79:1699–1708

Issawi B, Hinnawi M (1980) Contribution to the geology of the plain west of the Nile between Aswan and Kom Ombo. In: Close AE (ed) Loaves and fishes: the prehistory of Wadi Kubbaniya. Southern Methodist University Press, Dallas, pp 311–330

McCarthy TS (1993) The great inland deltas of Africa. J Afr Earth Sci (and the Middle East) 17:275–291

McCauley JF, Schaber GG, Breed CS, Grolier MJ, Haynes CV, Issawi B et al (1982) Subsurface valleys and geoarchaeology of the Eastern Sahara revealed by shuttle radar. Science 218:1004–1020

McCauley JF, Breed CS, Schaber GG (1995) SIR-C definition of the Serir-Kufra River system in SE Libya. EOS supplement, S196

Pachur H-J (1993) Palaeodrainage systeme im Sirte-Becken und seiner Umrahmung. Wurtzburger. Geogr Arbeiten 87:17–34

Paillou A, Schuster M, Tooth T, Farr T, Rosenqvist A, Lopez S et al (2009) Mapping of a major paleodrainage system in eastern Libya using orbital imaging radar: The Kufrah River. Earth Planet Sci Lett 277:327–333

Parashar S, Langham E, Mcnally J, Ahmed S (1993) Radarsat mission requirements and concepts. Can J Remote Sens 19:280–288

Pratt D, Ellyett CD (1979) The thermal inertia approach to mapping of soil moisture and geology. Remote Sens Environ 8(2):151–168

Robinson CA, El-Baz F, Singhroy V (1999a) Subsurface imaging by radarsat: comparison with Landsat TM data and implications to ground water in the Selima Area, Northwestern Sudan. Can J Remote Sens 25(3):268–277

Robinson CA, Kusky T, El-Baz F, El-Etr H (1999b) Using Radar data to assess structural controls from variable channel morphology: examples from the Eastern Sahara. In: Proceedings of the 13th international conference applied geologic remote sensing, vol II, p 381e386

Robinson CA, El-Baz F, Ozdogan M, Ledwith M, Blanco D, Oakley S et al (2000) Use of radar data to delineate palaeodrainage flow directions in the Selima Sand Sheet, Eastern Sahara. Photogram Eng Remote Sens 66(6):745–753

Robinson CA, El-Baz F, Al-Saud T, Jeon SB (2006) Use of radar data to delineate palaeodrainage leading into the Kufra Oasis in the Eastern Sahara. J Afr Earth Sci 44:229–240

Robinson C, El-Kaliouby H, Ghoneim E (2017) Influence of structures on drainage patterns in the Tushka Region, SW Egypt. J Afr Earth Sci (Special Issue) 136:262–271

Roth LE, Elachi C (1975) Coherent electromagnetic losses by scattering from volume inhomogeneities. IEEE Trans Antennas Propag 23:674–675

Schaber G, Mccauley J, Breed C (1997) The use of multifrequency and polarimetric SIR–C/X–SAR data in geologic studies of Bir Safsaf, Egypt. Remote Sens Environ 59:337–363

Seiffert RE, Bown MT, Clyde CW, Simons E (2008) Geology, paleoenvironment, and age of Birket Qarun Locality 2 (BQ-2), Fayum Depression, Egypt. In: Elwyn Simons: a search for origins. Springer, Berlin, pp 71–86

Thurmond AK, Stern RJ, Abdelsalam MG, Nielsen KC, Abdeen MM, Hinz E (2004) The Nubian Swell. J Afr Earth Sci 39:401–407

Wendorf F (1977) Late Pleistocene and recent climatic changes in the Egyptian Sahara. Geogr J 143(2):211–234

Wendorf F, Schild R (1980) Prehistory of the Eastern Sahara. Academic Press, New York

Wendorf F, Schild R (1986) The prehistory of Wadi Kubbaniya; the Wadi Kubbaniya skeleton: a Late Paleolithic burial from Southern Egypt, vol 1. Southern Methodist University Press, Dallas

Wendorf F, Close AE, Schild R (1987) Recent work on the Middle Paleolithic of the Eastern Sahara. Afr Archaeol Rev 5:49–63

Wolfe RW, El-Baz F (1979) The wind regime of the Western Desert of Egypt. Reports of planetary geology program, 1978–1979. NASA technical memorandum 80339, June 1979, pp 299–301

Chapter 9
Monitoring and Protection of Egyptian Northern Lakes Using Remote Sensing Technology

Sameh B. El Kafrawy and Mahmoud H. Ahmed

Abstract Coastal lakes in the Mediterranean region constitute a major aquatic resource, yet many in the Egyptian Mediterranean region (EMR) are severely degraded. Despite of the acute problems that sometimes stem from development activities, the lakes aquatic ecosystems are of high or potentially high value for local human populations as well as for regional biodiversity. Northern Egyptian lakes are all impacted by a variety of environmental change processes, but direct human activities have had the greatest effect during the 20th Century. This study is designed to promote these pressing environmental management issues through a variety of tasks that include monitoring, modeling and protecting. Management planning and policies are poorly supported by environmental science in the EMR. Satellite imagery is obtained for the study lakes and was subject to geometric correction and transformations. Ecological classification of satellite images are undertaken to establish the distribution of major ecological zones including distribution of major vegetation communities. For each lake, an array of different remote imagery and topographic maps from different time periods were employed to identify changes in the distribution of open water and aquatic vegetation as well as major changes in the configuration of these lakes. Geographic Information Systems (GIS) are used for all the data for each lake. Hydrodynamic models for some lakes are parameterized using the baseline historical data, bathymetric data, ancillary data, field survey data and monitoring data. The finite element two-dimensional modeling systems MIKE 21 and three-dimensional MIKE 3 were employed. Model results were used to examine ecosystem functioning including, the relative importance of freshwater inflows and exchange to the sea and their influence upon salinity, circulation patterns, and sources and distribution of nutrients. The principal results, achievements, and experience from this study enables to address some recommendations for future actions concerning coastal aquatic environments in the Egyptian Mediterranean Region including; furthering the need for an integrated approach to environmental assessment and management of Northern Egyptian lakes with special attention to harmonizing analytical skills by developing on-line facilities, Establishing continuity in the operation of integrated eco-hydrological investigations including remote sensing, GIS,

---

S. B. El Kafrawy (✉) · M. H. Ahmed
National Authority for Remote Sensing and Space Sciences (NARSS), Cairo, Egypt
e-mail: sameh@narss.sci.eg

S. F. Elbeih et al. (eds.), *Environmental Remote Sensing in Egypt*,
Springer Geophysics, https://doi.org/10.1007/978-3-030-39593-3_9

modeling at the coastal lakes that retain links with meaning stakeholders and experts, expanding impact assessment to include social and economic themes that include the societal implications of different scenarios, promoting links between environmental scientists, environmental managers, decision makers and wider society to increase transparency and communication of collected information using online facilities, and building upon the results of lakes studies by raising awareness of issues facing coastal aquatic ecosystems so that their sustainable management can be better integrated into water and land use policy nationally and internationally.

Keywords Northern lakes · Egyptian lakes · Remote sensing · Monitoring · Modeling · Biodiversity

9.1 Introduction

The Northern coastal zone of Egypt, including lakes environment, is of great socioeconomic and environmental significance. In Egypt, the lakes areas along the Mediterranean coast comprise five lakes. Four lakes are deltaic water bodies (Mariut, Edku, Burullus and Manzala) and the other is non-deltaic Lake Bardawil. The deltaic lakes are brackish, shallow (<2 m), with an average depth of ~1.0 m. Among these water bodies, Mariut Lake is artificially enclosed and has been without a major connection to the sea. The other three display typical lake characteristics separated from the Mediterranean by low-lying, long narrow coastal sandbarriers, and connected to the sea by protected inlets. These inlets are vital for a dozen of species of fish that depend on the lakes for at least a part of their life cycles, and they are the only inlet/outlet available for thousands of commercial fishing boats that use the lake. Delta lakes receive much of their freshwater input from irrigation drains that entering the lakes from the southern, eastern, and western margins. The non-deltaic lake is situated away from the delta region at north Sinai, Lake Bardawil. The ecosystem of these lakes has been controlled by the interaction of natural and man-induced factors (Fig. 9.1).

The southern and eastern margins of the Nile Delta lakes are bordered by extensive marshes of aquatic macrophytic plants. Surficial sediments of the delta lake composed of plant- and shell-rich muds and sandy silts, whereas Bardawil Lake is covered by sand-size sediments mixed with evaporates (Levy 1974a, b). The coarsest sediments are found near the inlet where current velocities are maximum, and the finest sediments is in the innermost reaches where current velocities approach zero and where drains are localized. Biological processes contribute significantly to the production of carbonate sediment through the formation of mollusk shells, ostracod and foraminiferal tests (Bernasconi and Stanley 1994).

For better access management of the Nile Delta coastal zone and adjacent coastal lakes, a coastal highway, is nearly completed, extending along the Mediterranean coast of Egypt from El Salum City in the west to Rafah City in the east. The road is more or less parallel to the present shoreline at a minimum distance of about 1 km

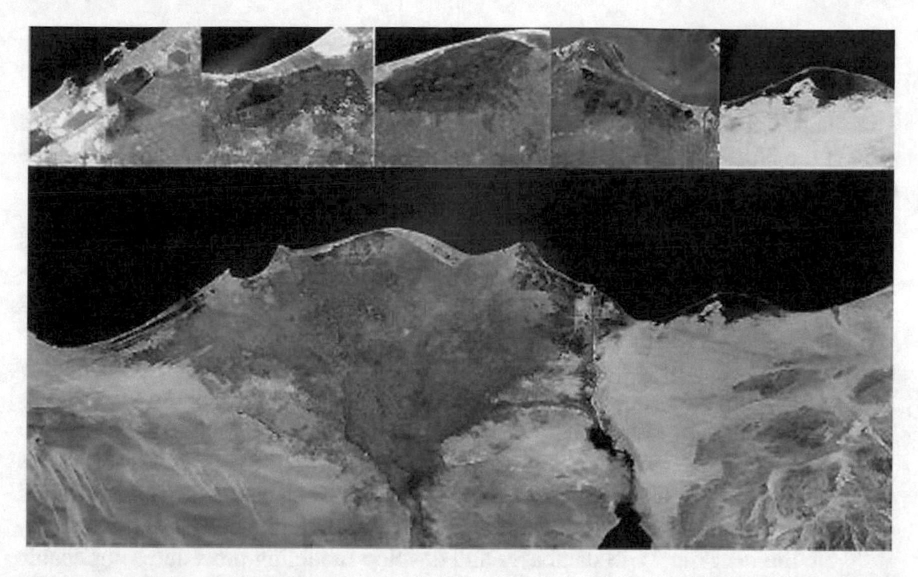

Fig. 9.1 Northern coastal lakes of Egypt (study area). From left to right (top of the figure), the lakes are: (Mariout Lake, Idku Lake, Burullus Lake, Manzala Lake, and Bardawil Lake) (Marine sciences department-NARSS)

from the shore (Mohsen 1992). The coastal road is crossing Burullus inlet by a pill bridge to permit navigation of fishing boats as well as the lake/sea water exchange. This highway will facilitate better access between lakes and other communities, agriculture and industrial centers across the northern coast of Egypt. This means that urban expansion is increasing around the coastal lakes and thus causing more pressure on the lake ecosystem (Ahmed et al. 2000a, b).

The primary objective of this study is to monitor the important problems that exist at the Northern coastal lakes of Egypt. Also, the authors attempt to use the environment of these lakes to be used as a lesson learned for any other lake management processes. The study is based largely on studies that used the satellite data, information technology and consultation of available literature. Most of the existing environmental problems of the Egyptian Northern lakes are derived from anthropogenic and natural influences. In contrast, the other is a hypersaline water body (Bardawil Lake), which is located at the northern coast of Sinai. Using earth observations and information technology (satellite data, GIS, etc.) enhance and update the understanding of the behaviours of lakes ecosystems.

On the other hand, ongoing natural factors have induced substantial changes in the lake environment. These changes include barrier erosion, inlet siltation, land subsidence and rising sea level, subsidence and prevailing sedimentary processes. Monitoring of the these problems of these lakes is discussed herein in view of identifying problems, causes, impacts and proposed actions for protecting and managing the northern Egyptian lakes.

9.2 Objectives

This chapter aims to review the role of satellite remote-sensing imagery to assess and protect the Northern lakes of Egypt; this can be achieved through addressing three major key issues:

(i) Monitoring the present status of the northern lakes of Egypt by providing the origin description and problems identification of the lakes ecosystems. Moreover, consulting the studies and reports that were used the remote sensing technology will be presented.
(ii) Addressing the developed techniques and procedures for using satellite imagery in studying the coastal lakes on a regional scale. These procedures are describing the satellite-based methods for implementation on a routine basis the lake natural resource management. Apply the satellite-based procedures to a series of images that span several decades to evaluate how lakes changed over time and space in relation to land-use and land-cover conditions. This issue is to assemble the necessary GIS databases and develop modeling procedures applicable at the small watershed scale to predict pollutant loadings and water quality responses in lakes based on size, drainage patterns, land-use/cover, and other related watershed information available in GIS format.
(iii) Protecting and conservations of lakes resources as well the management activities will discuss in the light of utilization of information technology in previous lakes studies.

9.3 State-of-the-Art Review

Although previous investigations have demonstrated reliable empirical relationships between satellite data and nearly contemporaneous ground observations, satellite imagery has not been incorporated into routine lake-monitoring programs conducted by institutes and agencies (Kloiber et al. 2002). This study examined the key issues involved in using satellite imagery in the regional assessment of lake, developed a procedure for the use of satellite imagery to assess lake water clarity and applied the procedure to a series of images of lakes taken during last four decades (Ahmed et al. 2000a, b, 2009).

The basic requirement for remote sensing environmental monitoring is the availability of different dates of imagery upon which the same area of land can be observed. Monitoring global, regional and local areas can be performed by restricting the analysis to a single sensor series or by using different satellite data. New multi-source satellites are creating data at higher spatial and temporal resolution than have been collected at any other time on earth. The selection of low cloud cover imagery with careful attention to selecting the dates through the year(s) is very important.

The unique characteristics of the Landsat data make it an ideal data source for monitoring. Its sensor (Thematic Mapper TM) covers a wide range in the electromagnetic wave spectrum with a ground (spatial) resolution of 30 m, which enable precision mapping. Its spectral range starts from the visible range (0.45–0.69 μm), to near Infrared (IR) (0.76–0.90 μm) and mid IR (1.55–2.35 μm), up to the thermal IR band (10.4–12.5 μm). Moreover, its temporal resolution (the ability to obtain repeated coverage of a specific geographic area) is 16 days, which is a considerably short period that permits continuous monitoring and environmental developments over time.

Remotely sensed data could be used as a tool to detect, monitor and evaluate changes in ecosystems to develop management strategies for ecosystem resources, satellite and airborne systems offer a plausible monitoring system for large-scale, earth surface viewing and provide a usable database for change detection studies. Remote sensing data can be used to span temporal and spatial scales ranging from local systems to aggregated global systems (Graetz 1990).

Milne (1988) noted that common types of detectable change in remotely sensed data are associated with the clearing of natural vegetation, increased cultivation, urban expansion, changing surface water levels, post-fire vegetation regeneration, and soil disturbanccs resulting from mining, landslides and animal overgrazing.

Milton and Mouat (1989), found that the use of remote sensing techniques allows ecologists and resource managers the opportunity to monitor vegetation condition, patterns, and trends in arid regions where rugged terrain, poor access, and extreme climate conditions make field investigations difficult. The use of satellite data is very suitable for inventorying the kind, quality, distribution, and condition of natural vegetation found on the range and forest lands (El-Kafrawy et al. 2017). Observations made by the Advanced Very High-Resolution Radiometer (AVHRR) on board the NOAA-7 platform, assisted in monitoring desertification in Africa (Choudhury 1990).

Dewidar and Khedr (2005), used Landsat Thematic Mapper TM data combined with surface measurements for mapping water quality parameters of Burullus Lake. They indicated that some of the water quality parameters were significantly correlated with TM radiance data.

Subsequently, multiple linear regression models were used to prepare digital cartographic products depicting the water quality over the entire study area (Kloiber et al. 2002).

Temporal pattern mapping of vegetation may further provide clues as to the causes of certain ecosystem changes. Briggs and Nellis (1991) developed a satellite-based textural gontrim which quantifies the temporal changes resulting from the influence of seasonal fires. It has also been demonstrated that it is possible to assess accumulated biomass (Fung 1990; Ringrose and Matheson 1987; Milne and O'Neill 1990; Franklin 1991), the degree of use and changes in the dominant species composition (Hall et al. 1991; Hobbs 1990), land use (Ambrose and Shah 1990; Teng 1990; Wharton 1987; Ramdani et al. 2009; Thompson et al. 2009; Ahmed and Donia 2007) and hydrological differences (Howman, 1988; Christensen et al. 1988; Jensen 1986) in various ecosystems using satellite and/or airborne data.

Recording land cover change over time is perhaps one of the most important applications of digital remote sensing data (Christensen et al. 1988). For example, the conversion of rural to urban land use can be detected using a temporal comparison of spatial change determined from satellite and airborne data. The value of utilizing remotely sensed data for change detection studies is limited only by the imagination of the investigators and potential users.

Satellite remote sensing using Landsat Thematic Mapper (TM) has been explored in several studies as a method of reducing the cost and labor of sampling water clarity in the field (Khorram and Cheshire 1985; Lathrop 1992; Dewider and Khedr 2001; Kloiber et al. 2002). An advantage of using remote sensing is that data for multiple lakes within a single image can be collected quickly and relatively inexpensively.

Remote sensing technology has been used for several decades in oceanography to measure chlorophyll, water-color, and suspended sediments over large areas, but has only recently been explored in lake studies in Egypt. Although sensors such as Landsat TM were primarily designed for detecting land features, recent improvements now provide better spatial and spectral resolutions for aquatic studies than previously available (Zilioli 2001; Kloiber et al. 2002). Recently, remote sensing has been shown to correlate well with Lake Secchi disk transparency (SDT) values (Khorram and Cheshire 1985; Lathrop 1992). However, to effectively implement remote sensing into a state monitoring program for inland lakes, there remain many unanswered questions.

Barale and Folving (1996) demonstrated the use of satellite images and remote sensing in studying the coastal interactions in the Mediterranean region. They showed that several interactions had occurred within the area in front of the Nile Delta during the last 50 years which led to detectable changes in the coastal area.

Ahmed et al. (2006) studied the case of eutrophication assessment of Lake Manzalah using the GIS technique. Their results indicated that the eutrophication index value could be calculated in combination with GIS overlay technique to give a relatively accurate prediction of the status of eutrophication of the lake. They also concluded that the water quality spatial distribution map could also produce using the same technique.

El-Raey et al. (2006), have applied change analyses techniques to TM images for identifying changes inside and outside Lake Burullus. Analysis of the results reveals the occurrence of erosion and accretion along both sides of the lake inlet. During the period (1995–2000), the lake area decreased by an average rate of 1.34 km^2 per year mainly due to illegal drying for land use.

9.4 Origin and Description of the Northern Egyptian Lakes

The Nile and other Mediterranean deltas contain a widespread and generally consistent late Pleistocene to Holocene stratigraphic succession. In respect to the Nile Delta, Stanley and Warne (1993) divided this section into three main sequences. Generally, this section consists of late Pleistocene to Holocene stratigraphic succession of a basal sequence I of late Pleistocene fluvial deposits; an overlying sequence II of late Pleistocene to early Holocene shallow marine transgressive sandy deposits; and an upper sequence III of Holocene deltaic of variable lithologies. Of importance of this study is sequence III which acquires the lake facies. This sequence was mostly formed by a change in sea level rather than by regional climate factors. The rate of sea-level rise decelerated markedly from about 7000–5000 BC to ~1 mm/yr. As sea level slowly was elevated and the gradient decreased to 1:5800. During this period, the rate of sediment accumulation matched or exceeded sea-level rise, shoreline position became relatively stable, and formation of the modern Nile Delta began (sequence III), locally included lake facies. With a slow rise of sea level, a series of smaller individualized lakes and marshes were developed in the northern delta, landward of coastal barrier beaches or dune ridges (Ahmed et al. 2000a, b). The other is hypersaline Lake Bardawil, probably have been developed in a process similar to the delta lakes combined with neotectonic land subsidence. The following is a brief description of these lakes (Ahmed et al. 2000a, b).

9.4.1 Lake Mariut

Lake Mariut has a strategic importance at the regional and local level. It plays an important role in the water balance of the Delta western region. Without it and with the direct drainage to the sea, the level of water would continue to rise, which would eventually flood wide areas of land. Also, due to the scarcity of land for new development in Alexandria, Lake Mariut and the surrounding area are now viewed as prime land for urban expansion as well as a significant economic resource for the city. Accordingly, Lake Mariut represents a vital economic resource to the governorate of Alexandria (Fig. 9.2) (World Bank 2005; Ahmed 2003; Ahmed and Elaa 2003).

Fishing is one of the major activities in the Mariut Lake and was characterized in the past by large fish harvest rates with reputed quality. According to available estimated statistics, there are over 2000 fishing boats owned and employed by about 5000 fishermen representing an estimated community of about 25,000 individual. This community relies solely on fishing as the only profession they know and practices for many years. Their adaptability and willingness to explore and engage in new earning venues are very remote. The current total estimated fish catch is about 4000 tons annually and are declining steadily due to the deterioration of water quality and the drying of vast areas for land acquisition (GAFRD 2008).

Around the lake, there is a wide area of reclaimed land which includes various residential, industrial, commercial, recreational, and other settlements activities such as the Mubarak sports city with a projected total area of 500 acres of which 130 have been already utilized, International garden, Mega market Carrefour. A variety of industrial activities also exist around Lake Mariut comprising a host of activities and including oil refining. The discharges of the combined activities, human and industrial, are the main culprit for Lake Mariut"s current and future vitality (Ahmed

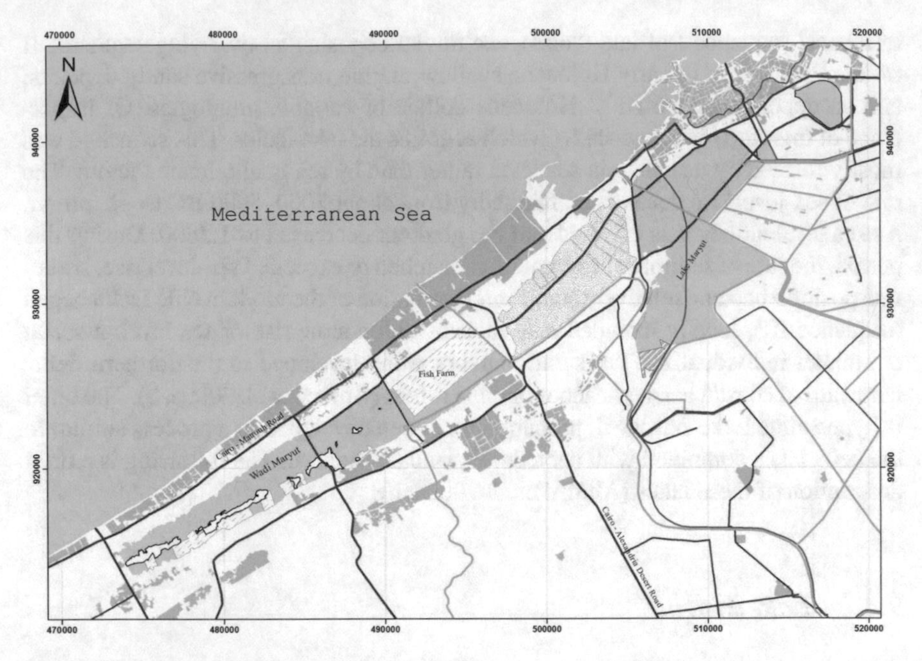

Fig. 9.2 Topographic map of Lake Mariut (Marine sciences department-NARSS)

and El-Leithy 2008). The assessment of the current concerns and challenges facing the sustainable development of Lake Mariut emphasized the following areas:

- Deterioration of water quality
- Continuous shrinkage of lake's area resulting from drying
- Declining fish production and biodiversity leading to deterioration of livelihood of Fishermen society
- The spread of water reeds and vegetation and emission of foul odours
- Negative impacts on public health and the high cost of medical care
- The spread of uncontrolled (squatter) settlements
- The occurrence of illegal, informal, and socially-threatening groups and racketeers.

Despite the many challenges facing the sustainable development of Lake Mariut however, the issues of Lake Mariut were never absent of the minds of officials, active civil society, the general public, and international development organizations. They all share the common interest in the proper development of this vital natural resource and its surroundings important to Alexandria"s future development and as part of the larger Mediterranean basin environmental sustainability.

9.4.2 Lake Edku

Lake Edku is a coastal lake in the eastern Mediterranean and is located about 40 km east of Alexandria City and 18 km west of Rosetta branch of the Nile River. It is located west of the Nile Delta between longitudes 30° 08′ 30″ and 30° 23′ E and latitudes 31° 10′ and 31° 18′ N (Montaser et al. 2017). The lake is connected to the adjacent Abu Qir Bay through the inlet channel Boughaz El Maadia. The actual surface area of the lake has decreased since 1964 due to the reclamation of a large area from the eastern side for cultivation purposes. Water depths in the lake vary from 10 to 140 cm, the maximum depths being in the central and eastern parts (Fig. 9.3). The lake receives large quantities of drainage water released from the agricultural lands of the Beheirah Province, through the Barzik, Edku and E1-Bosely drains, where the last two drains meet together before entering the lake and discharge their water through the extension of Edku drain. During the period of high discharge, there is an outflow of fresh water from the lake, and during the other period, seawater influx occurs. The marine water influence is limited to the areas near the Boghaz E1 Maadia.

There are two main drains discharge their wastes into the lake; namely El-Khayry drain and Barsik drain. The first drain is joined to three sources of drainage water coming from El-Bosely, Edku and Damanhour subdrains, which transport domestic, agriculture, and industrial wastewaters as well as the drainage water of more than 300 fish farms. The second drain transports mainly agricultural drainage water to the

Fig. 9.3 Lake Edku (Marine sciences department-NARSS-LANDSAT8-2018)

lake. This drainage water creates in most time water moves through the lake from both west and south to the north towards the sea (Fig. 9.3).

9.4.3 Lake Burullus

Burullus Lake is the second largest coastal lake in the northern lakes of Egypt and covers an area of about 410 km^2. Its long axis lies parallel to the coastline and separated from the Mediterranean by a sand barrier. The only connection of the lake with the Mediterranean Sea is the Burullus inlet. The lake margin is irregular and bordered by marshes. It is about 54 km long and has a maximum width of 12 km. Water depth ranges from 0.1 to 2.4 m, and salinity from 0.32 to 2.4 g/l (Ahmed et al. 2001; Saad 1976a). The lake receives fresh water from numerous drains along the southern and eastern margins (Fig. 9.4). Burullus Lake receives only agricultural runoff water not contaminated by industrial wastes. The southern margin is bordered by marshes. Lake water is brackish, salinity up to 3.51 g/l (Saad 1990). The lake is divided into several subbasins by natural and artificial barriers. Annual freshwater influx into the lake through drains is about 2.1×10^9 m^3 (Saad 1976b). The input runoff contains mainly agriculture waste and completely free from industrial wastes.

A huge amount of drainage water enters the Lake at the southern coast through several drains, causing dilution of water and rise in the Lake level above sea level. The Lake current towards the sea at certain seasons of the year is weak leading to accumulation of deposits at the Lake–sea connection area. To sustain fish life in the Lake, these deposits have to be removed periodically.

Burullus Lake receives 2.46×10^9 m^3/day of brackish water through drains. Compared with the present total size of the Lake, the residence time of water should be 2.5 months. This contrasts with the measured amount of water leaving the Lake annually, i.e. 446×10^6 m^3/year through Boughaz Al-Burg. This leaves about 2 ×

Fig. 9.4 Lake Burullus (Marine sciences department-NARSS, lANDSAT8, 2018)

10^9 m^3 in excess that has to find another pathway to leave the Lake. Part of this amount is lost through evaporation estimated to be about 0.71×10^9 m^3/y (Maiyza et al. 1991) and possibly the rest through the bottom and consumption by aquatic plants (Fig. 9.4).

Numerous islands characterize Lake Burullus. Most of these islands are elongated from south to north. Others are oriented either parallel or normal to the present coast. These islands consist mostly of mud; however, others are formed of sand (e.g. El Koom El Akhder). These islands are important paleo-geographic indicators of relict deltaic features such as beach ridges, dunes and riverbanks of former distributes (Ahmed et al. 2001).

The bottom sediments in Lake Burullus have a specific textural composition. Shells and shell fragments constitute a significant part of the sediments. Mostly shells, shell fragments, quartz, feldspar, Ostracoda and Foraminifera dominated sand. The fine fraction of the sediments is composed of silt and clay together with fine carbonate particles. Calcium carbonate content of the sediments is less than 30%. In the central and western region of the Lake, the carbonate contents reach higher values (up to 75%) due to the dominance of mollusca. The sediments in the eastern part of the Lake have the lowest carbonate content. The organic matter content of the sediments of Lake Burullus varies between 1 and 2% with an average of 1.8%. The organic matter content becomes higher near the southern eastern and western parts of the Lake (Fig. 9.4).

Several water plants like reeds (Phragmites, Typha, etc.) spreads all over the Lake affecting the movement of water. These plants play an important role to keep the internal coasts of the Lake from immolation. The decrease of the area of Elodea in Lake Brullus indicates the increase of the salinity in the Lake.

9.4.4 Lake Manzala

Manzala Lake is the largest of the northern coastal lakes. It is an important and valuable natural resource area for the fish caught, wildlife, the hydrologic and biologic regime and table salt production. It produces about 50% of the fish catch of the northern lakes and freshwater fisheries (Kriem et al. 2009).

Lake Manzala is not only the biggest one among the Egyptian lakes (area = 0.3 km^2) but also serves five provinces of Nile Delta, Namely Port Said, Ismailia, Sharkiya, Dakahliya and Damietta. It has been recognized as the most productive fishery ground of the country's lakes, since it contributed nearly 50% of the total country yield during early 1970s and about 35% during 1980s (Khalil 1990) (Fig. 9.5).

The lake is the largest brackish water body (~1000 km^2), and is located in the northeastern shoreline of the Nile Delta. The lake is about 50 km long and has a maximum width of 30 km. The lake is shallow (<2 m, with an average depth of ~1.0 m, and salinity varies from 0.77 to 11.67 g/l (Saad 1990). The western and southern sectors are supplied by drainage water from seven main sources. As in Edku and Burullus lakes, emergent and submerging marshes border the southern

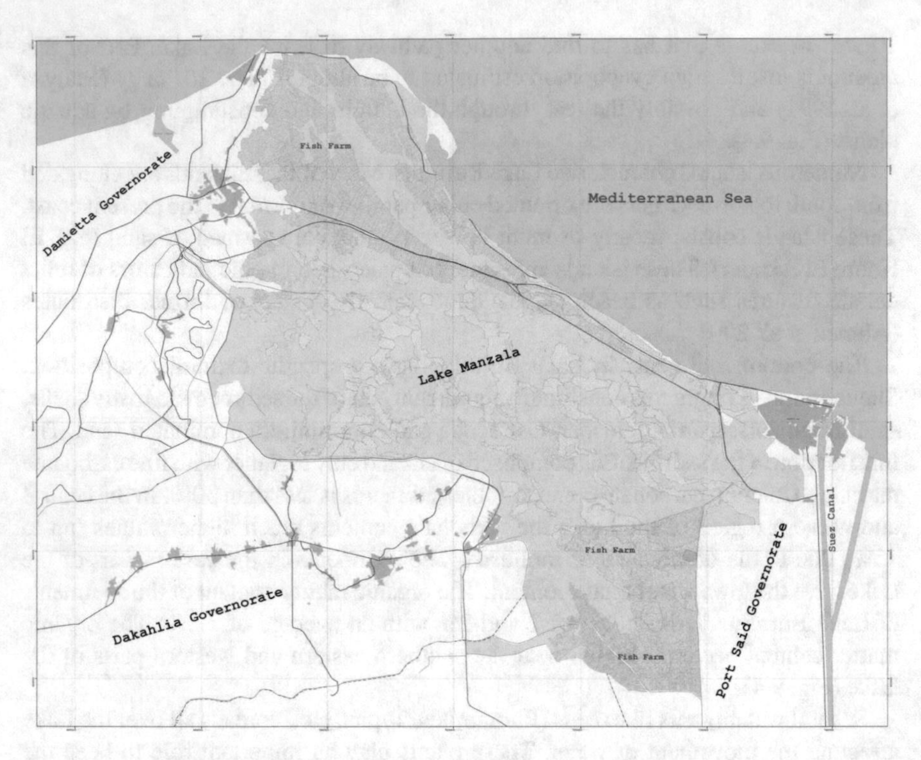

Fig. 9.5 Topographic map of Lake Manzala (Marine sciences department-NARSS)

margin. The sand barrier which separates the lake from the sea suffers from erosion. As a result of this erosion, the coastal road between Damietta and Port Said had been replaced by a new road farther inland connecting the lake islands. Manzala Lake supplies about 50% of the total Egyptian fish catch. El Gamil inlet is the only inlet for release of marine water inflow into the lake. Two jetties were constructed to protect this inlet from siltation and longshore migration (Donia and Ahmed 2006) (Fig. 9.5).

It is divided into sub-basins by natural and artificial barriers. These sub-basins are the sites of aquaculture development. The Ginka sub-basin in the southeast sector of the lake is identified as "black spot". This subbasin is heavily polluted by heavy metals and high nutrient discharging from Bar El Baqar drain (Global Environmental Facility 1992). This subbasin receives a discharge of municipal sewage, industrial effluent, and agricultural runoff. High concentrations for metal pollutants are recorded in the upper 20 cm of the Ginka subbasin, probably from industrial sources. High values for Hg (822 ppb), Pb (110 ppm), Zn (635 ppm), Cu (325 ppm) were recorded in bottom sediments of Ginka subbasin (Siegel et al. 1994; El-Kafrawy et al. 2006).

9.4.5 Lake Bardawil

Bardawil Lake is situated along the northern coast of Sinai, from a point about 45 km east of Port Said and extending to a point 20 km west of El-Arish. Its geographical boundaries extend from 32° 40′ to 33° 30′ east longitude and from 31° 03′ to 31° 14 north latitude. Lake Bardawil is mainly a flat low lying plain, bordered from the north by Sinai Mediterranean coast, from the south by a sand dune belt which extends inland to the region of fold and anticlinal hills, from the west by the Tineh Sabkha flat constituting eastern margin of the Nile Delta plain and from the east by Arish-Rafah sector. Its elliptical shape represents a major morphological feature in north Sinai coast (Fig. 9.6). This lake has an area of about 164,000 feddan (685 km^2), extends for a distance of about 80 km, with a maximum width of about 20 km and a maximum depth of about 3 m. It is separated from the Mediterranean by a long convex sand bar; the main water body of the lake lies towards the east occupying a section along the coast of about 30 km long ending with Zaranik Pond in the east (has an area of about 243.68 km^2), of which Zaranik Pond occupies about 42.02 km^2. The latter is now exploited for salt production (it is locally called Malahat Sebikah). The western part of the lake extends as a long narrow arm of about 50 km length (it has an area of about 445.36 km^2).

Historically, Bardawil Lake (also called Sabkhat El Bardawil or the Sirbonian Lake) is named after King Baldwin I, who took part in the Crusades and according to tradition was killed at El-Arish (Ben-Tuvia 1979). During the Roman period, the lake was called Port Sirbon. Many historians and archaeologists speculate that the Exodus of the tribes of Israel from Egypt passed through this area and the biblical

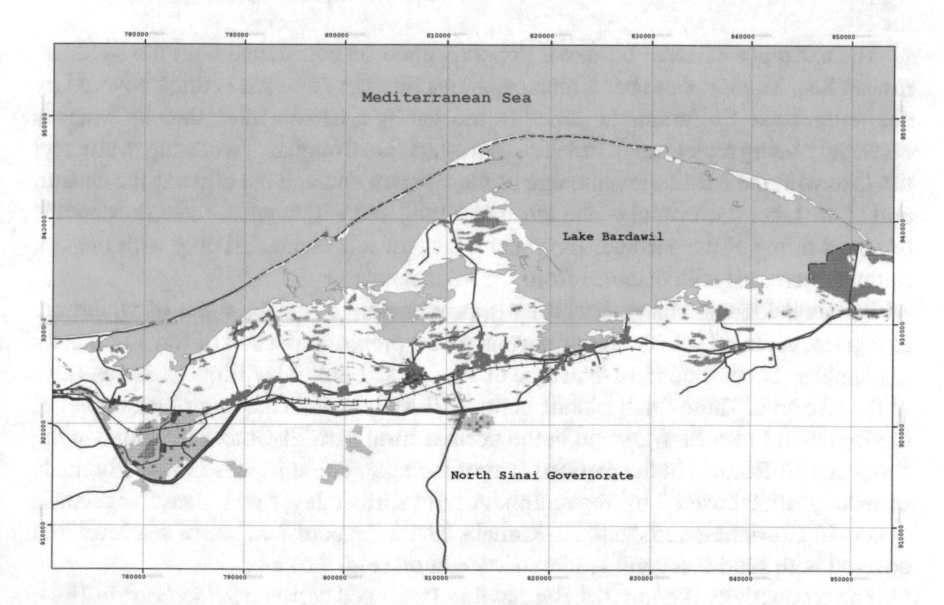

Fig. 9.6 Lake Bardawil (Marine sciences department-NARSS)

«Red Sea» or «Sea of Reeds» is the Bardawil Lake. Some recent excavations on the wide part of the sandy bar (Kals or Mount Cassius) have been reported and found evidence of settlements from the period of the Early Iron Age and a town, evidently Cassius, from the Hellenistic-Roman Period (El-Malky et al. 2003).

Seawater enters the lake at present through three inlets: two artificial tidal inlets (270 and 300 m wide and 4–7 m deep), they are maintained open by periodic dredging, and a natural eastern inlet of Zaranik which is now occasionally closed by silting (Fig. 9.6). Fish production of Bardawil Lake depends on the water exchange between the lake and sea, which regulates lake salinity. Joined to Bardawil Lake, are a number of bays (e.g. El-Telul and Misfiq) and a few restricted shallow water ponds (e.g. El-Rowaq and El-Marqab to the south) in which water depth is only a few centimetres. In the southern areas of the lake, extensive salt pans-sabkha complex occur mostly interrupted by a series of sand dunes (ridges) running mostly parallel to the coast and extending southwards. Thus sabkha of Bardawil may be coastal flat sabkha fringing the lake particularly at the extreme eastern margins or dunal sabkha south the lake (Shaheen 1998).

The bar separating the lake from the sea is arc-shaped, 300–1000 m wide. Its highest point is El-Kals (Mount Cassius), a 60 m high dune located about midway. The western part of the bar is an extension of a dune-covered higher ground which starts at Qantara on the Suez Canal. This ridge is part of the Pelusium Line, a compressional zone dividing the thin Mediterranean crust on the east from the oceanic-type crust on the west. However, there is little evidence that the lake was the estuary of the Pelusiac branch of the Nile. Pelusiac branch debouched at a site situated to the west of the above mentioned ridge (Neev and Friedman 1978), but undoubtedly the Nile supplied the quantities of sand that formed of the bar separating the depression from the sea.

The existence of Lake Bardawil depends upon its connective with the Mediterranean Sea. The low sand bar which separates the lake from sea is often covered by sea water. Lake El-Zaranik is joined to the sea by a narrow inlet; thus its water is constantly being replenished. In 1955, two inlets (i.e. Boughaze) were dug to connect the lake with the Mediterranean, one at the western end and the other at the eastern end of the lake. Each canal is one kilometer long and 150 m wide. Lake Bardawil is the most saline of the northern Egyptian lakes, for it is connected only with the sea. Salinity increases with distance from the inlet canals.

Bardawil Lake is characterized by approximately 51 islets, some of which are elongated, oriented subparallel or normal to the present coast and of few kilometers in diameter. They comprise a total area of about 13.31 km^2 which approximates 1.9% of the lake area. These small islands include El-Mahasnah in the north eastern part of the Bardawil Lake, El-Watawite in the north central part, El-Gouz El-Ashhab in the south and El-Romaia in the western arm of the lake. These islands are mostly made of muddy sand, covered by vegetation. A mud surface layer with dense vegetation occurs in El-Mahasnah Island. El-Romaia Island is about 2 m above sea level and covered with sand sheet and scattered vegetation (Fig. 9.6).

The productivity (kg/km^2) of Bardawil Lake ranged between 9.7 kg/km^2 in 1994, 24.38 kg/km^2 in 1999 and 21.4 kg/km^2 in 2006. It is clear that the productivity per feddan fluctuated over the years, based on the change in gross production of fish and the water area of the lake. Fishing gained significance for the Lake Bardawil. Although it is considered to be one of the best quality fishing areas in Egypt, its production was accounting to about 3534 tons and 0.7% of the national fish production in 2005.

9.5 Problems of Identification

There are some ecological problems invaded and impacted the northern lakes ecosystem due to the continuing and increasing water discharge, either domestic, agricultural or industrial. These problems are eutrophication, excessive growth of aquatic vegetation and the heavy metal pollution, reclamation, illegal fishing, etc.

9.5.1 Eutrophication

The increase of agricultural and domestic discharge into the lakes leads to increase of external nutrient load into the water body and finally eutrophication. This, in turn, changes the characteristics of the aquatic ecosystem. Ahmed et al. (2006) investigated the phytoplankton in the lake and reported that it is considered to be eutrophic as the levels of nutrient are significantly high. Accordingly, this lowered the diversity of phytoplankton population and affected the species composition that was dominated by species considered as indicator for water pollution such as *Euglina* spp. and *Phacus* spp. Ramdani et al. (2001) studied the seasonal chemical composition of Lake Edku water and reported that the lake receives high amounts of nutrient salts (nitrogen salts, phosphates and silicates) through the drainage water and became eutrophic (Ahmed et al. 2006).

9.5.2 Aquatic Vegetations

Aquatic vegetations represent all types of plants that grow in the aquatic ecosystems either rooted, floating or submerged. The undesired growth of aquatic vegetations takes place due to the increased nutrients, which stimulate and support plant growth. Samaan (1974) and Guerguess (1979) studied the primary production and hydrophytes of Lake Edku and found an inverse relationship between increased densities of the macrophytes and standing crops of epiphytes phytoplankton (El Nahry et al. 2006).

Table 9.1 Detection of size change of Lake Burullus during last 30 years using satellite images

Lake name	Year	Size (km^2)	Difference 1984–1997 (km^2)	Difference 1997–2000 (km^2)	Difference 2000–2003 (km^2)	Difference 2003–2006 (km^2)	Difference 2006–2008 (km^2)
Lake Burullus	1984	701.7	52.2	7.1	177.7	5.0	3.8
	1997	649.5					
	2000	642.4					
	2003	464.7					
	2006	459.7					
	2008	456.9					

9.5.3 Pollution

The delta lakes are influenced by fresh water runoff from the land via drains and canals. This water enriched the lakes with nutrients including phosphate, nitrate and silicate. Also, some of these drains discharging considerable amounts of sewage and industrial wastes directly into the lakes (Saad 1990; Mitwally 1982; Siegel et al. 1994). The high concentrations of phosphorus, nitrogen in the organic waste, pesticides as well as heavy metals in water and sediments have altered the ecosystem of these lakes. The lake areas adjacent to these drains have been deteriorated and subsequently eutrophicated (Ahmed and Elaa 2003).

Human activities were found to be the main source of ecosystem pollution such as untreated water discharge and sewage. Lake Edku received large amount of domestic sewage, agricultural and industrial effluents; hence, it is highly polluted with humic substances and dissolved organic carbon (DOC) (El-Sayed 1993; Aboul-Ezz and Soliman 2000). Nafea (2005) reported that high concentrations of heavy metals were detected in macrophytes of Lake Edku. Some other studies were carried out on the metal concentration of Lake Edku such as those of Abdel-Moati and El-Sammak (1997), Adham et al. (1999), Appleby et al. (2001).

Coastal lakes are generally a region of high biophysical and human activities. Given the socio- economic importance of the Mediterranean coastal zone of Egypt, diverse activities have increased pressure on the uses of the contiguous lakes. These activates include fishing, aquaculture (fish farm), dumping liquid and solid wastes, land reclamation, urbanization, and recreational uses. Common activities and their magnitudes in each of the study lakes are listed in Table 9.1 and discussed here after.

9.5.4 Fisheries and Aquaculture

Fish landing in Egypt (including aquaculture) reached 407,000 tons in 1995 (GAFRD 2008). Lakes fishing accounts for 33% of the Egyptian fish catch (33,000 tons) Manzala and Burullus lakes producing 119,000 tons (59,500 tons each) versus 27%

from fresh water resources, namely Lake Nasser produced 109,000 tons while the production of Mediterranean and Red seas reached 91,000 tons (22%). Fishing fleet consists of 46,269 ships and boats, 2897 of them were engine vessels, the rest were manual boats. Fishing activities sustain a large number of people: 35,000–40,000 fishermen in Manzala and about 47,000 in Burullus (Sestini 1992). Common fish species are tilapias, catfish (fresh water), mullets, seabream, seabass, grouper, pagrus, sardines (sea water). At Bardawil Lake the fish caught are mainly migratory; the main species are Sparus aurata, Dicentrarchus labrax and Mugil cephalus (Meininger and Atta 1994; Zakaria et al. 2007).

Aquaculture is concentrated around the southern margin of Edku and Manzala lakes and Mariut Valley. Aquaculture production accounted for about 15% (62,000 tons) of total fish production in 1995. Tilapia represents about 38% (129,000 tons) of total fish production in 1994 (CAPMS 1994). In the meantime, illegal harvesting of fish fry in Burullus and Manzala lakes will disturb the fish habitat and reduce the natural fish stocks. Moreover, over fishing in Burullus and Manzala lakes have an adverse effect on both fish and larval stocks (Ramadan and Ibrahim 1995).

The total areas of Northern lakes have recently been decreased from 2222.60 to 1936.90 km^2. This is due to land reclamation. The reclaimed lands are used mainly for aquaculture and urban activities. Euryhaline fish fry used for aquaculture especially Mullet, sea bream, sea bass and eel are collected from collection centers scattered along the discharge canals (lakes inlets) connected to the Mediterranean Sea. However, fresh water fish fry especially tilapia and carp are produced artificially.

9.5.5 Lakes Inlets

Lake inlets are capable of changing their cross-sectional dimensions (depth/width) quickly, migrating rapidly along the shore, and even closing completely (Leatherman 1991). In Egypt, siltation problems of the lake inlets generally taken place as a result of the combination of sand transport in the longshore and cross-shore directions characterize most of the coastal lakes (Fanos et al. 1993). Siltation causes shoaling or closing the lake inlets resulting in navigation hazards, decreasing water flow in and out in the inlet channel, as well as negative implications on fishing activities. Keeping the inlet open is important for the fishery to keep the salinity down and to allow migratory movements. The sand barriers of study lakes (Burullus, Manzala, Bardawil lakes) are being subjected to severe beach erosion. This erosion is mainly due to the effect of prevailing dynamic processes of waves and currents, and the absence of the sediment supply resulted from the construction of the High Aswan Dam in 1964. This problem causes shoreline retreat of the lake barrier and hence reducing its function as nature protective line from sea invasion (Fanos et al. 1993).

9.5.6 Reclamation

Large parts of the delta lakes have dried up due to land reclamation. Reclamation activities are due to gain new land for cultivation, aquaculture and also for building illegal houses. These activities have increased the pollution of the delta lakes. In this study, published data obtained from different sources are used to estimate the average rate of lake areas dried. These results show that some of the delta lakes have been intensively reclaimed being at present less than 50% of their former size. Among the Delta lakes, Manzala has been reclaimed rapidly at an average rate of 6.12 km^2/year. The others Edku and Burullus are being reclaimed at a lower rates being 3.42 and 1.99 km^2/year, respectively.

9.5.7 Sea-Lake Interactions

Low-lying littoral deltaic regions are highly vulnerable to even minor changes in sea level, particularly because most deltas are actively subsiding. Moreover, predicted global warming might accelerate sea-level rise, which would have a pronounced impact on low-lying deltas all over the world. In particular, the Egyptian Nile delta coast is expected to be severely affected by sea-level rise (El Fishawi and Fanos 1989; Frihy 1992). Sea level rise could have a serious impact on the lake ecosystem and the fertile land that has been reclaimed from the lakes. Sea-level rise might affect the ecosystem of the coastal lakes by eroding the lake barriers that protect the lakes from the sea and hence altering the water quality of the lake.

9.5.8 Man-Made Interference

Coastal lakes have been recently planned to attract investors to implement large-scale developing projects. Presently, a large-scale project has been implemented to modify coastal lakes to be used development area through construction of artificial inlets associated with long jetties, to connect the large lake with the Mediterranean Sea. These protruded jetties have created local erosion along adjacent beaches on the down drift side. Also, a large urban expansion was built around the lakes. These are presently undergoing extensive investment development. Experiencing, at the time being-potential problems resulting from increasing human pressure and man-made interventions. Bad management of the remarkable white calcareous sand beach has created some serious erosion environmental problems. As a consequence, rapid and uncontrolled development has created many human infringements on the coastal zone and subsequently impact on the coastal development projects (Rasmussen et al. 2006). This impact has changed the shoreline stability and expected to alter the water quality of the sea. Studies have shown that, in addition to the natural phenomena, man has affected physiographic and ecological transformations of the northern lake of Egypt. Anthropogenic alterations include pollution, land reclamation, agriculture, aquaculture, irrigation and industrial projects. Coastal lakes are undergoing very rapid change as a result of increased human activities resulting in pollution problem. Industrial, sewage and agriculture wastes are dumped into the Nile drain system which discharges locally into delta lakes (Mariut and Manzala) (Ahmed et al. 2009). The degree of pollution of these lakes varies from high in the eastern part of Mariut and in the southern, eastern part of Manzala lakes to very minor in the Burullus Lake. In view of increasing pressure in the coastal zone, a great amount of pesticides, insecticides, fertilizer are used for agricultural fields around the delta lakes. Pollutants are washed out from these fields to the canals and drains and then directly to the lakes. These runoff has been badly affected the water quality of the lakes.

9.6 Information Technology Challenge

Remote sensing of the environment is one of the subfields that has materialized in the past three decades as a foundational element in Egyptian lakes monitoring and in informing investigations into global environmental change and environmental resource management.

The use of remote sensing in Northern lakes studies in Egypt is not recent. Despite the commendable progress made in the field of using remote sensing in northern lakes, the adoption of a "scientific" view in the analysis of remote sensing images is quite limited. Little work has been carried out on automated estimation and mapping of lake variables, quantitative modeling and analysis of uncertainties and errors associated with estimated variables, or utilized remotely sensed measures in physical dynamic models of environmental phenomena. Several factors contributing to this limitation are documented below. It should be noted however that these factors are not exclusive to the Egyptian remote sensing community.

The first factor concerns the current-state-of-art in remote sensing analysis. Most remote sensing scholars in Egypt, still rely on tools that mix a variety of qualitative techniques of visual image interpretation with traditional computerized classification methods developed in the formative years of remote sensing when the quality and options of satellite imagery were limited. There are obvious restrictions to this mode of remote sensing practice that make it inefficient for environmental applications (Gong 2006). The most obvious is a great deal of human subjectivity that renders the results inconsistent and incomparable across spatial and temporal scales, and the inability of such techniques to handle large volumes of data or to produce useful inputs for quantitative models and/or simulations.

The second limiting factor relates to the nature of landscapes in which human/lakes interaction take place. Egyptian northern lakes are diverse and composed of a mosaic of many anthropogenic and natural surface materials interwoven with one another in irregular patch sizes and arrangements, and in varying densities that do not fit the idealized representation of image pixels (Fisher 1997; Forster 1985). Experiments have shown that even a 1-m spatial resolution image is likely to include mixed pixels that may produce misclassification errors due to the spectral “noise” from small features that become noticeable at this fine scale (Aplin 2003; Mesev 2003). Various methods of fuzzy classification (Foody 1999; Zhang 2001) and spectral mixture analysis (Phinn 1998; Rogan and Chen 2004; Ward et al. 2000) have been used to perform the sub-pixel mapping. However, studies have documented considerable degrees of uncertainty in the final results (Herold and Roberts 2006; Rashed et al. 2003; Small 2001a).

A third factor complicating quantitative remote sensing and the automatic identification of earth surface features relates to design tradeoffs between the spectral and spatial resolutions of a satellite sensor due to the cost/power/transmission considerations of data acquisition (Aplin 2003; Price 2001). Higher spatial resolution satellite sensors generally have a coarse spectral resolution (broad bandwidths) chosen to provide an optimal contrast between generalized land cover classes (e.g., urban versus vegetated) (Jensen and Cowen 1999). This imposes a considerable limitation on the extraction of features. Although experiments have shown that the majority of Built Surface Materials (BSM) are in fact spectrally separable (Chen and Hepner 2001; Herold and Roberts 2005, 2010; Small 2001b), the majority of operational satellite sensors are “blind” to the spectral regions at which BSM can be differentiated from one another, simply because they were not designed to detect subtle spectral differences in BSM reflectance.

A fourth factor limiting the wider application of quantitative remote sensing analysis in Egypt stems from the general lack of knowledge about the spectral properties of environmental materials. The near future project at National Authority of Remote Sensing and Space Sciences (NARSS) carried out to develop an understanding of the spectral response patterns of native materials in Egypt, i.e. Northern Egyptian Lakes. Contributing to this lack of knowledge is the limited availability of trained personnel as well as the sensors that can be used to provide a quantitative and qualitative characterization of environmental features for lake ecosystem. This situation has changed recently with the launch of the Earth Observing-1 (EO-1) satellite, carrying an onboard hyperspectral instrument, Hyperion, thus opening new horizons in the automated identification and mapping of feature spectra on a global basis.

This and other recent improvements, such as a large number of available and better-calibrated sensors and their higher measurement precision, has made room for the extensive use of quantitative algorithms for measuring and estimating biophysical variables from remotely sensed images, radiative modeling of the atmosphere, etc., and state-of-the-art techniques for sensor calibration and modeling. The potential of these developments will not be fully realized in Egypt, however, without developing a priori knowledge about the spectral response patterns of earth surface materials and improving upon quantitative techniques for utilizing such spectral measurements in image analysis.

It is in this capacity that remote sensing can be effectively used to generate information for use in the assessment, management, and monitoring of environmental risks in Egyptian northern lakes and to remedy environmental problems at various scales.

9.6.1 Lake Satellite Specifications

Five types of satellite data products are used in northern lakes studies: QuickBird, Aster, Landsat, Hyperion, and EgyptSat1. These sets of images acquired for every site. Aside from Landsat data, and their full potential, in terms of the automated mapping of lakes parameters or their utility for quantitative models of the lake environment, has not been assessed yet.

The first sensor, QuickBird, has a limited spectral range but produces the type of data typically welcomed in the remote sensing community due to its fine spatial resolution. The second, ASTER sensor, has a medium spatial resolution comparable to Landsat but its spectral range is considerably extended regarding shortwave infrared radiation. ASTER's spectral library has been limited to a generalized list of environmental materials, the majority of which are vegetative materials.

The third sensor, Hyperion, has a spatial resolution that is slightly coarser than ASTER (30 m) but provides 220 unique spectral bands with a 10-nm bandwidth. Recent studies have reported difficulties in applying spaceborne hyperspectral data, especially to lakes ecosystem, and highlighted the need for more investigation (Rogan and Chen 2004). To date, the literature provides little information about the utility of Hyperion data in automatic mapping and reliable quantitative estimation of environmental features (Ahmed et al. 2009).

Finally, EgyptSat1 is Egypt's first remote sensing that was successfully launched earlier this year (jointly built by NARSS and Yuzhnoye Design Bureau in Ukraine). The sensor produces "SPOT-like" data with a 50 day revisit cycle in the visible (Green and Red bands) and near infra-red with 7.8 m spatial resolution. A fourth band is produced in the mid-infrared but at a coarser spatial resolution (~40 m). EgyptSat1 is the first of a series of satellite sensors NARSS has planned to build before 2015. In this study, lakes ecosystems were monitored and examined the utility of EgyptSat1 in quantitative remote sensing analyses and provide spectral measurements for sensor calibrating, mentoring of sensor performance, and informing the design and implementation of the future sensors in Egypt. In addition to remote sensing data used in this study, several geospatial ancillary data layers (e.g., DEM, land use map, urban streets, geological maps, housing census, native vegetation species, crop yields, population) are used in the applied lakes studies and in guiding the fieldwork and sampling, and validating the results.

9.6.2 Spectral Analyses

This is the most critical and vital task in remote sensing studies. Field spectroscopy is a technique used to study the spectral response of features on the lakes components based on their emitted or reflected energy. Field spectrometry is produced using portable spectroradiometer (or spectrometers) to collect quantitative measurements of radiance, irradiance, reflectance and transmission of ground targets in the field. The quality of collected spectra depends on a variety of factors including the suitability of the field spectroradiometer for the materials of interest (e.g., spectral range, signal-to-noise ratio), the time of year/day for collecting spectra, spatial scale of the field measurement, target viewing, and illumination geometry (Goetz 1992; Rollin et al. 1997).

Using this spectroradiometer ensure that the spectra acquisition follow pre-defined standards that will be applied uniformly across all study lakes to ensure sufficient quality and comparability of the images Processing results. This spectrometer instrument represents state of the art in the technology with a fine spectral resolution of 3 nm possible spectral sampling interval and 10 spectra per second. Its light weight and ability to stay operational under severe weather condition will help ensure repeatability of results and thus better discrimination and analysis of spectra.

In addition to relying on the stratified spatial sampling procedures for spectra acquisition, imaging spectrometry data compiled in other contexts (such as those developed by ASTER images) will be examined to guide and complement the field surveys for some lakes and the help scaling between in-situ and satellite observations. All spectra and metadata are described and stored in a standard spectral database format.

9.7 Methodology

9.7.1 Image Processing (IP)

Image processing is a branch of computer graphics based on image data, which are pieces that make up a picture. In essence, image processing is a special form of two-dimensional (and sometimes three-dimensional) signal processing. Image processing is a powerful technique for uncovering information. The principle idea behind image processing is to make an image more informative. Computer system treats images as arrays or series of elements. The size of elements in an array determines the resolution of the image, and the number of bits available to any element of the array determines the number of colors that each element can have.

Images occur in various forms. Some visible and others not, some abstract and other physical, some suitable for computer analysis and others not, It is thus important to have an awareness of different types of images. A lake of this awareness can lead to considerable confusion, particularly when people are communicating ideas about images when they have different concepts of what an image is. An image is: "a representation, likeness, or imitation of an object or thing, a vivid or graphic description, something introduced to represent something else". A picture is a restricted type of image, it may be defined as "a representation made by painting, drawing or photography, a vivid or graphics, accurate description of an object or thing to suggest a mental image or give all an accurate of the thing itself". For our purposes, we take the word picture to mean a distribution of matter that is visible when properly illuminated. So the word picture is sometimes used as an equivalent to the word image.

Now digital image processing is subjecting a numerical representation of an object to a series of operations to obtain the desired result. Digital image processing starts with one image and produces a modified version of that image; it is, therefore, a process that takes an image into an image. Digital image analysis is taken to mean a process that takes a digital image into something other than digital images, such as a set of measurement data or a decision. The following are the steps used in the analysis of the satellite image.

9.7.2 Geo-referencing

It is the process of assigning map coordinates to image data. The image data may already be projected onto the desired plane, but not yet referenced to a proper coordinate system. Rectification involves geo-referencing as all map projection is associated with map coordinates. Image-to-image registration involves georeferencing only if the referencc image is already georeferenced. Georeferencing, by itself, involves changing only the map coordinate information in the image file, the grid of the image does not change. Geocoded data are images that have been rectified to a particular map projection and pixel size, and usually have radiometric correction applied.

9.7.3 Image Registration (Rectification)

The problem with using imagery from different dates, or from different sensors, is that, in order to be useful and to be able to compare separate images pixel by pixel, the pixel grid of each image must conform to the other images in the database, they have to be registered to a common coordinate system. This coordinate system can be one of the images or can be a map projection. The selection of appropriate map projection and the coordinate system is based upon the primary use of the data.

Any conversion to another coordinate system not only implies a loss in radiometric accuracy through resampling but is extremely time-consuming and error prone. Usually, it is done through the manual identification of ground control points "GCPs" of the same object in both coordinate systems. Control point identification is difficult, the difficulty increase with decreasing spatial resolution, and it is time-consuming, as it requires a matrix of a sufficient number of suitably spaced control points.

A transformation matrix is computed from the GCPs; the matrix consists of coefficients that are used in the polynomial equation to convert the coordinates. The size of the matrix depends on the order of transformation; linear transformation or first order transformation is used to drive the polynomial equations with the least error to be used to transform the reference coordinate of the GCPs to the UTM coordinates. Control points, which have large errors concerning the polynomial, are described, and a new fit is carried out. This procedure is repeated until an acceptable error is reached. After the polynomial is defined, it is used to transform the image into the desired coordinate system. To reduce the time taken, remove subjective human intervention, while achieving the highest accuracies, an automatic procedure has been developed. This procedure is to automatically register the imagery to a sub-pixel level of accuracy. It is assumed that gross distortions, rotations, and scale changes will not be encountered. The registration is carried out on the shapes of significant objects in the imagery.

9.7.4 Image Resampling

After completing the registration process, create the output file. Since the grid of pixels in the source image rarely matches the grid for the reference image, the pixels are resampled so that new data file values for the output file can be calculated (Fig. 9.7). The nearest neighbor is one method of resampling which uses the value of the closest pixel to assign the output pixel value. This method is the easiest one of all resampling methods, and it has a great advantage as it transfers original data values without averaging them as the other methods do; therefore, the extremes of the data values are not lost.

9.7.5 Image Mosaicking

After rectification and resampling of the desired images, mosaicking of every two images to obtain the area under study took place. Now the images multi-used are ready for the purpose applications.

9.7.6 Image Classifications

The classification process is described as the identification of the pattern associated with each pixel position in an image regarding characteristics of the objects or materials at the corresponding point on the Earth's surface (Mather 1978). The large area covered, the range of cover types and the paucity of ground data of TM series required the use of the classification method based on an interactive ISODATA (iterative self-organizing data algorithm) approach (ERDAS 2000). The ISODATA algorithm attempts to cluster the spectral (feature) space into a number of groups specified by the user. From an examination of the image data along with available ancillary data, both spectral and spatial, were separated initially. Statistics from the clustering were then used as input to the maximum likelihood classifier. With this classifier, it is assumed that the statistics of each class in each band are normally

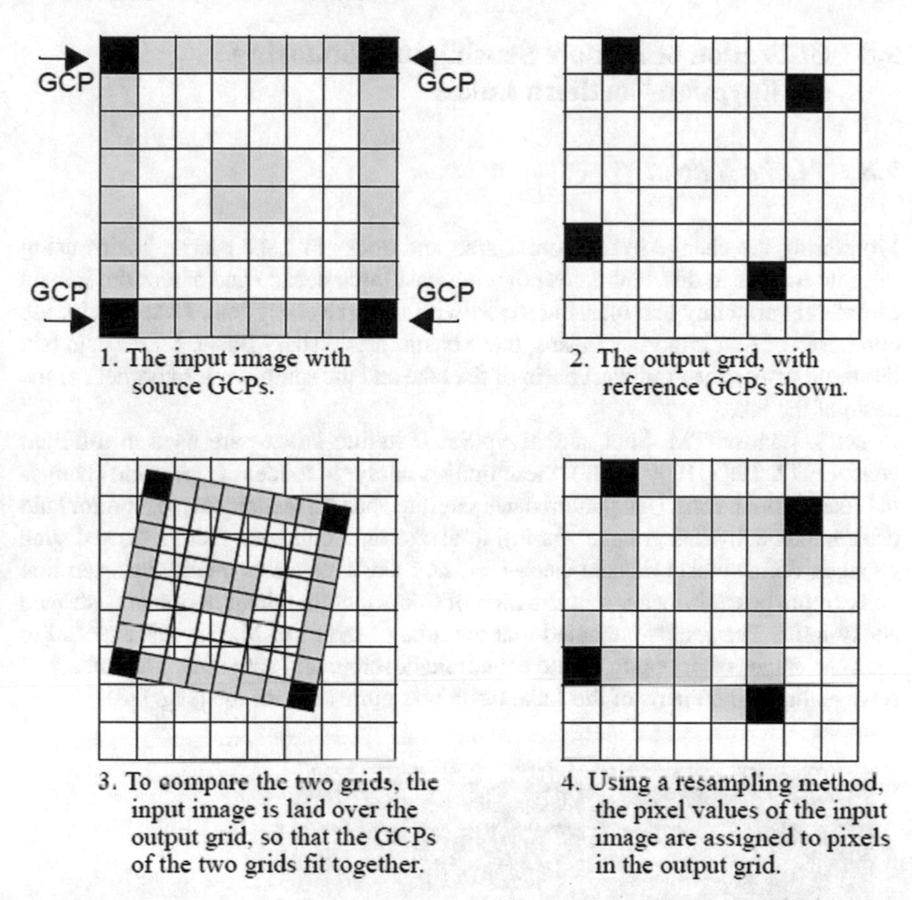

Fig. 9.7 Resampling methods

distributed, and that given pixel belongs to the class in which it has the highest probability (based on this distribution) of occurring. The results from the classification were interpreted with supporting data from topographic maps, aerial photography, local knowledge and the spectral and temporal properties of the classes themselves. Ten Classes were identified covering the lake and neighboring areas, which were too big or covered many types were further divided by clustering and reinterpreted.

9.8 Utilization of Remote Sensing in Monitoring the Egyptian Northern Lakes

9.8.1 Lake Mariut

Monitoring the change in the border area and docks in Lake Mariut basins using satellite images in different dates covering past three decade and assess the impact of natural variability and man-interfered with the lake ecosystem. Lake Mariut are composed of four important basins, these basins are; Fishery Basin, Main basin (six thousand acres), the northwest basin of the lake and the south-western borders of the basin of the lake.

MSS, Landsat-TM, Spot, and EgyptSat-1satellite images are used in different years (1972, 1985, 1999, 2007), these images analyzed to identify possible changes in Lake Mariut basins. Using multi dates satellite and GIS techniques, to monitor land uses/human activities in Lake Mariut, to assess environmental factors caused such changes, define most probable causes for the Lake's resources quality deterioration and explore possible means of mitigation of side effects or rehabilitation of destroyed ecosystems. The results indicated that the area of the Lake Mariut was exposed to multiple stages of drying to create economic development activities. The impact of sewage disposal on parts of the Lake basin was quite obvious too (Fig. 9.8).

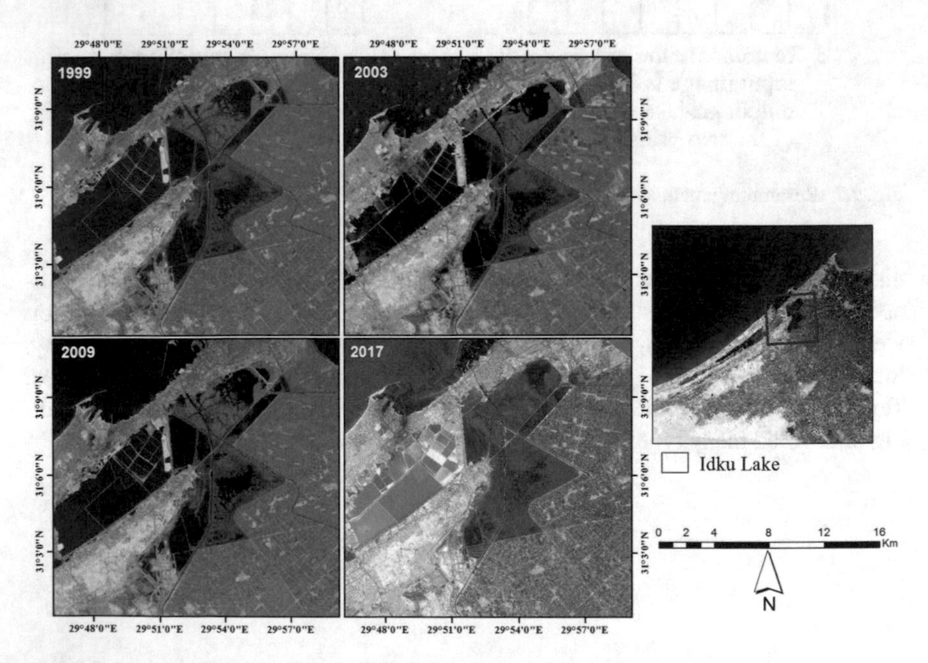

Fig. 9.8 Size change measurements from satellite images of Lake Mariut (Marine sciences department-NARSS)

The use of multidates satellite images of different sensors make the identification of historical changes of Lake Mariut available. This technology makes the quantitative and qualitative assessment for the Egyptian lakes is important to manage and take the correct decision. The monitoring and protecting of lakes ecosystems are considered as a priority for assessment and sustainability of Egyptian lakes.

The study environmental situation of Lake Mariut since Lake Mariut suffered during the past forty years to various types of pollution, whether the exchange polluter and ill-treatment resulted in the end to the pollution and the deterioration of their situation.

Most suffered from the pollution of the lake is dumping huge quantities and persistent remnants of industry, agricultural waste, sewage, and large areas were drying them randomly decreasing area from 248 to 61,414 km^2) (Figs. 9.9 and 9.10).

Monitoring the environmental situation of Lake Mariut indicated that the lake is complaining various types of pollution and anthropogenic inputs to the lake body; however, it is necessary to understand the potential actions lead to reduce negatively the change of environmental development of the lake using information technology. The use of such advanced techniques are useful in the identification of evolution incident lake area with large scale and lead to the availability of much information on the lake historical evolution since long periods up to now. Monitoring and assessment these changes in Lake Mariut basins using multidates images assist the decision makers to estimate the causes and rate of change, drying in the lake basins; to propose the solutions of identified problems, i.e. dynamic changes of vegetation covering percentage in the lake basins.

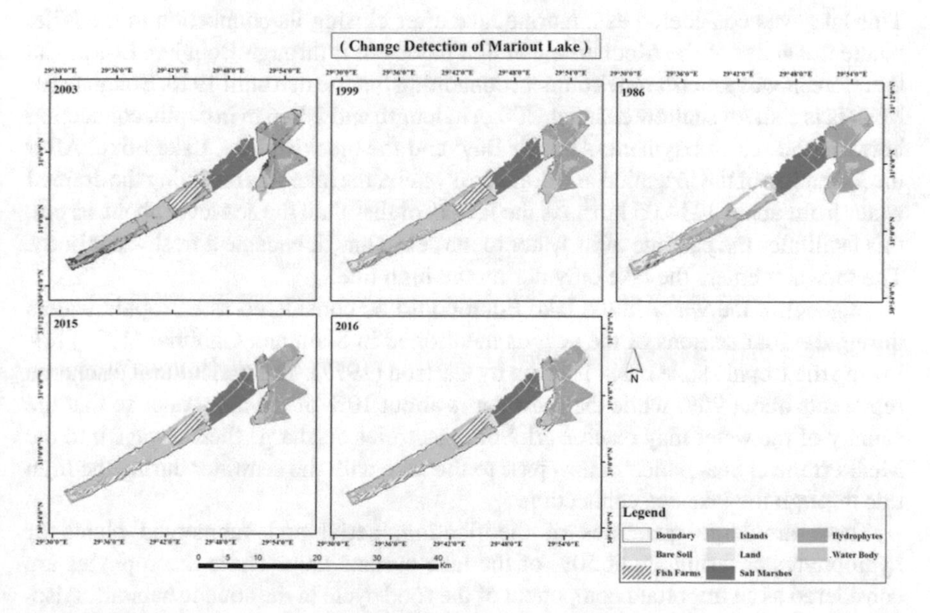

Fig. 9.9 Change detection of Mariut Lake

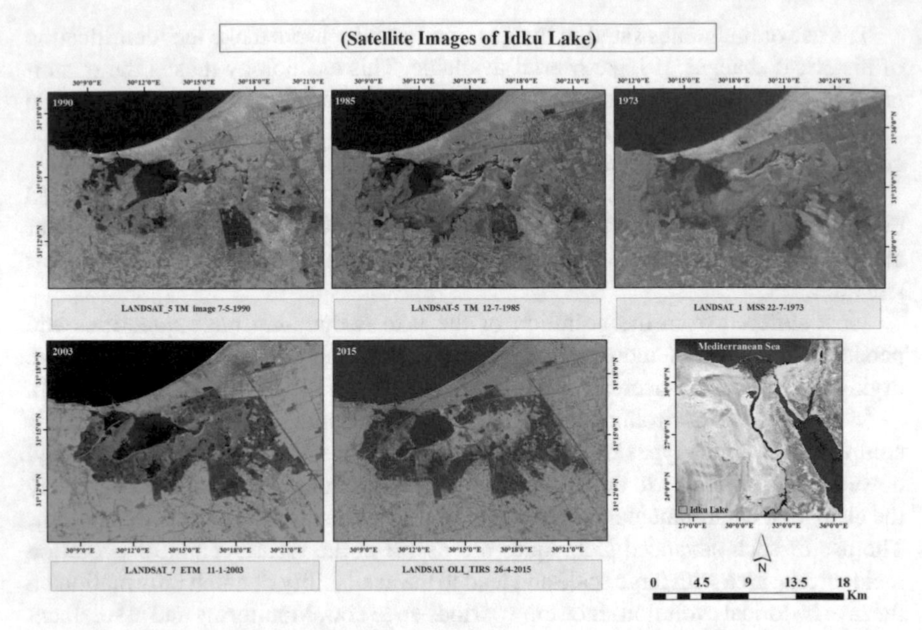

Fig. 9.10 Satellite images of Lake Edku (Marine sciences department-NARSS)

9.8.2 Lake Edku

This lake was considered as a marine lake after closing its connection to the Nile; where the water of the Mediterranean Sea passes in it through Boughaz El-Maadia therefore, it was a marine environment containing marine fish until 1920. Boughaz El-Maadia is a short, shallow channel, 300 m in length and 2.5–6 m in depth, connecting between the sea embayment, Abu Qir Bay; and the brackish lake, Lake Edku. After the formation of the irrigation and drainage system, the lake was receiving the drained water from about 1214.05 km^2. As the lake is higher than the sea level about 16 cm, this facilitates the passage of its water to the sea. Thus, it became a freshwater body. The seawater enters the lake only during the high tide.

In general, the water in the lake Edku could be considered as eutrophic waters during the four seasons of the year as mentioned in Siam and Ghobrial (1999) following the trophic state index is given by Carlson (1977). The agricultural discharge represents about 90% while the seawater is about 10% of the lake water so that the salinity of the water may reach 2 g/l. Some factories discharge their sewage into the Mediterranean Sea, which in turn pass to the lake with the seawater during the high tide through the lake-sea connection.

There are large quantities of the floating, aerial and submerged plants or hydrophytes covering about 50% of the lake surface area. These macrophytes are considered as an important component of the food cycle in the aquatic habitats. Also, this vegetation resists the movement of the fishing boats causes many problems for the fishermen and decreasing the surface area of the lake. The fish production of Lake

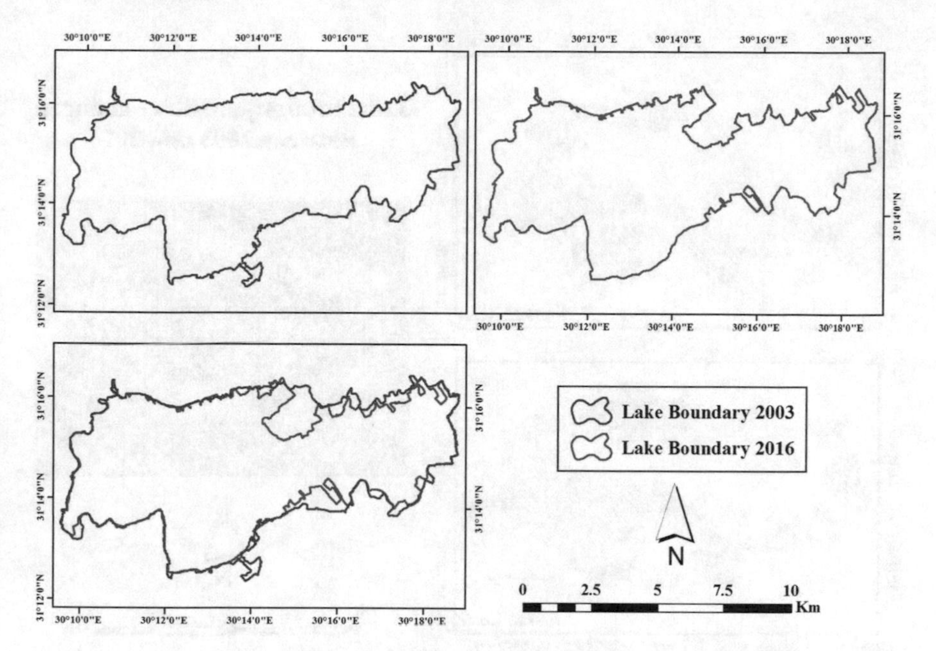

Fig. 9.11 Lake Edku area changes between 2003 and 2016. Explain better what is shown on every picture

Edku may reach 19,000 tons, which is approximately a value added to the Egyptian economy.

Lake Edku has complained dramatic changes in its boundary. The size of the lake recorded about 62 km^2 in 2003 while the area of the lake decreases to 55.88 km^2 in 2016. This difference reveals to the abuse of the lake management and drying of the lake (Fig. 9.11). The fish farm is increased from 61.43 to 67.79 km^2 (Fig. 9.12). This interprets the loss parts of the lake area replaced by the fish farm activities.

The Lake Edku vegetation is covering more than 70% of the lake area. This dynamic movement of the vegetation reflects the lake water interaction with the saline water (Fig. 9.13).

9.8.3 Lake Burullus

Multi-dates Images "Landsat TM" are used to monitor the characteristic of both water and land and determine the long-term changes taking place along the lake system and peripheries. The dates of the images are 1984, 1997 and 2000 with a resolution of 25 m/pixel. The multi- date images are also used for Burullus inlet change detection with time (Fig. 9.14).

The merging of multi-sensor image data is becoming a widely used procedure because of the complementary nature of many data sets. Merging information from

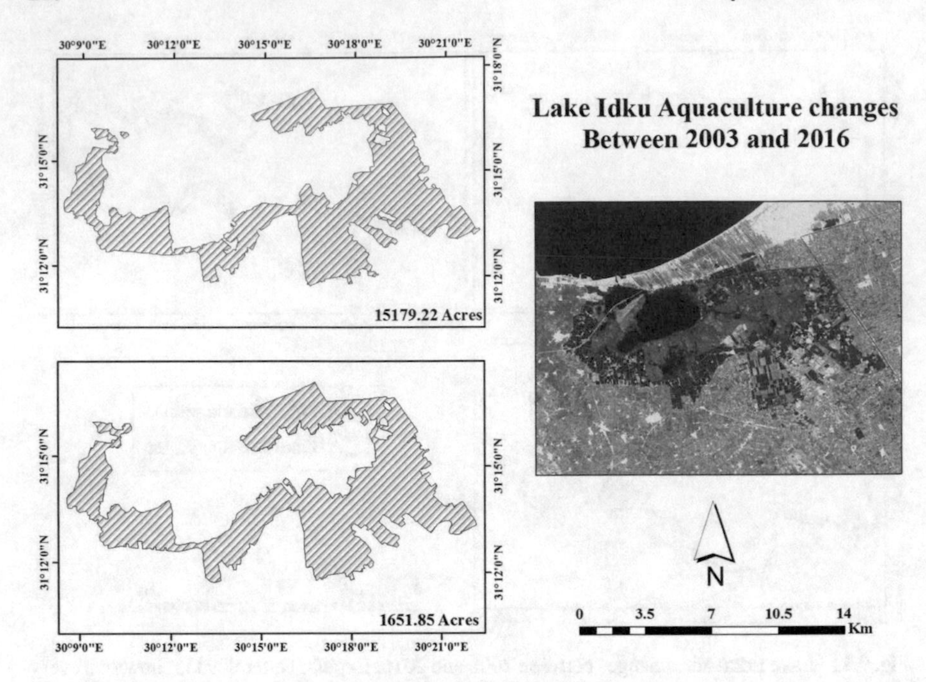

Fig. 9.12 Lake Edku fish farm changes between 2003 and 2016

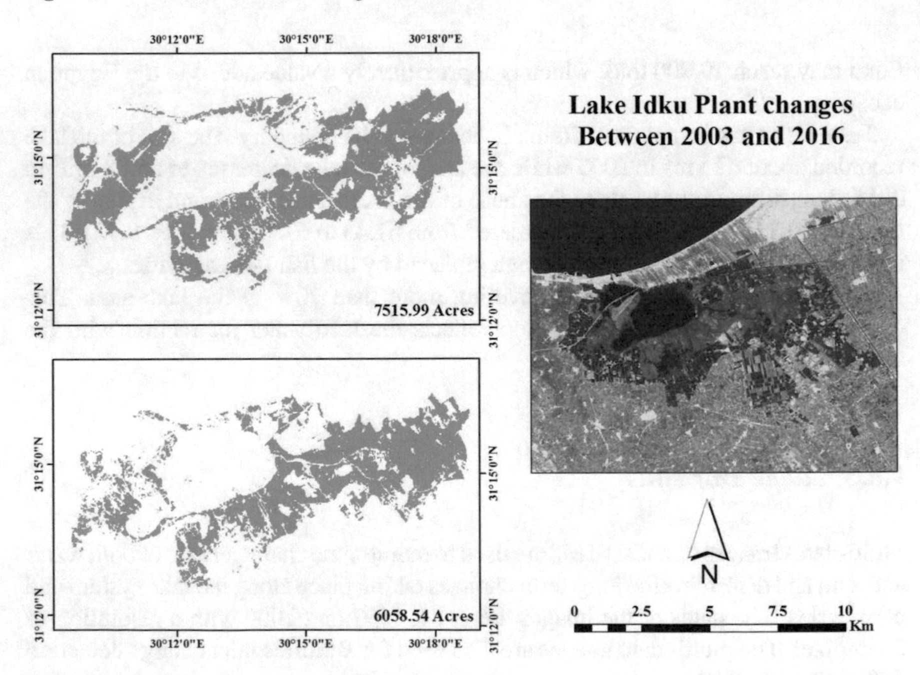

Fig. 9.13 Lake Edku vegetation changes between 2003 and 2009

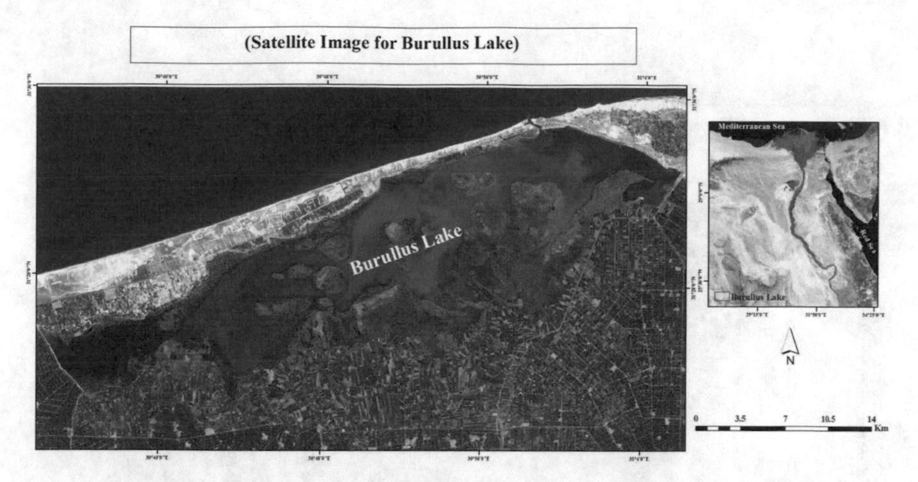

Fig. 9.14 Processed satellite image of Lake Burullus shows the vegetation and water of the lake

different imaging sensors involves two distinct steps. First, the digital images from both sensors are geometrically registered to one another. Next, the spectral and spatial information contents are mixed to generate a single data set by using various transformations such as Hue-Intensity-Saturation (HIS) transformation (Carper et al. 1990).

Reclamation projects in the southern and eastern margins have substantially decreased the size of the lake. To protect the lake inlet from siltation and beach erosion, two jetties were constructed on the western and eastern sides of the inlet. In order to combat siltation problem in the inlet channel, a new jetty was constructed in 1991 between the two jetties to reduce the equilibrium cross section. The identification classes of Lake Brullus are lake water, seawater, plants, marshes, islands, cultivated areas, urban areas, sand dunes and shallow water areas (Fig. 9.15).

The results reveal that the lake has been increasingly subjected to intensive and diverse human activities including fishing, aquaculture industry, dumping wastes, land reclamation, urbanization, saltpan, and recreational uses. Some of these activities have resulted in several environmental problems at a time when the population is expanding exponentially. Pollution, reclamation, fragmentation, over-fishing and illegal harvesting of fry fish are the major environmental issues threatening the fragile ecosystem of the lake. Also, ongoing natural factors have induced substantial changes in the lake environment. These changes include barrier erosion, inlet siltation and rising sea level as well as prevailing sedimentary processes (Table 9.1).

The outputs gained from this study can be considered as a practical lesson learned in future planning and management of Burullus Lake. Moreover, the information presented here is important for managing, improving and conservation of other coastal lakes of Egypt.

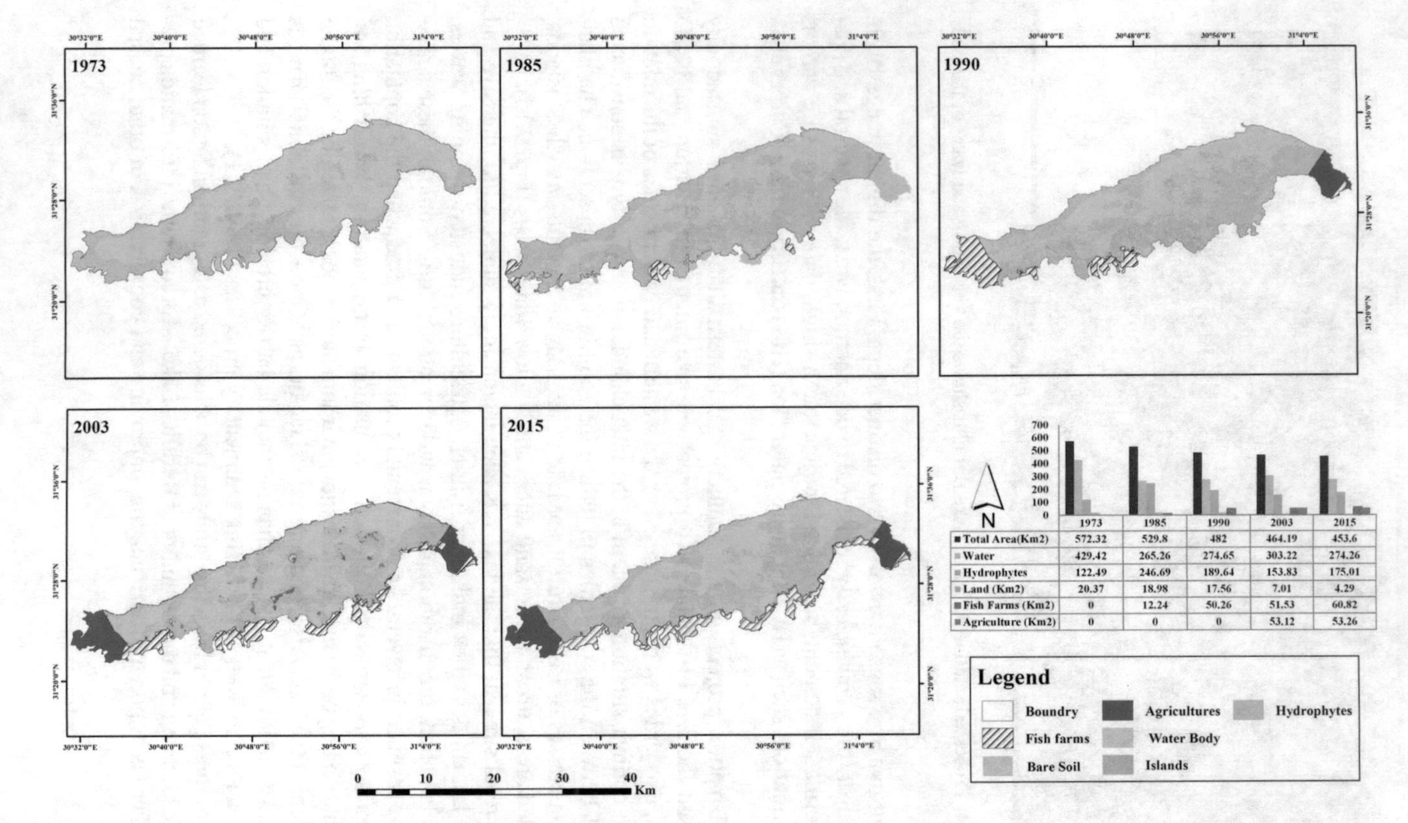

Fig. 9.15 Images classification of Lake Burullus between 1973 and 2015

9.8.4 Lake Manzala

Multidates satellite images have been used to detect the trends of environmental changes in Lake Manzala, during the past three decades. The results represent the way in which ecosystem degradation processes, affecting Lake Manzala, are increased due to external factors (human-induced). A study of ecological changes in the Lake Manzala over the period 1973–2003 has been undertaken to evaluate the effects of the human activities of drying and reclaiming some areas to establish development projects and urban settlements, and to estimate the consequential effects on the lake (Fig. 9.16). A combination of Landsat MSS, TM and ETM+ data were used to analyze temporal changes in the lake ecosystem. Different image processing techniques such as unsupervised classification and enhancement were used to determine the proportions of major ecological features within the lake. An arbitrary number of classes were assigned to represent all types of features. Classes were grouped into three Major classes (water, vegetation and land) within the lake boundary. Each feature class was identified spectrally, and consequently, the area of each major feature was calculated.

The change detection algorithm used is based upon the unsupervised classification technique. Details are provided by Olmanson et al. (2001). The basic premise in multi- spectral computer classification is that terrestrial objects manifest sufficiently

Fig. 9.16 Satellite image of Lake Manzala (Marine sciences department-NARSS)

different reflectance properties (digital values commonly known as spectral signatures) in different regions of the electromagnetic spectrum. Based on these spectral signatures, natural and cultural surface features can be discriminated and a new output image could be created having a specific number of classes or categories.

The results of the classification process revealed that five classes of vegetation showed different spectral patterns. Also, eleven water classes and five different land classes were identified. Changes that are analyzed and calculated from the extracted information explain how agricultural land-use practices interact with the characteristics of the study area. The results of the analysis indicate that Manzala Lake has been subjected to various physical and biological changes mainly due to the different human activities that have serious impacts on its quality and a subsequent deterioration in its ecological parameters.

This project is a research project designed to explore the application of a coastal lake changes based on satellite remote sensing and ground truth data (Fig. 9.17 and Table 9.2).

9.8.4.1 Vegetation of Lake Manzala

The aquatic plant classification methods consist of two major procedures: separation of image features into discrete units and classification of the pixels in each unit as described by Olmanson et al. (2002b). This was accomplished by performing an unsupervised classification. Vegetation classes were spectrally analyzed presenting 7 different species (plants). The different plants were identified/named according to the field survey information. The amount and percentage of each plant were identified. Manzala has clear water and an abundance of aquatic vegetation. Therefore, we assumed that aquatic vegetation was present throughout the lake. The next step thus was to stratify the lake into emergent and submergent vegetation features.

Terrestrial Change

The ecological change along the past two decades was investigated by qualitative and quantitative comparison between the historical images of the lake. The size of the lake was calculated, and as shown in Fig. 9.6, the overall size of the lake decreased about 50% from 1973 until 2003. Also, the water percentage area is decreasing whereas the vegetation and land percentage of areas overall the lake are increasing from 1973 until 2003 ass shown in Table 9.3. The decrease of the area is about 24% from 1978 comparing to 1973. The decrease of the area is changing in the next 5 years to 6%. Then, in the next 3 years the decrease in area is increasing to about 20%. In the next 3 years, the decrease in area is 7%. In the next decade, the area of the lake is decreasing about 8%. Finally, the decrease of the area in 2003 compared to 2000 is about 3%.

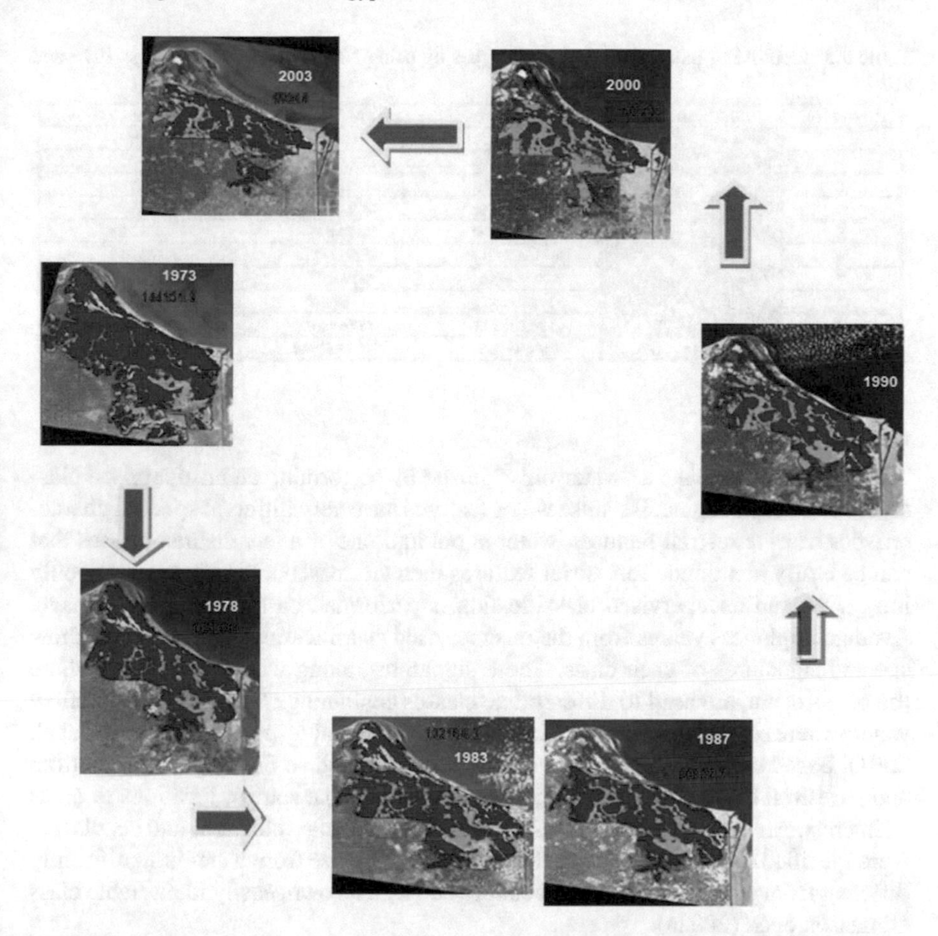

Fig. 9.17 Changes of water, vegetations and terrestrial components of Lake Manzala derived from satellite imagery during the last three decades (Marine sciences department-NARSS)

Table 9.2 Changes of water, vegetations and terrestrial components of Lake Manzala

Date	Water (%)	Land (%)	Vegetation (%)
03/07/1973	74.25	8.89	16.86
10/05/1978	74.68	7.85	17.46
16/05/1983	68.21	15.81	15.98
07/11/1987	71.68	8.01	20.31
04/08/1990	67.57	7.86	24.57
22/07/2000	55.35	6.46	38.19
05/02/2003	72.33	9.30	18.37

Table 9.3 Identifying Lake Manzala habitat types by using NDVI techniques between 1973 and 2017

Habitat type	NDVI Color index	1973 Area/Acres	2017 Area/Acres	changes in Area/Acres
1		0.872822	0.024	−0.849
2		709.917445	6677.489	5967.572
3		709.917445	19226.091	18516.174
4		8692.309893	18344.178	9651.868
5		32372.78213	15969.545	−16403.237
6		42264.24249	9095.747	−33168.496
7		10020.64646	3813.622	−6207.024
8		1463.91501	2488.836	1024.921
9		134.071722	1173.140	1039.068
	Habitat Total	96368.67541	76788.672	−19580.003

Water Classes

The first step is to make a "water-only" image by performing an unsupervised classification using Imagine. Because water features have very different spectral characteristics from terrestrial features, water is put into one or a few distinct classes that can be easily identified. Terrestrial features then are masked, creating a water-only image. Second unsupervised classification is performed on the water-only image. Average brightness values from the unsupervised classification are graphed to show spectral signatures of each class. These signatures, along with the location where the pixels occur, are used to differentiate classes containing clear, turbid, or shallow water (where sediment and/or macrophytes affect spectral response) (Olmanson et al. 2001). Based on this information, classes are re-colored so that vegetation, bottom and terrestrial effects can be avoided when selecting lake sample locations or areas of interest. All seven Landsat bands were used for the classification, and six classes were specified. Because the spectral-radiometric response from water is significantly different from terrestrial features, it should be given its own, easily identifiable class Olmanson et al. (2002a).

NDVI Classification of Lake Manzala Habitat

The Normalized Difference of Vegetation Index (NDVI) is a measure of the amount and vigor of vegetation at the surface. The magnitude of NDVI is related to the level of photosynthetic in the observed vegetation. In general, higher values of NDVI indicate greater vigor and amounts of vegetation. The NDVI was calculated from TM images 1973, and 2017 indicated that eight types of vegetative habitat could be recognized. Each of these habitats is recognized by certain colour index (Fig. 9.18). During the current study, two images were selected to perform this analysis belonging to the years 1973 and 2017, collecting of these two dates is due to the greening coverage on the surface of Lake body have to be detected at the same time. The data obtained from the two images were subjected to analysis including spatial vegetation patterns and assessment of vegetation dynamics. Furthermore, this biophysical parameter is a key remote sensing observation related to several important biospheric properties including the proportion of photosynthetically absorbed radiation and leaf area index were also used.

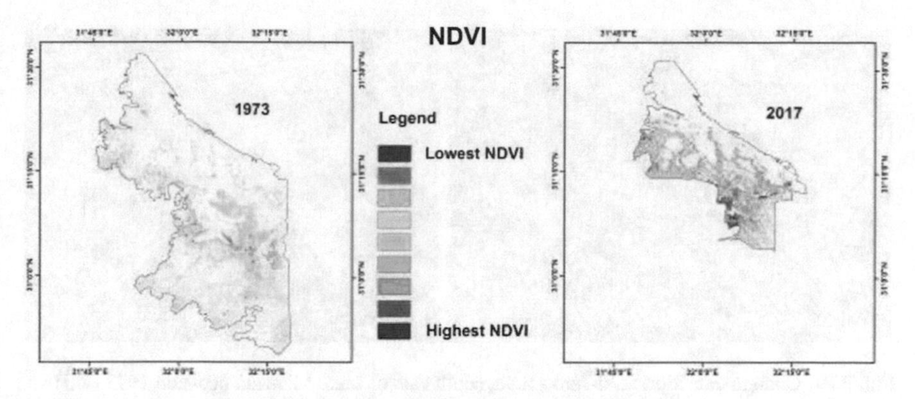

Fig. 9.18 Different types of vegetative habitat in Lake Manzala (Marine sciences department-NARSS)

The data calculated from both images indicated changes in the size of the different habitat within the lake between the two years. Table 9.3 listed the area of different habitat as obtained from the images and the value of changes between 1973 and 2017. The examining of these results showed that some of the lake habitats suffer a considerable decrease in area while others showed an increase in area, i.e. the decreasing percentage in greening habitat area ranged from −1.5% in habitat type 5 to −48.5 in habitat type 8, while the increasing percentage in the habitat area ranged from 10.03 in habitat type 4 to 28.95 in habitat type 2.

The data in Table 9.3 indicated that the Lake habitat which decreased in area has a total of 66% while those increased in area are represented by 55% in total. These means that a certain area of the habitat, with almost 11% value, was lost completely from the lake. However, it is safe to say that the change of certain habitat area is most probably connected to the faunal and floral composition of these habitats and the current rate of habitat loss will affect the biodiversity in the lake in the near future.

Detecting of Landfilling

EL Bashteer area is one of the fast developing areas in Lake Manzala. One of the most alarming practices in this area is land filling. In the current study images of 1973 and 2017 were analyzed for indicating the area lost from the lake due to landfilling. The results of quantitative analyses of Bashteer area are indicated the human interventions in the Lake ecosystem. These results for the Bahr El-Bshteer recorded the 100.726 km cutoff (dryed) the Lake area during 13 years. This is indicating that the developments of Lake are affected on the land reclamation with the rate of 7.75 km a year (Fig. 9.18).

Detection of Anthropogenic Effects

El Jenka area is one of those area in the lake subjected to different sorts of land filling for the establishment of a small fish farm named Hosha. The examination of the images at 1993 and 2017 proved that the human interventions "cultivated the land". This Hosha is used for fish fry catches which is illegal behavior inside the Lake. Results of such this illegal behavior are a loss of areas from the lake. Quantitative

Fig. 9.19 Change detection at El-Jenka area, south east of Lake Manzala between 1993 (left) and 2017 (right). *Source* Satellite, date?

analysis for El Genka area (black spot area) recorded about 8.45 km cutoff (drayed) the Lake area during 13 years, Fig. 9.19.

Detection of Pollutants Outflow

The out flow of Lake Manzala water into the Mediterranean Sea may cause certain problems connected to the water quality in the area. The analysis of the satellite images taken during the opening and the closing of the Lake outlet showed that a considerable amount of organic matters and phytoplankton bloom to be discharged into the Mediterranean waters. The extend of these blooms may reach a few kilometers then directed towards the east with the current. Such alarming situation may result in dramatic changes in the water quality to as far as El-Arish area (Fig. 9.20).

Detection of Lake Islands Change

Lake Manzala is famous for the presence of a large number of small Island, which represents part of the natural habitat of the lake. Due to the development of certain unsustainable activity within the lake, a large number of these islands disappeared

Fig. 9.20 Blooms of polluted water flowing from Manzala tidal inlet and dispersed by the eastern longshore current, during the opening of the outlet (Marine sciences department-NARSS)

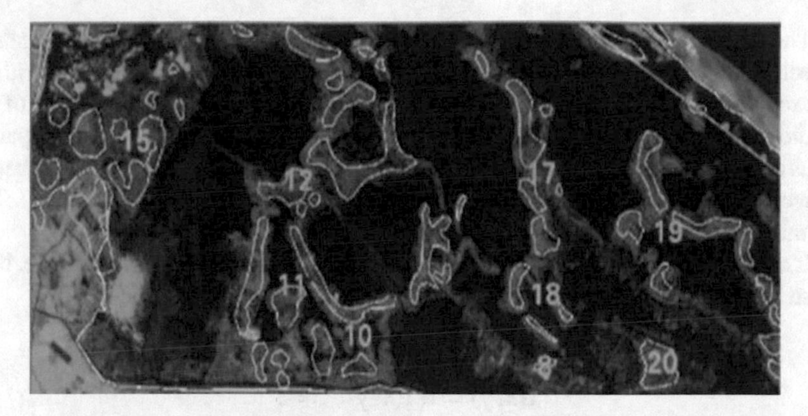

Fig. 9.21 Islands in 1987 and it dynamic change in 2000 (red) situation of the island in Lake Manzala (Marine sciences department-NARSS)

and/or replaced by another for of activity (fish bound, cultivated land, etc.). By comparing the 1987 and 2000 satellite images marking the boundary of these island detection of the amount of drifting islands that were occurred during this period (Fig. 9.21).

9.8.5 Lake Bardawil

The Lake is sustained by one natural strait called Zaranik and two artificial sea inlets. These inlets, locally called Boughaze, are silting up considerably; keeping them open is essential to keep Lake Salinities down and for fish recruitment from the sea. A salt plant has recently been established at Zaranik, with extensive evaporation ponds for salt harvesting.

Bardawil Lake is the greatest hypersaline lake in Egypt, Situated in the north of Sinai Peninsula. The 0.6 km^2 of Lake Bardawil yield some 1500–2500 tons of fish yearly; fisheries management and some in-lake aquaculture development could raise this production to a sustainable 3000 tons. The significance of Bardawil Lake fisheries is not only in the supply of fish for regional domestic consumption but especially in employment and export earnings.

Multi-date Images "Landsat TM" are used to monitor the characteristic of both water and land. The dates of the images are 1993 and 2000 with pixel resolution 25 m (Fig. 9.23). A Landsat TM image dated 1996 is used as geo-reference to correct the satellite data. The multi-date images are used for detecting the changes in the lake size as well as the inlet and barrier changes.

The accurate special registration of the two images is essential for all change detection methods. This necessitates the use of geometric rectification algorithms that register the images to each other or a standard map projection. Also, most of the methods require a decision as where to place the threshold boundaries to separate areas of change from those of non-change as it was done to identify the barrier change between the Bardawil Lake and the Mediterranean Sea (El-Asmar et al. 2015). The technique for change detection is briefly described:

If an image contains light objects, change, on a dark background, no change, then these light objects may be extracted by a simple thresholding

$$I(x, y) = 1 \ I(x, y) > T$$
$$I(x, y) = 0 \ I(x, y) <= T$$

where,

I (x, y)	represent the intensity value of the pixel at x, y.
T	is the threshold value supplied empirically or statistically by the analyst.
1	is the code of the pixel, which belongs to the object change.
0	is the code of the pixel, which belong to the no-change object "background".

This is exactly our case to define the changes in both inlets of Bardawil Lake where rationing is the technique of change detection used; this technique is considered a relatively rapid mean of identifying the areas of change. In rationing, two registered images from different dates (1993 and 2000) are radioed; the data is compared on a pixel-by-pixel basis. Figure 9.24 shows the areas subjected to deposition or accumulation and that subject to erosion or inundation in 1993 and 2000. The barrier is eroded in 2000 than that in 1993, and this is an indication that the lake is decreasing in size. Several dredging operations have been carried out since 1927, and the most recent one took place in 1987. To systematically investigate this siltation problem, the CRI conducted a comprehensive field monitoring program during the years 1985 and 1986 which was re-evaluated and updated in 1990, 1991, 1992 (Khafagy et al. 1988, 1990; Khafagy and Fanos 1990, 1992). Figure 9.22 shows the shoreline changes for both inlets (1986–1996).

Results of using the change detection techniques reveal that the size of Bardawil lake is decreasing with a rate of 0.71 km^2/year; (0.89% negative change in the size of the lake). The decrease in size of Bardawil Lake can be calculated by combination of the remote sensing results derived from present chapter with other results (see Hasan 2001; Ahmed et al. 2000a, b). The dataset of the yearly size changes indicated that the long-term size changes of Bardawil Lake decreased dramatically from 1922 to 2000 with a rate of 0.98 km^2/year whereas it decreases from 1935 to 2000 with a rate of 2.21 km^2/year, this is listed in Table 9.4. This means that the lake complains

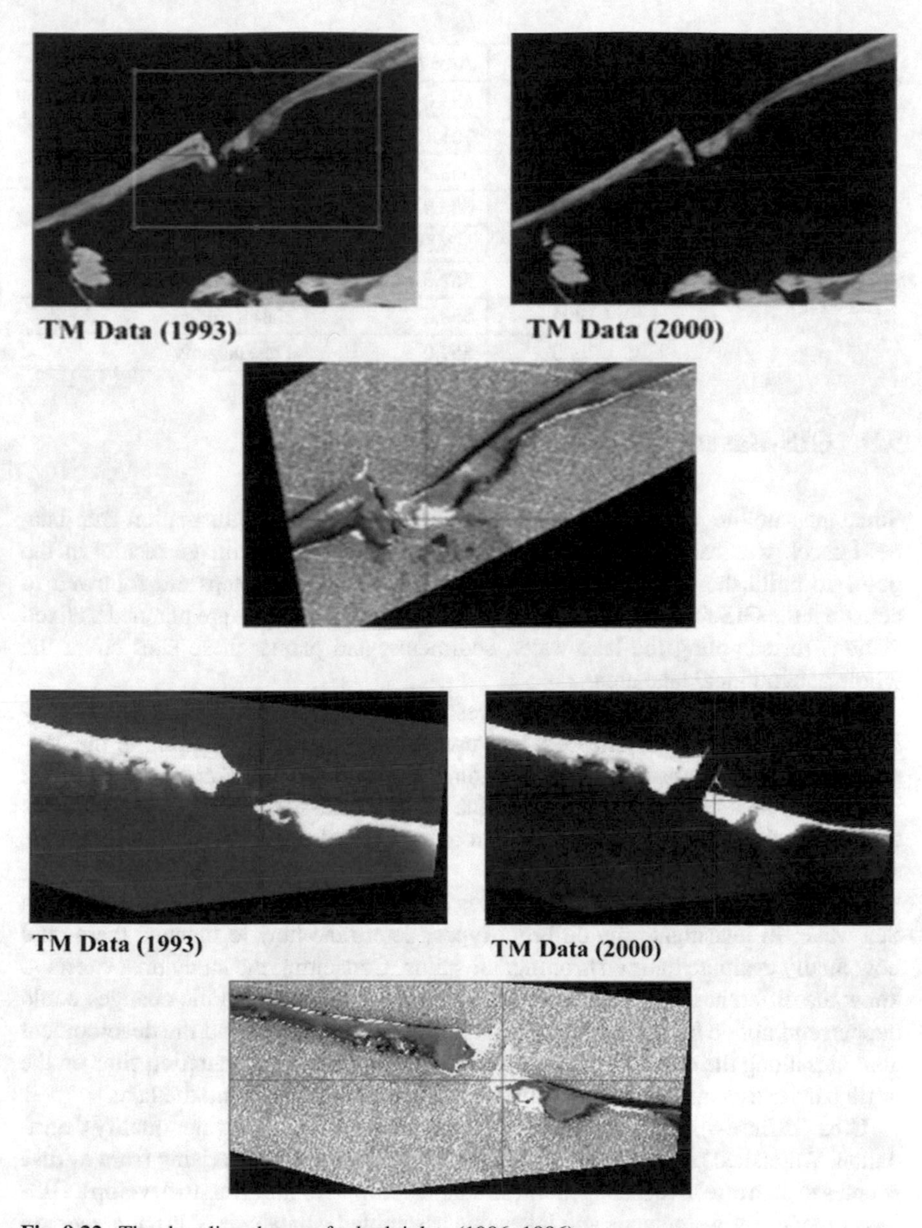

Fig. 9.22 The shoreline changes for both inlets (1986–1996)

of continuous natural and anthropogenic impacts affecting the lake development; therefore, the lake deterioration is becoming a national problem for the Egyptian Governorate.

Table 9.4 The size change in Bardawil Lake (1922–2000)

Year	Area (km^2)	References
1922	634.0	Ahmed et al. (2000a, b)
1935	701.15	Hasan (2001)
1955	663.23	Hasan (2001)
1969	685.9	Hasan (2001)
1984	598.89	Hasan (2001)
1993	562.0	Present study
1994	558.6	Hasan (2001)
2000	557.0	Present study

9.9 GIS-Based Lakes

Since the satellite image processing has been provided with information and data, field check was carried out to check the consistency of the image results in the field. To build the GIS system for the lakes, the following steps are followed to construct the GIS for northern Egyptian lakes. Selected samples are planned as fixed stations for sampling the lake water, sediments, and plants; these sites cover the whole geographical lake area.

Two GPSs were used in each site investigation on some of the sites specified by the used satellite images. The GPS was used for accurate determination of the geographical place and instrument calibration. Collecting the earth observations found in the study area help in recognition of the different classes of unsupervised images that reflect the types of Lake Ecosystem to supervised classification image for the study area.

Complete scanning of the surrounding lake features adjacent to Mediterranean Sea water, to investigate the different types, determine how to manage them, and how finally evaluate their environmental status. Consulting the study area exerts to know the different human activities in the area, as well the dynamic changes could happen and notice by the in-suite observer. Specify the places and the development activities along the coast by following up the erosion and sedimentation sites on the north barrier especially the border between the Mediterranean and the lake.

Urban diffuse-source pollution is a significant contributor to water quality degradation. Watershed planners need to be able to estimate the loads arising from diffuse sources to plan effective management strategies. As part of the effort to develop a GIS-based model for watersheds and lakes, we assembled a database of urban/suburban runoff data from field works on lakes drainage areas conducted over the past 40 years. Relationships between runoff variables and drainage characteristics were monitored and surveyed. The best regression equation to predict runoff volume for rain events was based on rainfall amount, drainage canals, and percent impervious area.

As an essential step in the development of the GIS of the lakes, i.e. GIS water quality, image processing procedure, a database consisting of growing-season measurements of water quality, these measurements were analyzed to determine the nature and consistency of seasonal patterns among lakes in the region.

Therefore, the following are the GIS of different northern coastal lakes and how it can help in understanding the lake behaviours and interaction between the lake and the sea. Figures 9.23, 9.24, 9.25 and 9.26 are demonstrate the Northern lakes GIS-based resources.

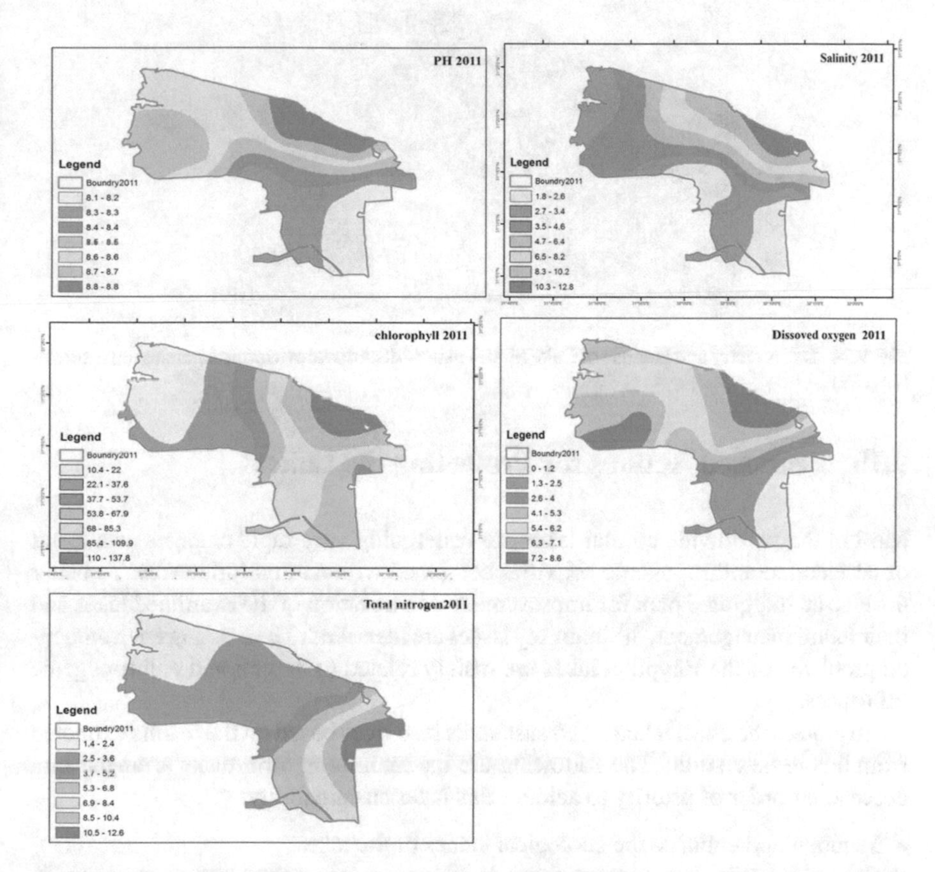

Fig. 9.23 Water quality parameters distribution using GIS software (Marine sciences department-NARSS)

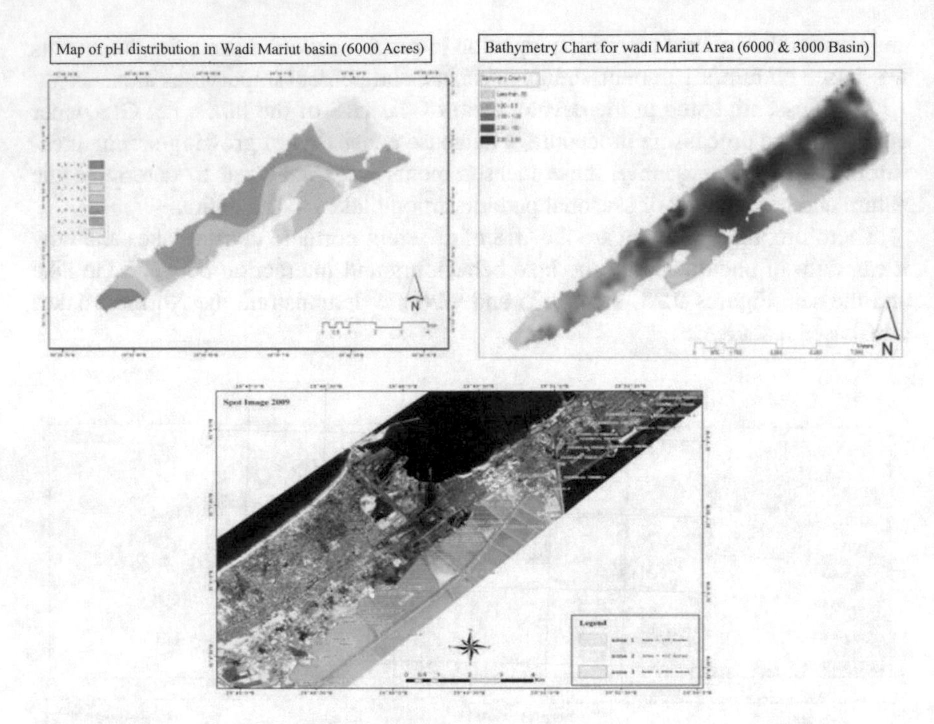

Fig. 9.24 Bathymetry and landuse of Lake Maruit using GIS software (Marine sciences department-NARSS)

9.10 Proposed Actions for Protecting the Lakes

Most of the worldwide coastal lakes are undergoing very rapid changes as a result of accelerated anthropogenic activities combined with natural influences. To better develop an integrated plan for improvement, conservation of the examined lakes and their future management, the main key issues are identified. The existing environmental problems of the Egyptian lakes are mainly related to natural and anthropogenic influences.

To protect the coastal lakes and sustain its resources based on the results extracted from this review study. The following are the main long-term tasks arranged in a descending order of priority to achieve this lakes sustainability:

- Maintain and enhance the ecological values of the lake,
- Conserve available resources through proper environmental management tools, i.e. Remote sensing technology, etc.
- Improve the socio-economic opportunities of the local population of the lake,
- Develop public awareness and participation in monitoring programs and nature conservation of the lake resources,
- Resolve existing legal conflicts, especially those of land ownership, and responsibilities.

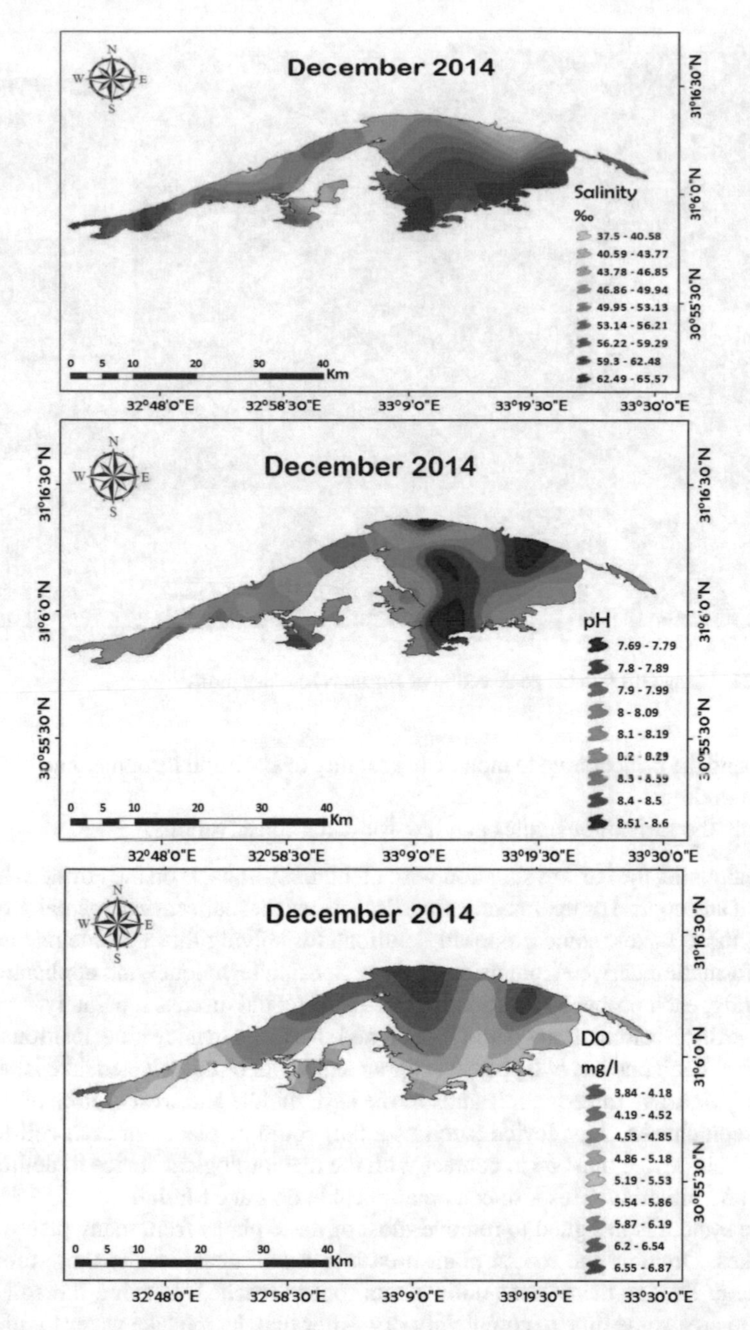

Fig. 9.25 Water quality parameters distribution of Lake Bardawil using GIS software (Marine sciences department-NARSS)

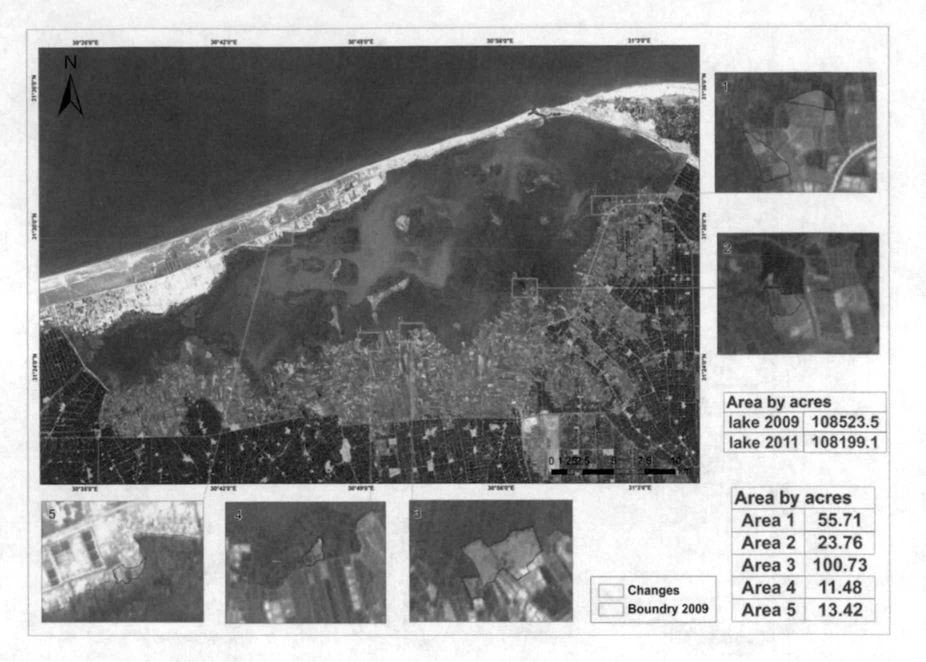

Fig. 9.26 Using GIS for change detection of Burullus lake fish farms

- Expand the protectorate to include the totality of Lake and its immediate terrestrial surroundings.
- Apply the legislations, rules and law for lakes conservations.

Meanwhile, the current situation with all humans impacts on the northern lakes is without any control or treatment. After discussions the main environmental problems facing these Lakes, some proposed solutions for solving those problems based on the information derived from using Remote Sensing techniques and applications. In following, each proposed solution for problems would discuss separately.

An artificial circulation system is designed such as to include nine positions inside the lakes. Each one has two positions upper and lower of the selected lake island. i.e. in Lake Manzala, three main islands in the east, middle and west sectors of the lake have been chosen. Any device (such as a fan) could be placed in each cell to push water. This device must be in contact with the meteorological device to define wind speed and direction. This also can be applicable on Lake Mariut.

The system is designed to remove most of these plants from many places inside the lakes. Dividing the rooted plant area into small areas, isolate these areas and dry them. Dredge these areas until plants root is reached. Leaving the soil of the dredged area some time to completely dry. After that, let the lake water to release to the cleaning area.

The harvesters units are paddle wheel propelled and have a large frame extending down into the water ahead of the bow. This frame is made up of vertical sickles on the sides and a sickle across the bottom connecting the two sides. A conveyer belt extends up from this frame to the boat and carries the cut macrophyte to the surface where they are collected (Klein 1997).

Duckweeds such as Lemna gibba have been utilized to treat municipal wastewater and removal of very high Biochemical Oxygen Demand (BOD). These plants must be removed from the lake (manually) with the timetable (depends on the season of growth).

The present situation of Lake Bardawil is rather precarious, vulnerable and could deteriorate in the future. Ill-advised management of the Lake plus the detrimental impact of major development projects (land reclamation for agricultural purposes, El-Salam Canal, new and expanding urban centers, the international highway, the new railway line, the fast proliferation of large-scale seaside resorts, etc.) will undoubtedly impact the current pristine status of the site, possibly beyond repair. All of these projects are in various stages of implementation, and it will not be long before their undesirable impact on Zaranik is felt.

Therefore, in order to keep the changes in the lakes under control and to minimize their impact on the lake ecosystem, the ideal concept envisaged for protectorate area inside the lake is that it should become a core conservation area for the totality, i.e. Lake Bardawil. In other words, the boundaries of the Protectorate should be expanded to include the entire Bardawil region. The newly added areas (aquatic and terrestrial) will become an associated zone, and the enlarged protectorate will be managed on the basis of sustainable development (i.e. environmental management).

The management plan of the protected areas located with the northern lakes should be a legal and technical instrument which helps to achieve the main long-term objectives as well as to maintain the delicate balance between the interests of the local population and the sustainable conservation and development of the lake's resources.

9.11 Conclusions

Results of this study indicated that these lakes had been increasingly subjected to intensive and diverse development activities including fishing, aquaculture industry, dumping wastes, land reclamation, lakes drying, urbanization, saltpan, and recreational uses. Some of these activities have derived to several environmental problems as well; the population is expanding exponentially. These problems are pollution, eutrophication, reclamation, fragmentation, over-fishing, illegal harvesting of fry fish, is considered as major environmental issues threatening the fragile ecosystem of these lakes.

The experiences gained from this study can be considered as a practical lesson learned in future planning and management of coastal lakes at the northern Egyptian lakes of the Mediterranean region. On a local scale, monitoring of such these ecosystems using Multidates satellite images presented here are important for managing, improving and conservation of the Northern coastal lakes of Egypt.

Remote sensing proves to be very successful in monitoring the water quality, vegetation types, and ecological changes along the Manzala Lake. Manzala Lake is characterized by special sensitive environments. However human activities including the discharge of sewage and industrial waste and the impact of canal and road networks have a serious impact on the Lake. Image processing techniques as enhancements were applied to help the identification and discrimination of the different features and classes in and around the Lake.

The ultimate goal of this remote sensing research is to provide resource managers with a useful tool to monitor natural resources such as Coastal Lakes and make better-informed decisions about their use and conservation. For example, fishermen need accurate forecasts of locations of good fishing waters, and for fish areas, managers need to monitor the water quality. Remote sensing is a valuable tool for studying Water Quality of the Manzala Lake. Integrating all available data into an easily accessible data system further improves remote sensing images potential as a tool for resource managers.

Steps which could facilitate improvement in the Lake Manzala ecosystem through reductions in nutrient loads include:

(i) treatment at source of the domestic and industrial wastewater which is at present responsible for the high BOD and nutrient loads within the Bahr El-Baqar drain;
(ii) treatment of wastewater from lakeside communities so that the use of the lake as a secondary and tertiary wastewater treatment unit declines and eventually ceases;
(iii) diversion of nutrient-rich wastewater from the Port Said wastewater treatment plant away from the lake and instead out to sea using a deep-water delivery system; and
(iv) improved control and regulation of fish farms within the lake.

These steps will require that Lake Manzala is managed sustainably as a large shallow lake ecosystem which supplies important goods and services. Restoration of Lake Manzala will undoubtedly bring about major ecosystem improvements that will benefit both people and biodiversity. Nevertheless, such changes will require political will, considerable financial resources and careful environmental planning.

The present-day coastal lakes host the Mediterranean coast of Egypt are mainly differ in shape, dimension, depositional environment, water quality, runoff of tributary streams, and inlet stabilization. The delta lakes (Mariut, Edku, Burullus, and Manzala) are brackishly influenced by freshwater runoff from drains and irregular canals. In contrast, the other lake is hypersaline and not connected to any freshwater flow (Bardawil). Like other Mediterranean lakes, they were generally originated as a result of changes in sea level combined with neotectonic land subsidence. These lakes are socially and economically important for the coastal population. The quality of these lakes is influenced to a large degree by various types of human activities and human habitats. Human activities differ throughout these lakes including fishing, wastewater through drains, lake reclamation, recreation, salt paning, hunting of water birds and engineering works at inlets.

Some of these lakes are presently experiencing environmental problems resulted from natural and human activities. Because these lakes are very valuable regarding natural resources and related economic activities, these problems have caused serious implication on the lake ecosystem, resources, fauna, flora, and human beings as well. Moreover, the lacks of effective and consolidated earlier management of these lakes have accelerated such implications.

Within the present task of finding practical solutions for the environmental problems of the lake, this study integrates the existing scientific knowledge in order to obtain an ecological understanding of the lake. It also implements a systematic decision support procedure in such a way that final judgments can be as objective as possible.

An enhanced model requires discharge time series measured daily or weekly (according to the type of study). These measurements should be carried out at the same points where the boundary conditions are placed. Ideally, a gauging station would be established at each known inflow to the lake. For an appropriate calibration and validation of the model, water levels referred to standard reference level are needed. At least one water level measurement point is needed in each basin. Water level and discharge measurements should be carried out simultaneously. Openings through the embankments of the drains should be surveyed as they define the interaction between canals and basins. In the presence of culverts connecting basins and drains, it is also necessary to model their hydraulics properly.

9.12 Recommendations

Based on this comprehensive study for the most important characteristics, problems and future development options of northern Lakes have been carried out. It is recommended that an integrated management program be implemented for the development of the lake and adjacent area taking into account the capability of recent remote sensing techniques in identifying and assessing changes and law enforcement.

Lake management is a dynamic process, the rules keep on changing, the problems keep on changing, and approaches of management also keep on changing. It is necessary to be able to make decisions at the proper time.

Periodic monitoring and assessment are therefore necessary. In the absence of main tools for management, (legal frame, corrective actions, technical and socioeconomic), conditions of the coastal zone will deteriorate without control. A decentralized remote sensing and geographic information system (GIS) capability must be developed to collect and upgrade available data and to help decision makers on local and national scales.

A sensitivity analysis must be carried out for all coastal lakes in Egypt to assess vulnerabilities to various problems including pollution, lake reclamation, erosion of Lake Barrier, illegal fish fry, etc. A contingency plan for protection and emergency measures must be developed.

The model built for some lakes is only indicative, and its results are only approximations of the flow pattern, water levels and velocities in the lake. Moreover, for an enhanced model that would permit a detailed analysis of the lake hydrodynamics, further detailed topographic and bathymetric data could be required, e.g. openings and interaction between drains and basins. Besides, the input data should be provided as time series instead of constant average values. Furthermore, it is essential to calibrate and validate the model to have reliable results.

The use of the ecological models to calculate the water flow velocities may give better results and understanding of the lake's ecosystem than using the increased roughness values for the vegetated areas, as done in this study. Therefore, it is recommended to do further research in this respect.

References

Abdel-Moati MAR, El-Sammak A (1997) Man-made impact on the geochemistry of the Nile Delta lakes. A study of metals concentrations in sediments. Water Air Soil Pollut 97:413–429

Aboul-Ezz SM, Soliman AM (2000) Zooplankton community in Lake Edku. Egypt J Aquat Res A.R.E. 26:71–99

Adham KG, Hassan IF, Taha N, Amin TH (1999) Impact of hazardous exposure to metals in the Nile and Delta lakes on the catfish, Clarias lazera. Environ Monit Assess 54:107e124

Ahmed MH (1991) Temporal shoreline and bottom changes of the inner continental shelf of the Nile Delta, Egypt. M.Sc. thesis, Alexandria University, 218 p

Ahmed MH (2000) Long-term changes along the Nile Delta coast: Rosetta promontory a case study. Egypt J Remote Sens Space Sci 3:125–134

Ahmed MH (2003) Erosion and accretion patterns along the coastal zone of Northern Sinai, Egypt. J Sedimentological Soc Egypt 11:281–290

Ahmed MH, Elaa AAA (2003) Study of molluscan shells and their enclosed bottom sediments in Manzala Lagoon, Nile Delta, Egypt. Bull Natl Inst Oceanogr Fish ARE 29:423–446

Ahmed MH, Donia NS (2007) Spatial investigation of water quality of Lake Manzala using GIS techniques. Egypt J Remote Sens Space Sci 10:63–86. ISSN: 1110-8923

Ahmed MH, El-Leithy B (2008) Utilization of satellite images for monitoring the environmental changes and development in Lake Mariout during the past four decades, Alexandria, Egypt. In: Proceeding of international conference "environment is a must", 10–12 June 2008, Alexandria

Ahmed MH, Frihy OE, Yehia MA (2000a). Environmental management of the Mediterranean coastal lagoons of Egypt. J Ann Geol Surv Egypt V(XXIII):491–508

Ahmed MH, Nicholls RJ, Yehia MA (2000b) Monitoring the Nile Delta: a key step in adaptation to long-term coastal change. In: The 2nd international conference on earth observations and environmental information (EOEI), 11–14 Nov, Cairo, Egypt

Ahmed MH, Abdel-Moati MAR, El-Bayomi G, Tawfik M, El-Kafrawy S (2001) Using geoinformation and remote sensing data for environmental assessment of Burullus Lagoon, Egypt. Bull Natl Inst Oceanogr Fish ARE 27:241–263

Ahmed MH, Noha D, Fahmy MA (2006) Eutrophication assessment of Lake Manzala Egypt using GIS techniques. J Hyrdroinformatics 8(2):101–109

Ahmed MH, El Leithy BM, Thompson JR, Flower RJ, Ramdani M, Ayache F et al (2009) Application of remote sensing to site characterisation and environmental change analysis of North African coastal lagoons. Hydrobiologia 622(1):147–171

Ambrose J, Shah P (1990) The importance of remote sensing and mapping for resource management: a case study of Nepal. Integrated Surveys Section Tomkcal Surveys Branch, Government of Nepal, pp 2161–2164

Aplin P (2003) Comparison of simulated IKONOS and SPOT HRV imagery for classifying urban areas. In: Mesev V (ed) Remotely sensed cities. Taylor & Francis, London, pp 23–45

Appleby PG, Birks HH, Flower RJ (2001) Radiometrically determined dates and sedimentation rates for recent sediments in nine North African wetland lakes (the CASSARINA project). Aquat Ecol 35:347e367

Barale V, Folving S (1996) Remote sensing of coastal interactions in the Mediterranean region. Ocean Coast Manag 30(2–3):217–233

Ben-Tuvia A (1979) Studies of the population and fisheries of Sparus aurata in the Bardawil Lagoon, Eastern Mediterranean. Invest Pesquera 43(1):43–67

Bernasconi MP, Stanley DJ (1994) Molluscan biofacies and their environmental implications, Nile Delta lakes, Egypt. J Coas Res 10:440–465

Briggs J, Nellis D (1991) Seasonal variation of heterogeneity in the tallgrass prairie: a quantitative measure using remote sensing. Photogram Eng Remote Sens 57(4):407–411

CAPMS (Central Agency for Public Mobilization & Statistics) (1994) Fish production statistics in ARE. # 71-12413/94

Carlson RE (1977) A trophic state index for lakes. Limnol Oceanogr 22(2):361–369

Carper WJ, Lillesand TM, Kiefer RW (1990) The use of intensity-hue-saturation transformations for merging SPOT panchromatic and multispectral images data. PERS 56(4):459–467

Chen J, Hepner GF (2001) Investigation of imaging spectroscopy for discriminating urban land covers and surface materials. Paper read at AVIRIS earth science and applications workshop, at Palo Alto, CA

Chen CM, Hepner GF, Forster RR (2003) Fusion of hyperspectral and radar data using the HIS transformation to enhance urban surface features. ISPRS J Photogrammetry Remote Sens 58:19–30

Choudhury BI (1990) Monitoring arid lands using A VHRR—observed visible reflectance and SMMR-37 GHZ polarization difference. Int J Remote Sens 11(10):1949–1956

Christensen E, Jensen J, Ramsey E, Mackey H Jr (1988) Aircraft MSS data registration and vegetation classification for wetland change detection. Int J Remote Sens 9(1):23–38

Dewider K, Khedr A (2001) Water quality assessment with simultaneous Landsat-5 TM at Manzala Lagoon, Egypt. Hydrobiologia 457:49–58

Dewidar K, Khedr AA (2005) Remote sensing of water quality for Burullus Lake, Egypt. Geocarto Int 20(3):43–49

Donia NS, Ahmed MH (2006) Remote sensing for water quality monitoring of lakes. Case study: Lake Manazla. In: 7th international conference of hydroinformatics (HIC 2006), Nice, France, 4–8 Sept 2006

El-Asmar HM, Ahmed MH, El-Kafrawy SB, Oubid-Allah AH, Mohamed TA, Khaled MA (2015) Monitoring and assessing the coastal ecosystem at Hurghada, Red Sea coast, Egypt. J Environ Earth Sci 5(6):144–160

El Fishawi NM, Fanos AM (1989) Prediction of sea level rise by 2100, Nile Delta coast. INQUA, Commission on Quaternary Shorelines, newsletter, vol 11, pp 43–47

El Kafrawy SB, Ahmed MH, Abu Zaied MM (2006) Environmental assessment of seasonal biological variations of Manzala Lagoon. In: The first international conference on environmental change of lakes, lagoons and wetlands in the Southern Mediterranean region, 4–7 Jan, Cairo, Egypt

El Kafrawy SB, Donia NS, Mohamed AM (2017) Monitoring the environmental changes of Mariout Lake during the last four decades using remote sensing and GIS. MOJ Ecol Environ Sci 2(5)

El-Malky MG, Ahmed MH, El-Marghany MA (2003) Heavy metals in water and sediments of Bardawil Lagoon, Northern Coast of Sinai, Egypt. J Environ Sci 7(3):753–785

El-Nahry AH, Ahmed MH, Madany SMA (2006) Land suitability modeling of Lake Manzala natural vegetation using remote sensing and GIS techniques. In: The first international conference on environmental change of lakes, lagoons and wetlands in the Southern Mediterranean region, 4–7 Jan 2006, Cairo, Egypt

El-Raey M, Hassan HM, Hussain MMA (2006) Remote sensing of environmental conditions and changes of Lake Brullus, Egypt. In: First international conference (ECOLAW, 06), 4–7 Jan, Cairo, Egypt

El-Sayed EM (1993) Comparative geological and geophysical studies on the origin of the northern Egyptian lakes and lagoons. Ph.D. thesis, Faculty of Science, Alexandria University

ERDAS (2000) Field guide, version 8.0. Erdas Inc., Atlanta, USA

Fanos A, Khafagy A, Anwar N, Naffaa M, Dean R (1993) Assessment and evaluation for the enhancement of the Brardawil Lake outlet, Egypt. In: Coastal dynamics conference, Spain, pp 189–203

Fisher P (1997) The pixel: a snare and a delusion. Int J Remote Sens 18(3):679–685

Foody GM (1999) Image classification with a neural network: from completely-crisp to fully-fuzzy situations. Adv Remote Sens GIS Anal 17–37

Forster BC (1985) An examination of some problems and solutions in monitoring urban areas from satellite platforms. Int J Remote Sens 6(1):139–151

Franklin J (1991) Land cover stratification using landsat thematic mapper data in Sahelian and Sudanian woodland and wooded grassland. J Arid Environ 20:141–163

Frihy OE (1992) Beach response to sea level rise along the Nile Delta coast of Egypt. In: Woodworth PL (ed) Sea level changes: determination and effects. Geophysical monograph 69, vol 2. American Geophysical Union, IUGG, pp 81–85

Fung T (1990) An assessment of TM imagery for land-cover change detection. IEEE Trans Geosci Remote Sens 28(4):681–684

GAFRD (General Authority for Fisheries Resources Development) (2008) Annual statistics year book, 195 p

George CJ (1972) The role of the Aswan High Dam in charging the fisheries of the Southeastern Mediterranean. Careless Technol 151–163

Goetz AFH (1992) Imaging spectrometry for earth remote sensing. In: Toselli F, Bodechtel J (eds) Imaging spectroscopy: fundamentals and prospective applications. Springer, The Netherlands, pp 1–19

Gong P (2006) Information extraction. In: Ridd MK, Hipple JD (eds) Remote sensing of human settlements

Graetz R (1990) Remote sensing of terrestrial ecosystem structure: an ecologist's pragmatic view. In: Hobbs R, Mooney H (eds) Remote sensing of biosphere functioning. Springer, New York, pp 5–30

Guerguess SK (1979) Ecological study of zooplankton distribution and macrofauna in Manzala Lake. Ph.D. thesis, Faculty of science, Alexandria University, 360 p

Hall F, Botkin D, Strebel D, Woods K, Goetz S (1991) Large-scale patterns of forest succession as determined by remote sensing. Ecology 72(2):628–640

Hassan MM, Khalil MT, Saad AEA, Shakir SH, El Shabrawy GM (2017) Zooplankton community structure of Lake Edku, Egypt. Egypt J Aquat Biol Fish 21(3):55–77. ISSN 1110-6131

Herold M, Roberts DA (2005) Spectral characteristics of asphalt road aging and deterioration: implications for remote-sensing applications. Appl Opt 44(20):4327–4334

Herold M, Roberts DA (2006) Multispectral satellites—imaging spectrometry—LIDAR: spatial—spectral tradeoffs in urban mapping. Int J Geoinformatics 2(1):1–13

Herold M, Roberts DA (2010) The spectral dimension in urban remote sensing. In: Remote sensing of urban and suburban areas. Springer, Dordrecht, pp 47–65

Hobbs R (1990) Remote sensing of spatial and temporal dynamics of vegetation. In: Hobbs R, Mooney H (eds) Remote sensing of biosphere functioning. Springer, New York, pp 203–219

Howman A (1988) The extrapolation of spectral signatures illustrates' landsat's potential to detect wetlands. In: Proceedings of lCARSS '88 symposium, Edinburgh, Scotland, 13–16 Sept, pp 537–539

Jensen JR (1986) Introductory digital image processing: a remote sensing perspective. Prentice-Hall, Englewood Cliffs, New Jersey, 379 p

Khalil MT (1990) Plankton and primary productivity of Lake Manzala, Egypt. Hydrobiologia 196:201–207
Khorram S, Cheshire HM (1985) Remote sensing of water quality in the Neuse River Estuary, North Carolina. Photogram Eng Remote Sens 51:329–341
Klein J (1997) Sediment dredging and macrophyte harvest as lake restoration techniques
Kloiber SM, Brezonik PL, Olmanson LG, Bauer ME (2002) A procedure for regional lake water clarity assessment using Landsat multispectral data. Remote Sens Environ 82(1):38–47
Lathrop RG (1992) Landsat thematic mapper monitoring of turbid inland water quality. Photogram Eng Remote Sens 58:465–470
Leatherman SP (1991) Coasts and beaches, Centennial special, vol 3, Chap 8. Geological Society of America, pp 183–200
Levy Y (1974a) Chemical changes in interstitial water from the Bardawil Lagoon, Northern Sinai. J Sediment Petrol 44(4):1296–1304
Levy Y (1974b) Sedimentary reflection of depositional environment in the Bardawil Lagoon, Northern Sinai. J Sediment Petrol 44(1):219–227
Maiyza IA, Beltagy AI, El-Mamoney M (1991) Heat balancet of Lake Burullus, Egypt. Bull Natl Inst Oceanogr Fish ARE 17(1):45–55
Mather J (1978) The climatic water budget in environmental analysis. Lexington Books, Toronto, Canada
Meininger PL, Atta GA (1994) Ornithological studies in Egyptian wetlands 1989/90. Foundation for Ornithological Research in Egypt (FORE), No 94-1, Report
Mesev V (ed) (2003) Remotely sensed cities: an introduction. In: Remotely sensed cities, pp 47–82
Milne A (1988) Change detection analysis using landsat imagery: a review of methodology. In: Proceedings of IGARSS '88 symposium. Edinburgh, Scotland, 23–16 Sept, pp 541–544
Milne A, O'Neill A (1990) Mapping and monitoring land cover in the Willandra Lakes World Heritage Region (New South Wales, Australia). Int J Remote Sens 11(11):2035–2049
Milton N, Mouat D (1989) Remote sensing of vegetation responses to natural and cultural environmental conditions. Photogram Eng Remote Sens 55(8):1167–1173
Mitwally H (1982) Review of industrial waste disposal in Alexandria. In: Proceedings of international symposium on management of industrial wastewater in developing nations, Alexandria, Egypt
Mohsen MA (1992) Descriptive analysis of the North Delta highway project. In: Workshop/roundtable discussion on the impact of the Nile Delta coastal road and its effectiveness as defense measure against the expected sea-level rise, 10 p
Nafea EMA (2005) On the ecology and sustainable development of the northern delta lakes, Egypt. Doctoral dissertation, Ph.D. thesis, Faculty of Science, Mansoura University
Neev D, Friedman GM (1978) Late Holocene tectonic activity along the margins of Sinai subplate. Science 202:427–429
Olmanson LG, Kloiber SM, Bauer ME, Brezonik PL (2001) Image processing protocol for regional assessment of lake water quality. Water resources center technical report # 14. University of Minnesota, St. Paul, 19 p
Olmanson LG, Bauer ME, Brezonik PL (2002b) Use of Landsat imagery to develop a water quality atlas of Minnesota's 10,000 lakes. In: Proceedings of Pecora 15 and land satellite information IV conference, Denver, CO, 10–15 Nov
Phinn SR (1998) Framework for selecting appropriate remotely sensed data dimensions for environmental monitoring and management. Int J Remote Sens 19(17):3457–3463
Price JC (2001) Spectral band selection for visible-near infrared remote sensing: spectral-spatial resolution tradeoffs. IEEE Trans Geosci Remote Sens 35(5):1277–1285
Ramdani M, Flower RJ, Elkhiati N, Kraïem MM, Fathi AA, Birks HH et al (2001) North African wetland lakes: characterization of nine sites included in the Cassarina project. Aquat Ecol 35: 281–301
Ramdani M, Elkhiati N, Flower RJ, Thompson JR, Chouba L, Kraiem MM et al (2009) Environmental influences on the qualitative and quantitative composition of phytoplankton and zooplankton in North African coastal lagoons. Hydrobiologia 622:113–131

Rashed T, Weeks J, Roberts DA, Rogan J, Powell P (2003) Measuring the physical composition of urban morphology using multiple endmember spectral mixture models. Photogram Eng Remote Sens 69(9):1011–1020

Rashed T, Weeks J, Couclelis H, Herold M (2007) An integrative GIS and remote sensing model for place-based urban vulnerability analysis. In: Integration of GIS and remote sensing

Rasmussen EK, Petersen OS, Ahmed MH (2006) A hydrodynamic-ecological model of the Manzala Lagoon, Egypt. In: The first international conference on environmental change of lakes, lagoons and wetlands in the Southern Mediterranean region, 4–7 Jan, Cairo, Egypt

Ringrose S, Matheson W (1987) Spectral assessment 50 of indicators of range degradation in the Botswana hardveld environment. Remote Sens Environ 23:379–396

Rogan J, Chen DM (2004) Remote sensing technology for mapping and monitoring land cover and land use change. In: Treitz P (ed) Progress in planning (forthcoming)

Rollin EM, Emery DR, Milton EJ (1997) The design of field spectroradiometer: a user's view

Saad MAH (1976a) Core sediments from Lake Brollus (Bahra el Burullus) Egypt. Acta Hydrochim Hydrobiologia 4:469–478

Saad MAH (1976b) Some limnological investigations of Lake Edku Egypt. Arch fur Hydrobiol 77:411–430

Saad AHM (1990) State of the Egyptian delta lakes, with particular reference to pollution problems. In: Regional symposium of environmental studies (UNARC), Alexandria

Samman AA (1974) Primary production of Lake Edku. Bull Inst Oceanogr Fish 4:259–317

Sestini G (1992) Implications of climate change for the Nile Delta. In: Jeftic L, Milliman JD, Sestini G (eds) Climate change and the Mediterranean. Environmental and social impacts of climate and sea-level rise in the Mediterranean sea. Edward Arnold, London, pp 533–601

Shaheen SE (1998) Geoenvironmental studies on El-Bardawil lagoon and its surroundings, North Sinai, Egypt. Ph.D. dissertation, Faculty of Science, Mansoura University, Mansoura

Siam EE, Ghobrial MG (1999) Pollution influence on bacterial abundance & chlorophyll-a concentration case study Idko Lagoon Egypt. J Arab Acad Sci Technol

Siegel FR, Slaboda ML, Stanley DJ (1994) Metal pollution loading, Manzala Lake, Nile Delta, Egypt: implications for aquaculture. Environ Geol 23:89–98

Small C (2001a) Estimation of urban vegetation abundance by spectral mixture analysis. Int J Remote Sens 22(7):1305–1334

Small C (2001b) Spectral dimensionality and scale of urban radiance. In: AVIRIS workshop, vol 27. Pasadena, CA, USA

Stanley DJ, Warne AG (1993) Nile Delta: recent geological evolution and human impact. Science 260:628–634

Teng W (1990) AVHRR monitoring of U.S. crops during the 1988 drought. Photogram Eng Remote Sens 56(8):1143–1146

Thompson JR, Flower RJ, Ramdani M, Ayache F, Ahmed MH, Rasmussen EK et al (2009) Hydrological characteristics of three North African coastal lagoons: insights from the MELMARINA project. Hydrobiologia 622:45–84

Ward D, Phinn SR, Murray AT (2000) Monitoring growth in rapidly urbanization areas using remotely sensed data. Prof Geogr 52(3):371–385

Wharton S (1987) Knowledge based recognition of urban land cover in high resolution multispectral data. In: Proceedings of IGARSS '87 symposium. Ann Arbor, Michigan, 18–21 May, pp 119–124

World Bank Study (2005) Report of Lake Maryut 2005

Zakaria HY, Ahmed MH, Flower R (2007) Environmental assessment of spatial distribution of zooplankton community in Lake Manzalah, Egypt. Acta Adriat 48(2):161–172. ISSN: 0001-5113

Zhang Y (2001) Detection of urban housing development by fusing multisensor satellite data and performing spatial feature post-classification

Zilioli E (2001) Lake water monitoring in Europe by means of remote sensing. Sci Total Environ 268:1–2

Chapter 10
An Overview of the Environmental and Anthropogenic Hazards Threatening Lake Nasser, Egypt National Water Reservoir

Nabil Sayed Embabi

Abstract Lake Nasser is the most significant man-made creature in Egypt built during its history. It was created by impounding of the Nile water by the Aswan High Dam, which was built in the 1960s and operated since the year 1971. From that date, it represents the national water storage for Egypt. Therefore, it is a must to keep the lake water unthreatened by any factor or feature of local/outside origin. Understanding the geomorphic aspects of the Lake is a pre-requisite for the analysis of the hazards threatening Lake Nasser as Egypt National Water Reservoir. Lake Nasser is threatened by several factors. The most dangerous threat comes from Ethiopia, where a new huge dam (El-Nahda/Renaissance) is under construction on the Blue Nile which provides Egypt with more than 50% of its water needs. This dam will deprive Egypt from some of its share in Nile water during filling of the dam and its operation, resulting in a decrease in the stored water, lowering the lake level, and reduction in electricity production. Locally, the Lake Nasser storage capacity is shrinking due to the continual filling of its basin by dune sands and Nile silt. High evaporation rates will deprive the Lake and consequently Egypt from 10 to 18 billion m^3/year.

Keywords Lake Nasser · Geomorphology · Nile discharge · Climate · Sand encroachment · Silting · Pollution

10.1 Geomorphology of the Lake Region

10.1.1 Introduction

Lake Nasser is a man-made landform, which was created after the construction of the Aswan High Dam at about 6.0 km to the south of the city of Aswan in the late1960s (Fig. 10.1). It extends from the High Dam in Egypt (23° 58′ N, 32° 58′ E)

N. S. Embabi (✉)
Department of Geography, Faculty of Arts, Ain Shams University, 20 Ibn Qotaiba Street, 7th District, Nasr City, Cairo 11471, Egypt
e-mail: nabilsayedembabi@gmail.com

S. F. Elbeih et al. (eds.), *Environmental Remote Sensing in Egypt*, Springer Geophysics, https://doi.org/10.1007/978-3-030-39593-3_10

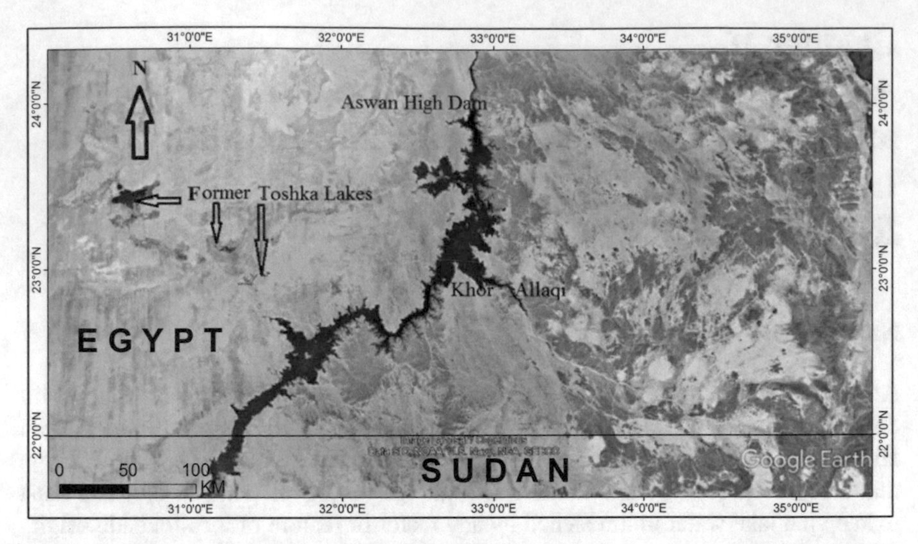

Fig. 10.1 Google earth image showing the general features of Lake Nasser and its environs. (*Source* Google Earth 2018)

southwards to Dal Cataract in Sudan (20° 30′ N, 30° 20′ E) when the lake level reaches 180 m asl. The lake is filled gradually by the excess in floodwater, which comes from the Ethiopian Highlands. Lake lies in a region characterized by a hyper-arid climate according to the UNESCO Index of Aridity (UNESCO 1977). The Lake level depends on the annual mean of both rainfall and evapo-transpiration. In Aswan Meteorological Station, the annual mean of rainfall is 0.7 mm (variable 1), and that of evapotranspiration (variable 2) is 7044 mm (Climatological Normals of Egypt, up to 1975). By dividing variable (1) by variable (2), it gives an Index of Aridity of 0.00001 indicating a hyper-arid climate according to the UNESCO classification. The Lake Region is of varied geologic and topographic characteristics. The geological structure is dominated by the basement rocks, which are exposed on the surface in some localities, and are overlain unconformable by the Sabaya, Lake Nasser and Abu Simbel Formations of the Nubian Sandstone (Fig. 10.2). The region is affected by a series of faults and fractures; some of which cut across the lake or run parallel to its shores (Conoco Coral and The Egyptian General Petroleum Company Corporation 1989). Before the lake, the Nubian Nile was a narrow river, which had no flood plain. Controlled by faults, the Nubian Nile changed its way several times. To the east and west, lies the great sandstone plain, which is best developed on the western side and is called by Butzer (1965) the Lower Nubian Pediplain. Generally, the sandstone plain rises on both sides of the Lake to about 200–300 m asl. In the low-lying areas, several hills rise up to 150–250 m a.s.l. of which Gabal El-Allaqi (218 m) and Gabal El-Maharaqah (199m) on the east side are examples. The plain is dissected by several wadis, which are well defined on the eastern side comparatively to those of the western side. Wadi El-Allaqi, Wadi Korosko, Wadi Kurkur, Wadi Kalabsha, and

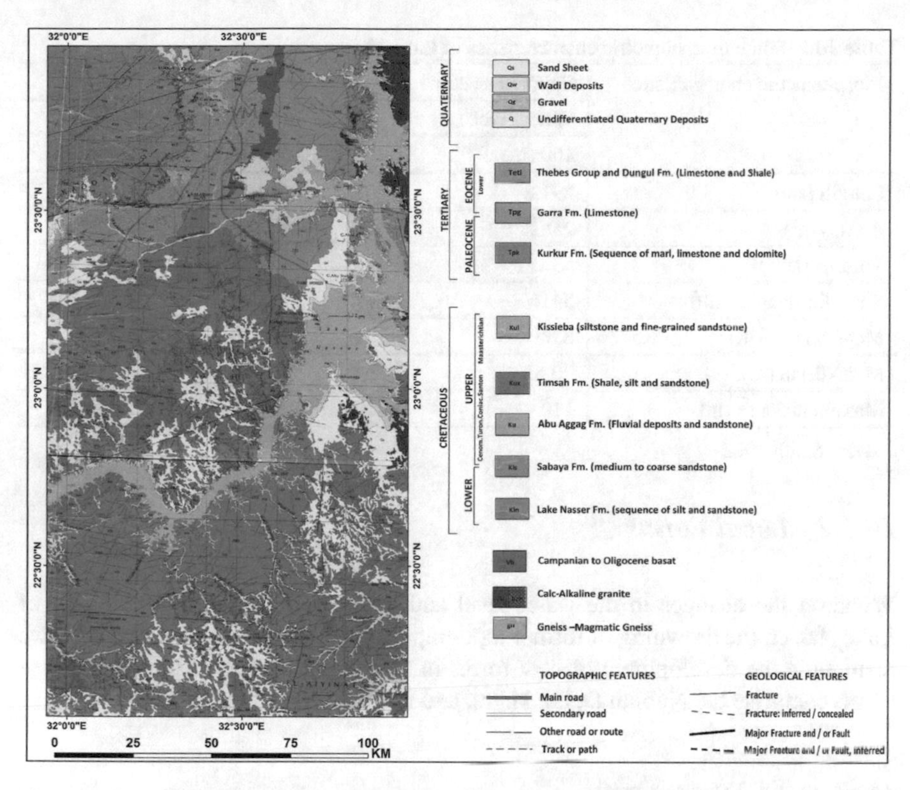

Fig. 10.2 Geological map of Lake Nasser area showing the distribution of formations and structural lines (Faults and fractures). (*Source* Conoco Coral and The Egyptian General Petroleum Company Corporation 1989)

Wadi Toshka are examples of those wadis on the east and west. Some wadis open up in semi-closed basins before reaching the Nile.

Previous topography and water level both control the morphometric characteristics of Lake Nasser. Firstly, the general shape and orientation of the lake follow the former Nubian Nile (Fig. 10.1). Secondly, surface area, shoreline length, mean width, mean depth and maximum depth vary greatly with water level changes (Table 10.1). Due to the flattening of the local topography, surface area at 160 m level is less than 50% of that at 180 m level. Shoreline length at 160 m level is approximately 68% of that at 180 m level, increasing rapidly in the rugged areas relative to the flat or undulating areas. Therefore, the shoreline length generally increases with a rise in water level on the more rugged eastern shores compared to the less rugged western shores. The mean width of Lake Nasser at 180 m level is twice that at 160 m level.

Table 10.1 Some morphometric characteristics of Lake Nasser

Morphometric characteristics	Egyptian sector		Total lake	
	Water level		Water level	
	160 (m)	180 (m)	160 (m)	180 (m)
Length (km)	291.8	291.8	419.3	481.8
Area (km^2)	2562	5237	3084	6276
Volume (km^3)	53	131	64	158
Shoreline length (km)	5416	7875	5860	8804
Mean width (km)	8.85	17.95	7.36	13.03
Mean depth (m)	20.53	25.01	20.75	25.23
Maximum depth (m)	110	130	110	130

Source Smith 1982

10.1.2 Local Forms

Whatever the changes in the water level and their effect on the morphometry of Lake Nasser, the drowning of former topography and the deposition of Nile silt have resulted in the development of new forms in the Nubian region of the Nile. These forms comprise the Nubian Delta, khors, and islands.

10.1.2.1 The Nubian Delta

The most important geomorphologic impact of the development of Lake Nasser is the water, and the deposited silt in the Lake. Since the Lake is acting as a local base level for the upper reaches of the Nile, the narrow Nubian Valley and its channel were filled gradually by Nile sediments starting from the south and advancing northwards. This filling is happening at present in Sudan and in the extreme south of Egypt (Fig. 10.3). It is already known that since1964 when the High Dam closed the Nile channel, virtually all the annual alluvium (about 13×106 m^3) carried out by the River Nile has been deposited in the lake. Most deposition occurs in a 135 km section extending to the south of Adindan at the Egyptian-Sudanese border (Smith 1982). The processing of Landsat-TM images shows that the southern half of the Lake bed is shallower than the northern one (Fig. 10.3). In time, these sediments extend northward as shown by the successive surveys which were carried out for the lakebed (Fig. 10.4). This sedimentation will lead, in the long term, to the formation of a new delta within the lake.

According to Smith (1982), the development of the new delta is governed by factors controlling the deposition of silt in Lake Nasser. A major factor is the sudden decrease in the velocity of the flow of Nile water as soon as the river reaches the lake in Sudan. It is expected that deposition should be greatest at the junction area and would decrease progressively downstream. This is confirmed by the measurements in

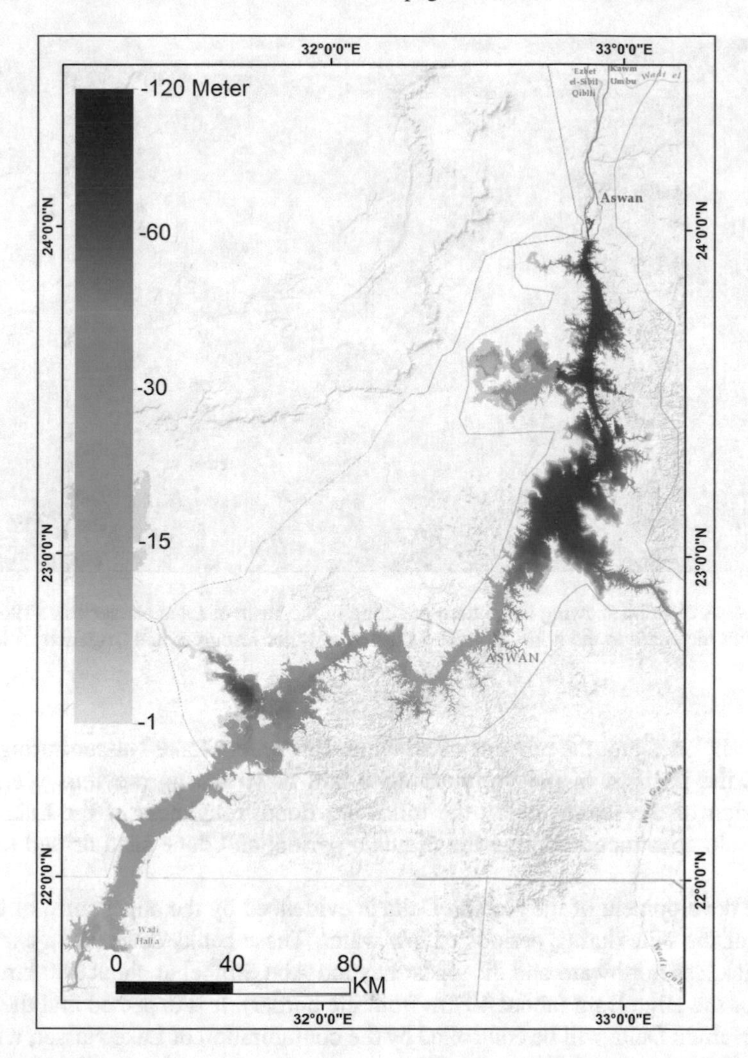

Fig. 10.3 Water depth in Lake Nasser (Processed by Mohamed El-Raei from a landsat-TM image taken in 2016)

the lakebed (Fig. 10.3). Since Lake Nasser acts as a local base level for the upstream reaches of the River Nile, the pre-flood level is the effective base level for the river as it enters the lake. At this level, the river enters the lake and deposition occurs. The pre-flood level changes annually and therefore base level and location of deposition changes accordingly. The rise of water level during the flood period is another factor affecting deposition in the lake. As a result of this rise, deposition will advance gradually upstream, and the sites of minimum and maximum water level of the lake will control annual deposition each year. Other factors affecting sediment deposition and the development of what may be called the Nubian Delta are several. The most

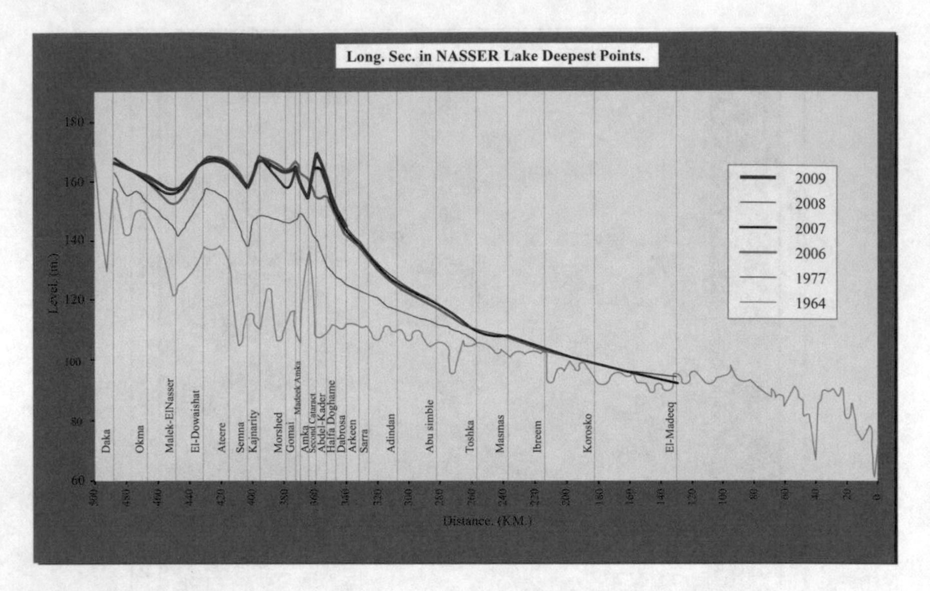

Fig. 10.4 A diagram showing the pattern of silting in the basin of Lake Nasser since 1964 up to 2009 from the south to the north. (*Source* Ministry of water resources and irrigation, Nile water sector)

significant ones are: the amount of alluvium that enters Lake Nasser during flood period, the position of the sediment mass laid down during previous years, rate of erosion of the sediments by the following flood, roughness of the Lake floor, man-made construction during the pre-dam period, and dune sand drifted into the Lake.

The development of the Nubian Delta is evidenced by the appearance of the silt banks of the Nile during periods of low water. These banks decrease in extension and thickness northward and disappear beyond Abu Simbel at about 280 km to the south of the High Dam (about 40 km from the border). It is expected that the shape of the Nubian Delta will be controlled by the configuration of Lake Nasser, which is longitudinal as a whole. When this delta appears on the Egyptian territory is an issue which is difficult to answer due to the complexity of the factors affecting deposition in the lake basin. Nevertheless, based upon three facts: 1—the annual rate of deposition in Sudan sector of the lake between 1971 and 1979 of about 1.0 m (Smith 1982), 2—the level of the lake bed is about 100 m asl in the southern part of Egypt, 3—the maximum water level in the Lake is 180 m asl, it can be inferred that the emergence of the Nubian Delta requires around 80–100 years from the present time; i.e. by the end of the 21st Century.

10.1.2.2 Embayments (Khors)

Embayments are the most common landforms along the sides of Lake Nasser and are responsible for irregularities of the shoreline (Figs. 10.1 and 10.5). Nearly all embayments are drowned wadis, which are known locally as Khors. Other embayments are drowned lowlands along the sides of the lake. Khors developed long before the construction of the High Dam. In fact, they started to develop after the old Aswan Dam had been built and heightened at the beginning of the 20th Century. Khor El-Allaqi, Khor Dehmit, Khor Sharaf El-Din and Khor Kalabsha are examples of pre-High Dam embayments. At that time, the size and number of the Khors were small relative to those of Lake Nasser because the maximum level of water storage in the old Aswan Dam was 121 m asl.

Khors are extremely variable in shape, length, width, depth, shoreline length and surface area. Although these characteristics change with water level, embayments can be classified into four types. Simple Khors are small, single drowned wadis, which extend inland not more than 1.0 km. Dendritic Khors are large and branching drowned wadis of which Khor El-Allaqi is the largest (about 40 km at 150 m asl). Irregular dendritic embayments are the result of the submergence of some wadis which open at adjacent lowlands. These two dendritic types dominate the shorelines of Lake Nasser since all large embayments belong to these types. The fourth type of embayment has an irregular shape which developed under the effect of the previous shape of lowlands around the lake. Although they are a few, some of the irregular embayments cover several square kilometers. The combined surface area of the largest 40 Khors would occupy about 60% of the total area of Lake Nasser at 180 m

Fig. 10.5 Space image for the northern part of Lake Nasser, showing the High Dam, the embayments (Khors), and sand encroachment along the eastern side of the Lake. (*Source* Google Earth 2014)

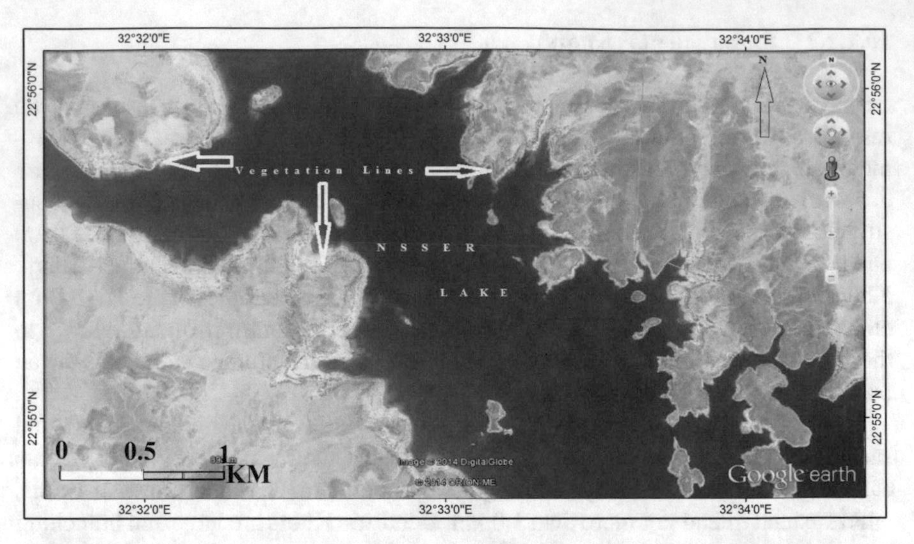

Fig. 10.6 Space image showing the growth of lines of vegetation at certain localities along Lake Nasser shores. (*Source* Google Earth 2018)

level (Entz 1973). High resolution space images (Google Earth 2018) show that the sides of Khors all around the lake are lined by two or three lines of small trees and shrubs (Fig. 10.6). Most probably, these tree lines came into being, and are kept alive, due to the underground water seepage along bed contact of the sandstone. The aquifers of the underground water were charged with water when the water level of the lake was higher—in the early eighties of the 20th Century—than that of the present day.

10.1.2.3 Islands

Islands represent submerged hills or high grounds along the sides of Lake Nasser. Similar to the two previous forms, islands change greatly in size and shape with changes in water level. With the rise in the level of water in Lake Nasser, some islands may disappear; others may become smaller, while others still might come into being as additional land is submerged. According to the topographic maps at scale 1:50,000 (1991) which were compiled from aerial photographs taken in 1988 when the water level was 155 m asl, there were about 2000 islands in the Egyptian sector of Lake Nasser. At that date (1988), there were 65 islands rising more than 180 m asl. Generally, small islands of areas less than 500 m^2 are dominant, whereas the large ones are very few. The largest island is 9 km long and 3 km wide in the sector of Abu Hur. Large islands are irregular in shape due to the presence of Khors and headlands, whereas small ones are circular in general. Only a few islands have names, which are in fact their former names when they were hills during the pre-lake period, such as Gabal El-Allaqi Island (218 m) and Gabal El-Miharraqah Island

(202 m). The number of islands changes according to the level of the lake. All islands whose level is less than 180 m asl appear and disappear according to changes that occur in the level of the lake due to water discharge and recharge. Islands disappear when submerged in water due to the rise in the level of the lake. Islands also disappear when the lake level decreases and some near-shore islands are reconnected to the mainland.

10.2 Environmental Hazards Threatening Lake Nasser

The main purpose of constructing the High Dam on the Nile in Upper Egypt was to control floods and provide water for irrigation, by storing water reaching Egypt from Ethiopian Highlands during times when the demands for economic activities do not need all the water that reaches it, such as during the summer season. At the same time, some of this excess water is needed to compensate the deficiency in water reaching Egypt from the Equatorial sources of the Nile during the winter season. Furthermore, water which is not consumed during the winter may be needed when there is a water deficit due to what is known in Egypt as low floods; or in other words, when the quantity of water is less than the normal quantity reaching Egypt during pre-dam times because of low rainfall on the Ethiopian Highlands.

However, in time, Lake Nasser grew gradually because of the annual excess water. It nearly reached its maximum capacity in the early 1980s. As can be seen from Table 10.1, the maximum capacity of water storage for the whole lake basin is in the range of 150×10^9 m^3, and in the Egyptian section it is 130×10^9 m^3. This water storage is known as century storage, since excess water could remain and be stored in the lake basin for a hundred years or more. Hence any decrease in the storage capacity would affect the quantity of water which could be stored in the lake. Three factors are currently threatening the capacity of water storage in the lake by decreasing the size of its basin, or by decreasing the water reaching Egypt from the Nile sources. First: present-day climatic change; second: the deposition of Nile silt; and third: the deposition of aeolian sand. The high evaporation rates represent another natural factor leading to deprive the Lake from some of the stored water.

10.2.1 Present-Day Climatic Changes

Present day climatic conditions throughout the world change significantly from one year to another. Some years, the changes may be slight, causing hardly any problems to economic activities; but in other years, the changes are drastic and cause serious problems to human activities. These year-to-year changes may continue for several consecutive years, resulting in severe problems to national economies. The most significant climatic element relevant to Lake Nasser is rainfall on the Ethiopian Highlands. This is because the Blue Nile water which is ~48 billion m^3, represents about 60% of the total water reaching Aswan (Embabi 1995). Rainfall may increase resulting in an exceptional rise in the level of water discharge reaching Egypt. These are considered locally high or very high floods, and the lake basin cannot accommodate this discharge. For those years of exceptional rainfall and water discharge, the Egyptian Irrigation Administration constructed a spillway through which the excess water is discharged into small basins in the Western Desert. Severe problems to the Egyptian economy occur when rainfall decreases to the extent where water discharge is much less than the demands of the economy. This has already occurred several times in Egypt's history from the times of the Ancient Egyptians resulting in famine and deterioration of the economy (Said 1993). The most recent of these critical conditions occurred during the years 1986 and 1987 (Embabi 1995) when the Nile discharge in Aswan was only 33×10^9 $m^3$and 35×10^9 m^3 respectively, whereas Egypt's annual demands are 55×10^9 m^3 causing an annual deficit ~20×10^9 m^3. That time, for the first time in Egyptian history, the country was not affected by economic harshness because water storage in Lake Nasser compensated the deficit in the Nile water discharge.

10.2.2 Silting in the Lake Basin

It is well known that the load of Nile water which reaches Egypt from the Ethiopian Highlands is in the range of fine sediments (sand, silt, clay), which are collectively called Nile silt. In pre-dam times and during the flood season, the silt was deposited on the Nile flood plain and along the northern coasts of the Nile Delta. By storing the flood water in the lake from the early 1970s, all the silt was deposited in the lake basin, starting in the extreme south and gradually extending northwards (Fig. 10.4). As mentioned in Sect. 10.1.2.1, the deposition of the silt resulted in the development of a local delta in the southern part of the Lake, and with the continuous deposition of silt, the delta will be growing so long as the lake is alive. This deposition of silt is filling the lake basin, resulting in a decrease in water depth, and consequently in water storage capacity. Figure 10.3 shows that the southern half of the Lake basin is shallower than the northern half. This is because silting is occurring since the Lake came into being, in the southern half of the basin of the Lake. Pre-dam plans estimated that it would take 500 years to fill the lake basin with silt, and the lake would then be obsolete. Due to the intervention of the threatening factors, those five hundred years of the working age of Lake Nasser are subject to change. It has been suggested that to increase the planned age, it is possible to dredge the Nile silt from the lake, and spread it on its shores to be used in several activities of which agriculture is a priority. Other researchers suggested extracting minerals of economic value from it such as gold (Hassaan et al. 1995).

10.2.3 Aeolian Sand Deposition in the Lake Basin

The second factor threatening the storage capacity of the Lake is the deposition of aeolian sand in the lake basin. It is a well-known fact that there are sand accumulations on the western side of the Nile. It is also a well-known fact that during pre-Dam period some of the sand was drifted into the Nile channel and was carried out northwards by flowing water in the Nile channel. From the time the lake came into being, all the deposited sand has accumulated in the lakebed, starting along the lake shores and extending into the lake basin. The result of this accumulation of sand is the same as that of the deposition of Nile silt in the lake basin; i.e. a decrease in the size of the lake basin and consequently a decrease in its water storage capacity. Although the aeolian sand could be dredged from the lakebed like Nile silt, this sand so far has no economic value. Until now, there have been no studies that evaluate the actual threat of aeolian sand, nor how to stop or minimize it.

10.2.4 High Evaporation Rates

The analysis of climatic data of Aswan Meteorological Station in Sect. 10.1.1 showed that Lake Nasser lies in a region characterized by a hyper-arid climate. This is because of the very high rates of expected evapotranspiration, which is ~7 m/year, and the very low rate of annual rainfall which is 0.7 mm. The question now is how much water evaporates every year from Lake Nasser at different levels? The following table shows the expected evaporated water per year at 160 and 180 levels.

This table shows that the estimated water loss from Lake Nasser is between 21.5 billion m^3 when the level of the Lake is 160 m, and 43.9 billion m^3 when the Lake level is 180 m.

These figures of water loss by evaporation from the surface of Lake Nasser exceed those calculated by previous studies as shown in the following table.

It is clear from this Table 10.3 how the estimated water loss differs from one study to another. This is due to the method used by each study in estimating evaporation rate from Lake Nasser. However, it is clear from Tables 10.2 and 10.3 how huge the estimated evaporated water according to any of the previous estimation. The High Dam Authority (1982) estimated the loss by evaporation is 10 billion m^3/year, and this loss would lead to a decrease of 2 m in the Lake level when it stands at 180 m and its surface area is 5000 m^2. Aboul-Ata (1978) estimation is 21.6×10^9 m^3. This loss is as twice as that of the High Dam Authority. This means that it will lead to a decrease in water level in the Lake basin by more than 4 m. The present study (Table 10.2), shows that the Lake reservoir will lose 43.9×10^9 m^3 representing about 33.1% of the storage capacity 131 billion m^3 when the water level is 180 m, and the loss will be about 33.7% when storage capacity is 53 billion m^3 and the water level is 160 m.

Table 10.2 Expected evaporated water per year from Lake Nasser (billion m^3)

Sector	Egyptian section		Total lake	
Level of lake (m)	160	180	160	180
Surface area (km^2)	2562	5237	3084	6276
Estimated water loss by evaporation (billion m^3)	17.9[a]	36.5[a]	21.5[a]	43.9[a]

[a]Calculated from Table 10.1 and the evapotranspiration rates in Sect. 10.1.1

Table 10.3 Estimated water loss by evaporation from the surface of Lake Nasser at 180 m level

Previous studies	Estimated loss (billion m^3)
Aboul-Ata (1978)	21.6
Gishler (1976)	12.5
Harb and El-Bakry (1979)	14.0
High Dam Authority (1982)	10.0

10.3 Anthropogenic Hazards

Anthropogenic hazards are those created by the people deliberately or accidentally. Lake Nasser is threatened by both types of hazards. The first is the construction of new dams on the Nile reaches, and the second is the pollution of Lake Nasser water by the new Nubian settlers and by the working touristic and fishing boats in the Lake.

10.3.1 Construction of New Dams on the Nile Reaches

Without a political agreement between the concerned governments, the construction of new dams on the upper Nile reaches especially in Ethiopia and Sudan, is the most serious threat to Lake Nasser. The first of these dams is the **Grand Ethiopian Renaissance Dam (El-Nahda in Arabic)** on the Blue Nile in Ethiopia. Although the precise impact of this dam is not yet known, there are three scenarios were suggested in some previous studies (Abdel Hamid 2017).

The first scenario suggests that Ethiopia is going to abide with the International water agreements which were held between the concerned states, but after filling the basin of the Ethiopian Dam to its full capacity. The filling of the dam basin will spread over several years (2–5 years). Therefore, there will be a loss of 11–29 × 10^9 m^3 of water per year, which would cause large sectors of farmers to lose their income while the reservoir is being filled. It will also affect Egypt's electricity supply by 20–40% while the dam is being built.

The second scenario presumes the destruction of the Ethiopian Dam by extraordinary accidental factor (s), such high floods, very strong earthquakes due to pressure from the stored water in the Lake basin of the Dam, or due to some fatal faults in the construction of the Dam. This will result in pouring of the stored water in the Lake of the Dam downstream in the Blue Nile towards Sudan and Egypt. On its way to Egypt, the pouring water will pass the Lakes of three dams in Sudan: El-Roseiras, Sennar, and Merwe. It is expected that Sudan will open the gates of these dams to drain the huge quantities of water coming from Ethiopia to avoid their destruction. This means that Lake Nasser will receive not only the pouring water the Ethiopian Dam, but also the drained water from these three Sudanese dams. The water journey from Ethiopia to Lake will take about 12 days, which is considered enough for the administration of Lake Nasser to take the precautions to avoid any negative accidents resulting from the arrival of these huge quantities of water to the basin of the Lake. The negative effects of destruction of the Renaissance Dam of Ethiopia cover several economic sides. The first is rise of the level of Lake Nasser more than 183 m asl which represent a real threat to the High Dam and the Electric Power Station. The second side is the submergence of all newly reclaimed land and settlement, and some touristic sites and some archeological locations along the sides of Lake Nasser and some islands.

The third scenario presumes a long period of drought in which there will be climatic change characterized by a drastic decrease in rainfall not only in Ethiopia, but also Sudan and East Africa like the one which happened during the Mid-Eighties and Early seventies of the 20th Century. If this happens, it will enforce Ethiopia to store all the water flowing in the Blue Nile in the basin of the Renaissance Dam, resulting in the deprive of Egypt of any drop of water coming from Ethiopia.

10.3.2 Pollution of the Lake Water

10.3.2.1 By New Settlers

Examination of high resolution images of Google Earth (2018) shows that there are several small human colonies established along the shores of Lake Nasser (Fig. 10.7a, b), where the main economic activity is agriculture. In each colony, a plot of land is reclaimed and cultivated by withdrawing water from Lake Nasser by several means of which canals are the most frequent ones as appears from images of Google Earth (2018), Fig. 10.7a, b. Most probably, all these colonies were established without official plans to settle down some Nubian families along the shores of Lake Nasser. The success of these colonies to establish viable communities, will encourage new settlers to establish new unplanned colonies. These colonies as such represent a contribution to the local and national economy. In the meantime, they represent a threat to the water of Lake Nasser since their waste will be drained or thrown into the Lake leading to the pollution of its water. By the increase in the number of colonies and settlers, pollution will increase accordingly. In an attempt to discover this assumption, the number of new human colonies on both sides of the Lake were counted from Google Earth, 2017 and 2007, it was found that the total area of the

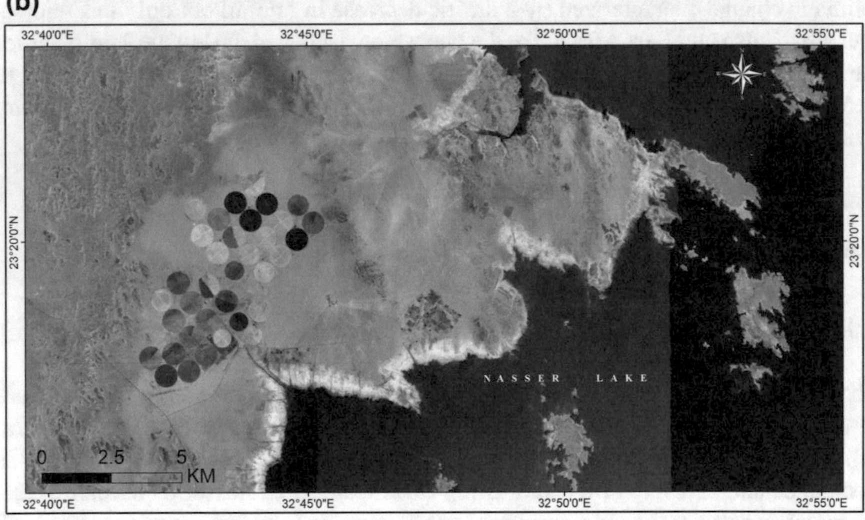

Fig. 10.7 **a** A human colony at the head of a khor in the western side of Lake Nasser, depending on water withdrawal from the Lake by a main canal and distributed by secondary canals. **b** Another locality on the western side of Lake Nasser, where several colonies of various sizes were established, the lagest of which is the one using the pivot irrigation system. All the colonies depend on water withdrawal by canals from Lake Nasser

colonies increased in ten years from 19.5 km^2 in July 2007 to 25.7 km^2 in July 2017, which is about 32% of the total area in 2007.

What is needed at the time being is that the Government of this country should prohibit any new human settling along the shores of Lake Nasser, and suggest a regional plan for economic development of the region of Lake Nasser. This plan is multi-sided including the establishment of small planned villages and towns, the suitable economic activities, and the number of people who will be living in this region. But, the most important side in the development of this region is to prohibit by law any drainage of waste water or solid material into the Lake basin.

10.3.2.2 By Fishing and Touristic Boats

Another source of pollution of the Lake Nasser water is the waste draining from fishing and touristic boats which cross the Lake from Egypt and Sudan and vice versa. Although there are some laws which prohibit draining waste water of boats, or throwing solid waste into the Lake, it is well-known that most of them do not abide with law, because no strict supervision by the Government administration. In fact this aspect needs a scientific study to determine the level of pollution of the water of the Lake.

10.4 Protection Measures

Protection measures cannot be applied to all hazards. Therefore, hazards threatening Lake Nasser can be classified into two groups. The first group comprises those hazards in which measures can be taken to prevent or minimize the effects of hazards on Lake Nasser. The second group comprises hazards in which measures are not applicable. This latter group originates by climatic accidents or tectonics. Rainfall on the Ethiopian highlands is the most significant climatic element as a potential hazard. Rainfall changes from to year, but rainfall becomes a hazard when it decreases to the extent that water flow reaching Lake Nasser is far less than Egypt's water needs for economic (Agriculture, Industry, Services) and domestic purposes.

Egypt faced such a situation in the mid-eighties of the 20th Century. At that time, water deficit was compensated from the stored water in Lake Nasser Basin. Luckily, this situation continued for a few years only, because water level of the Lake reached 125 m asl in 1986 and if it remained like that for another year, Egyptian economy will suffer severely because there is no enough water to cover Egyptian needs. The question now is there any other water source(s) to compensate the water deficit caused by the decrease in rainfall on the Ethiopian Highlands if it continues for a longer period than that of the 1980s? Other water sources in Egypt are underground water stored in the Nubian aquifers and in the Nile Valley aquifers, and the desalinated sea water. It is well-known that both water sources cannot produce enough water to compensate what is lost by rainfall decrease on Ethiopian Highlands.

Only when rainfall on the Ethiopian Highlands increases above normal, and water flow reaching Lake Nasser is above the storage capacity of the Lake Basin, some protection measures can be done to avoid the damage which might happen to the High Dam and other irrigation establishments by the excess water. In the 1990s, twice water reached Egypt from Ethiopia exceeded the capacity of Lake Nasser Basin. At that time the Egyptian Ministry of irrigation decided to diverge the excess water by a spillway (partially an ancient wadi, and partially artificially) called Toshka Spillway on the western side of Lake Nasser. This spillway took the excess water to several small closed depressions creating some lakes (Toshka Lakes) 'in the southern part of the Western Desert. Figure 10.1, shows Lake Nasser and the remains of some Toshka Lakes, since most of the lakes were dried up.

Evapotranspiration is a second climatic element representing a permanent hazard to the water storage in La6.ke Nasser. As indicated in Sect. 10.2.4, the Lake lies in a hyper-arid region where evapotranspiration is more than 10,000 times the expected rainfall on the region of the Lake. Several estimations were presented in some previous studies for the expected annual sum of water which could be evaporated from the surface of Lake Nasser, and it was found to vary between 10^9 m^3 and ~44^9 m^3 (~ one third of the total storage of the Lake. Since this hazard is inevitable, no protection measure can be applied.

The second potential hazard comes from earthquakes. Although quakes occurred every now and then, 83 events were reported to have occurred in and around Egypt, have caused damage of different degrees in various localities in Egypt since 2800 BC (Kebeasy 1990). Of these past earthquakes, only one affected Lake Nasser Region and occurred in Abu Simbil area in 1210 BC, with an intensity of 7° (Kebeasy 1990). However, the most serious earthquake which affected Lake Nasser region is that one which occurred in 14 November 1981 with a magnitude of 5.5. Its epicenter was in Kalabsha, 60 km SW of Aswan, and this why it called Kalabsha Earthquake. Although fears were raised about its effects on the Aswan High Dam and Lake Nasser, only minor effects occurred on both sides of the Lake (Kebeasy 1990).

The second group of hazards are those ones where protection measure can be applicable. They comprise physical hazards and man-made hazards. The physical one includes silting of the Nile silt in the southern section of the Lake, and deposition of aeolian sand mainly in the northern section of the Lake. As mentioned in Sects. 10.2.2 and 10.2.3, the deposition of these sediments will minimize the storage capacity of the Lake Basin, decreasing the sum of water which be stored. The only protection measure which can partially solve this problem is the dredging these sediments spreading on both sides of the Lake as suggested in some previous studies (Hassaan et al. 1995).

There are two man-made hazards. The first is the construction of dams on the upper reaches of the Nile to store water for the various purposes of which generating cheap electricity. The most dangerous Dam as a hazard for Egypt is the Ethiopian Renaissance Dam which be erected on the Blue. This hazard will deprive Egypt from some of its historical rights in the Nile waters, especially those arriving from the Blue Nile. The only protection measure which can be suggested is to make an agreement with Ethiopia and Sudan in which all partners share the Nile water according to their needs and historical rights with the maximum benefits and minimum harm.

The second man-made hazard is pollution of the Lake water. The potential pollution comes from two sources. The first is the fishing and touristic boats roaming in the lake, and the second is the new Nubian settlers on both sides of the Lake. In both cases, pollution comes from the wastes of the human beings and animals. Although there are some laws prohibiting draining waste water/throwing solid materials into the Lake Basin, most of boats or settlers do not abide with the laws. So far, no scientific studies were carried on determining the degree of pollution in the water of the Lake. At the time being, it is recommended to inspect all boats working in the Lake periodically to see what kind of measures they use to protect the water of the Lake from pollution. The same inspection could be applied on the settlers on both sides of the Lake.

10.5 Conclusions

With the aid of remote sensing, it was possible to discover many aspects of Lake Nasser or make them more accurate. Therefore, the analysis of the environmental and anthropological hazards that threaten Lake Nasser as Egypt's National Water Reservoir depended heavily on space images. Some hazards developed due to the environmental conditions of the region of the Lake, or due to the interaction between its geomorphologic aspects and water characteristics entering the Lake. Also, hazards are generated by humans not only in the Lake region, but also in the upper reaches of the Nile.

The Lake lies in a region with hyper-arid climate, with high temperature, and nearly no rain, resulting in very high evaporation rates, which deprive the Lake from a huge water amount stored in its basin. Estimates of this water loss vary between 10 and 43.8 billion m^3. Rainfall variability on the upper reaches of the Nile, mainly on the Ethiopian Highlands represents another environmental hazard, especially if there is a drastic decrease in rainfall. No protection measures can be taken against these two hazards, because no one can change their physical characteristics.

The size of Lake Nasser Basin is threatened by continuous silting of the Nile silt and deposition of aeolian sand carried out by wind blowing from NW direction. Since siltation starts in the southern section of the Lake, depth of the Lake floor is shallower here than the northern one. This characteristic is an indication that the deposited Nile silt and aeolian sand is minimizing the storage capacity of Lake Basin. Dredging of silt and sand from the floor of the Lake is suggested in previous studies to avoid filling the Lake with these sediments. It is also suggested that the dredged material should be spread out on desert land which could be reclaimed and cultivated in the future. This protection measure will solve the problem of filling the Lake basin by silt and sand only if dredging material is equal to the deposited one.

Humans generate some hazards that threaten Lake Nasser, of which the construction of the Great Ethiopian Dam (Ethiopian Renaissance Dam) on the Blue Nile is the most dangerous one. This dam is under construction -at present- on the Blue Nile which provides Egypt with more than 50% of its water needs. This dam will deprive Egypt from some of its share in Nile water during filling of the dam basin and during its operation, resulting in a decrease in the stored water, lowering the lake level, and reduction in electricity production. Under certain climatic conditions, a drastic decrease in rainfall will enforce Ethiopia to store every drop of water in the basin of the Dam, leading to a very critical economic situation to Egypt. In fact, the threat which results from the construction of this dam outweighs all the threats which come from all other environmental and other human threats. Therefore, a political agreement between Ethiopia, Sudan and Egypt is an urgent necessity to avoid a catastrophic situation in this region of Africa.

Other human threats are the direct pollution of the waste of the boats (fishing, touristic, commercial) crossing the Lake from Egypt to Sudan and vice-versa. Although there are some laws which inhibits the throw of the waste of the boats in the waters of the Lake, most of them do not abide, a matter which requires a strict permanent monitoring of the Lake and inspection of boats to minimize the pollution of the Lake waters. A second human hazard polluting the waters of Lake Nasser is the new settlers around the shores of the Lake. As the space images show agricultural colonies are present mostly on the western side of the Lake, because most of the plots which can be reclaimed and cultivated are present on this western side. The problem in this issue is not in cultivating newly reclaimed land whatever their size is, but it is in the waste of the settlers especially when their number increases to the extent it becomes a real threat for the quality of the water of the Lake. As far as these new societies became a fact in this region, the government administration is obliged to make a master plan for settling newcomers and building rural villages around Lake Nasser.

10.6 Recommendations

Lake Nasser is a new man-made feature created since about 50 years in a hyper-arid environment in the southern part of the Nubian Nile. The creation of the Lake transformed the region from a desert to aquatic environment. By time, the Lake is exposed to several physical and anthropogenic hazards, and new economic activities such as fishing, and tourism were introduced.

It became clear from the previous assessment of protection measure that all hazards are interrelated. Therefore, it is recommended to establish an authority to supervise all physical and anthropogenic aspects of the Lake and monitor the Lake region regularly from space and in-situ for protection purposes. It can be called **Lake Nasser Authority for Protection and Development**. It should be an independent authority, have the right to take measures to protect the Lake, and suggest new projects to develop new viable rural and urban centres in the Lake region. This Authority can be seen as an organizing corps to maximize the benefits of the available resources of Lake Nasser region.

Acknowledgements I am indebted to so many people who helped me writing this chapter. First, I would like to thank Dr. Salwa El-Beih of the National Authority of Remote Sensing and Space Sciences for nominating me to write this chapter. Second, my thanks also go to the Egyptian Editors and the Springer Editor who revised thoroughly the manuscript of this Chapter and for their valuable suggestion which added to the scientific value of this Chapter. I would like also to express my gratitude Mr. Mohamed El-Raei who works as an assistant lecturer in the Department of Geography, Faculty of Arts, Ain Shams University for processing the Landsat image of Fig. 10.3 and producing this figure which shows water depth in Lake Nasser, and for processing the Google Earth figures producing them with the geographic grids in tiff form. I acknowledge also Google Earth for providing high resolution space image on the internet free of charge, without which so many characteristics of Lake Nasser could not have been studied.

References

Abdel Hamid AKA (2017) Environmental impacts of the Renaissance Dam in Ethiopia on Lake Nasser using GIS and remote sensing. Unpublished Dissertation (in Arabic). Benha University

Aboul-Ata A (1978) Egypt and the Nile after the High Dam. Ministry of Irrigation and Land Reclamation (in Arabic). Cairo

Butzer K (1965) Desert landforms in the Kurkur Oasis, Egypt. Ann Assoc Am Geogr 55:578–591

Conoco Coral and the Egyptian General Petroleum Company Corporation (1989) Geological map of Egypt, NF 36 NW El Saad ElAli, scale 1:500,000

Embabi NS (1995) Water resources of Egypt: the Nile water (in Arabic). Middle East Research Center, Ain Shams University, Cairo

Entz B (1973) Morphometrics of Lake Nasser and Lake Nubia, Working paper no 2. Lake Nasser Development Center, Aswan

Gishler C (1976) Salt balance of Lake Nasser and the Nile. Second annual report. River Nile and Lake Nasser Project, Egypt, pp 50–57

Google Earth (2018) Egypt images: latitudes 22-32 N and longitudes 25-34 E. http://www.google.com

Harb MSED, El-Bakry M (1979) High Dam Lake and its impact on climate. Report prepared to River Nile and Lake Nasser project. Academy of Science Research and Technology, Cairo

Hassaan M, ElDardir M, Nassar Y (1995) Alluvial gold placer in main sedimentation area, High Dam Lake, Egypt: the first record. Sedimentol Egypt 3:125–129

High Dam Authority (1982) Report of the second research trip of the High Dam Lake. Aswan

Kebeasy R (1990) Seismicity. In: Said R (ed.) The geology of Egypt. Balkema, Roterrdam

Said R (1993) The River Nile: geology, hydrology and utilization. Pergamon Press, Oxford

Smith S (1982) Application of remote sensing techniques to the study of the impacts of the Aswan High Dam. Dissertation, University of Michigan

UNESCO (1977) Explanatory note on map of the world distribution of arid regions. MAB technical report 7, Paris

Chapter 11
Oil Pollution in the Mediterranean Waters of Egypt

Andrey Kostianoy, Evgeniia A. Kostianaia and Dmitry M. Soloviev

Abstract Oil pollution in the Mediterranean waters of Egypt is related to two main factors: (1) a huge ship traffic related to the passage through the Suez Canal, as well as operation of 6 ports located on the coastal zone of Egypt; and (2) significant increase of offshore activities related to exploration and development of several gas fields, which were discovered during the past decades. A review of the historical oil pollution in the Mediterranean Sea has not revealed serious cases of oil pollution in the Mediterranean waters of Egypt, but small and medium-size oil spills do exist as it is well seen on accumulated maps of oil spills detected by satellite remote sensing. A dozen of such cases have been found in records of the REMPEC MEDGIS-MAR tool. Unfortunately, Egypt has no satellite monitoring system for detection of oil pollution in the Mediterranean and Red Sea waters. This is why we provide satellite SAR imagery acquired in 2017–2019 to demonstrate detection of oil spills in the Mediterranean waters of Egypt. The presented cases of oil pollution show the necessity of establishing permanent operational integrated satellite monitoring of oil pollution in the Mediterranean and Red Sea waters of Egypt.

Keywords Egypt · Mediterranean Sea · Suez Canal · Shipping routes · Satellite monitoring · Oil pollution · Synthetic aperture radar

---

A. Kostianoy (✉) · E. A. Kostianaia (✉)
P.P. Shirshov Institute of Oceanology, Russian Academy of Sciences, 36, Nakhimovsky Pr., Moscow 117997, Russia
e-mail: kostianoy@gmail.com

E. A. Kostianaia
e-mail: evgeniia.kostianaia@gmail.com

A. Kostianoy
S.Yu. Witte Moscow University, 12, Build. 1, 2nd Kozhukhovsky Pr., Moscow 115432, Russia

D. M. Soloviev
Marine Hydrophysical Institute, Russian Academy of Sciences, 2, Kapitanskaya Str., Sevastopol 299011, Russia
e-mail: solmit@gmail.com

S. F. Elbeih et al. (eds.), *Environmental Remote Sensing in Egypt*,
Springer Geophysics, https://doi.org/10.1007/978-3-030-39593-3_11

11.1 Introduction

About 15% of all global maritime trade and 10% of global seaborne oil pass through the Suez Canal to/from the Mediterranean Sea. In 2017, the total number of ships passing through the Suez Canal was 17,550; most of which were container ships (31.73% of the total number of ships), followed by tankers (25.85%) and bulk carriers (18.74%) (Suez Canal Authority 2019). There is a little increase in the ship traffic related to the reconstruction of the Suez Canal in 2015. Besides, Egypt has 6 important ports on the coast of the Mediterranean Sea with heavy ship traffic. Our analysis of sources and volumes of oil pollution in the Mediterranean Sea shows that shipping activities are the main cause for oil pollution in the Mediterranean Sea because oil and gas production and exploration are still not so dramatic, unlike in the Gulf of Mexico, the North Sea or the Caspian Sea. Without major oil spill accidents from ships, which are very rare events in the Mediterranean, different expert reports and estimates provide total volumes of oil pollution ranging from 1600 to 1,000,000 tonnes per year. The other sources of oil pollution in the Mediterranean Sea include oil and gas platforms, ports and oil terminals, land-based sources, military conflicts, natural oil seeps, and the atmosphere (Kostianoy and Carpenter 2018a).

In 2011, Regional Marine Pollution Emergency Response Center for the Mediterranean Sea (REMPEC) in its Report states that in 1977–2010 geographically the most of oil releases of >100 tonnes occurred in Greece (30%), Italy (18%), Spain (14%), Egypt (8%), Algeria (6%), Israel (6%), Lebanon (6%), and others (12%) (REMPEC 2011). This can be explained by the distribution of the maritime traffic in the Mediterranean Sea. The most common type of accidents reported is grounding (21%), followed by collision/contact (17%), fire/explosion (14%), cargo transfer failure (11%), and, finally, sinking (9%). The other accidents related to acts of war, ruptured pipes on land, various mechanical or structural failures on board ships, etc. account for 28% of all accidents in 1977–2010. The largest accidents followed by a release of >700 tonnes of oil were caused by collisions of ships (32% of the accidents), fire/explosion (26%) and groundings (21%), cargo transfer failure (10%), and other (11%). Regarding accidents with released oil quantities between 101 and 700 tonnes, collisions of ships account for 28%, grounding—for 21%, and sinking—for 18%. For oil spills <100 tonnes cargo transfer failure accidents account for 26% of these accidents, collisions—13%, grounding—12%, sinking—9%, fire/explosion—2%, and other—38% (REMPEC 2011). As concerns the vessel type, REMPEC notes a gradual decrease in the proportion of oil tanker accidents from nearly 70% of all accidents in 1977–1984 to 23% in 2004–2010. However, the cargo proportion has increased from 17% in 1977–1984 to 30% in 2004–2010, as well as an increase for "other" vessel type (container, bulk, and chemical carriers) from 14 to 46% (REMPEC 2011).

Besides the very busy ship traffic in the Mediterranean waters of Egypt, in the last decade, there has been a substantial development of offshore gas and oil fields along the Mediterranean coasts of Egypt, where the most active companies are BP, BG, Eni, IEOC, EGAS, Total, RWE Dea and Dana Gas. BG is active in five concessions, Dana Gas, a regional company, has about 10 producing fields. In August 2015, Eni announced the discovery of the Zohr gas field 150 km from Egypt's north coast. The Zohr gas field reserve is estimated as 30 trillion standard cubic feet (trn scf) of natural gas, making it the largest in the Mediterranean. This discovery seems to increase the Egypt's total gas reserves by about 40% (Oxford Business Group 2018; Kostianoy and Carpenter 2018b).

The above mentioned numbers show that shipping activities and offshore oil and gas exploration and production represent a serious threat to the marine environment, coastal zones and beaches of tourist resorts in Egypt and other countries of the Eastern Mediterranean, in particular. The Egyptian coast of the Mediterranean Sea has no important sea resorts unlike in the Red Sea (Kostianaia et al. 2019), but any oil spills can seriously impact the vulnerable Nile Delta (Negm 2017), the coastal zone of Egypt and neighbouring countries. Fortunately, till present Egypt has avoided large oil pollution catasrophes like with the *MV Haven* which exploded off Genoa (Italy) in April 1991 and 144,000 tonnes of oil were spilled in the sea; with the *Irenes Serenade* (fire on board) in Navarino Bay (Greece) on 23 February 1980, when 100,000 tonnes of oil were spilled to the sea, and with grounding of *Juan A. Lavalleja* in Arzew Harbour, Algeria, in 1980, which caused a release of 39,000 tonnes of oil. However, such catastrophes can occur because the main shipping route in the Mediterranean Sea goes to the Suez Canal, unless precaution measures are considered and lessons are learnt from the past.

Moreover, the analysis of satellite imagery for different years has showed that operational oil spills of a size of about 1–10 tonnes released by ships of different types due to routine operations occur almost daily in the Mediterranean Sea, and these remain a major oil pollution challenge as their number may reach 1500–2500 every year (Kostianoy and Carpenter 2018a). Satellite monitoring of the Eastern Mediterranean around Cyprus during 5 years from 2007 to 2011 showed more than 1000 potential oil spills concentrated along the main shipping routes which are located not far from Egypt (Zodiatis et al. 2012; Kirkos et al. 2018).

We begin this chapter with a brief description of shipping activities in the South-eastern Mediterranean connected to passage via the Suez Canal (Sect. 11.2), offshore development of oil/gas fields (Sect. 11.3), and cases of oil pollution (Sect. 11.4). In Sect. 11.5, we provide satellite SAR imagery acquired in 2017–2019 to demonstrate detection of oil spills in waters of Egypt. In the Conclusions (Sect. 11.6), we discuss the necessity of establishing permanent operational integrated satellite monitoring of oil pollution in the Mediterranean waters of Egypt.

11.2 Shipping Activities in the Coastal Zone

The Mediterranean Sea is among the world's busiest waterways, accounting for 15% of global shipping activity by number of calls and 10% by vessel deadweight tonnes. More than 325,000 voyages occurred in the Mediterranean Sea in 2007, which have transported about 3800 million tonnes of oil and goods (Grid-Arendal 2019). The significant part of maritime transportation goes through the Suez Canal which is an artificial waterway 193 km long, connecting the Mediterranean Sea (Port Said) and the Red Sea (Port of Suez). It is one of the most vital and heavily used waterways in the world. For example, in 2017, the total number of ships passing through the Suez Canal was 17,550; most of which were container ships (31.73%), followed by tankers (25.85%) and bulk carriers (18.74%) (Suez Canal Authority 2019).

Egypt has six ports on the coast of the Mediterranean Sea: Alexandria, Dekheila, Damietta, Port Said, East Port Said, and El Arish (Fig. 11.1). The Port of Alexandria is one of the oldest ports in the world: its earliest port facilities were built in 1900 BC. The Port includes the Western Port of Alexandria, as well as small ports of Abu Qir, Sidi Krer, and Alexandria's old Eastern Port, which is no longer used for shipping. It is the country's largest port which handles over three quarters of Egypt's foreign trade (Jeffreys 2011). The Port of El-Dekheila was built in the 1980s seven miles west of Alexandria as an extension of the Port of Alexandria. It has facilities for container shipping and infrastructure to serve the nearby steel factory. In 2017, the Port of Alexandria received 4232 ships, handled 0,565 million tonnes of total goods, 1.6 million containers, and 25,480 passengers. The Port of Damietta received 2834 ships, handled 0,353 million tonnes of goods, and 1.13 million containers (Maritime Transport Sector 2019). The port of Port Said which is divided into the Port of Port Said and the Port of East Port Said, with 3,47 million containers transported in 2009, is the 28th busiest container seaport in the world and the first one in Egypt. Port Said is also an important harbour for exports of Egyptian cotton, rice, and other products, but also is a fueling station for ships that pass through the Suez Canal (Maritime Transport Sector 2019). El Arish, the largest city in Sinai Peninsula, the capital of Egyptian Governorate of North Sinai, has a very small port and marina.

The intensity of the marine traffic in the Southeastern part of the Mediterranean Sea for all ship types registered in 2017 is shown in Fig. 11.2.

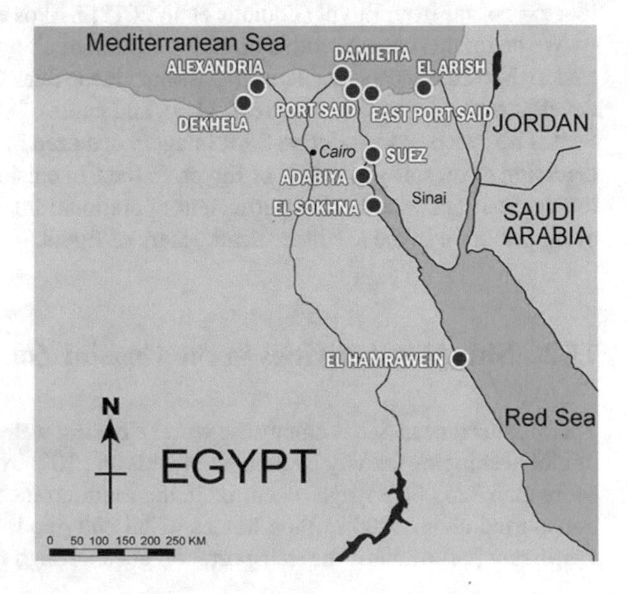

Fig. 11.1 The Mediterranean ports of Egypt. *Source* El Amira (2019)

Fig. 11.2 Marine traffic density in the Southeastern part of the Mediterranean Sea for all ship types registered in 2017. *Source* Marine Traffic (2019)

11.3 Offshore Oil/Gas Fields and Infrastructure

In the last decade, there has been a serious development of offshore gas and oil fields along the Mediterranean coasts of Egypt (Figs. 11.3 and 11.4), where the most active companies are British Petroleum (BP, UK), British Gas (BG, UK), Eni (Italy), Italian Egyptian Oil Company (IEOC), Egyptian Natural Gas Holding Company (EGAS), Total (France), DEA Deutsche Erdoel AG (Germany), and Dana Gas (UAE). BG is active in five concessions, Dana Gas has about 10 producing fields. In September 2013, BP announced a significant gas discovery in the East Nile Delta named "Salamat". The Salamat gas field is located around 75 km north of Damietta City and only 35 km to the north-west of the Temsah offshore facilities. The Atoll gas field was discovered by BP in March 2015 by drilling the Atoll-1 deepwater exploration discovery well 15 km north of the Salamat gas field. In August 2015, Eni announced the discovery of the Zohr offshore gas field 150 km from Egypt's north coast (Figs. 11.3 and 11.4). The reserve is estimated as 30 trillion standard cubic feet (trn scf) of natural gas, making it the largest in the Mediterranean Sea. This discovery seems to increase the Egypt's total gas reserves by about 40% (Oxford Business Group 2018; Kostianoy and Carpenter 2018b).

The Atoll gas field is a significant discovery lying in the North Damietta Concession offshore Egypt in the East Nile Delta (Fig. 11.4). BP announced the Atoll discovery in March 2015. The field development was approved by BP in collaboration with EGAS on 20 June 2016. It was decided that Atoll will be developed as a fast-track project after the agreement was signed in November 2015 between BP and the Egyptian Minister of Petroleum and Mineral Resources. The field was developed by BP, which has 100% equity in the discovery, and on 12 February 2018 BP announced

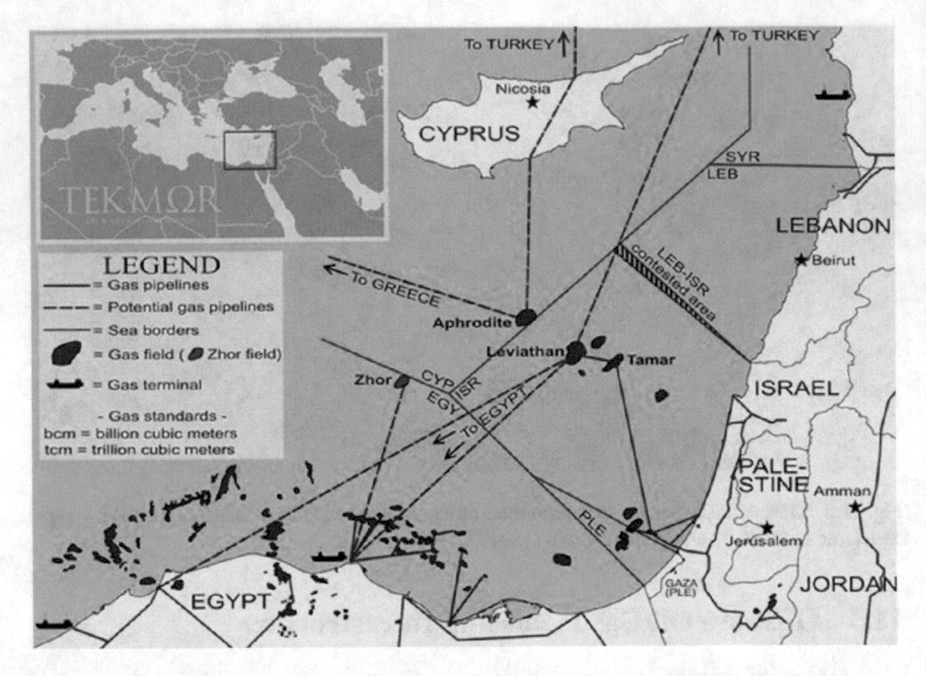

Fig. 11.3 Gas fields, active and potential gas pipelines, and gas terminals in the Southeastern Mediterranean. *Source* Vesti Finance (2019)

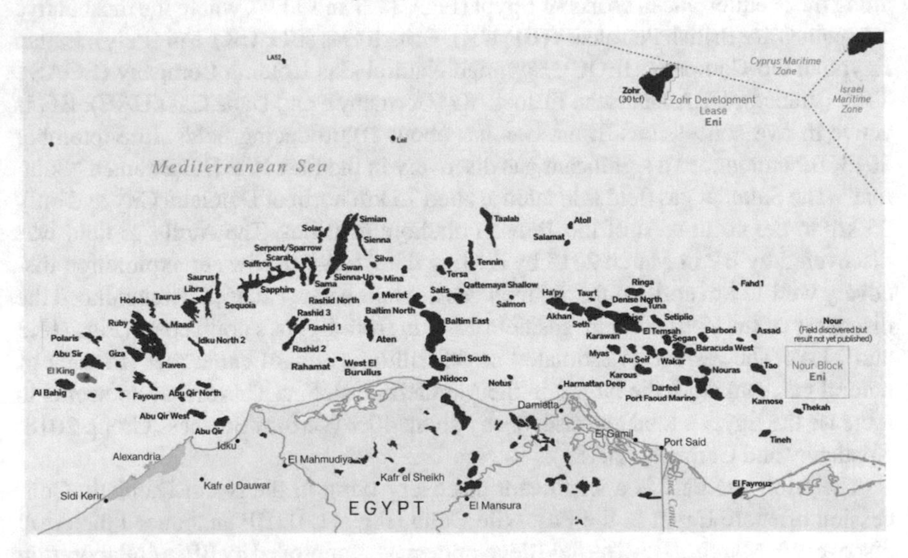

Fig. 11.4 Potential: Egypt's offshore gas fields. *Source* Stephen (2019)

the start of gas production. BP has also signed a number of agreements for transportation and processing arrangements related to the field development. Pharaonic Petroleum Co. (PhPC), BP's joint venture with EGAS and Eni, will execute and operate the project. The field is estimated to contain approximately 1.5 trillion cubic feet (tcf) of natural gas and 31 million metric barrels (mmbl) of condensates. The gas and the liquids produced from the field are processed onshore at the existing West Harbour gas processing facility, which currently processes 280 mmscfd from Ha'py and 265 mmscfd from Taurt fields (Offshore Technology 2018a).

The Zohr gas field (Figs. 11.3 and 11.4) is located within the 3,752 km^2 Shorouk Block, within the Egyptian Exclusive Economic Zone (EEZ), in the southeastern part of the Mediterranean Sea. The field is situated more than 150 km from the coast of Egypt, close to the Cyprus EEZ. Eni owns a 100% stake of the Shorouk license through IEOC Production, and the property is operated by Belayim Petroleum Company (Petrobel), a joint venture between IEOC and Egyptian General Petroleum Corporation (EGPC). Eni was granted approval for the Zohr Development Lease by the EGAS in February 2016. Production at the deepwater gas field started by the end of 2017 and will reach full production capacity of 2.7 bcf/d in 2019. The project will include the installation of an offshore control and production platform, which will further be connected to an onshore processing plant by subsea pipelines. The full field development plan envisages the drilling of 254 wells over the field's production life. The gas produced from the Zohr gas field is expected to be distributed within Egypt, while the excess will be exported to overseas markets (Offshore Technology 2018b).

In summer 2018, Eni announced the discovery of another offshore field "Nour", located about 50 km north of Sinai Peninsula, with the 30 trn scf of gas reserve (Fig. 11.4). The concession covers a total area of 739 km^2, with water depths ranging from 50–400 m. Eni operates the Nour North Sinai Concession in participation with EGAS, which has a 40% stake. Partners in the concession are BP (25%), Mubadala Petroleum (20%), and Tharwa Petroleum (15%). Consultancy Wood Mackenzie calls the last decade discoveries "Egypt's astonishing gas renaissance", estimating there are 61 trn scf of gas reserves in existing fields with another 45 trn scf waiting to be proved (Stephen 2019).

REMPEC MEDGIS-MAR tool (2019) allows to illustrate offshore activities related to gas exploration and production in the Mediterranean waters of Egypt, which are well seen by the intensity of the ship traffic from the ports of Alexandria and Port Said to the offshore gas installations and gas fields in the coastal zone offshore the Nile Delta (Fig. 11.5).

11.4 Oil Pollution Events

Our recent investigation of historical cases of oil pollution in the Mediterranean Sea has not revealed serious cases of oil pollution in the waters of Egypt in the past decades (Kostianoy and Carpenter 2018a). However, the REMPEC MEDGIS-MAR

Fig. 11.5 Offshore activities (ship traffic intensity) related to gas exploration and production in the Mediterranean waters of Egypt in 2013. *Source* REMPEC MEDGIS-MAR tool (2019)

tool (2019) on the map of the Mediterranean Sea shows several cases of oil pollution of the size 7–700 tonnes in the vicinity of the Nile Delta recorded in 1990–2013 (Fig. 11.6). We used the REMPEC MEDGIS-MAR tool (2019) to extract information about 22 cases of ship accidents in the Mediterranean waters of Egypt recorded from 1982 till present, which we gathered in Table 11.1. In these accidents, oil pollution of 60–350 tonnes was recorded only in 5 cases (all of them occurred with tankers), and in two cases it was about 1 ton. In 7 cases, there is no information about the resulted oil pollution. The reason for oil pollution was cargo transfer failure and collision of ships.

Fig. 11.6 Density of oil pollution accidents in the Eastern Mediterranean in 1990–2013. *Source* REMPEC MEDGIS-MAR tool (2019)

Table 11.1 Information about ship accidents and related oil pollution events in the Mediterranean waters of Egypt

Date	Coordinates	Ship/flag	Accident type	Oil spill volume
1 September 1982	31.15° N, 29.88° E, Alexandria	*Nortrans Enterprise*, bulk carrier, Japan	Grounding	Unknown
31 May 1987	31.10° N, 29.62° E, Sidi Kerir terminal	*Vergo* oil tanker Liberia	Cargo transfer failure	60 tonnes
8 August 1987	31.28° N, 32.37° E, Suez Canal	*Peaceventure L* oil tanker, Panama	Grounding	No
30 March 1989	31.10° N, 29.62° E, Sidi Kerir terminal	*Esso Picardie,* Oil tanker, France	Cargo transfer failure	300 tonnes
14 April 1992	31.10° N, 29.62° E, Sidi Kerir terminal	*Olympic Star,* Oil tanker, Panama	Cargo transfer failure	200 tonnes
27 October 1992	31.43° N, 32.37° E, offshore Port Said	*Soheir*, oil tanker, Egypt	Collision	350 tonnes
9 November 1992	31.13° N, 29.60° E, off Sidi Kerir	*Rosario Del Mar* oil tanker	Other	No
21 February 1993	31.15° N, 29.60° E, off Sidi Kerir	*Carlova,* oil tanker, Bahamas	Engine breakdown	No
2 July 1994	31.15° N, 29.88° E, Alexandria	*Seaoath,* oil tanker, Malta	Grounding	No
30 October 1995	31.15° N, 29.88° E, Alexandria	*Capo Argento,* LPG carrier Italy	Grounding	No
18 October 1997	31.28° N, 32.30° E, offshore Port Said	*Irving Galloway,* oil tanker, Barbados	Other	No
6 May 2000	31.27° N, 30.10° E Abo-Qir Bay, 6 km from Alexandria	*Dalia S.*, Syrian Arab Republic	Other	322 tonnes

(continued)

Table 11.1 (continued)

Date	Coordinates	Ship/flag	Accident type	Oil spill volume
17 June 2000	31.25° N, 29.00° E	*Captain Fouad*, Sao Tome and Principe	Other	No
10 May 2013	31.15° N, 32.305° E, Port Said	*CMA CGM Onyx*, container carrier, Singapore	Other	Unknown
21 June 2013	31.15° N, 32.305° E, Port Said	*MSC Perle*, Panama	Contact	Unknown
13 September 2013	31.19° N, 29.86° E, Alexandria	*Ahmad-M*, general cargo, Sierra Leone	Other	1 ton
12 March 2014	31.18° N, 29.86° E, Alexandria inner anch	*Genco Reliance*, China, Hong Kong	Engine or machinery breakdown	Unknown
16 March 2014	31.16° N, 29.83° E, Alexandria	*Ceylan*, Sierra Leone	Hull structural failure	Unknown
10 October 2014	31.17° N, 29.80° E 2.4 nm from El Dekheila	*Long Bright*, Panama	Grounding	Unknown
12 June 2015	31.18° N, 29.86° E, Alexandria Port	*Alex*, Bolivia	Other	1 on
26 April 2017	31.45° N, 30.04° E offshore Alexandria	*Rinella M.*, Italy	Fire or explosion	Unknown
21 May 2017	31.25° N, 29.00° E offshore Idku	*Schillplate General*, Gibraltar	Grounding	No

Source REMPEC MEDGIS-MAR tool (2019)

11.5 Satellite Monitoring of Oil Pollution

Satellite monitoring of oil pollution in the Mediterranean Sea performed in 1999–2004 by European Commission—Joint Research Centre in Ispra, Italy, shows that the REMPEC MEDGIS-MAR tool (2019) data are incomplete, because they contain only major cases of oil pollution and do not include small-size oil spills (Fig. 11.7). The analysis of satellite imagery for different years has showed that operational oil

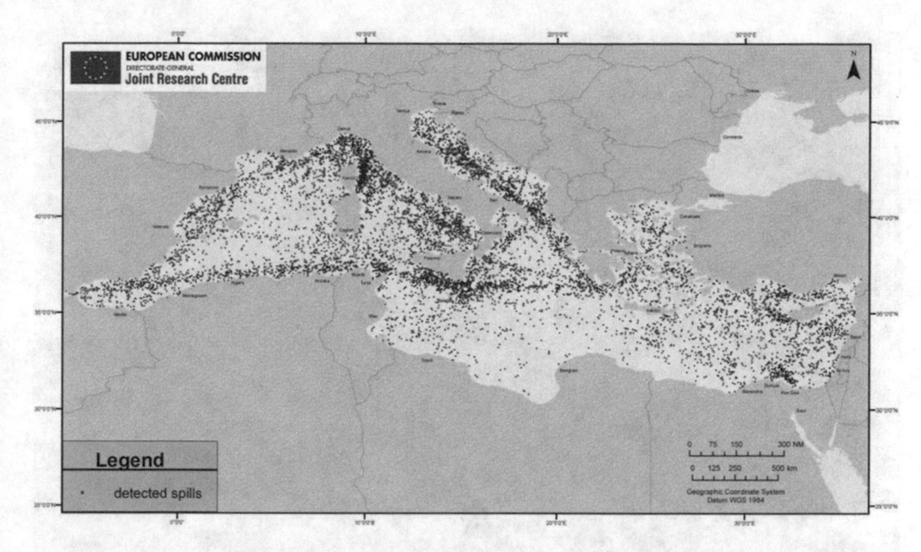

Fig. 11.7 Potential oil spills detected by satellite monitoring in 1999–2004 by European Commission—Joint Research Centre (Ispra, Italy). *Source* Ferraro et al. (2007)

spills of a size of about 1–10 tonnes released by ships of different types due to routine operations occur almost daily in the Mediterranean Sea, and these remain a major oil pollution problem as their number may reach 1500–2500 every year (Kostianoy and Carpenter 2018a). This is exactly what we see in Fig. 11.7, where oil spills detected from satellites are concentrated along the main shipping routes in the Mediterranean Sea, and in the waters of Egypt, in particular. We also observe a large patch of oil spills in the offshore area between Damietta and Port Said, which coincide both with maritime traffic related to the Suez Canal passage (Fig. 11.2) and offshore activities related to gas exploration and production (Figs. 11.4 and 11.5). It is evident that it is impossible to discriminate between them without near-real-time satellite monitoring of this area, very sensitive to oil pollution.

In 2007–2011, the Cyprus Oceanography Center (Nicosia, Cyprus) performed satellite monitoring of oil pollution in waters around Cyprus. It revealed more than 1000 potential oil spills concentrated along the main shipping routes in the Eastern Mediterranean Sea, some of them are directed to the coast of Egypt (Fig. 11.8) (Zodiatis et al. 2012; Kirkos et al. 2018). Unfortunately, the monitoring area was limited by latitude 32.5° N from the south, this is why we have no information about oil pollution in the waters of Egypt. To get a clear geographical view on the area which was outside of the monitoring area, we added the missing part of the sea to the accumulated oil spill map (Fig. 11.8). This map provides detailed information on oil pollution because it shows the real shape and size of the detected oil spills, which is not the case in the whole Mediterranean Sea map in Fig. 11.7. Different oil spill colors correspond to different years when they were detected. Based on Figs. 11.7 and 11.8, we can expect the same density of oil spills in the coastal waters of Egypt

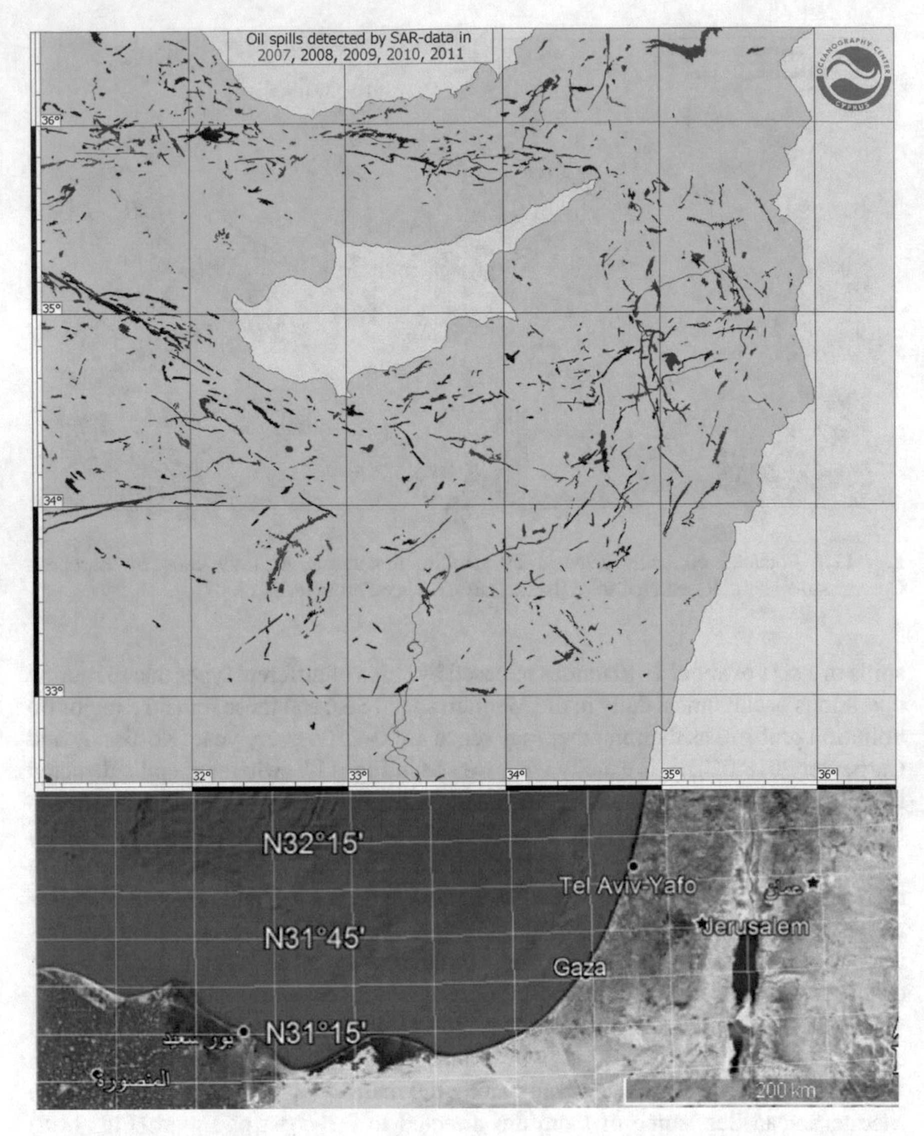

Fig. 11.8 Accumulated map of potential oil spills detected by ASAR Envisat imagery in 2007–2011. Courtesy by D.M. Soloviev. Adapted from Kirkos et al. (2018) (the upper part) and Google Earth (the lower part) to show Mediterranean waters of Egypt and Israel

as well, but still we have no statistical data on this area as satellite oil pollution monitoring is missing in Egypt.

Figure 11.8 shows one of the most spectacular and longest oil spill located in the southern part of the map. Below we provide the original ASAR Envisat satellite image where the spill was detected on 26 September 2011 (Fig. 11.9). In the left part of Fig. 11.9, the spill is shown in a zoom. The length of the oil spill is about 120 km

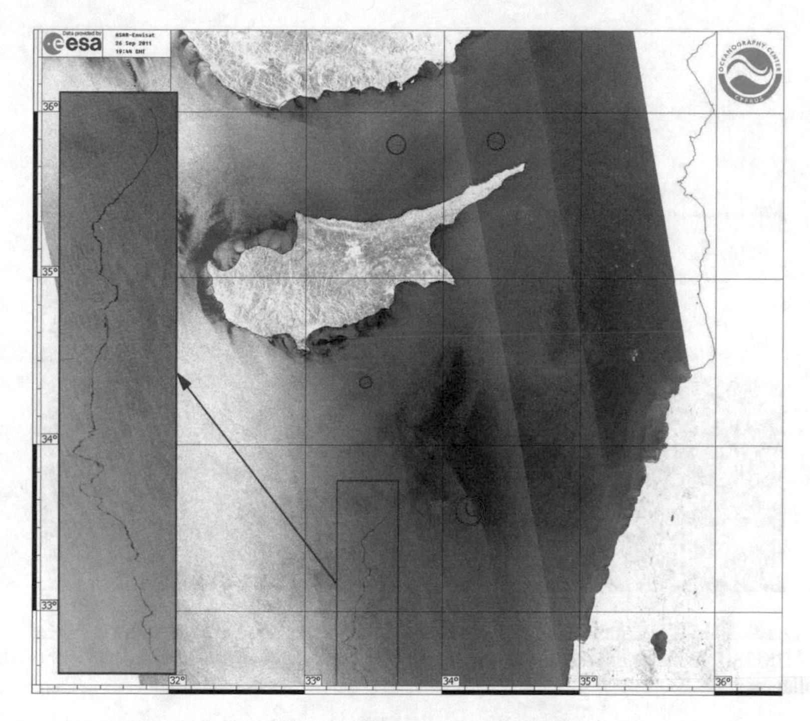

Fig. 11.9 ASAR Envisat image of oil pollution between Egypt and Cyprus on 26 September 2011, 19:44 GMT. Red circles and rectangle show cases of oil pollution. Courtesy by D.M. Soloviev

and it was released behind the ship on its way. The zig-zag shape of the oil spill is explained by the action of local currents and wind which modify the straight line of the ship trajectory with time. Besides, Fig. 11.9 shows other four small oil spills marked in red circles.

To fill the gap, below we provide satellite SAR imagery acquired in 2017–2019 to demonstrate detection of oil spills in the Mediterranean waters of Egypt (Figs. 11.10, 11.11, 11.12, 11.13, 11.14, 11.15, 11.16, 11.17, 11.18, 11.19, 11.20, 11.21, 11.22, 11.23, 11.24, 11.25 and 11.26). Oil spills displayed as black patches on SAR images are marked by red circles and are zoomed in a large circle to show details of the polluted area as well as ships and offshore oil/gas installations (platforms) that are normally well visible as bright white dots. Unfortunately, the presented oil spill gallery has confirmed our suggestion that the Mediterranean coastal waters of Egypt are seriously polluted due to the intense shipping traffic and rising offshore oil/gas exploration and production. For the reference, the Sentinel-1A and Sentinel-1B Interferometric Wide Swath (IW) Mode has 5-by-20 m spatial resolution and a 250 km swath.

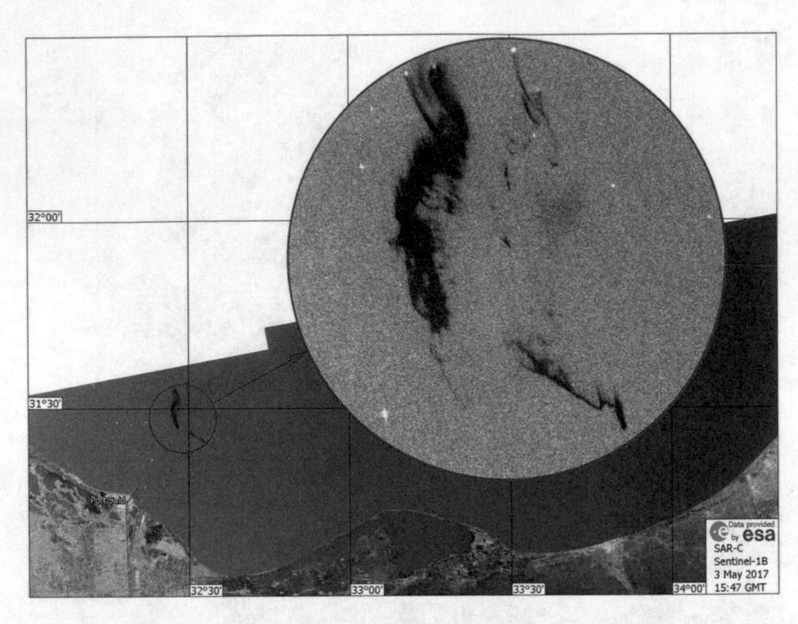

Fig. 11.10 SAR-C Sentinel-1B image of oil pollution northeastward Port Said on 3 May 2017, 15:47 GMT. Hereinafter red circles show cases of oil pollution and white dots are ships or offshore installations

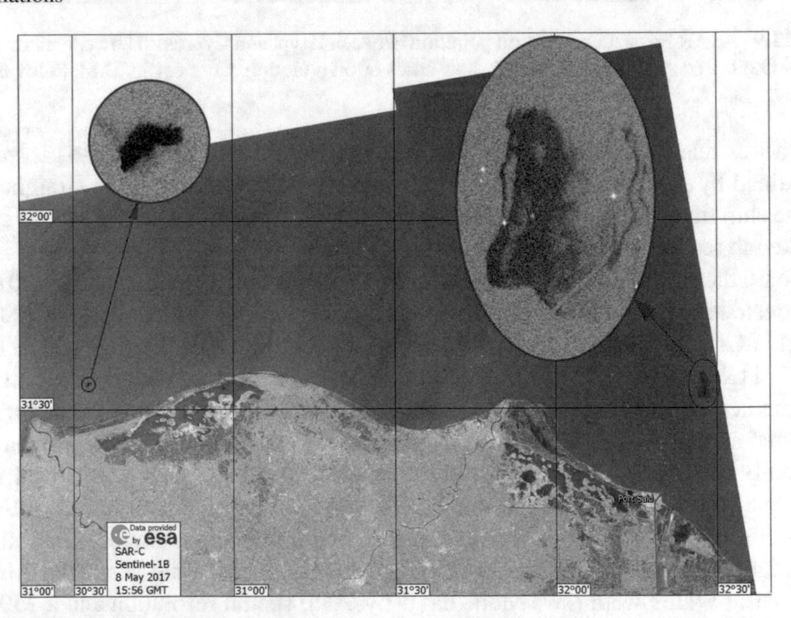

Fig. 11.11 SAR-C Sentinel-1B image of oil pollution northeastward Port Said and offshore the Nile Delta on 8 May 2017, 15:56 GMT

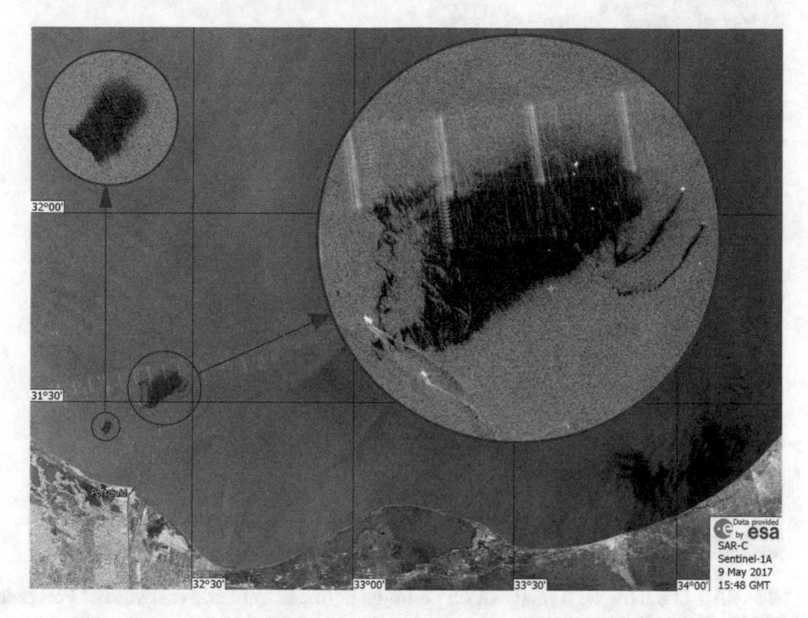

Fig. 11.12 SAR-C Sentinel-1A image of oil pollution offshore Port Said on 9 May 2017, 15:48 GMT

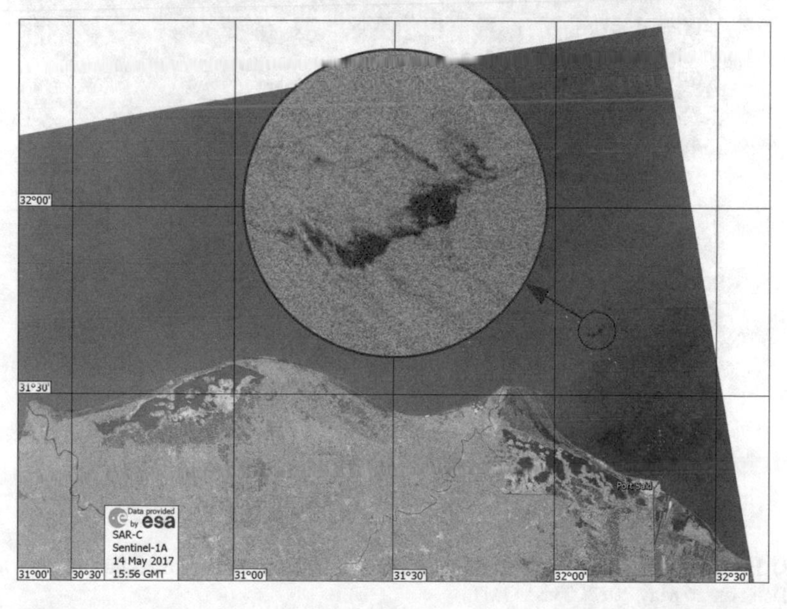

Fig. 11.13 SAR-C Sentinel-1A image of oil pollution northwestward of Port Said on 14 May 2017, 15:56 GMT

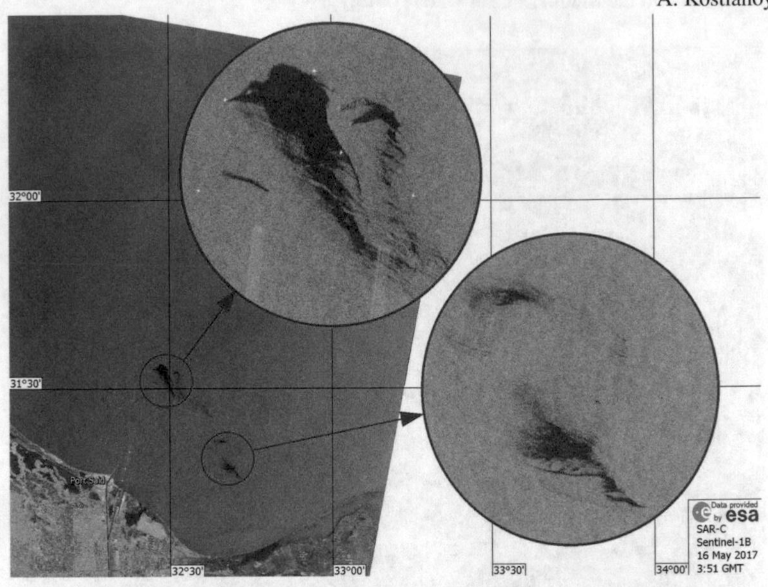

Fig. 11.14 SAR-C Sentinel-1B image of oil pollution northeastward and eastward of Port Said on 16 May 2017, 03:51 GMT

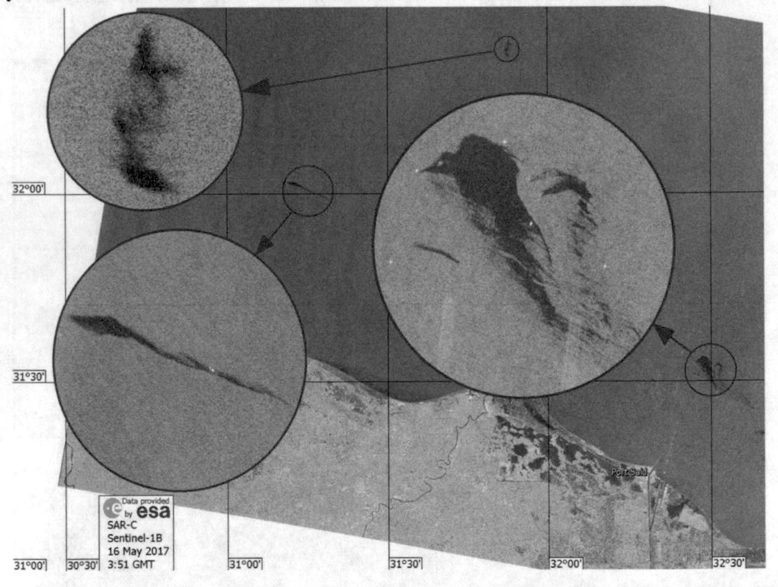

Fig. 11.15 SAR-C Sentinel-1B image of oil pollution northeastward of Port Said and offshore the Nile Delta on 16 May 2017, 03:51 GMT

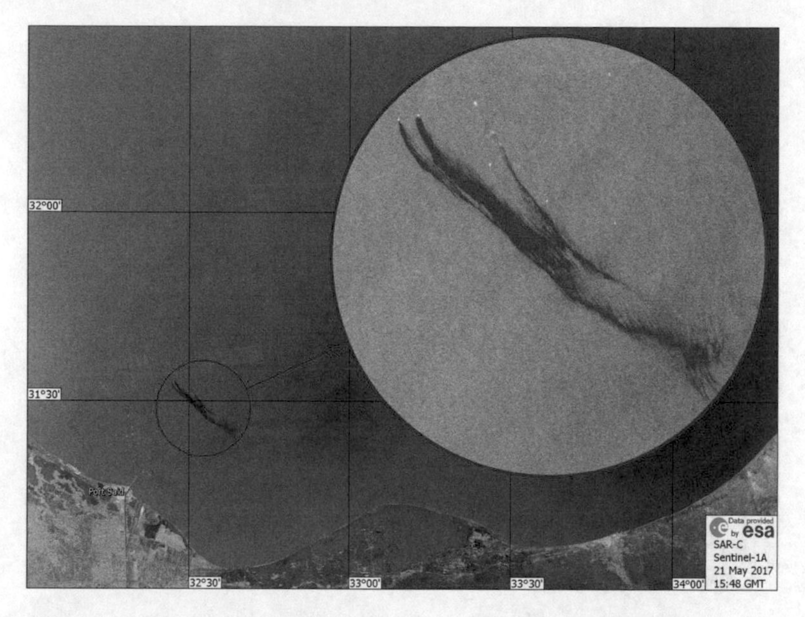

Fig. 11.16 SAR-C Sentinel-1A image of oil pollution northeastward of Port Said on 21 May 2017, 15:48 GMT

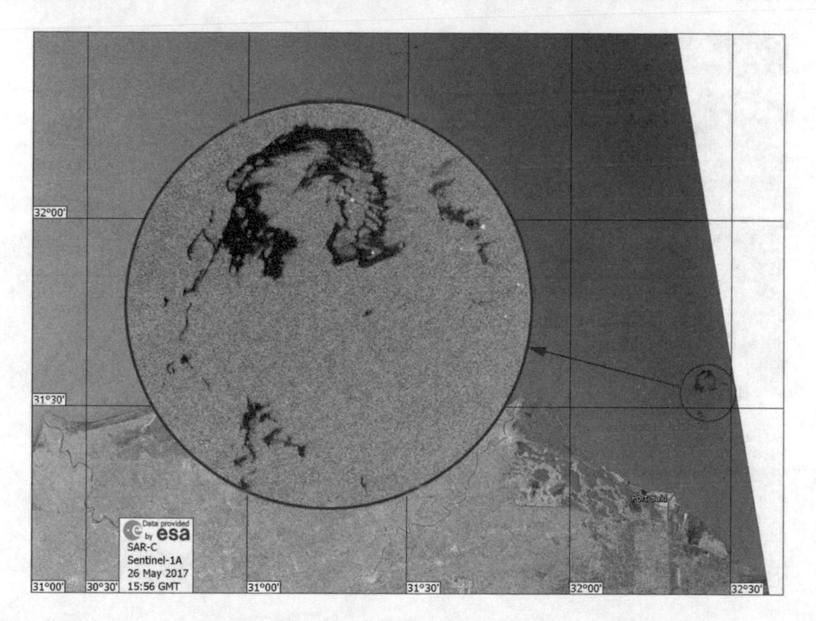

Fig. 11.17 SAR-C Sentinel-1A image of oil pollution northward of Port Said on 26 May 2017, 15:56 GMT

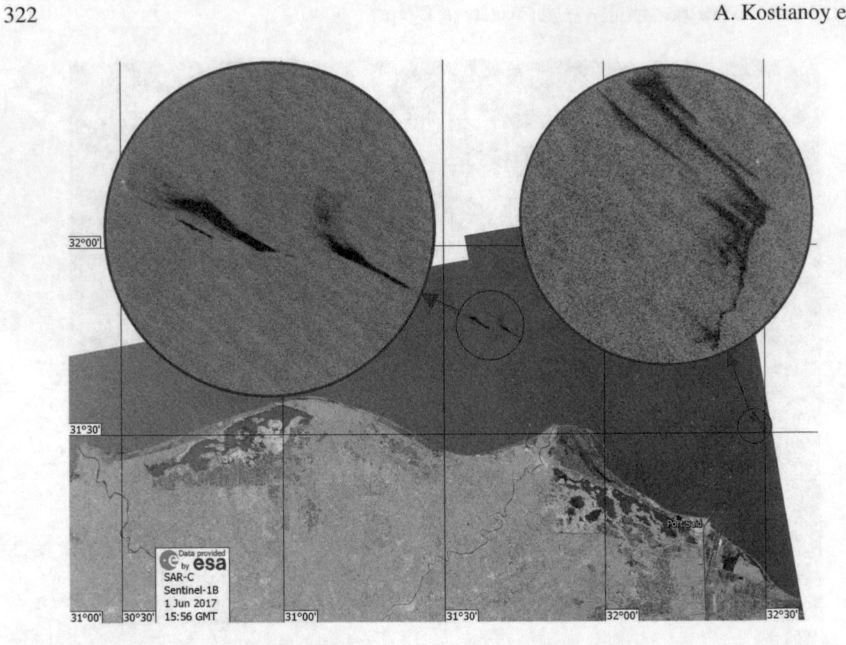

Fig. 11.18 SAR-C Sentinel-1B image of oil pollution northeastward of Port Said and offshore the Nile Delta on 1 June 2017, 15:56 GMT

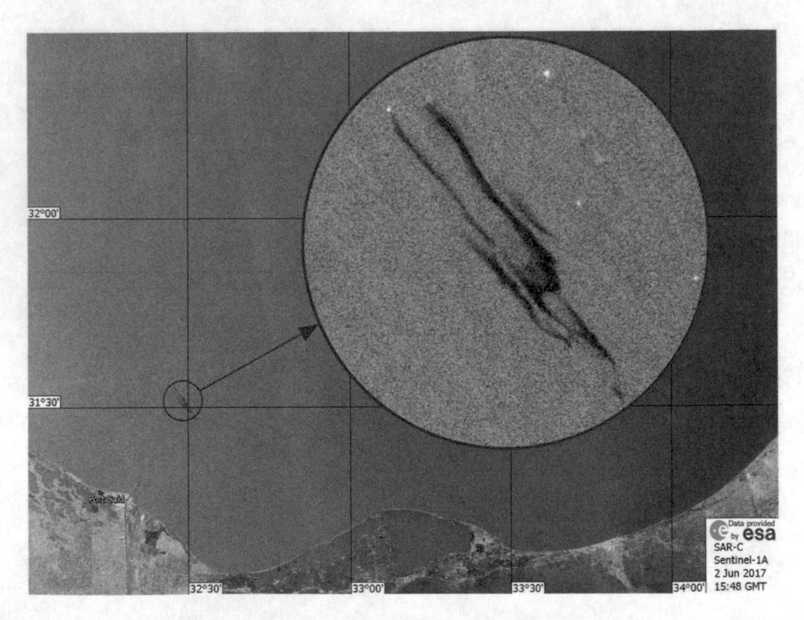

Fig. 11.19 SAR-C Sentinel-1A image of oil pollution northeastward of Port Said on 2 June 2017, 15:48 GMT

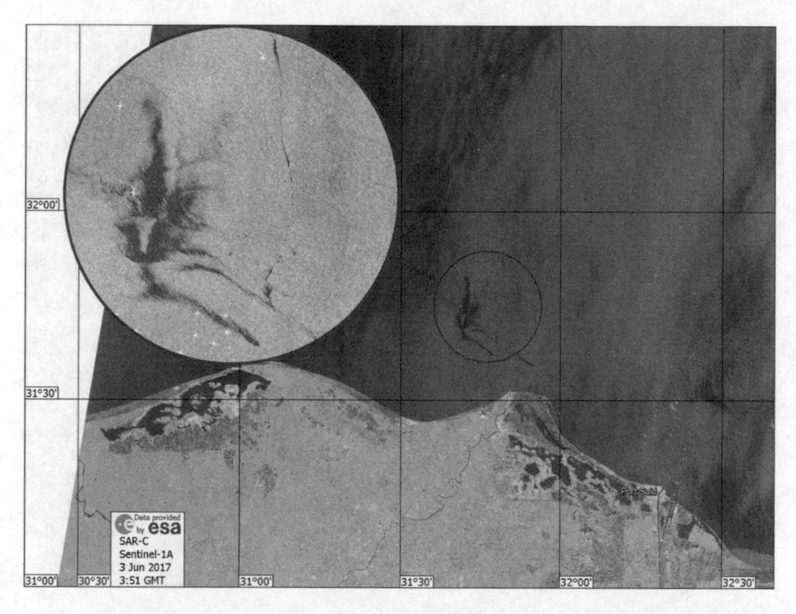

Fig. 11.20 SAR-C Sentinel-1A image of oil pollution northwestward of Port Said on 3 June 2017, 03:51 GMT

Fig. 11.21 OLI Landsat-8 optical image of oil pollution northeastward of Port Said on 27 June 2017, 08:22 GMT

Fig. 11.22 SAR-C Sentinel-1B image of oil pollution northwestward of Port Said on 15 June 2018, 15:48 GMT

Fig. 11.23 SAR-C Sentinel-1A image of oil pollution offshore of Port Said on 3 July 2018, 15:48 GMT

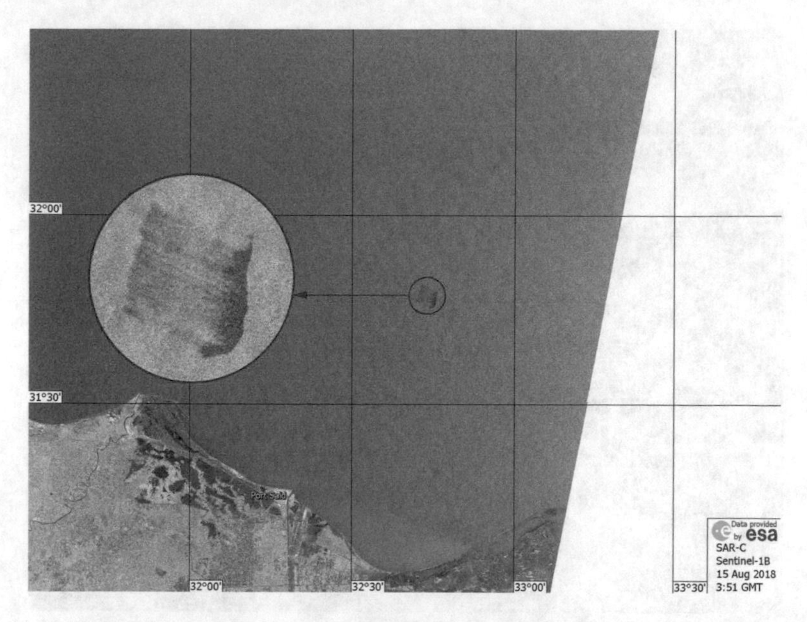

Fig. 11.24 SAR-C Sentinel-1B image of oil pollution northeastward of Port Said on 15 August 2018, 03:51 GMT

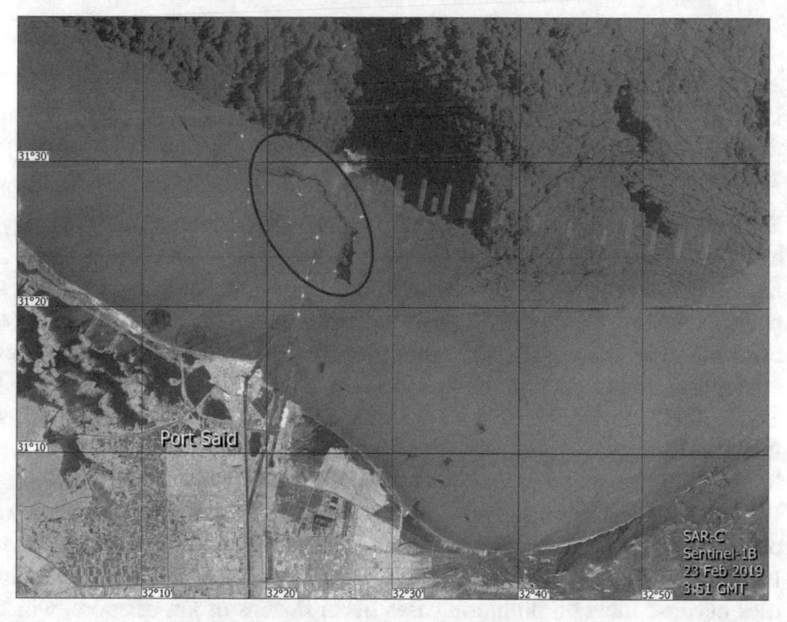

Fig. 11.25 SAR-C Sentinel-1B image of oil pollution northward of Port Said on 23 February 2019, 03:51 GMT

Fig. 11.26 SAR-C Sentinel-1B image of oil pollution northeastward of Port Said on 18 March 2019, 15:48 GMT

11.6 Conclusions

The above mentioned cases of oil pollution in the Mediterranean waters of Egypt led us to the following conclusions:

(1) We found that the most polluted area in the Mediterranean waters of Egypt is a region close to 31° 30′N and 32° 30′E located 18 n.m. northeastward of Port Said. This is explained by the ship traffic from Port Said (Suez Canal), shipping activities related to offshore installations (Fig. 11.5) and offshore gas/oil exploration in this region. The other places of oil pollution are mainly related to shipping activities. Figures 11.10, 11.11, 11.12, 11.13, 11.14, 11.15, 11.16, 11.17, 11.18, 11.19, 11.20, 11.21, 11.22, 11.23, 11.24, 11.25 and 11.26 show that almost in every satellite image there is one or several oil spills, as well as films of surface-active substances (oily and waste waters).
(2) Oil pollution is a serious environmental problem for the Mediterranean waters of Egypt because the number of detected oil spills and their size is very large.
(3) It seems that the real state of oil pollution in this region is unknown to the authorities because most oil pollution cases occur dozens of km offshore, which are not under control of the port authorities. The reason is the absence of permanent aerial surveillance and satellite monitoring of oil pollution over the region.
(4) Increased ship traffic and offshore gas/oil exploration and production in the region of the Nile Delta represent a serious potential threat to its environment, which one day can result in an environmental catastrophe as it already has

happened in different parts of the World Ocean, as well as in the Mediterranean Sea (Kostianoy and Carpenter 2018a, b).

(5) There is an urgent need to establish permanent daily satellite monitoring of the Mediterranean waters of Egypt in order to control oil pollution in the real or near real-time regime as it was done 15 years ago for the Southeastern Baltic Sea (Kostianoy et al. 2006, 2015; Lavrova et al. 2011, 2016). This is the only way to discriminate polluters between transit ships, shipping activities related to offshore gas/oil exploration and production, and gas/oil exploration/production at offshore platforms operated by different foreign and domestic companies.

11.7 Recommendations

The Mediterranean coastal zone of Egypt is not a widely known resort area like the Northern Red Sea due to different reasons. However, serious oil pollution in this region may deteriorate the fragile environment of the Nile Delta (Negm 2017) as well as the whole coastal zone of Egypt. Increased transit ship traffic as well as offshore activities related to the discovery of new gas fields in the waters of Egypt may also lead to transboundary water pollution, which in the future may constitute a considerable problem for the neighboring countries. Our analysis of oil pollution in the Northern Red Sea has led to a recommendation for the establishment of the International Satellite Monitoring Center to perform permanent satellite monitoring of oil pollution and water dynamics in the Northern Red Sea (Kostianaia et al. 2019). As for the Mediterranean coastal zone of Egypt, we came to the same conclusion. This Center should combine integrated satellite monitoring of oil pollution in both seas, as well as provide the authorities with other useful information for coastal zone management, tourism and port development, urban planning. It can also be used for monitoring of water resources, vegetation, desertification for the whole territory of Egypt subjected to climate change.

Acknowledgements The research was partially supported in the framework of the P.P. Shirshov Institute of Oceanology RAS budgetary financing (Project N 149-2019-0004).

References

El Amira (2019) Agency services. Available from https://www.elamira.com/egyptian_ports.php. Accessed 30 June 2019

Ferraro G, Bernardini A, David M, Meyer-Roux S, Muellenhoff O, Perkovic M et al (2007) Towards an operational use of space imagery for oil pollution monitoring in the Mediterranean basin: a demonstration in the Adriatic Sea. Mar Pollut Bull 54(4):403–422

Grid-Arendal (2019) Maritime transportation routes in the Mediterranean. Available from https://www.grida.no/resources/5920. Accessed 13 April 2019

Jeffreys A (ed) (2011) The report: Egypt 2011. Oxford Business Group, 103 p

Kirkos G, Zodiatis G, Loizides L, Ioannou M (2018) Oil pollution in the waters of cyprus. In: Carpenter A, Kostianoy AG (eds) Oil pollution in the Mediterranean Sea, part II—national case studies (Hdb Env Chem) vol 84. Springer International Publishing, pp 229–246 https://doi.org/10.1007/698_2017_49

Kostianoy AG, Carpenter A (2018a) History, sources and volumes of oil pollution in the Mediterranean Sea. In: Carpenter A, Kostianoy AG (eds) Oil pollution in the Mediterranean Sea: part I—the international context (Hdb Env Chem). Springer

Kostianoy AG, Carpenter A (2018b) Oil and gas exploration and production in the Mediterranean Sea. In: Carpenter A, Kostianoy AG (eds) Oil pollution in the Mediterranean Sea: part I—the international context (Hdb Env Chem). Springer

Kostianoy AG, Litovchenko KT, Lavrova OY, Mityagina MI, Bocharova TY, Lebedev SA, Stanichny SV, Soloviev DM, Sirota AM, Pichuzhkina OE (2006) Operational satellite monitoring of oil spill pollution in the southeastern Baltic Sea: 18 months experience. Environ Res Eng Manag 4(38):70–77

Kostianoy AG, Bulycheva EV, Semenov AV, Krainyukov AV (2015) Satellite monitoring systems for shipping, and offshore oil and gas industry in the Baltic Sea. Transp Telecommun 16(2):117–126

Kostianaia EA, Kostianoy AG, Lavrova OY, Soloviev DM (2019) Oil pollution in the Northern Red Sea: a threat to the marine environment and tourism development. In: Elbeih SF, Negm AM,Kostianoy A (eds) Environmental remote sensing in Egypt. Springer

Lavrova OY, Kostianoy AG, Lebedev SA, Mityagina MI, Ginzburg AI, Sheremet NA (2011) Complex satellite monitoring of the Russian seas. IKI RAN, Moscow, 470 pp (in Russian)

Lavrova OY, Mityagina MI, Kostianoy AG (2016) Satellite methods of detection and monitoring of marine zones of ecological risks. Space Research Institute, Moscow, 336 p (in Russian)

Marine Traffic (2019) Live map. Available from https://www.marinetraffic.com/en/ais/home/. Accessed 30 Mar 2019

Maritime Transport Sector (2019) Commercial ports. Available from https://www.mts.gov.eg/en/sections/10/1-10-Commercial-Ports. Accessed 30 Mar 2019

Negm AM (2017) The Nile Delta. Springer International Publishing, 537 p https://doi.org/10.1007/978-3-319-56124-0

Offshore Technology (2018a) Atoll gas field, North Damietta offshore concession, East Nile Delta. Available at https://www.offshore-technology.com/projects/atoll-gas-field-north-damietta-offshore-concession-east-nile-delta/. Accessed on 20 Oct 2018

Offshore Technology (2018b) Zohr gas field. Available at https://www.offshore-technology.com/projects/zohr-gas-field/. Accessed on 20 Oct 2018

Oxford Business Group (2018) New discoveries for Egyptian oil producers. Available at https://oxfordbusinessgroup.com/overview/fresh-ideas-new-discoveries-have-oil-producers-optimistic-about-future. Accessed on 13 Oct 2018

Regional Marine Pollution Emergency Response Center for the Mediterranean Sea (REMPEC) (2011) Statistical analysis. Alerts and accidents database, 33 p. Available from https://www.rempec.org/admin/store/wyswigImg/file/Tools/Operational%20tools/Alerts%20and%20accidents%20database/Statistics%20accidents%202011%20EN%20FINAL.pdf. Accessed 13 Apr 2019

REMPEC MEDGIS-MAR tool (2019) Available from https://medgismar.rempec.org/. Accessed13 Apr 2019

Stephen C (2019) Egypt's gas gold rush. In: Petroleum economist, vol 28. Available from https://www.petroleum-economist.com/articles/upstream/exploration-production/2019/egypts-gas-gold-rush. Accessed 13 Apr 2019

Suez Canal Authority (2019) Monthly number and net ton by ship type. Available from https://www.suezcanal.gov.eg/English/Navigation/Pages/NavigationStatistics.aspx. Accessed 31 Mar 2019

Vesti Finance (2019) The East-Mediterranean gas war begins. Available from https://www.vestifinance.ru/articles/97771. Accessed 30 June 2019

Zodiatis G, Lardner R, Solovyov D, Panayidou X, De Dominicis M (2012) Predictions for oil slicks detected from satellite images using MyOcean forecasting data. Ocean Sci 8:1105–1115

Chapter 12
Oil Pollution in the Northern Red Sea: A Threat to the Marine Environment and Tourism Development

Evgeniia A. Kostianaia, Andrey Kostianoy, Olga Yu. Lavrova and Dmitry M. Soloviev

Abstract The chapter briefly describes the development of the coastal tourism in Egypt, shipping activities connected to the passage using the Suez Canal, crude oil delivery to oil terminals, and development of oil fields in the northern part of the Red Sea. The main shipping routes going along the Egyptian coastline and tourist resort areas in the Red Sea with intense traffic, as well as drilling at offshore oil fields represent a threat to the unique marine environment. Since the 1970s, there have been a number of cases of oil pollution in the region, which seriously affected its coastline and coral reefs. Intensification of the ship traffic through the Suez Canal has led to an increase in the frequency of oil pollution events during the past decades. The Egyptian Environmental Affairs Agency and environmentalists state that today, at the shores from Ismailia to Hurghada, small-size oil spills occur almost monthly. Satellite Synthetic Aperture Radar (SAR) imagery for 2018 and 2019 clearly reveal cases of oil pollution in the Port of Suez, as well in the areas of major tourist resorts of Egypt. Satellite imagery (Sea Surface Temperature (SST), chlorophyll-a concentration, and water turbidity) of the northern part of the Red Sea for May–June 2017 showed the presence of intense cyclonic and anticyclonic eddies of 50–100 km, as well as dipoles and large-scale intrusions of relatively warm and cold waters propagating along and across the axis of the Red Sea. It is evident that this mesoscale water dynamics is

E. A. Kostianaia (✉) · A. Kostianoy
P.P. Shirshov Institute of Oceanology, Russian Academy of Sciences, 36, Nakhimovsky Pr., Moscow 117997, Russia
e-mail: evgeniia.kostianaia@gmail.com

A. Kostianoy
S.Yu. Witte Moscow University, 12, Build. 1, 2nd Kozhukhovsky Pr., Moscow 115432, Russia
e-mail: kostianoy@gmail.com

O. Yu. Lavrova
Space Research Institute, Russian Academy of Sciences, 84/32, Profsoyuznaya Str., Moscow 117997, Russia
e-mail: olavrova@iki.rssi.ru

D. M. Soloviev
Marine Hydrophysical Institute, Russian Academy of Sciences, 2, Kapitanskaya Str., Sevastopol 299011, Russia
e-mail: solmit@gmail.com

S. F. Elbeih et al. (eds.), *Environmental Remote Sensing in Egypt*,
Springer Geophysics, https://doi.org/10.1007/978-3-030-39593-3_12

a major driver for all kinds of pollution transfer in the sea, and in the regions of main tourist resorts of Egypt, in particular. Thus, the establishment of permanent satellite monitoring of oil pollution and water dynamics in the Northern Red Sea is of vital importance to avoid potential catastrophes resulted from large oil spills, and for sustainable development of coastal tourism business in Egypt.

Keywords Egypt · Northern Red Sea · Suez Canal · Tourist resorts · Shipping routes · Satellite monitoring · Oil pollution

12.1 Introduction

Paradoxically, but in many areas of the World Ocean and inland seas offshore oil and gas production, oil ports and terminals, as well as the main shipping routes are adjacent to major recreational areas and tourist resorts in different countries. There are many examples in the Baltic, Black and Mediterranean seas (Lavrova et al. 2011; Kostianoy and Lavrova 2014; Lavrova et al. 2016; Carpenter and Kostianoy 2018a, b). The Red Sea in Egypt is not an exception. Warm climate, sunny weather, warm and clean sea, beautiful beaches, rich marine life of coral reefs attract millions of tourists to the most popular resorts in Egypt, Sharm El-Sheikh, El Gouna, Hurghada, Safaga, Marsa Alam, almost all year round. Coastal tourist activities are not limited by the sea related ones (swimming, diving, etc.) but include visiting the rich natural environment (Egyptian pyramids, deserts, oases, the Nile River, etc.). For the last several decades, there has been a considerable growth in tourism development in Egypt, which is an important source of the national income, including foreign currency.

At the same time, about 15% of all global maritime trade and 10% of global seaborne oil passes through the Red Sea and the Suez Canal. The ship traffic goes along the coasts and marine resorts of Egypt, and there is an increase in the ship traffic related to the reconstruction of the Suez Canal in 2015. As gas and oil production are less important in the Mediterranean Sea than, for example, in the Caspian Sea, the Gulf of Mexico, or the North Sea, oil pollution here is mainly caused by ships of various types. According to various expert estimates, the total volume of oil pollution in the Mediterranean Sea ranges from 1600 to 1 million tonnes per year. This value excludes major oil spills from ships in the region, which is anyway very rare here (Kostianoy and Carpenter 2018). The other sources of oil pollution in the Mediterranean Sea include oil and gas platforms, ports and oil terminals, land-based sources, military conflicts, natural oil seeps, and the atmosphere. It seems that all these sources are valid for the Red Sea as well, because besides the heavy ship traffic, the Red Sea has several important ports and oil terminals with land oil pipelines, as well as offshore oil field exploration activities. These began in Egypt in the beginning of the 20th century with a discovery of the first oil in the Gulf of Suez in 1886, the Gemsa oil field in 1907, and the Hurghada oil field in 1911. Today, there are over 180 oil rigs operating in the Red Sea and the Gulf of Suez. These oil rigs account for a large percentage of the economy of Egypt (Alsharhan 2003).

Shipping activities and offshore oil and gas exploration and production represent a serious threat to the marine environment, coral reefs, islands, coastal zones and beaches of tourist resorts in Egypt and other countries. Egypt seems to be subjected to a greatest risk of oil pollution impact because it has a series of largest tourist resorts extended for dozens of km along the coast, and the extensive shipping traffic which goes to the Gulf of Suez and further via the Suez Canal to the Mediterranean Sea, as well as to the Gulf of Aqaba. Fortunately, till present Egypt avoided large oil pollution catastrophes like with the *MV Haven* off Genoa (Italy) in April 1991, when 144,000 tonnes of oil was spilled in the sea, or with the *Irenes Serenade* in Navarino Bay (Greece) on 23 February 1980, when 100,000 tonnes of oil was spilled to the sea. But the risk is still there as accidents with ships occur regularly in the Mediterranean Sea. The analysis of satellite imagery for different years showed that operational oil spills of a size of about 1–10 tonnes released by ships of different types due to routine operations occur almost daily in the Mediterranean Sea, and these remain a major oil pollution problem as their number may reach 1500–2500 every year (Kostianoy and Carpenter 2018).

It seems that this is the case in the Red Sea as well, because since the 1970s there have been a number of cases of oil pollution in the region, which seriously affected coastline and coral reefs. Intensification of ship traffic through the Suez Canal has led to an increase in the frequency of oil pollution events during the last 10 years. The Egyptian Environmental Affairs Agency and environmentalists state that today at the shores from Ismailia to Hurghada small-size oil spills occur almost monthly (Mayton 2009). In many cases, the culprit of the oil pollution cannot be identified because there is no permanent satellite monitoring of oil pollution of this area that should be established in Egypt.

We begin this chapter with a brief description of the following economic activities in Egypt in the northern part of the Red Sea: development of coastal tourism in Egypt (Sect. 12.2), shipping activities connected to passage via the Suez Canal (Sect. 12.3), crude oil delivery to oil terminals, and offshore development of oil fields (Sect. 12.4). In Sect. 12.5 we briefly point out cases of oil pollution recorded in the past several decades in the northern part of the Red Sea. To show capabilities of satellite remote sensing, in Sect. 12.6 we provide satellite imagery acquired with different sensors to demonstrate detection of oil spills, to show spatial and temporal variability of sea surface temperature (SST), Chlorophyll-a (Chl-a), and water leaving radiance (water turbidity) fields, as well as mesoscale water dynamics in the Northern Red Sea. In the Conclusions (Sect. 12.7) we briefly discuss the present state of application of satellite remote sensing methods to monitor the Red Sea and argue the necessity of establishing permanent operational integrated satellite monitoring of oil pollution in the northern Red Sea.

12.2 The Red Sea and Tourist Resorts

The Red Sea is a narrow, elongated body of water, which extends to the south-east from Suez, Egypt, for approximately 1930 km to the Bab el-Mandeb Strait, which connects the sea to the Arabian Sea and the Indian Ocean via the Gulf of Aden (Fig. 12.1). The Red Sea has a surface area of about 450,000 km^2, with a volume of approximately 250,000 km^3, and a maximum width of 355 km. About 40% of the Red Sea is considered to be shallow, with depths under 100 m; approximately 25% of the sea is under 50 m in depth. The Sinai Peninsula divides the Northern Red Sea into two parts: the Gulf of Suez and the Gulf of Aqaba. Through the Gulf of Suez, the Red Sea is connected to the Mediterranean Sea. The Gulf of Suez is 25–60 km wide and 300 km long. The Gulf of Aqaba is 19–25 km wide and 160–180 km long. It is narrow in the north and widens towards the south (Head 1987; Rasul et al. 2019; Ryan and Schreiber 2019). The following countries have access to the Red Sea: Djibouti, Eritrea, Sudan, Egypt, Israel, Jordan, Saudi Arabia, and Yemen. Figure 12.1 represents the political map of the Red Sea region.

The Red Sea region is a rich environment ranging from deserts with cultural heritage to the sea with widely known coral reefs, thus creating a serious potential for tourism development. Egypt has an abundance of natural environment activities that can be offered to tourists, for example, diving, snorkeling, windsurfing, water polo, and rowing. One of the main attractions is probably the famous coral reefs. Over 300 species of hard corals have been reported for the whole Red Sea, with about 200 species having been noted for Egypt alone (Gouda 2015).

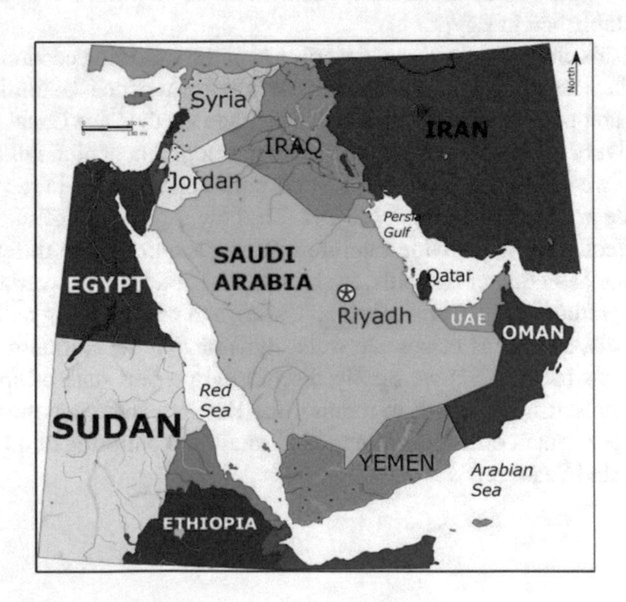

Fig. 12.1 The Red Sea political map. *Source* Africanglobe (2019)

For the last several decades, there has been a considerable growth in tourism development in Egypt. Tourism in general in Egypt is deemed to be one of the principal sources of the national income. It is also considered to be the main source of foreign currency (Hilmi et al. 2012). According to the 2018 report by the World Travel and Tourism Council (2018), the direct contribution of travel and tourism in Egypt to GDP in 2017 was 10.7 billion USD, which is 5.6% of total GDP. The forecast is for it to rise by 4.0% per annum from 2018 to 2028, to amount to 16.3 billion USD, or 5.3% of total GDP, in 2028. This primarily displays economic activities created by various linked industries, for example, travel agents, hotels, various passenger transportation services, including airlines, as well as activities of the restaurants and the leisure industry which are directly linked to tourism. The term "direct contribution of travel and tourism to GDP" indicates total spending within Egypt on travel and tourism both by residents and non-residents for leisure and business purposes, as well as spending by the government on travel and tourism services, which are directly related to visitors, such as recreational (for example, national parks) or cultural (for example, museums).

According to the same 2018 report by the World Travel and Tourism Council (2018), in 2017, travel and tourism directly generated 1,099,000 jobs, which represents 3.9% of the total employment in Egypt. The forecast is for this to rise by 1.9% per annum to amount to 1,383,000 jobs in 2028, which would represent 3.9% of the total employment. This comprises employment by travel agents, hotels, airlines, and various passenger transportation services without commuter services. This also comprises employment in restaurants and the leisure industry directly supported by tourism. Furthermore, the number of international tourist arrivals is forecast to total 21,315,000 by 2028, which means an increase of 3.8% per annum.

With the sunny climate, rich marine life, and a large variety of resorts, Egypt annually attracts a huge amount of visitors to its coastline. Some of the most popular resorts in Egypt are Sharm El-Sheikh, El Gouna, Hurghada, Safaga, Marsa Alam (Fig. 12.2).

12.3 The Suez Canal and Shipping in the Red Sea

The Suez Canal is an artificial waterway of the length of 193 km running from Port Said to the Gulf of Suez from the north to the south, thus connecting the Mediterranean and the Red Seas. It is one of the most vital and heavily used waterways in the world. Navigation started on the 17th of November, 1869, which was a cause for major celebrations. Since then, the Canal has been closed 5 times; it was reopened for navigation after the last closure on the 5th of June, 1975 (Suez Canal Authority 2017a, b).

At the time of its opening in 1869, the canal was about 8 meters deep, 22 m wide at the bottom and 61–91 m wide at the surface. Considerable changes started in 1876; due to these improvements, by the 1960s the canal had a minimum width of 55 m at a depth of 10 m along its banks. Its channel depth was now 12 m at low tide. The latest

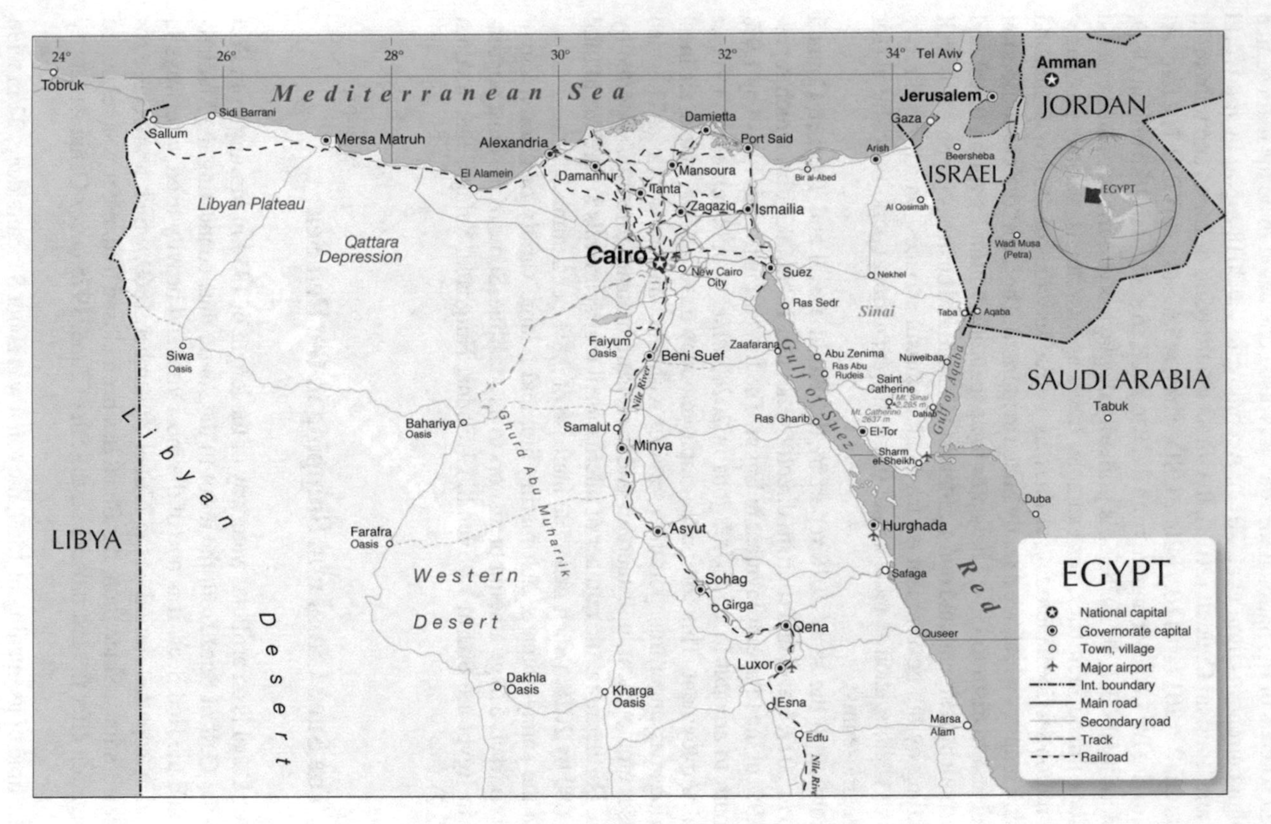

Fig. 12.2 Map of Egypt with the main touristic resorts. *Modified from* Nations Online (2019)

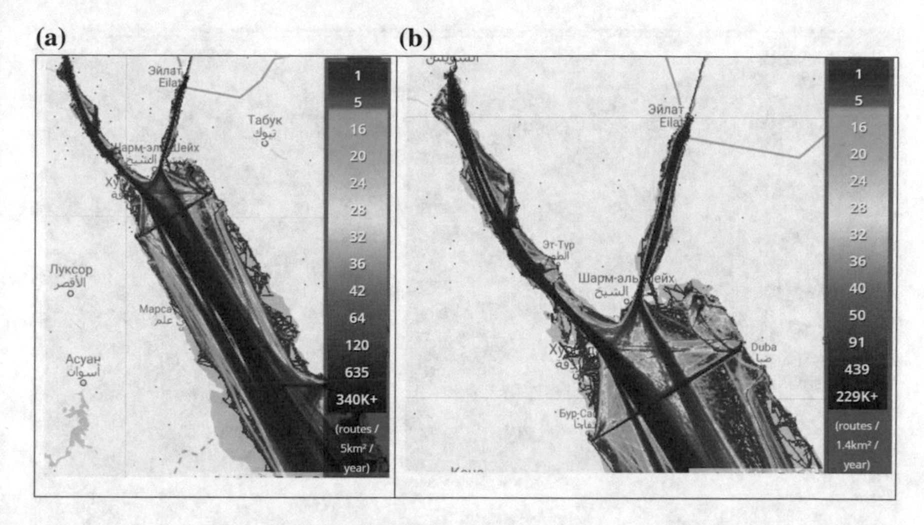

Fig. 12.3 **a** Marine traffic density in the northern part of the Red Sea for all ship types in 2017. **b** A zoom on the region around Sinai Peninsula. *Source* Marine Traffic (2019)

upgrade was completed in 2015, which allowed to significantly increase the capacity of the waterway: a new 35 km parallel route was opened, thus enabling two-way transit through the Suez Canal. Additionally, the existing western by-passes of the total length of 37 km were to be deepened to a depth of 24 m. The objective of this project was to: (a) shorten the transit time from 18 to 11 h for the southbound convoy; (b) minimize the waiting time for transiting ships down to 3 h instead of 8–11 h; (c) attract more ships to use the Suez Canal, and (d) overall increase the number of ships passing though the Canal every day, in response to the projected world trade growth (Suez Canal Authority 2017c; Fisher and Smith 2019).

The intensity of the marine traffic in the northern part of the Red Sea for all ship types for 2017 is shown in Fig. 12.3.

In 2017, the total number of ships passing through the Suez Canal was 17,550; most of which were container ships (31.73% of the total number of ships), followed by tankers (25.85%) and bulk carriers (18.74%) (Suez Canal Authority 2019).

Egypt has 15 commercial ports, 9 of which belong to the Red Sea Port Authority: Suez Port, Petroleum Dock Port, Adabia Port, El Sokhna Port, El Tor Port, Sharm El-Sheikh Port, Nuweiba Port, Safaga Port, and Hurghada Port (Maritime Transport Sector 2019; Red Sea Port Authority 2019). Figure 12.4 shows Egypt's commercial ports of the Mediterranean and Red Seas.

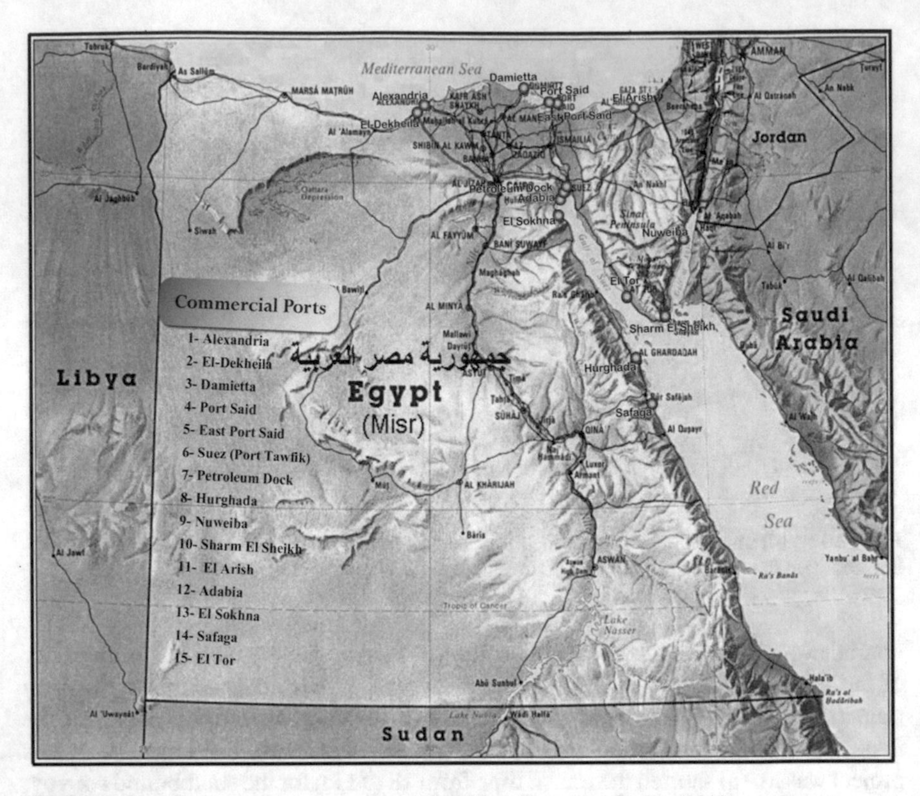

Fig. 12.4 Egypt's commercial ports. *Source* Maritime Transport Sector (2019)

12.4 Offshore Oil Fields and Infrastructure

The Gulf of Suez has a great hydrocarbon potential: its prospective sedimentary basin area totals about 19,000 km^2, and it is also regarded as the richest oil province rift basin of the Middle East and Africa. The basin contains more than 80 oil fields, from 1350 to less than 1 million bbl, in reservoirs of the Precambrian to Quaternary age. In the Gulf of Suez, oil was first found in 1886: crude oil then permeated into the tunnels which had been dug for extraction of sulphur in the Gemsa area on the western coast of the Gulf of Suez. Later, drilling was carried out near the surface oil seeps in the western coastal part of the southern Gulf of Suez. This resulted in the discovery of the Gemsa oil field in 1907, which was the first oil discovery in Africa and the Middle East. Another essential exploration was made in the Hurghada field in 1911, but unfortunately, the exploration activities had to be stopped for some time due to the World War I. The field though turned out to be a very productive one, with its production rate reaching 1.8 million barrels per year in 1931. In 1918, the Anglo Egyptian Oil Company drilled close to an oil seep on the eastern side of the Gulf of Suez and found the non-commercial Abu Durba oil field. The Standard Oil Company of Egypt discovered the Ras Gharib oil field, located between Suez and

Hurghada on the western side of the Gulf of Suez, in 1938. In 1946, oil was also found in Sudr, situated on the eastern coast of Sinai. During the next decades, more oil fields were discovered, which resulted in various petroleum agreements and laws (Alsharhan 2003; Khaled 2014).

There are several oil terminals located on the coasts of the Red Sea. Below is the overview of the oil terminals that are located near the territory of Egypt. Therefore, their operations might affect the environment or infrastructure of Egypt.

In the northernmost part of the Gulf of Aqaba, Eilat Ashkelon Pipeline Company (EAPC, Israel) operates a crude oil jetty at the Eilat Oil Port. The oil terminal with a water depth of 30 m alongside the jetty can receive tankers of up to 500,000 DWT (Fig. 12.5). The maximum discharge rate at this jetty is 20,000 m^3/h, and the maximum loading rate is 10,000 m^3/h. Eilat oil terminal receives oil from tankers calling at the Port of Eilat with an overall storage capacity of 160,000 m^3. From this site, the crude oil is pumped into the main tank farm at Ramat Yotam, which consists of 16 storage tanks with a total capacity of 1.2 million m^3 (EAPC 2019a).

Oil is then pumped via the Eilat-Ashkelon Pipeline to the coast of the Mediterranean Sea to be used in Israel and to be transported to Europe. The pipeline was built in 1968, it is 254 km long, and its capacity from a special pier in Ashkelon to the Port of Eilat is 400,000 barrels (64,000 m^3) per day and 1.2 million barrels (190,000 m^3) per day in the opposite direction. The pipeline is owned and operated by the EAPC, which also operates several other oil pipelines in Israel. The company set up a new transit capability in order to enhance the activity of the Eilat-Ashkelon pipeline: pumping of crude oil from the Mediterranean Oil Port of Ashkelon to Eilat on the Red Sea. The system is unique in its bidirectional feature as it complements the original south-to-north pumping direction. This pipeline route from Europe to Asia and back is shorter than the traditional one around Africa and cheaper than the one via the Suez Canal (EAPC 2019c).

Fig. 12.5 The Eilat oil terminal. *Source* EAPC (2019b)

Aqaba Oil & LPG Terminal started its commercial operations in October 2018. Together with the capacities of Aqaba's South Terminal, Aqaba Oil & LPG Terminal shall meet the present and future demands of petroleum products imports to Jordan. The terminal has 6 floating roof tanks with the size of 20,000 m^3 each, 3 LPG spheres with the size of 3760 m^3 each, and 15 loading bays. Aqaba Heavy Oil Terminal was built in 1978 as the country's biggest heavy oil storage terminal. It has 5 fixed roof tanks with the size of 42,000 m^3 each, and 4 loading bays (JOTC 2019a, b).

The location of the Oil terminal and the Port of Eilat (Israel), as well as Aqaba industrial port, Mota Berth (port), and Aqaba Port in Jordan in the northernmost part of the Gulf of Aqaba, can explain tremendous ship traffic in this part of the Red Sea (see Fig. 12.6).

The major terminal of Saudi Arabia on the Red Sea is the Yanbu King Fahd terminal with the total crude oil storage capacity of 12.5 million barrels. The terminal has 7 loading berths and can accommodate tankers of up to 500,000 dwt. With the launch of the Muajjiz oil terminal, incorporated in the Yanbu crude oil terminal, Saudi Arabia is expected to raise its total loading and export capacity to about 15 million b/d. The country has also smaller ports on the Red Sea, such as Jizan and Jeddah (US Energy Information Administration 2017).

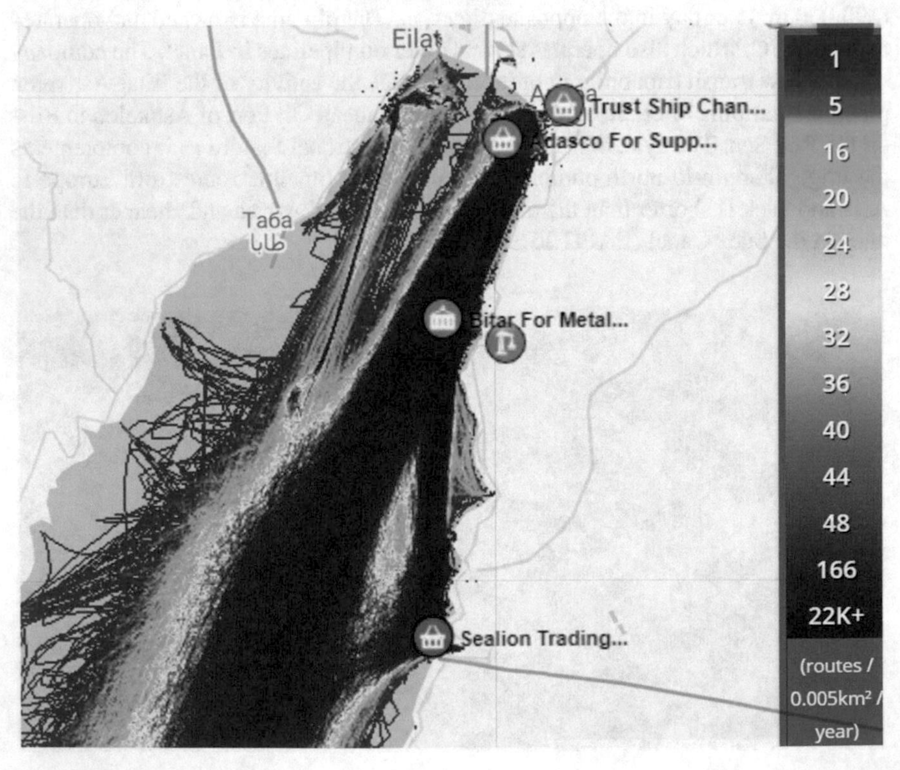

Fig. 12.6 Marine traffic density in the northern part of the Gulf of Aqaba for all ship types in 2017. *Source* Marine Traffic (2019)

12.5 Oil Pollution Events

Oil pollution and its consequences for the Red Sea have gained a lot of awareness in the recent years. Oil transportation and terminal development, the opening of the New Suez Canal increase the likelihood of a detrimental oil pollution event. Given the fragile nature of the Red Sea being a semi-enclosed water body, and the narrow character of the sea itself, a potential oil spill might have destructive ramifications on Egypt's coastal areas, damaging its marine environment and touristic areas.

The Red Sea has already seen a considerable number of oil pollution events. In 1970–1980s oil field exploration and intensification of ship traffic in the Red Sea led to a series of oil pollution events that have been reported in a variety of publications (see, for example, Wennink and Nelson-Smith 1979a, b; Hanna 1983; Dicks 1984; Awad 1989; Shaltout et al. 2005). In the recent years, also famous Egyptian resort areas, like Hurghada, very often become affected by oil spills. The scale and frequency of the recent events in the Red Sea region, covered by the international media, show that oil pollution represents a constant potential threat to the country's economy and ecosystems (see Mayton 2009; Al-Ghazawy 2010; Staff 2016; Egypt Independent 2017).

For example, Mayton (2009) informs that in 2006, there were two large oil spills in the Red Sea and inside the Suez Canal at Bitter Lake. On February 20, 2006, at least 3300 tons of heavy fuel oil were said to get spilled into the Red Sea, followed by 1100 tons of crude oil spilled into Bitter Lake in September. On May 20, 2009, oil got spilled onto the coastline to the north of Hurghada. Crude oil covered more than 2/3 of a mile of the sandy beach. 26 oil fields in the Red Sea negatively affect marine life due to almost permanent leakage from the old equipment used by oil companies. Frequent oil spills damage the marine environment despite the efficient clean ups. Mayton (2009) further indicated that similar small spills happened almost every month.

Another oil spill was detected on the Red Sea shore on June 16, 2010, which was labelled as the Jebel al-Zayt oil spill. It is said to have polluted about 160 km of the coastline, including several touristic places, and is considered to be the largest oil spill in the history of Egypt. There was no uniformity in the explanation of where the spill had originated. The Ministry of Petroleum of Egypt said the oil had leaked either from one of the tankers that were passing through the Gulf of Suez and cleaning their engines, or from some old oil deposits on the shores, that were melting because of the heat wave that was observed in Egypt at the same moment when the oil reached the shore. However, representatives from the Hurghada Environmental Protection and Conservation Association (HEPCA) insisted that the oil spill had originated from an oil rig about 50 nautical miles to the north of Hurghada (Al-Ghazawy 2010; Julian 2010; Leach 2014).

On June 24, 2011, 12 tons of oil and fuel got into the water near Almog Beach in Eilat from the ship "AVRAMIT" sailing under the Panamanian flag. The cleanup involved employees from the Society for the Protection of Nature in Israel, from the Nature and Parks Authority, from the Eilat municipality, as well as port and shipping workers, and volunteers (Israel Ministry of Environmental Protection 2013; Amir 2018).

On December 3, 2014, a breach happened in the Eilat-Ashkelon pipeline, south of Be'er Ora in Israel's southern Arava region, due to works of the relocation of Eilat's airport to a new site. Crude oil then traveled for several kilometers reaching the Evrona Nature Reserve, which has an extensive deer population and the northernmost Egyptian Doum palm trees in the world. The total estimation of the leaked crude oil was 5,000,000 l. Most of the crude oil was pumped up. About 55,000 tons of contaminated soil had to be removed to restore the damaged nature reserve (Israel Ministry of Environmental Protection 2014; EAPC 2019d).

Egypt Independent (2017) draw attention to the recurrent appearance of oil spills on the Red Sea's Ras Ghareb beach and the resort town of Hurghada since February 2016. Another incident happened in August 2016, when approximately 200 tons of crude oil got spilled from a burst pipe at the port of Aqaba in Jordan (Staff 2016).

Various research has identified the vulnerability of the coastal and marine ecosystems to pollution from massive tourism development and oil infrastructure, for example, declines in the coral cover due to the diving activity and oil pollution from dive boats (see Frihy et al. 1996; PERSGA 2006; Saleh 2007; Gladstone et al. 2013). Individual oil spills from tourist and dive boats are small in size but they are more frequent and are concentrated at coral reefs daily visited by divers.

12.6 Satellite Monitoring

The above mentioned cases of oil pollution in the Red Sea, and specifically in the regions of Egyptian resort areas led us to the following conclusions:

(1) oil pollution is a serious environmental problem for the Red Sea;
(2) oil pollution is caused mainly by offshore oil exploration and shipping activities;
(3) in many cases sources of oil pollution and the responsible for oil contamination is difficult to establish because there is no continuous aerial surveillance and satellite monitoring of oil pollution over the region;
(4) increased ship traffic in the narrow Red Sea, Gulf of Aqaba, and Gulf of Suez along the Egyptian resort areas represents a serious potential threat, which one day can result in an environmental catastrophe as it already has happened in different parts of the World Ocean, as well as in the Mediterranean Sea (Kostianoy and Carpenter 2018);
(5) there is an urgent need to establish permanent daily satellite monitoring of the northern part of the Red Sea in order to control oil pollution in the real or near real-time regime as it was done 15 years ago for the Southeastern Baltic Sea (Kostianoy et al. 2006; Lavrova et al. 2011, 2016; Kostianoy et al. 2015).

Working on this chapter, we found very little information on remote sensing of oil pollution in the Red Sea and on satellite monitoring of the Red Sea in general (e.g., Salem et al. 2001; Nasr et al. 2007; Eladawy et al. 2017). Salem et al. (2001), in particular, argued that satellite hyperspectral observations with high spectral and spatial resolution could be used to detect the level of oil contamination of polluted areas in the shoreline, which is necessary for cleaning processes. Finally, Salem et al. (2001) state that the high level of the environmental stress of the Red Sea increases the need for a comprehensive surveillance strategy to establish oil spill contingency plans and new coastal region regulations. Unfortunately, Salem et al. did not mention application of the Synthetic Aperture Radar (SAR) technology, which is the main remote sensing instrument for detection of oil spills and long-standing oil pollution monitoring (Kostianoy et al. 2006; Lavrova et al. 2011, 2016; Alpers 2014; Kostianoy et al. 2015).

Eladawy et al. (2017) examined the relationships between sea surface temperature (SST), sea wind (SW) and various oceanic variables in the northern Red Sea during the period of 2000–2014. The aim of the study was to identify the SST fronts and their relationship with the dominant circulation patterns. The analysis was performed with available remote sensing and reanalyzed data together with 1/128 HYbrid Coordinate Ocean Model (HYCOM) outputs. Also, the SST, SW speed and Chlorophyll-a (Chl-a) variability showed insignificant trends during the analyzed period.

More information on ocean color, chlorophyll-a, algal bloom, *Trichodesmium* bloom in summertime, coral reefs, mesoscale and submesoscale eddies observations in the Red Sea can be found in several chapters of the book "Remote sensing of the African Seas" edited by Vittorio Barale and Martin Gade (Barale and Gade 2014).

Below we show the capabilities of modern remote sensing for monitoring of the Red Sea oil pollution, SST, water turbidity, Chl-a concentration, and mesoscale water dynamics, as well as high-resolution imagery of the resort areas and the Suez Canal.

12.6.1 Oil Pollution

A Synthetic Aperture Radar (SAR) and Advanced SAR (ASAR) are active microwave instruments, very useful in detecting oil spills and surface films. Besides, they display a number of oceanic and atmospheric phenomena, like internal waves, underwater bottom topography, oceanic fronts, oceanic eddies, upwellings, river plumes, atmospheric gravity waves, atmospheric fronts, atmospheric eddies/cyclones, and coastal wind fields (Lavrova et al. 2011, 2016; Alpers 2014). Satellite SAR imagery has spatial resolutions ranging from few meters to several hundred meters depending on the type of SAR and on the data processing. European Space Agency (ESA) and several other space agencies have launched a series of satellites carrying SARs onboard the European ERS-1, ERS-2, and Envisat satellites, the Russian Almaz-1 satellite, the Canadian Radarsat-1 and Radarsat-2 satellites, the Japanese JERS-1 and ALOS satellites, the German TerraSAR-X and TanDEM-X satellites, the Italian COSMO-Skymed satellites, the Indian RISAT-1, and RISAT-2 satellites, and the Chinese HJ-1C satellite (Lavrova et al. 2011, 2016; Alpers 2014).

The above mentioned cases of oil pollution detected on the coasts of the Northern Red Sea give confidence that oil spills will be found on satellite images. To find out cases of oil pollution, the authors of this chapter used the toolkit of the "See The Sea" (STS) information service (geoportal) elaborated in the Space Research Institute (Moscow, Russia). STS employs technologies of automatic management of continuously renewed archives of satellite data (SAR, IR, VIS, etc.), provides access to the data and versatile analysis tools. The toolkit includes color composition based on one or several images, brightness correction, derivation of spectral and temporal profiles. For hyperspectral data, analysis of spectral profiles of different types of water surfaces helps to select the most informative spectral ranges for further image classification. STS provides access not only to various satellite data and derived products, but also to powerful specialized analysis tools. One can perform integrated joint analysis of data different in physical nature, spatial resolution, units of measurement and acquisition times. STS supports a comprehensive description of various phenomena and processes, estimation of their quantitative and qualitative characteristics, provides tools to analyze their origin and development, spatial and temporal parameters of phenomena distribution (Lavrova et al. 2011, 2016; Loupian et al. 2012; Mityagina et al. 2014; Kashnitskiy et al. 2015; Loupian et al. 2015).

STS allowed us to detect several cases of oil pollution based on the analysis of SAR Sentinel-1A and Sentinel-1B images in 2018 and 2019. For this publication, we did not analyze all SAR and high-resolution optical imagery archived in the STS, because our task was to show capabilities of satellite remote sensing to detect oil spills only. We found that the most polluted area in the Northern Red Sea is the Port of Suez where ships are waiting for the passage through the Suez Canal. Figures 12.7, 12.8, 12.9, 12.10 and 12.11 show that almost in every satellite image there is one or several oil spills, as well as films of surface-active substances (oily and waste waters). Interferometric Wide Swath (IW) Mode has 5-by-20-m spatial resolution and a 250 km swath.

Multi-Spectral Instrument (MSI) with 13 spectral channels in the visible/near infrared (VNIR) installed on Sentinel-2A and Sentinel-2B satellites operated by ESA are used to: (a) monitor Earth's forests and vegetation, plant growth; (b) determine various plant indices, such as leaf area chlorophyll and water content indices; (c) map changes in land cover, and (d) provide information on pollution in rivers, lakes, and coastal waters. The MSI imager features 10–20–60 m spatial resolution (depending on the spectral band) and a 290 km swath. Figure 12.12 shows films of oily waters which are displayed as black patches offshore the Port of Suez.

Also, we checked for oil pollution water areas close to the main resorts in Egypt—Hurghada and Sharm El-Sheikh. These regions are much cleaner than the area off the Port of Suez, but even here we detected several oil spills shown in Figs. 12.13, 12.14 and 12.15. For example, on 27 February 2018, two oil spills as large as 4–5 km are clearly displayed in the coastal zone southward of Hurghada (Fig. 12.13). On SAR

Fig. 12.7 SAR Sentinel-1B image of Suez Port on 22 July 2018, 03:52:19 GMT. Black patches are oil spills, and white dots are ships

image acquired on 14 August 2018, we detected more than 15 small oil spills in the area between Hurghada and the western coast of Sinai Peninsula (Fig. 12.14). Most of them are along the main shipping route going to the Gulf of Suez and then to the Suez Canal, but some of them are located in the coastal zone northward and southward of Hurghada. On 3 March 2019, in front of Sharm El-Sheikh—around Tiran Island—we observe a large dark area caused by natural films, probably originated from the coral reefs surrounding Tiran Island from the northern part (Fig. 12.15). The island is located at the entrance of the Strait of Tiran, which separates the Gulf of Aqaba from the main part of the Red Sea.

12.6.2 Mesoscale Water Dynamics

The following satellite images derived from MODIS-Aqua, MODIS-Terra and VIIRS-NPP in May and June 2017 show SST, Chl-a concentration, water turbidity (water leaving radiance at wavelength 551 nm), and water dynamics in the northern part of the Red Sea (Figs. 12.16, 12.17, 12.18, 12.19, 12.20 and 12.21).

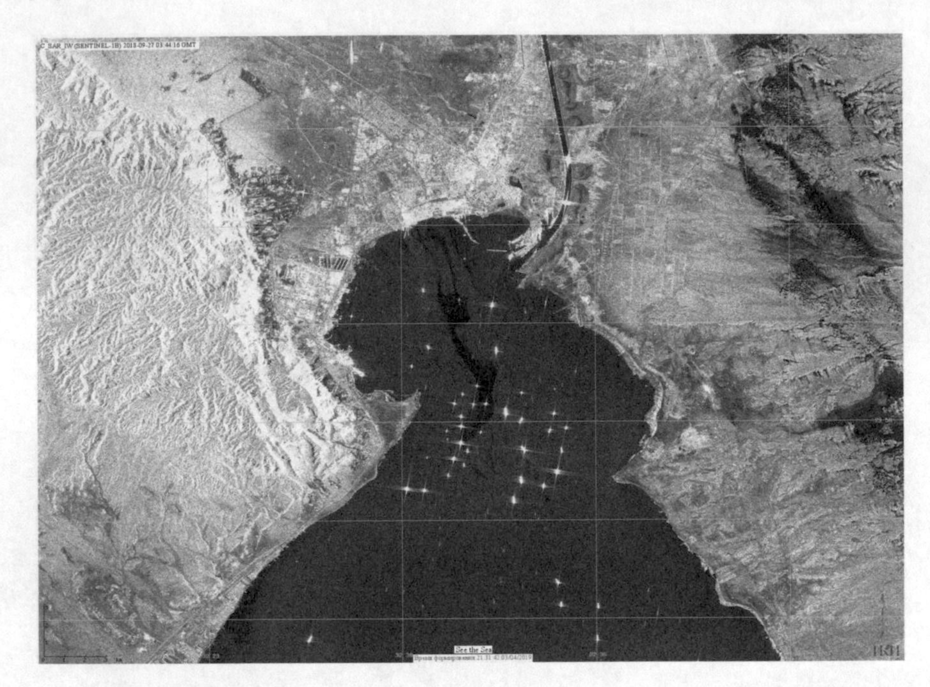

Fig. 12.8 SAR Sentinel-1B image of Suez Port on 27 September 2018, 03:44:16 GMT. Black patches are oil spills, and white dots are ships

12.6.3 The Suez Canal

We used high-resolution optical imagery of Operational Land Imager (OLI) Landsat-8 with 30 m spatial resolution and MSI Sentinel-2A to show changes in the Suez Canal structure related to the reconstruction of the canal in 2015. To demonstrate the difference, we compared satellite images of 2013 and 2017 in Fig. 12.22. The new double structure of the Suez Canal is clearly seen in front of Ismailia City situated on the west bank of the Suez Canal. The Canal splits right after the Bitter Lakes and merges again northward of Ismailia. High-resolution optical imagery allows monitoring land cover changes in details, which can have a spatial resolution up to 60 cm for some satellites.

12.6.4 Resort Areas

High-resolution optical imagery allows monitoring land changes in the areas of sea resorts of Egypt, as well as changes in coral reefs surrounding them as the main sightseeing for tourists and divers. As an example, on the MSI Sentinel-2A optical image of Hurghada area on 31 May 2017, we observe the City of Hurghada,

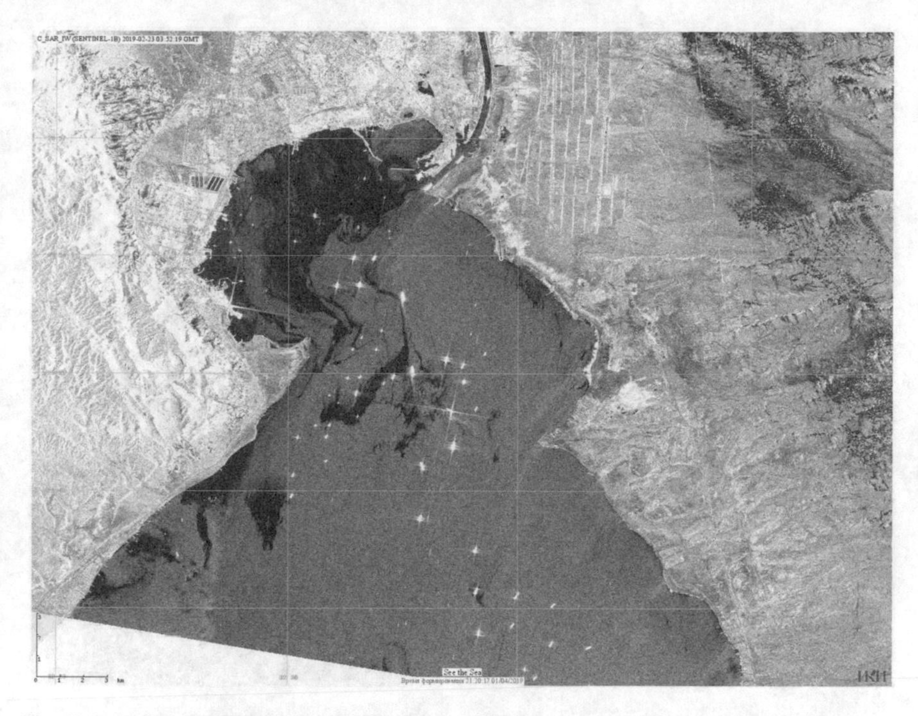

Fig. 12.9 SAR Sentinel-1B image of Suez Port on 23 February 2019, 03:52:19 GMT. Black patches are oil spills, and white dots are ships

coastal infrastructure, roads, Hurghada International Airport, and Giftun Islands with surrounding coral reefs displayed by light blue colors (Fig. 12.23). Figure 12.24 shows an MSI Sentinel-2A optical image of Sharm El-Sheikh area on 31 May 2017. We see Sharm El-Sheikh International Airport, the City of Sharm El-Sheikh, sea resorts and other objects of coastal infrastructure, roads, and the desert. Analysis of high-resolution optical imagery for the past several decades will allow to calculate the surface of coral reefs and detect areas of their degradation. This type of satellite imagery is very useful for urban and coastal infrastructure planning.

12.7 Conclusions

Analysis of scientific publications on satellite remote sensing of the Red Sea showed that there is very little information on satellite monitoring of oil pollution in the sea (El-Raey et al. 1996; Abdel-Kader et al. 1998). This is strange because about 15% of all global maritime trade and 10% of global seaborne oil passes through the Red Sea and the Suez Canal. From other regions of the World Ocean, we know that shipping activities and offshore oil and gas exploration and production represent a serious threat to the marine environment, coral reefs, islands, coastal zones, and beaches.

Fig. 12.10 SAR Sentinel-1B image of Suez Port on 11 March 2019, 15:56:09 GMT. Black patches are oil spills, and white dots are ships

Egypt seems to be subjected to a greatest risk of oil pollution impacts because it has a series of largest tourist resorts stretched for dozens of km along the coast, and the extensive shipping traffic which goes to the Gulf of Suez and further via the Suez Canal to the Mediterranean Sea, as well as to the Gulf of Aqaba. Fortunately, till present Egypt avoided large oil pollution catasrophes like with the *MV Haven* off Genoa (Italy) in April 1991, and the *Irenes Serenade* in Navarino Bay (Greece) on 23 February 1980 (Kostianoy and Carpenter 2018).

It seems that, as well as in the Mediterranean Sea, the routine oil pollution occur almost daily in different parts of the Red Sea. According to the Egyptian Environmental Affairs Agency and environmentalists, today at the shores from Ismailia to Hurghada small-size oil spills occur almost monthly. Moreover, since the 1970s there have been a number of cases of oil pollution in the region, which seriously affected coastline and coral reefs. The Gulf of Suez as a semi-enclosed area represents the most vulnerable site for oil pollution in the Red Sea, due to pollution from ships, including leisure boats; sewage wastes, offshore oil exploration at oil platforms. As a result, the level of environmental stress in the Red Sea is of raising concern. The Northern Red Sea area contains many natural resources and habitats, which could be negatively impacted by oil spills. They may affect the most important coral reef islands and diving areas in Hurghada south of the Gulf of Suez and Ras Mohamed

Fig. 12.11 SAR Sentinel-1A image of Suez Port on 01 April 2019, 03:45:02 GMT. Black patches are oil spills, and white dots are ships

south of the Gulf of Aqaba. All this increases the need for a comprehensive surveillance strategy to establish oil spill satellite monitoring over the coastal zone of Egypt and along the main shipping routes in the Red Sea.

To show the capabilities of satellite remote sensing techniques, we have received, processed and analyzed a series of satellite images acquired in 2017–2019 from different SAR, optical and infrared sensors. We detected a number of oil spills in the Northern Red Sea and found that the most polluted area is the Port of Suez, where almost every satellite image displays oil spills. The water around Hurghada and Sharm El-Sheikh resorts looks quite clean, but even here small-size oil spills occur as well. Medium resolution optical and infrared imagery revealed mesoscale water dynamics in the form of cyclonic and anticyclonic eddies, dipoles and intrusions, which can redistribute oil pollution across and along the sea, and contribute to coastal pollution. High-resolution optical imagery showed its importance for monitoring of coral reefs, land changes, coastal infrastructure, as well as for urban planning.

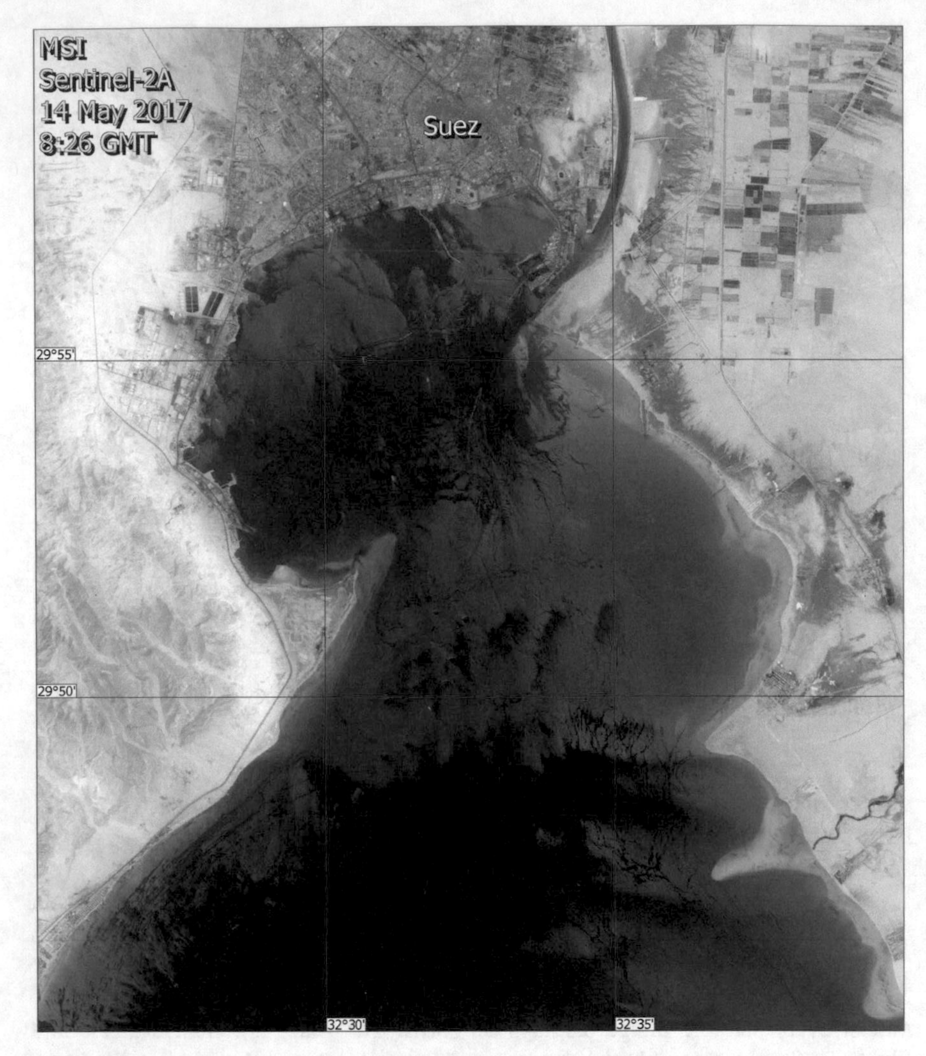

Fig. 12.12 MSI Sentinel-2A optical image of the water area near the Port of Suez on 14 May 2017, 08:26 GMT. Spatial resolution 10 m

12.8 Recommendations

It is a strange coincidence but in many places of the world ocean and inland seas, tourist places and resorts are located in close proximity to offshore oil/gas production, ports, oil terminals, and shipping routes. The Gulf of Mexico, the Mediterranean Sea, the southern part of the Baltic Sea, the Black Sea, and the Caspian Sea are typical examples. The Northern Red Sea with a dozen of large resort areas and beaches is not an exception. In-situ observations regularly reveal cases of oil pollution in the coastal zone, but satellite remote sensing is the most effective tool to monitor the coastal

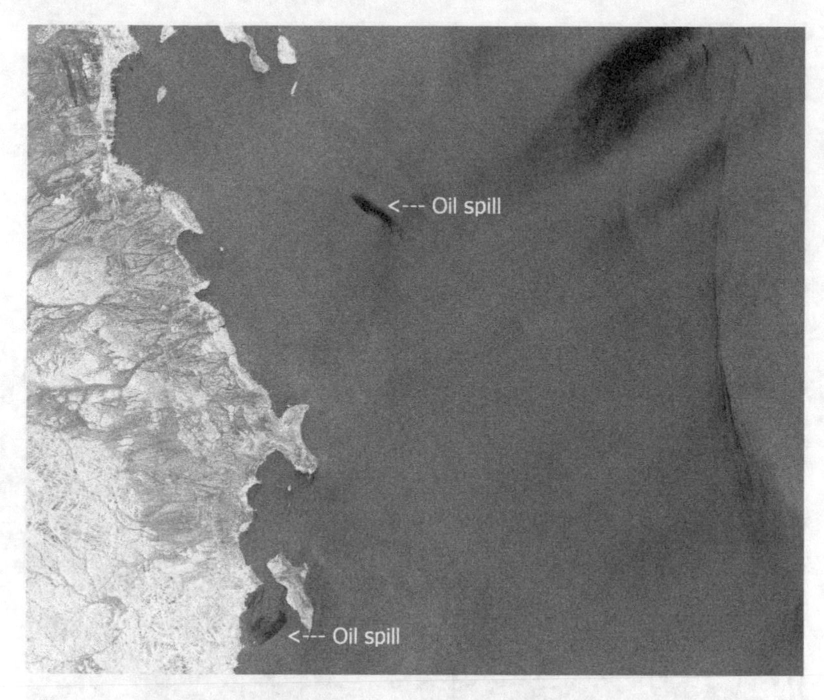

Fig. 12.13 SAR Sentinel-1B image of the coastal zone southward of Hurghada on 27 February 2018, 15:46:44 GMT. Oil spills (black spots) are indicated by arrows

zone and the open sea and to detect cases of oil pollution, their number, extension, shape, advection, and behavior. In many cases, the culprit of the oil pollution can be also identified. Numerous cases of oil pollution detected on satellite imagery in the Northern Red Sea and presented in this paper have proved the efficiency of satellite monitoring systems for environmental protection. This task could be productively solved in the International Satellite Monitoring Center that should be established in Egypt. The establishment of permanent satellite monitoring of oil pollution and water dynamics in the Northern Red Sea and the Southeastern Mediterranean Sea is of vital importance to avoid potential catastrophes resulting from large oil spills, and for sustainable development of the coastal tourism business in Egypt.

Fig. 12.14 SAR Sentinel-1B image of the water area between Hurghada and the southern tip of Sinai Peninsula on 14 August 2018, 15:46:52 GMT. Oil spills (black spots) are indicated by red arrows

Fig. 12.15 SAR Sentinel-1A image of the coastal zone of Sharm El-Sheikh on 3 March 2019, 03:36:59 GMT

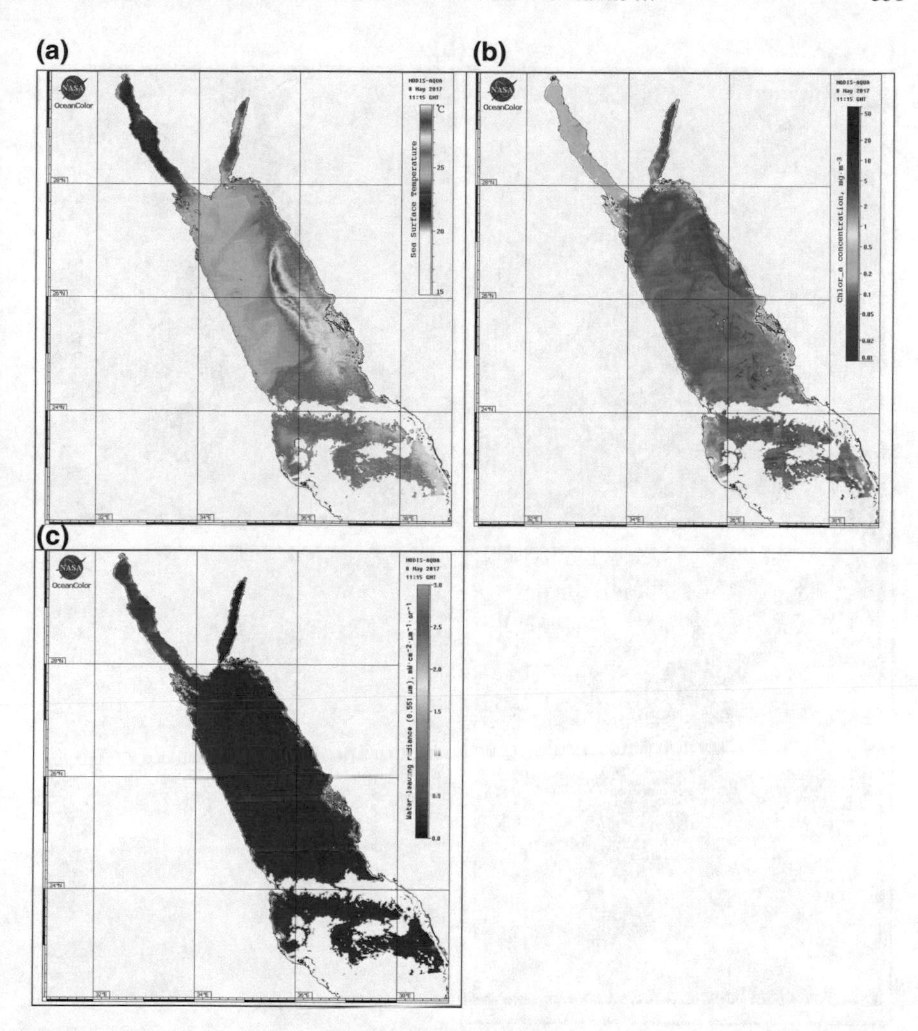

Fig. 12.16 **a** SST (°C); **b** Chl-a concentration (mg/m^3); **c** Water leaving radiance derived from MODIS-Aqua on 8 May 2017, 11:15 GMT. Spatial resolution—1 km. SST map shows the presence of mesoscale eddies along the Egyptian coast of the Red Sea as large as 50–100 km, which transfer relatively cold water to the central part of the sea, and warm waters from the eastern coast to the western one. There is a strong intrusion of warm water propagating from 25° to 27° N in the central part of the sea. This mesoscale water dynamics may redistribute potential pollutants (oil and plastic pollution) in the sea and impact the coastal zone of Egypt, including resort areas of Hurghada and Sharm El-Sheikh. Chl-a concentration and water turbidity show that water in the northern part of the sea is very clean, except from the very narrow bands along the coastal zone and around islands. White patches in the southern part of the image are clouds

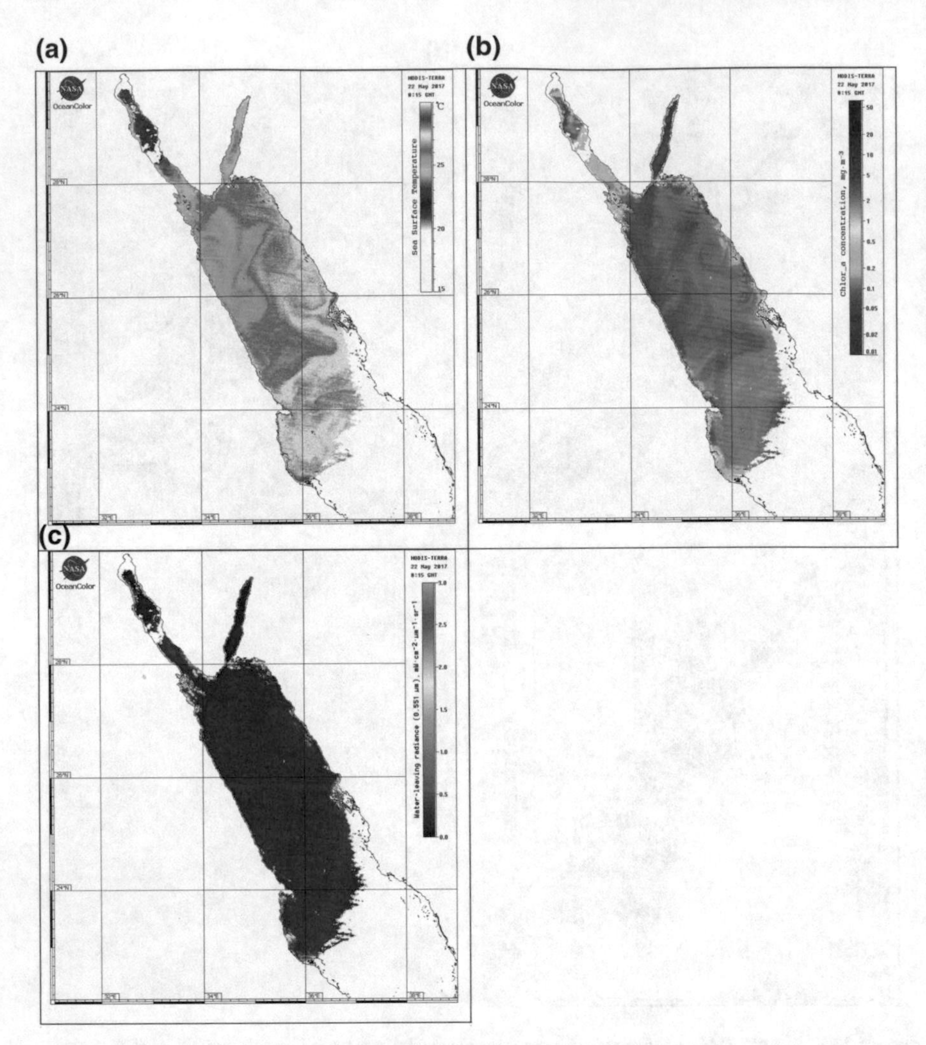

Fig. 12.17 **a** SST (°C); **b** Chl-a concentration (mg/m^3); **c** Water leaving radiance derived from MODIS-Terra on 22 May 2017, 08:15 GMT. Spatial resolution—1 km. SST map shows the presence of 3 mesoscale anticyclonic eddies along the coast of Saudi Arabia as large as 100 km, which transfer relatively warm waters northward along the coast. Offshore Hurghada we observe a large cyclonic eddy which transfers relatively cold water northward along the Egyptian coast. Southward of 25° N, there is a pool of relatively warm water, which fills the whole basin of the Red Sea. White patches in the southern part of the image are clouds. Chl-a concentration and water turbidity show that water in the northern part of the sea is very clean, except from the very narrow bands along the coastal zone and around islands

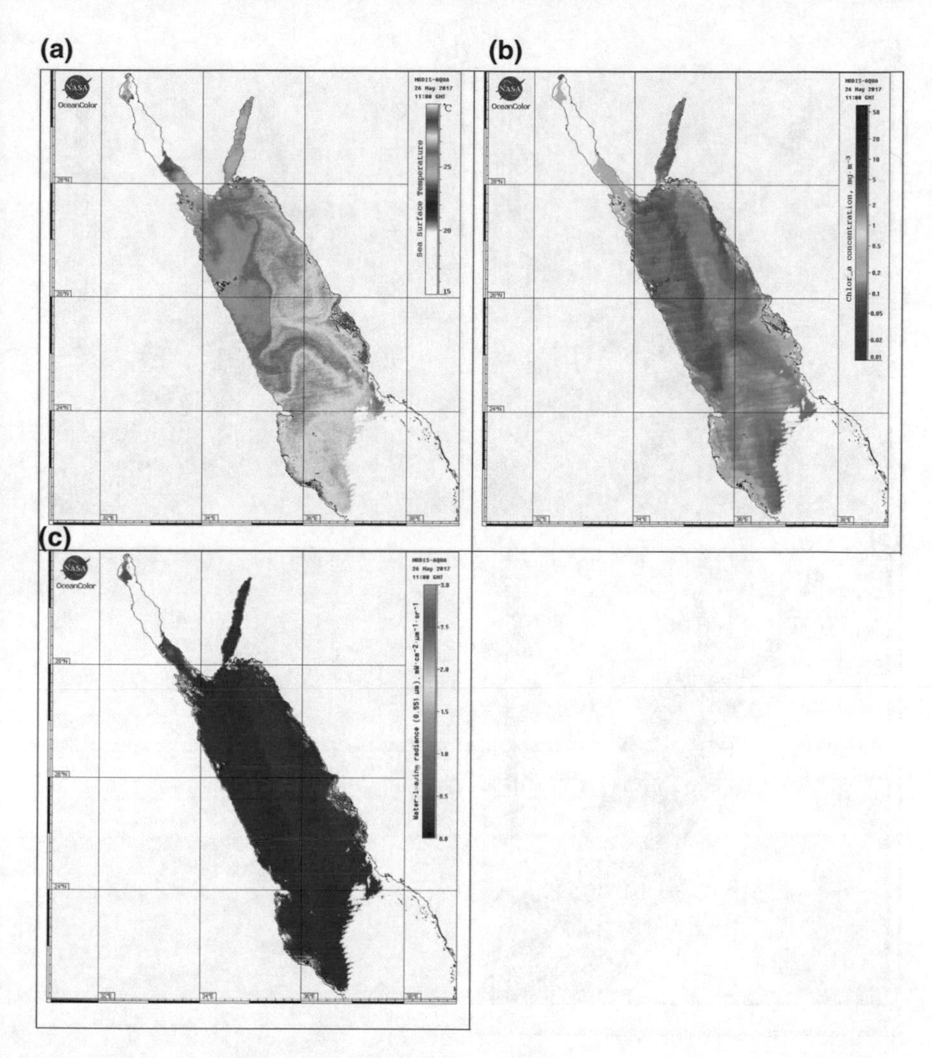

Fig. 12.18 **a** SST (°C); **b** Chl-a concentration (mg/m^3); **c** Water leaving radiance derived from MODIS-Aqua on 26 May 2017, 11:00 GMT. Spatial resolution—1 km. Four days later the SST pattern significantly changed. In the middle of the image (25–26° N) we observe a warm dipole structure 150 km large propagating from the east to the west, where it is blocked by a cold dipole structure moving from the north to the south. Southward of 25° N, there is a pool of relatively warm water, which fills the whole basin of the Red Sea. Chl-a concentration and water turbidity maps show that water in the northern part of the sea is very clean, except from the very narrow bands along the coastal zone and around islands. White patches in the southern part of the image are clouds

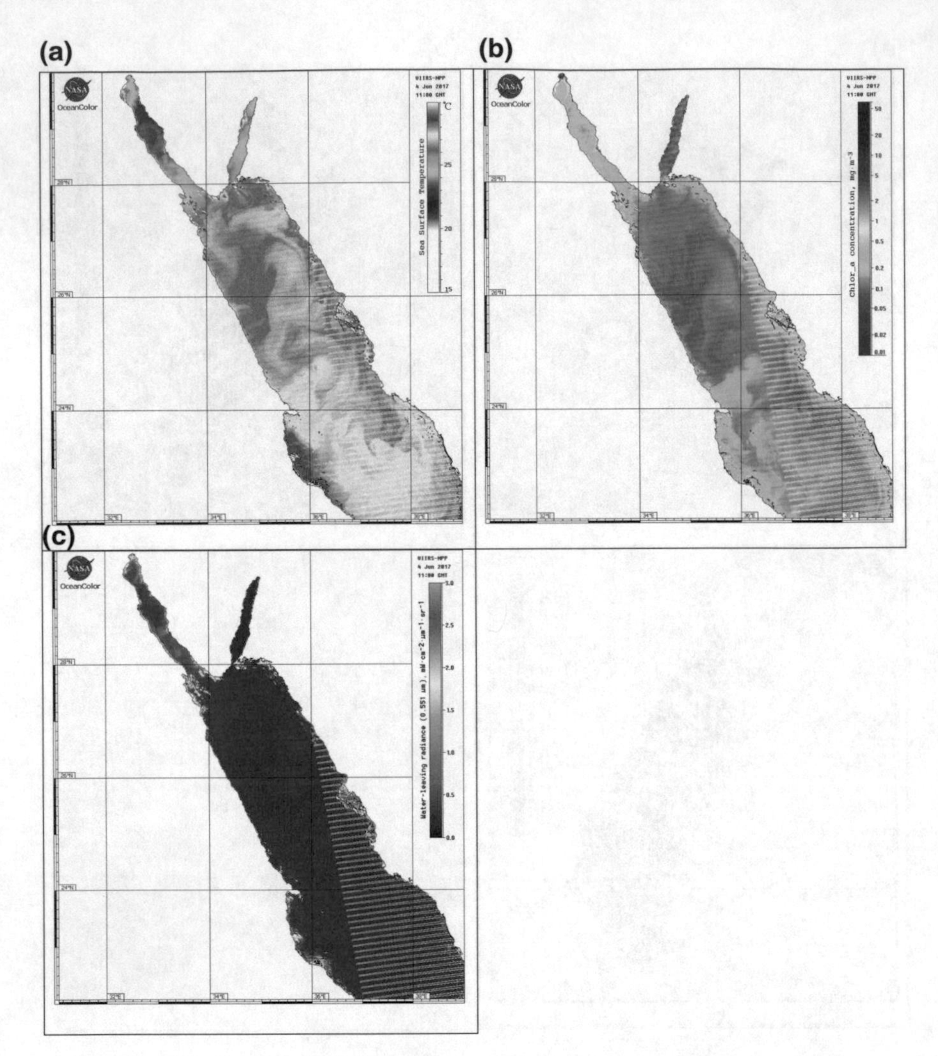

Fig. 12.19 **a** SST (°C); **b** Chl-a concentration (mg/m^3); **c** Water leaving radiance derived from VIIRS-NPP on 4 June 2017, 11:00 GMT. Spatial resolution—1 km. It seems that changes in wind force and direction led to permanent changes in the SST field structure. For example, a week later, on 4 June, we observe a northward propagation of warm water along the western coast till 24.5° N, while in May it moved along the eastern coast or in the middle of the sea. Then it goes in the northeastward direction in the form of a warm dipole, which meets a relatively colder dipole going from the east to the west at about 25.5° N. In its turn, its motion to the west is blocked by another much colder dipole moving northward to about 27.5° N, where it meets cold water moving southward from the Gulf of Aqaba. Chl-a concentration map in its southern part shows higher concentrations of Chl-a, which correlates with a location of a relatively warm pool of water. This can be probably explained by the beginning of algal bloom in this region of the Red Sea

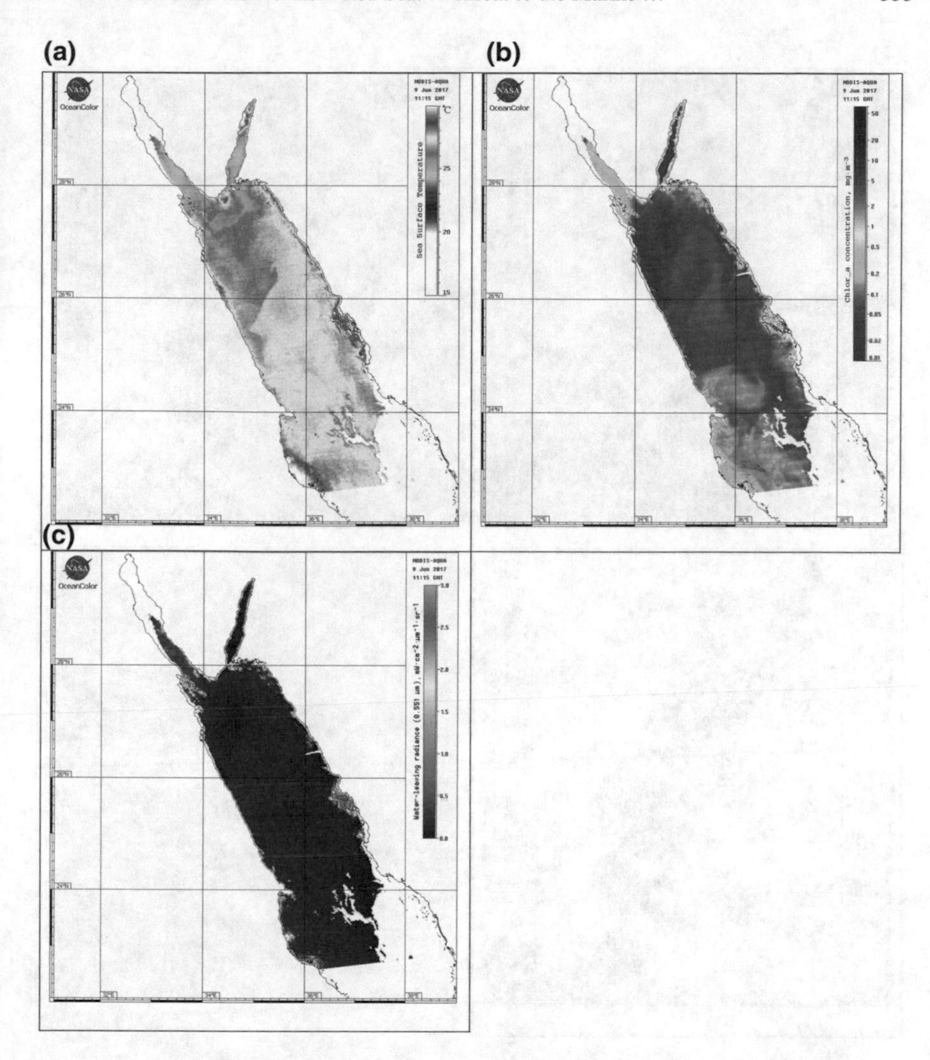

Fig. 12.20 **a** SST (°C); **b** Chl-a concentration (mg/m^3); **c** Water leaving radiance derived from MODIS-Aqua on 9 June 2017, 11:15 GMT. Spatial resolution—1 km. In the past five days, the situation again changed significantly. The dipoles disappeared, and the SST pattern became more uniform displaying a basin scale intrusion of relatively warm water propagating northward along the eastern coast of the sea till the gate to the Gulf of Aqaba. From the Gulf of Aqaba, we observe an intrusion of relatively cold waters moving exactly towards the Hurghada coast. Northward of 24° N, the water temperature increased by several degrees. Southward of 25° N, we observe patches of high Chl-a concentration, which can probably be explained by the beginning of algal bloom. White areas in the southeastern part of the image are clouds

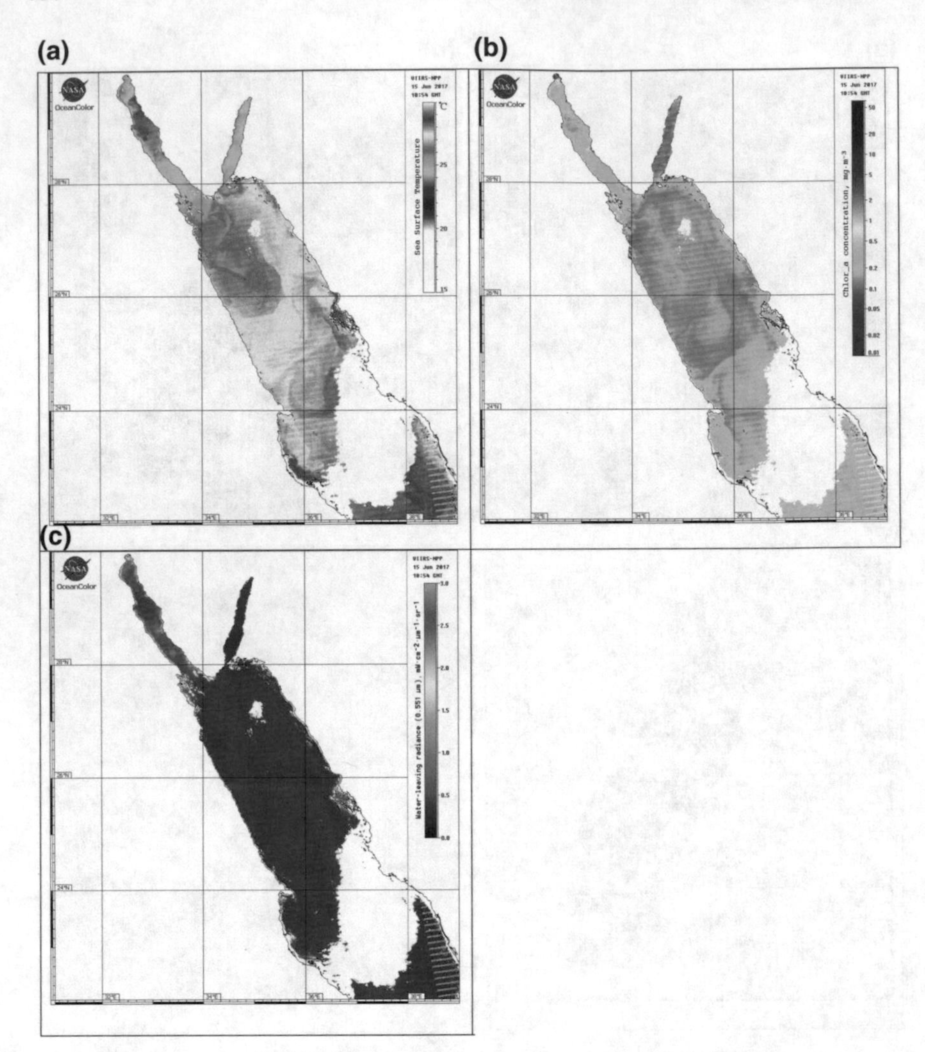

Fig. 12.21 **a** SST (°C); **b** Chl-a concentration (mg/m^3); **c** Water leaving radiance derived from VIIRS-NPP on 15 June 2017, 10:54 GMT. Spatial resolution—1 km. A week later, again we don't see vortical dynamics in the northern part of the Red Sea. In the middle of the image (at about 26° N), we observe a thermal front generated by a collision of a warm intrusion moving from the south to the north along the eastern coast, and the relatively cold intrusion propagating along the western coast in the opposite direction. These relatively cold waters are originated from the Gulf of Suez, which is evident from the SST image. Propagation of relatively cold waters from the Gulf of Aqaba is restricted by the warm intrusion from the south. This is why we observe only a small patch of colder water nearby the gate to the Gulf. It seems that algal bloom is propagating more and more northward of 25° N as we observe patches of high Chl-a concentration till Sinai Peninsula. Water turbidity patterns did not change because water-leaving radiance at wavelength 551 nm is responsible more for suspended matter (there are no rivers and other sources of suspended sediments in the area) than for Chl-a concentration. White patches in the southern part of the image are clouds

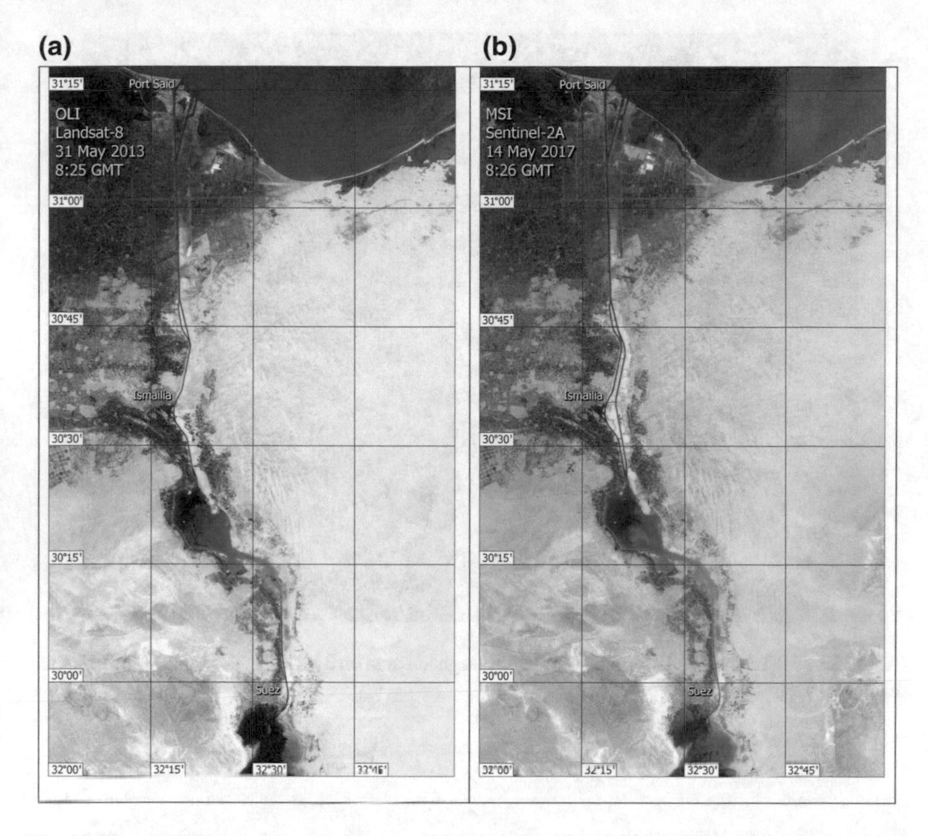

Fig. 12.22 **a** OLI Landsat-8 optical image of the Suez Canal on 31 May 2013, 08:25 GMT; **b** MSI Sentinel-2A optical image on 14 May 2017, 08:26 GMT

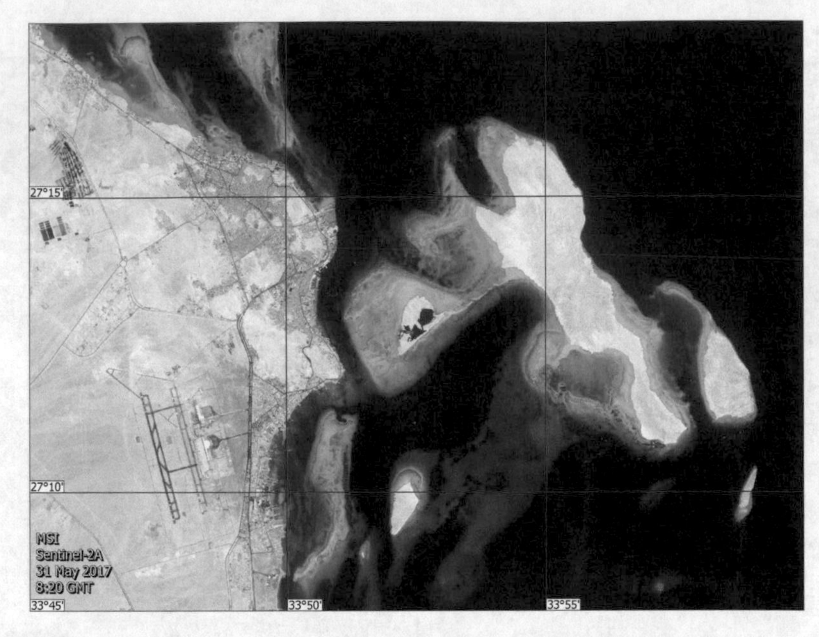

Fig. 12.23 MSI Sentinel-2A optical image of Hurghada area on 31 May 2017, 08:20 GMT. Spatial resolution 10 m

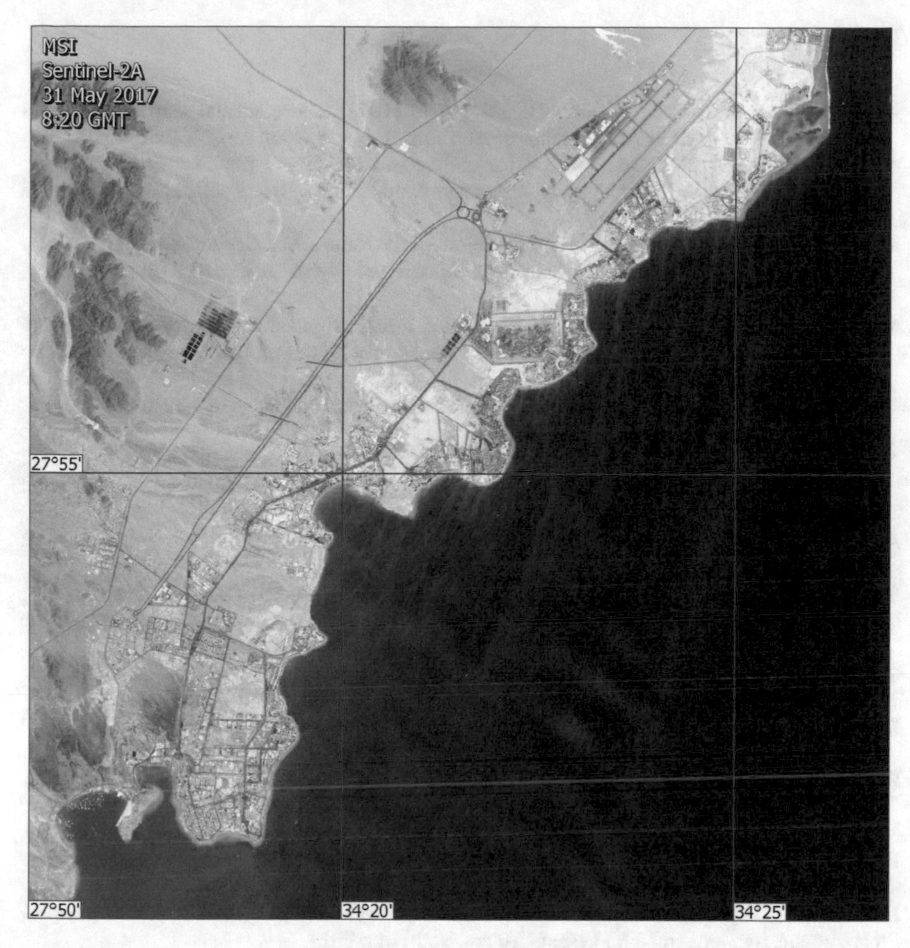

Fig. 12.24 MSI Sentinel-2A optical image of Sharm El-Sheikh area on 31 May 2017, 08:20 GMT. Spatial resolution 10 m

Acknowledgements E. A. Kostianaia analyzed the sources, cases, and impact of oil pollution on tourism development in Egypt, and she was supported in the framework of the P.P. Shirshov Institute of Oceanology RAS budgetary financing (Project N 149-2019-0004). A. Kostianoy was partially supported in the framework of the P.P. Shirshov Institute of Oceanology RAS budgetary financing (Project N 149-2019-0004) for the analysis of optical and infrared satellite imagery. O. Yu. Lavrova tested the capabilities of the "See The Sea" satellite monitoring system (SAR imagery) for the area of the Northern Red Sea in the framework of the Russian Science Foundation Project N 19-77-20060 «Assessing ecological variability of the Caspian Sea in the current century using satellite remote sensing data».

References

Abdel-Kader AF, Nasr SM, El-Gamily HI, El-Raey M (1998) Environmental sensitivity analysis of potential oil spill for Ras-Mohammed coastal zone, Egypt. J Coast Res 14(2):502–510

Africanglobe (2019) Red Sea politics [online]. Available from: https://www.africanglobe.net/africa/eritrea-red-sea-slipping-total-arab-control/attachment/red-sea-politics-2/. Accessed 15 June 2019

Al-Ghazawy O (2010) Red Sea suffers crude-oil slick. Nature Middle East [online], 14 July 2010. Available from: https://www.natureasia.com/en/nmiddleeast/article/10.1038/nmiddleeast.2010.175. Accessed 30 Mar 2019

Alpers W (2014) Remote sensing of African coastal waters using active microwaves instrument. In: Barale V, Gade M (eds) Remote sensing of the African Seas. Springer, The Netherlands, pp 75–94

Alsharhan AS (2003) Petroleum geology and potential hydrocarbon plays in the Gulf of Suez rift basin, Egypt. AAPG Bull 87(1):143–180

Amir R (2018) Oil pollution in the marine waters of Israel. In: Carpenter A, Kostianoy AG (eds) Oil pollution in the Mediterranean Sea: part II—national case studies. Handbook of environmental chemistry, vol 84. Springer, Berlin, pp 199–228

Awad H (1989) Oil contamination in the Red Sea environment. Water Air Soil Pollut 45(3–4): 235–242

Barale V, Gade M (2014) Remote sensing of the African Seas. Springer, The Netherlands, p 428

Carpenter A, Kostianoy AG (eds) (2018a) Oil pollution in the Mediterranean Sea. Part I: the international context. Springer, Berlin, Heidelberg

Carpenter A, Kostianoy AG (eds) (2018b) Oil pollution in the Mediterranean Sea. Part II: national case studies. Springer, Berlin, Heidelberg, 291 p. https://doi.org/10.1007/978-3-030-11138-0

Dicks B (1984) Oil pollution in the Red Sea—environmental monitoring of an oilfield in a coral area, Gulf of Suez. Deep Sea Res Part A. Oceanogr Res Pap 31(6–8):833–854

Egypt Independent (2017) Repeated Red Sea oil spills forces environment minister to take action. Egypt Independent [online], 13 Dec 2017. Available from: https://ww.egyptindependent.com/repeated-red-sea-oil-spills-forces-environment-minister-to-take-action/. Accessed 30 Mar 2019

Eilat Ashkelon Pipeline Co. Ltd. (2019a) Oil terminals [online]. Available from: https://eapc.com/the-crude-oil-system/oil-terminals/. Accessed 30 Mar 2019

Eilat Ashkelon Pipeline Co. Ltd. (2019b) The crude oil system [online]. Available from: https://eapc.com/cat-gallery/crude-oil-system/. Accessed 30 Mar 2019

Eilat Ashkelon Pipeline Co. Ltd. (2019c) Reverse flow project [online]. Available from: https://eapc.com/the-crude-oil-system/reverse-flow-project/. Accessed 30 Mar 2019

Eilat Ashkelon Pipeline Co. Ltd. (2019d) The leak incident [online]. Available from: https://eapc.com/evrona-incident/the-leak-incident/. Accessed 2 Apr 2019

Eladawy A, Nadaoka K, Negm A, Abdel-Fattah S, Hanafy M, Shaltout M (2017) Characterization of the northern Red Sea's oceanic features with remote sensing data and outputs from a global circulation model. Oceanologia 59(3):213–237

El-Raey M, Farid Abdel-Kader A, Nasr SM, El-Gamily HI (1996) Remote sensing and GIS for an oil spill contingency plan, Ras-Mohammed, Egypt. Int J Remote Sens 17(11):2013–2026

Fisher WB, Smith CG (2019) Suez Canal [online]. Available from https://www.britannica.com/topic/Suez-Canal. Accessed 30 Mar 2019

Frihy OE, Fanos AM, Khafagy AA, Aesha KA (1996) Human impacts on the coastal zone of Hurghada, Northern Red Sea, Egypt. Geo-Mar Lett 16(4):324–329

Gladstone W, Curley B, Shokri MR (2013) Environmental impacts of tourism in the Gulf and the Red Sea. Mar Pollut Bull 72(2):375–388

Gouda EA (2015) Obstacles to sustainable tourism development on the Red Sea Coast. Int J Innov Educ Res 3(3)

Hanna RG (1983) Oil pollution on the Egyptian Red Sea Coast. Mar Pollut Bull 14(7):268–271

Head SM (1987) Red Sea fisheries. In: Edwards AJ, Head SM (eds) Red Sea: key environments. Pergamon Press, Oxford, pp 363–382

Hilmi N, Safa A, Reynaud S (2012) Coral reefs and tourism in Egypt's Red Sea. Top Middle East N Afr Econ 14:416–434

Israel Ministry of Environmental Protection (2013) Shipping Co. ordered to pay NIS 750,000 for oil spill in Gulf of Eilat [online]. Available from: https://www.sviva.gov.il/English/ResourcesandServices/NewsAndEvents/NewsAndMessageDover/Pages/2013/4_April/AvramitRuling.aspx. Accessed 2 Apr 2019

Israel Ministry of Environmental Protection (2014) Arava oil spill latest [online]. Available from: https://www.sviva.gov.il/English/ResourcesandServices/NewsAndEvents/NewsAndMessageDover/Pages/2014/y%20December/Arava-Oil-Spill.aspx. Accessed 2 Apr 2019

Jordan Oil Terminals Company (2019a) Aqaba Oil & LPG Terminal [online]. Available from: https://www.jotc.com.jo/DetailsPage/JOTIC_En/OurTerminalsEn.aspx?ID=33. Accessed 31 Mar 2019

Jordan Oil Terminals Company (2019b) Aqaba Heavy Oil Terminal [online]. Available from: https://www.jotc.com.jo/DetailsPage/JOTIC_En/OurTerminalsEn.aspx?ID=34. Accessed 31 Mar 2019

Julian HL (2010) Egyptian oil spill in Red Sea. Arutz Sheva [online], 24 June 2010. Available from: https://www.israelnationalnews.com/News/News.aspx/138250#.UgsEWNK-ods. Accessed 2 Apr 2019

Kashnitskiy AV, Balashov IV, Loupian EA, Tolpin VA, Uvarov IA (2015) Development of software tools for satellite data remote processing in contemporary information systems. Sovremennye problemy distantsionnogo zondirovaniya Zemli iz kosmosa 12(1):156–170 (in Russian)

Khaled N (2014) More than a 100-year journey [online]. Available from: https://egyptoil-gas.com/features/more-than-a-100-year-journey/. Accessed 31 Mar 2019

Kostianoy AG, Carpenter A (2018) History, sources and volumes of oil pollution in the Mediterranean Sea. In: Carpenter A, Kostianoy AG (eds) Oil pollution in the Mediterranean Sea: part I—the international context. Handbook of environmental chemistry. Springer, Berlin

Kostianoy AG, Lavrova OY (eds) (2014) Oil pollution in the Baltic Sea. The handbook of environmental chemistry, vol 27. Springer, Berlin, Heidelberg, New York, 268 pp

Kostianoy AG, Litovchenko KT, Lavrova OY, Mityagina MI, Bocharova TY, Lebedev SA et al (2006) Operational satellite monitoring of oil spill pollution in the southeastern Baltic Sea: 18 months experience. Environ Res Eng Manage 4(38):70–77

Kostianoy AG, Bulycheva EV, Semenov AV, Krainyukov AV (2015) Satellite monitoring systems for shipping, and offshore oil and gas industry in the Baltic Sea. Transp Telecommun 16(2):117–126

Lavrova OY, Kostianoy AG, Lebedev SA, Mityagina MI, Ginzburg AI, Sheremet NA (2011) Complex satellite monitoring of the Russian seas. IKI RAN, Moscow, 470 pp (in Russian)

Lavrova OY, Mityagina MI, Kostianoy AG (2016) Satellite methods of detection and monitoring of marine zones of ecological risks. Space Research Institute, Moscow, 336 p (in Russian)

Leach L (2014) Egypt's oil spill preparedness. Egypt Oil and Gas Newspaper [online], 31 Dec 2014. Available from: https://egyptoil-gas.com/features/egypts-oil-spill-preparedness/9968/. Accessed 2 Apr 2019

Loupian EA, Matveev AA, Uvarov IA, Bocharova TY, Lavrova OY, Mityagina MI (2012) Satellite service See the Sea—a tool for instigation of processes and phenomena at the sea surface. Sovremennye problemy distantsionnogo zondirovaniya Zemli iz kosmosa 9(2):251–261 (in Russian)

Loupian EA, Proshin AA, Burtsev MA, Balashov IV, Bartalev SA, Efremov VY et al (2015) IKI center for collective use of satellite data archiving, processing and analysis systems aimed at solving the problems of environmental study and monitoring. Sovremennye problemy distantsionnogo zondirovaniya Zemli iz kosmosa 12(5):263–284 (in Russian)

Marine Traffic (2019) Live map [online]. Available from: https://www.marinetraffic.com/en/ais/home/. Accessed 30 Mar 2019

Maritime Transport Sector (2019) Commercial ports [online]. Available from: https://www.mts.gov.eg/en/sections/10/1-10-Commercial-Ports. Accessed 30 Mar 2019

Mayton J (2009) Oil spills poison the Red Sea. The Christian Science Monitor [online], 5 Nov 2009. Available from: https://www.csmonitor.com/Environment/2009/1105/oil-spills-poison-the-red-sea. Accessed 30 Mar 2019

Mityagina MI, Lavrova OY, Uvarov IA (2014) "See the Sea": multi-user information system for investigating processes and phenomena in coastal zones via satellite remotely sensed data, particularly hyperspectral data. In: Proceedings of SPIE 9240, remote sensing of the ocean, sea ice, coastal waters, and large water regions, 2014, 92401C. https://doi.org/10.1117/12.2067300

Nasr AH, El Leithy BM, Helmy AK (2007) Assessment of some water quality parameters using MODIS data along the Red Sea Coast, Egypt. ICGST-GVIP J 7(3):29–34

Nations Online (2019) Map of Egypt, Middle East [online]. Available from: https://www.nationsonline.org/oneworld/map/egypt_map.htm. Accessed 30 Mar 2019

PERSGA (2006) State of the marine environment, report for the Red Sea and Gulf of Aden. PERSGA, Jeddah

Rasul NM, Stewart IC, Vine P, Nawab ZA (2019) Introduction to oceanographic and biological aspects of the Red Sea. In: Oceanographic and biological aspects of the Red Sea. Springer, Cham, pp 1–9

Red Sea Port Authority (2019) Services [online]. Available from: https://rspa-eg.com/index.php/en/. Accessed 30 Mar 2019

Ryan WBF, Schreiber BC (2019) Red Sea [online]. Available from: https://www.britannica.com/place/Red-Sea. Accessed 30 Mar 2019

Saleh MA (2007) Assessment of mangrove vegetation on Abu Minqar Island of the Red Sea. J Arid Environ 68(2):331–336

Salem F, El-Ghazawi T, Kafatos M (2001) Remote sensing and image analysis for oil spill mitigation in the Red Sea [online]. In: Proceedings of the 2nd biennial coastal GeoTools conference, Charleston, SC 8–11 January 2001. Available from: https://pdfs.semanticscholar.org/c4c3/9b2bf769b5b4c64e469bfdf5d9404ef80608.pdf. Accessed 04 Apr 2019

Shaltout KH, Khalaf-Allah A, El-Bana M (2005) Environmental characteristics of the mangrove sites along the Egyptian Red Sea Coast. Assessment and management of mangrove forests in Egypt for sustainable utilization and development: a project funded by ITTO (Japan) and supervised by MALR/MSEA—EEAA, Cairo

Staff T (2016) Aqaba spill dumps 200 tons of crude oil into Red Sea. The Times of Israel [online], 24 Aug 2016. Available from: https://www.timesofisrael.com/aqaba-spill-dumps-200-tons-of-crude-oil-into-red-sea/. Accessed 30 Mar 2019

Suez Canal Authority (2017a) About Suez Canal [online]. Available from: https://www.suezcanal.gov.eg/English/About/SuezCanal/Pages/AboutSuezCanal.aspx. Accessed 30 Mar 2019

Suez Canal Authority (2017b) Canal history [online]. Available from: https://www.suezcanal.gov.eg/English/About/SuezCanal/Pages/CanalHistory.aspx. Accessed 20 Mar 2019

Suez Canal Authority (2017c) New Suez Canal [online]. Available from: https://www.suezcanal.gov.eg/English/About/SuezCanal/Pages/NewSuezCanal.aspx. Accessed 30 Mar 2019

Suez Canal Authority (2019) Monthly number and net ton by ship type [online]. Available from: https://www.suezcanal.gov.eg/English/Navigation/Pages/NavigationStatistics.aspx. Accessed 31 Mar 2019

US Energy Information Administration (2017) Country analysis brief: Saudi Arabia [online]. Available from: https://www.eia.gov/beta/international/analysis.php?iso=SAU. Accessed 31 Mar 2019

Wennink CJ, Nelson-Smith A (1979a) Coastal oil pollution study for the Gulf of Suez and the Red Sea Coast of the Republic of Egypt, Part A. IMCO, London, p 69

Wennink CJ, Nelson-Smith A (1979b) Coastal oil pollution evaluation study for the Gulf of Aqaba Coast of the Hashemite Kingdom of Jordan, Part A. IMCO, London, p 18

World Travel and Tourism Council (2018) Travel and tourism economic impact 2018, Egypt [online]. World Travel and Tourism Council, London

Chapter 13
Satellite Monitoring of the Nile River and the Surrounding Vegetation

Andrey Kostianoy, Evgeniia A. Kostianaia, Dmitry M. Soloviev and Abdelazim M. Negm

Abstract The chapter shows capabilities of high-resolution satellite optical imagery to monitor in details natural vegetation and agricultural fields around the Nile River. The course of the Nile was divided into 19 scenes of OLI Landsat-8 images acquired in July–September 2018 to show spatial and vegetation peculiarities of every sub-region. Besides, the NASA Giovanni online data system v.4.30, developed and maintained by the NASA Goddard Earth Sciences Data and Information Services Center (GES DISC), was used to show seasonal and interannual (2000–2018) variability of NDVI as a measure of vegetation health. We did this analysis for the whole Egypt, the area around the Nile Valley, the Nile Delta, and two important agricultural areas of Egypt located along the Nile in the southern and northern part of the country. The obtained results show that during the past two decades, there is a general growth of NDVI in all observed areas, but with different rates. This is likely explained by the growth of arable and irrigated lands. For larger areas, like Egypt or the whole Nile River, two clear cycles of about 10 years were observed, which is likely explained by regional climate change oscillations.

A. Kostianoy (✉) · E. A. Kostianaia
P.P. Shirshov Institute of Oceanology, Russian Academy of Sciences, 36, Nakhimovsky Pr., Moscow 117997, Russia
e-mail: kostianoy@gmail.com

E. A. Kostianaia
e-mail: evgeniia.kostianaia@gmail.com

A. Kostianoy
S.Yu. Witte Moscow University, 12, Build. 1, 2nd Kozhukhovsky Pr., Moscow 115432, Russia

D. M. Soloviev
Marine Hydrophysical Institute, Russian Academy of Sciences, 2, Kapitanskaya Str., Sevastopol 299011, Russia
e-mail: solmit@gmail.com

A. M. Negm
Water and Water Structures Engineering Department, Faculty of Engineering, Zagazig University, Zagazig 44519, Egypt
e-mail: amnegm85@yahoo.com; amnegm@zu.edu.eg

S. F. Elbeih et al. (eds.), *Environmental Remote Sensing in Egypt*,
Springer Geophysics, https://doi.org/10.1007/978-3-030-39593-3_13

Keywords Egypt · Nile River · Nile Delta · Nile Valley · High-resolution · Optical imagery · Vegetation · NDVI

13.1 Introduction

The Nile River is the major and longest river in Africa. Its overall length is about 6800 km, and during its course, the Nile crosses Tanzania, Uganda, Rwanda, Burundi, the Democratic Republic of Congo, Kenya, Ethiopia, Eritrea, South Sudan, Republic of Sudan, and Egypt. The Nile River enters the Egyptian territory near Adindan on the Egyptian-Sudanese border and flows for a distance of about 1536 km until emptying into the Mediterranean Sea via the Nile Delta (Negm 2017a, b). In Egypt, the Nile River is surrounded by the Egyptian Western and Eastern deserts, which occupy nearly 96% of the country's territory. The cultivated areas in Egypt are limited to the Nile Valley, the Nile Delta, oases, and desert depressions, which altogether account for about 4% of the total area of the country (Gad 2020). From ancient times, the Egyptian civilization depended on the Nile River, and till present most of the population live along the river banks.

From the ancient times, the Nile River was used by the local population for navigation, trading, transportation of goods and people, fishery, irrigation, food and building materials production, etc. Since 1970, with the Aswan High Dam construction, the Nile became the main source of electricity production. Besides, the Nile River is a source for 95% of water resources in Egypt. Thus it is of vital importance for people and Egypt's economy.

Global warming and regional climate change significantly affected Egypt's territory and the Nile River, in particular (Egypt. Second National Communication 2010). The situation is aggravated by the fact that more than 95% of Egypt's water resources are generated outside its territory. UNDP Climate Change Adaptation states that: "Although we can not yet predict the impact of climate change on the Nile Basin, there are indications that the impacts will be significant. The studies have indicated that the following areas [in Egypt] are the most vulnerable in order of severity and certainty of results: agricultural areas, coastal zones, aqua-culture and fisheries, water resources, human habitat and settlements, and human health" (UNDP. Climate Change Adaptation. Egypt 2019).

These are serious arguments in favor of permanent satellite monitoring of the Nile River and the Nile Valley. Normalized Difference Vegetation Index (NDVI) as well as other land cover and water surface indices can be used to follow seasonal and interannual variability of the state of vegetation or even desertification along the whole length of the Nile. We already demonstrated such capabilities of high-resolution satellite remote sensing on the example of the Amu Darya River, the largest river in Central Asia, which length is about 1450 km from the confluence of the Pyandzh and Vakhsh Rivers to the Aral Sea (Kostianoy et al. 2013). High resolution satellite imagery allows to monitor the areas around the Amu Darya River, to control desertification processes, to calculate the normalized difference vegetation index (NDVI) and the normalized difference water index (NDWI) for further seasonal and interannual analysis and estimates of the state of vegetation and water capacity in the Amu Darya River area. This information is of vital importance for agriculture and water management in Turkmenistan and Uzbekistan—the countries where deserts occupy most of the territory as in Egypt.

In this chapter, we show capabilities of high resolution optical imagery to monitor the whole area of the Nile Valley and seasonal/interannual variability of NDVI as a measure of vegetation health. We did this analysis for the whole Egypt, the area around the Nile Valley, the Nile Delta, and two important agricultural areas of Egypt located along the Nile in the southern and northern part of the country.

13.2 Data and Methods

To show peculiarities of vegetation along the Nile River on the territory of Egypt we divided the area of the Nile Delta into 6 frames (NN 1-6), the fame N 7 covers the area of the southern part of the Suez Canal, and the main course of the Nile is divided into 12 frames (NN 8-19) from the Nile Delta to the border with Sudan (see Fig. 13.1). This division is explained by high-resolution optical imagery acquired by OLI Landsat-8 that we used for the analysis. Landsat-8 was launched in February 2013 with a 16-day repeat cycle. Operational Land Imager (OLI) installed at Landsat-8 has a swath of 185 km and collects data from 9 spectral bands with a spatial resolution of 30 m. The approximate scene size is 170 km in the north-south direction and 183 km in the east-west direction. Landsat data are used in many disciplines, especially: land cover, agriculture, forest management, natural disasters, water management, ecosystems, biodiversity, ocean/marine studies, urban development, and climate change. For the present study, most of the OLI Landsat-8 imagery were collected in July–September 2018, and in many cases, two neighboring scenes were used to build frames shown below. In these cases, two dates are indicated in the figure caption.

To trace seasonal and interannual variability of the vegetation state in the Nile Valley and the Nile Delta, we used Normalized Difference Vegetation Index (NDVI) which quantifies vegetation by measuring the difference between near-infrared (which vegetation strongly reflects) and red light (which vegetation absorbs). NDVI is a dimensionless parameter which ranges from -1 to $+1$. It is very easy to analyze because if an NDVI value is close to $+1$, there is a high possibility that the land cover has dense green leaves. However, when NDVI is close to zero, there are no green leaves on the land cover, and it can be characterized a desert, or it can even be an urbanized area. Thus, an increase or a decrease of the NDVI value with time may be caused, on one hand, by the state of the vegetation, but on the other hand, by an increase or a decrease of the cultivated or irrigated area; desertification or urbanization processes.

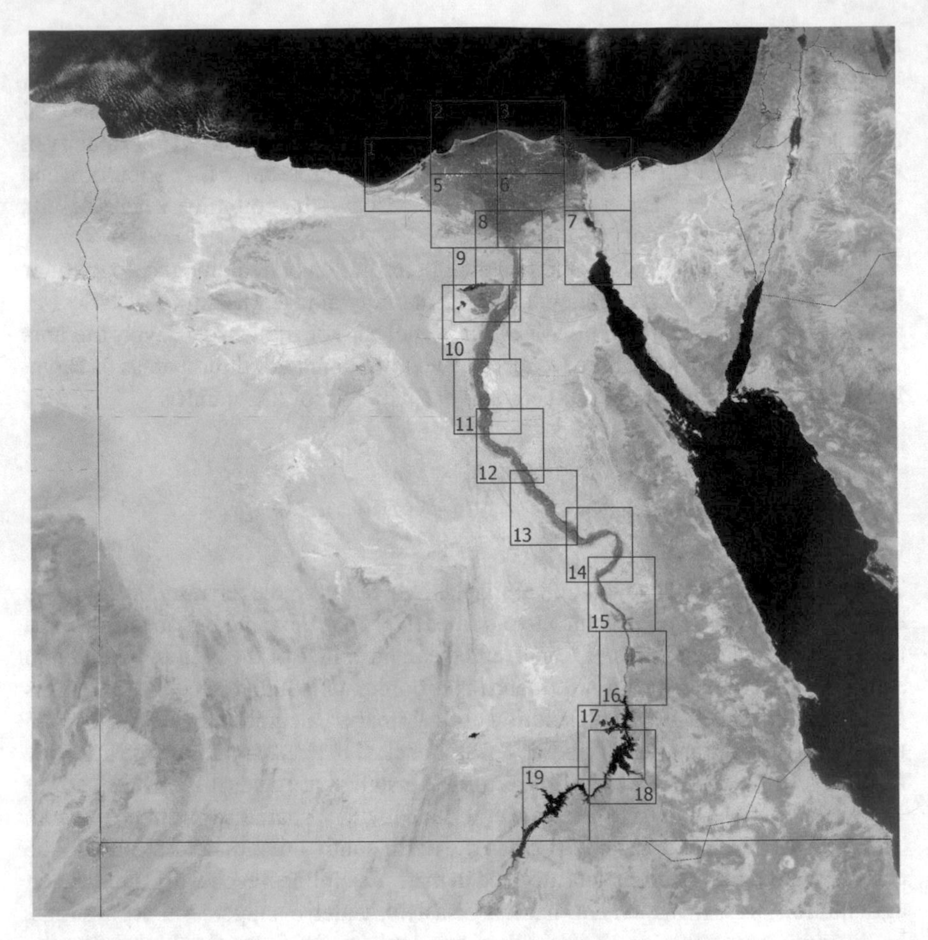

Fig. 13.1 Satellite view of Egypt, VIIRS-SNPP, 8 July 2018, RGB composite, channels 1-4-3

Analyses and visualization of the NDVI monthly data obtained from MODIS-Terra from February 2000 to December 2018 were produced with the NASA Giovanni online data system v.4.30, developed and maintained by the NASA Goddard Earth Sciences Data and Information Services Center (GES DISC) (Acker and Leptoukh 2007). The spatial resolution of these data is 0.05°. To compare seasonal and interannual variability of vegetation in different parts of Egypt, we calculated NDVI for the whole Egypt, the area around the Nile Valley, the Nile Delta, and for two important agricultural areas in Egypt located along the Nile in the southern and northern part of the country. The first one is located eastward of the Nile and the line between cities of Faris and Nagaa Al Hajar, the second one is located westward of the Nile and the line between cities of Al Maqatfiyyah and Beni Suef.

13.3 High Resolution Optical Imagery

Figures 13.2, 13.3, 13.4, 13.5, 13.6, 13.7, 13.8, 13.9, 13.10, 13.11, 13.12, 13.13, 13.14, 13.15, 13.16, 13.17, 13.18, 13.19 and 13.20 show the OLI Landsat-8 satellite images frames NN 1-19 focused on the Nile River, which cover the whole length of the river from the Mediterranean Sea to the border with Sudan (Fig. 13.1). The Nile, a "green belt" of vegetation with a various width along the river, a system of canals, lakes, irrigated fields and deserts are clearly visible in the gallery of high-resolution satellite imagery.

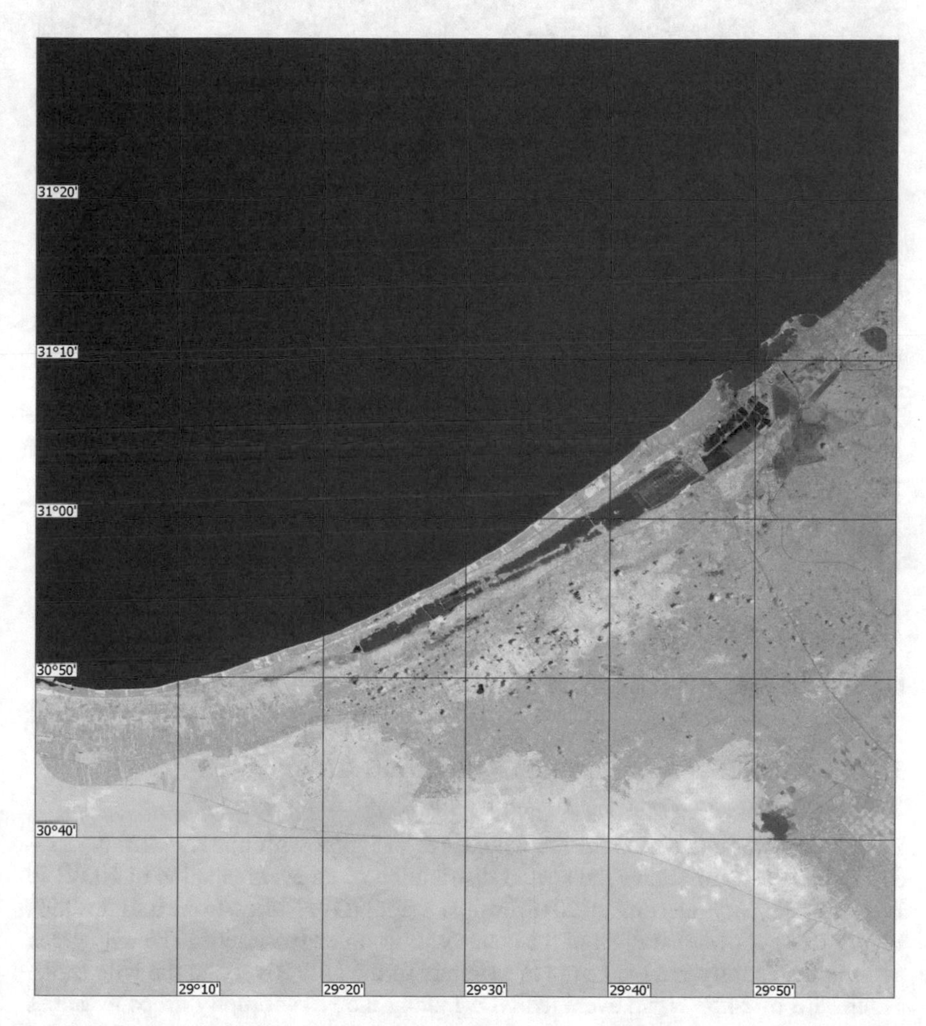

Fig. 13.2 Frame #1, OLI Landsat-8, 14 and 23 July 2018 Hereinafter for location, see Fig. 13.1

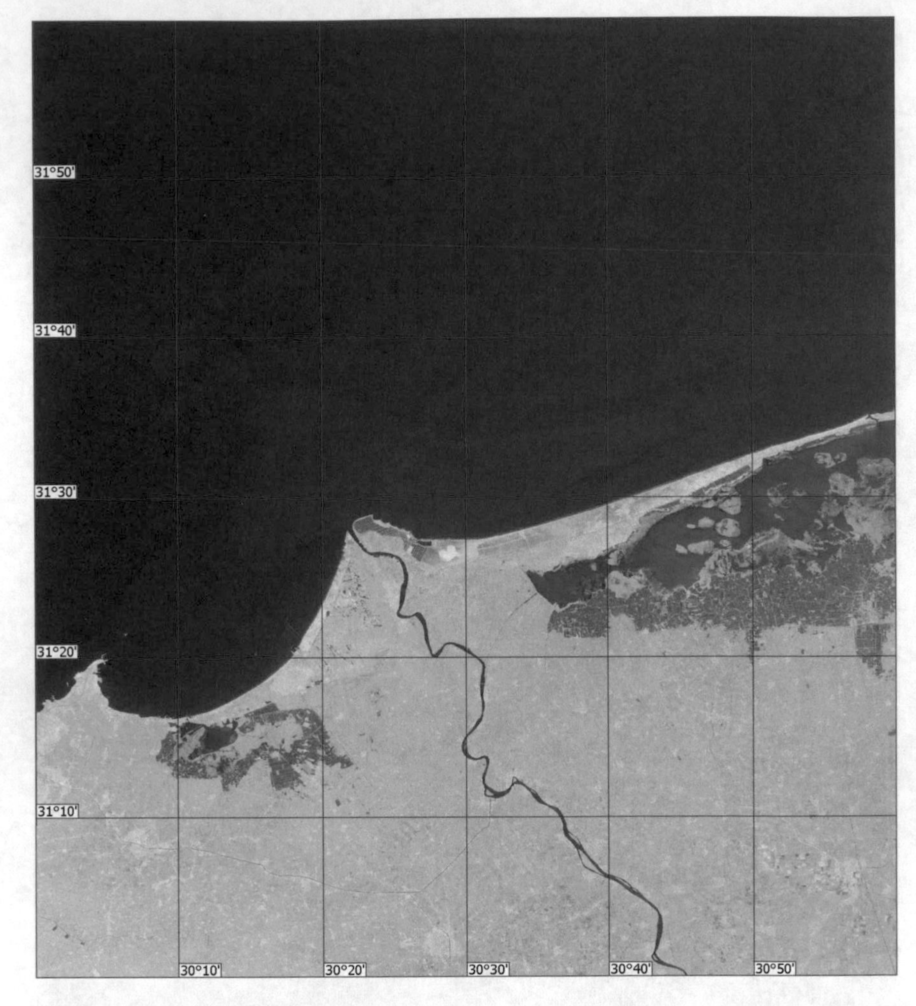

Fig. 13.3 Frame #2, OLI Landsat-8, 23 July 2018

13.4 Normalized Difference Vegetation Index

First, NDVI was calculated for the whole area of Egypt within; 31.5°–22° N, 25°–37° E. Figure 13.21 shows the spatial distribution of an average value of NDVI in Egypt for January–December 2018. In this map, NDVI varies from 0.01 to 0.99. Figure 13.21 shows that most of the country is occupied by deserts. The only green areas in the country are located in the vicinity of the Nile River and the Nile Delta. Southward of 24.5° N, an averaged NDVI along the Nile abruptly drops to values which are characteristic for desert areas.

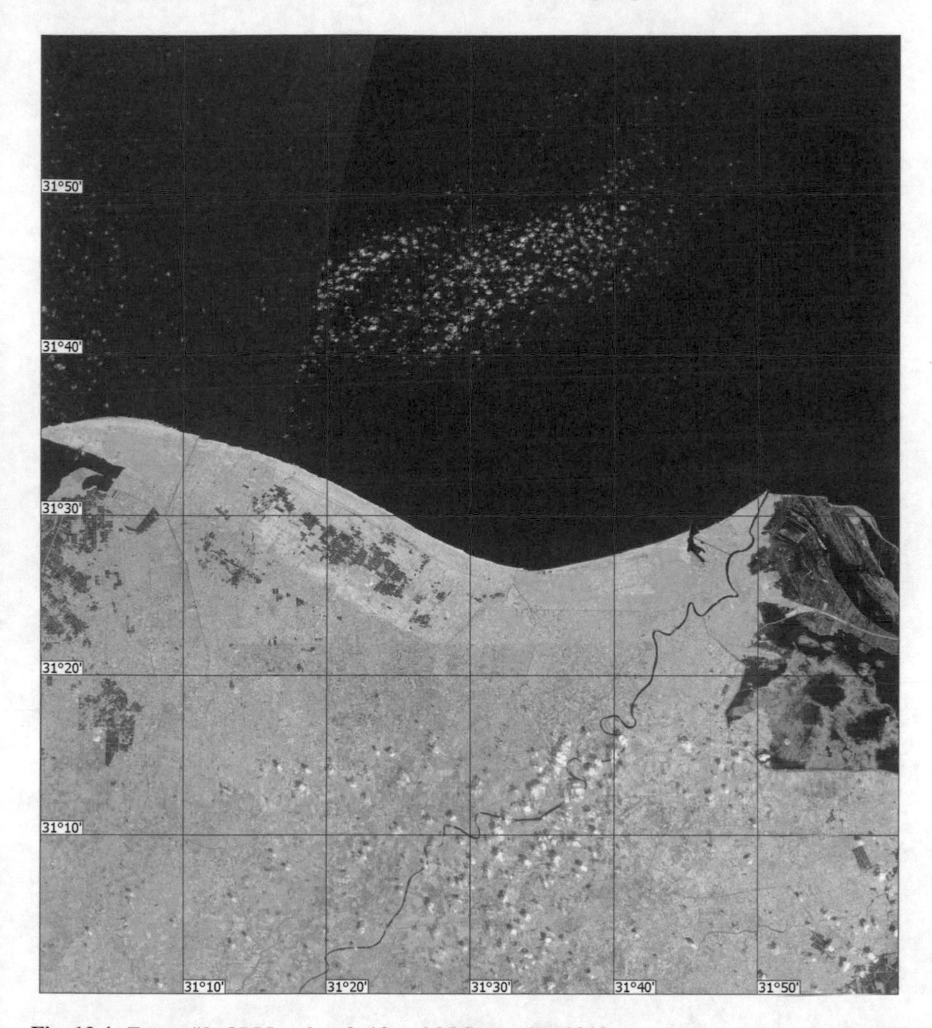

Fig. 13.4 Frame #3, OLI Landsat-8, 18 and 25 September 2018

Figure 13.22 shows the seasonal and interannual variability of NDVI for the whole territory of Egypt for 2000–2018. NDVI varies in the limits between 0.106 and 0.124, which is explained by deserts which occupy most of the country. In general, NDVI slowly increases with a linear rate of 0.00018/year. Almost two decades of NDVI observations allow to detect two cycles of about ten years which are visible on the graph, the first one ending by about 2010, and the second one ending likely by 2020. A characteristic feature of a seasonal cycle of NDVI for the territory of Egypt is a double minimum in May and October in the dry season and a single maximum in February in a wet season (Fig. 13.23).

Fig. 13.5 Frame #4, OLI Landsat-8, 27 June 2017

The same procedure was repeated for the Nile River within 31.5°–22° N, 30°–33.5° E. Figure 13.24 shows the spatial distribution of an average value of NDVI for the Nile River area for January–December 2018. On this map, the NDVI varies from 0.01 to 0.99. Figure 13.24 shows that most of the area around the Nile is occupied by deserts. The green areas are associated with the Nile River and the Nile Delta. Southward of 24.5° N an averaged NDVI along the Nile abruptly drops to values which are characteristic for desert areas.

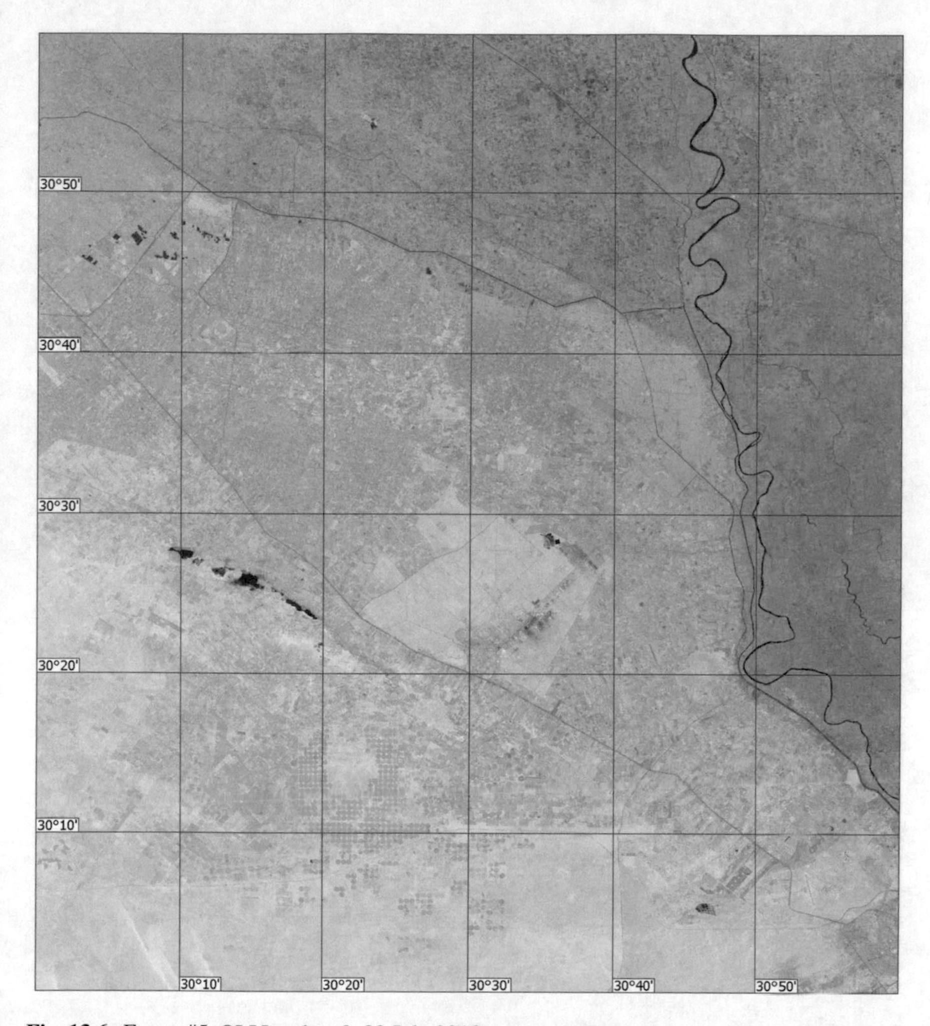

Fig. 13.6 Frame #5, OLI Landsat-8, 23 July 2018

Figure 13.25 shows the seasonal and interannual variability of NDVI for the area of the Nile River for 2000–2018. NDVI varies in the limits between a bit higher values of 0.123 and 0.161, which is explained by a reduced area of deserts included in the analysis. In general, NDVI slowly increases with a linear rate of 0.00025/year. Two decades of NDVI observations allow to detect the same cycles of about ten years which are visible on the graph, the first one ending by about 2010, and the

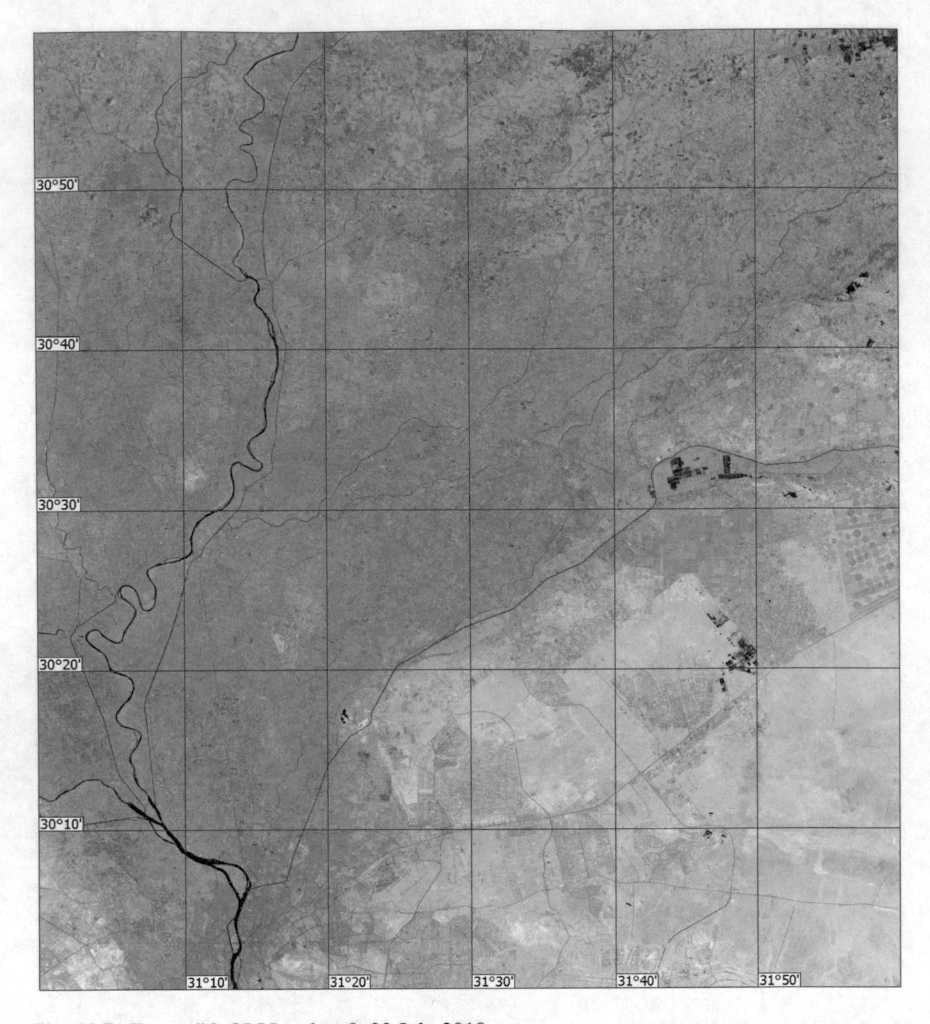

Fig. 13.7 Frame #6, OLI Landsat-8, 23 July 2018

second one ending likely by 2020. However, in this case, both are modified by intradecade variability, which results in additional minimums around 2004 and 2016. A characteristic feature of a seasonal cycle of NDVI for the area around the Nile is a double minimum in May and October in the dry season and a single maximum in February in a wet season (Fig. 13.26). In this case, the double minimum structure is more pronounced as in July and August NDVI values rise notably again.

Fig. 13.8 Frame #7, OLI Landsat-8, 10 and 17 August 2018

The same procedure was applied to the Nile Delta within 32°–30° N, 29.5°–32.5° E. Figure 13.27 shows the spatial distribution of an average value of NDVI for the Nile Delta for January–December 2018. On this map, NDVI varies from 0.01 to 0.99, but most of the region has high values of NDVI that is related to natural vegetation and highly developed agricultural fields.

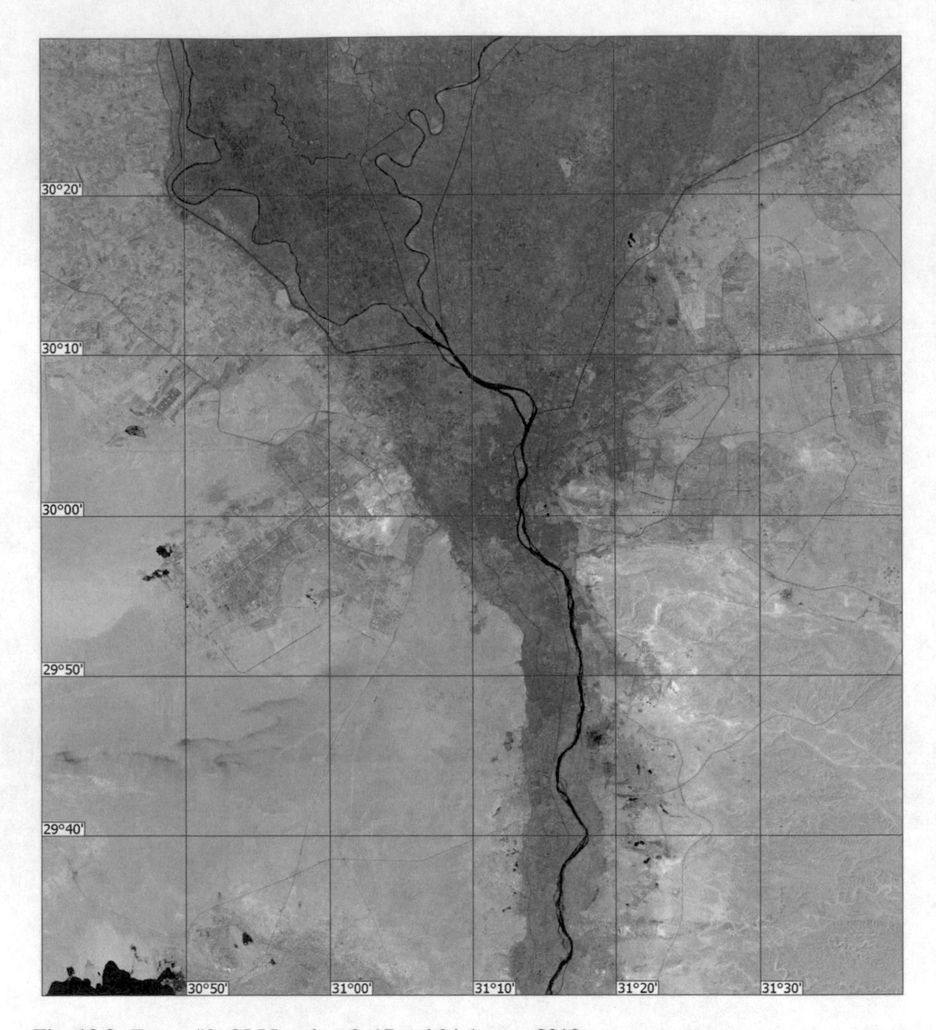

Fig. 13.9 Frame #8, OLI Landsat-8, 17 and 24 August 2018

Figure 13.28 shows seasonal and interannual variability of NDVI for the area of the Nile Delta for 2000–2018. NDVI varies in the limits between (a much higher values) 0.226 and 0.423, which is explained by a highly developed vegetation in the Nile Delta. In general, NDVI slowly increases with a linear rate of 0.0015/year, which is an order higher than in previous cases. In total during 20 years, we have an increase of NDVI by about 10%. Decadal cycles are still visible in the interannual variability of NDVI, but both are modified by intradecade variability. Again, we

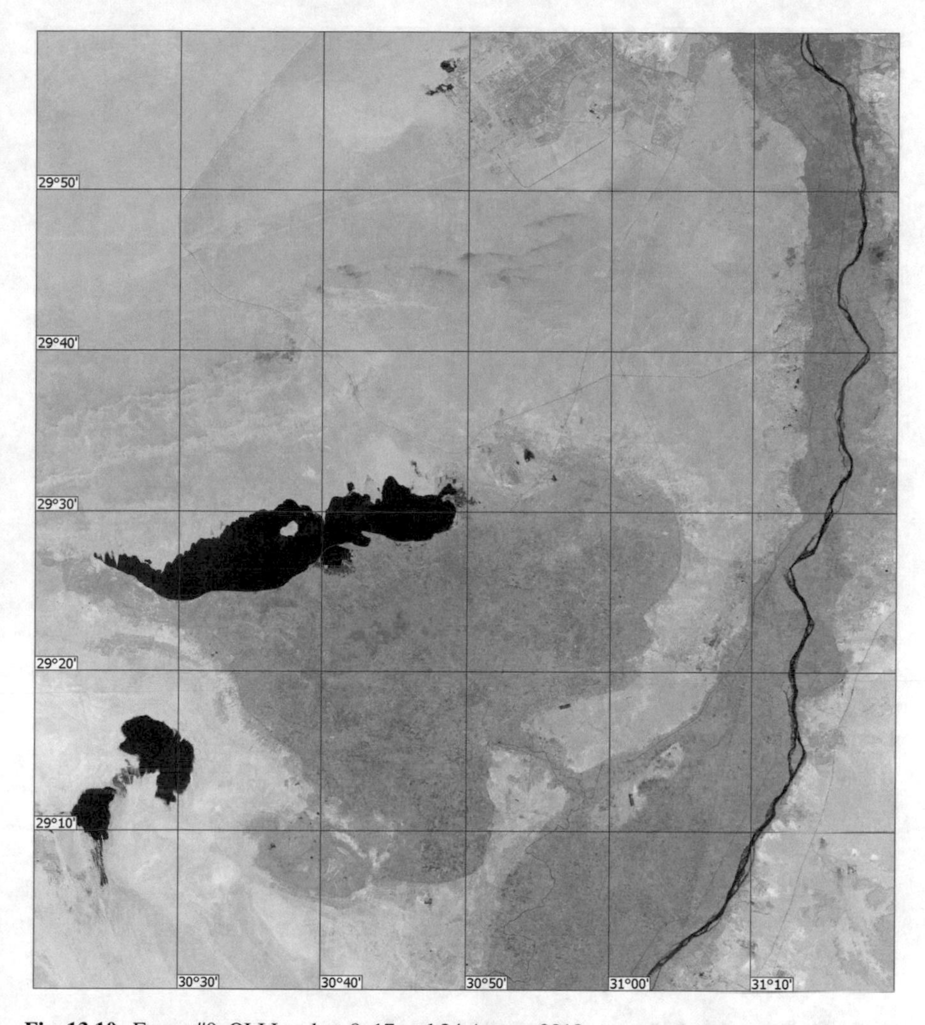

Fig. 13.10 Frame #9, OLI Landsat-8, 17 and 24 August 2018

observe a double minimum in May and September–October in the dry season and a single prolonged maximum in January–February in a wet season (Fig. 13.29). In this case, the double minimum structure is also very pronounced as in July and August NDVI values rise by a half of a seasonal amplitude of NDVI variability.

The same procedure was applied to an important agricultural area westward of the Nile and the line between cities of Al Maqatfiyyah and Beni Suef within 29.6°–29° N, 30.3°–31.3° E (see Fig. 13.10 for location). Figure 13.30 shows the seasonal and interannual variability of the averaged value of NDVI for this agricultural area for

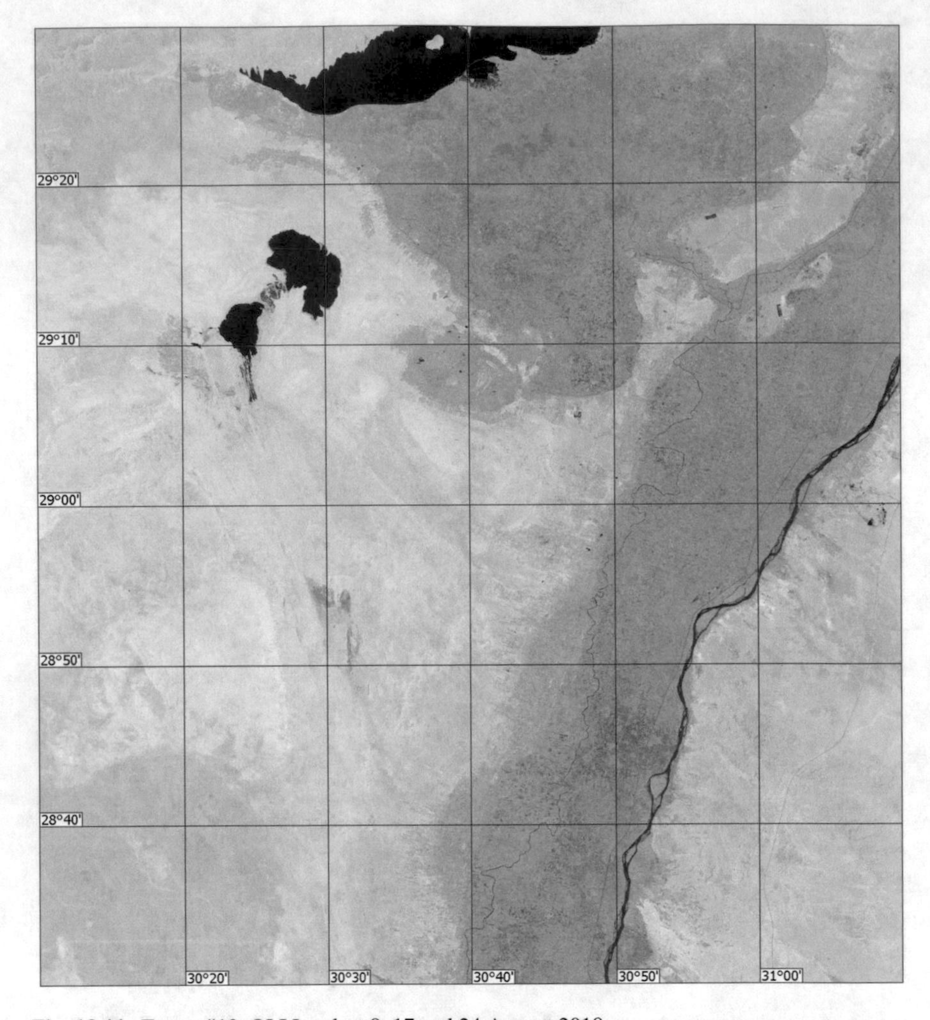

Fig. 13.11 Frame #10, OLI Landsat-8, 17 and 24 August 2018

2000–2018. NDVI varies in the limits between a bit lower values than in the Nile Delta: 0.186 and 0.339. High NDVI values are explained by agricultural fields located in this region. In this case, NDVI increases with a linear rate of 0.0010/year, which is comparable with the Nile Delta. In total during 20 years, we have an increase of NDVI by about 10%. Decadal cycles are less visible in the interannual variability of NDVI,

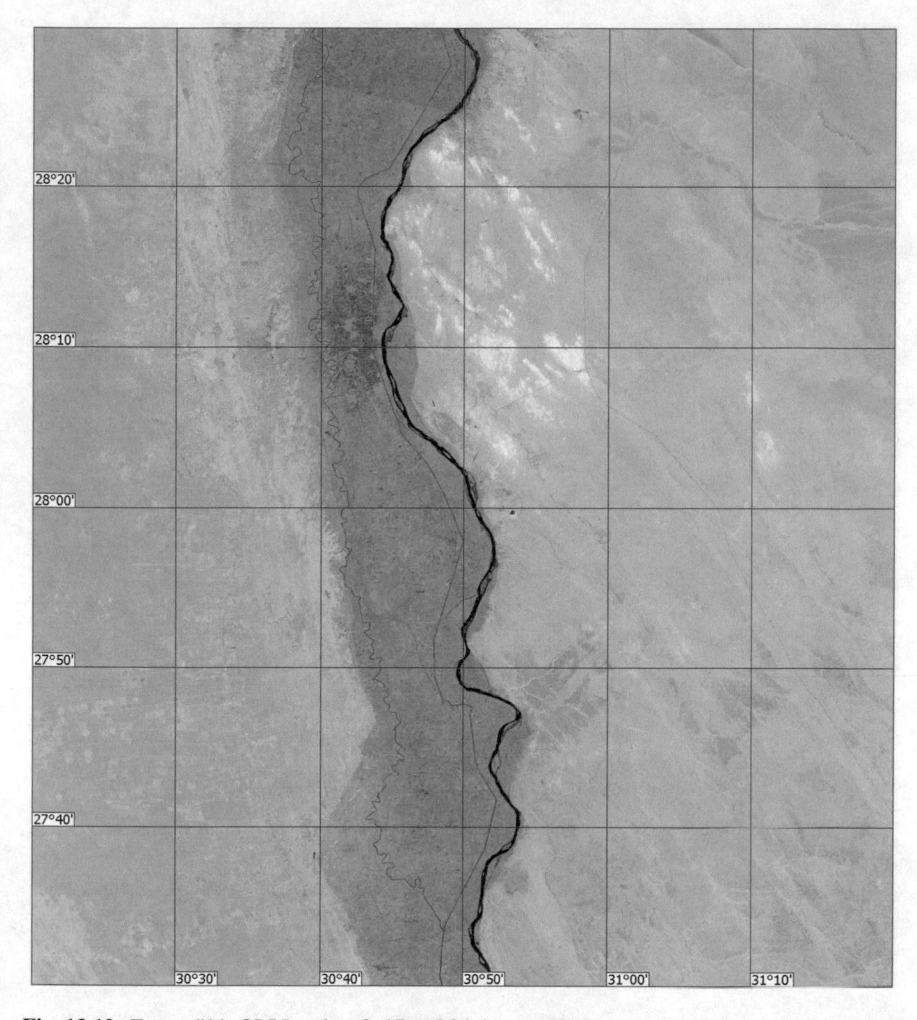

Fig. 13.12 Frame #11, OLI Landsat-8, 17 and 24 August 2018

and linear regression fits better the NDVI variability. In the seasonal variability, we observe a double minimum in May and September in the dry season and a single maximum in February in a wet season (Fig. 13.31). In this case, the double minimum structure is less pronounced than in the Nile Delta because of a little single maximum in July.

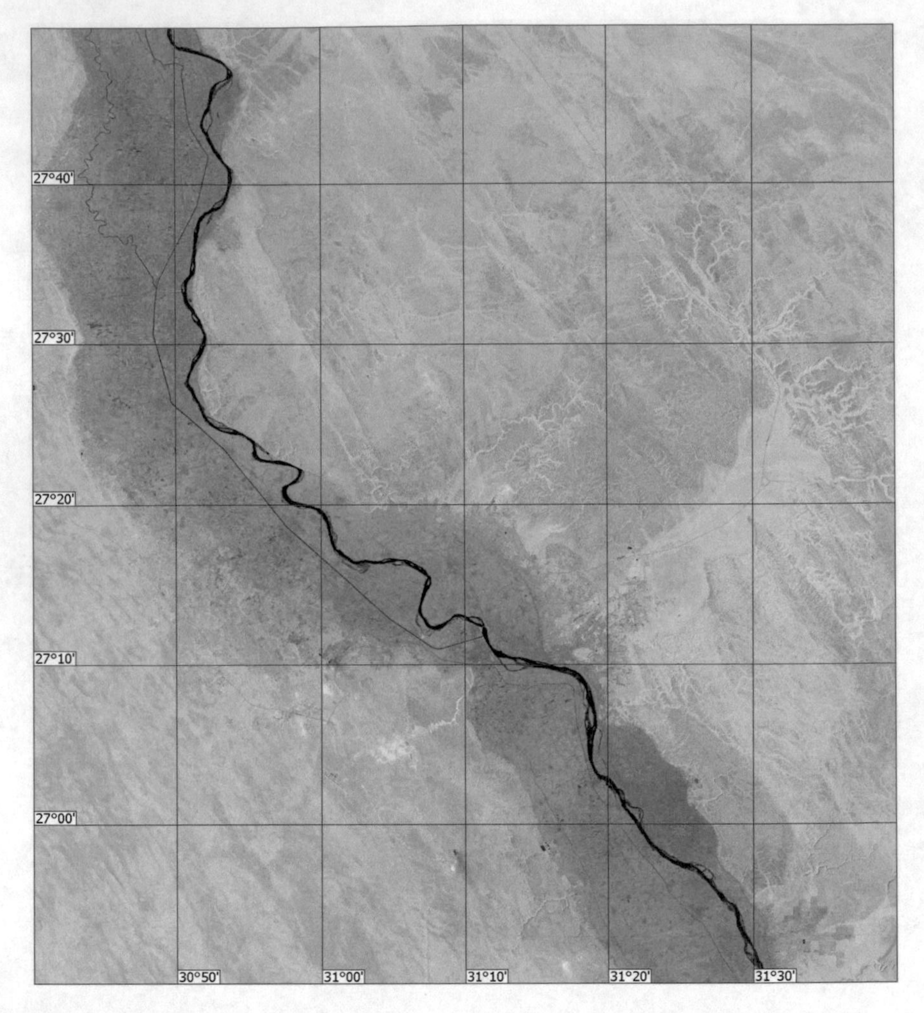

Fig. 13.13 Frame #12, OLI Landsat-8, 17 August 2018

For the Beni Suef agricultural area we tried to understand the reason for a 10% rise of NDVI from 2000 to 2018. The observed increase may be caused by two factors: an increase in the area of agricultural fields and an increase of NDVI in the same agricultural fields, or by a combination of both factors. We calculated the spatial distribution of an average value of NDVI for this region for 2000 (Fig. 13.32a) and

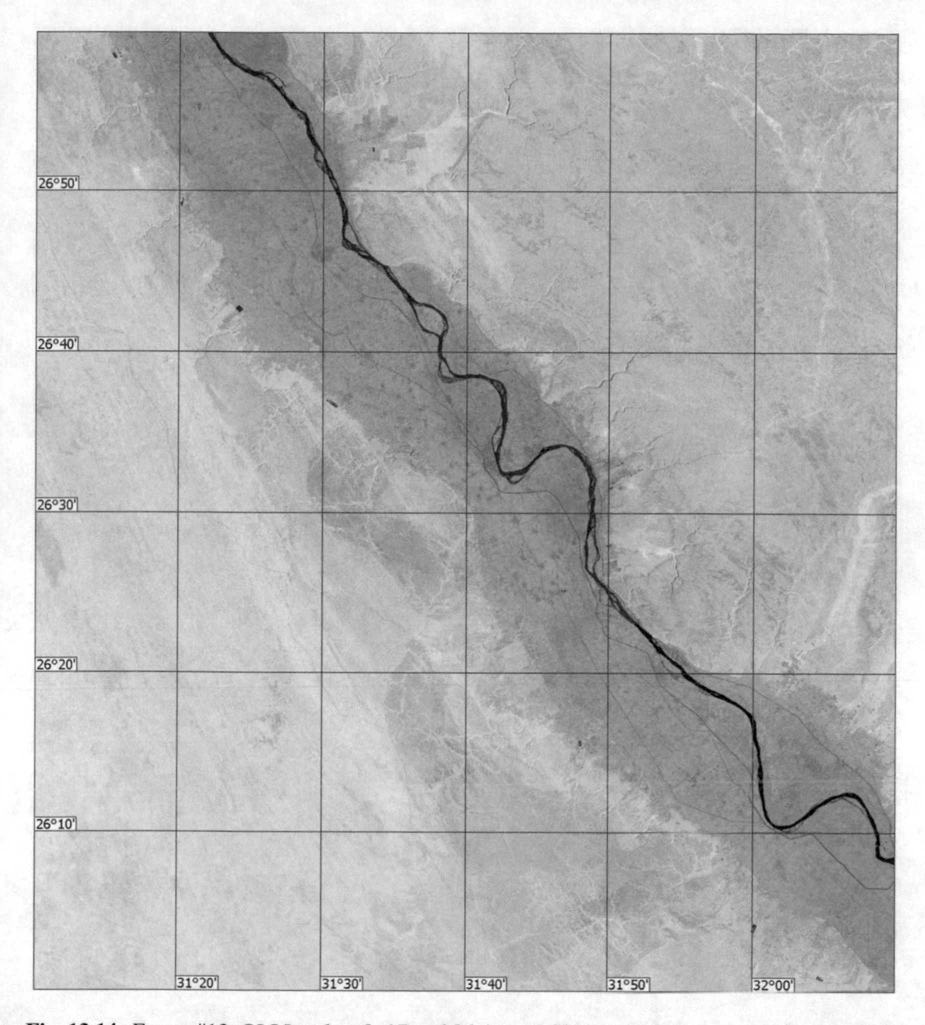

Fig. 13.14 Frame #13, OLI Landsat-8, 17 and 26 August 2018

compared it with the year 2018 (Fig. 13.32b). A pixel by pixel comparison of NDVI values for both years showed that there is a little expansion of the area of agricultural fields and a little increase of NDVI values in the fields that were already arable in 2000. Thus, both factors caused a 10% rise of NDVI from 2000 to 2018 in this region.

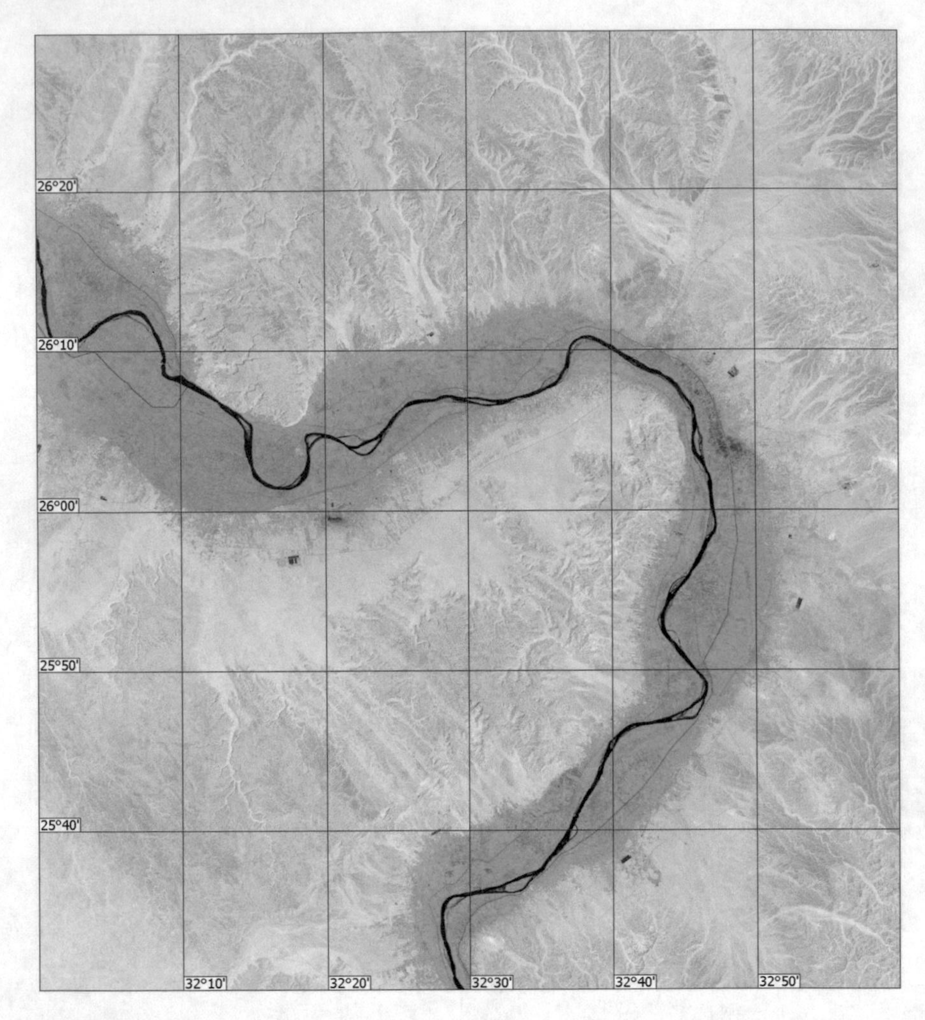

Fig. 13.15 Frame #14, OLI Landsat-8, 26 August 2018

The same procedure was applied to another important agricultural area located southward from the Beni Suef agricultural zone, eastward of the Nile and the line between cities of Faris and Nagaa Al Hajar within 24.7°–24.3° N, 32.8°–33.5° E (see Fig. 13.17 for location). Figure 13.33 shows the seasonal and interannual variability of the averaged value of NDVI for this agricultural area for 2000–2018. NDVI varies in the limits between a bit lower values than in the Beni Suef region: 0.140 and 0.212. Relatively high NDVI values are explained by agricultural fields located in

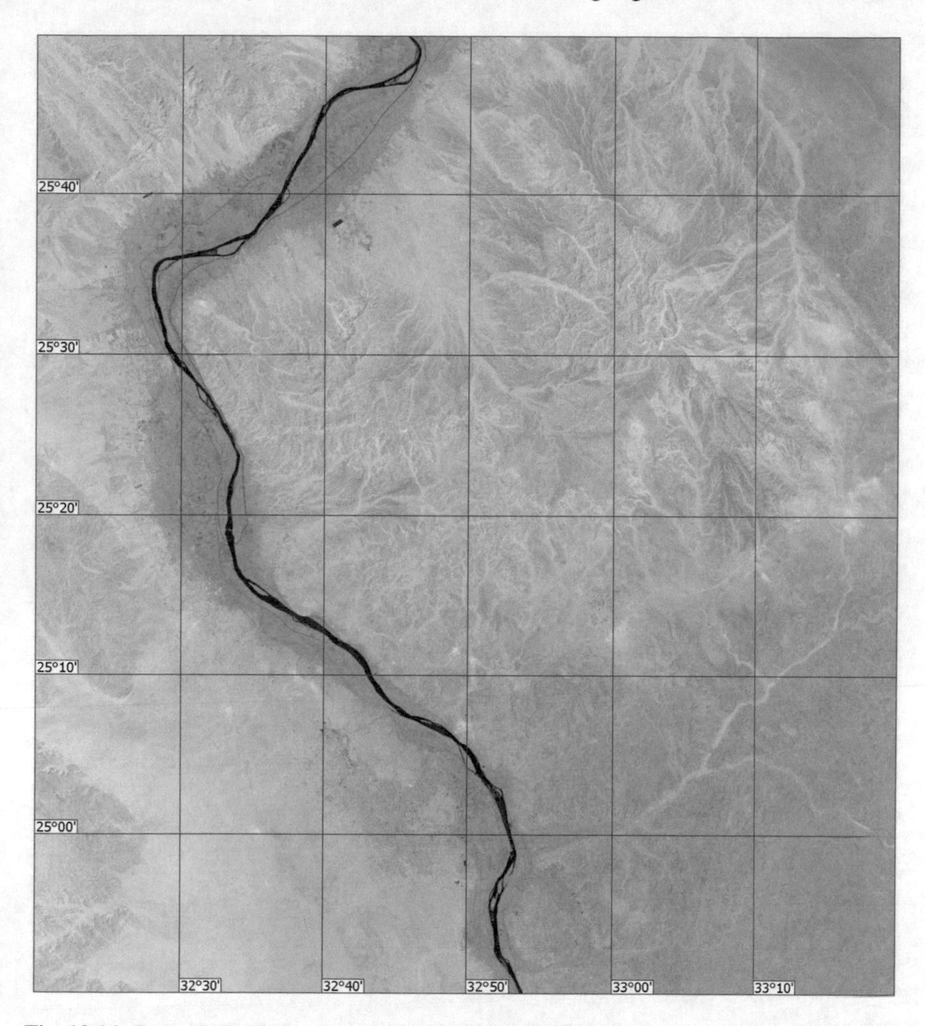

Fig. 13.16 Frame #15, OLI Landsat-8, 19 and 26 August 2018

this region. In this case, NDVI increases with a linear rate of 0.0022/year, which is the highest rate for the five cases under investigation. In total during 20 years, we have an increase of NDVI by more than 25%. Decadal cycles are not visible in the interannual variability of NDVI because linear regression dominates in the NDVI variability (Fig. 13.33). In the seasonal variability, we observe an absolute different behavior of NDVI with one minimum in April and a very prolonged (7 months) period of high NDVI values from August to February (Fig. 13.34).

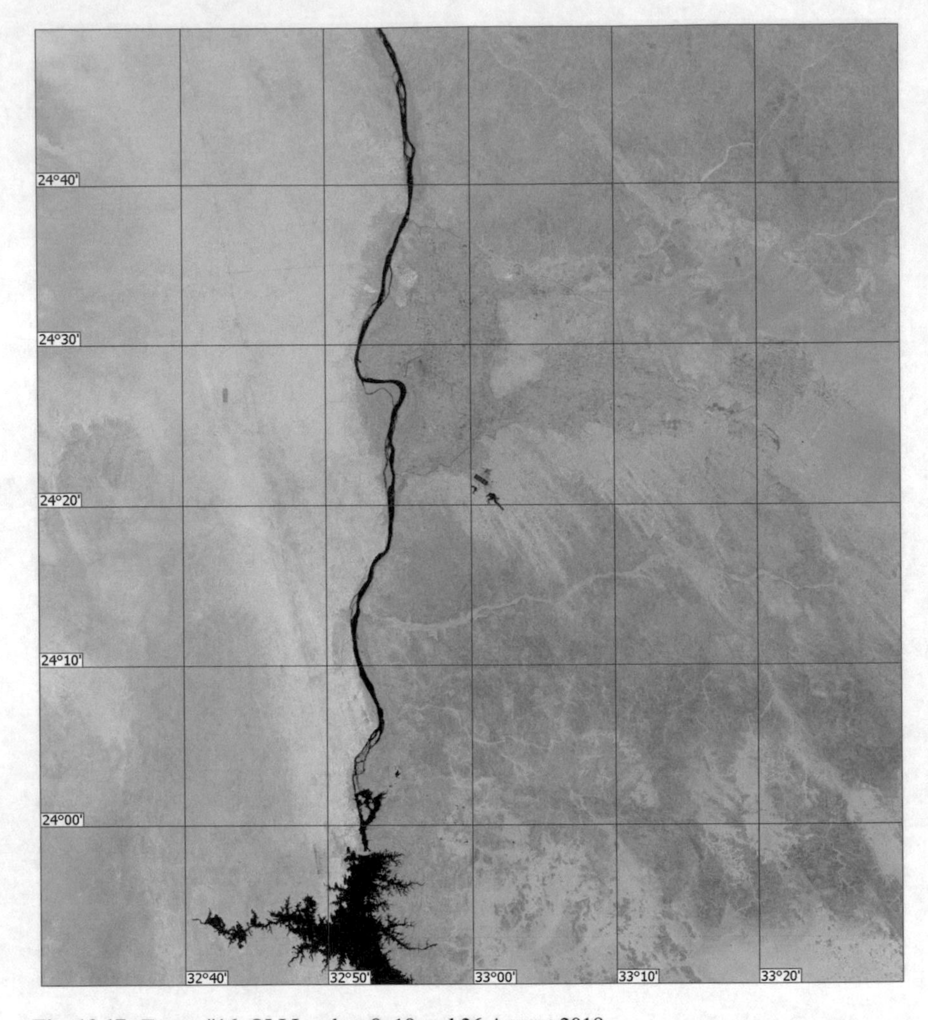

Fig. 13.17 Frame #16, OLI Landsat-8, 19 and 26 August 2018

For this agricultural area, we also tried to understand the reason for a 25% rise of NDVI from 2000 to 2018. As it was mentioned before, the observed increase may be caused by a combination of two factors: an increase in the area of agricultural fields and an increase of NDVI. We calculated the spatial distribution of an average value of NDVI for this region for 2000 (Fig. 13.35a) and compared it with the year 2018

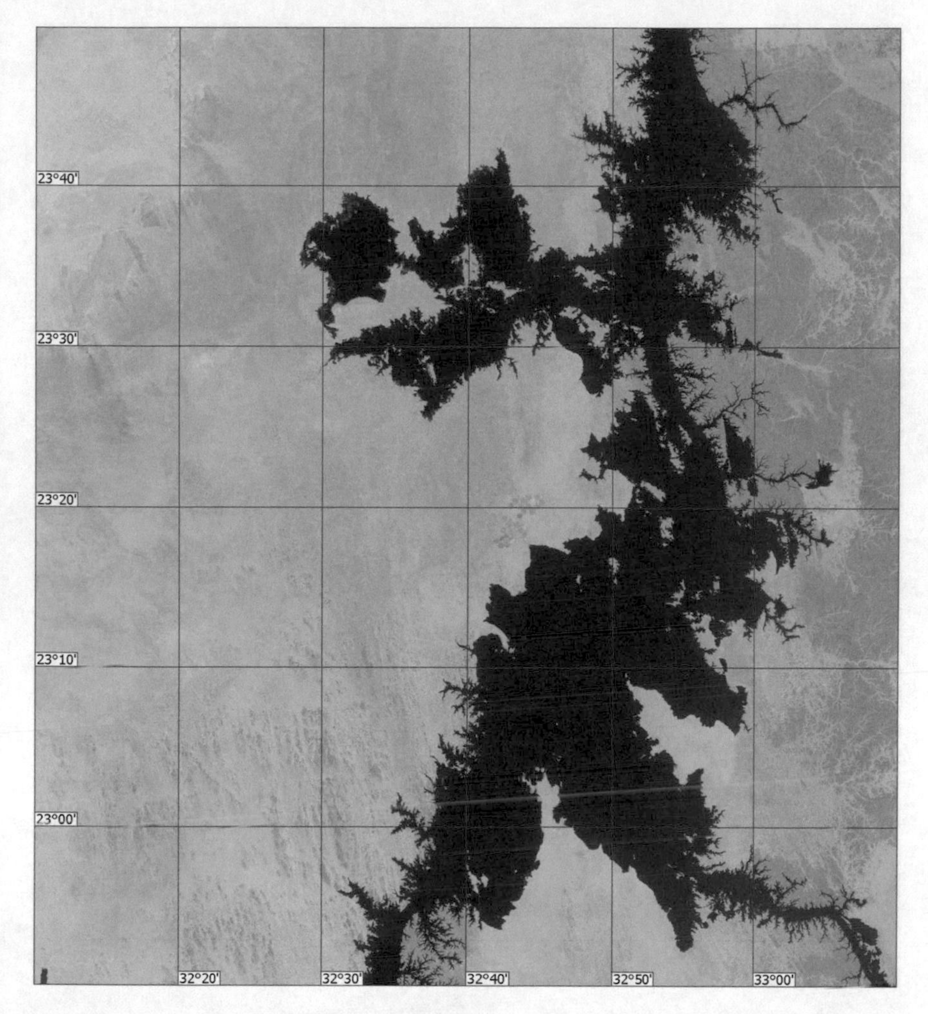

Fig. 13.18 Frame #17, OLI Landsat-8, 19 and 26 August 2018

(Fig. 13.35b). A pixel by pixel comparison of NDVI values for both years showed that there is a notable expansion of the area of agricultural fields directed to the east, and a little increase of NDVI values in the fields that were already arable in 2000. Thus, both factors caused a 25% rise of NDVI from 2000 to 2018 in this region, but the expansion of agricultural fields dominates.

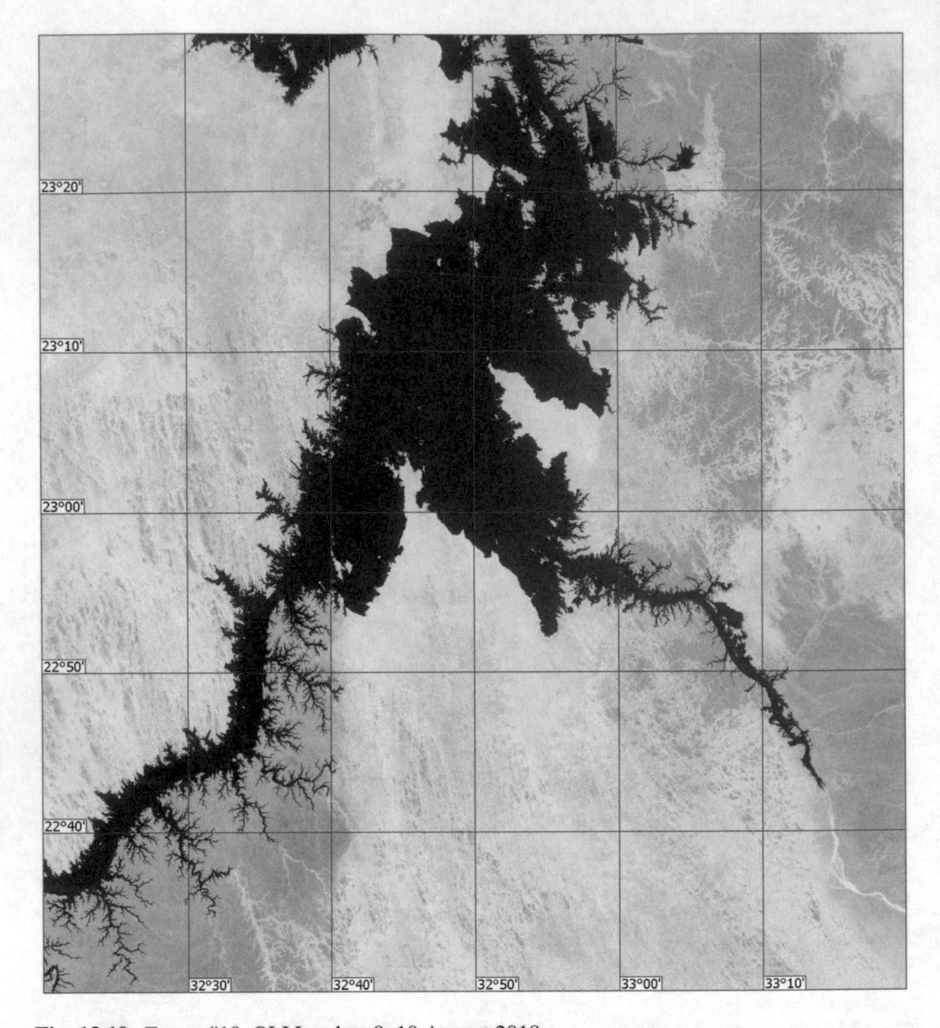

Fig. 13.19 Frame #18, OLI Landsat-8, 19 August 2018

13.5 Conclusions

In this chapter, the authors demonstrated modern capabilities of satellite remote sensing technologies in environmental monitoring of the River Nile Valley, the Nile Delta, and important agricultural zones. High-resolution satellite imagery is very effective in monitoring of the whole Nile River area because today we can use,

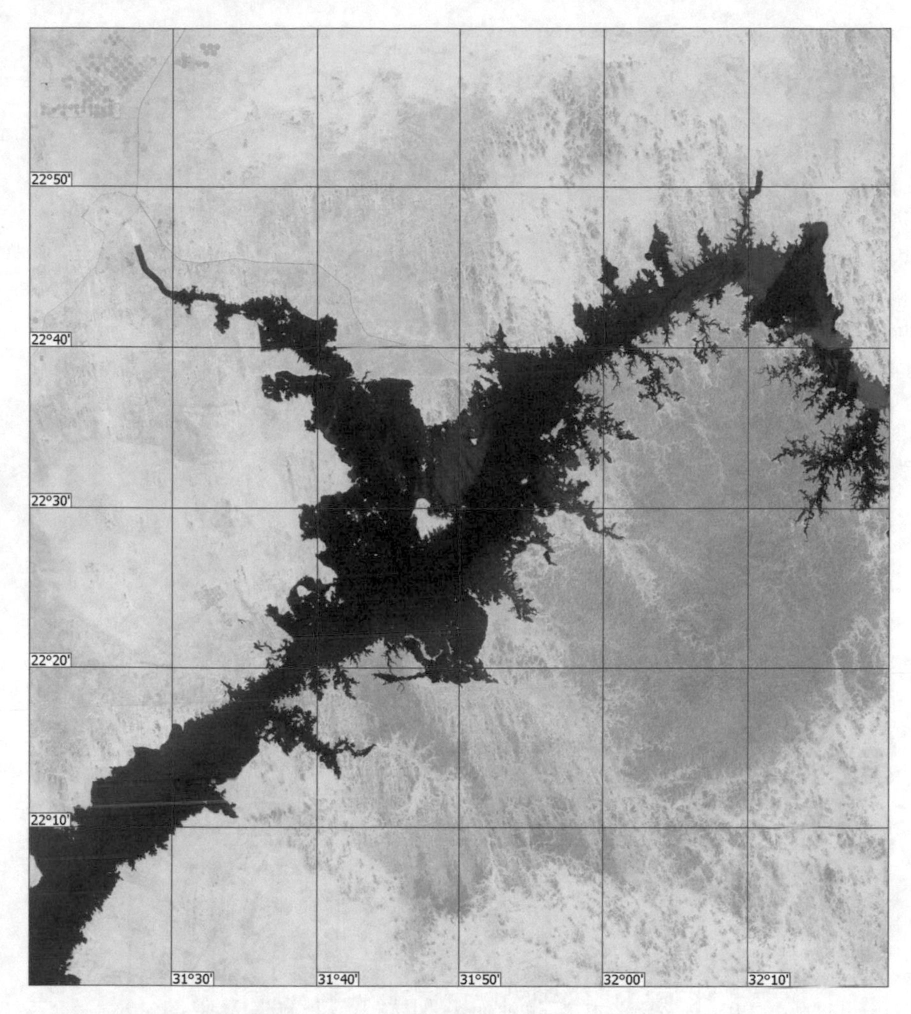

Fig. 13.20 Frame #19, OLI Landsat-8, 26 August 2018

for example, freely available OLI Landsat-8 optical imagery to trace seasonal and interannual variability of the vegetation or desertification processes in different parts of the river with a spatial resolution of 30 m. This information can be used to calculate different vegetation and water indices which are widely used for assessment of the state of the agricultural fields and natural vegetation along the Nile and in other parts of Egypt. To show such kind of analysis we used NDVI monthly data obtained from

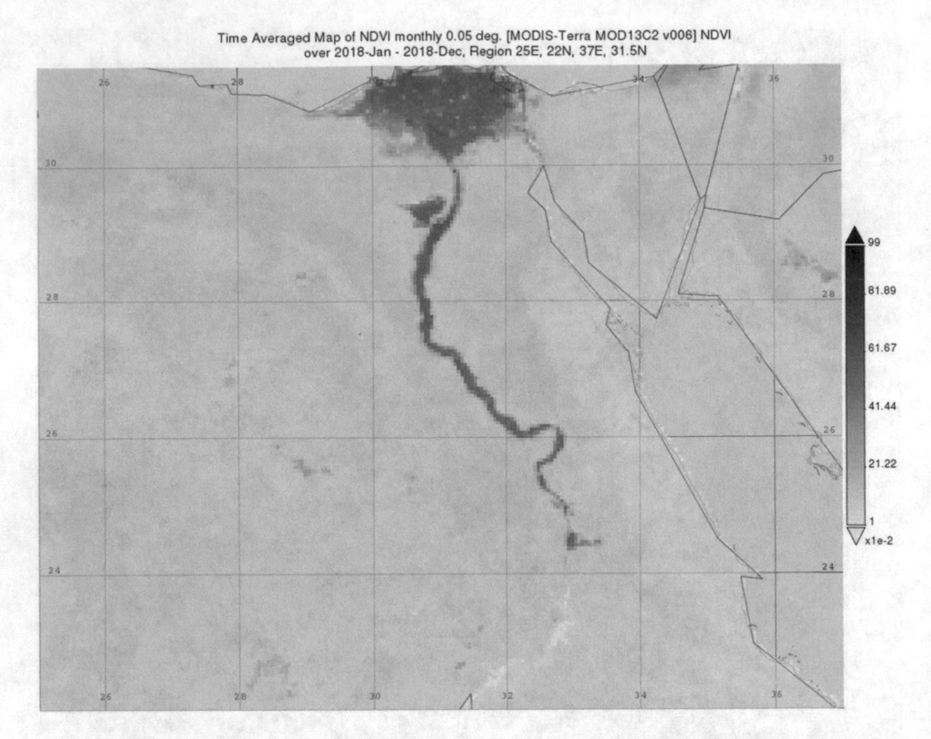

Fig. 13.21 Spatial distribution of NDVI in Egypt with an average value of NDVI for 2018 (Giovanni NASA GES DISC)

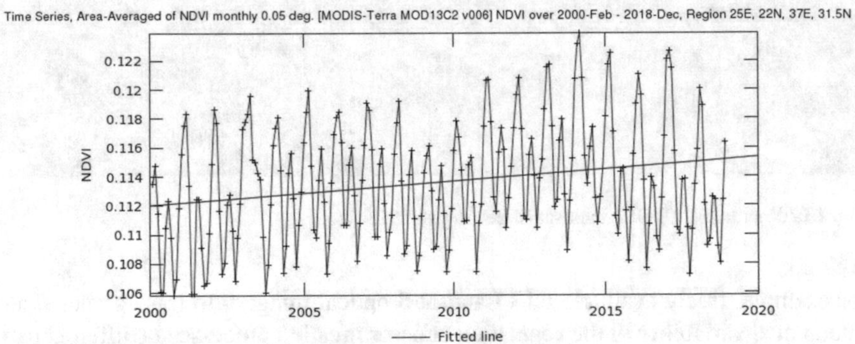

Fig. 13.22 Seasonal and interannual variability of Egypt's territory averaged value of NDVI for 2000–2018. Linear regression is shown by the blue line (Giovanni NASA GES DISC)

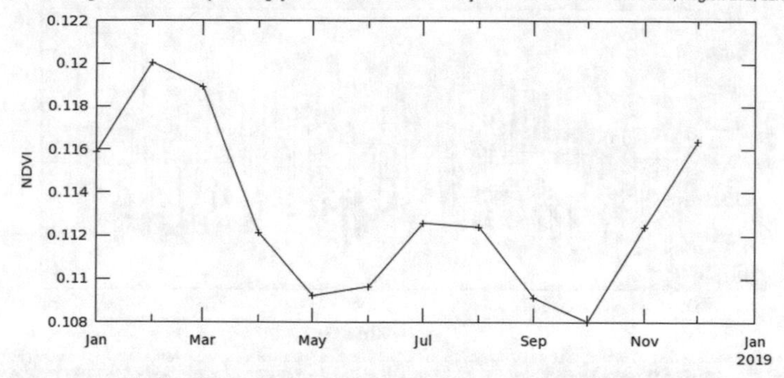

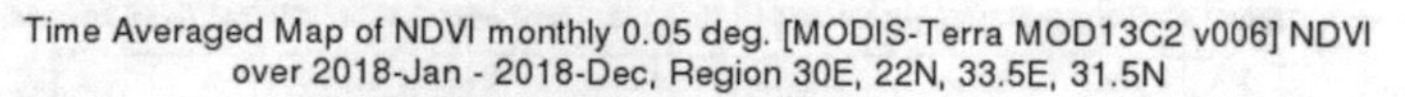

Fig. 13.23 Seasonal variability of Egypt's territory averaged value of NDVI for 2018 (Giovanni NASA GES DISC)

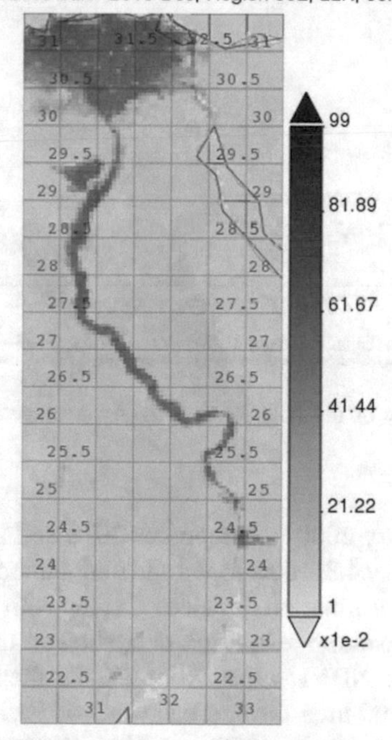

Fig. 13.24 The spatial distribution of NDVI along the Nile River with an average value of NDVI for 2018 (Giovanni NASA GES DISC)

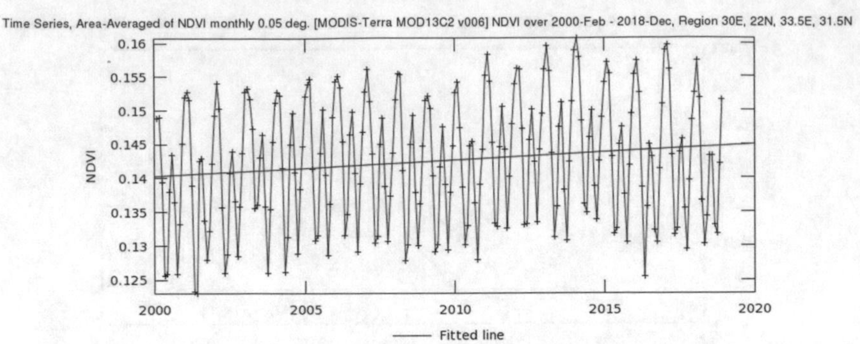

Fig. 13.25 Seasonal and interannual variability of the averaged value of NDVI for the Nile River area for 2000–2018. Linear regression is shown by the blue line (Giovanni NASA GES DISC)

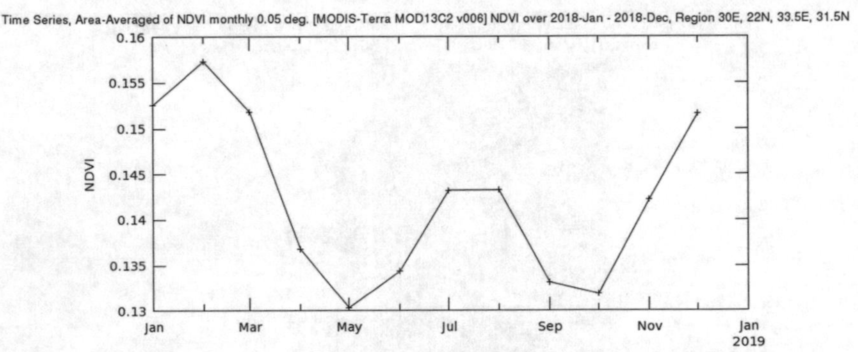

Fig. 13.26 Seasonal variability of the Nile River area. An averaged value of NDVI for 2018 (Giovanni NASA GES DISC)

MODIS-Terra from February 2000 to December 2018 and provided by the NASA Giovanni online data system v.4.30, developed and maintained by the NASA Goddard Earth Sciences Data and Information Services Center with a spatial resolution of about 5 km. Even these moderate resolution data showed that during two decades there is a steady rise in the NDVI values from 2% for the whole country to 25% for the important agricultural area located eastward of Faris and Nagaa Al Hajar. A comparison of irrigated lands in 2000 and 2018 showed that the reason for a

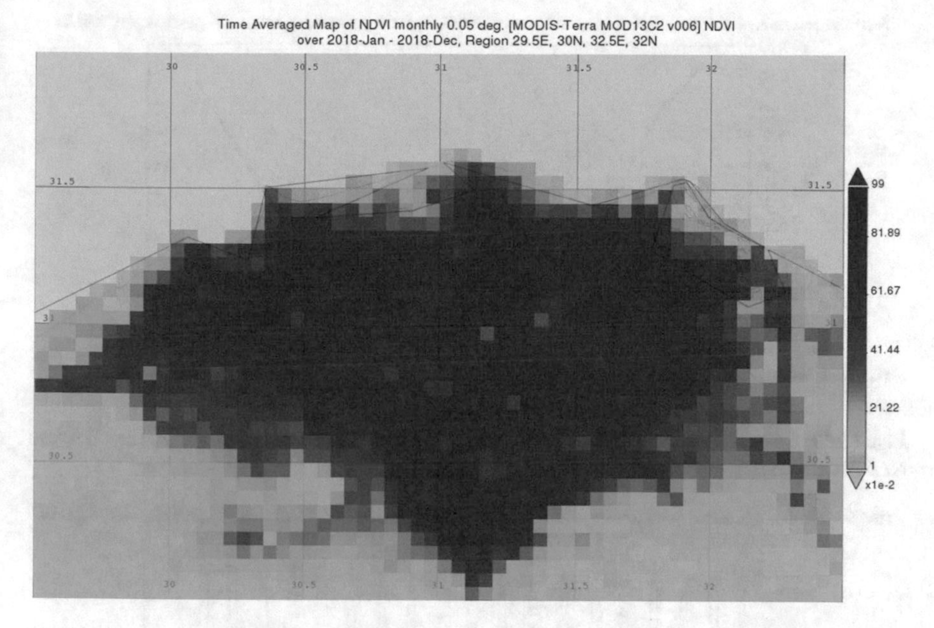

Fig. 13.27 The spatial distribution of NDVI in the Nile Delta. An average value of NDVI for 2018 is presented (Giovanni NASA GES DISC)

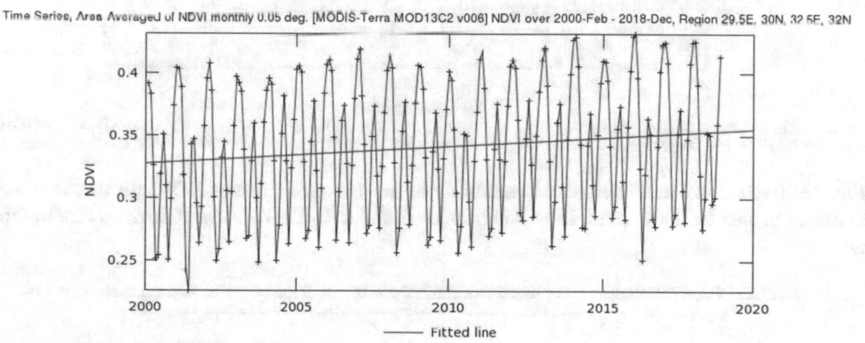

Fig. 13.28 Seasonal and interannual variability of the averaged value of NDVI for the Nile Delta for 2000–2018. Linear regression is shown by the blue line (Giovanni NASA GES DISC)

substantial growth of NDVI in the Beni Suef and Faris–Nagaa Al Hajar agricultural zones is a combination of two factors. The first one is an increase of the area of arable lands and the second one is an increase in the NDVI values in the fields which existed in 2000, which is likely a signature of a higher quality of agriculture.

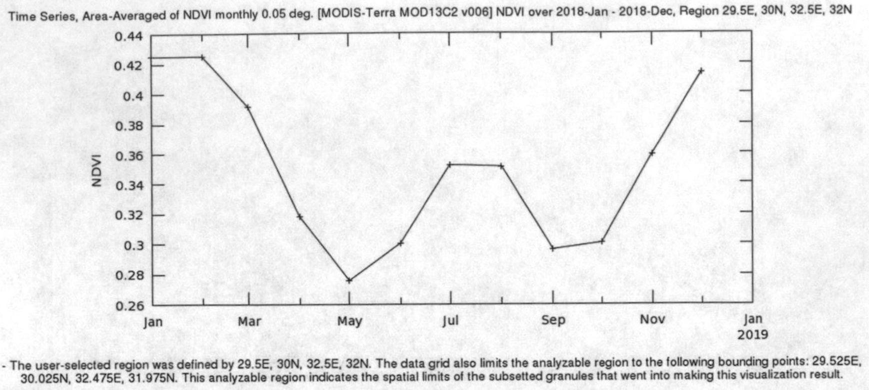

Time Series, Area-Averaged of NDVI monthly 0.05 deg. [MODIS-Terra MOD13C2 v006] NDVI over 2018-Jan - 2018-Dec, Region 29.5E, 30N, 32.5E, 32N
NDVI
0.44
0.42
0.4
0.38
0.36
0.34
0.32
0.3
0.28
0.26
Jan
Mar
May
Jul
Sep
Nov
Jan 2019
- The user-selected region was defined by 29.5E, 30N, 32.5E, 32N. The data grid also limits the analyzable region to the following bounding points: 29.525E, 30.025N, 32.475E, 31.975N. This analyzable region indicates the spatial limits of the subsetted granules that went into making this visualization result.

Fig. 13.29 Seasonal variability of the Nile Delta averaged value of NDVI for 2018 (Giovanni NASA GES DISC)

Time Series, Area-Averaged of NDVI monthly 0.05 deg. [MODIS-Terra MOD13C2 v006] NDVI over 2000-Feb - 2018-Dec, Region 30.3E, 29N, 31.3E, 29.6N
NDVI
0.32
0.3
0.28
0.26
0.24
0.22
0.2
2000
2005
2010
2015
2020
Fitted line
- The user-selected region was defined by 30.3E, 29N, 31.3E, 29.6N. The data grid also limits the analyzable region to the following bounding points: 30.325E, 29.025N, 31.275E, 29.575N. This analyzable region indicates the spatial limits of the subsetted granules that went into making this visualization result.

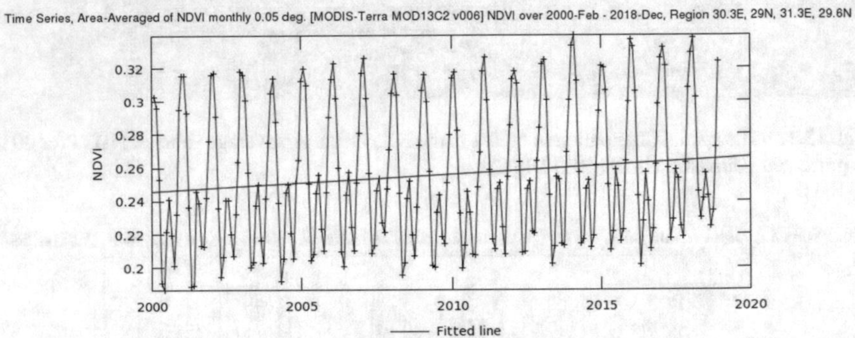

Fig. 13.30 Seasonal and interannual variability of the averaged value of NDVI for the Beni Suef agricultural area for 2000–2018. Linear regression is shown by the blue line (Giovanni NASA GES DISC)

Time Series, Area-Averaged of NDVI monthly 0.05 deg. [MODIS-Terra MOD13C2 v006] NDVI over 2018-Jan - 2018-Dec, Region 30.3E, 29N, 31.3E, 29.6N
NDVI
0.34
0.32
0.3
0.28
0.26
0.24
0.22
Jan
Mar
May
Jul
Sep
Nov
Jan 2019
- The user-selected region was defined by 30.3E, 29N, 31.3E, 29.6N. The data grid also limits the analyzable region to the following bounding points: 30.325E, 29.025N, 31.275E, 29.575N. This analyzable region indicates the spatial limits of the subsetted granules that went into making this visualization result.

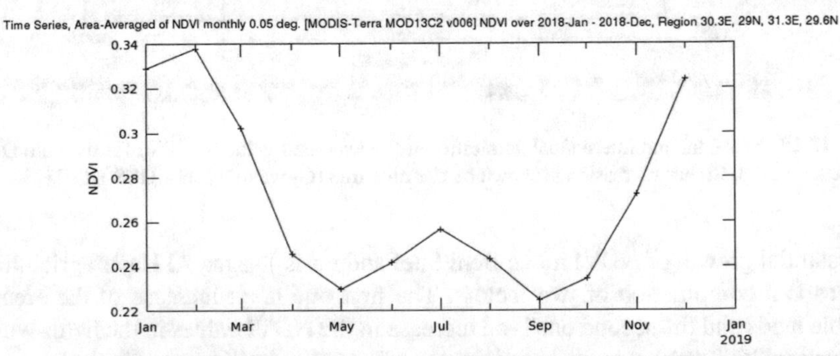

Fig. 13.31 Seasonal variability of the Beni Suef agricultural area averaged value of NDVI for 2018 (Giovanni NASA GES DISC)

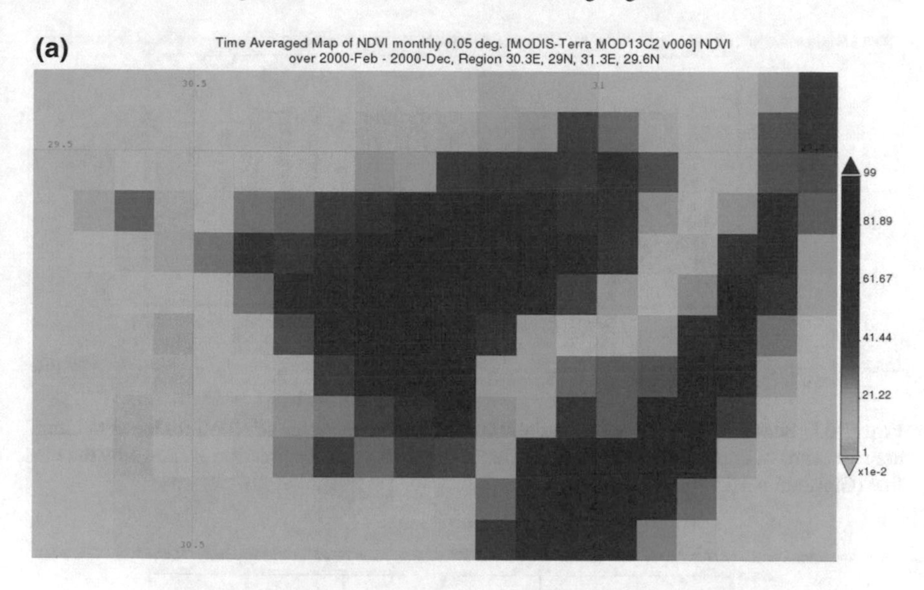

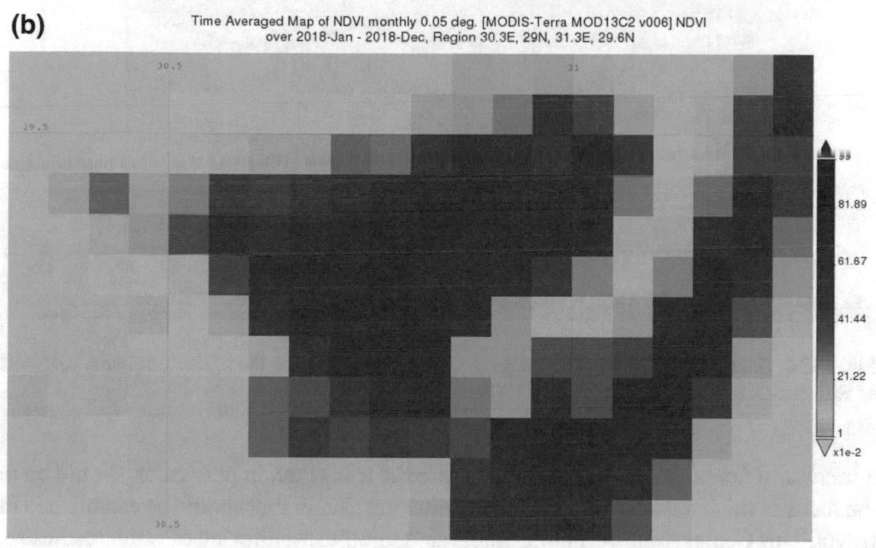

Fig. 13.32 Spatial distribution of NDVI in the Beni Suef agricultural area in 2000 (**a**) and 2018 (**b**) (Giovanni NASA GES DISC)

13.6 Recommendations

The performed analysis of high and moderate resolution optical imagery of the Nile River and surrounding natural and agricultural vegetation shows the advantages of both methods in the investigation of land cover change due to natural phenomena like regional climate change and man-made impacts like irrigation and expansion of

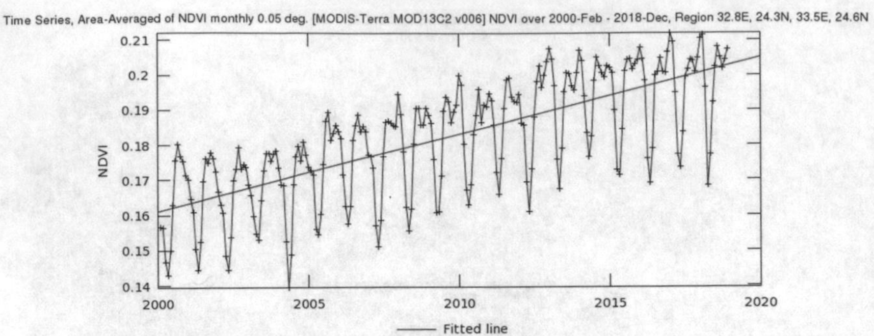

Fig. 13.33 Seasonal and interannual variability of the averaged value of NDVI for the agricultural area eastward of Faris and Nagaa Al Hajar for 2000–2018. Linear regression is shown by the blue line (Giovanni NASA GES DISC)

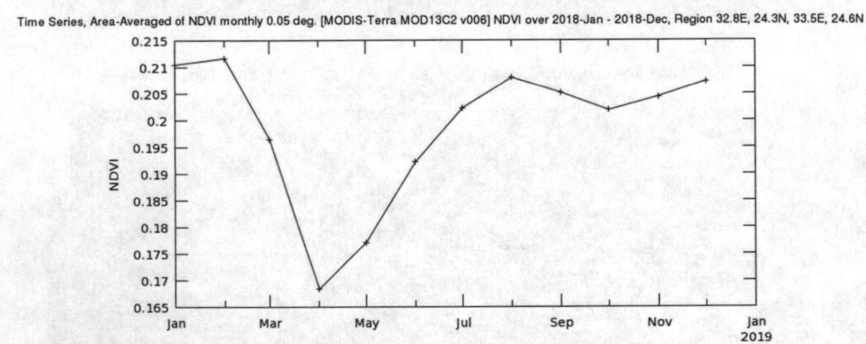

Fig. 13.34 Seasonal variability of the averaged value of NDVI for the agricultural area eastward of Faris and Nagaa Al Hajar in 2018 (Giovanni NASA GES DISC)

agricultural fields. It is evident that these are the tasks that, in particular, should be in the focus of the International Satellite Monitoring Center that should be established in Egypt. This Center should combine integrated satellite monitoring of water resources,

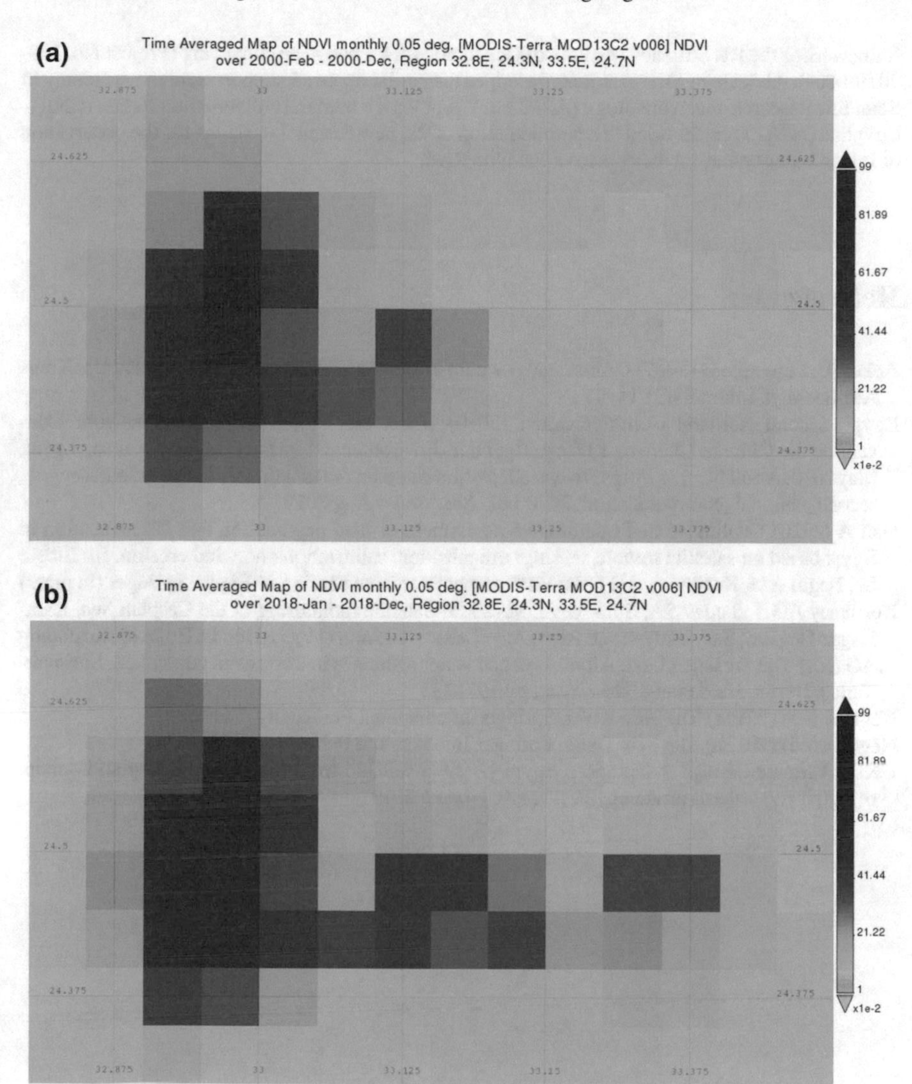

Fig. 13.35 Spatial distribution of NDVI in the agricultural area eastward of Faris and Nagaa Al Hajar in 2000 (**a**) and 2018 (**b**) (Giovanni NASA GES DISC)

vegetation, desertification for the whole territory of Egypt subjected to climate change as well as operational satellite monitoring of oil pollution in the coastal zone of the Red and Mediterranean seas.

Acknowledgements Analyses and visualizations used in this study were produced with the Giovanni online data system v.4.30, developed and maintained by the NASA Goddard Earth Sciences Data and Information Services Center (GES DISC). A. G. Kostianoy was partially supported in the framework of the P.P. Shirshov Institute of Oceanology RAS budgetary financing (Project No. 149-2019-0004). Abdelazim M. Negm acknowledges the partial financial support from the Academy of Scientific Research and Technology (ASRT) of Egypt via the bilateral collaboration Italian (CNR)–Egyptian (ASRT) project titled "Experimentation of the new Sentinel missions for the observation of inland water bodies on the course of the Nile River".

References

Acker JG, Leptoukh G (2007) Online analysis enhances use of NASA earth science data. Eos Trans Am Geophys Union 88(2):14, 17

Egypt. Second National Communication (2010) Under the United Nations Framework Convention on Climate Change, 137 pp. Egyptian Environmental Affairs Agency, Cairo, Egypt, May 2010. Available from https://www.adaptation-undp.org/sites/default/files/downloads/egypt_second_national_communication_2010.pdf. Accessed 4 Aug 2019

Gad A (2020) Qualitative and quantitative assessment of land degradation and desertification in Egypt based on satellite remote sensing: urbanization, salinization and wind erosion. In: Elbeih SF, Negm AM, Kostianoy AG (eds) Environmental remote sensing in Egypt. Springer (in press)

Kostianoy AG, Lebedev SA, Solovyov DM (2013) Satellite monitoring of the Caspian Sea, Kara-Bogaz-Gol Bay, Sarykamysh and Altyn Asyr Lakes, and Amu Darya River. In: Zonn IS, Kostianoy AG (eds) The Turkmen Lake Altyn Asyr and water resources in Turkmenistan, vol 28. Springer-Verlag, Berlin, Heidelberg, New York, pp 197–232

Negm A (ed) (2017a) The Nile River. Springer International Publishing, 741 pp

Negm A (ed) (2017b) The Nile Delta. Springer International Publishing, 537 pp

UNDP. Climate Change Adaptation. Egypt (2019) Available from https://www.adaptation-undp.org/explore/northern-africa/egypt. Accessed 4 Aug 2019

Part IV
Environment and Climate Change

Chapter 14
Remote Sensing and Modeling of Climate Changes in Egypt

Mohamed El Raey and Hesham El Askary

Abstract Profound impacts of climate change occur at regional levels, affecting, among others, ecosystems, agriculture, hydrology, and carbon cycle at global levels. These changes will have signficant impacts on all aspects of human societies, including food, water, energy, and, not least, the economy itself. In particular, major uncertainties exist for natural and managed ecosystems. We will start by demonstrating the physical fundamentals of the global phenomenon of climate change, its origin and greenhouse gasses emissions, lifetimes, its global impact on critical sectors and resources of sustainable development. Then, we emphasize the vulnerability of Egypt to climate changes and the need for large-scale, global systems for monitoring, modeling, assessment, and follow-up of mitigation and adaptation measures. An outline of the capabilities of remote sensing and GIS techniques for monitoring, mitigating, assessing vulnerabilities to impacts, success stories, modeling, and early warning of extreme events in Egypt, is presented. We end up with an assessment of the needs of Egypt to fulfill its strategic development goals (SDGs) in harmony with reducing the risk of climate change.

Keywords Remote sensing · Modeling · Climate change · Sustainable development

M. E. Raey (✉)
Institute of Graduate Studies and Research, Alexandria University, Alexandria, Egypt
e-mail: melraey@Alexu.edu.eg

H. E. Askary
Center of Excellence in Earth Systems Modeling & Observations, Chapman University, Orange, CA 92866, USA
e-mail: elaskary@chapman.edu

Schmid College of Science and Technology, Chapman University, Orange, CA 92866, USA

Department of Environmental Sciences, Faculty of Science, Alexandria University, Alexandria 21522, Egypt

S. F. Elbeih et al. (eds.), *Environmental Remote Sensing in Egypt*,
Springer Geophysics, https://doi.org/10.1007/978-3-030-39593-3_14

Abbreviations

AATSR	Advanced Along Track Scanning Radiometer
AMSRE	Advanced Microwave Scanning Radiometer-EOS
AVHRR	Advanced Very High Resolution Radiometer
CDM	Cleaner Development Mechanism
COP	Conference of Parties
EEAA	Egyptian Environmental Affairs Agency
ESA	European Space Agency
GHRSST	Group for High Resolution Sea Surface Temperature
GMES	Global Monitoring for Environment and Sustainability
GOES	Geostationary Operational Environmental Satellite Imager
GOSAT	Greenhouse Gas Satellite
ICZM	Integrated Coastal Zone Management Plan
IPCC	Intergovernmental Panel of Climatic Changes
JAXA	Japan Aerospace Exploration Agency
LST	Land Surface Temperature
MODIS	Moderate Resolution Imaging Spectroradiometer
MOEJ	The Ministry of the Environment, Japan
MTSAT-1R	Multi-functional Transport Satellite 1R
NASA	National Aeronautics and Space Administration
NIES	National Institute for Environmental Studies, Japan
NOAA	National Oceanic and Atmospheric Administration
SDG	Sustainable (Strategic) Development Goals
SEVIRI	Spinning Enhanced Visible and Infrared Imager
SLR	Sea Level Rise
SRTM	Shuttle Radar Topography Mission
SST	Sea Surface Temperature
TMI	Tropical Rainfall Measuring Mission Microwave Imager
TNC	Third National Communication
TRMM	Tropical Rainfall Monitoring Mission
TSP	Total Suspended Particulates
UNFCCC	UN Framework Convention on Climate Changes
WMO	World Meteorological Organization

14.1 Introduction to Climate Change and Its Potential Impacts

One of the essential global phenomena that affect the quality of human life, economy, and welfare in recent years, is the climatic changes. This phenomenon is due to the gradual increase of long-lived, greenhouse gases molecules (CO_2, CH_4, NO, O_3, water vapor HO_2 and the chlorofluorocarbons (CFC)) in the atmosphere, due to the high rates of growth of energy generation, human overconsumption, and industrial development, in most countries. These so-called greenhouse gases allow the solar radiation in the visible, to reach the earth's surface but do not allow the longwave emitted radiation from the warmed earth's surface, in the infrared, to escape to space. This imbalance of radiation reflectance is causing the observed increase in the average global temperature. Almost all of these gases are gradually increasing in the atmosphere, except for CFC, which are controlled by Montreal Protocol. In particular, CO_2 has increased dramatically from 380 ppm in 2006 to over 400 ppm in 2018 (Fig. 14.1 top). Figure 14.1 also illustrates the change in global surface temperature relative to 1951–1980 average temperatures. Seventeen of the 18 warmest years in the 136-year record all have occurred since 2001, except for of 1998 (NASA 2011).

So far, the average global temperature increase did not exceed 0.85 Degree Celsius over the past 100 years (IPCC-AR4 2014a). The observed increase in temperature is associated with an increase in sea level due to the melting of the polar caps. The accelerated temperature is also associated with a correspondingly

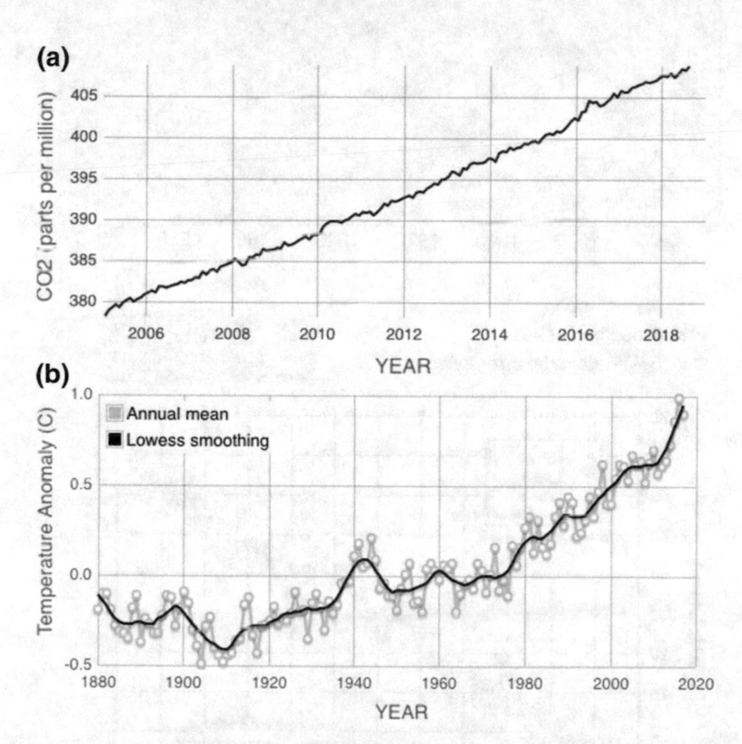

Fig. 14.1 Top: Monthly CO_2 concentration since 2005 (average seasonal cycle removed). Bottom: Global land-ocean temperature anomaly since 1880 (NASA 2011)

accelerated sea level rise. The phenomenon is evident in the accelerated temperature over the past 15 years and the corresponding acceleration in the rate of sea level rise (Fig. 14.2a, b).

Moreover, the rise in average global temperature is also associated with an increase of the frequency, severity, and duration of extreme events (IPCC-AR4 2014a). This increased rates of occurrence, severity, and duration of hazardous events such as heat waves, flash floods, dust storms, and droughts. It is also noted that the increase of temperature is mostly manifested in the lower troposphere layer, which becomes

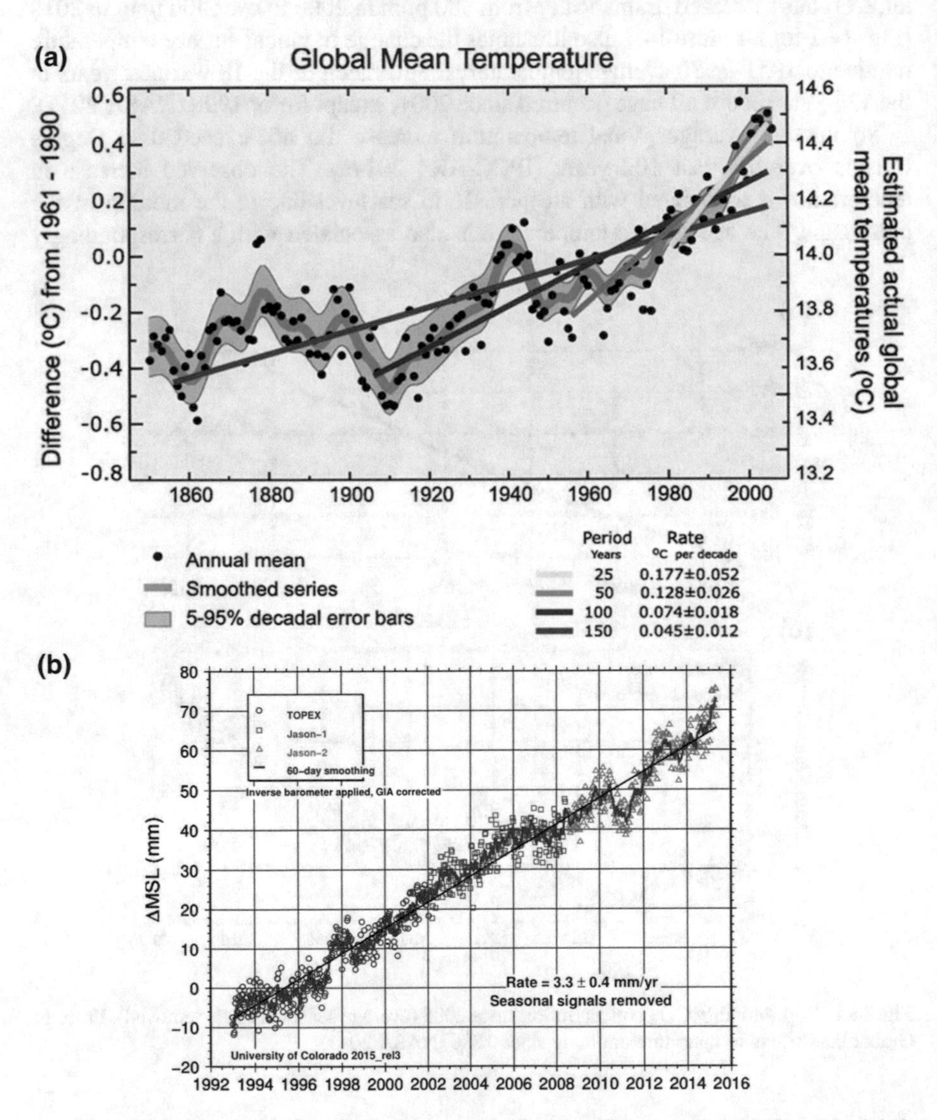

Fig. 14.2 Global representation of **a** temperature acceleration (Solomon et al. 2007) and **b** global-mean sea level from satellite altimetry (Nerem 2016)

responsible for holding air pollution, causing health impacts to humans. It is also estimated that, if no action is taken to reduce greenhouse gases, this increase of average global temperature may reach 4–5 °C by the end of this century. If this is not stopped, the sooner, the better, polar caps will completely melt, and sea level rises. The expected rise in sea level may reach over 100 cm due to water from polar caps as well as the expansion of deep cold water in oceans, shift of bioclimatic zones, and the precipitation rates changes.

Severe impacts on all sectors of the development in all countries are therefore expected. Water and coastal resources, agriculture resources, food security, health and, socioeconomic conditions, in all developing countries, are projected to suffer a setback, and they have to carry out mitigation and adaptation measures. So, in addition to the responsibilities of developing countries to achieve development according to SDG, developing countries have to mitigate and adapt. However, it should be noticed that the UN Strategic Development Goals (SDG) go hand in hand with the reduction of the impact of climatic changes.

So, to reduce the risk associated with the increasing temperature, it is necessary to mitigate emissions of these long-lived gases, estimate and assess vulnerability to impacts on water resources, agricultural resources, coastal areas, and cities and adapt to changes in the climate. Recently in COP22, the world (194 countries) realized these facts and came to a remarkable agreement in Paris (2015), not to allow the global temperature to increase more than 2 °C above present time mean temperature. Egypt, because of its environmental vulnerability, has presented the case on behalf of 50 African countries to the conference of parties. Egypt signed the Paris Agreement in 2015 and ratified it in 2016. The scarcity of available information on emissions of various countries, the shortage of capabilities of developing countries, and lack of awareness among diverse communities are the main obstacles facing the immediate solution to this problem.

Satellites remote sensing observation can close the gap of the scarcity of information by complementing the sparse data available from ground-based observations. The global nature and broad spatial scale of remote sensing are essential characteristics that make satellites an ideal platform for the mitigation and control of greenhouse gas. Also, the spatial, spectral, temporal, and radiometric characteristics of new satellite sensors are most suitable for features identification and its final temperature time series makes monitoring of changes feasible. The advantage of satellite observation is the long-time stability, global calibration, and broad spatial view. Therefore, building an accurate geographic database of emissions, vulnerability assessment of climate changes on earth, and modeling of adaptation measures becomes necessary. We shall give several examples of used techniques of remote sensing and GIS in vulnerability assessment and the potential uses of mitigation and adaptation measures.

14.1.1 Monitoring of Gases

Remote sensing of atmospheric particles, molecules, and gases is one of the primary goals of satellites and ground-based measurements. Figure 14.3 shows an example of a basic remote sensing observation and analysis system. The use of satellite observation in weather forecasting is an early application of satellite systems for persistent impacts of climate change, and it acted as an early warning system. Even though the spatial resolution was low, its time resolution was high enough to check some indices. However, many networks are now capable of detecting fast impacts, such as the increase of frequency, duration, and severity of extreme events associated with an increase in global temperatures. The low spatial resolution, sun-synchronous satellite systems with high time resolution are still needed for climate change research. It has been used for assessing levels of atmospheric CO_2, CH_4, and aerosol gasses. The methodology involves using multiple measurements at some spectral wavelengths of these gases in either the visible, infrared or the microwave region to invert an integral equation of the type (Dubovik and King 2000; Wurl et al. 2010)

$$X(h) = \text{Constant} \times \int K(h, y) \cdot S(y) \cdot dy \tag{14.1}$$

where X(h) is a set of measurements at fixed wavelengths (e.g., measurements of Aerosol Optical Depth (AOD) at some wavelengths), or, (measurements of atmospheric brightness temperatures in oxygen lines for temperature profile inversion).

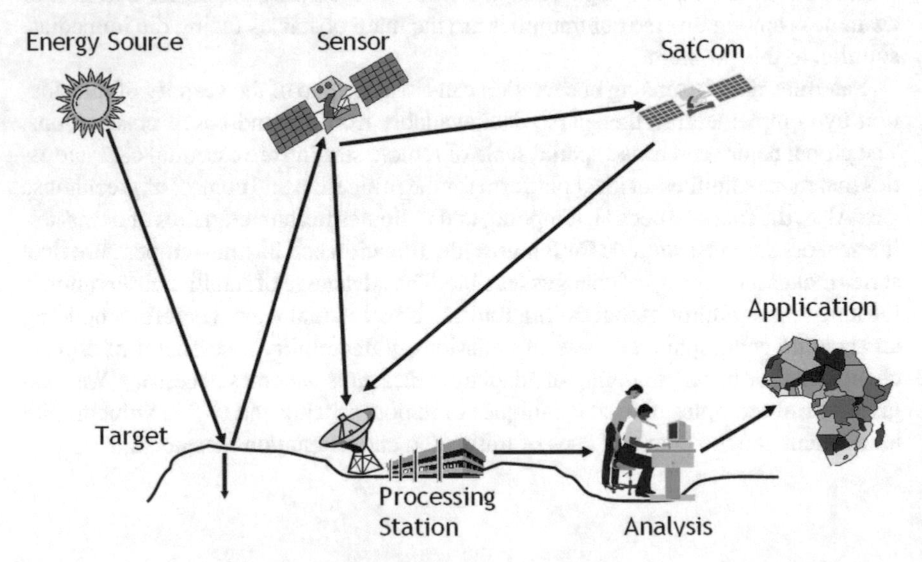

Fig. 14.3 A diagram illustrating the basic remote sensing platform and processing station for a ground target (Natural Resources Canada 2016)

The kernel K(h,y) is a function dependent on both h and y, and the function S(y) is the sought function, e.g., the inverted profile of the density of the gas, or particle size distribution (Nakajima et al. 1996). Results indicated the success of the methodology (e.g., Dubovik et al. 2000, 2002; Lauvaux et al. 2016). This method has been applied to check greenhouse gas emissions from sources on country scales and can be utilized to check the emission of sources and build models for mitigation.

Monitoring of atmospheric parameters such as density profiles, temperature profiles, and optical depth parameters are carried out by NASA AURA satellite, AIRS, and MODIS as well as by CALIPSO satellite (Cloud-aerosol infrared Lidar pathfinder Satellite) and most recently by ESA Copernicus Sentinel-5p. An inversion technique is necessary to obtain temperature and density profiles in the atmosphere, and the particle size concentration.

For instance, MODIS (The MODerate-resolution Imaging Spectroradiometer) is a radiometer sensor launched by NASA, 1999 on board of Terra satellite platform (a second series on board Aqua satellite platform, launched 2002) to study global dynamics of the earth's atmosphere, land, and oceans. MODIS captures data in 36 bands ranging in wavelengths from 0.4 to 14.4 μm at varying spatial resolutions (2 bands at 250 m, five bands at 500 m and 29 bands at 1.0 km). The Aqua platform is a sun-synchronous near polar at 705 km altitude. MODIS Aqua instrument images the entire Earth every one to two days. The level 3 Standard Mapped Images (SMI) Chlorophyll-a dataset (https://oceancolor.gsfc.nasa.gov), has a daily temporal resolution and 4.6 km spatial resolution at the equator. The MODIS Aqua instrument provides quantitative data on ocean bio-optical properties to examine oceanic factors that may affect global climate change as well as other biogeochemical cycles. Subtle changes in chlorophyll-a signify various types and quantities of microscopic marine plants (phytoplankton), rather like a pump, phytoplankton transport gases, and nutrients from the ocean surface to the deep. Their role in the carbon cycle is quite different from that of trees and other land plants, which actually absorb CO_2 and serve as a storehouse, or "sink" of carbon. The NASA Ocean Biology Processing Group (OBPG) was constituted for this knowledge, which has both scientific and practical applications.

14.1.2 Solutions

14.1.2.1 Mitigation and Adaptation

Climate change is starting to be factored into a variety of development plans: how to manage the increasingly extreme disasters we are seeing and their associated risks, how to protect coastlines and deal with sea-level encroachment, how to best manage land and forests, how to deal with and plan for reduced water availability, how to develop resilient crop varieties and how to protect energy and public infrastructure. Earth's climate stability has been crucial for the development of our modern civilization and life as we know it. Modern life is tailored to the stable climate we have become accustomed to. As our climate changes, we will have to learn to adapt. The faster the climate changes, the harder it could be.

Responding to climate change involves two possible approaches: reducing and stabilizing the levels of heat-trapping greenhouse gases in the atmosphere ("mitigation") and/or adapting to the climate change already in the pipeline ("adaptation"). Mitigation—reducing climate change—involves reducing the flow of heat-trapping greenhouse gases into the atmosphere, either by reducing sources of these gases (for example, the burning of fossil fuels for electricity, heat or transport) or enhancing the "sinks" that accumulate and store these gases (such as the oceans, forests, and soil). The goal of mitigation is to avoid significant human interference with the climate system, and "stabilize greenhouse gas levels in a timeframe sufficient to allow ecosystems to adapt naturally to climate change, ensure that food production is not threatened and to enable economic development to proceed in a sustainable manner" (IPCC-AR4 2014a). Adaptation—adapting to life in a changing climate—involves adjusting to actual or expected future climate. The goal is to reduce our vulnerability to the harmful effects of climate change (like sea-level encroachment, more intense extreme weather events or food insecurity). It also encompasses making the most of any potential beneficial opportunities associated with climate change (for example, longer growing seasons or increased yields in some regions).

Since the average global temperature has already been increased by about 0.8 degrees Celsius over the past one hundred years and still increasing at accelerated rates, we have to carry out the adaptation in addition to the mitigation. It is also well recognized (e.g., Stern 2008) that, no matter what we do for mitigation, we still have to carry out the adaptation.

14.1.2.2 Risk Reduction

One of the most important consequences of increased global temperatures, is the increasing frequency, duration, and severity of extreme events such as; droughts, flash floods, heat waves, dust storms and sea surges (IPCC-AR5 2014b). This constitutes a large-scale hazard to various sectors of development for all countries and constitutes a risk to human life and welfare. The estimated Risk is:

$$\text{Risk} = \text{Probability of Hazard} \times \text{Vulnerability to Impacts/Resilience of Communities}. \quad (14.2)$$

To reduce these risks through mitigation and proactive adaptation measures, we have to upgrade community resilience. The community resilience depends on vulnerable population size, unemployment rate, demographic conditions and human development index (HDI) which in turn depends on health, education, income levels in the community. The resilience could be upgraded by raising education level of the community, and use of technology. For estimating the risk associated with extreme events of changes of temperature on vulnerable communities, we rely on

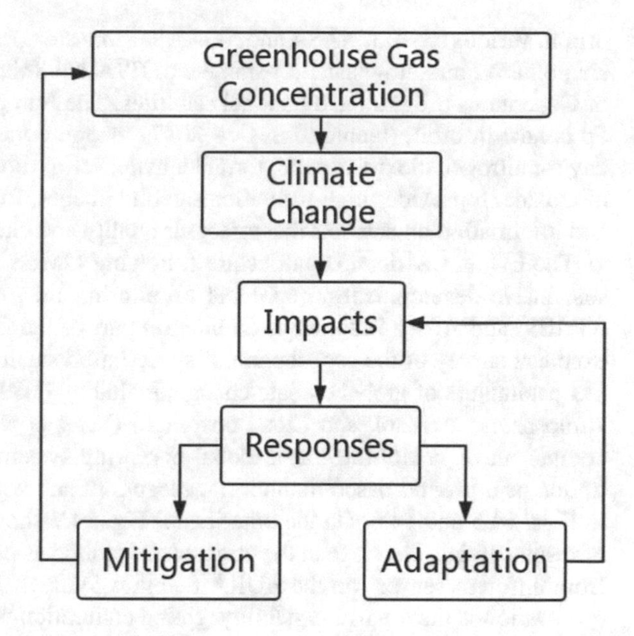

Fig. 14.4 A block diagram illustrating various phases of activities of combating the climatic changes

remote sensing and GIS technologies. A block diagram illustrating various phases of activities of combating the climate changes is shown in Fig. 14.4.

14.2 Role of Remote Sensing and GIS

Note that satellite remote sensing is an essential tool for monitoring of gaseous greenhouse emissions, vulnerability assessment, mitigation and adaptation measures. Also, early warning systems of flash floods in or around cities could be established based on analysis of satellite images (LANDSAT, SPOT, SRTM, InSAR) and GIS to analyze and check land cover and land use over vulnerable areas and vulnerable cities to increase community resilience to risk reduction.

The Ministry of the Environment, Japan (MOEJ), together with the National Institute for Environmental Studies (NIES), and Japan Aerospace Exploration Agency (JAXA) have launched sensors for carbon dioxide (CO_2) and methane (CH_4) monitoring. The Observing Satellite "IBUKI" (GOSAT) has revealed that the whole-atmosphere monthly mean CO_2 concentration detrended with average seasonal variation (CO_2 trend) has exceeded 400 ppm in February 2016, for the first time since GOSAT was launched in 2009. These satellites and others can be used for monitoring, testing compliance, enforcing agreements and modeling greenhouse gases emissions of each country and each source.

NASA and NOAA have started a program of Earth Observation that involves measurements of many climatic and weather parameters, which cover aspects of extreme events and provide an early warning system of hazards of atmospheric origin. Various types of NASA and NOAA sensors are now monitoring and assessing air pollution and atmospheric parameters. ESA has followed by the development of Copernicus program and Sentinel satellites. The European Copernicus Sentinel-5p is now in orbit, (launched in Oct. 2017) mapping the global atmosphere every day for nitrogen dioxide, ozone, formaldehyde, sulfur dioxide, methane, and carbon monoxide. It provides high-resolution satellite images, for free, on these trace gases and information on aerosols that affect air quality and climate.

The EU and African countries are following GMES (2017–2020) program for sustainable development. The Global Monitoring Environment and Sustainability (GMES) and Africa initiative is an integral part of the EU-Africa Partnership. An excellent survey of the contribution of space-based satellite remote sensing to various parameters of global climate change, including (LST, SST, Snow Cover, SLR, Atmospheric Aerosol, and Cloud coverage) (Yang et al. 2013). The space-based component of WMO integrated global observing system along with another illustration of a ground-based monitoring, reception, and analysis platform are shown in Figs. 14.5 and 14.6. On the other hand, Fig. 14.7 shows an example of the type of results of data coverage in the atmosphere plotted as earth's atmospheric profiles from different sensors on the AURA Satellite *Systems.* The advantage of satellite observation is the long-time stability, global calibration, and wide spatial view.

A geographic information system (GIS) is a computer system designed to capture, store, manipulate, analyze, manage, and restore all types of geospatial data. The implementation stage of GIS consisted of the creation of a relational database,

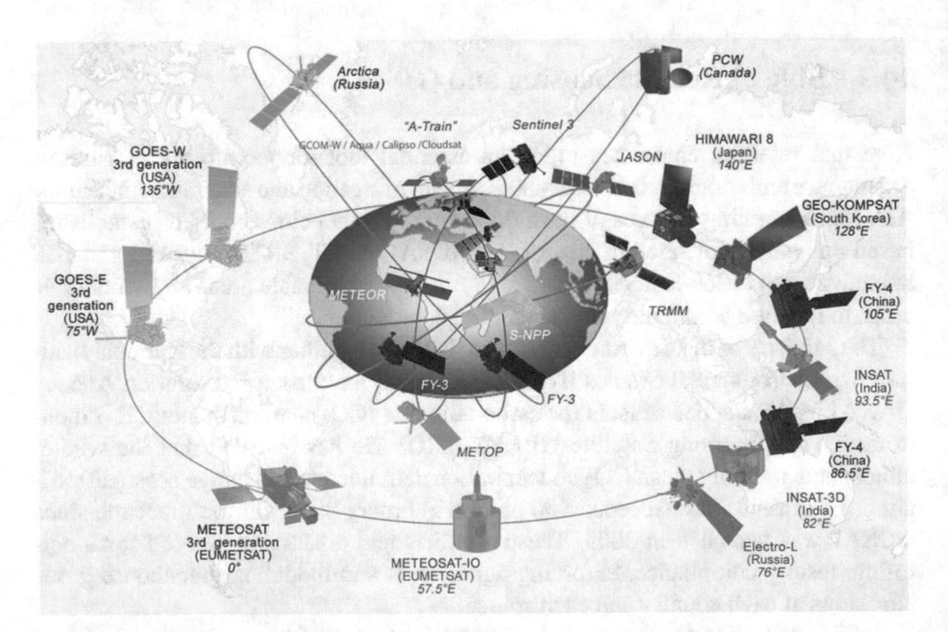

Fig. 14.5 WMO global observing geosynchronous and polar orbital satellites, each carrying some sensors (Deutscher Wetterdienst 2018)

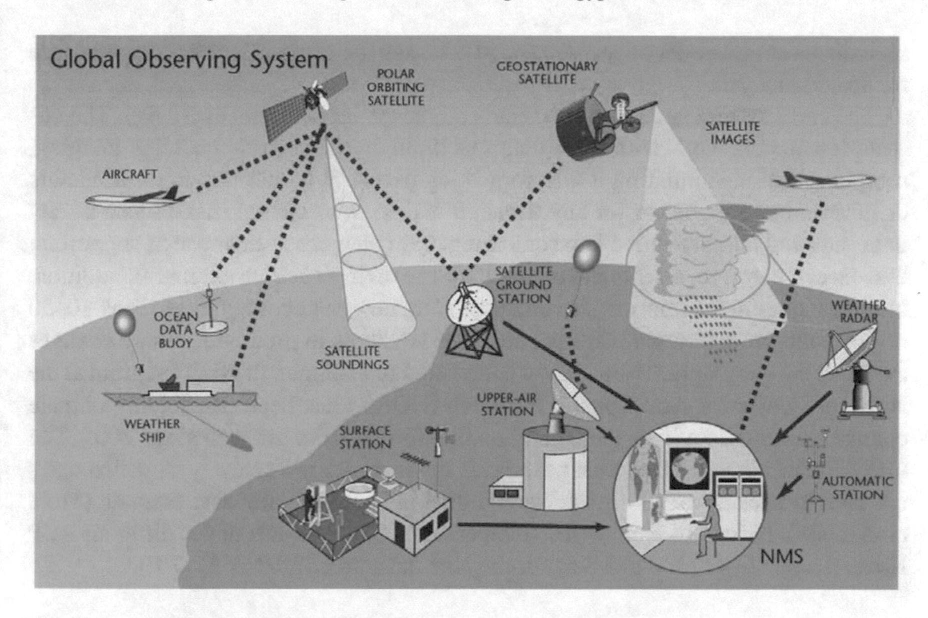

Fig. 14.6 A diagram of a global observing station

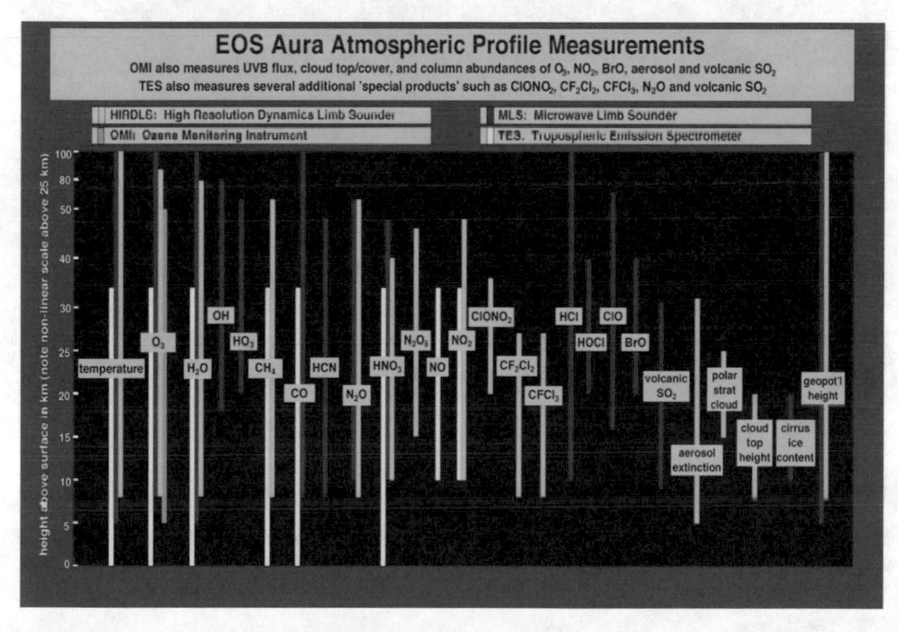

Fig. 14.7 Four sensors on board of EOS Aura Satellite and their measurements profiles. Example of what can be measured by different sensors (NASA 2018)

including a complete set of georeferenced data, and the elaboration of a user interface for spatial analysis.

The earth system generates thousands of geospatial data sets every day. The climate researchers work with these data sets through often cross-checking, merging, morphing and manipulating it into something useful. Whether that is for academic or government reports, or for any decision makers who cannot make sense of raw data and need it transformed into something they can work with or use as supporting evidence. Therefore, GIS is the technology of solving such problems. In addition, climate modelling is important to understanding how our ecology might look 10, 50 or 100 years from now and GIS is one of the key tools in predicting future changes based on the geographic data already collected. For example, the GIS Program at the National Center for Atmospheric Research (NCAR) has been distributing climate change scenario data in shapefile and textfile format from the web since 2005. The GIS Climate Change Scenarios portal was the first internet gateway in which users are able to access global climate model data in standard GIS data formats (Moss et al. 2008). Figure 14.8 shows the temperature change trends under different RCP levels (RCP 2.6, 4.5, 6.0 and 8.5) for the Nile delta area (30.9°N 31.0°E).

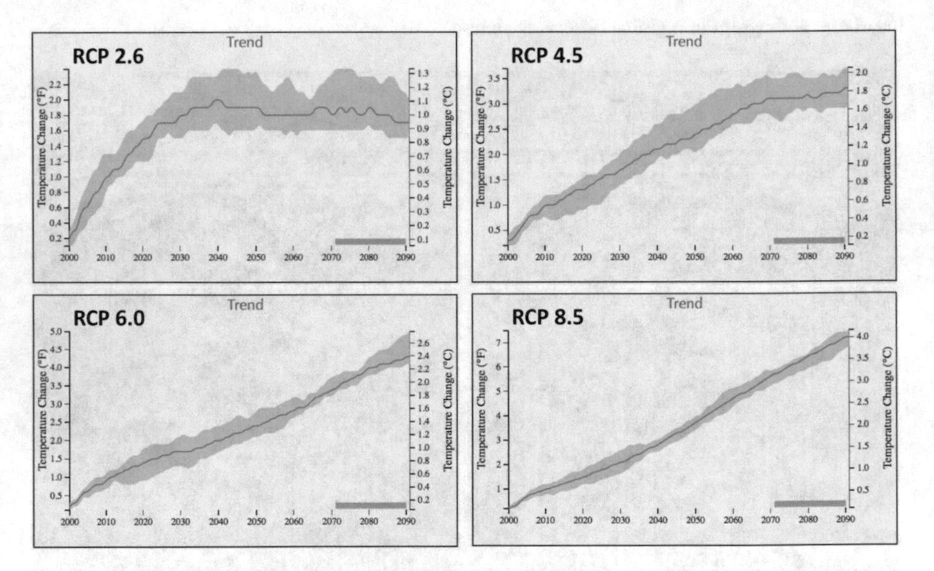

Fig. 14.8 20-year running mean of annual temperature change for emission trajectory: for (RCP 2.6, 4.5, 6.0 and 8.5) ensemble average (dark line) and spread of ensemble members (shaded area). Values are for the model grid cell containing: 30.9°N 31.0°E (Moss et al. 2008)

14.2.1 Early Warning Systems

It is well recognized practically and approved theoretically (IPCC-AR4 2014a) that associated with the increase of temperatures is an increase in the frequency and severity of extreme events. So that the rates of flash floods, heat waves, droughts and dust storms will be increased together with their severity and we have to warn against this new phenomenon.

Early warning systems are systems based on satellites sensors, made to predict that a hazardous situation may be approaching a specific area or region on earth so that a prepared contingency plan must be activated. Tornadoes and Hurricanes, severe heat waves and droughts are good examples. These early warning systems are based on remote sensing images monitoring and fast analysis of changes as shown in Fig. 14.9. The formation and movement of tornadoes and Hurricanes can be monitored by satellites and the predicted tracks over urbanized or populated areas could be estimated. Figure 14.10 shows the JAXA platform early warning system for precipitation based on TRMM sensors. It demonstrates the high congestion that was later manifested by the heavy precipitation encountered in Alexandria, Egypt in October 2015. The JAXA Global Rainfall Watch based on MSMAP offers a google earth map that, presents the cloud map which can be used to estimate rainfall area and how many mm/h are expected over the region.

Many satellites launched in recent years have been comprising satellite constellations to meet the growing needs of hazards warning (Fig. 14.11). Among the satellite observations are the ability to provide information to determine the Normalized Difference Vegetation Index NDVI index defined as:

$$\text{NDVI} = (\text{NIR} - \text{IR})/(\text{NIR} + \text{IR}) \tag{14.3}$$

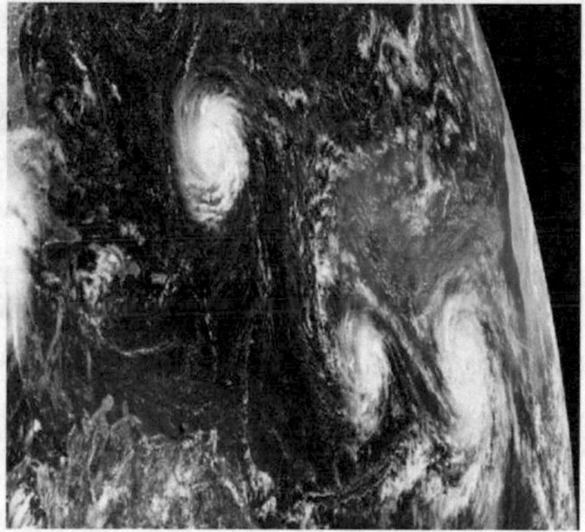

Fig. 14.9 Low spatial resolution images provide an early warning system for Hurricanes and tornadoes, GOES Image of Hurricane Fran (left) and 2018 Frances (right) (NOAA 1996; AerosolWatch 2018)

Fig. 14.10 Global rainfall watch platform using google earth for monitoring of expected rainfall over the Mediterranean (JAXA 2015)

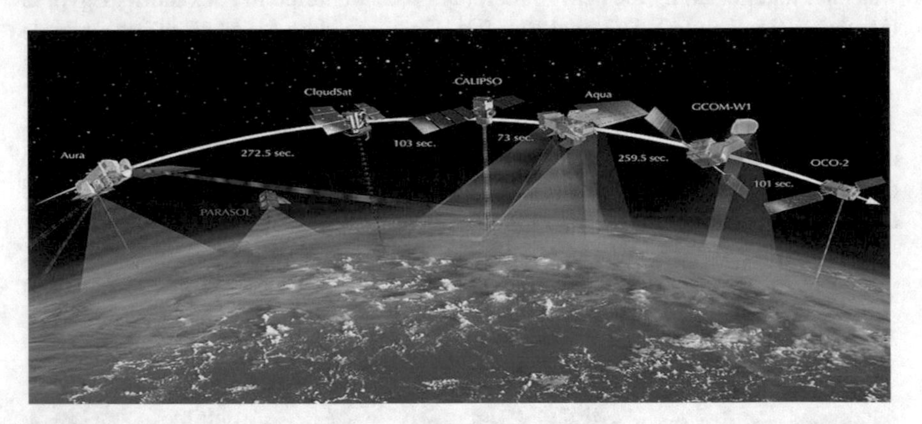

Fig. 14.11 Six observation satellites comprising the A-train satellite constellation as of 2014. Aura, Cloud sat, Calipso, Aqua, GCOM w1 and OCO-2 Satellites (NASA 2014)

where NIR is the near infrared channel and the IR is the infrared channel, and which varies between -1 and $+1$ could also be used to indicate the quality of agricultural productivity and soil characteristics. Vegetation indices like NDVI would be very useful for monitoring agriculture changes and other applications (Kim et al. 2014; Whitney et al. 2018; Li et al. 2019).

14.3 Vulnerability of Egypt to the Impacts of Climate Changes

According to the official numbers released by the Central Agency for Public Mobilization and Statistics (CAPMAS) in 2012, Egypt's population has reached 82 million living on just 5.3% of the country's area. The Egyptian gross national product (GDP) was 107.43 billion US$ dollars in 2006, reached 336.3 billion US$ in 2016, with growth rate around 14%, and GDP-per capita is 6.600$ in 2013. The GDP composition by end use is 78.6% household consumption, 11.8% government consumption, 14.3% investment in fixed capital, 0.4% investment in inventories, 18% exports of goods and services and −23.2% imports of goods and services. GDP composition by sector of origin in 2013 shows 14.5% agriculture, 37.5% industry, and 48% services. The labor force in 2013 is 27.69 million, 29% in agriculture, 24% in industry and 47% on services. The unemployment rate reached 13.4% in 2013 and inflation rate 9% (TNC 2016).

Historically, Egypt is one of few countries which has recognized its vulnerability to the impacts of climate changes as early as 1989 (Sestini 1989). Figure 14.12 shows the global land and ocean temperature departure from the average of month of June. Northern region of Egypt along the Mediterranean Sea experienced warmest June on record. According to the NOAA (Fig. 14.13), the global land and ocean surface temperatures of four major cities of Egypt all reached the warmest record of Jan-June temperature anomalies since the meteorological record, followed by other fourth highest records during the years of 2010, 2016 and 2012.

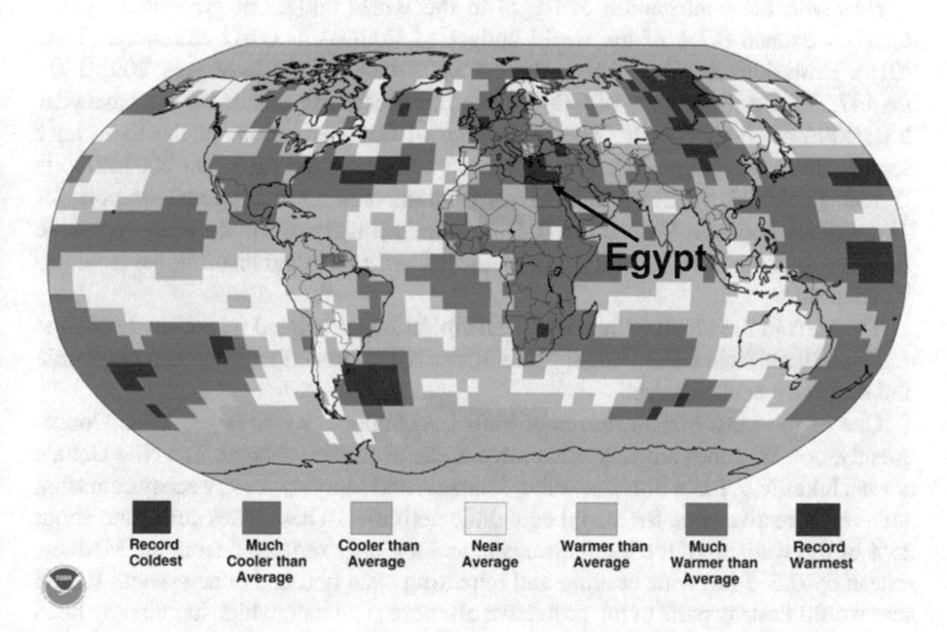

Fig. 14.12 Land and ocean temperature percentiles of June 2018 (NOAA NCEI 2018a)

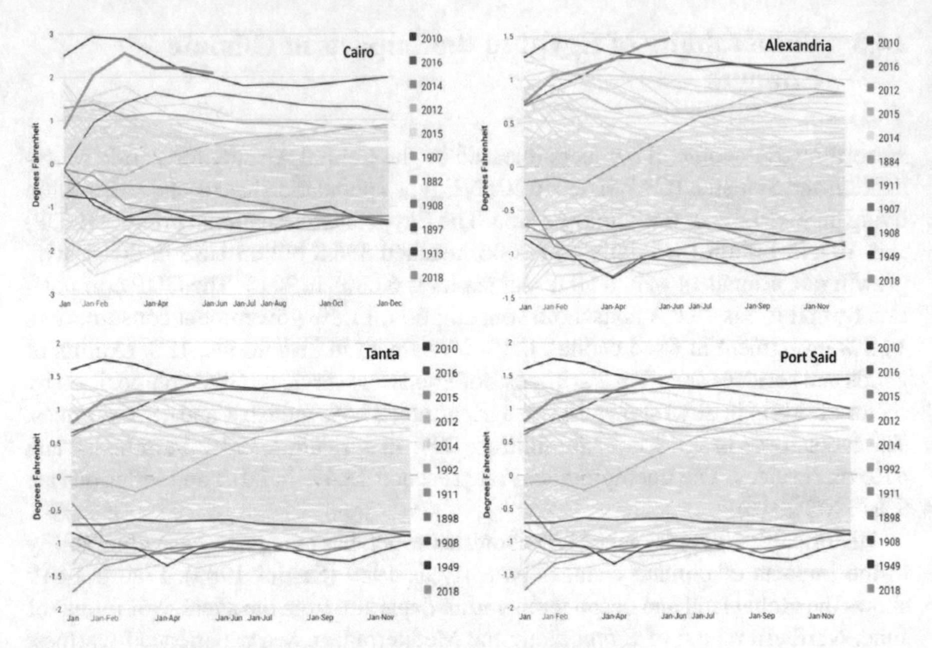

Fig. 14.13 Year to date temperature anomalies for four major cities of Egypt, the 5 warmest years, 5 coolest years, and current year 2018 are highlighted to stand out, all other years are in light gray (NOAA NCEI 2018b)

However, the contribution of Egypt to the world budget of greenhouse gases does not exceed 0.7% of the world budget of Equivalent GHG emissions (TNC 2016). Emissions of CO_2 e from the energy sector for the base year 2005/2006 are 147,324 Gg. Emissions of CO_2 e from the Electricity Sector for the base year 2005/2006 are 54,845.6 Gg. The percentage of GHG emissions of the Electricity Sector relative to all energy consumption for CO_2 e for the base year 2005/2006 is 37.23% (TNC 2016). Egypt has already implemented an institutional structure of Cleaner Development Mechanism (CDM) at the Ministry of Environment to reduce global emissions, but it has not yet implemented an equivalent institutional structure for adaptation.

The shift in the bioclimatic zone of precipitation associated with climate change undoubtedly affects water budget of the River Nile. This affects all socio-economic and health aspects of Egypt.

One of the most certain results of global warming is a rise in sea level. Consequently, coastal zones are very vulnerable areas to climate change. The Nile Delta's coastal lakes (e.g. Lake Burullus, Idku, Manzala and Maryut) are key ecosystems that act as a protective zone for inland economic activities. These lakes constitute about 25% of the wetland of the Mediterranean, and are only separated from the Mediterranean by 0.5–3 km wide eroding and retreating sand belt and dune system. Rising seas would destroy parts of the protective offshore sand belt, which has already been weakened by the reduction of sediment flows after the construction of the Aswan

Dam in 1964. The sediment belt protects the lagoons and the low-lying lands. Without this sediment belt, recreational tourism and beach facilities will be inundated, water quality will be altered in coastal freshwater lagoons and groundwater will be salted.

In addition to the impact of sea level rise, the low coastal area of the Nile delta is subsiding (Stanley and Warne 1993; Becker and Sultan 2009; Sušnik et al. 2013), which means more inundation and saltwater intrusion. It drastically affects the agricultural land productivity, as well as the health of people and animals.

Since 2017, the Egypt Ministry of Water Resources and Irrigation, Green Climate Fund of United Nations Development Programme (UNDP) have collaborated to support the 'Enhancing Climate Change Adaptation in the North Coast of Egypt' with US $73.8 million fund. The project aims to protect the densely populated low-lying lands in the Nile Delta, which have been identified as highly vulnerable to climate change induced sea-level rise (By adopting ICZM).

The Ministry of Environment in cooperation with UNFCCC and UNDP has worked out three National Communications and a Strategic Action Plan for adaptation. A National Strategy for adaptation to climate changes was advanced by Information Decision Support Centre (IDSC) of Egypt in 2011. Climate changes were also considered in the Egyptian Strategy 2030 as one of the priority issues.

The Nile Delta of Egypt hosts a population of 60% of the country's 95 million people in the coastal zone (CAPMAS 2016), most of whom do not realize that the low topography is affecting their life. Figure 14.14 shows the general environmental vulnerability index of the Nile delta region to various impacts of climate change. The northern coast is vulnerable to the impacts of SLR due to its low elevation (<1.0 m) with relative to mean sea level of the Mediterranean Sea. The region representing Greater Cairo Region (>100.0 m), is socioeconomically vulnerable due to an increase of frequency and severity of heat waves and dust storms, which increases health impacts and mortality rates by heat stresses and heat strokes and favors the mobility of micro-organisms carrying microbes of Malaria diseases. Also, it negatively affects the land productivity and socioeconomic conditions in the region. Upper Egypt is also vulnerable because of poverty and the very limited socioeconomic resilience to temperature changes, heat stress, and socioeconomic implications.

All sectors of development of Egypt including its water resources, agricultural resources, coastal, and touristic resources are vulnerable, one way or another. Hence, the socioeconomic and the health sectors are also vulnerable. Monitoring, assessment, and modeling have checked this vulnerability in at least the following sectors:

14.3.1 Water Resources

Nile flows are very sensitive to small changes in average basin rainfall or temperature. It consists of some distinct sub-basins that each respond quite differently to possible climatic variations. Although the impacts of a global warming trend are uncertain at

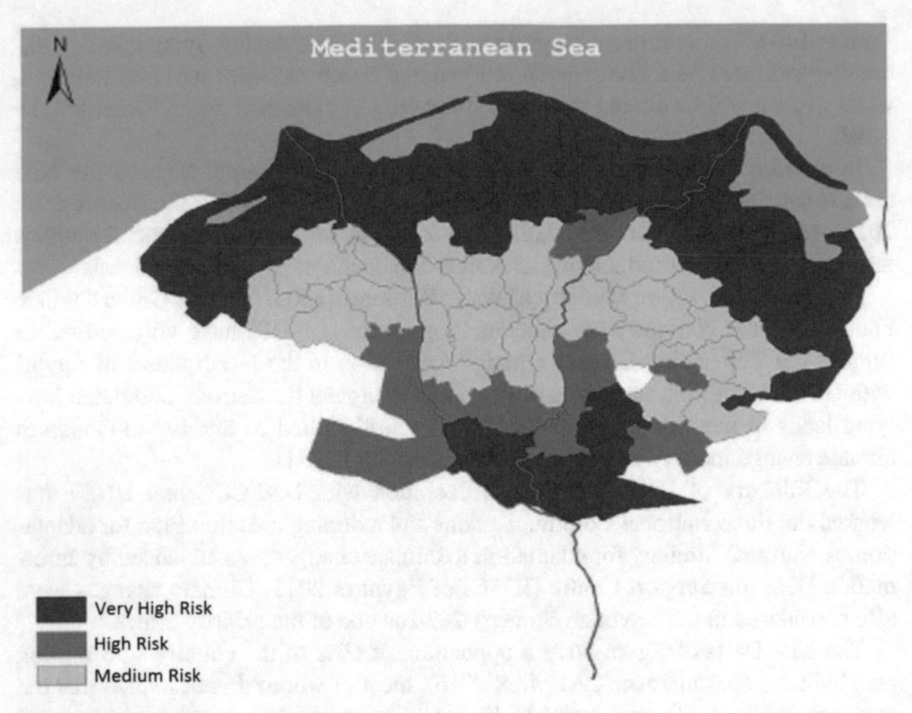

Fig. 14.14 Vulnerability index map of the Nile Delta

the regional and local level, the basin countries would do well to implement some 'no-regret' or proactive measures aimed at building resilience to current climate variability while enhancing adaptive capacity for future threats (Nile Basin Initiative 2012). A sensible approach is to be prepared for more variable conditions than currently recorded.

The water resources of Egypt are due to 95% from the River Nile, 3.5% rainfall in the northern coastal zone and 1.5% groundwater. The assessment of the vulnerability of Egypt water resources to impacts of climate changes stems from the fact that a shift of bioclimatic zone and precipitation pattern may have its origin in climate changes. The River Nile basin that is being filled each year by flood, 65% of which to precipitation on the Ethiopian hills may change due to the shift of bioclimatic zones. This necessitated to carry out analysis of the waterfalls on the Ethiopian hills, in addition to the modeling of the water that reaches Lake Nasser each year through the Blue Nile. Global circulation models have been downscaled to gain details of the picture (Strzepek et al. 1995). The analysis of global and regional circulation models indicates a high uncertainty of the results. One model showed that the water budget of the River Nile might increase by 25–30%, while all other models showed that it might decrease by as much as 70% over the average, in the next few decades (Strzepek et al. 1996).

Even though the Reconnaissance Dam that is being built in Ethiopia will affect the budget of the river Nile drastically, the impact of climate changes cannot be overlooked. Together with the rapid growth of population, urbanization, and needs for clean energy generation for development of industry and food security. This places Egypt under severe stress of uncertainty of water resources. Rain harvesting and groundwater resources in Egypt may exist in some areas of the country. However, wastewater treatment must be carried out, and utilization of treated wastewater for afforestation must be of primary priority. Integrated remote sensing and GIS techniques have been used recently to draw new maps of the hydrogeology of Egypt (e.g., Elbeih 2015).

One of the important parameters of coastal waters is its content of phytoplankton measured by Chlorophyll-*a*. It indicates the fishing productivity of the coast. MODIS data were collected (https://oceancolor.gsfc.nasa.gov) over the Nile Delta and its seasonal variation of Chlorophyll indicating fishing productivity averaged over the past seven years (2011–2017) are shown in Fig. 14.15. It shows that the index varies drastically over different parts of the delta and that it is higher in the winter (DJF). The long-term analysis may reveal changes with time. On the Red Sea, the variation of areas and concentration Chlorophyll-*a* could be an important indicator of regional climate changes, which contribute the favorable water temperature and nutrient sources (such as ocean upwelling, atmosphere deposition) for the algae blooms (Li et al. 2017, 2018).

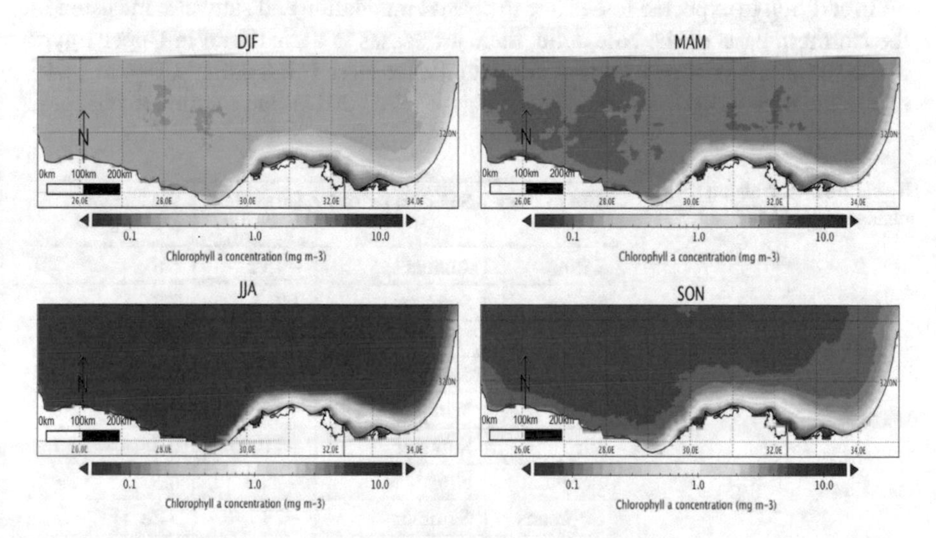

Fig. 14.15 Average Chlorophyll-a over the Northern coast of Egypt (2011–2017)

14.3.2 *Agricultural Resources*

The land area of Egypt is around 995,000 km^2 (3% of cultivated land) and the population is 83 million in 2014, with estimated 55% of the labor force in Egypt is engaged in agricultural activities and 7.5% engaged in agricultural production activities, a sector which consumes about 80% of the freshwater resources and contributes about 13.5% to the GDP, in 2012/2013. Although it is still behind the sugar cane in terms of value and yield, wheat, corn and rice are still the main food crop in Egypt. The cultivated land base of Egypt is about 3.5 million hectares, with a total annual cropping area of about 6.2 million hectares, representing 176% of the total cultivated land area (SADS 2010). Egypt is currently the world's leading wheat importer, requiring approximately 10 million tons of wheat per year, far exceeding its domestic wheat production. In addition, Egypt needs to import large quantities of corn (6 million tons) and soybeans (2 million tons) each year. Egypt mainly exports fruits, tomatoes and vegetables.

The agricultural resources are the most impacted since Egypt is used to be considered an agricultural country, in the first place. The water resources of Egypt are consumed as 80% for surface irrigation, and 20% for domestic use and industrial applications. The temperature rise has proved to be negatively impacting the agricultural productivity of most products, except the cotton. The cotton productivity shows an increase by as much as 20% in case of global temperature rise (Table 14.1).

In addition to expected losses due to coastal inundation and saltwater intrusion in the Northern parts of the Nile delta, there are losses in agricultural in Upper Egypt due to the shortage of resilience of communities to already hot climate. Figure 14.16 shows the agricultural vulnerability of Egypt (FAO 2011) due to climate changes.

Table 14.1 Estimated change in yield and water

Use of crop	Selected season	Crop yield	% Change water use
Citrus	Annual	−15.2	6.6
Cotton	Summer	+19.8	7.2
Lentil	Winter	−28	7.8
Maize	Summer	−15.2	6.6
Onion	Winter	−1.53	7.28
Rice	Summer	−11	6.6
Sorghum	Nili[a]	−15.2	6.6
Soybeans	Summer	−28	7.28
Sugarcane	Annual	−15.2	6.6
Tomato	Winter	−28	8.16
Vegetable	Summer	−28	8.16
Wheat	Winter	−19.2	7.2

[a]Nili is during fall months between summer and winter

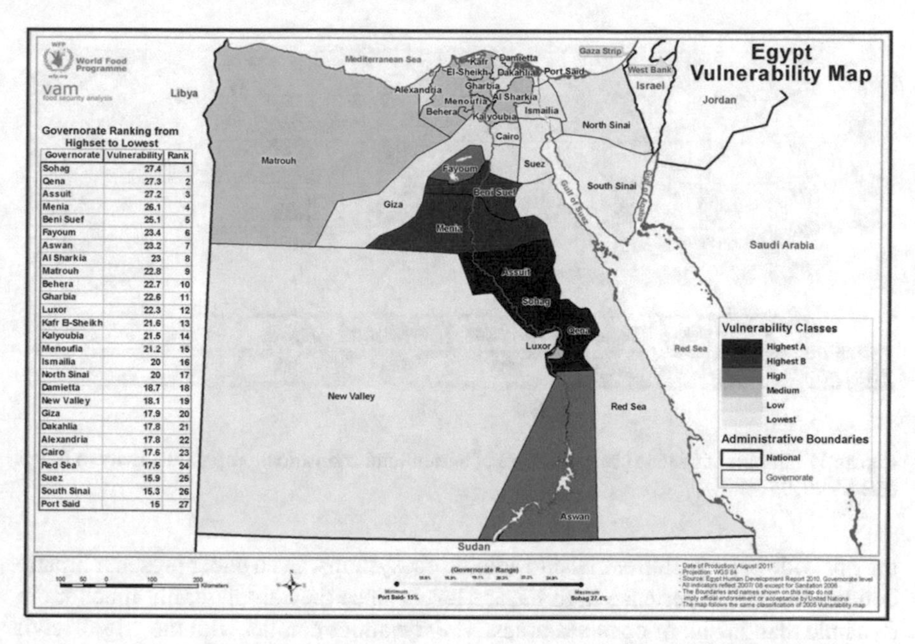

Fig. 14.16 FAO vulnerability map for Egypt. It indicates the vulnerability of Upper Egypt to climate change due to hot temperature, severe poverty, and low resilience of communities (FAO 2011)

It again emphasizes the vulnerability of Upper Egypt to impacts of heat and heat stress and the very low resilience of the communities due to lack of technology and shortage of education. In addition, since Egypt heavily relies on the food import, the climate changes in the areas of major suppliers such as Black Sea and Europe, could also threaten food security of Egypt through the increasing import price.

The agricultural vulnerability of the country causes a food deficit due to the shortages of water resources, increased contamination of already existing groundwater due to saltwater intrusion, and an increase of evaporation. Also, it was found that the crop productivities decrease drastically (15–20%) due to increase in temperature and urban encroachment in agricultural land. Figure 14.17 shows how these productivities are changed given changes in temperatures, which shows that a reduction of productivities of all crops, except for Cotton, where it increases by as much as 20% over the same period. It is worth to mention here that remote sensing techniques could monitor the land productivity of crops.

14.3.3 Coastal Resources

Coastal zones experience one of the highest population densities in the Mediterranean area. They support the concentration of uses and infrastructures like no other are in the region. They are also critical for their high biological productivity and the link

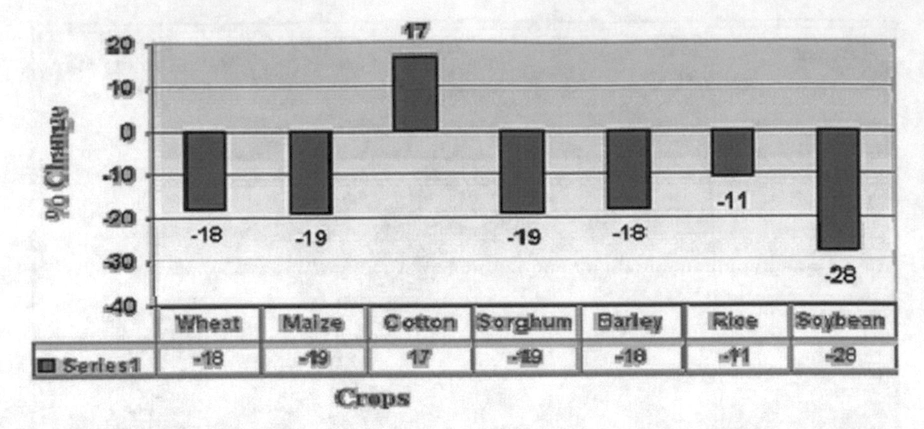

Fig. 14.17 Impact of climate change on loss of agricultural and various crop productivity in Egypt (EEAA 2010)

they provide between terrestrial and aquatic ecosystems. Even under present climatic conditions, they experience high hazard levels since they are naturally much more dynamic than inland or open sea areas. The dynamics conflict with the "rigid" civil engineering structures usually built or projected in coastal zones to help to support the high number of uses and activities. Because of this, coastal zones are excellent case studies to evaluate cross-sectorial impacts of climate variability. The primary hazards for coastal zones arise from the fact that coastal zones are in a dynamic and fragile equilibrium between terrestrial, marine, riverine and atmospheric factors. Any change in any of these factors due to natural climatic variability or an accelerated climatic change will lead to a change of status in the coastal zone, usually not compatible with present uses and infrastructures. Among the many climate threats experienced by present coastal zones of Egypt and likely to be aggravated in the future are:

1. Sea level rise
2. Excessive saltwater intrusion in freshwater resources
3. Increase of frequency and severity of storm surges.

Egypt has long coasts on the Mediterranean Sea extend for over 1200 km and over 2300 km on the Red Sea. Also, the wetland represented by the five northern lakes and Lake Qarun represent huge coasts. Also, freshwater (Lake Nasser) to the south. In addition to marine resources of these lakes, it offers fantastic touristic sites. The low elevation beaches of the Mediterranean, especially in Alexandria city and the northern Nile delta beaches and lakes, represent an essential central domestic tourist resource, while the beaches of the Red Sea are mainly for foreign and domestic tourism. Coral reefs, as well as mangrove trees in the Red Sea, represent a unique resource that attracts tourists from all over the world. These are also vulnerable to the increase of temperature and whitening of the coral reef has already been observed using remote sensing observations (e.g., El-Askary et al. 2014).

The relatively low elevation Nile Delta region on the Mediterranean is the most fertile region for agricultural land of Egypt, and it hosts over 50% of the population in the delta and over 70% of the industry of the country. It also hosts the most important five northern lakes which were producing about 66% of the fish production of the country, in addition to its touristic values. This region is vulnerable to SLR, inundation of beaches and low urban areas and agricultural land and saltwater intrusion into the soil and groundwater resources. It is also vulnerable to the impacts of extremes of hot waves, flash floods, droughts and severe storms. The vulnerability includes a loss of productivity of the soil due to saltwater intrusion, and inundation due to sea level rise both of which could be monitored by remote sensing. Early warning systems based on remote sensing must be established to provide information on flash floods, rainfall, marine storms, and dust storms.

Figure 14.18 shows a map of the delta with the levels of contour elevation indicating that in case of SLR by 1.0 m, about 15–20% of the land in the delta and vicinity will be saltwater intruded or inundated. The red areas represent a dry and inhabited land below sea level.

The FAO has facilitated the piezometric network throughout the Nile delta (Fig. 14.19). The network of piezometers is monitored by satellites to determine rates of saltwater intrusions.

Extensive work has been carried out on this area using remote sensing and GIS techniques, for assessment of the vulnerability of cities overlooking the Mediterranean (Alexandria, Abu Qir, Rosetta, Port Said, Marina, Damietta, and Matruh)—(El-Raey 1997, 2011; El-Raey et al. 1995, 1997; Smith et al. 2014; Masria et al. 2014). The idea is to develop an accurate geographic database of several layers. A

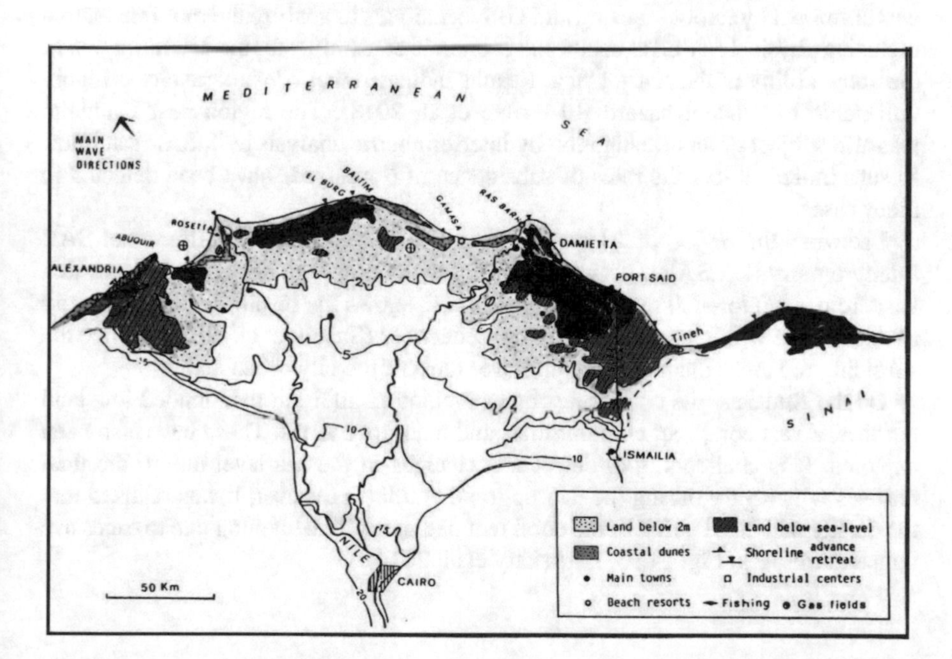

Fig. 14.18 Contour maps of the Nile Delta low land areas indicating, the lakes of Egypt (in dark blue), the vulnerable areas (in Red) and the coastal land below 2 m (yellow) (El-Raey 2011)

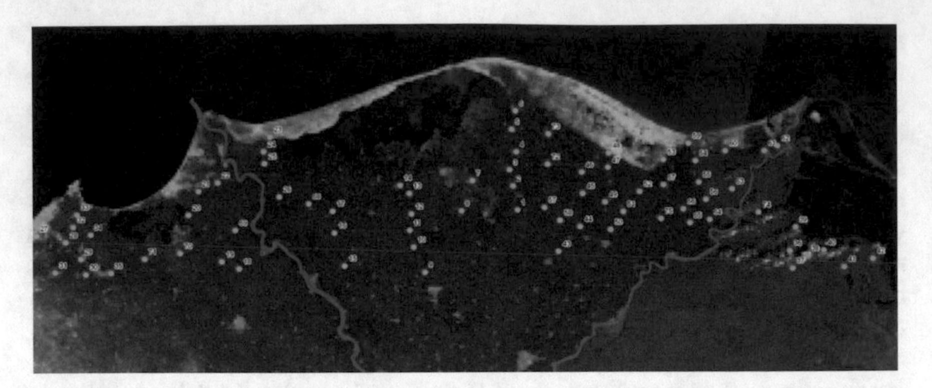

Fig. 14.19 Piezometric network survey of the Nile delta

layer of topography obtained from SRTM (Shuttle Radar Topography Mission) and Digital Elevation Mode (e.g., Fig. 14.20), land cover and land use (obtained from classified imagery) (e.g., Fig. 14.21), and population density, distributions of jobs, and assuming SLR scenarios, to find % vulnerability to inundation of each locality. The GIS is used to estimate the % area of the vulnerable population, jobs lost and vulnerable land. This may also relate to excessive rainfall flooding in case of concave structures of various scenarios of precipitation.

For example, potential tsunami risk assessment to the city of Alexandria, Egypt was examined by remote sensing and GIS techniques to analyze the multiple factors including physical as well as social and economic constraints for the determination of the vulnerability of the coastal area. Results indicated that Alexandria city is highly vulnerable to tsunami hazard (El-Hattab et al. 2018). The region near Damietta was also subject of an investigation by interferometric analysis by InSAR satellites. Results indicated that the rates of subsidence of 9 mm/year have been detected in many cases.

Likewise, the region of Alexandria was also investigated by Differential SAR interferometry (DInSAR). It was found that many areas are subsiding at rates that vary from −4.0 to −6.0 mm/year. Figure 14.22 shows the resulting map of the city of Alexandria with pictures of urban consequences (Sušnik et al. 2015). Notice the subsiding red areas that are spreading over most of the city urban areas.

On the Red Sea, the coastal resources include in addition to extended low land beaches, a vast coral reef communities, and mangrove areas. These have also been explored. It is challenging to detect any changes in the sea level due to the tidal changes which vary during the day up to 90 cm/day. However, it was realized that the quality and distribution of the coral reef decreases by whitening due to excessive temperatures (e.g. Fig. 14.23, El-Askary et al. 2014).

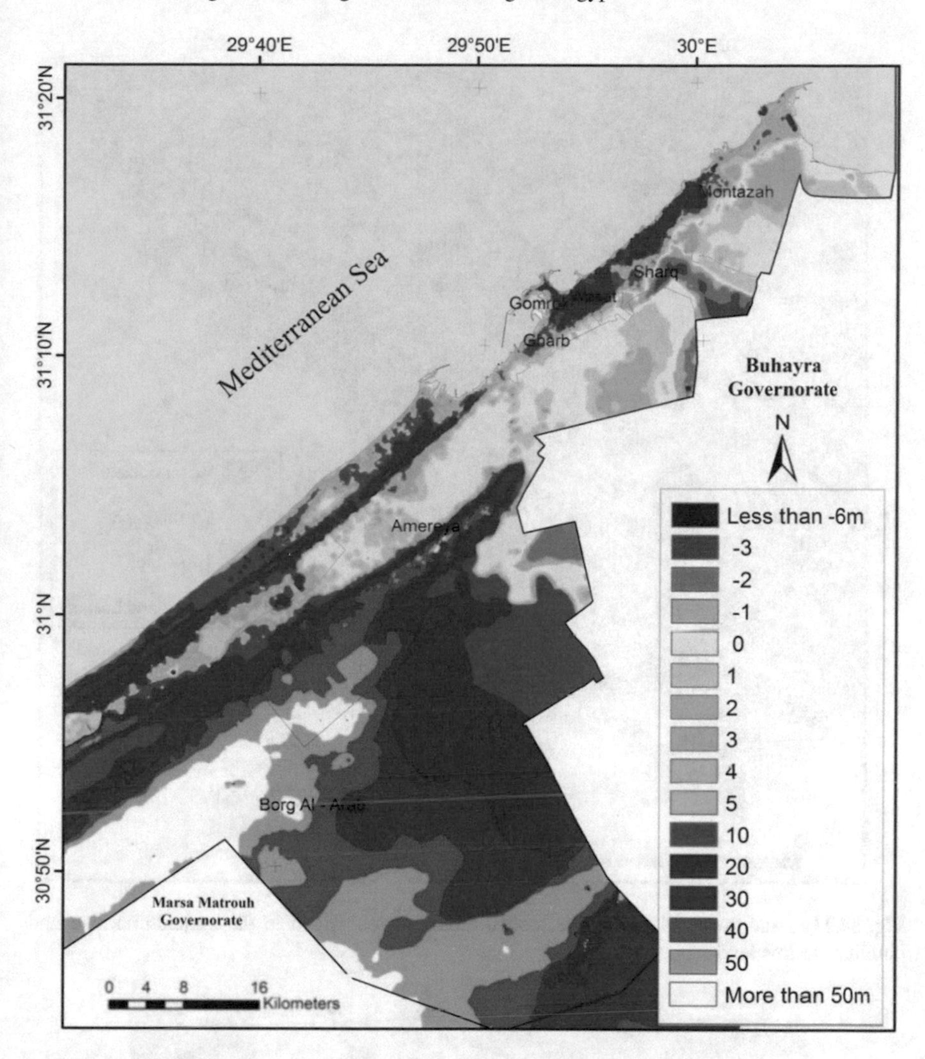

Fig. 14.20 Digital elevation model of the governorate of Alexandria. Notice the low elevation area to the East of Alexandria, in Montaza District

14.3.4 Urban Vulnerability

In addition to the high vulnerability of coastal zone, the high rates of population growth in Egypt have severe implications on urbanization rates. The urban encroachment on agricultural land in Egypt is shown in Fig. 14.24 for some towns in the Delta and Cairo and Alexandria, as well as many other cities. It shows the rapid urbanization, hence the rapid formation of urban heat islands of higher temperatures, which places more stress on human beings.

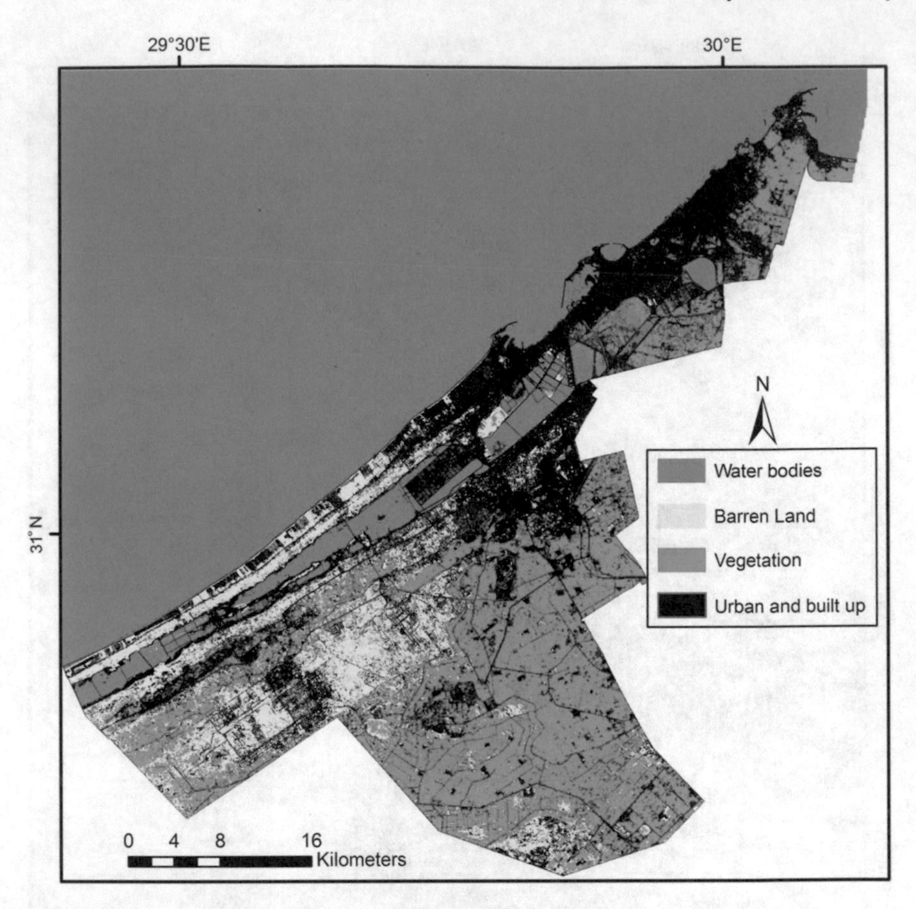

Fig. 14.21 Land cover of Alexandria, notice urban encroachment in the agricultural land and buildings in low land

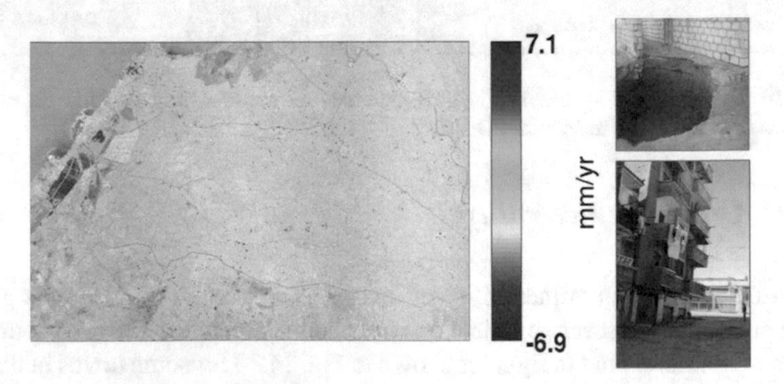

Fig. 14.22 Topographic deformation in the Alexandria Governorate from 2006 to 2010 using Differential SAR-interferometry (left). At right, examples of the damage caused in Alexandria due to local land subsidence (Sušnik et al. 2015)

Fig. 14.23 Coral reef along the coast of Hurghada, Egypt the Red Sea identifying vulnerable beaches (El-Askary et al. 2014)

Figure 14.25a, b show a satellite image of the urban areas of Giza and Cairo in 1984 and 2002. The impact of the encroachment of urbanization in the surrounding agricultural area without planning is evident. The problem here is that the troposphere of urban areas are heat storage areas and tend to increase temperature due to Urban Heat Islands (UHI) effect, also, to increase of temperature due to climate change. This is manifested in prevailing temperature inversions which keeps pollution down and cause severe episodes of pollution.

Remote sensing of urban pollution in the cities is a crucial aspect of revealing this effect in the delta and Greater Cairo region.

14.3.5 Resilience

To estimate and assess risks associated with an increase in frequency and severity of extreme events, we have to estimate the resilience of the community. Data collected from six districts in Alexandria are used to calculate the community resilience of each district. The vulnerability involved physical vulnerability parameters (including topographic slope, number of floors, building height, height above the sea level, and distance from the shoreline), and socio-economic vulnerability parameters (including building type and land use). Table 14.2 shows the numerical values of community resilience of each of the six districts of Alexandria estimated based on mapping of

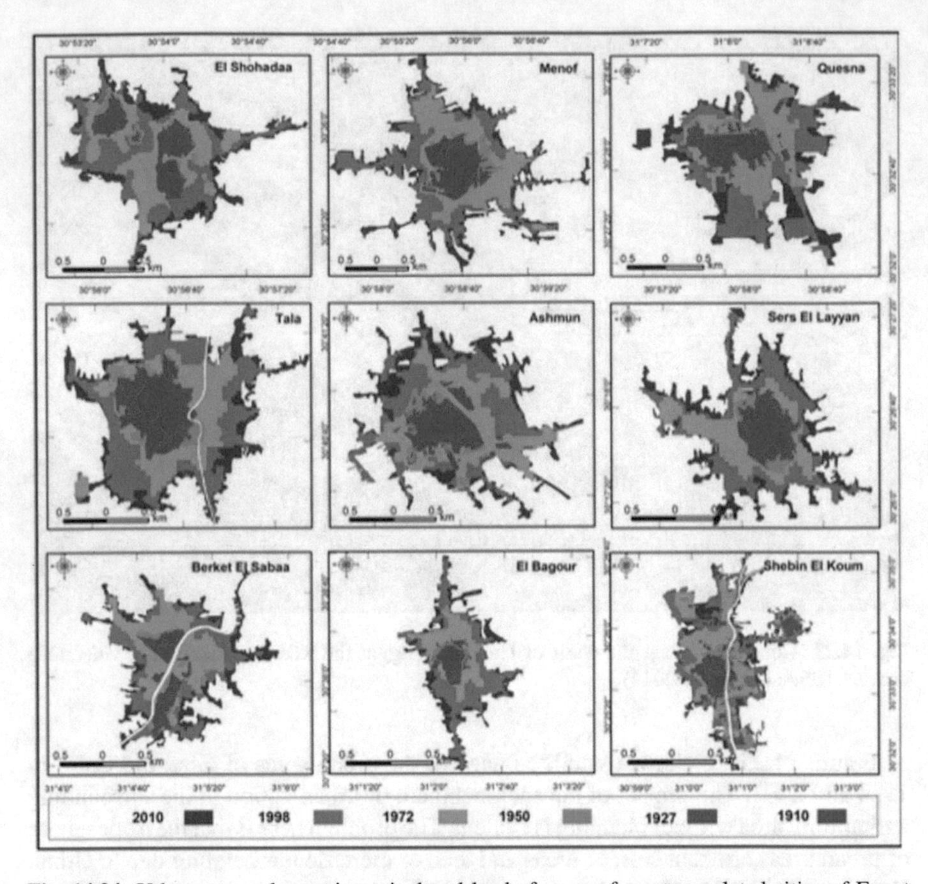

Fig. 14.24 Urban encroachment in agricultural land of some of most populated cities of Egypt (El-Magd et al. 2015)

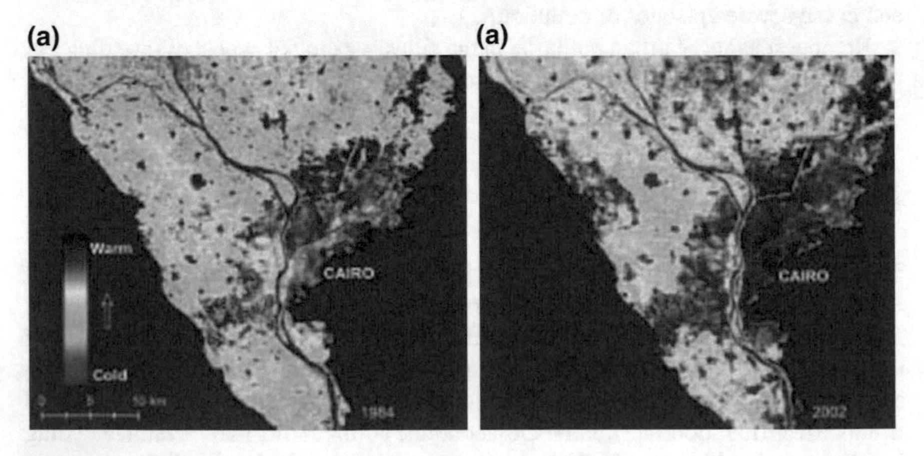

Fig. 14.25 Change of Cairo and Giza urban areas **a** 1984 **b** 2002. Notice changes in heat storage (Tolba and Saab 2009)

Table 14.2 GIS data from the six districts of Alexandria are used to retrieve resilience of the community for various scenarios of wave heights and extremes (Ahmed 2017)

Resilience index					
District	SLR Scenarios				Extreme event
	0.5 m Scenario	1 m Scenario	1.5 m Scenario	2 m Scenario	5 m Scenario
Montazah	0.4652	0.4600	0.4599	0.4405	0.14
Sharq	0.4790	0.4706	0.4558	0.4344	0.19
Wasat	0.3798	0.3798	0.3793	0.3769	0.34
Gomrok	0.2940	0.2938	0.2938	0.2932	0.28
Gharb	0.3632	0.3632	0.3615	0.3601	0.29
Amereya	0.4219	0.4218	0.4198	0.4132	0.05
Borg Al-Arab	0.2774	0.2753	0.2727	0.2683	0.25
Total	0.383	0.381	0.378	0.369	0.22

each, GIS values of population density, education, health, and technology capabilities (Ahmed 2017). Results indicate that the resilience of Alexandria City communities to impacts of surges is very low, which means that the risk is high.

14.3.6 Socioeconomic Vulnerability

Many researchers such as (Agrawala et al. 2004; Smith et al. 2014) addressed problems of socioeconomic implications of climate changes over Egypt. The main impact of climate change on the Nile Delta region would be deterioration of living conditions of people due to inundation of low elevation beaches on the Mediterranean. A half-meter rise in sea level would force 1.5 million people to evacuate and lose nearly 200,000 jobs. This will also have a disastrous impact on tourism, which accounts for 12.6% of the employed population. Reducing the area of beaches, saltwater intrusion in the groundwater of the northern delta contaminating groundwater reduces productivity and forces farmers and their dependents to abandon the land and move away to look for other jobs. These lead to the estimated immigration of 8 million people by 2100. It was estimated (El-Raey et al. 1997) that a loss of land in Alexandria exceeds $30 billion and a loss of jobs of almost 295,000 job over this century.

14.4 Remote Sensing for Mitigation Measures in Egypt

It was also recognized that mitigation and adaptation measures must be carried out, simultaneously. An internal network of most emitting factories is being monitored at EEAA Headquarters, day and night, by sensors placed in smokestacks, for compliance with the Egyptian regulation. The high need for adaptation was advocated as early as 1999 (El-Raey et al. 1999a). It is now recognized that (Stern 2008; IPCC-AR5 2014b) that no matter what we do for mitigation, we will still encounter some temperature rise because of the long lifetimes of gases, and we have to mitigate and adapt.

Assessing the impact of climate change on Egypt requires up-to-date, accurate and high-quality information. Due to the availability of historical data, reducing costs and increased resolution of satellite platforms, remote sensing technology seems to be expected to have a major impact on decision-making agencies and providing better understanding and prediction of the dynamics of the climate system, in order to mitigate the expected impact on Egypt.

Remote sensing of vegetation change is one of the widely used approaches for providing scientific evidence of climate change. For example, Fig. 14.26 shows monitored land cover type using MODIS yearly average dataset in Nile delta between 2001 and 2012. Classification and change detection was carried out to show the changes from grass or shrub (colored in orange) to wetland (color in navy) in the coastal area (areas in blue squares) due to accelerated inundation as a result of both climate change and anthropogenic activities. The increased cropland (colored in green) could also be found in the areas located in two wings of the southern Nile delta.

Remote sensing of atmospheric gases is being used for estimation of densities of atmospheric gasses in recognition of its long lifetime. Atmospheric Optical depth has been monitored by sensors on board of satellites including MODIS, MISR, ENVISAT, CALIPSO, and most recently the European Satellite Copernicus Sentinel-5p. Figure 14.27 shows an example of observing a dust storm across the Red Sea using

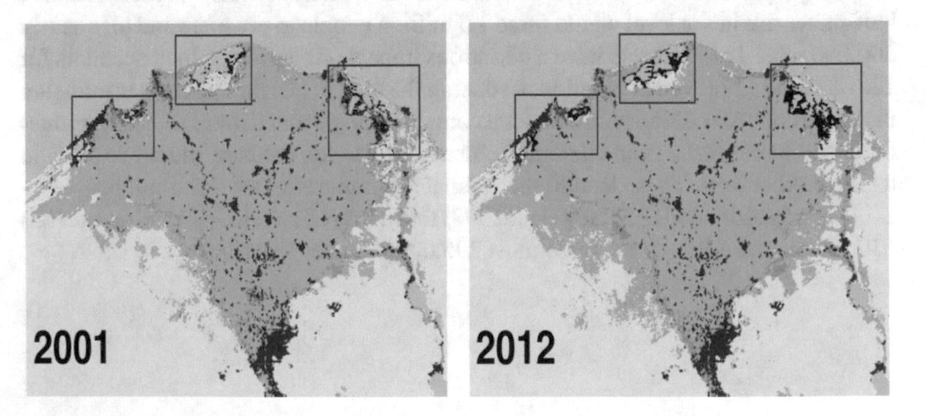

Fig. 14.26 Land cover changes in coastal areas of Nile delta from 2001 to 2012

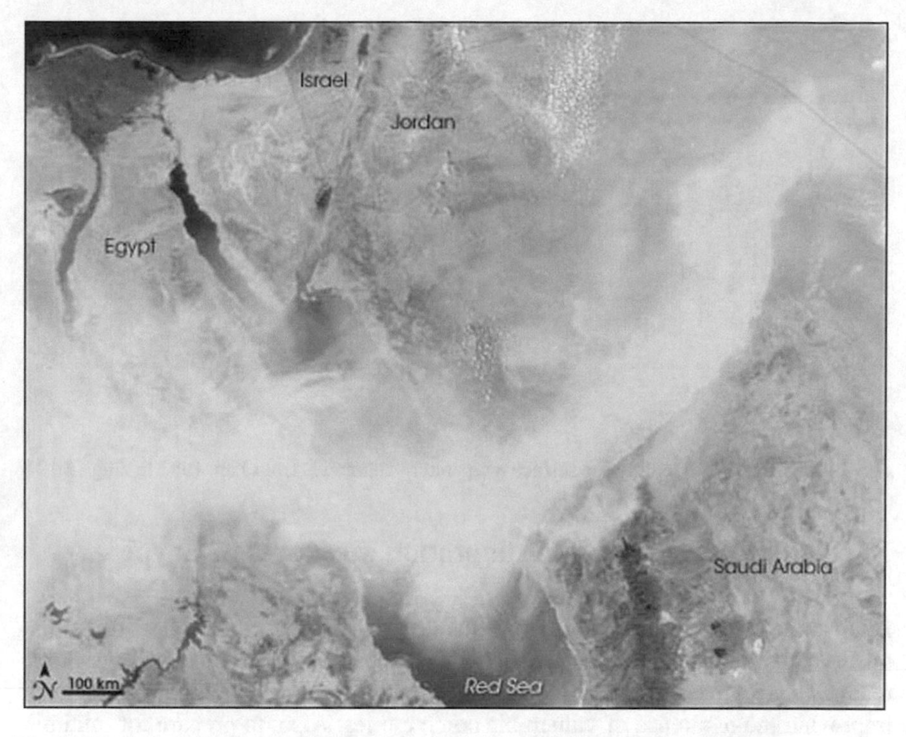

Fig. 14.27 A dust storm across the Red Sea, the frequency, and severity of which increases as the temperature increases (NASA 2005)

satellite image. The dust storm event has global impact which could reach as further as to even North America (El-Askary et al. 2018). It also influences meteorology parameters such as precipitation in the Nile delta region (El-Askary et al. 2019). Analysis of data from these satellites and others are necessary for better control of emissions of greenhouse gases, noting that Egypt contributes only 0.7% of world greenhouse gases.

Besides, ocean climate variability modifies both oceanic and terrestrial surface heat and CO_2 flux, resulted in strong impacts on the land surface temperature and soil moisture in Egypt. Oceanographic parameters such as sea surface temperature (SST), biomass, surface wind and sea surface height are important indicators of global weather conditions. A Group for High Resolution Sea Surface Temperature (GHRSST) Level 4 sea surface temperature analysis dataset uses satellite data from sensors that include AVHRR, AATSR, SEVIRI, AMSRE, TMI, MODIS, GOES, MTSAT-1R radiometer, and in situ data. Figure 14.28 shows the SST anomalies from the Aug 1981 to Feb 2018 near coast of northern Egypt using GHRSST dataset. The accelerated increase of SST is presented after the year 2008, with almost every month above the seasonal average.

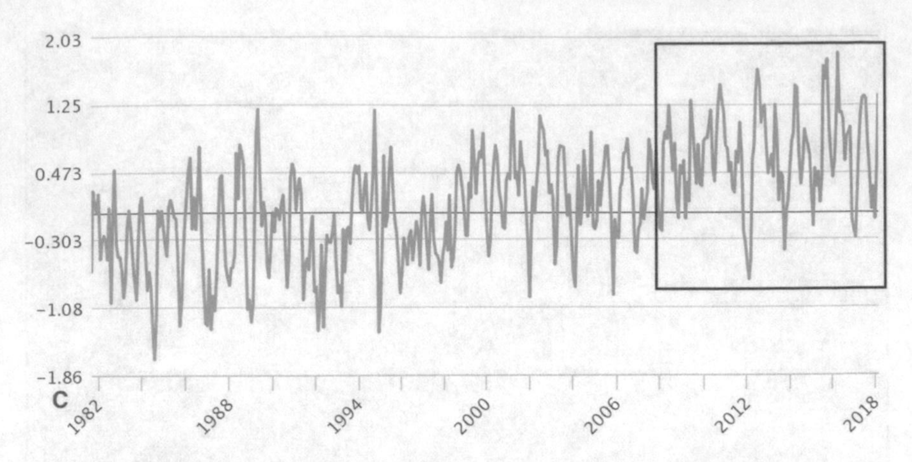

Fig. 14.28 GHRSST Monthly sea surface temperature anomalies from Feb. 1982 to Aug. 2018

14.5 Remote Sensing for Adaptation Measures in Egypt

Adaptation measures involve many separate and integrated activities, depending on the vulnerable sector. The primary objectives are to minimize impacts of climate changes through proactive planning, helping decision makers prioritizing and improving the resilience of vulnerable communities. Also, to prepare for minimizing risks associated with extreme events of hazards. Current technologies such as remote sensing (RS) and geographic information system (GIS) provide a cost effective and accurate alternative solution to make a great impact on planning agencies and providing better understanding the dynamics of the climate system, predict, adapt and mitigate the expected global changes and the effects on human civilization. For example, remotes sensing has been intensively applied in Land Use Land Cover (LU/LC) studies at a variety of spatial scales. Coupled with the accessible historical remote sensing data, the recent advances of remote sensing technologies with reduction in data cost and increased resolution from different satellite platforms, which improve our understanding of landscape dynamics and human-environment interaction. Furthermore, remote sensing has given a rise to the advent of more accurate and referenced data, which in turn have created opportunities for improved assessments and linking between the cover and the use of land.

Adaptation options in Egypt must go in harmony with the general strategy of development of the country's Sustainable Development Goals (SDG) (e.g., El-Raey 2011). One or more of these options can be selected depending on the case under consideration. The following adaptation options can be reactive or proactive, and we must reject the no action option.

1. Soft Measures (incentives for vulnerable stakeholders)
2. Hard Measures (Engineering structures)
3. Upgrading resilience of vulnerable communities
4. Proactive planning, follow up and Early warning systems

5. ICZM and stakeholders participation
6. Upgrading Research opportunities for vulnerability assessment and adaptation.

All of these options need to make use of remote sensing and GIS techniques.

14.6 Success Stories

There are many examples of success stories that have been developed based on the needs of the communities and to a great extent satisfy the criteria mentioned above. We select the following four case examples:

14.6.1 Aerosol and Dust Storms

In the atmosphere of Greater Cairo region (17 million people during the day), a persistent Black Cloud of Carbon appeared starting during September and October of each year, for about the past 12 years. It was realized that this is due to the burning of the waste of rice in the delta region, which causes this pollution that one cannot miss its smell. The prevailing temperature inversions in the troposphere caused by heating of the lower troposphere and UHI and climate changes cause air pollution to increase and may cause a black cloud. A study of the black cloud over Cairo has been carried out by (e.g., Aboel Fetouh et al. 2013; El-Askary and Kafatos 2008; El-Askary 2006; Marey et al. 2010, 2011; Prasad et al. 2010). The Egyptian Ministry of Environment has used satellites to control burning of rice ashes in the Delta through using daily images by MODIS sensors and has succeeded to a great extent.

14.6.2 Mohamed Ali Sea Wall

At the time of Mohamed Ali (1820), the people of eastern Alexandria complained that water that they drink through the canal from the River Nile reaches them saline. Mohamed Ali engineers thought to separate Abu Qir Lagoon and a part of Lake Maryut, by building Mohamed Ali Seawall (1830) on the eastern side of Alexandria city. After building the seawall, the land was dried, and people cultivated the dry land, built their houses and later established industries on an area of about 600 km^2, at 3–5 m below mean sea level. This area extends to Behaira Governorate. This was the first adaptation method in Egypt to get rid of saltwater intrusion.

Recently, due to the realization of adverse impacts of saltwater intrusion on agricultural productivity and human health, a network of sensors has been established to monitor changes on groundwater salinity through satellites to assess conditions, and estimate rate of saltwater intrusion. Also, the erosion rates encountered in this area is very high at the Rosetta promontory. These and adjacent shorelines have been well enforced by the shoreline authorities to reduce rates and minimize the risk of collapse and SLR is monitored by tide gauges and satellites.

14.6.3 Adaptation to Extreme Flash Flood (FLAFLOM-EWS)

An adaptation success story involves the use of modeling techniques to develop an early warning system against flash floods in eastern Egypt, particularly in Sinai. This early warning system, known as (FlaFloM) has been established in Sinai.

It is well known that the roads in Sinai are highly vulnerable to the impacts of flash floods almost every year. This is particularly risky for tourists and people living in these areas. As flash floods are expected to increase in the frequency of occurrence and severity with climate changes in the future, it becomes necessary to realize some dams to collect fresh water from stream torrents identified over mountains by remote sensing. Assessment of drainage pattern and stream torrents are carried out by remote sensing.

A project was developed primarily to be applied to Wadi Water Valley in Sinai (www.flaflom.org), to monitor, model, and issue an early warning message. The steps for the development of an EWS involves is shown in Fig. 14.29 (Cools et al. 2012):

14.6.4 Regional Circulation Models

Remote sensing can contribute to the climatic modeling of temperature and precipitation. As remote sensing monitors the sea surface temperature (SST) and Land surface temperature (LST), the needed data for input of models are usually provided by remote sensing or ground-based observation. Global Circulation Models (GCMs) have been used at the beginning and were downscaled to obtain RCMs. Then the Regional Circulation Models (RCMs) were used. However, RCM still needs regional information for verification.

The Regional Circulation Model RegC4 predicts future time series of temperature and precipitation and has proved accurate and reliable. Depending on the needed domain size and resolution, RCM models are computationally demanding, which limits the length and applicability of many experiments to date. Figure 14.30 shows the results of the application of Model RegCM4 for Egypt and the Gulf region indicating that the region is heading toward higher temperatures and less precipitation than today (Hemming et al. 2007).

According to (Almazroui 2011) the outcomes of the application of 28 regional models to Saudi Arabia indicate that the region should be subject to higher temperatures and less precipitation, over the next few decades, and we have to be prepared

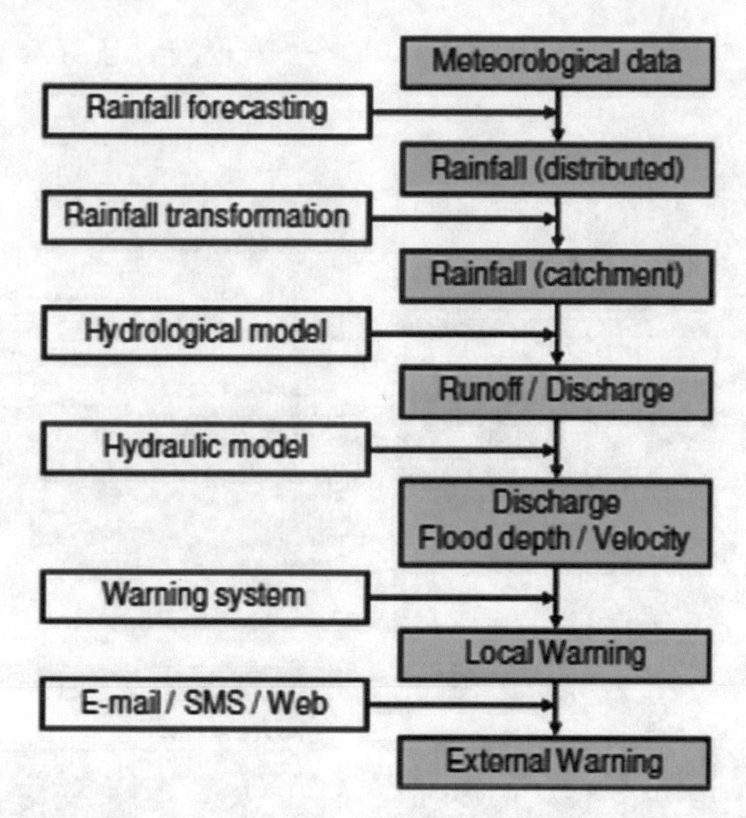

Fig. 14.29 The chain of components that forms the early warning system of Wadi Water, Sinai, Egypt (Cools et al. 2012)

to face that. To conclude this section, it is clear that satellite remote sensing and GIS have significantly contributed to the development of our understanding of the phenomenon of climate changes in Egypt.

14.7 Needs for Adaptation in Egypt

According to the most recent assessment by the team of Third National Communication, the priority is building an infrastructure for monitoring in the northern coastal zone of Egypt (TNC 2016). In general, it is necessary to carry out proactive planning, a strategic environmental assessment (SEA), and an Environmental impact assessment (EIA) before any development, taking care of renewable energy resources, wastewater treatment and utilization and recent technologies of transportation. It has also stressed the necessity to consider reducing the overconsumption of all resources. In general, it is necessary to consider transferring to a green economy and green technology.

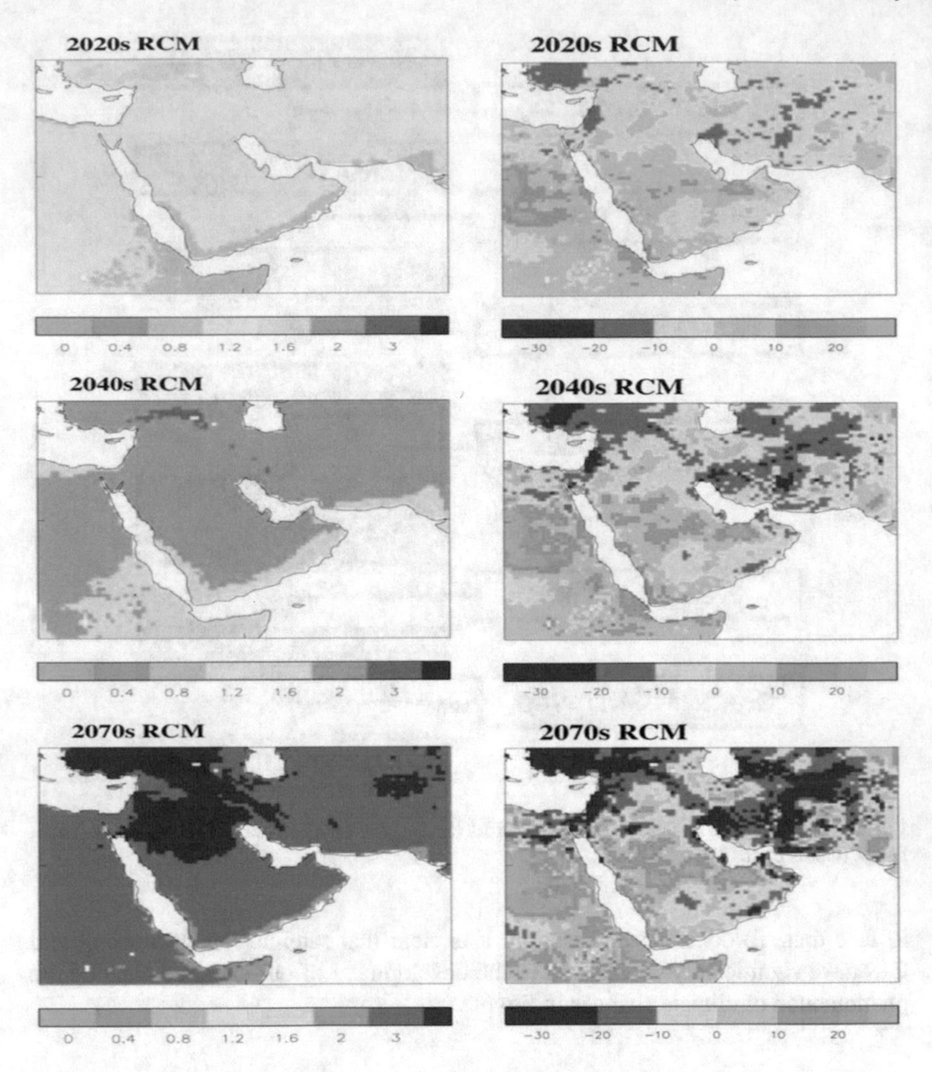

Fig. 14.30 Results of regional climate circulation model RegMC4 projections over Egypt and Gulf region for **a** average temperature changes, **b** precipitation change % for 2020s, 2040s, and 2070s relative to 1990s (Hemming et al. 2007)

The steps followed for implementation of the adaptation strategy follows the guidelines published by El-Raey et al. (1999b). A broad framework for the evaluation of adaptation strategies to cope with climate change can be identified. This comprises the following steps:

1. Defining the objectives. Some overall goals and evaluation principles must guide any analysis of adaptation. Two examples of general goals commonly propounded are (i) the promotion of sustainable development, and (ii) the reduction of vulnerability.

2. Specifying the climatic impacts of importance. This step involves an assessment, following the methods outlined elsewhere above, of the possible impacts of climate variability or change on the exposure unit. Where climatic events are expected that will cause damage, these need to be specified in detail so that the most appropriate adaptation options can be identified. A complete vulnerability assessment must be carried out so as to identify as accurately as possible the extent of the damage expected.
3. Identifying the adaptation options. The main task of assessment involves the compilation of a detailed list of possible adaptive responses that might be employed to cope with the effects of climate. The list can be compiled by field survey and by interviews with relevant experts, and should consider all practices currently or previously used, as well as possible alternative strategies that have not been used, and newly created or invented strategies. Six types of strategies for adaptation to the effects of climate have been identified: (i) prevention of loss (ii) tolerating loss (iii) spreading or sharing loss (iv) changing use or activity (v) changing location (vi) restoration.
4. Examining the constraints. Many of the adaptation options identified in the previous step are likely to be subject to legislation or be influenced by prevailing social norms, which may encourage. restrict or totally prohibit their use. Thus, it is important to examine closely, possibly in a separate study, what these constraints are and how they might affect the range of feasible choices available. Two important criteria should be taken into consideration when considering adaptation strategies: (i) Flexibility (ii) Benefits must exceed costs.
5. Quantifying measures and formulating alternative strategies. The next step is to assess the performance of each adaptation measure with respect to the stated objectives. It may be possible, if appropriate data and analytical tools exist, to use simulation models to test the effectiveness of different measures under different climatic scenarios. Historical and documentary evidence and survey material or expert judgments are alternative sources of this information. Uncertainty analysis and risk assessments are also considered at this stage. This step is a prelude to developing strategies, which maximize the level of achievement of some objectives while maintaining baseline levels of progress towards the remaining objectives.
6. Weighting objectives and evaluating trade-offs. This is the key evaluation step, where objectives must be weighted according to assigned preferences and then comparisons made between the effectiveness of different strategies in meeting these objectives. Standard impact accounting systems can be used in the evaluation. For example, a four-category system might consider: (i) national economic development; (ii) environmental quality; (iii) regional economic development; and (iv) other social effects. Selection of preferred strategies then requires the determination of trade- offs between the categories.
7. Recommending adaptation measures. The results of the evaluation process should be compiled in a form that provides policy advisers and decision-makers with information on the best available adaptation strategies. This should include some

indication of the assumptions and uncertainties involved in the evaluation procedure, and the rationale used (e.g., decision rules, key evaluation principles, national and international support, institutional feasibility, technical feasibility) to narrow the choices.

An institutional capability in the form of a multidisciplinary National Center for Climate Change is highly needed and is recommended. It should have components of.

14.7.1 Water Resources Needs for Adaptation

Development of unconventional water resources and adaptation measures is highly needed and timely yet requires a great deal of raising awareness along with other actions. These actions include shifting to modern agriculture and adopting water conservation and new irrigation techniques. It also involves a thorough analysis of precipitation and implementation of early warning systems for better preparedness for flood water saving. Periodic fixing of stream torrents, and proper management of groundwater should fall on the priority list as well. All of the aforementioned actions could benefit from earth observations, namely remote sensing and GIS as an essential component, for site selection, EIA, optimization and follow up. The following actions are recommended for water resources adaptation to climate changes:

1. Improve capacity in government departments and universities for training and implementation and management of water.
2. Develop a program for groundwater management to protect reservoirs from overexploitation and pollution.
3. A recent survey of infrastructure, along with improved drainage systems, is a crucial adaptation measure. Encouragement of Information exchange among researchers and governorates officials.
4. Build new catchments, identify and clean stream torrents and construct dams for water collection from flash flood areas such as Sinai, and east of the country.
5. Reactivate old rainwater catchments (Karma systems) detected by satellites on the northwestern coast and use for rainfed cultivation (to serve tourists especially during the summer). This will also reduce transport of food supplies and water from Alexandria city centers to coastal resorts.
6. Encouraging farmers of Egypt to switch the surface irrigation into modern irrigation techniques or/and aquaponics.
7. Protecting development from flooding during heavy rain such as that encountered in Alexandria, 2015. This involves mapping of topography and use of remote sensing.
8. These EWS are especially crucial for coastal cities, which need to manage freshwater and need to pump treated water up to natural watercourses. Encouraging rainwater harvesting and separation of the network for stormwater from the

general wastewater, in coastal cities. Accurate geographic Information systems are necessary which could be obtained only through remote sensing and GIS:

9. Develop renewable solar energy and wind energy for applications in desalinization of water and groundwater extraction.
10. Develop new methodologies for wastewater treatment and use for safe afforestation.

14.7.2 Agricultural Resources Needs for Adaptation

The following actions are recommended for agricultural resources adaptation to climate changes:

1. Immediate change of crop distribution. Change of water irrigation techniques to water-efficient agriculture and aquaponics to decrease net losses in agricultural distribution systems. Currently the irrigation efficiency is assumed to be 80% across most of the country with the upper and middle Egypt regions at 75% and the Delta at 77% (McCarl et al. 2015).
2. Development of a market shared cropping pattern. Development of genetically high yield crops, tolerant to high temperatures, and water efficient. Development of modern techniques of irrigation and cultivation.
3. Diversification in time and space, increased drought tolerance, and flood resilience.
4. Better control of rice cultivation and less burning of rice ashes.
5. Research on plantations that are tolerant of heat, droughts and extreme events.
6. These are all monitored by satellites and estimated accurately by satellites and GIS.
7. Improved municipal and industrial distribution efficiency. Current distribution efficiency results in 50% of the water distributed to municipal and industrial sector being lost through network leakage or other forms of inefficiency (McCarl et al. 2015).
8. Involve increasing technical progress in the form of faster crop yield growth.
9. Develop programs to upgrade awareness of farmers on water conservation and modern irrigation techniques. Population growth rate reduction is also possible to assess its impact on the agricultural sector, which slow population growth leads to a reduction in water and agricultural demands and thus lower stress placed on the land and water base (McCarl et al. 2015).

14.7.3 Coastal Resources Needs for Adaptation

To consider options of adaptation measures in coastal areas, it is necessary to consider large-scale structures. This can be selected and monitored for effectiveness by remote sensing techniques. Measures and action should include Integrated Coastal Management (ICZM), aquaculture and living shoreline.

ICZM

Integrated Coastal Management (ICZM) is a coastal management process that uses an integrated approach, regarding all aspects of the coastal zone, including geographic and political boundaries, to achieve sustainability. Periodic nourishment, in a package of ICZM, may be the best solution to some of the eroding and subsiding beaches. Natural water barriers (coastal dunes, natural dikes), can be combined by artificial nourishment and walls (dikes, levees, and super levees) and temporary storm surge barriers to prevent coastal or river flooding. The construction of permanent flood walls needs to take into consideration historical and current flood trends, future sea-level rise and higher frequency of extreme weather events after consultation with vulnerable stakeholders. Underwater breakers may be designed to protect beaches from strong marine surges. Additionally, flexible storm surge barriers require apriori comprehensive analysis. Strengthening a network of permanent dikes and levees, building flexible, extensive levees on river banks that allow for urban development on top of the barriers as well as direct public access to riverside recreation areas. Remote sensing can monitor all parameters.

The project "Adaptation to climate change in the Nile Delta through Integrated Coastal Zone Management in Egypt", is to integrate the management of SLR risks into the development of Egypt's Low Elevation Coastal Zone (LECZ) in the Nile Delta. This project was co-financed US$12,840,000 (as of December 1, 2010), by UNDP Special Climate Change Fund (SCCF), Egypt Ministry of Water Resources and Irrigation and Coastal Research Institute. The ICZM plan was developed for the North Coast of Egypt that links the plan for shore protection from sea-level rise with the national development plan of the coastal zones. The ICZM plan will be associated with the establishment of a systematic observation system, including in situ surveys and remote sensing observations to monitor Oceanographic parameters changes under a changing climate as well as the impact of the different shore protection scenarios on the coastal erosion and shore stability.

Aquaculture

Aquaculture in the coastal zone is a top priority option to be implemented on vulnerable coastal areas, in fact, the implementation of aquaculture at Kafr El Shiekh is considered one of the most crucial adaptation measures, of Egypt. These can be monitored and assessed by remote sensing for water quality, depth of water in each basin, adjacent basin productivity. The infrastructure needed for transport and management of the products can be monitored by satellites.

Some recommended adaptive options to climate change and SLR along the Mediterranean coast of Egypt are listed below:

1. Setting up regulations and guidelines for coastal development incorporating the adaptation of the impacts of the SLR. Incorporating the effect of the SLR in the context of the environmental impact assessment (EIA) and strategic

impact assessment (SEA). Setting limits and binding requirements when issuing permits to drill wells for the withdrawal of groundwater in the coastal zone.
2. Designing engineering structures to protect high risk vulnerable areas to inundation. Introducing appropriate measures to mitigate penetration of seawater toward the cultivated Delta soil.
3. Strengthening the protection works which are basically functioning to mitigate beach erosion and wave overtopping to be an integrated element of the protection system.
4. Rehabilitation and strengthening the existed engineering coastal structures, particularly on the low-lying lands and areas at risk, to adapt with the likely impacts of the SLR. Examples of the coastal structures are: the international coastal road, Mohamed Aly Seawall wall, and the banks of Al-Salam Canal that connect the Damietta branch to Sinai.
5. Encouraging the application of nonstructural erosion protection techniques as part of a Living Shorelines approach applied by the UNDP's "Living with the Sea" Project in the Nile Delta of Egypt.
6. Preserving the natural protection environments, this can be realized by stabilizing dunes, creating or restoring wetlands, prohibiting land infill of wetlands and preventing quarrying and destruction of the shore-parallel carbonate ridges.
7. Integrating the adaptation plans and strategies to climate change and SLR in the national development plans.
8. Enhancing public and scientific awareness and capacity building.
9. Enhancing institutional capabilities for monitoring and assessment.
10. Considering human and fund resources mobilization plans and developing an early warning system.

Living Shoreline
Establishment of a living shoreline on islands along the shoreline can provide scenic views and is meant to protect the urban areas from severe storms such as that associated with extreme surge events and Tsunamis. Figure 14.31, illustrates a simulation of a living shoreline for Alexandria.

14.7.4 *Urban Needs for Adaptation*

Proactive planning of urban areas is necessary. Monitoring of development structure for compliance with planning and enforcement of regulations which also could be carried out by remote sensing.

1. Modeling of temperature and precipitation, collecting rainwater and wind flow. Advancing proper contingency plans for flash floods and heat waves.

Fig. 14.31 A living shoreline for the coastal area of Alexandria. It illustrates protection of beaches against Tsunamis and severe storms

2. Considering renewable energy sources, wastewater treatment and upgrading of awareness of population.
3. Upgrading awareness of hazardous occasions among children in schools.
4. Reducing risks associated with Tsunamis by building underwater breakers in open coastal areas.

14.7.5 Socioeconomic Measures of Adaptation

Removal of slum areas and upgrading of awareness is the first step to upgrading the resilience of the community for identifying any changes in climatic conditions. This again can be checked by remote sensing and GIS techniques. For example, NASA Socioeconomic Data and Applications Center (SEDAC) developed products and applications through synthesizing Earth science and socioeconomic data, to provide useful information to decision makers and other applied users.

14.8 Summary

Climate change is a global phenomenon that has impacts on a local scale; it could be monitored and assessed by a global monitoring system and local ones. Remote sensing by satellites offers the best available platform to monitor, assess vulnerability to impacts, mitigate, and adapt to this phenomenon. Advantages include periodic coverage, universal calibration, and wide area view.

Egyptian resources, despite being vulnerable to climate changes, yet it has not considered adaptation measures for institutional structure. It has an immense amount of renewable solar energy, yet it has not exploited it. Egypt has one of the most significant rates of population growth and poverty in the world, yet it has not enforced control of population growth, and urban encroachment in agricultural land, at least for food security.

Satellite monitoring of emissions of greenhouse gases is an essential tool for control of emissions in case of many developed countries, and remote sensing is the primary tool for vulnerable developing countries to carry out adaptation applications:

1. Satellite remote sensing and GIS systems with image processing can be a handy large-scale tool, to monitor, mitigate atmospheric impacts and assess vulnerability, to adapt and to follow up its implementation. It can also monitor and assess critical early warning of extreme climatic impacts such as heat waves, droughts, flash floods, forest fires and many other anthropogenic impacts.
2. High-resolution satellite data can help proactive planning and provide adaptation measures clues involving large-scale infrastructures and models such as afforestation that can be monitored by high-resolution satellites.
3. Early warning based on satellites systems can forecast and predict extreme events of flash flooding, droughts, heat waves and fires and can save lives. A program with a contingency plan must be developed.
4. Advanced and fast techniques of change detection could be easily interpreted with the availability of satellite images to identify and assess unplanned changes to enforce regulations.
5. Remote sensing can follow up adaptation measures and provide useful indicators of success and failure. Proper exchange of data and information is a prerequisite. It can help finding renewable resources of groundwater and identify conditions of stream torrents.
6. GIS data retrieved for various districts can be used to estimate the resilience of each district which facilitates the immediate action of the community.

References

Aboel Fetouh Y, El Askary II, El-Raey M, Allali M, Sprigg WA, Kafatos M (2013) Annual patterns of atmospheric pollutions and episodes over Cairo Egypt. Adv Meteorol 2013:1–11

AerosolWatch (2018) GOES-16 image of Hurricane Frances on September 10, 2018 at 1600 UTC. Retrieved from https://www.star.nesdis.noaa.gov/smcd/spb/aq/AerosolWatch

Agrawala S, Moehner A, El-Raey M, Conway D, Van Aalst M, Hagenstad M et al (2004) Development and climate change in Egypt: focus on coastal resources and the Nile. Organisation for Economic Co-operation and Development

Ahmed S (2017) Environmental risk assessment in the city of Alexandria using remote sensing and GIS. Ph.D. thesis

Almazroui M (2011) Sensitivity of a regional climate model on the simulation of high intensity rainfall events over the Arabian Peninsula and around Jeddah (Saudi Arabia). Theor Appl Climatol 104(1–2):261–276

Becker RH, Sultan M (2009) Land subsidence in the Nile Delta: inferences from radar interferometry. Holocene 19(6):949–954

Central Agency for Public Mobilization and Statistics (CAPMAS) (2016) Statistical year book

Cools J, Vanderkimpen P, Afandi GE, Abdelkhalek A, Fockedey S, Sammany ME et al (2012) An early warning system for flash floods in hyper-arid Egypt. Nat Hazards Earth Syst Sci 12(2):443–457

Deutscher Wetterdienst (2018) Weather satellites around the world. Retrieved from https://www.dwd.de/EN/research/observing_atmosphere/satellites/weather_satellites_node.html

Dubovik O, King MD (2000) A flexible inversion algorithm for retrieval of aerosol optical properties from sun and sky radiance measurements. J Geophys Res Atmos 105(D16):20673–20696

Dubovik O, Smirnov A, Holben BN, King MD, Kaufman YJ, Eck TF et al (2000) Accuracy assessments of aerosol optical properties retrieved from Aerosol Robotic Network (AERONET) sun and sky radiance measurements. J Geophys Res Atmos 105(D8):9791–9806

Dubovik O, Holben B, Eck TF, Smirnov A, Kaufman YJ, King MD et al (2002) Variability of absorption and optical properties of key aerosol types observed in worldwide locations. J Atmos Sci 59(3):590–608

Egyptian Environmental Affairs Agency (EEAA) (2010) Egypt second national communication under the United Nations Framework Convention on Climate Change (UNFCCC). Egypt Environmental Affairs Agency, Cairo

El-Askary H (2006) Air pollution impact on aerosol variability over mega cities using remote sensing technology: case study, Cairo Egypt. Egypt J Remote Sens Space Sci 9:31–40

El-Askary H, Kafatos M (2008) Dust storm and black cloud influence on aerosol optical properties over Cairo and the Greater Delta region Egypt. Int J Remote Sens 29(24):7199–7211

El-Askary H, Abd El-Mawla SH, Li J, El-Hattab MM, El-Raey M (2014) Change detection of coral reef habitat using Landsat-5 TM, Landsat 7 ETM+ and Landsat 8 OLI data in the Red Sea (Hurghada, Egypt). Int J Remote Sens 35(6):2327–2346

El-Askary H, Li J, Li W, Piechota T, Ta T, Jong A et al (2018) Impacts of aerosols on the retreat of the Sierra Nevada Glaciers in California. Aerosol Air Qual Res 18(5):1317–1330

El-Askary H, Li W, El-Nadry M, Awad M, Mostafa AR (2019) Strong interactions indicated between dust aerosols and precipitation related clouds in the Nile Delta. In: El-Askary HM et al (eds) Advances in remote sensing and geo informatics applications. Springer International Publishing, Cham, pp 1–4

Elbeih SF (2015) An overview of integrated remote sensing and GIS for groundwater mapping in Egypt. Ain Shams Eng J 6(1):1–15

El-Hattab MM, Mohamed SA, El-Raey M (2018) Potential tsunami risk assessment to the city of Alexandria, Egypt. Environ Monit Assess 190(9)

El-Magd IA, Hasan A, El Sayed A (2015) A century of monitoring urban growth in Menofya Governorate, Egypt, using remote sensing and geographic information analysis. JGIS 07:402–414

El-Raey M (1997) Vulnerability assessment of the coastal zone of the Nile Delta of Egypt, to the impacts of sea level rise. Ocean Coast Manage 37(1):29–40

El-Raey M (2011) Mapping areas affected by sea-level rise due to climate change in the Nile Delta until 2100. In: Coping with global environmental change, disasters and security. Springer, Berlin, Heidelberg, pp 773–788

El-Raey M, Fouda Y, Nasr S (1997) GIS assessment of the vulnerability of the Rosetta area, Egypt to impacts of sea rise. Environ Monit Assess 47(1):59–77

El-Raey M, Frihy O, Nasr SM, Dewidar KH (1999a) Vulnerability assessment of sea level rise over Port Said Governorate, Egypt. Environ Monit Assess 56(2):113–128

El-Raey M, Dewidar KR, El-Hattab M (1999b) Adaptation to the impacts of sea level rise in Egypt. Mitig Adapt Strat Glob Change 4(3–4):343–361

El-Raey M, Nasr S, Frihy O, Desouki S, Dewidar K (1995) Potential impacts of accelerated sea-level rise on Alexandria Governorate, Egypt. J Coast Res 190–204

Food and Agriculture Organization (2011) FAO vulnerability map for Egypt. Retrieved from https://www1.wfp.org

Hemming D, Iowe J, Biginton M, Betts R, Ryall D (2007) Impacts of mean sea level rise based on current state-of-the-art modeling. Hadley Centre for Climate Prediction and Research, Exeter, UK

Intergovernmental Panel on Climate Changes (2014a) Fourth assessment report (IPCC-AR4)
Intergovernmental Panel on Climate Changes (2014b) Fifth assessment report (IPCC-AR5)
Japan Aerospace Exploration Agency (JAXA) (2015, Oct 31) JAXA global rainfall watch. Retrieved from https://sharaku.eorc.jaxa.jp/GSMaP
Kim S-R, Prasad AK, El-Askary H, Lee W-K, Kwak D-A, Lee S-H et al (2014) Application of the Savitzky-Golay filter to land cover classification using temporal MODIS vegetation indices. Photogrammetric Eng Remote Sens 80(7):675–685
Lauvaux T, Miles NL, Deng A, Richardson SJ, Cambaliza MO, Davis KJ et al (2016) High-resolution atmospheric inversion of urban CO_2 emissions during the dormant season of the Indianapolis Flux Experiment (INFLUX). J Geophys Res Atmos 121(10):5213–5236
Li W, El-Askary H, ManiKandan K, Qurban M, Garay M, Kalashnikova O (2017) Synergistic use of remote sensing and modeling to assess an anomalously high chlorophyll-a event during summer 2015 in the South Central Red Sea. Remote Sens 9(8):778
Li W, El-Askary H, Qurban M, Proestakis E, Garay M, Kalashnikova O et al (2018) An assessment of atmospheric and meteorological factors regulating Red Sea phytoplankton growth. Remote Sens 10(5):673
Li W, El-Askary H, Qurban M, Allali M, Manikandan KP (2019) On the drying trends over the MENA countries using harmonic analysis of the enhanced vegetation index. In: El-Askary HM et al (eds) Advances in remote sensing and geo informatics applications. Springer International Publishing, Cham, pp 181–183
Marey HS, Gille JC, El-Askary HM, Shalaby EA, El-Raey ME (2010) Study of the formation of the "black cloud" and its dynamics over Cairo, Egypt, using MODIS and MISR sensors. J Geophys Res Atmos 115(D21)
Marey HS, Gille JC, El-Askary HM, Shalaby EA, El-Raey ME (2011) Aerosol climatology over Nile Delta based on MODIS, MISR and OMI satellite data. Atmos Chem Phys 11(20):10637–10648
Masria A, Negm A, Iskander M, Saavedra O (2014) Coastal zone issues: a case study (Egypt). Procedia Eng 70:1102–1111
McCarl BA, Musumba M, Smith JB, Kirshen P, Jones R, El-Ganzori A et al (2015) Climate change vulnerability and adaptation strategies in Egypt's agricultural sector. Mitig Adapt Strat Glob Change 20(7):1097–1109. https://doi.org/10.1007/s11027-013-9520-9
Moss R, Babiker M, Brinkman S, Calvo E, Carter T, Edmonds J et al (2008) Towards new scenarios for analysis of emissions, climate change, impacts, and response strategies. Intergovernmental Panel on Climate Change, Geneva, p 132
Nakajima T, Tonna G, Rao R, Boi P, Kaufman Y, Holben B (1996) Use of sky brightness measurements from ground for remote sensing of particulate polydispersions. Appl Opt 35(15):2672–2686
Nile Basin Initiative (2012) State of the River Nile basin. Nile Basin Initiative Secretariat, Entebbe, Uganda
National Aeronautics and Space Administration (NASA) (2005) Dust storm across the Red Sea. Retrieved from https://visibleearth.nasa.gov/view.php?id=72897
National Aeronautics and Space Administration (NASA) (2011) Global climate change: vital signs of the planet. Retrieved from https://climate.nasa.gov/
National Aeronautics and Space Administration (NASA) (2014) A-train constellation with details. Retrieved from https://oco.jpl.nasa.gov/galleries/galleryspacecraft/
National Aeronautics and Space Administration (NASA) (2018) EOS Aura atmosphere profile measurements. Retrieved from https://aura.gsfc.nasa.gov/images/instruments/eoschemchart_big.jpg
National Oceanic and Atmospheric Administration (NOAA) (1996) Hurricane Fran near peak intensity on September 4, 1996 at 1700Z. Retrieved from class.ncdc.noaa.gov
Natural Resources Canada (2016) Fundamentals of remote sensing tutorial [PDF file]. Retrieved from https://www.nrcan.gc.ca/sites/www.nrcan.gc.ca/files/earthsciences/pdf/resource/tutor/fundam/pdf/fundamentals_e.pdf

Nerem RS, National Center for Atmospheric Research Staff (eds) (2016, Jan 19). The climate data guide: global mean sea level from TOPEX & Jason Altimetry. Retrieved from https://climatedataguide.ucar.edu/climate-data/global-mean-sea-level-topex-jason-altimetry

NOAA National Centers for Environmental Information (2018a) State of the climate: global climate report for June 2018. Retrieved from https://www.ncdc.noaa.gov/sotc/global/201806

NOAA National Centers for Environmental information (2018b) Climate at a glance: global time series. Retrieved from https://www.ncdc.noaa.gov/cag/

Prasad AK, El-Askary H, Kafatos M (2010) Implications of high altitude desert dust transport from Western Sahara to Nile Delta during biomass burning season. Environ Pollut 158(11):3385–3391

Sestini G (1989) Implications of climate change for the Nile Delta. Report WG 2/14. UNEP/OCA, Nairobi

Smith JB, McCarl BA, Kirshen P, Jones R, Deck L, Abdrabo MA et al (2014) Egypt's economic vulnerability to climate change. Clim Res 62(1):59–70

Solomon S, Qin D, Manning M, Averyt K, Marquis M (eds) (2007) Climate change 2007-the physical science basis: working group I contribution to the fourth assessment report of the IPCC, vol 4. Cambridge University Press, Cambridge

Stanley DJ, Warne AG (1993) Nile Delta: recent geological evolution and human impact. Science 260(5108):628–634

Stern N (2008) The economics of climate change. Am Econ Rev 98(2):1–37

Strzepek KM, Onyeji SC, Saleh M, Yates D (1995) An assessment of integrated climate change impacts on Egypt. Cambridge University Press, Cambridge, UK and New York, NY, USA, pp 180–200

Strzepek KM, Yates DN, El Quosy DED (1996) Vulnerability assessment of water resources in Egypt to climatic change in the Nile basin. Clim Res 6(2):89–95

Sušnik J, Vamvakeridou-Lyroudia LS, Savić DA, Kapelan Z (2013) Integrated modelling of a coupled water-agricultural system using system dynamics. J Water Clim Change 4(3):209–231

Sušnik J, Vamvakeridou-Lyroudia LS, Baumert N, Kloos J, Renaud FG, La Jeunesse I et al (2015) Interdisciplinary assessment of sea-level rise and climate change impacts on the lower Nile Delta, Egypt. Sci Total Environ 503:279–288

The Sustainable Agricultural Development Strategy towards 2030 (SADS) (2010) Agricultural Research & Development Council (ARDC)

Third National Communication (TNC) (2016) Egypt third national communication under the United Nations Framework Convention on Climate Change (UNFCCC). Egypt Ministry of Environment

Tolba MK, Saab NW (2009) Arab environment: climate change. Beirut, Arab forum for environment and development

Whitney K, Scudiero E, El-Askary HM, Skaggs TH, Allali M, Corwin DL (2018) Validating the use of MODIS time series for salinity assessment over agricultural soils in California, USA. Ecol Ind 93:889–898

Wurl D, Grainger RG, McDonald AJ, Deshler T (2010) Optimal estimation retrieval of aerosol microphysical properties from SAGE~II satellite observations in the volcanically unperturbed lower stratosphere. Atmos Chem Phys 10(9):4295–4317

Yang J, Gong P, Fu R, Zhang M, Chen J, Liang S et al (2013) The role of satellite remote sensing in climate change studies. Nat Clim Change 3(10):875

Chapter 15
Qualitative and Quantitative Assessment of Land Degradation and Desertification in Egypt Based on Satellite Remote Sensing: Urbanization, Salinization and Wind Erosion

A. Gad

Abstract This chapter analyzes, generally, the problems of desertification and land degradation, highlighting the difference between the two terms and aspects to be assessed. The common land degradation/desertification processes, in Egypt will be highlighted, however the current chapter, as a start of an articles series focusing on desertification, will only consider three important processes (i.e. urban encroachment, salinization and wind erosion) to be detailed studied. Both the descriptive and quantitative approaches will be followed and merged, showing advantages of combining both approaches in assessment, sizing and combating preparedness. Regional assessment scale for the whole territory of Egypt, in addition to some detailed case studies will be introduced, with adaptation of indicators scale. Remote sensing, in addition to thematic maps, may supply valuable information concerning landscape features, vegetation type and quality and land use/cover, as inputs of the FAO-UNEP provisional methodology to assess aspects of different desertification processes. Multi scale and multi spectral satellite sensors supply reliable data sources to point out variable indicators needed to evaluate the present status and risk of different degradation/desertification processes. Multi temporal satellite imageries and thematic data make it possible to detect temporal land use/cover changes, hence compute the annual rate of a desertification process. The EU-MEDLUS methodology assessing the environmental sensitivity to desertification is rather due to the use of remote sensing data in computing the Soil Quality Index (SQI), Vegetation Quality Index (VQI), and Management Quality Index (MQI). Climate Quality Index (CQI) may be computed, using meteorology satellites data. The Geographic Information System (GIS) is a valuable tool to store, retrieve, update and manipulate the huge amount of data needed to map aspects if each desertification process. The system also facilitates computation and mapping environmental parameters of different quality indices, hence determining Desertification Sensitivity Areas (DSA's). The Egyptian territory is susceptible to very high-to-high desertification sensitivity. However the

A. Gad (✉)
National Authority for Remote Sensing and Space Sciences (NARSS), 23 Joseph Tito Street, El-Nozha El-Gedida, P.O. Box: 1564 Alf-Maskan, Cairo, Egypt
e-mail: agad@narss.sci.eg

S. F. Elbeih et al. (eds.), *Environmental Remote Sensing in Egypt*,
Springer Geophysics, https://doi.org/10.1007/978-3-030-39593-3_15

Nile Valley is moderately sensitive due to cultivated vegetation cover. Combating desertification measures are essential for the sustainable agriculture. Special concerns have to taken at the desert oases, wadis and interference zone due to their rule in decreasing food gaps and accelerating agriculture expansion. Operational innovative monitoring is recommended to an early control of desertification sensitivity. Defining and followup the Environmentally Sensitive Areas (ESA's) are needed to point out the risk, magnitude and causes of land degradation and desertification processes. Combining both descriptive and quantitative desertification approaches may grantee full sizing of desertification impacts.

Keywords Land degradation · Desertification · Erosion · Urbanization · Salinization · Remote sensing · Egypt

15.1 Introduction

The studied topic is a combination of different processes resulting in the exposure of the soil to other agents of deterioration (FAO 1986). It includes the clearing of ground cover, bush fire, over grazing and destruction of woody species. The progressive disappearance of favorable pasture is also included in this category. Numerous definitions of deserts were developed, where none got common agreement about what constitute the desert. However, the most widely accepted definitions are those using climatic factors. Desert regions are not necessarily characterized by great heat, nor do they always consist of vast expanses of shifting sand dunes. Polar deserts rarely experience temperature in excess of 10 °C. The common characteristic to all desert types is their aridity throughout most or all year (Thompson 1977). The arid regions of the world fall into three classes; Semi-arid, arid and hyper-arid. The semi-arid areas have an annual rain fall less than 2/3 of the potential evapotranspiration. Arid areas have less than 1/3 and hyper-arid areas have, additionally, at least one 12-month period without rainfall. The hyper-arid Saharan areas, including Egypt, hardly sustain life based on cultivation. Irrigation is the essential way to any kind of intensive agriculture.

Desertification, the extension of desert like conditions, is a severe problem in the arid and semi-arid regions. Younes (1982) identified desertification as the diminution or destruction of the biological potential of the land which can lead ultimately to desert like conditions. It occurs through different processes (e.g. wind erosion, water erosion, vegetation degradation, salinization…etc.). The United Nation Environmental Program—UNEP (1977) stated that 20 million square kilometers have recently reverted to desert or desert-like conditions. Human life has always been closely linked with environment and habitat. The pressure created by industrialization and urbanization has affected the environment's capacity to absorb the results of man's activities and preserve the balance in the natural equilibrium which has existed so far. Climate has not, during the past millennia, changed drastically. Therefore, the pollution that exists today and the irrational exploitation of natural resources are the direct products of man's activities.

Desertification conditions are to a certain extend man-made. They are caused by one or a combination of the following: transient (patchy) tillage (agriculture), overgrazing, lumbering and deforestation of perennials; drought, agriculture mismanagement practices, fire, Aeolian sand deposition and dune formation; soil erosion, water logging and soil salinization (Helldén 1991).

The Egyptian deserts occupy nearly 96% of the country's surface areas. Parts of these deserts were different in the not too distant past. Egypt was a great field for cultivating wheat in the time of the Roman civilization. Also, history tells about the oases where Qampis lost his army under the creeping sand dunes. The paleo botanists, looking at the fossils embedded in the sedimentary rocks, refer that most of these deserts can be older than the late Cenozoic (Leopold 1963) and El-Baz (1982a, b), in a trip to the Egyptian Western Desert, referred to the findings of his companions Drs. Haynes and Mchugh of eggshell pieces, which were later carbon dated as being more than 8000 years old. Droughts and crop failures have always threatened arid lands, as Joseph recognized when he advised Pharon to set aside grain reserves in ancient Egypt.

15.1.1 Scope of Desertification Concept

The word desertification was apparently introduced by Aubreville (1949) while he was observing the progressive replacement of tropical and sub-tropical forests by savannas, a process he called savanization. He identified both fire and deforestation as the main factors that allowed more arid conditions to set in, and used the term desertification to indicate extreme cases of savanization. His prediction indicates that desertification is characterized by severe soil erosion, changes in the physical and chemical soil properties, and invasion of more xeric plant species. Le Houerou (1969) introduced the term "desertization" to emphasize the role of human activities in arid and semi-arid regions bordering actual deserts. Neither word was really popular, as that no dictionary included either of them until very recently. Meanwhile, scientific research on the problem of these regions continued, in particular under the supervision of the United Nations Educational Scientific and Cultural Organization (UNESCO).

A review of literature reveals that desertification is not a new phenomenon. The Bible and the Quran referred to seven lean and seven fat years in Egypt and Kana'n (Buttrick 1996). In the sixth century B.C. Napata, the capital of the Kingdom of Kush, near the fourth cataract (Sudan), grazing around the city was abandoned because it resulted in erosion and advancement of the desert (Dixon 1959). Leopold (1963) attributed the decay of some ancient empires and town in southwestern and central Asia to a general decrease of the annual rainfall. In 1922, Lord Lugard reported how he had seen the advances of the Sahara to Sokoto, Nigeria.

Fig. 15.1 Sahel region (*source* https://www.eshailsat.qa/en/DynamicPages/channels)

Afican Sahel (Fig. 15.1), is a semiarid region of western and north-central Africa extending from Senegal eastward to Sudan. It forms a transitional zone between the arid Sahara (desert) to the north and the belt of humid savannas to the south. The West African Sahel is well known for the severe droughts that dominated the region in the 1970s and 1980s. Plenty of Meteorological investigations on the region has prospered during the last decade as a result of several major field experiments. Recently, increasing concern has been expressed about desert and desertification in a reaction against the disaster of the Sahel. Actually the word "Sahel" is a geographic description of the narrow strip land bordering the sahara, and is derived from the Arabic meaning "border". This disaster included six west African countries that faced drought from 1968 to 1973; Mauritania, Senegal, Mali, Upper Volta, Niger and Chad. The drought has also affected the southern border of Sudan and parts of many other countries as Somalia, Ethiopia, Kenya and Tanzania.

15.1.2 Magnitude of Desertification in the Globe

The tropical and subtropical deserts and semi-deserts represent 18% of earth's land surface. These areas are expanding annually due to the misuse of the environment. The FAO and UNEP (1984) estimated that desertification threatens more than 785 million people or 17.7% of the world population who lives in the dry lands. Out of this number 60–100 million people are affected directly by decreases in productivity associated with current desertification processes. Falk (2019) refered that between 50,000 to 70,000 square kilometers of useful land are lost for production every year, through desertification.

Ministry of Environment, India (2018) and Dregne (1987) referred that about 30% of Spain's arid land are severely decertified, 27% in North, 22% in Asia, 18% in Africa and 6% in Australia. The moderately decertified land ranges from 11% in Africa to 70% in Spain.

Approximately 80% of the agricultural land in the arid regions of the world has experienced moderate to severe desertification. Mabbutt (1985) referred that following indicators are related to desertification danger:

1. Encroachment and growth of dunes and sand sheets;
2. Deterioration of rain fed croplands;
3. Water-logging and salinization of irrigated lands;
4. Deforestation and destruction of woody vegetation;
5. Declining availability and quality of ground water and surface water supplies.

According to Horowitz and Potter (1971) "The social factors leading to desertification are of recent origin, as evidenced by the fact that affected areas have previously sustained human population for long periods of time without serious degradation". The two most important factors are viewed to be the rapid increase of human population and the massive inclusion of external demands in the local economy.

15.1.3 Desertification Processes

It is a combination of different processes resulting in the exposure of the soil to other agents of desertification (FAO 1986). It includes the clearing of ground cover, bush fire, over grazing and destruction of woody species. The progressive disappearance of favorable pasture is also included in this category. Different scientific opinions deal with the causes of desertification, some are emphasizing upon the role of climatic factors, and others palming the human action. The first group, because of lack in meteorological and hydrological data, is relying upon the archeology, geomorphology, geology and history of the flora and fauna. These studies concluded major changes of climate in the last 20,000 years, through which the Sahara regions passed several times pluvial and dry periods (Grove 1969, 1996; RAPP 1974).

The second group considers that the climate is only a supporting factor. The indication of this opinion is the creation of desert away from the desert margins. Over-cultivation can lead to a sharp decrease in land productivity. Overgrazing also results in the removal of vegetation cover and thus leaving the soil exposed to the wind action. In addition, bush fire is an important factor for the desertification in some regions. All these factors resulted in the destruction of the natural vegetation on which livestock and wild animals depend, destruction of organic matter and hence the formation of infertile and fragile soils. The bad quality or misuse of irrigation water may lead to the accumulation of salts; possibly increase of exchangeable Sodium and the starting of alkalization and sodification, which may result in a decrease of crop production.

Some studies, compromising between the two previous opinions, claim that the level of rainfall in some important arid zones may be decreased partly by human mismanagement, Bryson and Baerreis (1967) referred that the increased atmospheric dust produced by over grazing, rangeland burning and over cropping can reduce local rainfall or can encourage global climatic shift.

The desert area is characterized by a very scarce rain fall. However, flash rains are frequently occurring. This phenomenon results in the removal of soil on sloping areas and may lead to the exposure of the bedrock. Hagedorn et. al. (1977) stated that sporadic rains can be very intensive, and run off is usually not retarded by soil or vegetation cover. The low average annual runoff factor of 5% conceals the fact that even in higher-arid deserts; the run off factor can become as high as 30 or 40% in a single flash-flood.

According to Cooke and Warren (1973), surface runoff will start when precipitation is higher than 6.6 mm during single rain. Water erosion in three forms (i.e. sheet erosion, rills and gullies formation and mass movements) will occur. During the occasional flood rains, the water transports the erodible top soil, which is accumulated at the downslope. The water erosion in the arid areas is of natural type, man influence is not frequent due to the sparse vegetation cover.

The FAO/UNEP (1984) listed the responsible processes of desertification as follows;

1. Degradation of vegetation cover
2. Water erosion
3. Wind erosion
4. Salinization
5. Reduction in soil organic matter
6. Soil crusting and compaction
7. Accumulation of substances toxic to plants and animals.

15.1.4 Active Desertification Processes and Their Extent in Egypt

It is true that Egypt is a gift of the Nile. However the investment of this gift, its conservation and development would have never been realized without the keen efforts exerted by the Egyptians, throughout history (Hereher 2017). The cultivated area in Egypt (Fig. 15.2) is limited to the Nile Valley, Delta and desert depressions which all together present less than 4% of the total area. On the other hand about 96% of the country is occupied by almost barren land representing a part of Africa's dry desert region. Apart from cultivated land, Egypt is divided into three typical desert areas namely, Western Desert, Eastern Desert and Sinai Peninsula.

Fig. 15.2 Egypt cultivated area limited to the Nile basin and desert depressions (*source* Hereher 2017)

The Western Desert is a Northern dipping plain of sedimentary rocks, composed of sand stone in the south and limestone in the north (Said 1962). It occupies 681,000 km^2, or more than two thirds of Egyptian territories. The Landscape is interrupted by low situated oases, and by granitic mountains in the south-west intersection between Egypt, Libya and the Sudan. It is characterized by extensive bodies of sand including long chains of sand dunes. Some of these dunes extend in N-S direction of several hundreds of kilometers. The moving sand dunes overwhelm cultivations and inhabited oases. El-Baz (1978a) found, by using Apollo-Soyuz photographs that some longitudinal dunes are encroaching on fertile lands of the Nile Valley, especially after the completion of Nile alluvium. El-Baz (1978b) managed to use Low altitude aerial photographs (Fig. 15.3) in defining Aeolian submerged areas at the western borders of Nile Valley, Egypt.

Water erosion processes by occasional thunderstorms are becoming frequent in Egypt. Intensive rains fall on bare desert rocks and drains rapidly into the nearest wadi. The United Nation report (1980–1982) refers that thunderstorms became more frequent during last years; El-Menia Governorate was exposed to destructive effects

Fig. 15.3 Low altitude aerial photographs of Aeolian submerged areas (*source* El-Bazb 1978a, b)

of rain storms between the years 1965 and 1975. Studies were later performed to protect new El-Menia town against the run-off caused by the thunderstorms (Salem et al. 1982).

Desertification, caused by salinization and water logging, is an ecological hazard which faced almost all large scale projects of irrigated land in the world. In Egypt, uncontrolled floods and flooding irrigation in the absence of efficient drainage system have resulted in such problems. In west Nubaria schemes, the calcareous silty soil responded to irrigation and early vigor of biomass productivity seemed most promising, but it didn't take many years to notice that the ground water was rising at an alarming rate of 2–3 m annually (Remote Sensing Center and ISPRA 1982).

Other desertification problems which exist in the arable land of Egypt have to be closely studied. Urbanization and bricks industry are influencing the arable coverage and land fertility. Also, water logging occurs as a result of non-efficient drainage systems. Vegetation degradation is common in many areas of the arable land; however the demand for animal products in Egypt is rising much faster than the increase in the output. As a result there is a high pressure (over grazing) on the natural vegetation in the grazing areas, as Sinai and the Northern Coastal region. Ayyad (1977) stated that the northwestern coast region is the richest part of Egypt in flowering plants. The flora of this region comprises a stock of genetically growing resources for quite a number of cultivars. However, some species in the fauna and flora of the area are in urgent need of immediate protection. It is worthy to highlight that Mariota's area, along the north Mediterranean coast, which flourished during the Roman history and supplied vines and olives, has now changed into nomadic pastoral life. In addition, non-palatable species are lumbered for fuel, thus adding to and aggravating the problem of overgrazing; a problem which would become irreversible unless sufficient attention is given to it.

The soil biota simultaneously interacts and exerts their effects on vegetation cover (Ragan Shrestha et al. 2005). Among the biota, insects are considered the most detrimental in reducing the number of viable seeds substantially per unit area. The least palatable species gain preponderance. However, if plants in good years are not given the chance to produce seeds then the number of potential seed stock in soil germinate in the following season will be substantially reduced.

15.1.5 Monitoring of Desertification

In order to assess and map desertification processes, it is necessary to define, describe, quantify and codify the various aspects of each process, starting with the current condition. Since desertification is a continuous process which proceeds, if not halted in time, through several stages before reaching the final stage, its rate of progress in terms of time is an important aspect. The combined effects of climatic conditions, animal and human impacts need to be well defined in order to determine and forecast the socio-economic reflections of desertification problems. The provisional methodology of FAO/UNEP (1981, 1983, 1984) listed the following aspects to be considered:

i. The status, meaning the present situation as it is in the time of studying compared to conditions which existed in the past.
ii. The rate, meaning the spread and progress of the problem in relation with time.
iii. The inherent risk, depending on the vulnerability of landscape to desertification process, natural conditions (e.g. climate, physiography, soil erodibility, water quality and water table depth).
iv. The hazard, reference has to be made to natural susceptibility of the land to desertification (i.e. status, rate and inherent risk) and to manmade factors (e. g. livestock and population pressures on environment).

Furthermore, it is necessary to identify specific indications that could be used to monitor different desertification processes. These indicators should ideally be quantitative, sensitive to changes in the factor being measured, easy to measure and few in number. Criteria proposed by (FAO/UNEP) are shown in Table 15.1.

15.1.6 Utilization of Remote Sensing in Studying Desertification

The importance of using remote sensing technologies in the study of desertification has been confirmed in many scientific and technical events by several authors. It proves to be highly effective in determining the physical, biological and human processes leading to desertification. The fast progress in remote sensing technology, added more facilities to monitor the physical structure and dynamic nature of desertification indicators (De Pina Tavares et al. 2015). At large scale combined with stereoscopic coverages, certain degradation phenomena can be highly accurately measured. At all scale, however, remote sensing has important advantages. Shrestha et. al. (2005) applied imaging spectrometer data to detect and map desert-like surface features. The authors referred that absorption feature parameters in the spectral region between 0.4 and 2.5 μm wavelengths were analyzed and correlated with soil properties, such as soil color, soil salinity, gypsum content, etc. Some types of remote sensing data are relatively inexpensive and readily available. The multispectral nature of Landsat imagery has many advantages in the natural resources evaluation, planning and management program. In addition to multispectral capabilities, Landsat image offers the advantage of synoptic, multi-data coverage which allow mapping on a near-real time basis and performing comparative studies of dynamic surface conditions. Seasonal changes in vegetation pattern can be discerned. For the concept of desertification mapping, remote sensing can identify many of the boundaries and mapping units needed, as it gives a very exact delineation of certain physiographic features.

It is often, in the studies of desertification and land degradation, to combine satellite remote sensing information with other spatial data, like topography, soils, and

Table 15.1 Criteria indicating aspects of desertification processes *(FAO/UNEP 1981)

Desertification process	Desertification aspect	Regarded criteria
Urbanization	Status	• Loss of fertile and within ten years (% of inhabited area)
	Rate	• Increase in urban area per year
	Risk	• Population rowth
Vegetation degradation	Status	• Canopy cover of perennial plants (%) • Biomass in kg of Dm/ha/year • Production of fodder in unit/ha/year • % of potential productivity • Biomass in kg of Dm/mm of rain
	Rate	• Degradation increment (% per year) • Range trend line for last 10 years (in absence of drought • Woodland trend line/year • Cereal yield trend line/year
	Risk	• Increase of dryland (rainfed) arable land • Increase in livestock production (in % per year) • Overgrazing (Au/year) • Animal unit (Au) (growth rate in % per year) • Human population (growth rate in % per year) • Animal unit carrying capacities • Length of growing period (Less 180 days) • Climatic index for biological degradation • Moisture index (z) • Bush fire (detrimental effects) • Potential for reclamation

(continued)

Table 15.1 (continued)

Desertification process	Desertification aspect	Regarded criteria
Water erosion	Status	• Density of rills and gullied per km • Surface status • Type of erosion • Loss of top soil and subsoil % • Soil deposits in cm • Sequence of horizons • Thickness of soil (A + B) in % of the original thickness • % of yield compared with non-eroded soils • Decrease of organic matter content in % of non-eroded soils
	Rate	• Removal or deposition above normal in Mt/ha/year or in mm/year • Sediment deposits in dams in % of retention per year
	Risk	• Slope • Precipitation in mm • Weight of soil lost in t/ha/year • Rainfall factor • Soil erodibility factor • Topography factor • Biotir factor • Rainfall erosion index per type of climate • Summer drought (two rainy and winter precipitations) and winter drought (two rainy and winter precipitations) • Transition regimes

(continued)

Table 15.1 (continued)

Desertification process	Desertification aspect	Regarded criteria
Wind erosion	Status	• Loss of top soil (%) • Aeolian formations • Total surface covered by Aeolian formations (%) • Surface covered by veneer of Aeolian deposits • Surface covered by hummocks (%) above normal surface covered by moved dunes (%) • Relative concentration of gravels and or stones on ground surface
	Rate	• Removal above geologic rate (kc/ha/year) • Sand blasting of fine material • Depth of soil removed per year (cm) • Growth rate of the area invaded by sand in % of affected area • Depth (thickness • Wind erosivity index • Sand storm frequency (times per year per ten years period). • Number of days per sandstorm per year • Number of hours of wind storm per year • Number of days of wind storm in Spring (March–April) most dangerous in sub-tropical areas) • Maximum wind speed at 2 m height (m/s) of soil transported by wind (cm/year)
Salinization	Status	• Maximum Ece * 10 in mmhos/cm in upper 75 cm of soil of the soil • Maximum ESP in mmhos/cm in upper 75 cm of soil of the soil • Soil: plant yields (% of yields of similar non saline field soil) • New formations • Morphological Observation • Salt in t/ha/ 1.5 m

(continued)

Table 15.1 (continued)

Desertification process	Desertification aspect	Regarded criteria
	Rate	• EC increase in mmhos in upper 75 cm of soil • ESP increase in upper 75 cm of soil, in % per year • Yields in % per year • Surface affected by soluble salt in %
	Risk	• Climatic index for salinization, number of dry months (in absence of the critical depth of ground water table) • Average depth of ground water (cm) salt concentration of irrigated water
Soil crusting and compaction	Status	• Calcic accumulation and cementation form (depth in cm) • Gypsic accumulation and cementation form (depth in cm) • Ferric accumulation and cementation form (depth in cm)
Reduction in soil organic matter	Status	• Actual situation in % of optimum natural level
	Rate	• Reduction in organic matter in surface layer in % per year for last three years
Excess toxic substances in the soil	Status	• Lead, Zinc and copper

land use, into a Geographic Information System (GIS). A GIS provide the possibilities of analyzing combinations of different data layers, which may result in understanding the rule of different environmental parameters in causes and consequences of a desertification process. Also, The calculation function of the GIS may assist in developing simple empirical and semi-empirical models to determine the risk of land degradation or desertification in the studied areas. Oktoth (2003) used a simple logit regression equation to integrate the parameters of slope and ground cover, derived from remote sensing, to determine water erosion, confirmed by field validation in Kiambu, Kenya. A similar approach was used by (Vrieling et al. 2002) to determine water erosion risk in the Colombian Eastern Plains.

15.2 Methodologies of Monitoring, Analyzing and Mapping Desertification

15.2.1 FAO and UNEP Provisional Methodology for Desertification Assessment and Mapping

The provisional methodology of FAO/UNEP (1984) for assessment and mapping desertification was adopted to investigate the active desertification processes in the arable land of Egypt. In this chapter, three most important desertification processes (urban sprawl, salinization and wind erosion) will be analyzed and mapped from three aspects point of view (i.e. status, rate and risk). For mapping desertification aspects, the categorization guidelines, shown in Table 15.2 were followed.

Table 15.2 Bases of desertification aspect categorization

Desertification categories	Present area in various desert categories			
	Slight	Moderate	Severe	V. Severe
Slight	>30%	<30%	<40%	
Moderate	<30%		<40%	
Severe			>40%	
			<30%	
Very severe	<20%		>40%	
				>30%

15.2.2 Mapping Sensitivity to Desertification (DISMED)

Desertification is the consequence of important processes that turns productive land into non-productive one, due to poor land-management. The United Nation Convention for Combating Desertification (UNCCD) defined desertification as "land degradation in arid, semi-arid and sub-humid areas resulting from various factors, including climatic variations and human activities" (https://sites.google.com/a/owu.edu/land-degradation-and-desertification/home/gen). Batterbury and Warren (2001) postulated that where land use performance, in the ecosystems, is the main limiting degradation factor, desertification refers to active degradation in arid and semi-arid environment.

The EU funded project "Mediterranean Desertification and Land Use—MEDLUS" distinguished between degradation processes in European Mediterranean environments and the more arid areas. Physical soil loss associated with nutrient, caused by water erosion dominant problems in the European Mediterranean region. On the other side, wind erosion and salinization processes are most active ones in the arid Mediterranean areas (Glantz 1977; Quintanilla 1981; Zonn 1981).

Desertification assessment methodology is base upon the fact that environmental systems are generally in a state of dynamic equilibrium with external driving forces. Small changes in the driving forces, such as climate or imposed land use tend to be accommodated partially by a small change in the equilibrium and partially by being absorbed or buffered by the system. Desertification of an area will proceed if certain land components are brought beyond specific threshold, where further change produces irreversible alteration (Tucker et al. 1991; Nicholson et al. 1996, 1998). Environmentally Sensitive Areas (ESA's) to desertification around the Mediterranean region exhibit different sensitivity status to desertification for various reasons. For example there are areas presenting high sensitivity to low rainfall and extreme events due to low vegetation cover, low resistance of vegetation to drought, steep slopes and highly erodible parent material (Ferrara et al. 1999).

Mapping of environmental phenomena (e.g. desertification, degradation) needs to monitor and collect information indicating the assessed aspect. Desertification indicators are those, which indicate the potential risk of desertification while there still time and scope for remedial action. Regional indicators should be based on available international source materials, including remotely sensed images, topographic data, climate, soil and geologic data (Woodcock et al. 1994; Pax-Lenney et al. 1996). At the semi-detailed scale (ranging 1:25,000 to 1:1,000,000) the impact of socio-economic drivers is expressed mainly through pattern of land use. Each regional indicator or group of associated indicators should be focused on a single desertification process. The key indicators for defining ESA's to desertification, which can be used at regional or national level, can be divided into four broad categories defining the qualities of soil, climate, vegetation, and land management (Kosmas et al. 1999). This approach includes common measurable parameters, found in existing soil, vegetation and climate reports.

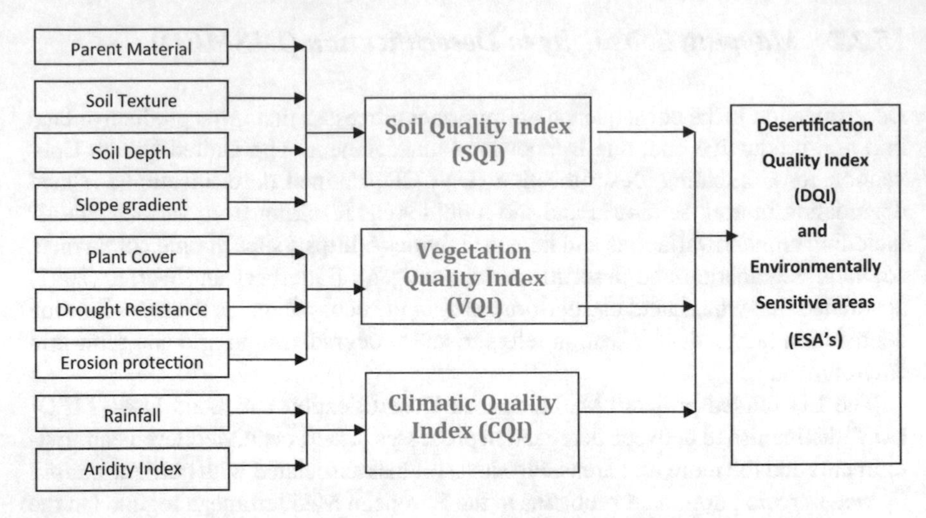

Fig. 15.4 Flow chart of mapping Environmentally Sensitive Areas (ESA's)

The following three quality indices were computed;

(a) Soil Quality Index (SQI)
(b) Vegetation Quality Index (VQI)
(c) Climatic Quality Index (CQI).

Figure 15.4 illustrates the main flow chart of concepts and steps followed to deduce desertification sensitivity indices (DSI) and define the environmental sensitive Areas (ESA's). The main input data for calculating theses indices include a mosaic of LANDSAT ETM+ image, geologic map of Egypt, produced by CONOCO (1989), climatic data derived from the Ministry of Agriculture. An image processing system (i.e. ERDAS IMAGINE 8.3) and a GIS system (i.e. Arc GIS 9) were the main tools in indices computations and ESA's mapping.

15.2.3 Mapping Soil Quality Index (SQI)

As vegetation and water are scares in arid and semi arid areas, soil is the dominant factor in such ecosystems. Soil quality indicators for mapping ESA's can be related to water availability and erosion resistance (Briggs et al. 1992; Basso et al. 1998). In the current work, a number of four measurable oil parameters (i.e. parent material, soil texture, soil depth and slope gradient) were determined at the field investigation. Weighting factors were assigned to each category of the considered parameters, on basis of OSS (2004) he fyors were adapted from MEDALUS project methodology (European Commission 1999). Table 15.3 demonstrates the assigned indexes for different categories of each parameter. The Soil Quality Index (SQI) is calculated on basis of the following equation, and classified according to categories shown in Table 15.4.

Table 15.3 Classes, and assigned weighting index for soil quality indicators (*source* European Commission 1999)

Soil quality parameters	Class of indicators	Description	Score	
			Wind	Water
Parent material	**Coherent**: Limestone, dolomite, non-friable sandstone, hard limestone layer	Good	1.0	
	Moderately coherent: Marine limestone, friable sandstone	Moderate	1.5	
	Soft to friable: Calcareous clay, clay, sandy formation, alluvium and colluvium	Poor	2	
Soil thickness	Very deep	thickness is more than 1 m	1	
	Moderately deep	thickness ranges from <1 m to 0.5 m	1.33	
	Not deep	thickness range <0.5 m to 0.25 m	1.66	
	Very thin	Soil thickness 0.15 m	2.00	
Texture classes	Not very light to average	Loamy sand, sandy loam, Balanced	1	1.66
	Fine to average	Loamy clay, clayey sand, sandy clay	1.33	2
	Fine	Clayey, clay loam	1.66	2
	Coarse	Sandy to very sandy	2	1.66
Slope gradient	<6%	Gentle	1	
	6–18%	Not very gentle	1.33	
	19–35%	Abrupt	1.66	
	>35%	Very abrupt	2	

Table 15.4 Classification of soil quality index (*source* European Commission 1999)

Class	Description	Range
1	High quality	>1.13
2	Moderate quality	1.13–1.45
3	Low quality	>1.46

$$SQI = (Ip * It * Id * Is)^{1/4} \tag{15.1}$$

Ip index of parent material, **It** index of soil texture, Id index of soil depth, **Is** index of slope gradient.

The erosion agent (i.e. water erosion/wind erosion) is considered in rating the impact of soil texture.

15.2.4 Mapping Vegetation Quality Index (VQI)

Vegetation quality, according to Basso et al. (2000) is assessed in terms of three aspects (i.e. erosion protection to the soils, drought resistance and plant cover). The Landsat-5 TM satellite images mosaic covering Egypt (Fig. 15.5) is the main material used to map vegetation and plant cover classes. Adapted rating values for each of

Fig. 15.5 Landsat-5 TM satellite images mosaic covering Egypt

Table 15.5 Ranges and classes of desertification sensitivity index (DSI) (*source* European Commission 1999)

Classes	DSI	Description
1	>1.2	Non affected areas or very low sensitive areas to desertification
2	1.2 < DSI < 1.3	Low sensitive areas to desertification
3	1.3 < DSI < 1.4	Medium sensitive areas to desertification
4	1.3 > DSI < 1.6	Sensitive areas to desertification
5	DSI > 1.6	Very sensitive areas to desertification

erosion protection, drought resistance and vegetal cover classes were adapted on basis of OSS (2004) as shown in Table 15.6. Vegetation Quality Index was calculated according the following equation, while VQI was classified on basis of the ranges indicated in Table 15.5.

$$\mathbf{VQI} = (\mathbf{IEp} * \mathbf{IDr} * \mathbf{IVc})^{1/3} \tag{15.2}$$

where: **IEp** index of erosion protection, **IDr** index of drought resistance and **IVc** index of vegetation cover.

15.2.5 *Mapping Climatic Quality Index (CQI)*

Climatic quality is assessed by using parameters that influence water availability to plants such as the amount of rainfall, air temperature and aridity, as well as climate hazards, which might inhibit plant growth (Thornes 1995). Table 15.5 reveals the classification categories of climatic quality index according to OSS (2003). The Climate quality index is evaluated through the Aridity Index (AI), using the methodology developed by FMA in accordance with the following formula In the current study, rainfall and evapotranspiration data on a number of 33 metrological stations were used to calculate the CSI as follows;

$$\mathbf{CQI} = \mathbf{P/PET3} \tag{15.3}$$

where: **P** is average annual precipitation and **ETP** is average annual Potential Evapotranspiration.

ArcGIS10.2 software was used to map ESA's to desertification Kosmas et al. (1999) were followed by integrating all data concerning the soil, vegetation and climate in a GIS model. Different quality indices were calculated and displayed as GIS ready maps from which class areas are deduced. The Desertification Sensitivity Index (DSI) was calculated in the polygonal attribute tables linked with the geographic coverage according to the following equation;

$$\mathbf{DSI} = (\mathbf{SQI} * \mathbf{VQI} * \mathbf{CQI})^{1/3} \quad (15.4)$$

15.3 Results of Case Studies

15.3.1 Assessment of Desertification Processes, According to Adapted FAO/UNEP

15.3.1.1 Urban Expansion Process

Investigating and evaluating urbanization may be elaborated by using multi temporal aerial photographs, satellite images and survey maps. Urban encroachment process is only dangerous in the arable land, as the Nile Delta and Valley, Study areas were selected within these regions. They represent geographical locations characterized by variable environmental conditions, including large cities or towns and some small villages. Figures 15.6 and 15.7 show location of study cases. The employed remote sensing means, for studying each site, was in accordance with the site areas, studying scale and resolution of remote sensing data (Table 15.6). The study sites comprise the following locations in northern Nile Delta (i.e. Greater Cairo), South Nile-Delta (Kalub El Balad) Middle Nile Delta (i.e. Tokh and Banha), and Middle Nile Valley (i.e. Mallawi and Maghagha). Moreover 7 villages were also studied (Table 15.7).

Monitoring of urban areas was mainly based on remote sensing, in addition to validation by ground truth data. The multi-concept of remote sensing was used to Coincide with different study site conditions. It is found that small scale satellite

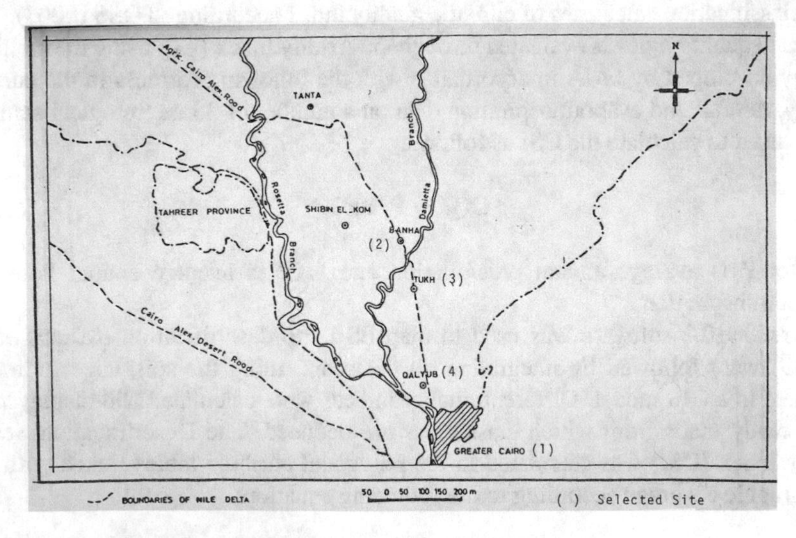

Fig. 15.6 General location map of study areas in the Nile Delta

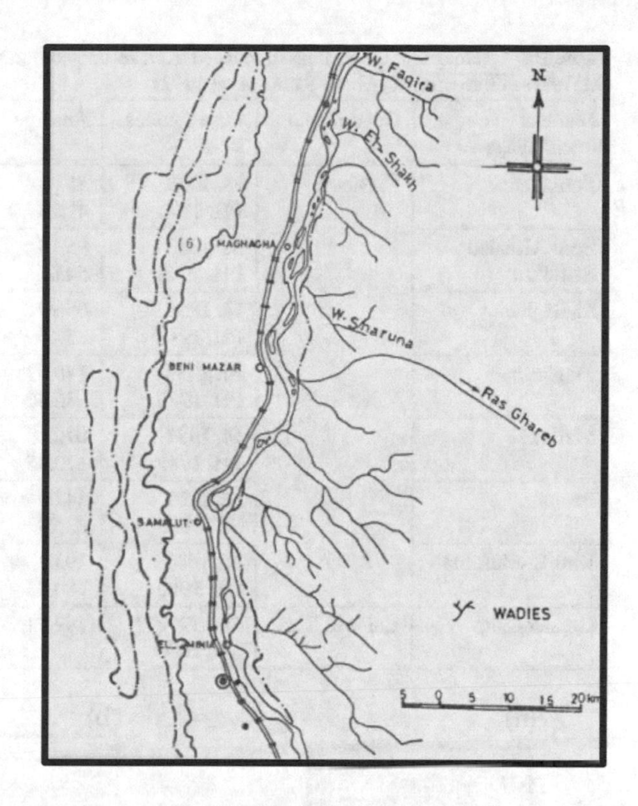

Fig. 15.7 General location map of study areas in the Nile Valley

Table 15.6 Chosen urbanization study sites and used Remote sensing Facilities

Townes			Villages		
Serial no.	Name	Used RS	Serial no.	Name	Used RS
1	Greater Cairo	Landsat images	7	Beni Amir	Aerial photos
2	Banha		8	Beni Khalid El Baharia	
3	Tokh				
4	Kalub El Balad		9	El-Baghur	
5	Mallawi	Aerial photos	10	Kafr El Museilha	
6	Maghagha		11	Kom El-Ahmar	
			12	Basus	

images are suitable to monitor large areas. The single scenes have the advantage of large area coverages. The use of large scale aerial photographs permits the delineation and analyses of smaller features.

The use of relatively recent aerial photographs, compared with earlier dated topographic maps (Figs. 15.7 and 15.8), made it possible to delineate boundaries of urban areas, using ArcGIS 3.3. The GIS provided the multi-temporal urban area acreages, hence calculating change rates. The same trend is approved, using multi temporal Landsat TM 5 images of Greater Cairo (Fig. 15.9).

Table 15.7 Acreages of urban areas monitored from aerial photographs (PH) and topographic maps (M) and satellite images (Abdel-Samie et. al 1992)

Name of town/village	Governorate	Data sources	Acreage	Increase (Acres)	Rate
Beni Amir	El-Menia	M. 1953 PH. 1948	75.89 41.28	34.61	16.74
Beni AKhaled El-Bahria		M. 1953 PH. 1948	95.15 54.62	40.53	09.28
El-Baghur		M. 1953 PH. 1948	05.19 25.46	20.27	12.60
Maghagha		M. 1953 PH. 1985	146.07 386.06	239.99	05.30
Mallawi		M. 1981 PH. 1985	076.13 170.05	93.92	30.84
Basus		M. 1961 PH. 1985	14.09 83.29	69.20	20.46
Kafr El-Muselha		M. 1945 PH. 1985	19.03 180.43	161.40	21.20
Greater Cairo	Cairo & Giza	Sat. 1978 Sat. 2006	149.11		

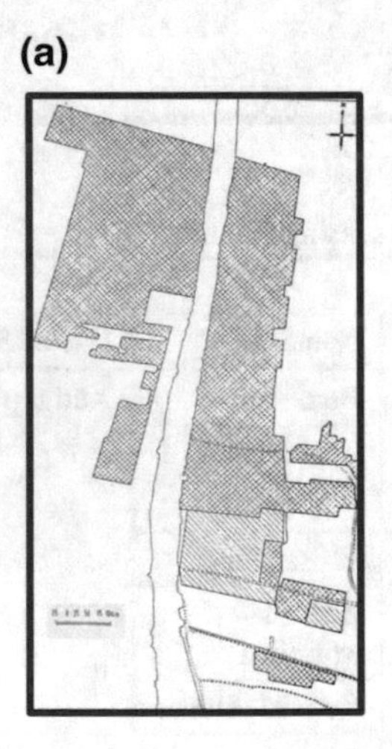

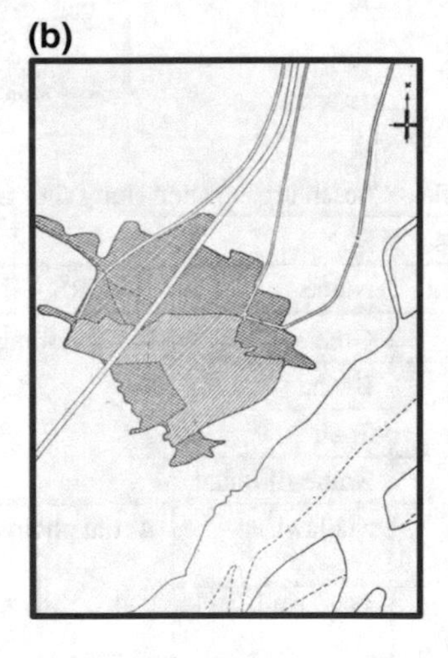

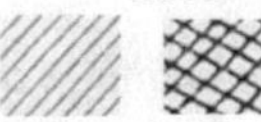

Fig. 15.8 Boundaries of urban areas of **a** Kafr El-Museliha, El-Munfya Governorate, Nile Delta **b** Maghagha, El Menya Governorate, Nile Valley. Indicate **a** and **b** in figures or below figures

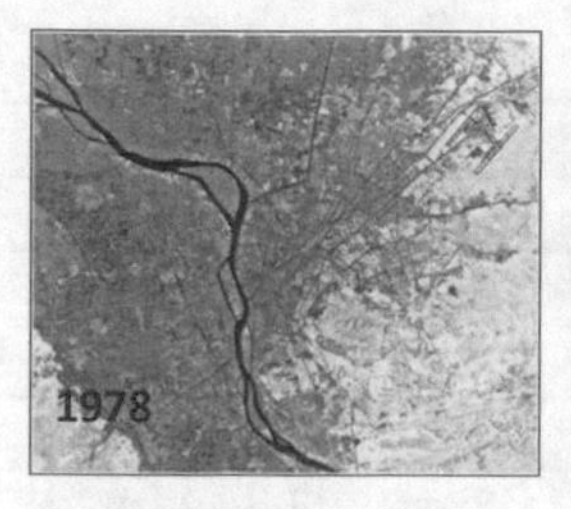

Fig. 15.9 Multi-temporal Landsat TM 5 images showing Greater Cairo areas in 1978, 1984 and 2006

The provisional methodology of FAO/UNEP and UNISCO (1984) was taken as a guideline to evaluate the problem of urbanization in Egypt. The methodology was modified on the assumption that accepted average increase of urban areas is <1% per year. This assumption is based on the usual rate of population growth in most developing countries as 1:2% per year. Thus, the suggested rating values for different aspects of urbanization are proposed in Table 15.8. The area of greater Cairo increased from 149.11 to 192.1 and 557.87 km^2 during 1978, 1984 and 2006 respectively. The increased area represents 28.85% of the original coverage.

The results (Table 15.9) show that the study areas are almost exposed to a very severe to severe status and rate of urbanization; all the study areas face severe risk. These measures are based on the assumption that all urban expansion happened on cultivated fertile lands. Indeed, this assumption is mostly true because the creation of new urban communities in the desert is a recent policy started 1990's as an attempt to save the remaining fertile lands from being lost. Also, constructing new urban areas away from the cultivated land will protect more fertile soils to be scraped out for bricks industry.

The estimation of risk is based on population increase of 1,200,000 persons per year, which represent 2–3% in a population of 52 million people then. Following a more civilized population policy may reduce the future risk.

Table 15.8 Class limits suggested for assessing different aspects of urbanization

Desertification aspect	Assessment indicator	Class limits			
		Slight	Moderate	Severe	Very severe
Status	Loss of fertile land within ten years (% of inhabited area)	<20	20–30	30–50	>50
Rate	Increase of urban area per year (%)	1–1.5	1.5–2.5	2.5–4.5	>5
Risk	Population growth (%)	<1.5	1.5–2	2–3	>3

Table 15.9 Evaluation of different aspects of urbanization

Name of town or village	Status		Rate		Risk	
	(1)	Class	(2)	Class	(3)	Class
Village Beni Amir	170.0	V. Severe	16.74	Severe	2–3	Severe
Village Beni Khaled El-Bahria	93.00	V. Severe	09.28	Severe	2–3	Severe
Village Baghur	126.0	V. Severe	12.60	Severe	2–3	Severe
Town Maghagha	53.00	V. Severe	05.03	Severe	2–3	Severe
Town Mallawi	308.0	V. Severe	30.84	Severe	2–3	Severe
Town Basus	205.00	V. Severe	20.46	Severe	2–3	Severe
Village Kafr Museilha	212.00	V. Severe	21.20	Severe	2–3	Severe
Capital city Greater Cairo	48.10	Severe	04.80	Severe	2–3	Severe
Town Banha	46.40	Severe	04.64	Severe	2–3	Severe
Town Tokh	31.78	Severe	03.17	Severe	2–3	Severe
Village El-Kom El-Ahmar	43.70	Severe	04.37	Severe	2–3	Severe
Town Kalub El-Balad	67.73	V. Severe	06.77	Severe	2–3	Severe
Greater Cairo	67.06	V. Severe	06.71	Severe	2–3	Severe

(1) Loss of fertile land within ten years (% of inhabited area)
(2) Loss of fertile land within ten years (% of inhabited area)
(3) Population growth (%)

15.3.1.2 Assessment of Salinization and Water Logging Processes Case Studies

The origin of soil salinity is commonly related to evaporation of water solution and accumulation of salts in different soil profile layers and also at same solution. Generally, a number of four processes take place in this cycle; migration, accumulation, deposition and redisposition of salts in one or more of the following media;

(a) Marine solution
(b) Surface and ground water
(c) Artesian deep waters and
(d) Irrigation waters.

In Egypt, marine cycle of salinization is apparently connected with the soils of low-lying areas near the Mediterranean Sea coast at northern Nile Delta. The land is directly or indirectly influenced by sea water of Mediterranean area during periods of tide and storms. Consequently, ground water of very high salinity exists. Sodium chloride constitutes most of the soluble salts, together with much smaller concentrations of magnesium chloride. Continental mass of salt migrate with river water to the Nile Valley and delta regions, where they accumulate as a result of evaporation and water transportation.

Artesian water cycle salinization is best illustrated in the oases. The irrigation water salinization results from incorrect activities of man particularly through

Table 15.10 Areas classified under various soil-productivity classes based on soil surveys undertaken from 1957 to 1974

Soil Rating	Nile Delta		Nile Valley		Total	
	Area, Fed	%	Area, Fed	%	Area, Fed	%
Cultivated						
Class I	192,157	3.5	0,167,460	5.98	359,617	4.34
Class II	1,252,334	22.8	1,381,023	49.25	2,633,357	31.79
Class III	1,718,622	31.3	0,573,024	20.48	2,291,464	27.66
Class IV	0,429,025	7.8	169,655	6.06	598,680	7.22
Subtotal cultivated	3,592,138	65.48	2,291,162	81.87	5,883,300	71.01
Uncultivated						
Class V	1,431,046	26.11	194,809	6.97	1,625,855	19.61
Class VI	0,462,953	8.41	312,638	11.18	775,591	9.38
Subtotal cultivated	1,893,999	24.52	507,447	15.13	401,446	28.99
Total surveyed	5,486,137	100	2,798,609	100	8,284,746	100

impeded drainage, insufficient leaching and/or incorrect use of saline water. In his studies on the soils of Egypt, Kovda (1958) defined theses cycles and their existence in Egypt with specific reference to certain visited locations where saline soils prevail.

The first national soil salinity survey completed by the soil survey department, Ministry of agriculture, was carried out 1957–1974. During this period, an area of about 8.285 feddans was surveyed. The system adopted for that survey and classification into different classes was that which is formulated and used by the U.S. Bureau of Reclamation. Details of the areas classified under different ratings are presented in Table 15.10.

It is noteworthy that soil salinity condition is the main factor in rating soils under classes III and IV. Therefore, the figures for areas under these two classes (Table 15.10) are indications of the magnitude of saline and alkaline soil distribution.

A. **Monitoring of soil salinity**

Soil salinity, refers to the state of the soluble salts accumulation in soil. It can be determined by measuring the electrical conductivity of a solution extracted from a water-saturated soil paste. The electric conductivity as ECe (Electrical Conductivity of the extract) with units of decisiemens per meter (dS/m^{-1}) or milliohms per centimeter (mmhos/cm) is an expression for the anions and cations in the soil (Al-Khaier 2003). Since mid-fifties intensive research has been carried out on soil salinity and water salinity connected with irrigation and drainage. However, most of this work, whether in the universities or Agricultural research institutes, was not systematic. Hence, no clear period assessment on changes in salinity of soils, at the national level is available. Nevertheless, the soil and water research institute compiled two often used documents; (a) a soil salinity map of the Delta, (Fig. 15.10) which was based on the data collected from the soil survey, using aerial photographs, and analyses

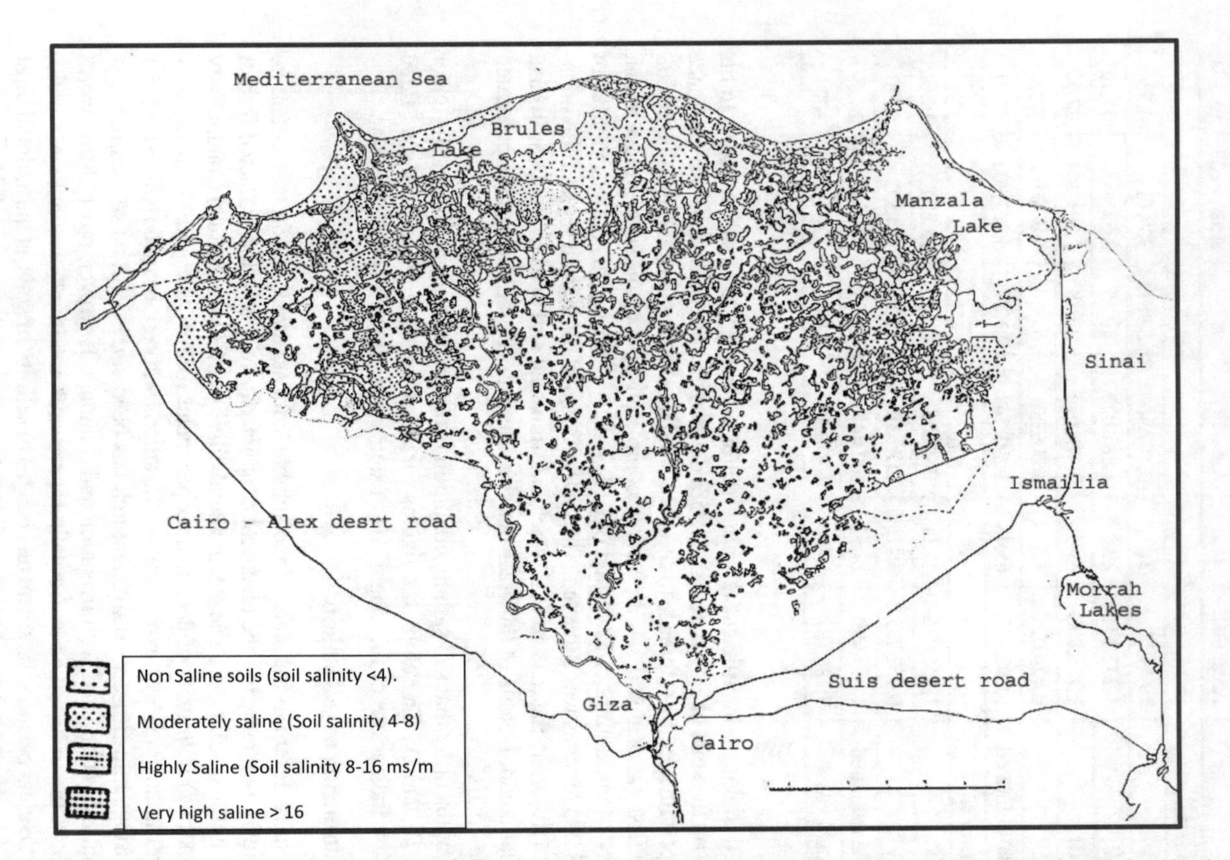

Fig. 15.10 Classification of soil salinity, according to salinity of top soil, based on aerial photography Compiled by: Soil survey and classification Dept., Soil and Water Use Research Institute, Agr. Res. Center 1975, was based on the data collected from the soil survey, using aerial photographs

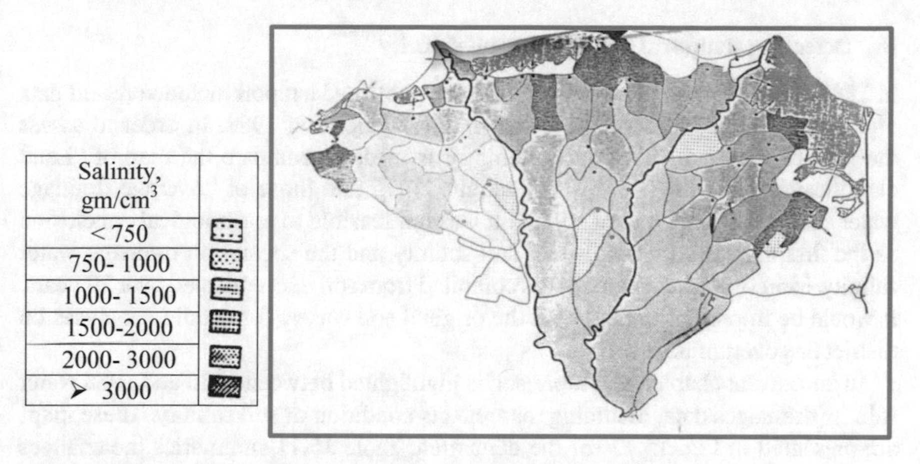

Fig. 15.11 Map showing salinity of drainage water in the Nile Delta in 1986

conducted in 1957–1974 period and; (b) a comparative study on soil changes in the Fayoum Governorate first survey (1960) and second survey (1980).

It may be stated that there is no systematic program for monitoring changes in salinity of irrigated soils on the national scale. However, there are hundreds of localized activities and investigations which collect data on soil characteristics.

Recently, some research institute collected accurate data on volumes and salinities of drainage water. It was aimed to compile a map showing salinity of drain water of the delta in 1986 (Fig. 15.11). The range of salinity categories used were as follows; 700, 750–1000, 1000–1500, 1500–2000, 2000–3000 and 3000 gm/cm. Taking drainage water salinity as an index of soil salinity, then this map indicates the following;

1. Non saline soils occupy small areas especially in south delta between the two Nile branches, Midwestern delta adjacent to Rosetta branch and north eastern delta around Mansoura.
2. Category of drainage water salinities in the range of 750–1000 gm/m^3 (ppm) dominates all south, middle and south east deltas. It covers the largest area compared with other categories.
3. Salinity range 1000–1500 is found in the northern half of the delta distributed in the middle, east and west of delta. It constitutes a large area second to the area covered by the 750–1000 gm/m^3 category.
4. The other categories represent soils located towards the edges of the delta. They include some of the newly reclaimed areas.

It should be highlighted that remote sensing and GIS became the key tool in most of the survey activities analyzing physiography, landscape, hyperspectral data… etc. These large numbers of activities do not follow the same systems of sampling or data recording that makes it difficult to compile tables or maps of soil salinity, particularly when specific study date is not mentioned.

B. Detecting temporal changes in soil salinity

In 1989, the drainage water research institute published a report included recent data on average salinity of drainage water in the delta during 1988. In order to assess the present status of salinity, an attempt was made to compare the map of "Land classification according to top soil salinity, 1975 and those of "Average drainage water salinity of same region, 1988". It was not feasible to reach logical conclusion as the first map is based on actual soil salinity and the second on drainage water salinity. Moreover, the first map was compiled from soil data collected over 20 years. It would be important to deal with the original soil survey for detailed analyses on district or governorate level.

In the current chapter, a comparison is highlighted between 1986 and 1988 water salinity drainages data, assuming that reflects condition of soil salinity. These maps are presented in Fig. 15.12 for the delta area. Table 15.11 summaries the changes occurred between 1968 and 1988.

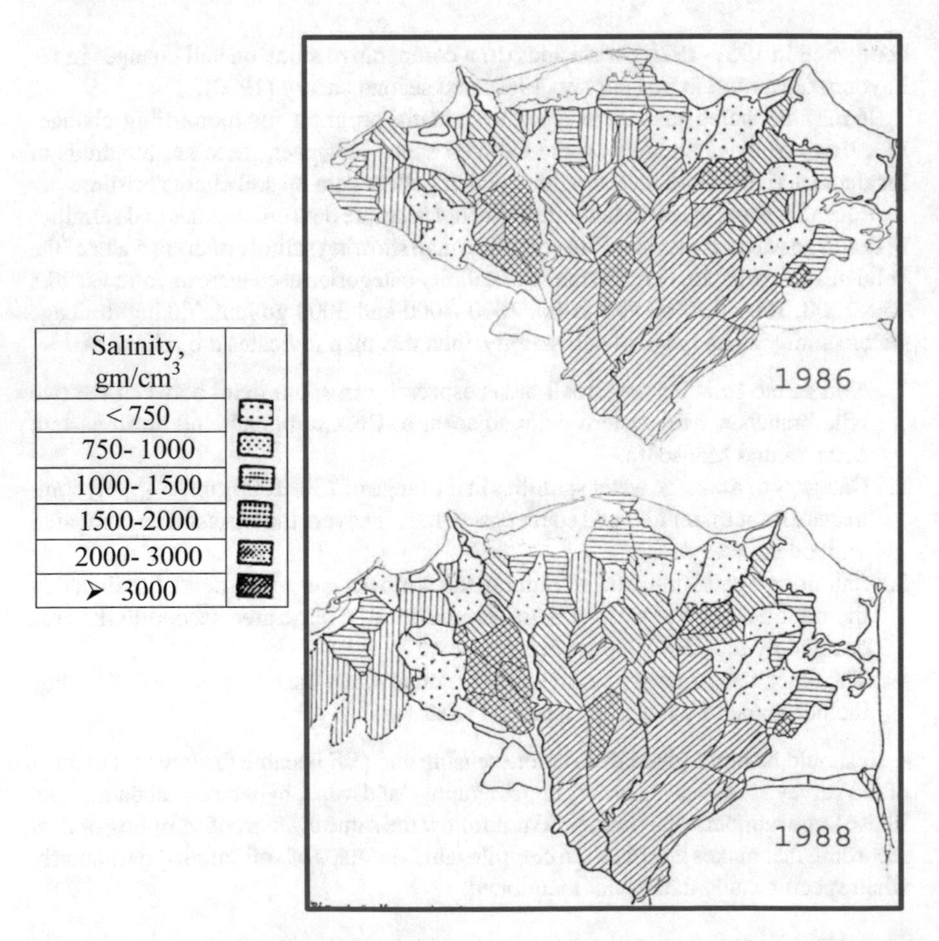

Fig. 15.12 Maps showing changes detected in drain wafer salinity, 1986 and 1988 for the Nile Delta area

Table 15.11 Changes in soil salinity in 1988 as compared with 1986 (drawn from drainage water salinities in the two years)

Catg. no.	Category of drainage water (soil) salinity gm/m^3	Quantitative changes in 1988 compared with 1986
1	<750	Area increased slightly on the expense of category 2
2	750–1000	Area decreased due improvement (i.e. changes to category 1)
3	1000–1500	Area decreased due improvement (i.e. changes to category 2)
4	1500–2000	Area decreased due improvement (i.e. changes to category 3)
5	2000–3000	Area decreased in delta but additional area added to newly reclaimed land west of the delta
6	>3000	Area increased on account of category 5 g

It can be stated that there is no systematic program for monitoring changes in salinity of irrigated soils on the national level. However, there are hundreds of localized activities and investigations which collect data on soil characteristics including salinity and alkalinity. These large numbers of activities do not follow the same sampling systems f or data recording and thus isn't convenient to be compiled. Recently, some research institutes started collecting accurate data on volumes and salinities of drainage water. The collected data were used to compile a GIS digital map showing the salinity of drain water for the Nile Delta in 1986.

C. **Use of remote sensing in assessment of soil salinity in Egyptian soils**

Soil salinity is a dynamic phenomenon, thus needs to be monitored regularly to secure it's up to update status, extent, degree of severity, spatial distribution, nature and magnitude. Remote sensing data provide great potentiality in analyzing and monitoring soil salinity through aerial photography or multi sensors satellite images (Allbed and Kumar 2013). There are plenty of satellites and sensors useful in detecting and monitoring the saline soil. Spatial resolution of the satellite images varies according to satellite platform, from 120 m (Landsat-5MSS/TM thermal bands) and 100 m (Landsat 8-Aster TIRS) to a meter-resolution IKONOS (4.0 m) or QuickBird (2.44 m), however there are some obstacles for the wide use of such data. In particular, optical waves back-scattered from the terrain and registered by the sensor are sensitive to weather conditions (dense clouds, rain, etc.) and, additionally as being passive sensors, images can be obtained only in daylight. To avoid these problems, radar images from satellites (as ERS-1 and ERS-2, JERS-1, RADARSAT-1 and ENVISAT) can be used for erosion research. The main advantages of active radar systems are weather independence and both day/night acquisition possibility. Multispectral data such as LANDSAT, SPOT, IKONOS, EO-1, IRS, and Terra-ASTER with variable resolution values, in addition to as hyperspectral sensors.

In the context of using remote sensing in studying salinization, mapping of salt affected soils is based on satellite image classification. Moreover, modelling of soil salinity is done in a GIS using a detailed digital elevation model and information about the impeded clay layer as the most important data layers. The results of the modelling correspond with the obtained image classification. This model is elaborated for the desert Delta fringes of Egypt, based on geomorphology, remote sensing and GIS. From the agricultural point of view, saline soils are the soils which contain sufficient neutral soluble salts in the root zone to adversely affect the growth of most crops (see Table 15.11).

Iqbal (2011) delineated surface soil salinity in the prime rice-wheat cropping area of Pakistan. The study employed an index-based approach of using optical remote sensing data in combination with geographic information system. An index-based approach of using optical remote sensing data, in combination with geographic information system, was tested. Variable satellite imagery indicators were examined, using different spectral band combinations.. The author found that Near infrared and thermal IR spectral bands proved to be most effective as this combination helped easy detection of salt affected area from the non-saline ones.

"For the purpose of definition, saline soils have an electrical conductivity of saturation extracts of more than 4 dS m^{-1} at 25 °C" (Richards 1954).

The study was based on recently collected accurate data on volumes and salinities of drainage waters and official data on available irrigation waters. Studying the change in soil salinity between years 1986 and 1988 revealed, in general, improvements in areas characterized by drainage water salinity of 750–2000 g/m^3. However, areas whose drainage water salinities ranged from 2000 to >3000 g/m^3 recorded an increase. Salt balance data were calculated in four separate pilot areas with a total area of 150,000 feddans. Deposition of salts was associated with low volumes of applied irrigation waters. Calculating the salt balance on the national level revealed a removal of 13.3 million tons of salts in 1990, from an irrigated area of 7.2 million feddans. Future prospects for the year 2017 were analyzed in view of the plans to increase the cultivated area by 3.4 million feddans. The estimated salt balance in the year 2017 indicates an addition of 62.2 million tons of salt every year to the irrigated soils, not including salts which are added through other sources such as chemical fertilizers and manure. These figures stress the necessity of a tight control on water use for irrigation combined with adoption of modern irrigation and efficient drainage systems. An ultimate necessity in face of such a situation is to breed and grow high salt-tolerant crops [1 feddan = 0.42 ha].

15.3.1.3 Assessment of Wind Erosion and Aeolian Activities

The agriculture in Egypt is concentrated in the Nile Valley and Delta (4% of total Egyptian geographical surface) where 95% of population is concentrated. Most of Egypt, except the narrow Mediterranean semi-desert belt, is characterized by arid to hyper arid environment. A Landsat-TM-5 satellite image mosaic of Egypt (Fig. 15.13) shows the distribution of different geomorphological units (i.e. Hamda Desert, Gravel, and land forms of aeolian and fluvial natures indicate desert environments).

The Western Desert (Western side of Nile Valley) is dominated by aeolian land forms, where longitudinal sand dunes indicate main wind direction from northwest to southeast. Moreover, Barchan dunes move, according to occasional sand storms from east to west. Sand sheets and rocky to gravely surface inhibit the dune corridors and plateau.

A. **Factors influencing wind erosion processes**

Concerning desertification by wind erosion process, Aeolian deposits present serious worldwide problems. The extensive Aeolian deposits from past geologic eras give evidence that it is a phenomenon of long history. Skidmore (1986) indicated the following conditions that allow wind erosion wind erosion to activate:

1. Strong wind.
2. Loose, dry and finely divided soil.
3. Relatively large area of open land.
4. Smooth soil surface devoid of vegetation cover.

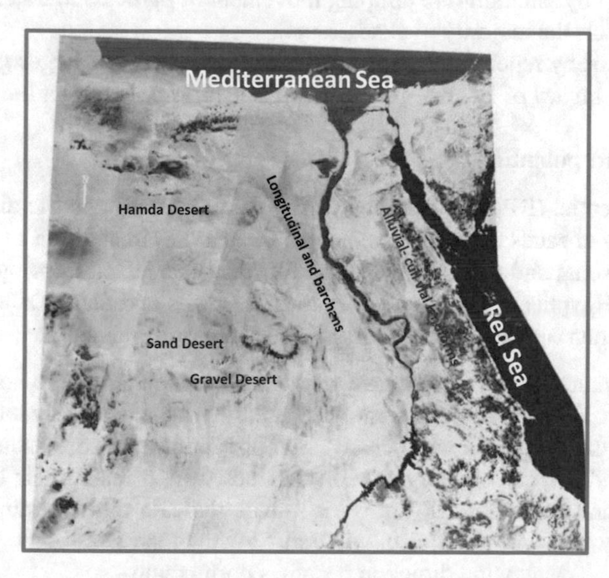

Fig. 15.13 Landsa TM 5 satellite image mosaic of Egypt, showing distribution of different geomorphological units of the true desert. (Aeolian and fluvial/colluvial landforms indicating variable desertification processes

These factors greatly characterize the Egyptian territory at western desert, northern parts of the eastern desert and Sinai. In these regions, Aeolian activity has its effect on the cultivated lands in different degrees. Climatic conditions affect greatly the formation of Aeolian landforms such as precipitation, rainfall variability and intensity, air and soil temperature, air humidity and wind. The most important climatic element for dune and sand sheet formation are constant wind direction. Wind speed at the ground has to be sufficiently high for sand grains transport. According to Bagnold (1954), the threshold wind speed ranges 12–19 km/h, lesser wind have no morphological effects. Also, consistency of wind direction and wind duration are of great influence on sand movement. A wind of rather low intensity, but blowing for a long time, will have the same effect of sand fields and sand dunes as strong wind blowing for short times only. Moreover, the mechanical effect of transported sand will increase ten times where speed rises from 24 to 34 km/h and 100 times when it rises from 24 to 56 km/h (Bagnold 1954). This shows that high speed sand-storm are especially important for the formation and movement of sand dunes.

The lower limit of particle sizes that can be formed into Aeolian bed forms is controlled by a number of factors. The primary control is through the settling velocities of particles in air. Particles below a certain size travel in suspension in the wind, whereas above that the movement is mainly saltation. The small particles is lifted by turbulence, and are therefore diffused to great heights. The fine particles are thus carried away more quickly and therefore separate from the coarser grains. Three possible types of grain motion during transportation of sand by wind as based by Hagedorn et al. (1977) are as follows;

i. Transport by suspension, relevant for clay and silt only.
ii. Transport by saltation (i.e. jumping movement of particles in trajectory curve). This affects the majority of sand grains).
iii. Movement by repetition (i.e. the creeping and rolling of larger grains by the saltation impact of smaller grains, creep is usually a quarter of the total load.

B. Sand drift potential

The drift potential (DP), the resultant drift potential (RDP) and the direction variability (VAR) of sands have been calculated (Gad 1988) in addition to drawing and mapping seasonal and annual sand roses for a number of meteorological stations covering all Egyptian lands. According to those maps and data (Table 15.12), the following results can be expected with respect to seasonal changes:

1. During winter, winds nearly have west-east direction along the northern coast, with high DP values toward the east, and shows moderate wind variability. Towards the south, the wind changes its direction to be south-southeast at El-Minya and Asyut with very low drift values, which reach their lowest at El-Faiyum and El-Minya. Hover, VAR values indicate high variability of wind especially at Asyut and Cairo, while at El-Minya the wind direction is nearly stable. At Aswan, wind direction becomes north-south.

Table 15.12 Resultant Drift Potential (RDP), Drift Potential (DP) and Drift Potential Variability (VAR) in the studied climatic stations (After Gad 1988)

Region	Winter			Spring			Summer			Autumn		
	DP	RDP	VAR	DP	RDP	VAR	RDP	DP	VAR	RDP	DP	VAR
Mediterranean Coast												
Salum	474.3	536.6	0.123	178.1	239.3	0.256	128.0	135.8	0.057	45.1	65.7	0.314
Marsa Matruh	299.5	533.5	0.439	161.0	422.1	0.619	74.3	109.5	0.321	37.4	71.0	0.473
Alex	189.0	251.0	0.247	111.8	189.8	0.411	62.9	68.7	0.084	31.5	39.1	0.194
Damietta	60.7	80.3	0.244	79.8	96.0	0.169	95.2	185.8	0.488	3.6	7.3	0.507
Pot Said	500.2	822.5	0.392	370.9	617.3	0.399	526.5	589.4	0.107	339.6	386.1	0.120
Lower Egypt												
Tanta	13.6	16.7	0.814	32.0	35.3	0.093	28.6	28.6	0.000	0.0	0.0	0.000
Cairo	2.8	3.4	0.824	18.8	22.0	0.145	0.9	0.9	0.000	10.0	11.5	0.130
Middle Egypt												
Fayoum	1.3	1.3	0.000	19.3	11.9	0.383	4.6	4.7	0.021	20.4	22.7	0.101
Beni Suef	0.0	0.0	0.000	0.0	0.0	0.000	3.0	3.0	0.000	0.000	0.0	0.000
Upper Egypt												
Menia	7.3	7.5	0.027	102.9	80.3	0.220	61.9	62.4	0.008	14.9	15.5	0.039
Assiut	20.0	55.0	0.622	154.8	75.9	0.510	18.6	21.3	0.127	25.3	30.0	0.157
Luxor	1.6	4.8	0.667	23.9	15.2	0.364	8.5	9.6	0.115	6.2	9.4	0.340
Aswan	154.1	159.6	0.034	592.7	551.4	0.070	276.3	296.8	0.069	159.9	180.7	0.115
Western Desert												
W. El Natrun	558.3	753.6	0.259	955.4	672.6	0.296	100.0	169.8	0.409	600.8	702.1	0.144
Siwa	54.6	65.4	0.165	137.2	79.9	0.418	56.5	60.5	0.066	19.9	30.4	0.345

(continued)

Table 15.12 (continued)

Region	Winter				Spring			Summer			Autumn		
	DP		RDP	VAR	DP	RDP	VAR	RDP	DP	VAR	RDP	DP	VAR
Baharia	30.3		40.8	0.257	71.3	45.5	0.362	145.2	159.2	0.088	9.5	9.5	0.04
Dakhla	1.1		3.5	0.686	71.8	65.3	0.091	28.8	30.8	0.065	1.8	4.6	0.609
Kharga	105.1		116.4	0.097	535.3	497.4	0.071	327.4	335.0	0.023	127.0	138.5	0.063
Red Sea Coast													
Ismaillia	105.0	186.3		0.436	176.0	100.5	0.429	10.9	49.0	0.165	25.3	25.3	0.000
Hurghada	1752.9	1946.9		0.100	26324	2247.7	0.146	2331.4	24460	0.047	1158.2	1219.2	0.050
Quseir	188.7	198.2		0.048	342.5	235.0	0.314	104.5	118.8	0.120	75.7	94.5	1.199

2. During summer, the wind has a remarkable north-south direction especially along the Nile Valley and Nile Delta, with slight change towards the south-south east at Alexandria, Cairo, Port Said and Asyut. The wind possesses low variability during this season allover Egyptian territory. The maximum drift values occur at Port Said. While the minimum values are found at Cairo, El-Fayum and Luxor, and the moderate values at El-Minya, Asyut and Alexandria.
3. During spring, the wind blows mainly towards the southeast along the northern coast. El-Fayum and El-Minya. Along the lands of upper Egypt and Delta, wind becomes of north-south trend. But at Asyut it blow up toward the east. The greatest values of wind variability are found at Asyut. Indeed, spring season is characterized by a maximum in the relative drift potentials compared with the other months or the year. This fact reflects the unstable character of wind direction in this season.
4. During autumn, the wind is mainly blown towards the south along the coasts, Nile delta and Nile Valley. Its direction slightly changes toward the south-south east at Alexandria and toward the east at Cairo. The wind variability shows that Luxor is characterized by unstable wind direction than the other regions along the Egyptian territories.

With respect to regional differences in wind dynamics, in general, the Egyptian lands especially the coastal area, delta and the Nile Valley are characterized by the following;

i. The Mediterranean coast, especially at Port Said shows the greatest values of wind variability and the (RDP). Wind is mainly blown up towards the east in winter and changing clockwise towards the southwest in spring, south-southeast in summer and to the south in autumn.
ii. In lower Egypt, the wind is blown up from northwest in winter, northeast in spring, north in summer and from the west in autumn. In Cairo, wind variability possesses the greatest values during winter and the lowest values during summer as compared with the other Egyptian regions. Generally, the relative drift potential over Lower Egypt is small during all seasons of the year.
iii. In Middle Egypt, the prevailing winds blow toward the east south east in winter, southeast in spring, south in summer and east in autumn. Wind variability and the relative drift potential value are nearly similar to those of Lower Egypt.
iv. In Upper Egypt, the main wind trend is nearly stable at Aswan and blows up towards the south. In Minya, the wind blows up towards south-southeast in winter and spring and to the south in the summer and autumn. In Asyut, it blows up towards the north east in winter, to the east in the spring, to south east in the summer and autumn. The highest relative drift potential value in Upper Egypt is found in Aswan, while the lowest value is in Luxor. Moreover, Asyut possesses higher value than El-Minya, except in the spring. The wind variability values show that in Aswan the wind possess an unstable character of wind direction, while in El-Minya the wind changes its direction only slightly.
v. Spring is apparently the preferential season for high wind and sand storms, and the wind variability is generally high in this season, and the anomalous wind

direction is from the south (Khamasin) and from the west (especially in the western regions of Egypt).

C. **Sand deposits sedimentation**

There are three ways by which sand grains rest deposition. On Egyptian territories, the three types of sediments are represented as follow;

- **True sedimentation**, takes place where particles falling through slowly moving air strike the surface with insufficient velocity for further movement. For grains carried in suspension, this is the mode of deposition where sand grains deposited over cultivated lands of the Nile Valley and Delta.
- **Accelerations** occur either when surface roughness increase or when velocity decreases, such as sand deposited on highly weathered land of the plateau bordering the Nile Valley. Also, when velocity decreases by external causes, such as the sand deposited on the flood plain bordering the Nile Valley, where wind velocity decreases by cultivated lands, or that deposited at the foot scarps of the plateau area. The difference between both true and accelerated sedimentation is that the first deposits sink to the ground and no more moves, while the other are still moved for some distance to a secure hollow.
- **Encroachment**: This deposition mode occurs without changes of wind speed, surface creep is held up by some obstacles over which saltation may pass on. A common case is the steep slope of the dunes and plateaus slopes.

The sand patches and dunes on the desert surface originate in gentle dip, behind small change irregularities on surface or where convergent secondary occurs. An increase in velocity or divergence in the wind leads to erosion, such as that which occurs during winter and spring. Steady velocity means neither deposition nor erosion and decreasing velocity or convergence means depositions such as that which occurs along the western boundary of the cultivated lands between Beni Suef and El Fashn (Landsat-TM 5 image at Fig. 15.14). Since sand particles rebound more easily off a hard surface of rock or desert pavement than off sand, any wind that can initiate sand movement over, it will be larger than over a sandy surface.

Aeolian activity is highest in the lower plains. This is shown in the area extending between the cultivated lands and eastern scarp of the western plateau between Beni Suef and Assyut. The large difference in altitude between the flood plain and its surrounding high plateau leads to development a regional circulation pattern with strong winds of high turbulence and frequent shifts of directions. The area taken up by sand fields has increased in the process of desertification.

D. **Morphology of the dune origin areas**

The following types of sand cover were delineated from Landsat images covering the Egyptian lands;

1. Beach areas along the Mediterranean coast, passing through Alexandria, Rosetta and North Sinai (Fig. 15.15): These areas represent the rule lead to development of coastal dunes only. The sand furnishing areas were larger during the cold

Fig. 15.14 LandsatTM-5 image of western Beni Suef and El Fashn, characterized by decreasing wind velocity or convergence and Aeolian depositions

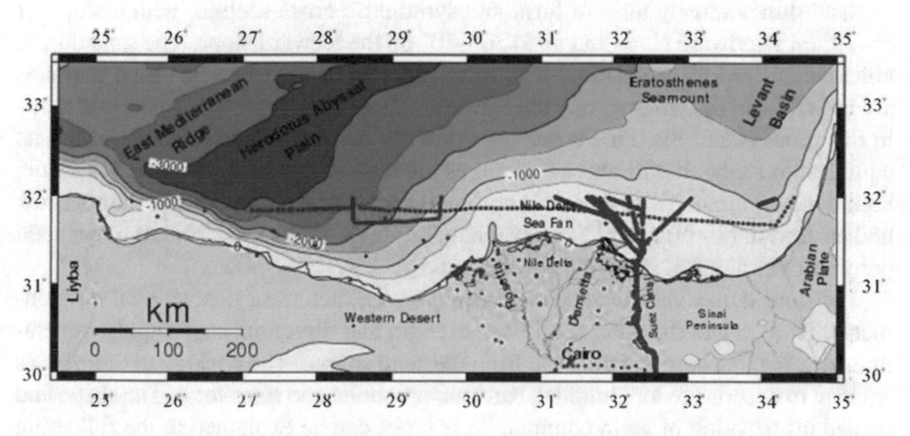

Fig. 15.15 Beach areas along the Mediterranean coast, passing through Alexandria, Rosetta and North Sinai, representing the rule lead to develop coastal dunes only

periods of Pleistocene due to related eustatic sea-level depression (Hagedorn et. al. 1977).

2. Flood plain of the river Nile in the western desert fringes between Asyut and El Fashn: Their dunes are mostly of hilly and hummocky types.

3. Low plains located in western desert and the northern parts of Egyptian lands as the Great Sand Sea, wadi El Rayan depression, the low plain east of, northern Sinai low lands and the west delta law plain the low plain west of the flood plain between Asyut and Beni Suef. Dunes formed over these plains are mostly of longitudinal with other types.
4. Terminal basins in internal drainage areas of Qattara, Sittra- El Baharia playas and El Baharia, Frafra, El Kharga and El Dakhla oasis. Their dunes are mostly longitudinal and barkhan types.

Sand blown out of beaches or flood and low plains have largely been reworked from older sandstone deposits either by weathering or by fluvial activity. Sand may also stem from Aeolian erosion of sandstone.

E. **Sand dune shapes and movement**

Sand dunes are mounds, hills or ridges of windblown sand. They are found where is a source of sand, a wind strong enough to erode and transports particles and a land surface on which to deposit the sand. The initial form of a dune is the sand patch likely to occur wherever topography leads to reduction in wind speed plus resultant deposition. Since particles rebound more easily off a hard rock surface than off sand, any wind that can initiate sand movement over a hard surface will be able to carry less sand per unit of time over a beginning sand patch. The later, therefore, act as a barrier to sand flow and induces further accumulation such as that found at El-Menya, Asyut and Ismaillya-Cairo cultivated lands.

Sand dunes mostly tend to form an asymmetric cross section, with a slope of 5°–10° on windward slope and up to 30°–40° on the leeward slope. The sand dunes, unless stabilized by vegetation, tend to migrate downwind because sand particles are blown from the windward to the leeward side of the dunes. The developed slops in this process and the dune crests are normally convex. When a dune reached its equilibrium shape, it will move forward as a whole conserving its shape. As a rule, high dunes migrate faster than low ones and broad dunes move faster than narrow bodies. Lettau (1969) calculated a barchan 3 m high transported 20.550 m^3 of sand per year over a line at right angle to the wind.

The sand dunes vary in shape according to the factors of depositional environments. They reflect the wind condition, strength and direction, sand supply, vegetation, physical barriers and distance from the sand source. The basic sand dune types existing in deserts are longitudinal, barchan, parabolic and transverse. The shape and related information of main common dune types can be explained in the following paragraphs;

1. **Longitudinal Dunes**

They form large, narrow parallel ridges of sand whose equal side slope indicate wind directions parallel to their long axes. This pattern is a result of very strong winds, small supply of sand and some variation in the prevailing wind direction. They show, on satellite images (Fig. 15.16), a remarkable straightness in the direction of the prevailing wind, only crests developing some sinuosity. The longitudinal dunes are

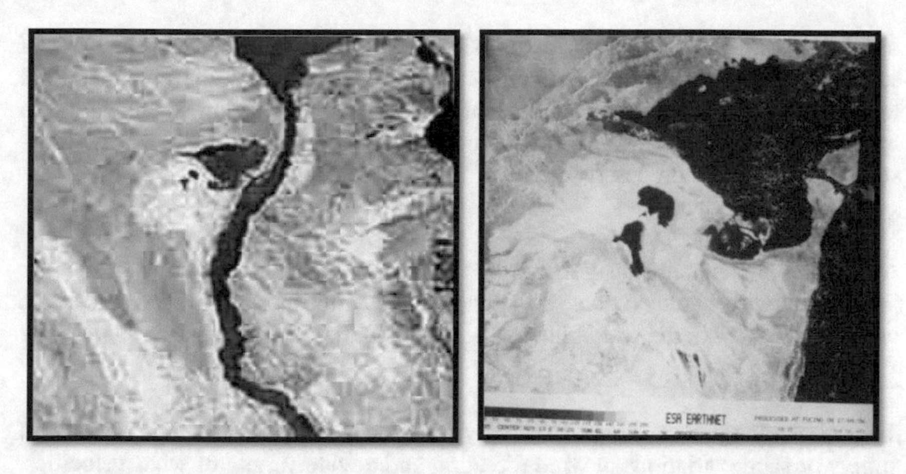

Fig. 15.16 Landsat-ETM 7, 2014 imageS mosaic showing longitudinal dunes, Western Elmenya, and Southern Fayoum Depression, Egypt. Longitudinal dunes are large, narrow parallel ridges of sand whose equal side slope indicate wind directions parallel to their long axes

locally called "Scif", meaning "Sword" in Arabic, as they appear looking like. This dune type is dense along the southern flank of Qattara Depression, El Khanka and east Bitter lakes. The most eastern one in Qattara is named Hathathiat dune belt which seems to be the source of sand that formed the longitudinal dune at Wadi El Rayan and that which is accumulated in the western low plain west El Fashn-Asyut region.

2. **Barchan dunes**

The Barchan dunes develop crescent-shaped outline with horns pointing downwind (Fig. 15.17). The crest line is toward the inside of crescent, indicating that the pre-

Fig. 15.17 Aerial photograph (Left) and a terrestrial one (Right) showing barchan dunes field, western Elmenya, Egypt. Horns pointing downwind

dominant wind directions from the outside face. These dunes develop in areas of strong wind, limited sand supply, constant winds from one major direction and flat topography. These conditions are present at El-Menya barchan arms, elongating in one direction than the other. Such difference may be to an asymmetry in wind pattern, sand supply or slope in desert surfaces.

3. **Transverse dunes**

Abundant barchan dunes may merge into barchanoid ridges, which then grade into linear (or slightly sinuous) transverse dunes, however the crest is oriented across, the wind direction, with the wind blowing perpendicular to the ridge crest.

4. **Beach dunes**

They occur along coastal areas. Their very hummocky form is the result of a sufficient supply of sand, variations in wind direction and a wide variety of wind velocities. Further inland, beach dunes may be stabilized by vegetation. Mostly this dune type contains ground water floating over the salty water of the sea. In Egypt, beach dunes are highly dense along the coasts of Sinai and Nile Delta, especially between Roseta and Port Said. They also cover separate areas along the coasts of western desert, especially in areas where beaches are not of cliff type.

F. Dune forms influenced by topography

Cooke and Warren (1973) describe echo-dunes developed at the foot of escarpment with fairly steep slopes presented toward the wind as being due to the effect of long trolled vertices, comparable to the gap between a sand fence and the accumulation sand behind it. Increased wind speed due to the topography of large escarpments results in the interruption of migrating dune fields crossing them. Dunes disintegrate a short distance before the rim, and the sand transported in streams over the rocky surface is again collected in dunes of the same type at one or even several kilometer's distance from the foot of escarpment. This pattern is well developed along the escarpment bordering El Fayoum Depression and east of Bitter Lakes.

G. Dune forms influenced by vegetation

Distinctive sand accumulation forms develop where sand comes into contact with vegetation, such as that in contact with western desert boundary of cultivated land between El Fashn and Asyut. Vegetation lowers the wind speed and thereby induces and accumulation.

H. Relation between trends of the cultivated lands sand dunes

The careful study of space imageries and wind patterns in Egypt resulted in pointing the relation at Fig. 15.18 between location trends of cultivated lands and Aeolian activities;

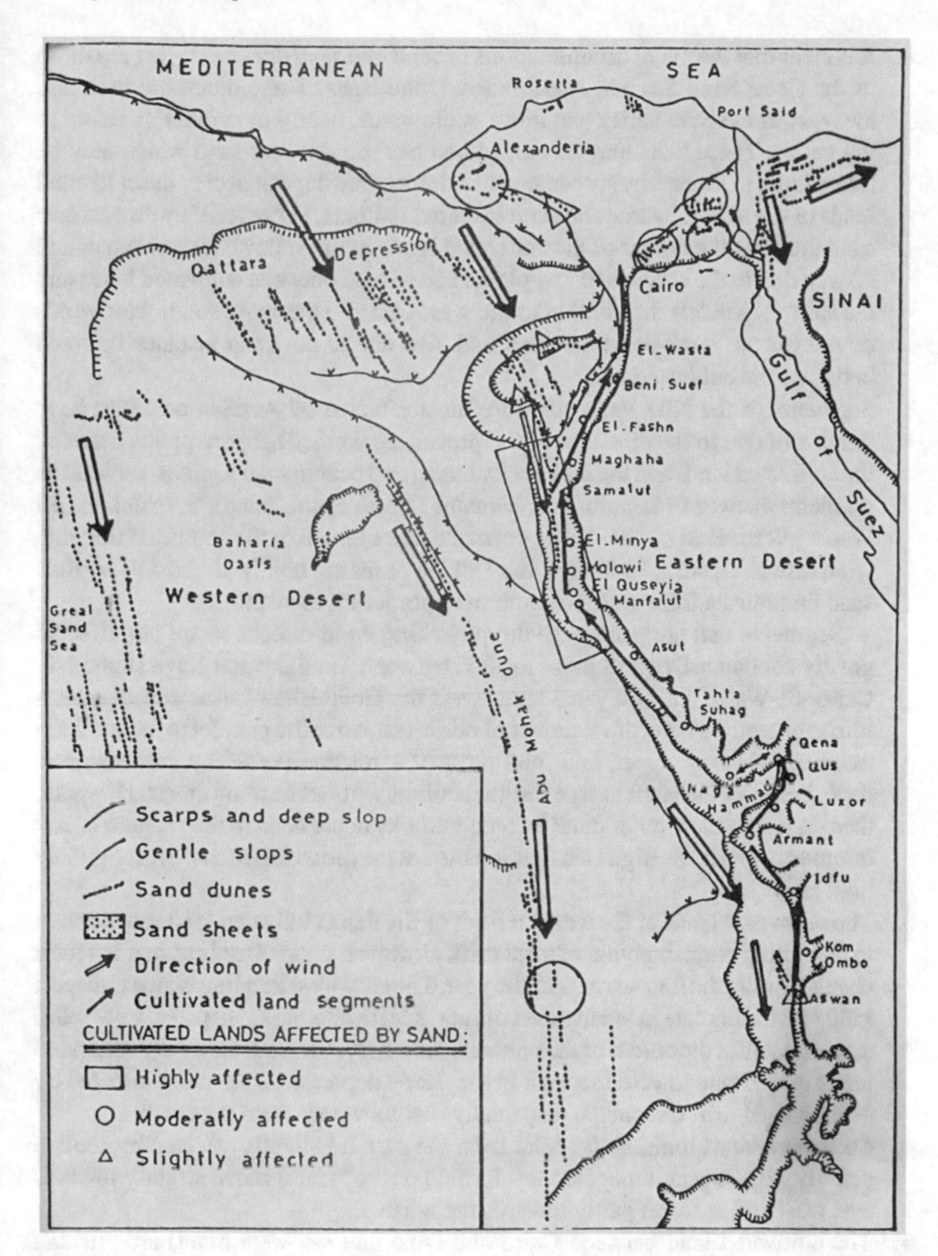

Fig. 15.18 Map of Egypt showing the cultivated land segments, directions of wind and trends and distribution of sand dunes

1. It is clear that the major accumulations of sand and sand dunes in Egypt are those of the Great Sand Sea and Abu Mahrik Dune Belts. These dunes are extended away from the Nile Valley and move southwards, nearly in parallel direction to the valley. These belts may be considered as reservoirs for sand which may be transported to the east by suspension or saltation and deposited over the cultivated lands of the valley, by true and accretions depositions. However, they do not form sand dunes at the border of the cultivated lands in the valley between Asyut and Aswan due to the absence of low plains that extend between cultivated lands and the stony highlands. In addition to the west-east and the south south-west winds do not prevail throughout the year, and also due to the great distance between belts and the cultivated lands.
2. Segments of the Nile Valley that are most effected by Aeolian activities have trends oblique to the direction of the prevailing wind. There are plains between the cultivated lands and the high rocky lands, and near to sand sources, such as the segments between Manfalut and Samalut (North-South trend), Samalut and El Wasta (North-East trend). These segments are oblique to the southeast trending wind direction. Also, El-kharga and Dakhla oases are highly affected by aeolian sand that moves from north to south over bordering highland.

   Segments that are parallel to the prevailing wind directions are not affected greatly aeolian activity, as those located between Manfalut and Naga Hammadi. Cairo–El-Wasta, the cultivated lands west the Great Bitter Lakes and Aramant-Idfu segment. Aswan-Idfu segment of north-south trend is parallel to the southern trending winds in upper, and thus most of sand transported by suspension is deposited over the lands as true sedimentations without forming dunes. However, there is a small low relief dune located in rocky depression in the western desert oriented towards the High Dam but is halted at the foot scarp of east-west trending high land.
3. The cultivated lands at the western limit of the delta cultivated lands are parallel to prevailing wind direction of sand drift. However, wind directions in this sector change during the four seasons of the year. The east-west trending winds transport sands to accumulate in small areas of sand sheets into the small sizes. Low relief areas located at the border of the cultivated lands. Northward, the newly cultivated lands that extend toward the west in the sandy depressions are more affected by aeolian sand drift directions, -especially that blown up from the north.
4. Cultivated lands limiting the delta from the east are slightly affected by aeolian activity. El Khanka dunes that are formed east of Cairo move slightly towards east west and east and partly towards the north.
5. The cultivated land between Cairo and Ismailia (east-west trend) are affected slightly by dunes that were formed at their southern part due to the effect of the southerly and westerly winds.

15.3.2 *Assessment and Mapping Environmental Sensitivity to Desertification (DISMED)*

15.3.2.1 Soil Quality Index (SQI)

The geologic map (CONCO 1989) was used to deduce the nature of parent material, which is demonstrated in Figs. 15.19 and 15.20. Table 15.13 demonstrates the areas of various parent materials classes, as deduced, from the geologic map, using ArcGIS

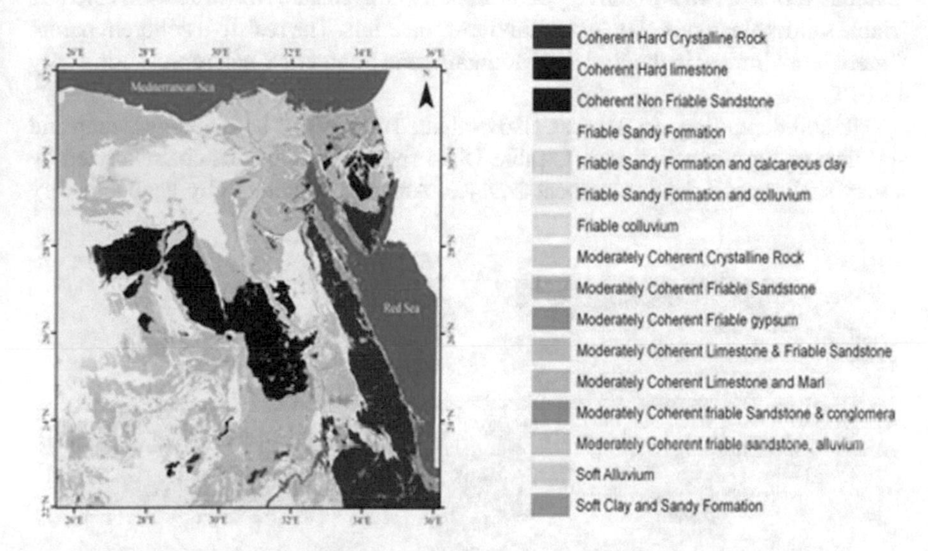

Fig. 15.19 Nature of parent material in the Egyptian Territory

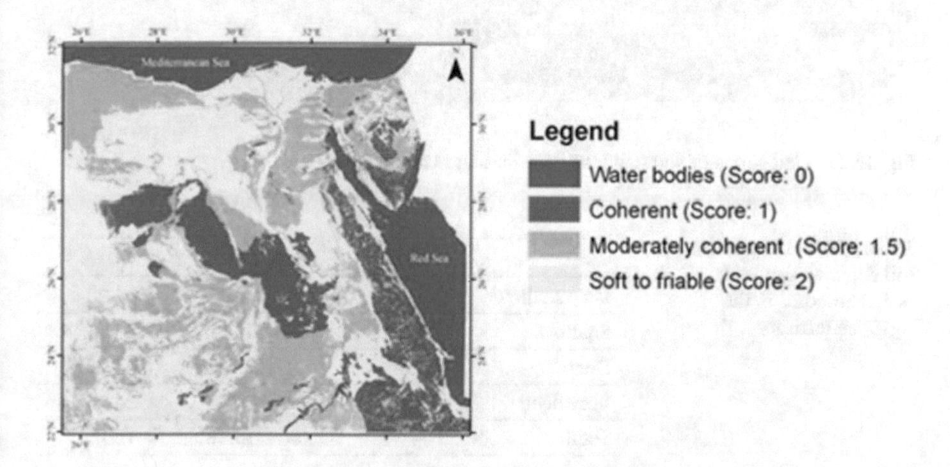

Fig. 15.20 Re-categorization of parent material natures in the Egyptian Territory

Table 15.13 Nature of parent material classes of Egyptian territory and assigned scores

Class	Score	Area (km^2)	%
Coherent	1	179,616.39	18.01
Moderately coherent	1.5	338,890.46	33.97
Soft to friable	2	479,009.13	48.02
Total	–	997,515.98	100

system. It is found thar the area exhibited by sois, originated by soft to friable parent material, represent 48% of the Egyptian territory is originated from Such soils include friable sand, calcareous clay and colluviums materials. The resistive coherent parent materials are limited in the Red Sea Mountains and Southern Sinai, representing only 18.01%.

The soil depth (Fig. 15.21) was also evaluated on basis of both geologic map and soil map of Egypt (ASRT 1982). Table 15.14 shows that the soils characterized by a very shallow soil depth represent 24.44%. Areas characterized by deep and very

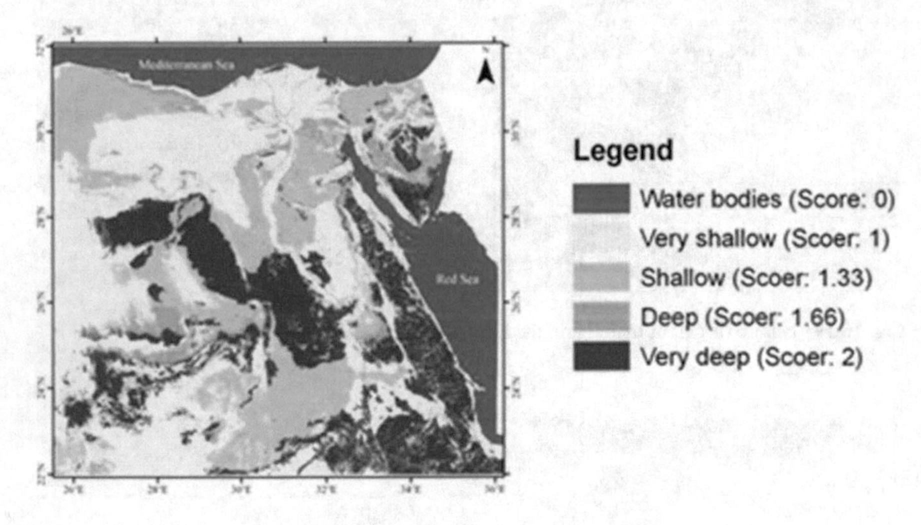

Fig. 15.21 Categories of soil depth as contributing in soil quality index

Table 15.14 Distribution of soil depth classes and assigned scores in the Egyptian territory

Class	Score	Area (km^2)	%
Very shallow	1.00	441,126.17	44.22
Shallow	1.33	265,446.21	26.61
Deep	1.66	47,103.87	4.72
Very deep	2.00	243,839.73	24.44
Total	–	997,515.98	100

deep soils do not exceed 30% of the whole Egyptian territory, located mainly in the Nile Valley and Delta and areas of sandy plains.

The soil texture was assessed on basis of the geomorphology, deduced from the ETM+ Landsat-7 satellite mosaic. Table 15.15 and Fig. 15.22 show that the most sensitive coarse textured soils amount 81.5% of whole territory. The alluvial the Nile Valley is exhibited by average textured soils, covering 8.25% of all soils. The colluviums (16.7%), brought by the alluvial fans and ravines, at the desert fringes, are exhibited by very light to average textured soils. The wadi soils are characterized by fine to average textured soils, covering 1.7% of all soils.

The slope gradient (Fig. 15.23 and Table 15.16) was classified, on basis of topographic maps and Digital Elevation Model (DEM).

Calculating the soil quality index (Table 15.17 and Fig. 15.24) show that 64.84% of Egyptian soils are characterized by very low soil quality. On the other hand, an area of 21% of the Nile Valley soils are characterized by moderate quality due to its capability to sustain soil structure.

Table 15.15 Distribution of soil texture classes and assigned scores in Egyptian territory

Class	Description	Score	Area (km^2)	%
Very light to average	Loamy, sandy, sandy-loam, balanced	1.00	167,425.65	16.78
Fine to average	Loamy clay, clayey-sand, sandy clay	1.33	16,994.83	1.70
Average	Clay, Clay-Loam	1.66	82,299.74	8.25
Coarse	Sandy to very sandy	2.00	730,795.76	73.26
Total		–	997,515.98	100

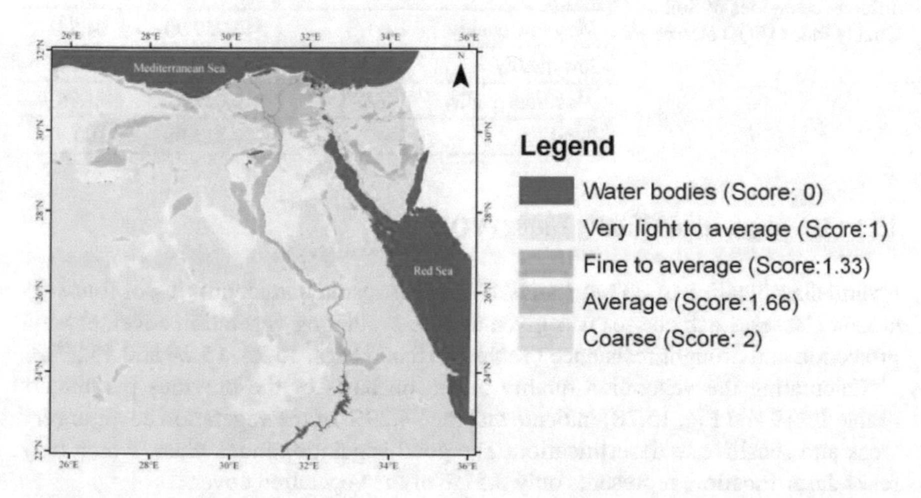

Fig. 15.22 Categories of soil texture as contributing in soil quality index

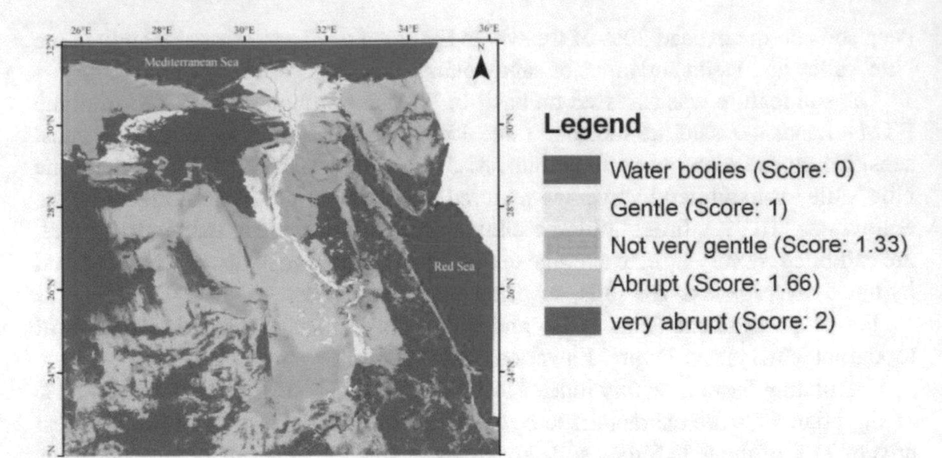

Fig. 15.23 Categories of slope gradient as contributing in soil quality index

Table 15.16 Distribution of slope classes and assigned scores in Egyptian territory

Class	Score	Area (km^2)	%
Gentle	1.00	57,134.61	5.73
Not very gentle	1.33	217,333.01	21.79
Abrupt	1.66	276,935.89	27.76
Very abrupt	2.00	446,043.05	44.72
Total	–	997,515.98	100

Table 15.17 Areas of different categories of Soil Quality Index (SQI) classes

Class	Score	Area (km^2)	%
Very low quality	>1.6	646,757.90	64.84
low quality	1.4–1.6	131,656.25	13.20
Moderate quality	1.2–1.4	219,032.41	21.96
Total	–	997,515.98	100

15.3.2.2 Vegetation Quality Index (VQI)

Hybrid classification of ETM+ Landsat-7 images resulted in identifying of four vegetation classes. Each classes was given a score evaluating vegetation cover, erosion protection and drought resistance (Table 15.18 and Figs. 15.25, 15.26 and 15.27).

Calculating the vegetation quality index, on basis of the previous parameters (Table 15.19 and Fig. 15.28) indicate that the 94.29% of the vegetation cover is very weak and sensitive to desertification. The good vegetation index class, which may resist desertification, represents only 3.51% of the vegetation cover.

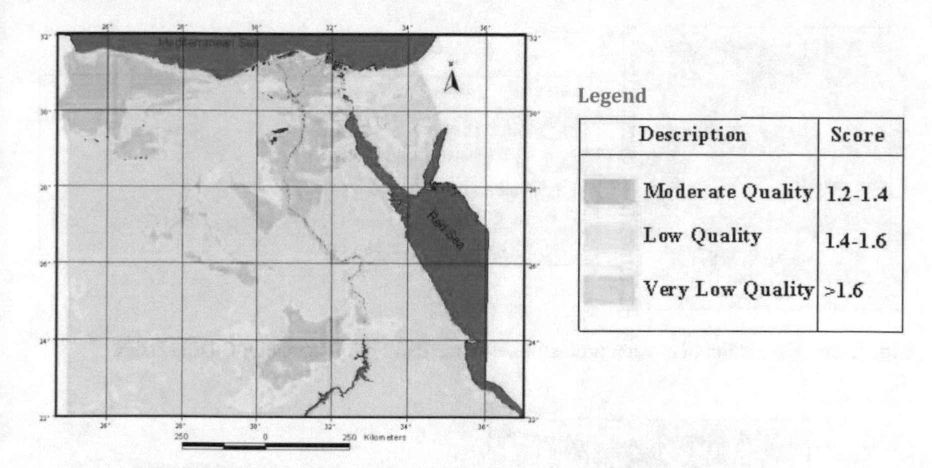

Fig. 15.24 Soil Quality Index map (SQI)

Table 15.18 Vegetation cover classes and assigned scores for different elements

Class	Area (km^2)	Drought resistance scores	Erosion protection scores	Vegetation cover scores
Cultivated land	45,536.36	1.00	1.00	1.00
Halophytes	13,851.56	1.00	1.33	1.33
Orchards mixed with crop land	9388.44	1.33	1.66	1.66
Saharan vegetation <40%	904,024.57	1.66	2	2
Saharan vegetation >40%	24,645.63	2.00		
Total	997,515.98			

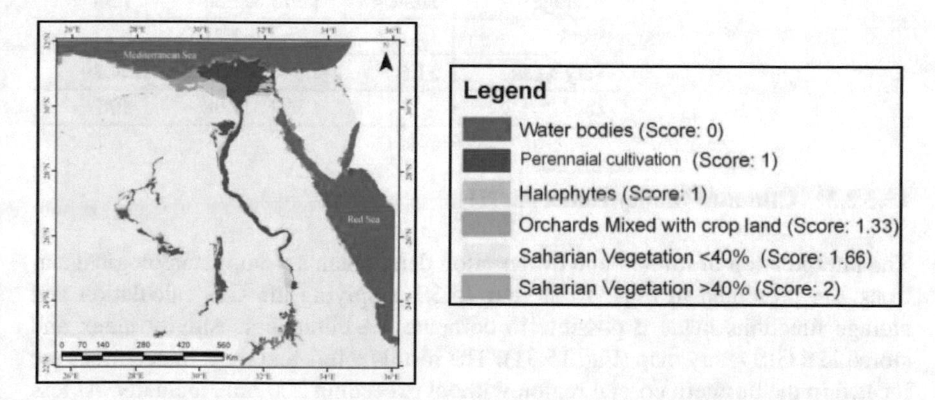

Fig. 15.25 Evaluation of drought resistance, as contributing in Vegetation Quality Index

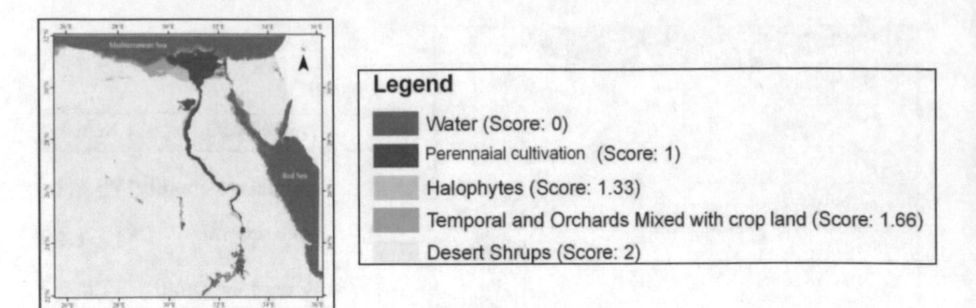

Fig. 15.26 Evaluation of erosion protection, as contributing in Vegetation Quality Index

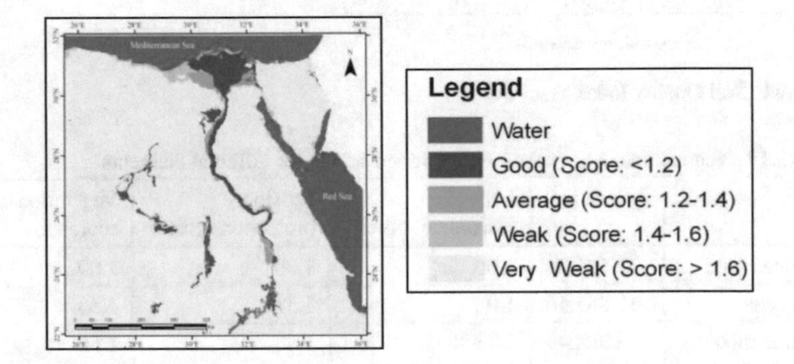

Fig. 15.27 Evaluation of vegetation cover, as contributing in vegetation

Table 15.19 Areas of different vegetation quality index classes

Class	Score	Area (km^2)	%
Good	<1.2	34,974.9	3.51
Average	1.2–1.4	13,851.56	1.39
Week	1.4–1.6	8142.71	0.82
Very week	>1.6	940,477.39	94.29
Total	–	997,515.98	100

15.3.2.3 Climate Quality Index (CQI)

The interpolation of rainfall and evaporation data, obtains from metereological stations, are presented in Figs. 15.29 and 15.30. Applying the GIS calculation and storage functions made it possible to compute the climatic sensitivity index and stored in a GIS ready map (Fig. 15.31). The results refers that most rained areas are located in the northern coastal region without exceeding 200 mm. annually. At less that 50–150 km southward from marine coast, annual rainfall drops down to almost zero, while average annual potential evapotranspiration remain high in whole country. The average annual potential evapo-transpiration is relatively high in the whole

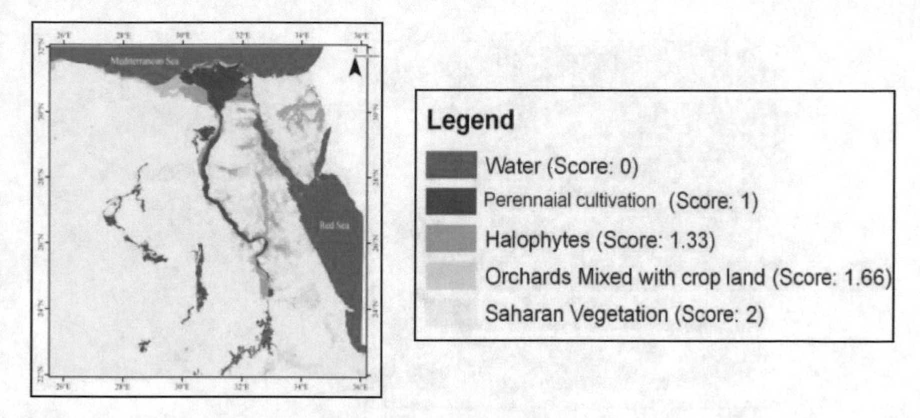

Fig. 15.28 Vegetation Quality Index (VQI), as contributing in desertification sensitivity

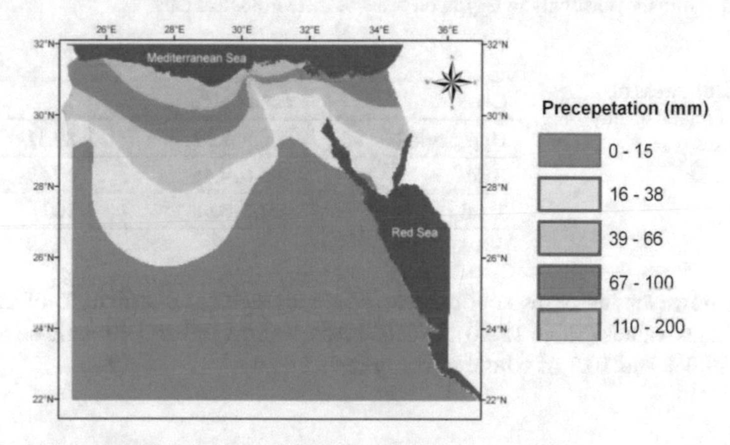

Fig. 15.29 Average annual precipitation in Egypt, on basis of meteorological data

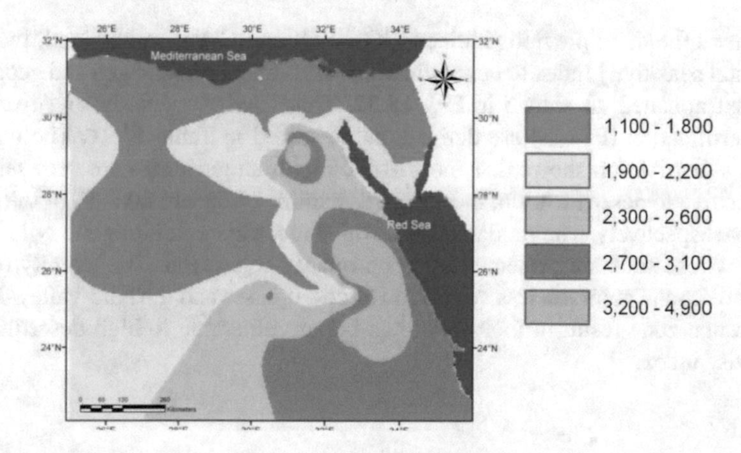

Fig. 15.30 Average annual potential evapotranspiration in Egypt, on basis of meteorological data

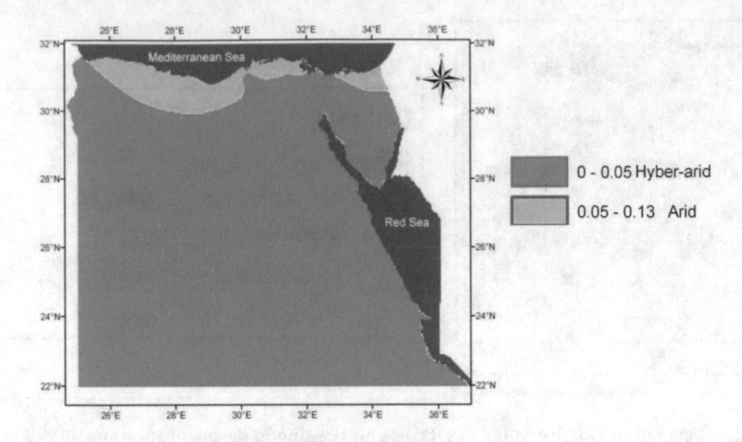

Fig. 15.31 Climatic sensitivity in Egypt, on basis of meteorological data

Table 15.20 Areas of different climatic quality index classes

Class	Area (km^2)	%
Hyper-arid	890,881.52	89.31
Arid	106,634.45	10.69
Total	997,515.98	100

country, however increases southwards. The geographical distribution of climatic quality index values (Table 15.20) reveals the hyper arid and arid climatic conditions exhibit 89.3% and 10.7 of whole territory respectively.

15.3.2.4 Environmentally Sensitive Areas (ESA's) to Desertification

As soon as the three previous indices (SQI, VQI and CQI) are obtained, the environmental sensitivity index to desertification (DSI) may be calculated and geographically extrapolated, as shown in Fig. 15.32. Thus Environmentally sensitive areas for desertification (ESA'S) are defined, as presented in Table 15.21. The out pots of ESA's distribution shows thar most of the Egyptian territories are very sensitive and sensitive to desertification; these classes exhibit 74.39 and 20.27% of the whole territory respectively. The moderately sensitive areas are exhibiting the Nile Valley region, whose soils are protected by good quality vegetation. The vicinity of aeolian deposits and common less vegetation at desert oases and the Nile Valley–Desert interference zone result in their nature as being vulnerable to high desertification sensitivity index.

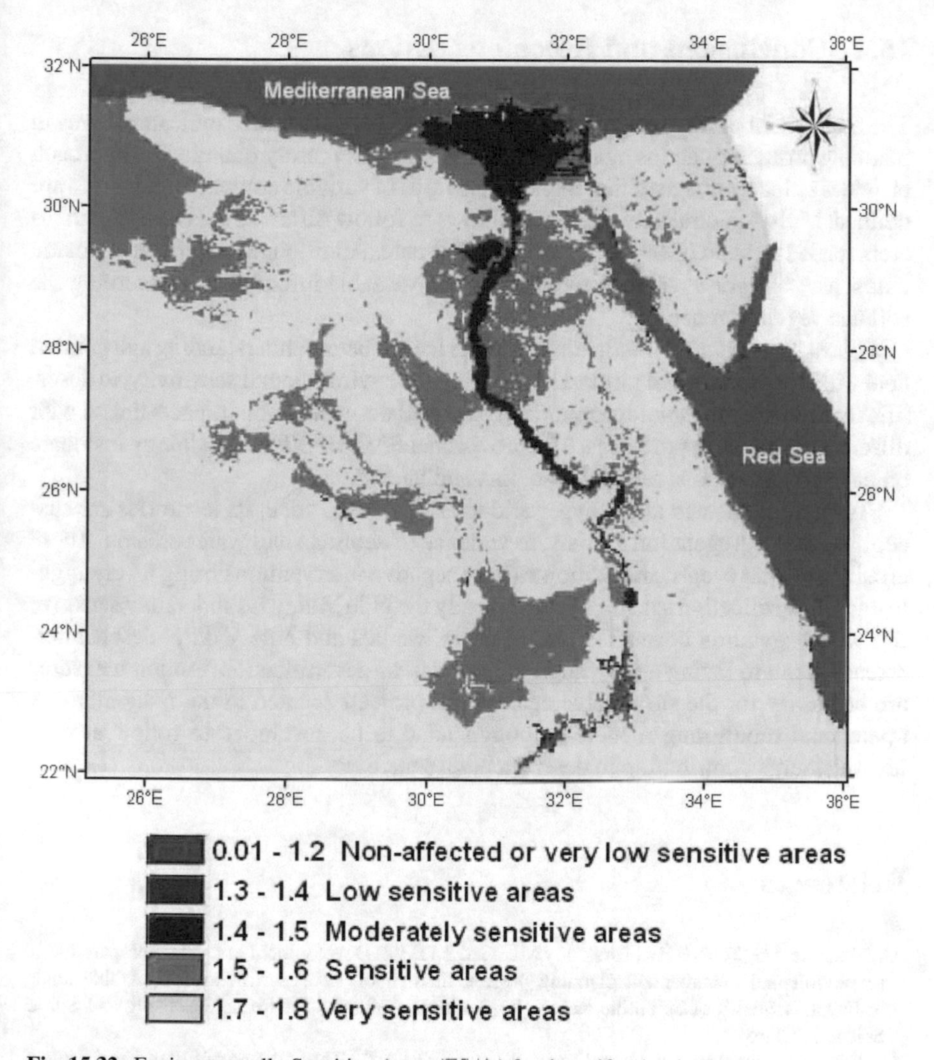

Fig. 15.32 Environmentally Sensitive Areas (ESA's) for desertification in Egypt

Table 15.21 Occurrence of Environmentally Sensitive Areas (ESA's)

Class	Score	Area (km^2)	%
Non affected or very low sensitive areas	0.01–1.2	798.01	0.08
Low sensitive areas	1.3–1.4	11,072.43	1.11
Moderately sensitive areas	1.4–1.5	41,396.91	4.15
Sensitive areas	1.5–1.6	202,196.49	20.27
Very sensitive areas	1.7–1.8	742,052.14	74.39
Total	–	997,515.98	100

15.4 Conclusions and Recommendations

The assessment of desertification sensitivity provides quantiative indicator serves in planning protection actins against desertification. The merely quantitative approach provides a clearer image of the risk state, thus ares of variable combating priorities are defined. Multi-spectral satlite imaging allows to follow different measurable parameters related to state of soil, vegetation and climate. Additionally, available thematic maps, may integrate with space data to obtain valuable information concerning the soil and vegetation quality.

Following multiple investigation schemes lead to better understanding and granted deal with the investigated target. Determining the environmental sensitivity, to desertification, defines the total magnitude of expected desertification impacts linked with different dominating processes. The provisional FAO/UNEP methodology insinuate on each process status, rate, inherent risk and hazard.

As Egypt is located at the hyper arid to arid climatic zone, its territories are susceptible, due to vegetation sacristy, to violence of aeolinan and water erosion. These erosive external events, in addition to man negative interventions bring a very high-to-high desertification sensitivity. Fortunately the Nile Valley is moderately sensitive due toits vegetation cover. The desert oases, wadies and Nile Valley- desert interference area are facing a very hugh to high risk of desertification. Action measures are necessary for the sustainable agricultural projects located in the risky areas. A operational monitoring model is recommended to be developed to follow at field level elements contributing to desertification sensitivity.

References

Abdel-Samie AG, Younes HA, Elrakaiby ML, Gad A (1992) Development of disaster preparedness, prevention and management planning project, final report on desertification of Arable lands in Egypt. Ministry of Scientific Research—National Authority for Remote Sensing and Space Sciences, Cairo

Academy of Scientific Research and Technology, ASRT (1982) The soil map of Egypt. Final project report, Cairo, Academy of Scientific Research and Technology, 379 p

Al-Khaier F (2003) Soil salinity detection using satellite remote sensing. Thesis submitted to the International Institute for Geo-information Science and Earth Observation in partial fulfillment of requirements for Master of Science in Geo-Information Science and Earth Observation. https://webapps.itc.utwente.nl/librarywww/papers_2003/msc/wrem/kh, 23 June 2019

Allbed A, Kumar L (2013) Soil salinity mapping and monitoring in arid and semi-arid regions using remote sensing technology: a review. Adv Remote Sens

Aubreville A (1949) Climate, forests et desertification d l'Afrique tropical, Paris

Ayyad MA (1977) System analyses of Mediterranean desert ecosystem of Northern Egypt "SAMDENE", Progress report No. 4, vol 11, Animal and socio economic studies. University of Alexandria, Alexandria

Bagnold RA (1954) The physics of blown sand and desert dunes. Methuen & Co. Ltd., London, p 265

Basso F, Bellotti A, Bove E, Faretta S, Ferrara A, Mancino G, Pisante M, Quaranta G, Taberner M (1998) Degradation processes in the Agri Basin: evaluating environmental sensitivity to desertification at basin scale. In: Enne G, D'Angelo M, Zanolla C (eds) Proceedings international seminar on "Indicator for assessing desertification in the Mediterranean, Porto Torres, Italy, Sept 18–20, pp 131–145, Supported by ANPA via Brancati 48, 00144 Roma

Basso F, Bove E, Dumontet A, Ferrara A, Pisante M, Quaranta G, Taberner M (2000) Evaluating environmental sensitivity at the basin scale through the use of geographic information systems and remotely sensed data: an example covering the Agri basin (Southern Italy). CATENA 40:19–35

Batterbury SPJ, Warren A (2001) Desertification. In: Smelser N, Baltes P (eds) International encyclopedia of the social and behavioral sciences. Elsevier Press, pp 3526–3529

Briggs D, Giordano A, Cornaert M, Peter D, Maef J (1992) CORINE soil erosion risk and important land resources in thc southern regions of the European Community. EUR 13233, 10 Luxembourg, 97 p

Bryson RA, Baerreis DA (1967) Desert expansion. Bull Am Meteorol Soc 48:141

Buttrick GA (1996) Prayer. Abingdon-Cokesbury Press, 333 p. This old Presbyterian pastor gives us a great treatise on the subject of prayer. Contents include: JESUS AND PRAYER

CONOCO Inc. (1989) Startigraphic Lexicon and explanatory notes to the geological amp of Egypt 1–500,000 (Hermina M, Klitzsch E, List FK, eds). CONOCO Inc., Cairo, 263 p. ISBN 3-927541-09-5.

Cooke RU, Warren A (1973) Geomorphology in deserts, 100 maps and diagrams, 80 pls. B. T. Batsford, London, xii + 374 pp

Dixon KL (1959) Ecological and distributional relations of desert scrub birds in Western Texas. Condor 61:397–409

Dregne HE (1987) Ecological stability of desert soils. In: Proceedings of a post graduate course, 31 Aug–25 Sept 1987. Riksumiversiteit, Ghent

De Pina Tavares J, Baptista I, Ferreira AJD, Amiotte-Suchet P, Coelho C, Gomes S, Amoros R, Dos Reis EA, Mendes AF, Costa L, Bentub J (2015) Assessment and mapping the sensitive areas to desertification in an insular Sahelian mountain region Case study of the Ribeira Seca Watershed, Santiago Island, Cabo Verde. CATENA 128:214–223

El-Baz F (1978a) The meaning of desert colour in earth orbit photographs. Photogram Eng Remote Sens 44(1):7–11

El-Baz F (1978b) Egypt as seen by landsat. Dar Al-Maaref, Cairo, 65 p

El-Baz F (1982a) Journey to Egypt farthest corner. In: El-Baz F, Maxwell A (eds) Desert landforms of southeast Egypt: a basis for comparison with Mars. NASA C.R 3611, Washington DC

El-Baz F (1982b) Egypt's desert of promise. Nati Geogr 161:197–220

European Commission (1999) The MEDALUS project Mediterranean desertification and land use, manual on key indicators of desertification and mapping environmentally sensitive areas to desertification. Directorate-General Science, Research and Development, p 94

Falk R (2019) Toward a just world order. Taylor & Francis Group, p 652. ISBN 9780367211912. https://books.google.com.eg/books?id=TgKaDwAAQBAJ&pg=PT458&lpg.

FAO (1986) The land resources base. FAO (ARC/86/3), Rome

FAO/UNEP (1981) Provisional methodology for desertification assessment and mapping. FAO, Rome, 57 p

FAO/UNEP (1983) Provisional methodology for assessment and mapping of desertification. FAO, Rome, 104 p

FAO/UNEP (1984) Provisional methodology for assessment and mapping of desertification, 84 p

Ferrara A, Bianchi S, Cimatti A, Giovanardi C (1999) Astrophys J Suppl Ser 123:437–445. 1999 August © 1999. The American Astronomical Society. All rights reserved. Printed in U.S.A.

Gad A (1988) The study of desertification processes and soil conditions in the transition zones between the desert and the Nile Valley, using remote sensing (Lower Nile, Egypt). Ph.D. thesis, Faculty of Sciences, State University of Ghent

Glantz MH (1977) Environmental and social impacts group, National Center for Atmospheric Research Boulder, CO 80307, USA (reproduced by: Glantz MH, Orlovsky NS (1983) Desertification: a review of the concept. Desertification Control Bull 9:15–22)
Grove AT (1969) Landforms and climate changes in the Kalabari and Ngamiland. Geograph J 138(2):191–212
Grove AT (1996) Physical, biological and human aspects of environmental change. In: MEDALUS II Project 3, Managing desertification. EV5V-CT92–016, pp 39–64
Hagedorn H, Giebner K, Weise O, Busche D, Grunert G (1977) Dune stabilization—a survey of literature on dune formation and dun formation and dune stabilization. Geographies, Institute University of Wurzburg, 400 p
Helldén U (1991) Desertification—time for an assessment? Ambio 20(8):372–383
Hereher E (2017) Geomorphology and drift potential of major aeolian sand deposits in Egypt. Geomorphology 304:113–120
Horowitz AS, Potter EF (1971) Introductory petrography of fossils. Springer, Berlin, 302 p
https://www.eshailsat.qa/en/DynamicPages/channels Provider, Channels, Freq. Pol. FEC, Encr, satellite. Al Jazeera Arabic, 11604. H, 3/4 DVBS QPSK, No, Es'hail-1. Al Jazeera Mubasher, 11604. H, 3/4 DVBS
Iqbal F (2011) Detection of salt affected soil in rice-wheat area using satellite image. Afr J Agric Res 6(21):4973–4982. https://doi.org/10.5897/AJAR11.634
Kosmas C, Ferrara A, Briasouli H, Imeson A (1999) Methodology for mapping Environmentally Sensitive Areas (ESAs) to desertification. In: Kosmas C, Kirkby M, Geeson N (eds) The Medalus project: Mediterranean desertification and land use. Manual on key indicators of desertification and mapping environmentally sensitive areas to desertification. European Union 18882, pp 31–47, ISBN 92-828-6349-2.
Kovda VA (1958) Studies on the soils of Egypt
Le Houerou N (1969) La vegetation de la Tunisie steppique these, Tunisie
Leopold AS (1963) The desert. Time Inc., The Hague, 191 p
Lettau H (1969) Evapotranspiration climatology: monthly evapotranspiration, run-off and soil moisture storage. Mon Weather Rev 97:691–699
Mabbutt JA (1985) Desertification of the world's rangelands. Desertification control Pull 12:1–11
Ministry of Environment, India (2018) Order of the National Green Tribunal regarding new road projects obstructing tiger corridors
Nicholson JWG, McQueen RE, Allen JG, Bush RS (1996) Composition, digestibility and rumen degradability of crab meal. Canad J Anim Sci 76(1)
Nicholson SE, Tucker CJ, Ba MB (1998) Desertification, drought and surface vegetation: an example from the West African Sahel. Bull Am Meteorol Soc 79(5):815–829
Okoth PF (2003) A hierarchical method for soil erosion assessment and spatial risk modelling. Thesis, Wageningen University, 232 p
OSS (2003) Map of sensitivity to desertification in the Mediterranean basin—proposal for the methodology for the final map. Observatory of the Sahara and Sahel (OSS), Rome
OSS—Roselt Sahara and Sahel Observatory (2004) Long term ecological monitoring observatories network, a common device for the monitoring of desertification in circum—Saharan Africa
Pax Lenney M, Woodcock CE, Collins JB, Hamdi H (1996) The status of agricultural lands in Egypt: the use of multitemporal NDVI features derived from Landsat TM. Remote Sens Environ 56:8–20
Quintanilla EG (1981) Regional aspects of desertification in Peru, in combating desertification through integrated development. In: United Nations Environment Program/USSR commission for UNEP international scientific symposium, abstract of papers, Tashkent, USSR, pp 114–115
Ragan Shrestha DS, Margate F, Van der Meer (2005) Analysis and classification of hyperspectral data for mapping land degradation: an application in southern Spain. Int J Appl Earth Obs Geoformation 7(2):85–96
Rapp A (1974) A review of desertification in Africa-water, vegetation and man. Lunds Universities Geographisca Institution (Report No. 39), Lund, 17 p

Remote Sensing Center and ISPRA (1982) Monitoring of desertification problems in Egypt, using remote sensing. A research proposal, March 1982
Richards LA (1954) Diagnosis and improvement of saline and alkali soils. Lippincott Williams & Wilkins 78(2):154
Said R (1962) The geology of Egypt. Elsevier, Amsterdam, New York, 349 p
Salem, MH, Dorah HT, Fatehelbab A (1982) Design of the New l-Menia town—thunderstorms study. Rep. Mins. cons. (Egypt), 74 p
Shrestha DP, Margate DE, Van der Meer F, Anh HV (2005) Analysis and classification of hyper-spectral data for mapping land degradation: an application in southern Spain. Int J Appl Earth Obs Geoformation 7(2):85–96
Skidmore EL (1986) Wind erosion climatic erosivity. Clim Change 9.1–2(1986):195–208
Thompson (1977) The desert. Orbis Publishing Limited, London, 128 p
Thornes JB (1995) Mediterranean desertification and the vegetation cover. In: Fantechi R, Peter D, Balabanis P, Rubio JL (eds) EUR 15415—"Desertification in a European context: physical and socio-economic aspects". Office for Official Publications of the European Communities, Brussels, Luxembourg, pp 169–194
Tucker CJ, Dregne HE, Newcomb WW (1991) Expansion and contraction of the Sahara Desert from 1980 to 1990. Sci New Sci 253(5017):299–301
UNEP (1977) United Nations conference on desertification—round-up, Plan of action and resolutions. United Nations, New York, 43 p
Vrieling A, Sterk G, Beaulieu N (2002) Erosion risk mapping: a methodological case study in the Colombian eastern plains. J Soil Water Conserv 57:158–163
Woodcock CE, El-Baz F, Hamdi H et al (1994) Desertification of agricultural lands in Egypt by remote sensing. Final Report
Younes HA (1982) Application of remote sensing techniques for detection of desertification in Saudi Arabia and what has been done to combat desertification. Rep of Ministry of Higher Education, King Abdulaziz University, Kingdom of Saudi Arabia, 29 p
Zonn IS (ed) (1981) USSR/UNEP projects to combat desertification. Moscow Centre of International Projects GKNT, 33

Part V
Hydrology and Geomorphology

Chapter 16
Landscapes of Egypt

Nabil Sayed Embabi

Abstract Landscape may be defined as a stretch of country with environmental characteristics and dominated by certain landforms. Landscape analysis is the subdivision of landscape for some purpose or another. It may be for a scientific study such as the environmental conditions that prevails at present-day or in the past. By using satellite images of different types, the analysis of Egypt's landmass made it possible to recognize various types of landscapes according to the dominant surface features. Examples of Egyptian landscapes are karst landscape on the carbonate plateaus, riverine landscape in the Nile Valley and the Nile Delta, and dune landscape in the sand seas of the Western Desert and Sinai. Since most of the Egyptian landscapes, especially the physical ones, do not develop under the present arid environmental conditions, it can be said that they were inherited from past environments.

Keywords Egypt · Mountain landscape · Structural domes landscape · Karst landscape · Fluvial landscape · Dune landscape · Ridge-depression landscape · Fluvio-marine landscape · Urban landscape · Rural landscapes

16.1 Introduction

16.1.1 Definitions

"*Landscape may be defined as a stretch of country as seen from a vantage point*" (Fairbridge 1968). The landscape is made up of rock with its cover of weathered material, and any other surface features such as vegetation, rivers and lakes. Examination of space images of Egypt shows a tremendous range of variations from one locality to another. These variations are due to local differences in their physical characteristics and historical development. A brief account on geographic regions, geologic aspects, and climatic conditions (Present and Past) of Egypt is a necessity

N. S. Embabi (✉)
Department of Geography, Faculty of Arts, Ain Shams University, 20 Ibn Qotaiba Street, 7th District, Nasr City, Cairo 11471, Egypt
e-mail: nabilsayedembabi@gmail.com

S. F. Elbeih et al. (eds.), *Environmental Remote Sensing in Egypt*,
Springer Geophysics, https://doi.org/10.1007/978-3-030-39593-3_16

for understanding the variations in the landscapes of Egypt. This account will be concluded by an attempt to classify the land of Egypt into several landscape types.

16.1.2 Geographic Regions of Egypt

Since the development of the Nile Valley 6–7 million years ago, Egypt is divided into 4 geographic regions: the Nile Valley and its Delta, the Western Desert, the Eastern Desert, and Sinai Peninsula (Fig. 16.1).

16.1.2.1 The Nile Valley and Its Delta

This is the smallest geographic region, covering about 3.5% of the total area of Egypt. Running in a desert region, the River Nile is a salient feature in the geomorphology

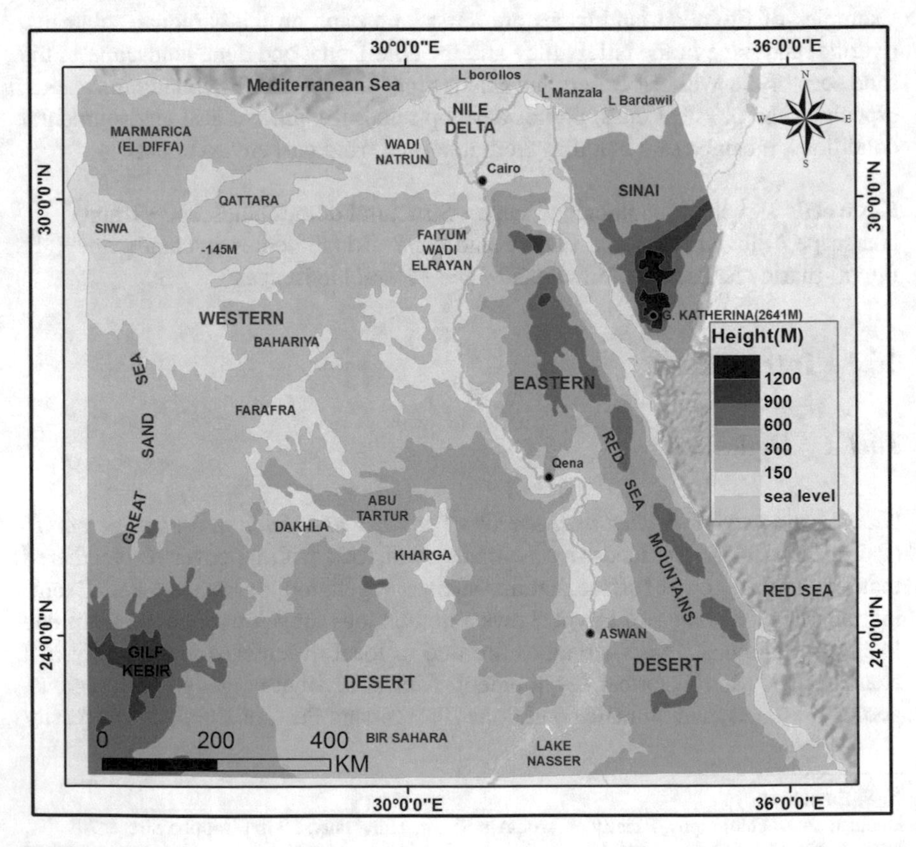

Fig. 16.1 Topographic map of Egypt

of Egypt (Fig. 16.1). Not only does it divide the country into two distinct physical regions (the Western and Eastern Deserts), it has also shaped Egypt history and human occupations. During its geological history, the Nile developed a valley and formed a delta. The Nile Valley widens gradually northwards from several hundreds of meters in the south to 23 km at the latitude of Beni Suef City (~122 km to the south of Cairo). To the south of the First Cataract, about 6 km south of Aswan, the Nubian Nile used to pass through a very narrow valley, which was surrounded by cliffs from both sides. At present, this reach is drowned by the waters of Lake Nasser, which came into being after the construction of the High Dam in the late 1960s of the twentieth century. Therefore, the Nile Valley in Egypt is divided at present into two distinct sub-regions connected by the River Nile. The Nile has two branches in both the Nile Valley (El-Sohagiya Canal and Bahr Youssef), and the Delta (Damietta and Rosetta Branches). However, in the old days, the number of Nile branches were more than that in both sub-regions, but they disappeared due to various reasons. Under population pressure and the increasing needs for more food crops, several irrigation projects were constructed on the Nile and its branches (canals, barrages, dams) to irrigate the Delta plain and the flood plain in the Valley and the newly reclaimed land at their peripheries. Some of these new canals follow the path of ancient branches. Therefore, the landscape of these two sub-regions is composed of a network of water canals to irrigate the cultivated fields and drain excess water.

16.1.2.2 The Western Desert

This region covers all the area lying to the west of the Nile Valley and its Delta. It is the largest geographic region since it covers more than two thirds of the total area of Egypt, which is about a million km^2. It is the region of plateaus, depressions, sand dunes and vast plains. Apart from the plains which dominate the landscape of the southern part of this Desert, other forms are repeated all over this Desert (see Fig. 16.1). Plateaus developed in the north (Marmarica/El-Diffa plateau), in the central part (the carbonate Eocene plateau and Abu Tartur), and in SW of this Desert (El-Gilf Kebir, Abu Ras, and the Peneplaned Plateau) Embabi (2004), Richter and Schandelmeier (1990), Clayton (1933). All the plateaus in the Western Desert rise between 200 and 500 m asl except for the plateaus lying in the SW of this Desert which rise more than 1000 m asl.

Depressions spread everywhere in the Western Desert. They range in size from the minute ones (a few square meters) to mega-depressions where length or width is more than 100 km and depth is 100–200 m. The largest one is the Qattara Depression which is about 45,000 km^2. All mega depressions are well-known since ancient times except for Qattara Depression which is discovered in 1926 by Ball (1927). Except for Qattara, prehistoric sites were discovered indicating that humans settled in these depressions depending on underground water discharged naturally or artificially. The floors of the five depressions which lie in the northern half of this region are below sea level (Qattara—145 m, Siwa—18.5 m, Wadi El-Natrun—24 m, El-Fayum—45 m, Wadi El-Rayan—64 m). The Fayum Depression is distinguished from other depressions in that the Nile water transformed to it by Bahr Yousef which carried out huge quantities of the Nile silt that was laid down on its floor, converting it to a land with Nile affinities. Therefore, the Fayum landscape is different from other depressions.

Although a huge drainage system preceded that of the Nile in the Western Desert, only its delta exists till the present-day, and only some of its tributaries can be traced as inverted wadis on the surface of the limestone plateaus of the Western Desert. Relics of other drainage nets can be traced in the SW in the area Gilf Kebir and Abu Ras plateaus. Recently, wadis were discovered by Radar Space shuttles buried below the Selima Sand Sheet, and this why they were called in some previous studies "Radar Rivers" (McCauley et al. 1986).

Sand dunes are distributed all-over the Western Desert, but in separate accumulations called sand seas and dune fields. Sand seas cover larger areas ($\geq$5000 km^2) than dune fields (<5000 km^2), with higher dune coverage of more than 50% (Embabi 1998). There are 5 sand seas and 4 dune fields in the Western Desert, of which the Great Sand Sea is the largest (>100,000 km^2 in Egypt only). In all sand accumulations, predominant forms are the crescent (barchans) and linear dunes. In the Great Sand Sea, linear dunes are aligned side by side in some instances so that there are no inter-dunes corridors, and are extending for more than 100 km, with heights exceeding 100 m. Other sand seas in the Western Desert are the Selima Sand Sheet, Ghard Abu Moharik, Farafra Sand Sea, and South Qattara Sand Sea (Fig. 16.2). Under the effect of wind, barchans move in the downwind direction which is—in general—the southward. Also, the linear dunes extend and grow in the same direction. In both

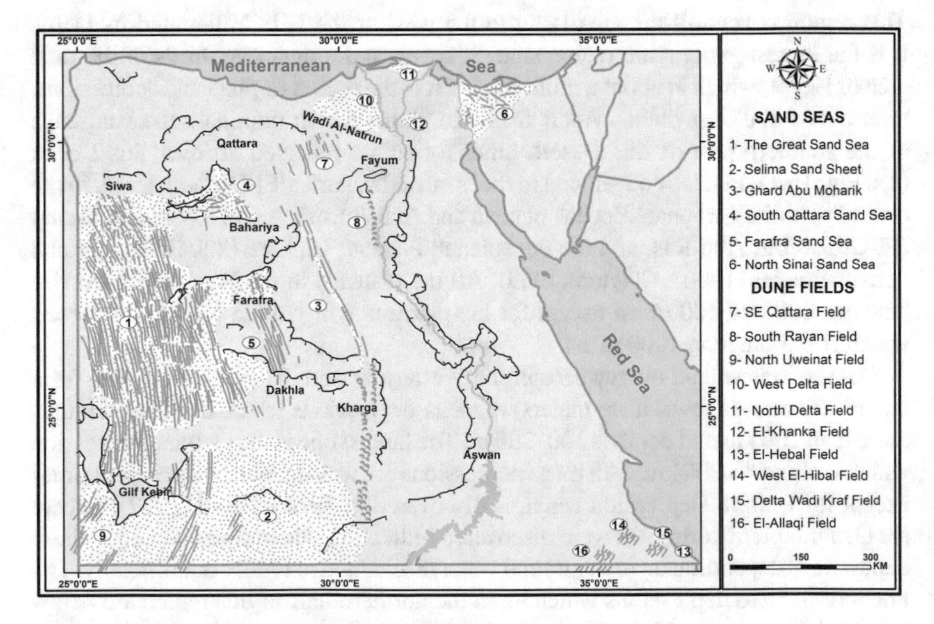

Fig. 16.2 Distribution of sand seas and dune fields in Egypt

cases, the dunes during this action might encroach on cultivated land, villages, and cross highways and railways.

In some sand seas, sand sheets dominate the landscape, which is exemplified by the Selima Sand Sheet in the southern part of the Western Desert. Here, the original landscape is a vast peneplane developed in the past by fluvial action of several networks of streams and wadis, which are hidden below the sand sheet in the present-time (McCauley et al. 1986).

16.1.2.3 The Eastern Desert

This is the region that lies between the Nile Valley and its Delta in the west, and the Red Sea and the Isthmus of Suez in the east. It covers about 25% of the total of Egypt. This is the region of mountains, plateaus and large drainage nets. Here, both the mountains and plateaus are dissected by wadis which drain either towards the Nile Valley or towards the Red Sea. Heights of some peaks in the mountains exceed 2000 m asl, of which Gabal El-Shayeb is the highest peak (2187 m asl.). A characteristic feature in the Eastern Desert is the circular forms known as "Ring Complexes". They are distributed all-over the Mountains, with some concentration in the southern region of the Mountains. Heights of the plateaus range between 1000 m and several hundreds of meters. Only in the depression known as El-Mallaha that lies between the small range or Aish El-Mallaha and the coastline of the Gulf of Suez, heights are a few meters below sea level (2–9 m below sea level).

The Red Sea Mountains extend southwards from Latitude 28° 30′ N parallel to the coastline of the Red Sea to and beyond the Egyptian-Sudanese borders (Latitude 22° N). To the west and the north of the Mountains, vast plateaus extend to the Nile Valley. The southern sandstone plateau is called "El-Ababdah" which extends from the borders to Qena Bend of the Nile Valley. To the north of this bend, there are the limestone plateaus of El-Maaza, El-Galalah El-Qibliah, El-Galalah El-Bahariyah and Gabal Ataqa. A narrow coastal plain developed between the Mountains and plateaus, and the coastline of the Red Sea. Along with this plain, several groups of forms of various origins developed such as small deltas and fans, sabkhas, marine terraces, raised beaches and small dune fields, of which El-Hebal Dune Field is the largest and most known. Drainage nets dissect the mountains and plateaus and drain either easterly in the Red Sea and the Gulf of Suez, westerly to the River Nile, and northwards to the plains of east Delta and hide beneath the Nile silt of the Delta. An example of the large wadis characterizing the Eastern Desert is Wadi Araba which separates El-Galalah El-Bahariya and El-Galalah El-Qibliah and drains into the Gulf of Suez. Wadi Qena which extends for about 250 km in a nearly N-S direction to the town of Qena in the Nile Valley is another example. Wadi El-Allaqi, with some of its tributaries come from the Sudan drains nowadays in Lake Nasser is a third example.

16.1.2.4 Sinai Peninsula

Though considered a sub-region of the Eastern Desert, the Sinai Peninsula with its distinct triangular shape possesses its own characteristics. It is bounded from the north by the Mediterranean Sea, from the west by the Gulf and Isthmus of Suez, and from the east by the Gulf of Aqaba and the Palestinian-Israeli boundary. Sinai covers about 6% (61,000 km^2) of the total area of Egypt. Its coasts extend for about 700 km; a characteristic which makes Sinai less continental compared to other regions of Egypt. Sinai can be divided into four regions: the southern mountainous region, the central tableland region, the region of the Syrian-Arcs domes, and the northern region of sand dunes (Embabi 2004).

The southern region is dominated by a rugged mountainous landscape, which rise to more than 2000 m asl in some localities such as mount St. Kathrine (2641 m) and Gabal Mousa (2285 m), and is dissected by a high-density drainage nets, which drain either in the Gulf of Aqaba or in the Gulf of Suez. The central region is composed of two vast carbonate tablelands called El-Tih Plateau lying just to the north Sinai Mountains and El-Egma Plateau which follows El-Tih from the north. Both plateaus are characterized by the whitish color of their calcareous rocks, which contrasts with the dark color of the igneous and metamorphic mountains of southern Sinai. Steep escarpments bound both plateaus from the south, west and east. The escarpments are bold and not dissected, except for a few localities where wadis were able to cut through its high walls. The relative heights of the escarpment of El-Tih and that of El-Egma are 300–700 m and 300–500 m respectively. Both escarpments rise more than 1000 m asl (Gabal El-Tih 1187 m; Gabal El-Gunna 1583 m; Gabal El-Egma 1626 m). Drainage lines of El-Tih form the tributaries of the main wadis which drain either into the Gulf of Aqaba or the Gulf of Suez. Wadis draining the surface of El-Egma form the upper tributaries of Wadi El-Arish, which crosses the northern region of Sinai and drains into the Mediterranean Sea.

The northern part of Sinai is divided into two distinct regions. The first is the region of the folded hills and mountains in the south, and the second is dominated by sand plains in the north. The folded forms are organized into parallel lines trending ENE-WSW, and rise between 1090 m asl (Gabal Yelleq) and 368 m asl (Gabal Risan Ainaza). They are elliptical and exhibit structural domal morphology. Their length varies between 10 and 50 km, and width between 3 and 20 km. Some of the drainage lines dissecting the slopes of these hills and mountains drain internally forming fans, bajadas and playa plains, whereas some other drainage lines form the northern tributaries of Wadi El-El-Arish. To the north of these domal forms, the sandy plains extend to the Mediterranean coastline. Dunes of different types, patterns and age spread from the eastern borders of Egypt to the Suez Canal, forming a sand sea, which is called the North Sinai Sand Sea (Embabi 1998). In this plain, there are other two significant forms. The first is the lower reaches of Wadi El-Arish and the second is the Bardawil Lake. To the west of this Lake, the sandy plain changes to muddy flats that are composed of ancient Nile silt and locally known as El-Tinah plain.

16.1.3 The Geologic Aspects of Egypt

16.1.3.1 Rock Formations

It is already well-known that rock formations in Egypt are classified into two groups according to their mode of development (Hume 1925; Said 1962, 1990a, b). The first group is the Crystalline Basement Complex rocks, composed of igneous and metamorphic rocks and developed during Precambrian times. The second group comprises all the sedimentary rocks which developed during the Phanerozoic period and cover most of the basement complex.

The geologic map of Egypt, scale 1: 2,000,000 (Survey of Egypt) shows that the basement rocks cover about one tenth of the total land surface of Egypt. They are exposed in the Red Sea Mountains of the Eastern Desert, the southern part of Sinai and Gabal Uweinat area (Fig. 16.3). Small exposures also occur in the southern part of the Western Desert and the Cataract region of Aswan and Nubia.

Said (1962) divided the sedimentary cover of Egypt into three divisions (lower, middle, and upper) according to the prevailing type of rocks and age. The strata of the Lower Division are primarily clastic with a few calcareous beds. The geological map of Egypt, scale 1:2,000,000, shows that the strata of this division are exposed in the southern part of Egypt to the south of Kharga—Dakhla Depression and further south into the Sudan, in Nubia around Lake Nasser and Aswan, and in Bahariya

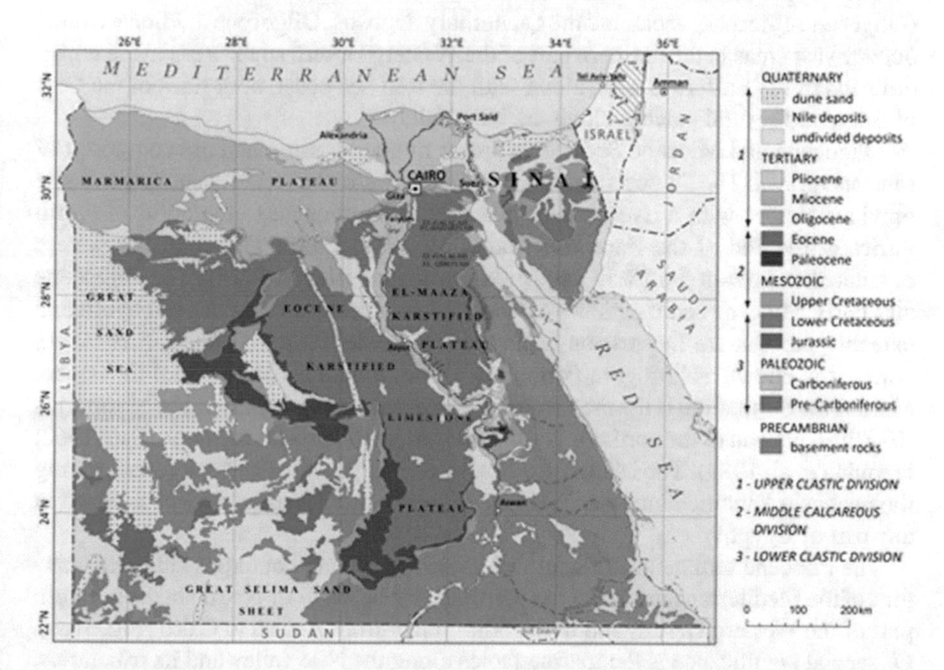

Fig. 16.3 Simplified geologic map of Egypt. *Source* Embabi (2018); simplified from the Geological map of Egypt, Scale 1: 2,000,000

Depression. They are also exposed in Wadi Qena, central and northern Sinai such as Gabal Maghara and other folded hills, and Wadi Araba between the two Galalah plateaus (Fig. 16.3). One of the most significant formations in this division from the economic point of view is the Nubia Sandstone. Although Said (1962) did not include this formation in the Lower Division, the Author found it would be better to include in that division since it is all clastics. This Formation is characterized by the presence of a huge reservoir of underground water. This reservoir is composed of several aquifers of which the deepest reaches more than 1000 m below the surface in some localities and the nearest is only 100–150 m deep. All the oases in the Western Desert depend on this underground reservoir since prehistory. Also, nearly all the recent reclamation and development projects since the forties of the 20th Century depended on the water extracted from this underground reservoir.

The Middle Sedimentary Division is composed of a series of beds, which are dominantly calcareous and is deposited during the period from Upper Cretaceous to the end of the Eocene (Said 1962). The geological map (1: 2,000,000) shows that these strata of this division cover wide areas in the central parts of the Western and Eastern Deserts on both sides of the Nile Valley and central and northern parts of Sinai (Fig. 16.3). They also extend to the southernmost parts of the Western Desert and along the Red Sea coast. These calcareous beds are the most relevant formations in this group, since they exhibit karst landscapes with different surface and subsurface forms of karstic origin.

The Upper Clastic Division is divided into two sub-divisions: Cenozoic (Oligocene-Pliocene) rocks and the Quaternary deposits. Oligocene to Pliocene units occupy vast areas in the northern part of the Western Desert, small areas in the northern parts of the Eastern Desert, along with the Red Sea coast, both sides of the Gulf of Suez, and several patches along the Nile Valley.

Oligocene and Miocene Sediments are of fluviatile origin and are composed of sand and gravel. The Oligocene fluvial sediments were connected according to some previous studies with a river system that flowed consequent to the uplift of North Africa at the end of the Eocene period (Ball 1939, 1900). This drainage system continued to flow in the Early and Middle Miocene (Salem 1976). The Oligocene and Early Miocene sediments were deposited as a delta in the north of Fayum and extending to Qattara Depression (Salem 1976; Abdel-Rahman and El-Baz 1979), where it is known as "Moghra Formation". The Middle Miocene reefal limestone, Marmarica Formation, is the most extensive exposure, covering the entire Marmarica (El-Diffa) plateau of the northern Western Desert, and Cairo-Suez stretch (Said 1962; Hermina et al. 1989). The Marmarica limestone exhibits various karst forms, giving the plateau a karstified landscape. Upper Miocene marine strata are not exposed in any part of Egypt.

The Pliocene sediments are scattered as relatively small outcrops along the margins of the Mediterranean coastal plain, in and around Wadi El-Natrun in the northern part of the Western Desert, and in the Nile Valley from Aswan to Cairo (Fig. 16.3). Of special significance is the marine facies along the Nile valley and its tributaries, which are taken as an indication of the rise of the Mediterranean Sea level forming a marine gulf that submerged the Nile Valley and the mouths of the tributaries.

This marine gulf is known as the Pliocene Gulf of the Nile Valley (Said 1993; Ball 1939). The sediments that were laid down in the Pliocene Gulf are significant to the development of the flood plain of the Nile Valley.

The Quaternary deposits are spread all over Egypt and lie unconformable over the Pliocene or older rocks. These deposits are closely connected with climatic changes and sea level fluctuations that occurred during the Pleistocene period. They are classified according to their mode of origin into aeolian, fluvial, and marine sediments.

Aeolian sands are the most widespread and most recent, covering more than 20% of the total area of Egypt (Embabi 1998). They occupy vast areas in the south and west of the Western Desert and North Sinai, and small areas in the north and east of the Western Desert, and the north and south of the Eastern Desert (Fig. 16.2). They are formed as dunes of different types and sand sheets shaped as sand seas and dune fields.

Fluvial deposits include Nile sediments, wadi gravels, playa silt, sand and gravel clastics, calcareous tufa and travertine. Of these fluvial sediments comes the Nile silt as the most significant one from the point of view of landscapes in Egypt. The Nile silt spreads on the floor of the Nile flood plain, the Delta and Fayum Depression, developing a type of landscape not repeated anywhere else.

Marine and coastal deposits include the Mediterranean coastal oolitic limestone in the form of parallel linear ridges separated by longitudinal depressions covered by evaporites, marl and gypsum. These ridges were formed as offshore bars in the receding Mediterranean Sea during the Pleistocene (Shukri et al. 1956; Selim 1974). Sabkha deposits developed along the Mediterranean coasts of the Nile Delta, various stretches to the west of Alexandria, Sinai and in Siwa, Qattara, Wadi El-Natrun and small patches in other depressions and along the Red Sea coast. Shelly-coralline, sandy and silty sediments spread along the Red Sea coast, which indicate a shallow reefal marine environment during the Pleistocene epoch.

16.1.3.2 Tectonic Framework

Although many tectonic events occurred during the geological history of Egypt, this section will deal with those features which are relevant to the landscapes of Egypt. Studies of Said (1962) have divided Egypt tectonically into four units: The Arabo-Nubian Shield, The Stable Shelf, The Unstable Shelf, and The Gulf of Suez-Red Sea Graben.

The rocks of the Arabo-Nubian unit (the Basement rocks) are exposed, as previously mentioned, over large areas in the Red Sea and Sinai Mountains) and small patches in the extreme southwestern part of the Western Desert and along both sides of the Nile Valley in the Aswan environs. Also, evidence indicates that these rocks extend northwards beneath the sedimentary cover of the Stable and Unstable Shelves (Said 1962). This tectonic unit is characterized by certain features which gives it a characteristic landscape. The first feature is the development of mountain ranges in the Eastern Desert and southern Sinai during the opening of the Red Sea. The second feature is the development of drainage systems dissecting the mountain ranges during past pluvial periods. The third feature is the presence of a high-density network of joints and fissures which resulted in a basin-range landscape. The basins were the locus of fluvial sediments which were carried out by active streams during past Pluvial Periods.

The second tectonic feature relevant to the landscapes is the "**Syrian Arcs**" **anticlines** which are concentrated in north Sinai (Fig. 16.4). All anticlinal axes take an NNE-SSW orientation and are plunging in both sides. Most of these anticlines are small ones (>1 km in length or breadth and a few hundred meters high). Only Gabal Magharah, G. Halal and G. Yeleq are relatively large anticlines (40–45 km long and rise 800–1000 m or more asl). Due to their folding origin, the beds are dipping with various degrees in opposite directions, developing hogbacks. Therefore, this region

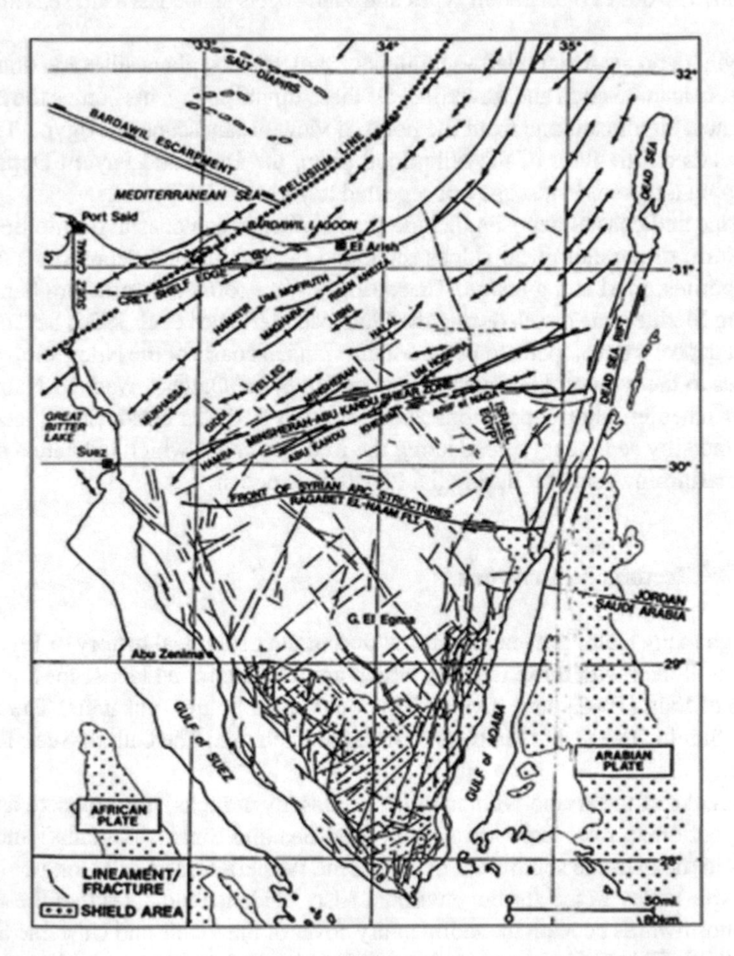

Fig. 16.4 Structural map of Sinai Peninsula showing Domal Landscape of North Sinai and the adjacent areas in the Negev Desert and Eastern Mediterranean Sea. *Source* Jenkins (1990)

is considered as a special type of landscape composed of anticlinal hills and small mountains and modeled by fluvial action that dissected the anticlines into circular ridges of hogbacks.

The third tectonic feature relevant to the landscapes of Egypt is the Hing Zone which separates the Unstable Shelf from the Miocene-Geosyncline of the Mediterranean Basin. According to Sigaev (1959) and Schlumberger (1984), the Nile Delta occupies a large trough in the Hinge Zone. This trough which is called by Said (1981) the North Delta Embayment is present within the continental slope of the Mediterranean. It is delimited in the south by a fault running along the northern boundary of the Cairo-Suez horst and in the north by structures consisting of parallel, elongated tilted blocks. This embayment has been filled with the sediments of the Nile Delta Cone since the Upper Miocene times. This sedimentary cone extends below the Mediterranean and covers an area which equals three times that of the present Nile Delta.

16.1.4 Climatic Conditions: Present and Past

16.1.4.1 Present Day Climatic Conditions

Although Present Day arid conditions are known to prevail over the whole of Egypt, variations in the degree of aridity are largely unknown. To investigate such aspects in a previous study (Embabi et al. 2012) the UNESCO Index of Aridity (1987) as a ratio of (Pmm) Annual Precipitation (mm)/(Eptmm) Annual Evapotranspiration (mm) is applied to all meteorological stations (60 Stations) shown in the published Climatological Normals of Egypt (Meteorological Authority of Egypt 1979). The calculated indexes were classified according to the UNESCO (1977) into four categories as follows:

A. Hyper-arid: <0.03
B. Arid: 0.03–0.2
C. Semi-arid: 0.2–0.5
D. Sub-humid: 0.5–0.75.

The calculated indexes of all meteorological stations (60 stations) were plotted according to the designated degree of aridity in Fig. 16.5. It can be seen from this figure that Egypt is dominated by hyper-arid climate. Only one station (Rafah) in the extreme northeast corner belongs to the semi-arid category. The map also shows that the arid climate prevails only along a narrow strip parallel to the Mediterranean coast from Sallum to El-Arish. If Rafah, which belongs to category C (semi-arid) is added, it can be said that present-day climatic conditions along the whole Mediterranean coast are hyper-arid to arid.

Although Fig. 16.5 shows that hyper-arid conditions prevail to the south of the arid Mediterranean coast, the actual indexes indicate that the southern half of the Western Desert (south of the latitude of Qena-Farafra) is extremely arid, since their indexes

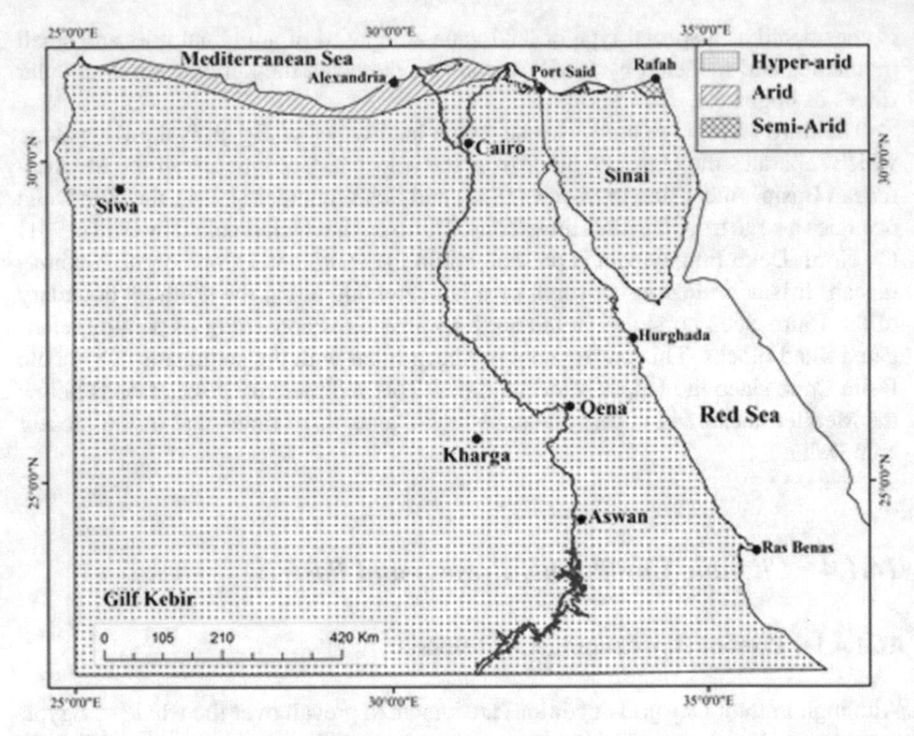

Fig. 16.5 Types of arid climates in Egypt. *Source* Embabi et al. (2012)

are zero. This suggests a new category of aridity, which can be called extremely arid, the index of which is zero. Therefore, the southern part of the Western Desert is known as the driest part in the Sahara, and probably the driest region on Earth (Kehl and Bornkamm 1993).

16.1.4.2 Past Climates

"Present is the Key to the Past" (Hutton 1785). This fundamental concept is based upon the fact that "*The same physical processes and laws that operate today operated throughout geologic time, although not necessarily always with the same intensity as now*" (Thornbury 1954). Thus, past climates will be recalled depending upon this fundamental concept. The analysis of past climates of Egypt will be divided into three sections according to geologic periods: the Paleogene, and the Neogene.

The Paleogene: is the first period in the Cenozoic era. It is divided into three epochs: Paleocene, Eocene and Oligocene. All evidence found by Rolfs (1873), Blankenhorn (1900), Ball (1939), Salem (1976), Abdel-Rahman and El-Baz (1979) and Said (1990a) points to the presence of a great river that flowed in a northern direction to enter the sea immediately to the north of what is now the Fayum, forming a huge delta in this region. With such a delta and tributaries and the fossil content of the sediments, the conclusion that Egypt enjoyed a rainy tropical climate during the Paleogene, is confirmed.

The Neogene: The climate during the first two epochs of Miocene and Pliocene does not differ much from that of the Paleogene. This is because the large drainage system which was dominating the scene during the Paleogene extended geologically to Lower Miocene (Salem 1976). Said (1962, 1990a) assigned an early Miocene age for Moghra Formation in northwest Egypt. Said (1990a) also found that the in-situ biota points to a tropical climate and African affinities.

Evidence found by Pickford et al. (2015) in the area between Bahariya and Farafra indicated that the Western Desert was covered by woodland during Late Miocene (11–12 Million years). Also, fauna found in this area indicated a mean annual rainfall more than 500 mm and perhaps as much as 1200 mm (Pickford et al. 2015). The formation of the Nile Valley in Egypt before the end of the Miocene indicates that a rainy (probably also tropical) climate prevailed during this period. The climate became arid during the early and middle Pliocene since no evidence was found to prove the contrary. So far, evidence indicates that it is only in the late Pliocene that Egypt witnessed the onset of more humid conditions. These conditions converted the marine gulf of the Nile into a veritable river channel, named Palaeonile by Said (1981).

Buried wadis in the southern plain of the Western Desert are other fluvial forms indicating the prevalence of a nationwide humid climate. This hidden drainage system below the Selima Sand Sheet, was discovered by radar imaging in 1981 (SIR/A), 1984 (SIR/B) and 1994 (SIR/C) during flights of the United States space shuttles. This palaeo-drainage includes wide wadis (10–30 km), a hundred kilometers long wadis and narrow anastomosing channels that developed at the floor of large wadis (McCauley et al. 1986). Regional reconstruction of these previously unknown palaeo-drainage systems was the subject of several studies (McCauley et al. 1982, 1986; McHugh et al. 1988; Schaber et al. 1997; El-Baz et al. 1998; Ghoneim et al. 2007).

The conclusion reached by most of these studies is that the large wadis are relics of a Cenozoic system that drained the Eastern Sahara, and that its narrow channels developed during the Quaternary Pluvial Periods (McCauley et al. 1982). Also, previous studies postulated that the Red Sea Mountains were the source area of the large wadis long before the development of the Nile Valley (McCauley et al. 1986). It can therefore be said that the buried wadis in southern Egypt reveal new evidence indicating that the climate was rainy during the Cenozoic not only in the northern parts, but also in the southern region, and consequently all over Egypt.

The advent of the Pleistocene (from ~1.84 Ma) brought a pattern of aridity that set the tone of the climate prevailing in Egypt, with minor fluctuations throughout the Pleistocene (Said 1990b). During the earliest Pleistocene, a long period of aridity (~one million years) dominated Egypt and the country was converted into a desert (Said 1990b). During the Middle and Late Pleistocene period, the climate changed several times from arid to pluvial during which rainfall was diminishing gradually until it reached the present arid conditions ~5000 years ago (Said 1990b).

16.2 Landscapes of Egypt

The preceding analysis of the physical characteristics of the Egyptian landmass showed vast variations from one locality to another, indicating that there is no one landscape overwhelming Egypt but rather there are a variety of landscapes. In the meantime, human activities on the Egyptian landmass contributed some new characteristics to previously physical ones modifying some of their aspects or creating new anthropogenic landscapes after obliterating the existing physical features. Therefore, the types of landscapes which can be recognized on the Egyptian landmass are grouped into two types: Physical Landscapes and Anthropogenic Landscapes.

16.2.1 Physical Landscapes

A wide variety of physical landscapes are developed in Egypt. Most of them are inherited from past environmental conditions, mainly from past wetter climates or from tectonics.

16.2.1.1 Mountain Landscape

This type of landscape is present on the Red Sea and Sinai Mountains. It is characterized by certain rock formations, morphological characteristics and certain landforms not repeated on any other landscape. Here, the mountains are formed mostly of igneous and metamorphic rocks. The highest areas in Egypt are recorded in this landscape, such as Mount St. Katherine (2641 m asl) which is the highest not only in Sinai, but also nationwide (Fig. 16.1). Other conspicuous peaks in Sinai are Gabal Um Shomar (2586 m), Gabal Mousa (2285 m) and Gabal Serbal (2070 m). Along the Red Sea range, about 30 peaks rise more than 1000 m asl, such as, Gabal Shayeb Al-Banat (2187 m), Gabal Hamata (1977 m), Gabal Qattar (1963 m), Gabal Shendib (1911 m) Gabal Abu Abid (1900 m).

The analysis of the Google Earth images revealed that this landscape is characterized by certain structural and erosional landforms (Google Earth 2017). They include Dyke Swarms (Fig. 16.6), Intermountain Basins, Ring Complex Circular Mounts (Fig. 16.7), Ancient Erosion Surfaces (Fig. 16.8), and sharp Peaks (Fig. 16.9). Dykes, in most cases, rise a few meters forming low-ridges that run parallel to or cross each other. Intermountain basins are small low-lying depressions surrounded by high ground and are filled with fluvial sediments brought by water flow during past pluvial periods. Ring complex are circular mounts that are structurally developed; therefore, the form is conformable with the structure. They are formed of one or more circular ridges (Fig. 16.7), with/without a core representing the peak of the mount. Ancient erosion surfaces developed on the basement rocks and are covered by the Nubian Sandstone Formation in Sinai (Awad 1950). In some localities,

Fig. 16.6 Parallel NE-SW dykes in the northern section of the Red Sea Mountains. *Source* Google Earth (2017)

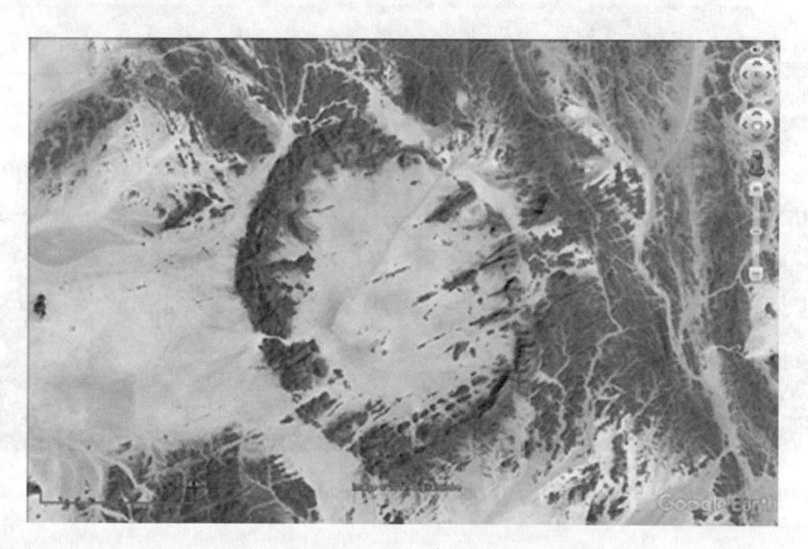

Fig. 16.7 A Google Earth image of a ring complex developed in the NE corner of an intermountain basin in the Red Sea Mountains. *Source* Google Earth (2017)

the sedimentary cover is removed resulting in the resurrection of these flat surfaces (Fig. 16.8), and this why they are called locally "Farsh". Sharp peaks developed due to the dissection of the steep-sided mountains by deep incised wadis during the past pluvial periods (Fig. 16.9).

Fig. 16.8 A field photo for the old erosion surface in the NE corner of Sinai Mountains, topped with some sandstone remnants

Fig. 16.9 A field photograph for sharp peaks of Red Sea Mountains in Halayeb Region

16.2.1.2 Structural Domes Landscape

North Sinai is characterized by the abundance of northeast to east-southeast oriented doubly plunged anticlines. They are known in geologic literature as "the Syrian Arcs" (Said 1990a, b) as mentioned in Sect. 16.1.3.2. As can be seen from Fig. 16.4, "the Syrian Arcs" are not restricted to Sinai, but extend eastwards into the Negev region, and northwards into the Mediterranean basin. They form a distinctive tectonic and geomorphic region in North Sinai. Although there are—similar forms in various areas in Egypt, domes are clustered in this region of Sinai to form a specific landscape entity (Fig. 16.4). Small or large, all hills/mountains are conformable with their domal structure. In some cases, the core of the domes is eroded so that it becomes lower than the outer ridges. Since the domes are formed of alternating sedimentary beds of varying resistance to erosion, and with varying outward dips, circular hogback ridges dominate the landscape.

16.2.1.3 Fluvial Landscape

The analysis of climatic conditions in Sect. 16.1.3 indicated two facts: the first is that the present-day rainfall does not develop permanent running water that can work on Egypt landmass and produce fluvial landscape. The second fact is that the geologic and geomorphologic evidences indicated that Egypt enjoyed rainy climates in the past during which surface water flow was able to develop fluvial forms on the surfaces of Egyptian landmass. From these two facts, it can be concluded that the fluvial landscape in Egypt is a product of past humid climate prevailed in Egypt during its geological history. In other words, this type of landscape is inherited from past humid environmental conditions, and it is not the product of present arid conditions.

The fluvial landscape is dominated by forms/features produced by surface water flow, of which drainage nets and wadis are still existed since they were developed during past pluvial periods. Nearly all the surfaces of the landmass of Egypt are affected by the action of water flow and the resulting forms/features can be seen here and there, but fluvial landscape applies only on areas where fluvial forms and features are overwhelming the surfaces of the landscape. Areas with these characteristics are present in various parts of Egypt, such as Gilf Kebir and Abu Ras Plateaus in SW the Western Desert, El-Maaza and El-Ababdah Plateau in the Eastern Desert between the Nile Valley and Red Sea Mountains, and Egma Plateau in Central Sinai (Fig. 16.10).

Relatively recently (in the early 1980s), radar imaging discovered fluvial forms, such as drainage nets, wide wadis and channels below the Great Selima Sand Sheet and are known as buried wadis (McCauley et al. 1986). Here, the landscape is composed of two types of landscapes. The first is the fluvial landscape which developed in the Cenozoic (McCauley et al. 1986) on bedrocks of southern Western Desert, and the second is the aeolian landscape in which aeolian sand and some dunes developed atop the fluvial landscape and buried it.

16.2.1.4 Karst Landscape

Karst landscape developed on areas where carbonate rocks are predominant on the surface and subsurface. Although much of rainwater percolates in the subsurface of carbonate formations, the surface flow has the energy to develop some fluvial forms side by side surface and subsurface karstic forms. Therefore, both groups of forms: karst and fluvial develop contemporaneously and can be found in the same area. However, the landscape can be coined karstic only when the karst forms and features dominate the scene. This indicates that the karst landscape is inherited from

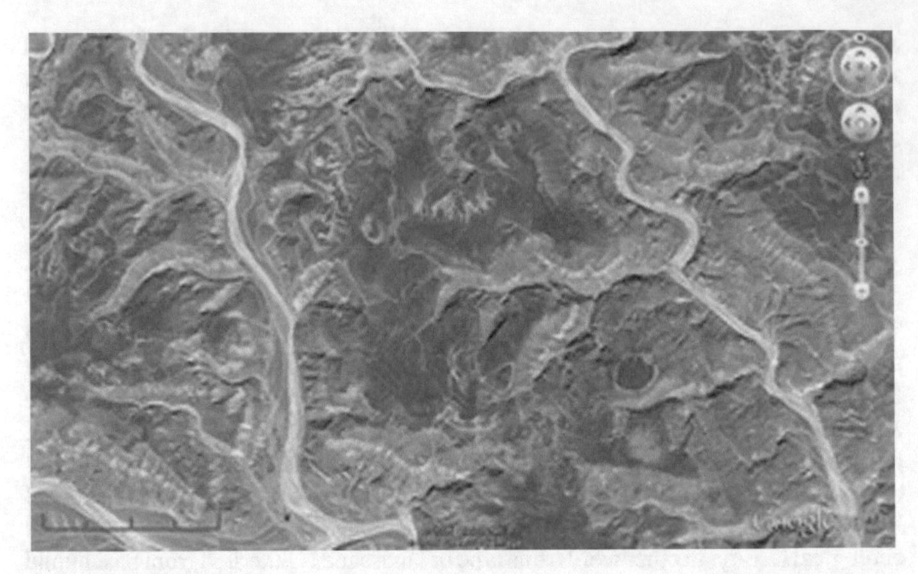

Fig. 16.10 Fluvial landscape atop El-Egma plateau in Central Sinai. *Source* Google Earth (2017)

past pluvial periods side by side fluvial landscape. The most recurrent surface forms which are detected on high resolution space images (Google Earth 2017) are cone-karst (Fig. 16.11), minute dissolution basins, giant flutes (Fig. 16.12). From field investigations, other features and forms were recorded such Terra Rosa, residues of carbonate fines, disintegrated angular flint gravel, and degraded caves. These caves keep some remnants of their original forms and sediments, such as stalactite and stalagmite column, flowstones precipitated on the walls of caves, and breccia. In the

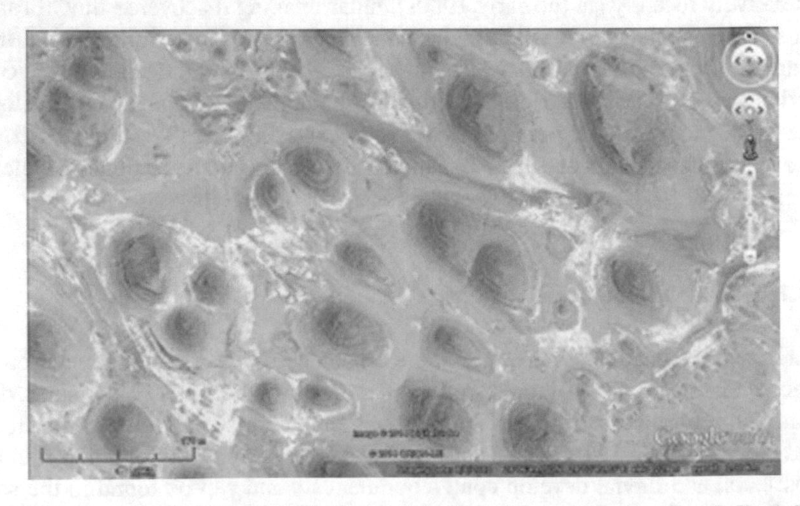

Fig. 16.11 Cone-karst landscape on the Eocene karstified carbonate plateau (Google Earth 2017)

Fig. 16.12 Giant flutes separated by giant ridges developed on the Eocene carbonate plateau in the Western Desert

subsurface, so many caves were discovered, and some of them were studied in detail such as Sannur cave in the Eastern Desert 70 km to the east of Beni Suef City (Dabous and Osmond 2000). Another example is the Djara cave in the Western Desert, that lies beside the camel track between Farafra and the Nile Valley and 10 km west Ghard Abu Moharik (Brook et al. 2002). St. Antony cave and some others developed in El-Galalah El-Qibliyah Plateau are other examples of caves that were found in the Eastern Desert (Haliday 2004). Other caves were known since the Pharaonic times but were not studied as the previous ones. Landscape with these karst characteristics is found on Marmarica plateau, the Eocene plateau, and the two Galalahs in the NE of the Eastern Desert.

16.2.1.5 Fluvio-Marine Landscape

This type of landscape developed along the coastal margins where fluvial forms and features intermingle with those of marine origin. Along these margins, not only these two groups of forms interact, but also rise, and fall of sea level gives this coastal strip some additional morphological aspects. Predominant forms and features are: Deltas, pro-deltas and drowned deltas, coastal lagoons with mangrove agglomerations in some localities, sabkhas, sand barriers, coral marine terraces, and sand beaches (Moawad 2008, 2013; Embabi 2017). They characterize the coastal margins of the Red Sea with their extension along the western side of the Gulf of Suez, and Sinai coasts on the Gulf of Aqaba and the Gulf of Suez. Although, this landscape extends for several hundreds of kilometers, its width does not exceed several tens of kilometers in most cases.

16.2.1.6 Coastal Ridge-Depression Landscape

Along the Mediterranean coast from Alexandria westwards to Sallum City at the Egyptian-Libyan borders, linear parallel low-lying convex ridges separated by depressions are developed forming a unique landscape (Fig. 16.13). The number of ridges and accordingly the number of depressions varies from one locality to another, between just one ridge (the coastal ridge) such as Sidi Barrani locality and eight ridges as in the Arab Gulf Region. Butzer (1960) and Shukri et al. (1956) mapped eight lines of these ridges with varying length and width in the Arab Gulf region. In Sallum basin, the examination of Google Earth images (Google Earth 2017) revealed eight ridges (Embabi 2004). Relative height does not exceed a few tens of meters, but due to the rise of the landmass inland, its height increases in this direction till reaches 110 m asl in the eighth ridges in Burg El-Arab locality. Using space images (Sundborg and Nilsson 1985; Lindell et al. 1991; Moussa et al. 1998; Moawad 2003) recognized submerged ridges in several localities along the Mediterranean coasts.

Previous studies showed that the ridges are mainly composed of alterations of marine, aeolian, fluvial sediments, and palaeosoils. Intra-depressions are filled with fine materials derived mainly from the ridges and some fluvial materials or occupied by some coastal lagoons where there is a connection with the sea such as the lagoons of Matrouh City. Due to water seepage from the sea and rainfall, the floors of the depressions are covered with sabkhas as can be seen in Fig. 16.13 (Hasouba 1980; El-Asmer 1991; Stanley and Hamza 1992; Wali et al. 1993; El-Asmer 1994a, b; Mahmoud 1995; El-Asmer 1998; El-Asmer and Wood 2000; Shata 2000; Embabi 2004). These studies found evidence shows that these ridges are multi-genetic in origin, where wind, marine and continental processes were responsible for their

Fig. 16.13 Ridge-depression landscape in Sallum Basin, where depressions are occupied by sabkhas. *Source* Google Earth (2017)

Fig. 16.14 Aquatic landscape of Borollos Lake and its environs along the northern of the central Nile Delta. *Source* Google Earth (2017)

development. Those studies which were concerned with age found that their age ranges between Lower Pleistocene (Wali et al. 1993; Shukri et al. 1956) for ridges 6–8, and a few hundred years ago (800 years) for the top layers of some submarine bars (Lindell et al. 1991), and 600 years for the top layer of El-Omayid (4th) Ridge in Alamain area (El-Asmer and Wood 2000).

16.2.1.7 Aquatic Landscape

This landscape can be detected in areas where wetlands are dominating the scene, either as water bodies (lakes), or damp soils (sabkhas/marshes). The principal region with such features extends uninterrupted along the Mediterranean coasts from Bardawil Lake in northern Sinai to Lake Maryut in the west. This region of aquatic landscape accommodates five lagoons (Bardawil, Manzalah, Borollos, Edko and Maryut), which are surrounded by vast tracts of sabkhas and salt marshes (Fig. 16.14). Under population pressure and the increasing needs for land to produce crops to feed the increasing population, the area of this landscape is shrinking due to land reclamation, and the transformation of land characteristics from sabkhas and salt marshes to cultivated land.

16.2.1.8 Sand/Dune Landscape

Figure 16.2 shows that sand dunes spread in all regions of Egypt. Previous studies found that about 20% of the landmass of Egypt is covered with sand seas and dune

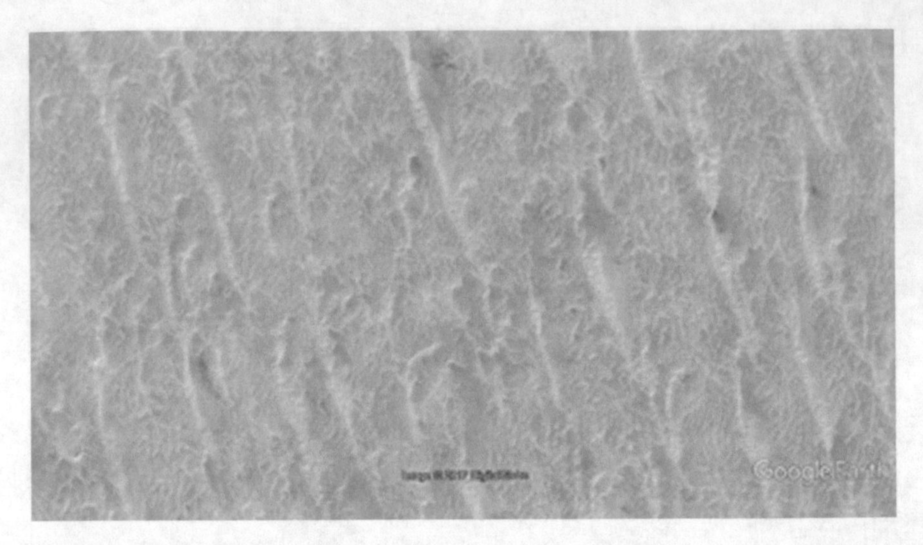

Fig. 16.15 Google Earth image showing a sample of a linear dune type/pattern landscape in the northern part of the Great Sand Sea. *Source* Google Earth (2017)

fields (Embabi 1998, 2017). Some sand seas extend for hundreds of kilometers without any break (The Great Sand Sea and The Selima Sand Sheet), forming a unique landscape of coalesced dunes in the case of the Great Sand Sea, or a continuous cover of sand as sand sheets. Although the dune landscape is undulating due to variations in dune height and sloping sides, sand sheets are nearly flat without any change in their surface characteristics. As recorded by Bagnold (1931), while travelling across the Selima Sand Sheet: *mirage deceives the vision making the surface as if it is covered with water sheet looking like a sea*. In the Great Sand Sea, dune types and patterns change from North to South, and from East to West (Figs. 16.15, 16.16 and 16.17). Other sand/dune landscape that covers smaller areas than the Great Sand Sea or the Selima Sand Sheet are present in South Qattara, East Farafra and North Sinai.

16.2.2 Anthropogenic Landscapes

16.2.2.1 Urban Landscape

As mentioned in Sect. 16.1.5, there are some areas of the landmass of Egypt where people changed the physical face of the landscape by obliterating the physical features and constructing new features for population agglomeration. The new features are mainly houses, roads and other ones to make life viable for people who will be living in (Fig. 16.18). These are the centers which are called cities where people are doing activities other than plowing the land. The best example for urban landscape is Cairo Conurbation and Alexandria City, where a house is built beside another

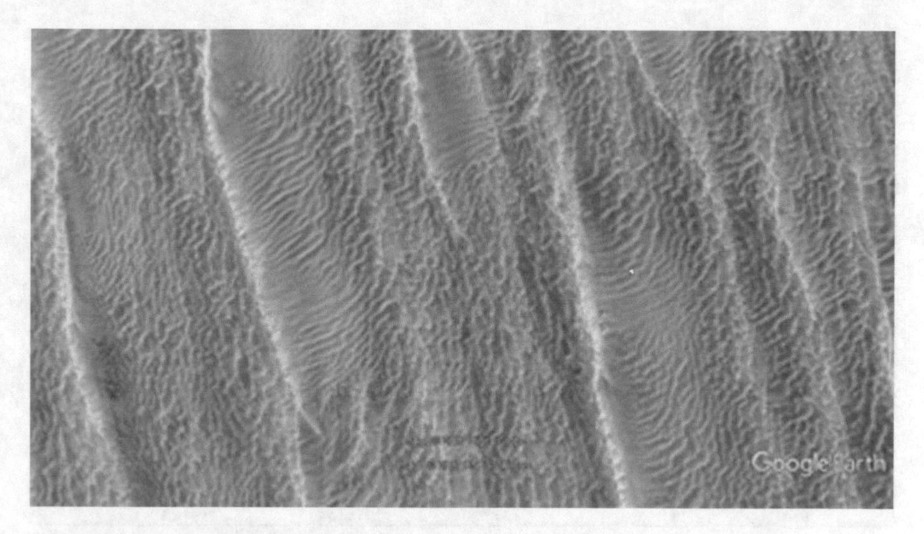

Fig. 16.16 A Google Earth image for a second sample of different dune types and patterns landscape in the central part of the Great Sand Sea. *Source* Google Earth (2017)

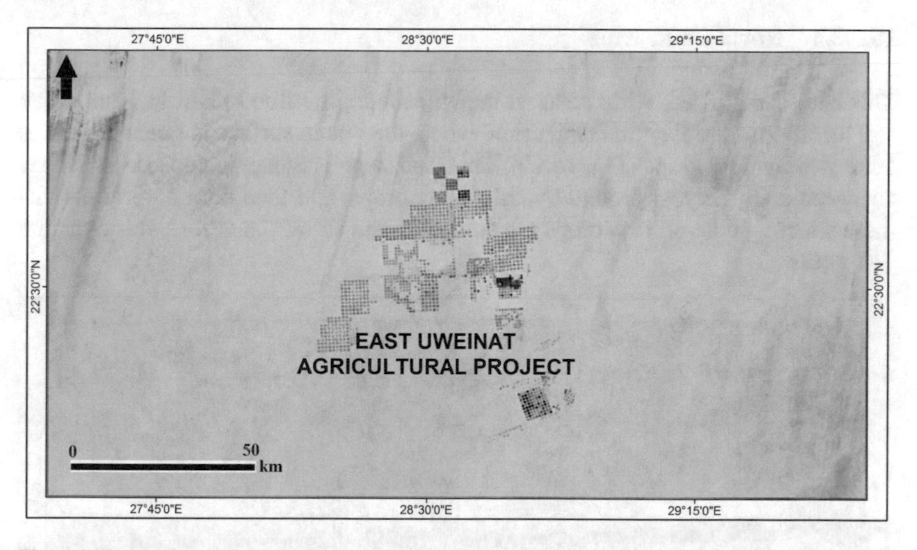

Fig. 16.17 Another sub-type of dune/sand landscape in Selima Sand Sheet, with some newly cultivated land of East Uwienat Project. *Source* Google Earth (2017)

house, separated only by roads, railway lines, and other public utilities. Other urban landscape can be seen in other cities of Egypt.

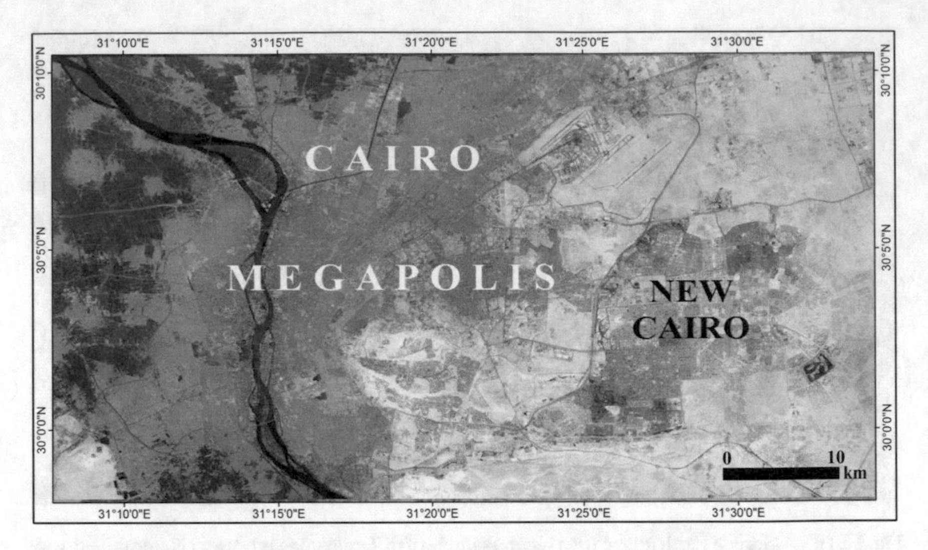

Fig. 16.18 Urban landscape represented by Cairo Megapolis

16.2.2.2 Rural Landscape

This landscape covers wide areas in the Nile Delta, the flood plain in Nile Valley and the floor of the Fayum Depression, where the whole surface is green since it is cultivated by field crops (Fig. 16.19). This landscape is dynamic because its face is changeable all the year around by cultivating crops, and then after several months these are cropped, and new crops are cultivated and so on. Therefore, this dynamic

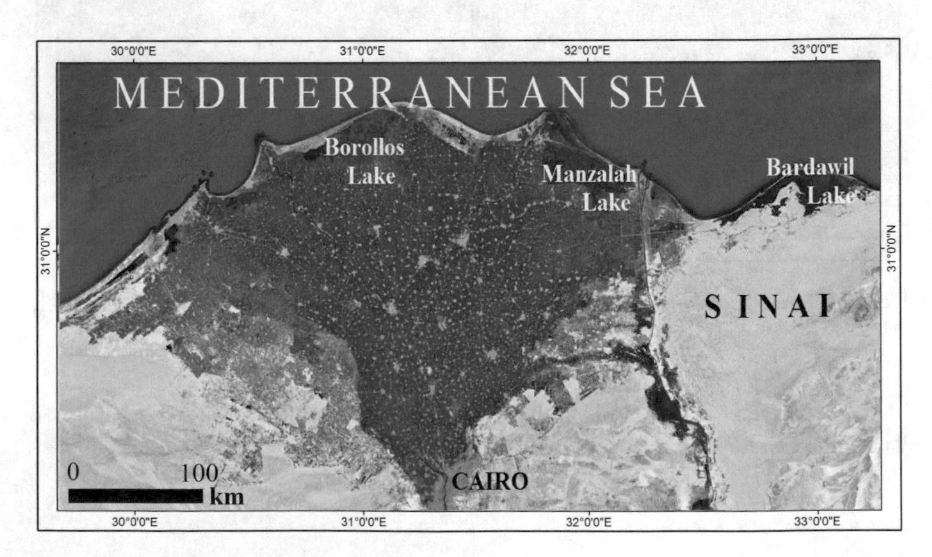

Fig. 16.19 Rural landscape in the Nile Delta

landscape is distinguished from other landscapes. Beside the cultivable land, there are several features that are necessary to keep this landscape dynamic. On the surface of this landscape, a very sophisticated network of canals which carry water to irrigate the fields were established gradually depending on the water of the Nile and its branches in the Delta. Irrigation is essential to cultivate this landscape since as mentioned in Sect. 16.1.3.1 present-day climate is arid/hyper-arid with rain which is not enough to cultivate crops. Some of the irrigation canals are—in fact—previous Nile branches such as Bahr Youssef which carry water to irrigate the floor of the Fayum Depression, and El-Sohagiyah Canal which takes its water directly from the Nile near the City of Sohag to irrigate lands in the extreme western areas of the flood plain. That this network of irrigation canals depends on sophisticated engineering establishments constructed on the Nile Channel and its branches to guarantee efficient system to distribute water according to the needs of crops all the year around. On the top of these irrigation establishments, comes the Aswan High Dam which was constructed in the late sixties of the 20th Century, and formed the largest artificial lake named "Lake Nasser" after the name of the President of Egypt at that time. This lake is representing at the time being the "National Water Reservoir", since all water comes from the upper reaches of the Nile are stored in it.

In addition to the irrigation canals system, there is another system of canals which is called "Drainage Canals". This drainage system is a necessity to drain all the excess water which is accumulated in the subsoil horizon. This is because the low gradient of the cultivated lands does not help to drain naturally the subsoil excess water. Because the drainage canals occupy a part of the cultivable land, the administration in the Ministry of Agriculture responsible for the development of agriculture in Egypt is planning to replace the surface drainage canal with a system of pottery pipes. These pottery pipes can absorb the subsoil water and drain the water into the main drainage canals.

A third feature characterizing this rural landscape is the villages which accommodate the rural population. Although these villages are built of houses, which represent a different land use type from that of the cultivated lands, they are a part of the rural landscape. During pre-Aswan High Dam times when the Nile discharge was high, and the Nile channel cannot accommodate it, water overflows and inundates the flood plain and submerge some of the villages, and this why the natives were building the houses of the villages on the relatively high ground. After the completion of the High Dam, and the water passing through the Dam is regulated according to the needs of crops, and no excess water can flow in the Nile or its distributaries, extensions of the villages in low-lying areas can be done.

A sub-type of the rural landscape is that one which can be called **The Oases Rural Landscape**, where natives are cultivating land on underground water discharged from underground aquifers. In the old days of prehistory, water was discharged naturally as springs (A'yun), but with advances in technology, water is discharged by digging wells tapping the near surface aquifer (150–100 m below the surface). The discovery of artifacts and tools in the oases of Kharga (Caton-Thompson 1952) and Dakhla (Kleindienst et al. 1999) represents evidence that the oases rural landscape started not only in prehistory, but also is older than that of the Nile Valley and the Delta.

Quite recently, attempts started in the forties of the 20th Century to dig deeper than the shallow aquifer to discharge more water from deeper aquifers (~1000 m). In the early sixties of the 20th Century, a new reclamation project was launched to reclaim vast areas mainly in the Depressions of Kharga and Dakhla and was called the "New Valley" project. Due to the limitations (Quantity and salinity) of ground water, the area of this sub-type of the landscape is relatively small comparatively to that one in the Nile Valley and its Delta. The main problem in this sub-type of the landscape is that the underground water discharge diminishes by time in most localities. Therefore, the area of this landscape will be decreasing; otherwise, to overcome this problem it is necessary to dig new wells to produce water to compensate for the drop-in water discharge in old wells. Although there several crops are cultivated in this landscape, palm trees are the most successful crop in all oases (Siwa, Bahariya, Farafra, Dakhla, and Kharga) for two reasons. First, water requirements are much lower than any other crop, and second palm trees can tolerate relatively high- salinity water, since the discharged underground water is saline.

16.3 Conclusions

As a vast country (~1000 km from north to south and ~1000 km from east to west), bordered by two seas from the north (the Mediterranean) and east (the Red Sea), Egypt is characterized by several physical and human aspects. It is covered by nearly all types of rocks, although the present climate is hyper-arid and arid, it changed from arid to wet several times. Also, it accommodates a large variety of landforms, and was inhabited by humans since the times of the Paleolithic. Therefore, it is expected that no single landscape will develop and dominate the landmass of this country. On the contrary, it was found a large variety of landscapes developed all-over the country due to variations in the characteristics of surface features. It was also found that most of landforms which characterize several landscape types were inherited from past wet environmental conditions. Therefore, several types of landscapes were recognized, and were classified into two groups: physical and anthropogenic types. The physical landscapes are classified into subtypes according to the dominant physical aspects. Also, the anthropogenic landscapes were classified into subtypes according to the dominance of human activity. Of the physical landscapes, it was found that three of them dominate the scene on the landmass of Egypt: the mountain landscape, the karst landscape, and the dune/sand landscape, whereas the ridge-depression landscape is the smallest. Of the anthropogenic landscape, Rural Landscape, Cairo Conurbation and Alexandria City are the biggest ones, and the smallest is the Oases Rural Landscape.

16.4 Recommendations

The diversification of landscapes is a unique characteristic of Egypt. This characteristic makes this country has diversified resources that can be used for the welfare of its people. Some of these resources were used since prehistoric times, but still there are some other resources are undiscovered, and still some others can be developed to increase its value-added.

The most precious resource in all landscapes is water, rain falling nowadays in arid and hyper-arid climate cover a very low percentage of the water needs of the population of Egypt. In the old days, the Nile water had the full attention of the Egyptian administration. This is because most of Egypt's population is concentrated in the Nile Valley. Under population pressure in the Nile Valley, attention is given to other landscapes to develop so that they can absorb some of the population increase. This policy requires to develop the local water resources or to find untraditional new water resources in these new frontiers.

Three different landscapes are going under development nowadays, depending on three different water resources. They are: the Rural/Oases landscape in the New Valley Governorate, the Red Sea Marine Front, and the NW coastal Fringe, and. The Rural System of the New Valley depends on the underground water extracted from the Nubian Aquifers. Locals in the oases of this rural landscape were depending on the near-surface aquifer, which is about 100–150 m deep since prehistory (Ezzat 1964), because they do not have the technology to dig deeper to reach deep water aquifers. With the beginning of the 20th Century and the advances in industry, four development attempts to extract underground water from deeper aquifers for the purposes of land reclamation and cultivation in the Oases of the Western Desert. The first attempt was carried out in the northern part of Kharga Depression by a British Firm in the early years of the 20th Century, using the near-surface aquifer and reclaimed and cultivated small plots of land (Beadnell 1905). The second attempt was carried out by the Ministry of Public Works (called nowadays Ministry of Irrigation and Water Resources) in the period from late forties to the early fifties of the 20th Century. In this attempt, 14 deep wells (between 220 and 507 m) were drilled in Dakhla, Kharga (7 each) (Paver and Pretorius 1954). Some of these wells gave a daily discharge of several thousands of cubic meters, while some others gave nothing or a few hundreds of cubic meters. According to the discharged water of each well, land with different areas were reclaimed and cultivated. One common feature characterized all the working wells is that water discharge decreased with varying rates (Paver and Pretorius 1954). The third attempt is a gigantic reclamation project depending on the huge quantities of water discharged from deep wells which were drilled to tap the deep aquifers. This project was called by President Gamal Abdel-Nasser the "New Valley Project". The plan in this project is to reclaim and cultivate several hundreds of thousands of acres and to establish new villages to settle down the new comers from the Nile Valley. This Project faced the same problem of the previous one which is the diminishing of water discharge of wells, resulting in the shrinkage of cultivable land. To keep the cultivable area unchanged needs to drill new wells every now and then at a high cost to compensate water decrease in the old wells. The fourth attempt is undergoing at present as a part of a nation-wide project to reclaim and cultivate 1.5 million feddans, which is initiated by President Abdel-Fattah El-Sisi. The project is undergoing in several localities in Egypt, and one of these localities is the Farafra Depression in the New Valley Governorate. No study is carried out to evaluating its results.

As far as water is the most precious element in the Rural Oases Landscape, two matters are recommended for the maximum value added of water in this desert landscape. Since the extraction of water is costly, water should not be provided for free for the consumers either for farmers, industry or householders. Before the beginning of the New Valley Project, the propriety was for water and not for land as in the Nile Valley. Drilling a well needs a collaborative work from those willing to cultivate a plot of land. According to the share in the costs of drilling a well, a similar share of water will be assigned to each of the shareholders. In this matter, the discharged water is divided into 24 Karat (Karat equals an hour/day). Therefore, every farmer will be provided by some Karats according to his share in the costs of drilling the well. By this way, every famer was making the maximum use of each drop of water. When the rule of local administration is applied to desert governorates, water was provided for free for all consumers, and land became the real propriety of farmers. Therefore, when the discharge of wells decreases, the cultivated land decreases accordingly till a new well is drilled, and this might take a time.

The second recommendation concerns the cultivated crops. It is well-known that the main crop cultivated by locals is palm trees. This is because this crop consumes less water and tolerates high-salinity water. Therefore, although other crops are needed, it is recommended to take the experience of locals into consideration when cultivating a new land. Beside palm trees, there are several crops of high value-added which do not consume much water were recommended in a previous study on the utilization of underground in the New Valley (Embabi 1977).

The second landscape where water represents a critical element in its development is the Fluvio-Marine Landscape along the Red Sea, Gulf of Suez, and Gulf of Aqaba coasts. Here, water resources are very limited, since only some water could be carried out by wadis from the inner highlands to this coastal fringe. Therefore, the quantity and quality of water cannot cover the needs of development projects such as tourism and new settlements or urban expansion. It was thought that transferring some Nile water across the Red Sea Mountains will solve the problem of water shortage, but after extending a pipe line from the River Nile in late seventies of the 20th Century, it was found that it will not solve the water shortage problem. Later, it was found that desalinating sea water is the only way to overcome the problem. Most probably, this landscape is the only one in Egypt which depends heavily on desalinated water to cover its water needs. Since Egypt's water needs is increasing, it is recommended that the technology of water desalination should be adopted as necessity to produce more water needed for future development in landscapes where there is no alternative or as a supplement for small local water resources.

The third landscape is the Ridge-Depression of the NW coast of Egypt, which is going under development since 3–4 decades. This stretch of the Mediterranean coast is developing as the main Egyptian touristic region in Egypt. Tens of summer resorts were established all along this coast from Alexandria to Matruh. Nearly 100% of the water needs in these resorts is covered by the Nile water, which arrives this long coastal stretch by a pipe line. Concomitant with the development of these resorts, development occurred also in another two elements in this region. The first is that infrastructures (highways, electricity, communications and others) were improved significantly. The second is the growth of ancient towns and cities. These service centers (El-Agami, El-Hammam, and Matruh) provided services for the new comers in summer holiday. Another activity which represents a unique characteristic of this landscape is the expansion in fig horticulture. Fig crop continues for the four summer months, and with the increase in production, it is exported to other parts of Egypt. Since this landscape is still undergoing development, water demands will increase in the future. Here, as in the coastal fringe of the Red Sea, shortage in water can be covered only by desalinated sea water. There are some desalination stations in some resorts and in some service centers like Matruh City since the sixties of the 20th Century, but with the increase of water demands, the source which can cover the future needs is the desalinated water.

Other landscapes can also be developed, according to the potentialities of each of them. This needs very detailed studies for the elements of each landscape, recommending the development of the best sides of each element in each landscape to give the maximum value added.

Acknowledgements I owe many thanks to many people who helped me to bring this Chapter into being. First, I would like to thank Dr. Salwa El-Beih of the National Authority of Remote Sensing and Space Sciences for nominating me to write this chapter. Special thanks goes to the Egyptian editors and the Springer editor who revised thoroughly the manuscript of this Chapter and for their valuable suggestion which added to the scientific value of this Chapter. I would like also to express my gratitude Mr. Mohamed El-Raei who works as an assistant lecturer in the Department of Geography, Faculty of Arts, Ain Shams University for processing Fig. 16.1 and producing it in a color form. I acknowledge also Google Earth for providing high resolution space image on the internet free of charge, without which so many characteristics of the landscapes and landforms could not have been studied.

References

Abdel-Rahman M, El-Baz F (1979) Detection of a probable ancestral delta of the Nile River. In: El-Baz F, Warner DM (eds) Appollo-Soyuz test project II, Earth observations and photography. NASA, Washington DC, pp 511–520

Awad H (1950) La Montabgne du Sinai Central. Etude morphologique, Spec Public. Societe Royale de Geographie d'Egypte, Le Caire 243 p

Bagnold RA (1931) Journeys in the Libyan Desert, 1929, 1930. Geogr J 78:13–39

Ball J (1900) Kharga Oasis: its topography and geology. Geological Survey Department, Cairo, Egypt, 308 p

Ball J (1927) Problems of the Libyan Desert. Geogr J 70:21–38, 105–128, 209–224

Ball J (1939) Contribution to the geography of Egypt, Cairo. Survey and Mines Department, Ministry of Finance, Government Press

Beadnell HJL (1905) Topography and geology of Fayum Province. Survey of Egypt, Cairo, 101 p
Blankenhorn M (1900) Neues zur Geologie and palaeontologie Aegyptens. II: Das Palaeogen (Eosin und Oligozaen). Zeitschrift der Deutschen Geologischen Gesellschaft 52:403–479
Brook GA, Embabi NS, Ashour M, Edwards M, Cheng H, Cowart JB et al (2002) Djara Cave in the Western Desert of Egypt: morphology and evidence of quaternary climatic change. Cave Karst Sci 29:57–66
Butzer K (1960) On the Pleistocene shore lines of the Arab's Gulf, Egypt. J Geol 68:626–637
Caton-Thompson G (1952) Kharga Oasis in prehistory. Athelon Press, London
Clayton PA (1933) The western side of the Gilf Kebir. In: A reconnaissance of the Gilf Kebir by the Late Sir Robert Clayton East Clayton. Geogr J 81(3):254–259
Dabous A, Osmond JK (2000) U/TH study of speleothems from the Wadi Sannur Cavern, Eastern Desert of Egypt. Carbonates Evaporites 15:1–16
Donner J, Ashour M, Brook GA, Embabi NS (2015) The quaternary history of the Western Desert as recorded in Abu El-Egl Playa. Bulletin de la Société de Géographie d'Égypte 88:1–18
El-Asmer HM (1991) Old shorelines of the Mediterranean coastal zone of Egypt in relation with sea level changes. Dissertation, Mansourah University
El-Asmer HM (1994a) Aeolianite sedimentation along the northwestern coast of Egypt: evidence for Middle to Late Quaternary aridity. Quat Sci Rev 13:699–708
El-Asmer HM (1994b) Recognition of diagenetic environments of the Peistocene carbonate rocks at El-Hammam, NW Mediterranean coast of Egypt, using stable isotopes and electron microprope. Neues Jahrbuch für Geologie und Paläontologie Monatshefte 1:7–22
El-Asmer HM (1998) Middle and late Quaternary Palaeoclimatic evolution, northern Mediterranean coast of Egypt. In: Alsharhan S, Glennie KW, Whittle GL, Kendall CG (eds) Quaternary deserts and climatic changes. Proceedings of the international conference on quaternary deserts and climatic changes. AlAin-UAE, 9–11 Apr 1995. A A Balkema, Roterrdam, pp 261–271
El-Asmer HM, Wood P (2000) Quaternary shorelines development: the northwestern coast of Egypt. Quat Sci Rev 19:1139–1149
El-Baz F, Robinson C, Maxwell TA, Hemida IH (1998) Palaochannels of the great selima sand sheet in the Eastern Sahara and implications to ground water. Palaoecology Afr 26
Embabi NS (1977) Problems of utilization of underground water in the Oases of the Western Desert of Egypt, with a special reference to the Kharga and Dakhla Oases (in Arabic). Bull Arab Res Stud Arab Educ Cult Sci Organ Arab League 8:149–184
Embabi NS (1998) Sand seas of the Western Desert of Egypt. In: Al-Sharhan A, Glennie K, Whittle G, Kendall C (eds) Proceedings of the international conference on quaternary deserts and climatic change. Balkema, Al-Ain, Rotterdam, 1995
Embabi NS (2004) The geomorphology of Egypt, vol. I: the Nile Valley and the Western Desert. Special Publication of the Egyptian Geographical Society, Cairo
Embabi NS (2018) Landscapes and landforms of Egypt. Springer International Publishing AG, Switzerland
Embabi N, Mostafa A, Mahmoud A, Azab M (2012) Geomorphology of Ghard Abu Moharik, Western Desert, Egypt. Bull Soc Géog Égypte 85:1–28
Ezzat MA (1964) New valley project: ground water conditions (in Arabic). Egyptian General Desert Development Organization, Tunis
Fairbridge RW (1968) The encyclopedia of geomorphology. Dowden, Hutchinson & Ross. Inc., Halsted Press: A division of John Wiley & Sons, Inc., Stroudsburg, Pennsylvania
Ghoneim E, Robinson C, El-Baz F (2007) Radar topography data reveal drainage relics in the Eastern Sahara. Int J Remote Sens 28:1759–1772
Google Earth (2017) Egypt images: latitudes 22-33 N and longitudes 25-33 E. http://www.Google.com
Haliday W (2004) Caves and karsts of Northeast Africa. Int J Speleol 32(1/4):19–32
Hasouba ABH (1980) Quaternary sedimentsfrom the coastal plain of NW Egypt (from Alexandria to El-O'mayed), Dissertation. University of London, UK

Hermina M, Klitzsch E, List F (eds) (1989) Stratigraphic Lexcon and explanatory notes to the geological map of Egypt, scale 1: 500,000. Conoco Inc., Cairo
Hume WF (1925) Geology of Egypt, vol I. Survey of Egypt. Government Press, Cairo, 408 p
Jenkins DA (1990) North and Central Sinai. In: Said R (ed) The geology of Egypt. Balkema, Rotterdam, pp 361–380
Kebeasy R (1990) Seismicity. In: Said R (ed) The geology of Egypt. Balkema, Rotterdam, pp 51–59
Kehl H, Bornkamm X (1993) Landscape ecology and vegetation units of the Western Desert of Egypt. In: Meissner B, Wycisk P (eds) Geopotential ecology analysis of a desert region, vol 26. Catena supplement, pp 155–178
Kleindienst MR, Churcher CS, McDonald MMA, Schwarcz HP (1999) Geography, geology, geochronology and geoarchaeology of the Dakhleh Oasis region: an interim report 1. In: Churcher CS, Mills AJ (eds) Dakhleh Oasis project: monograph 2. Reports from the Survey of the Dakhleh Oasis, Western Desert of Egypt, 1977–1987. Oxbow Books, Oxford
Lindell LT, Alexandersson ET, Norrman JO (1991) Satellite mapping of oolitic ridges in the Arab's Gulf, Egypt. Geocarto Int 1:49–59
Mahmoud AA (1995) Evidence of Late Holocene subsidence, NW Alexandria. Mediterranean coast. Egypt Bull Soc Geog Egypte 68:99–112
McCauley J, Schaber C, Breed M, Grolier C, Haynes V, Issawi B et al (1982) Subsurface valleys and geoarchaeology of the Eastern Sahara revealed by shuttle radar. Science 218:1004–1020
McCauley J, Breed M, Schaber G, Haynes V, Issawi B, El-Kilani A (1986) Palaeodrainages of the Eastern Sahara, the radar rivers revisited (SIR-A/B implications for a mid-tertiary trans-African drainage system). IEEE Trans Geosci Remote Sens 24:624–648.
McHugh WP, McCauley JF, Haynes CV, Breed CS, Schaber GG (1988) Paleorivers and geoachaeology in the southern Egyptian Sahara. Geoarchaeology 3:1–40
Meteorological Authority of Egypt (1979) Climatological normals of the Arab Republic for Egypt up to 1975, Cairo
Moawad M (2003) The geomorphology of the coastal zone between Ras Abu Girab and Ras Alam El-Roum, the NW coast of Egypt. Dissertation, Ain Shams Unversity
Moawad M (2008) Application of remote sensing and geographic information systems in geomorphological studies: Safaga-El-Quseir Area, Red Sea Coast, Egypt as an example. Dissertation, Johannes Gutenberg Universität, Mainz.
Moawad M (2013) Geomorphological characteristics of the submerged topography along the Egyptian Red Sea Coast. In: Scalon L, Ranieri J (eds) Continental shelf: geographical distribution, biota and ecological significance (Earth sciences in the 21st century). Nova Science Publishers, Inc., New York
Moussa A et al (1998) Carbonate sedimentation in coastal hypersaline lagoons of Alamain, Egypt. In: Alsharhan AS, Glennie WW, Whittle GL, Kindall CG (eds) Quaternary deserts and climatic changes. Proceedings of the international conference on quaternary deserts and climatic changes, Al-Ain, UAE, 9–11 Dec 1995. A A Balkema, Rotterdam, pp 43–56
Paver RM, Pretorius DA (1954) Report on hydrogeological investigations in Kharga and Dakhla Oases. Report no 4, Special Publication de l'Institut du Desert d'Égypte, 108 p
Pickford M, Wanasb H, Soliman H (2006) Indications for a humid climate in the Western Desert of Egypt 11–10 Myr ago: evidence from Galagidae (Primates, Mammalia). Palaeoecology 5:935–943
Richter A, Schandelmeier H (1990) Precambrian basement inliers of the Western Desert, geology, petrology and structural evolution. In: Said R (ed) The geology of Egypt, Chap 11. Balkema, Rotterdam, pp 185–200
Rohlfs G (1875) Drei Monate in der libyschen Wüste. Kassel (reprint Köln 1996)
Said R (1962) The geology of Egypt. Elsevier Publishing Co., Amsterdam
Said R (1981) Geological evolution of the River Nile. Springer, New York
Said R (1990a) Cenozoic. In: Said R (ed) The geology of Egypt. Balkema, Rotterdam, pp 451–486
Said R (1990b) Quaternary. In Said R (ed) The geology of Egypt. Balkema, Rotterdam, pp 487–507
Said R (1993) The River Nile: geology hydrology and utilization. Pergamon Press, Oxford, 320 p

Salem R (1976) Evolution of Eocene-Miocene sedimentation patterns in parts of northern Egypt. Am Assoc Pet Geol Bull 60:34–64
Schaber GG, McCauley JF, Breed CS (1997) The use of multifrequency and polarimetric SIR-C/X-SAR data in the geologic studies of Bir Safsaf, Egypt. Remote Sens Environ 59:337–3631
Schlumberger L (1984) Well evaluation conference, Cairo, Egypt
Selim A (1974) Origin and lithification of the Pleistocene carbonates of the Sallum Area, Western Coastal Plain of Egypt. J Sediment Petrol 44:70–78
Shata MA (2000) Sedimentological and mineralogical characteristics of subsurface sediments in Borg El-Arab region, NW coast of Egypt. Bull Nat Inst Oceanogr Fish 26:1–26
Shukri NM, Philip G, Said R (1956) The geology of the Mediterranean coast between Rosetta and Bardia, II Pleistocene sediments: geomorphology and microfacies. Inst Égypte Bull 37:395–427
Sigaev N (1959) The main tectonic features of Egypt: an explanatory note to the tectonic map of Egypt, scale 1: 2,000,000. Paper 39. Geological Survey of Egypt, Cairo
Soliman K (1972) The climate of the United Arab Republic. In: Griffiths J, Soliman K (eds) The Northern Desert (Sahara), climates of Africa. World survey of climatology, vol 10. Elsevier Publ. Co., Amsterdam, pp 79–92
Sundborg A, Nilsson B (eds) (1985) Qattara hydrosolar power project: environmental assessment. UNGI report no 62. Department of Physical Geography, Uppsala University, Uppsala, 194 p
Survey of Egypt (1980) Geologic map of Egypt, scale 1: 2,000,000, Cairo
Thornbury WD (1954) Principles of geomorphology. Wiley, New York
UNESCO (1977) World distribution of arid regions, Paris
Wali AMA, Brookfield ME, Schreiber BC (1993) The depositional diagenetic evolution of the coastal ridges of NW Egypt. Sediment Geol 90:113–136

Chapter 17
Integrated Watershed Management of Grand Ethiopian Renaissance Dam via Watershed Modeling System and Remote Sensing

Mohamed E. Dandrawy and El-Sayed E. Omran

Abstract For water resource technicians and decision-makers, long-term hydrological simulation is of great significance. The presence of input information from the long-term model does not always ensure effective and reliable results. In this respect, water resource technicians face some technical issues in modeling the watersheds of the Nile basin. Some of these technical issues are linked to the hydrological simulation program-fortran (HSPF) and its Watershed Modeling System (WMS) interface. The aim of this research is to use the HEC-1 model to create a hydrological modeling rain-flow form on the Grand Ethiopian Renaissance Dam (GERD) watershed. This research is an attempt to help to address and estimate the upper Blue Nile (UBN) basin's water budget using a fresh hydrological modeling structure and remote sensing information. This model is then used to forecast the basin's hydrological reaction to climate change and land-use situations. First, the Renaissance Dam's GIS-simulated reservoir at 606 m asl (i.e. 100 m maximum water depth) showed that the lake would cover roughly 745 km^2 and that the quantity of water stored would reach 74 billion cubic meters. Second, the islands will be submerged in the water when the Renaissance dam is turned on and the lake position is filled. The analysis of the geographical location of the Ethiopian GERD Dam Lake after its completion showed that the current riverine and riverine islands had been submerged and new islands, were originally emerged, that located within the new lake, with different sizes and dimensions depending on the water level in the lake. In the center of the GERD Dam Lake, a variety of small new islands between 18393 m^2 and 6 km^2 will be found in the center of the lake, about 77 islands except for one which will be about 27 km^2 and will appear in the middle of the lake in front of the Renaissance dam. Third, according to the EM-DAT Database, many earthquakes occurred between 1900 and 2013, affecting 585 individuals. Final, safety surveys on dams have shown that the Ethiopian GERD Dam's safety factor is only 1.5 degrees from 9 degrees. It is more probable that the GERD Dam will collapse. Experts said the dam was created

M. E. Dandrawy · E.-S. E. Omran (✉)
Department of Natural Resources, Institute of African Research and Studies and Nile Basin Countries, Aswan University, Aswan, Egypt
e-mail: ee.omran@gmail.com

E.-S. E. Omran
Soil and Water Department, Faculty of Agriculture, Suez Canal University, Ismailia 41522, Egypt

S. F. Elbeih et al. (eds.), *Environmental Remote Sensing in Egypt*,
Springer Geophysics, https://doi.org/10.1007/978-3-030-39593-3_17

to collapse. The safety of the dam is very low. This implies that owing to the weight of the water behind it and the weakness of the region in particular, any earthquake will damage this dam and it is a seismic zone adjacent to the African groove. In a case of GERD failure, it demonstrates the regions that would be flooded by the Blue Nile River and its tributaries flowing from the Ethiopian Plateau. In this scenario, the water would flow from a high altitude region of 520 m to the plains of Sudan at an altitude of 480 m, destroying the Renaissance Dam and flooding the capital of Sudan and the White Nile River under the influence of the topographical slope. The findings of this research show that approximately 667,228 km^2 was the highest extent of the flood region estimated from a DEM using a flood basin model. These regions are situated between the rivers of the Blue Nile and White Nile and cover all urban and agricultural regions.

Keywords Remote sensing · Water budget · Upper Blue Nile · Modeling · GERD · WMS · Integrated management

17.1 Introduction

In many parts of the globe, freshwater is a rare resource. Increasing populations and significant rises in demand for agricultural and industrial reasons continue to aggravate the issue. One such region is the Nile River basin, with comparatively arid climate due to elevated temperatures and solar radiation that fosters fast evapotranspiration. Most nations in the basin, like Egypt, Sudan, Kenya, and Tanzania, are not getting enough freshwater (Herbert 2014). Exceptions are the small regions on the equators and the Upper Blue Nile (hereafter UBN) basin in the Ethiopian highlands, which gets up to 2000 mm annually (Johnston and McCartney 2010). The Upper Blue Nile basin, in particular, is the region's primary water sources.

Management of the basin's water resources faces many pressures and difficulties. Ethiopia announced in April 2011 and during the Egyptian Revolution that it intended to build the Grand Ethiopian Renaissance Dam (GERD), one of the largest dams in the world (Omran and Negm 2019). This GERD will flood 1700 km^2 of forest in northwestern Ethiopia—about 811 km from Addis Ababa and 14 km from the Sudan frontier—producing a dam reservoir that will hold up to 67 billion m^3 of water, possibly taking up to seven years to achieve ability and eventually almost twice the size of Lake Tana, the biggest natural lake in Ethiopia (Mohamed and Elmahdy 2017).

Dam defects may lead from any of the following causes or combinations (IPoE 2013; Veilleux 2013; Omran and Negm 2019): (1) Long precipitation and flooding periods (the cause of most failures). (2) The presence of geological structures in the rocks like a fault, joint areas and folds. (3) Silting reservoir and accumulation of sediments in the dam reservoir owing to inadequate dam spills. (4) Landslides into reservoirs resulting in overtopping surges. (5) Earthquakes, typically causing longitudinal cracks at the tops of embankments, resulting in structural failure.

Only by collecting quantitative data can we tackle all these facts and difficulties and develop better water resource development policies (Hall et al. 2014). Therefore, understanding UBN's hydrological procedures is the foundation for both the trans-boundary water resource sharing agreements and the sustainability assessment of the region's farming systems. Indeed, due to the absence of hydrometeorological information and an adequate modeling structure, latest modeling attempts in the basin have obvious constraints in solving these issues. Regional studies are restricted to small reservoirs, especially in the Lake Tana basin where hydrometeorological information is comparatively better (Teferi et al. 2010; Uhlenbrook et al. 2010; Rientjes et al. 2011; Tekleab et al. 2011), or on the entire basin scale, but where spatial variability data is generally ignored (Tekleab et al. 2011). Other studies are restricted to a specific hydrological method, such as variation of rainfall (Abtew et al. 2009), time series and statistical analysis of in situ discharge/rainfall data (Teferi et al. 2010; Taye and Willems 2011) or perform modeling at very low temporal resolutions (e.g. monthly) (Tekleab et al. 2011).

There is no spatially distributed data on all parts of the water budget, and basin modeling methods tailored to a single element do not provide an efficient image of the dynamics of the basin's water resources. To tackle data scarcity, remote sensing (RS) products can support large-scale hydrological modeling, which fills the information gaps in estimating water equilibrium dynamics (Sheffield et al. 2012). Different scientists used hydrodynamic modeling methods to map flood susceptibility (Matkan et al. 2009). Hydrological techniques, however, involve fieldwork and enormous information collection budget (Fenicia et al. 2013). Despite the availability and accuracy of geographic information system (GIS) techniques and RS products at different (spatial and temporal) resolutions, their use is a new paradigm in estimates of the water budget closure (Andrew et al. 2014; Wang et al. 2014). The hydrologic study is an important study of the knowledge, tracing, utilization or protection against water hazards. Combination of HEC-1, WMS, and GIS has, capability in simulation of flood hazard zoning (Qafari 2004; Echogdali et al. 2018).

This chapter provides a way to integrate RS, and GIS with hydrological models HEC-1 of the Watershed modeling system. Where a case study of these models was presented on the basins that are in the storage of the Ethiopian GERD Dam. This study also aims to map and simulate the flooded area in Sudan and Egypt in the case of GERD failure. The integration of remote sensing and GIS is of great assistance in predicting and describing potential geohazard events as a result of dam failure. An attempt is presented to help addressing the quantitative problems associated with the above-mentioned problems and seeks to solve the UBN basin's water budget using a fresh hydrological modeling structure and remote sensing information to improve estimates from past research. It is also methodological research, in that it delineates various methodologies to overcome the data scarcity.

17.2 Methodology

17.2.1 The Study Basin

The Blue Nile catchment and its tributaries on the Ethiopian Plateau include roughly 250,000 km^2 and contain additional Sudanese tributaries, equating 324,000 km^2 at its confluence with the Khartoum White Nile. The Blue Nile is carved into the Ethiopian Plateau, rising at 2000–3000 m asl with several peaks up to or above 4000 m. Geologically, the Ethiopian Plateau is a significant component of the East African Rift System, which started at the end of the Cretaceous, and led to the creation of the Red Sea and the Main Ethiopian Rift (McConnell 1972). The earliest rocks in the Blue Nile basin are the Precambrian basement rocks, which are predominantly acidic to basic rocks, which include quartzite, granites, granodiorite genisses, diorite, metasediments and metavolcanics (Wolela 2012).

The Upper Blue Nile (UBN) river flows southeast through a series of cataracts from Lake Tana at Bahir Dar. The river reaches a profound canyon after approximately 150 km and shifts path to the south. After flowing for another 120 km, the river shifts its path toward the El Diem (Ethiopia-Sudan border) west and northwest again. The main river along its course is joined by many tributaries draining from many parts of the Ethiopian highlands. Ethiopia's complete river range is about 1000 km. The UBN reservoir accounts for up to 60% of the contribution of Ethiopia's hills to the flows of the Nile, which is 85% of the total (Conway 2000). The river basin region surrounded by a segment at the boundary between Ethiopia and Sudan is approximately 175,315 km^2 (Fig. 17.1), comprising approximately 17% of the country's total area.

17.2.2 Geography and Hydrological Setting of the GERD

The GERD is situated in northwestern Ethiopia, approximately 14 km from the Sudanese border, and stretches from latitude 11° 10′ 30″ N and 11° 17′ 15″ N, to longitude 35° 01′ 04″ E and 35° 07′ 48″ E (Fig. 17.2). It is situated on the northwest part of the Benishangul-Gumuz area, which between May and October is characterized by high precipitation. The estimated annual rainfall at the dam site is about 860 mm and at the adjacent hilly regions is about 2250 mm. The rainwater quantity is an agent of weather erosion that impacts the fill slopes, resulting in gully erosion and landslides. The Nubian Block, including the place of the GERD, is drained by extensive dendritic streams that flow under the gravitational influence from a high-altitude region to low-altitude areas. These dendritic streams represent steep slope topography. Rainwater speeds up in regions with a steep slope and the likelihood of rock erosion and landslides rises (Elmahdy et al. 2016; Mohamed 2016).

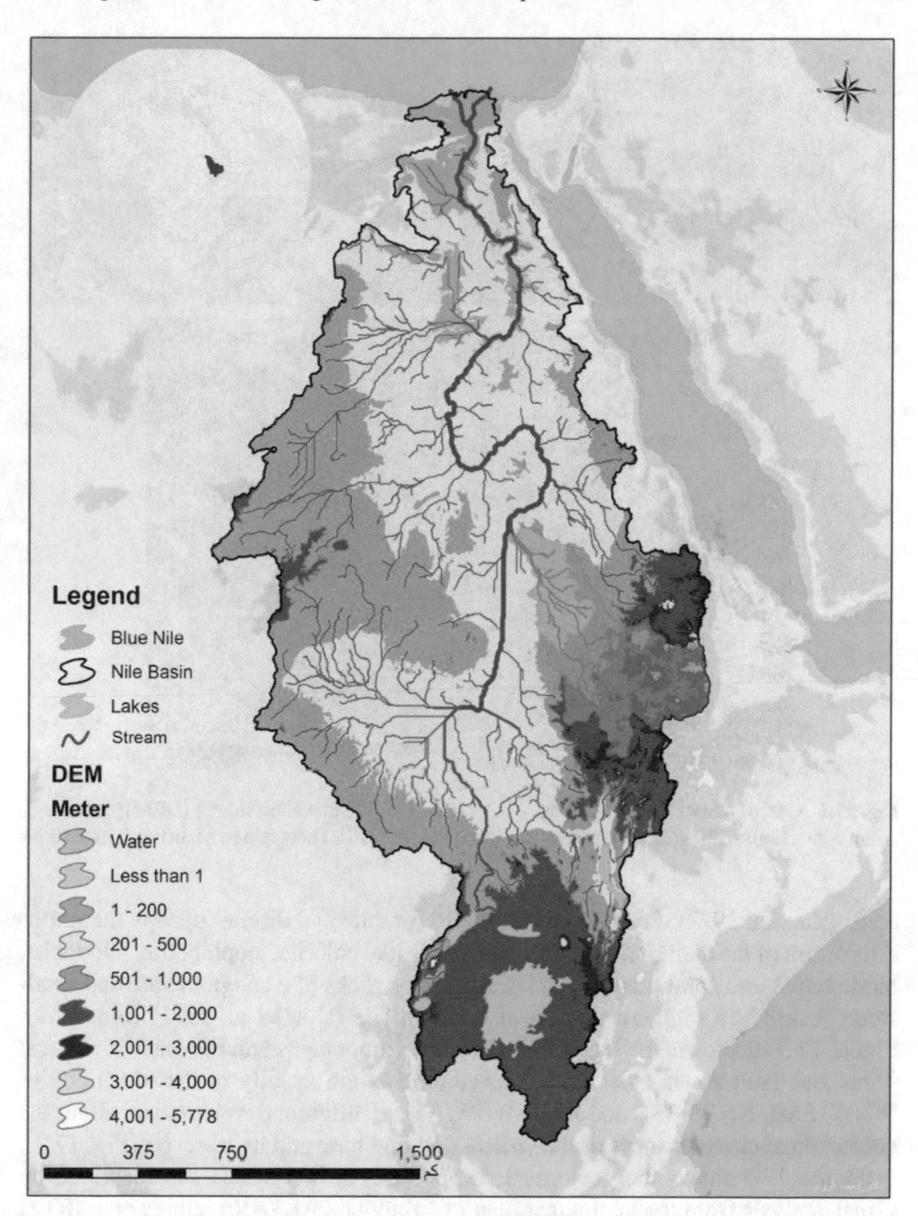

Fig. 17.1 Location map of the catchment of the Nile River Basin

17.2.3 Geological Setting of the GERD

Archaean basement rocks underline the Nubian Block and the dam site. Massive and extremely fractured igneous and metamorphic rocks are in high mountainous

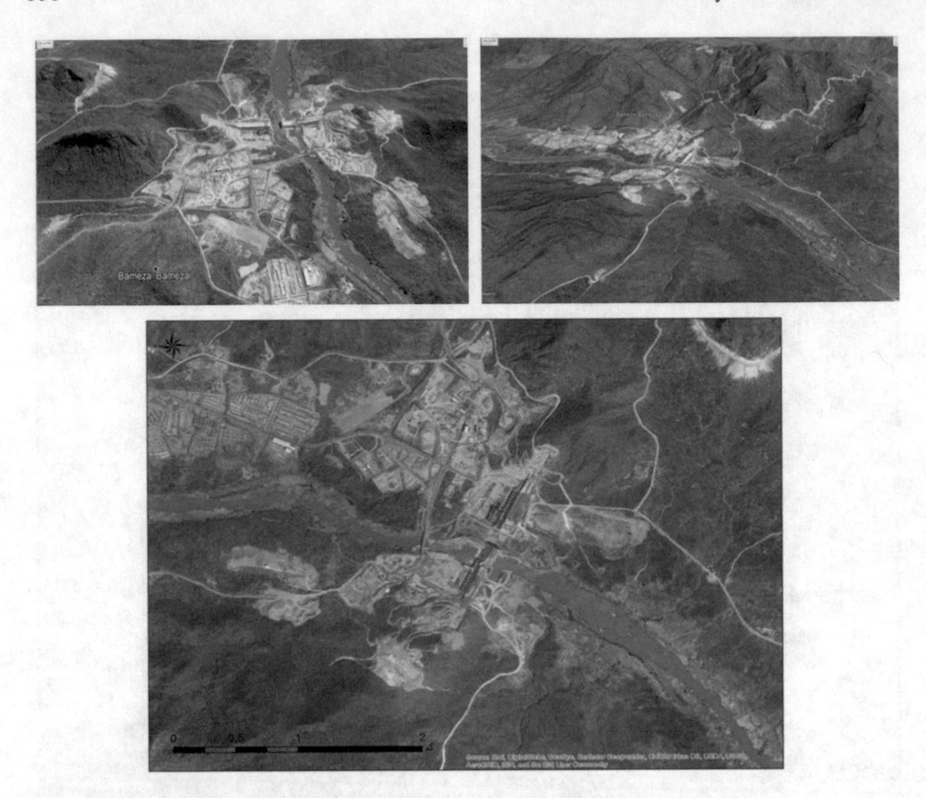

Fig. 17.2 Geography of the GERD basin: 3D view of digital elevation model (DEM); and image zoom with a high spatial resolution (lower) displaying the GERD between two hard rock mountains

areas (Kazmin 1972). Posttectonic granitoid (granite and diorite) occupy the northern portion of the dam site. Metamorphic rocks like chlorite, amphibolite, talc schist, biotic schist and coral graphite and sedimentary rocks like conglomerate and sandstone occupy the southern portion of the dam site (Dow et al. 1971) (Fig. 17.3). Figure 17.3 illustrates the GERD's geological components and lineaments. Several joints and fault zones exist. Their developments are usually in the directions of WNW–ESE, NNW–SSE and ENE–WSW. These different developments affect the geotechnical circumstances of the GERD building base and its reservoir (Fig. 17.3).

Figure 17.3 shows the geological components of the GERD site identified by visual analysis from the geological map of Ethiopia, PALSAR pictures and SRTM DEM (Mohamed and Elmahdy 2017). The map demonstrates that GERD site lies between two mountains. The right hill (1400 m) consists primarily of intrusive acidic rocks such as post-tectonic granitoid and granite, ranging from 24 to 42. While the left hill (1100 m) consists of metamorphic rocks such as chlorite slate, amphibolite, talc schist, biotic schist and quartz graphite, and conglomerate and sandstone sedimentary rocks with shear maximum intensity varying from 31 to 38. Cleavage and foliation characterize these rocks, causing water leakage and sand and clay forming rock

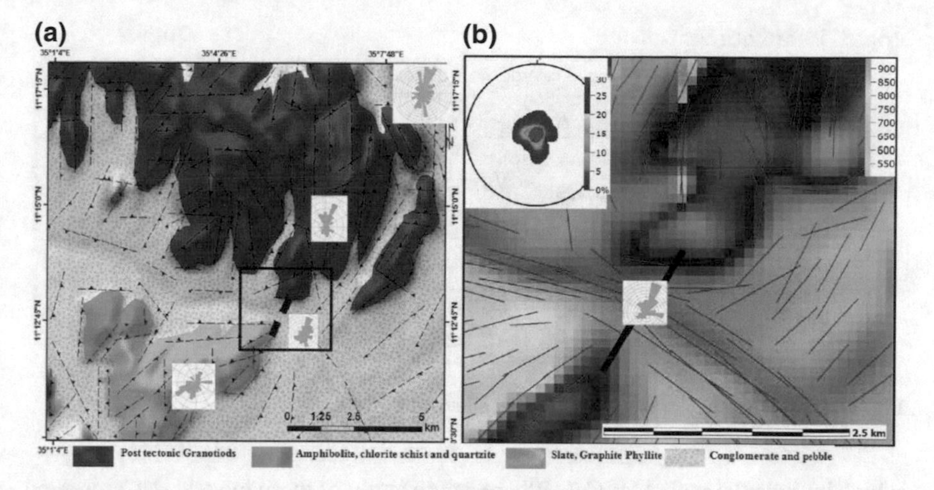

Fig. 17.3 Geological components (**a**), and lineaments extracted from a DEM (**b**) of the GERD control GERD site. Rose and earth normal diagrams highlight the trends of the geological structures that control the GERD site. All the normals either point to the NE or SW of bearing of 26.6° (Mohamed and Elmahdy 2017)

erosion. Thus, most of the GERD site's lithology is readily erodible, undesirable, and a stable basis is not available. The GERD site is regarded as less appropriate than others because it has a basis for softly dipping granite and slate and is very near to high-fracture areas (Mohamed and Elmahdy 2017).

Statistical assessment using rose diagrams indicates that the hilly regions, including the GERD location, are structurally regulated by trend fault zones NNE–SSW, NNW–SSE and ENE–WSW (Mohamed and Elmahdy 2017). It agrees well with the developments crossing the Nubian Block and the African Rift. Their fault trends are parallel to the GERD and their displacement is perpendicular to the walls of the dam, producing some alarm (Fig. 17.3). Building such a dam on extensive junctions of geological fractures can trigger tremendous water leakage through the conglomerates and boost the time it takes to fill the reservoir. Although all of these faults have previously suffered significant earthquakes, there have been several earthquakes on the brink of the Nubian Block and African Rift.

17.2.4 Overall Methodology

The overall methodology comprises of several steps (Fig. 17.4): data acquisition, HEC-1 import, centerline stream display, georeferencing cross-sections, terrain modeling, and floodplain mapping.

The hydrologic study is an important study of the knowledge, tracing, utilization or protection against water hazards. This chapter provides a way to integrate GIS software with hydrological models such as Watershed modeling system (WMS),

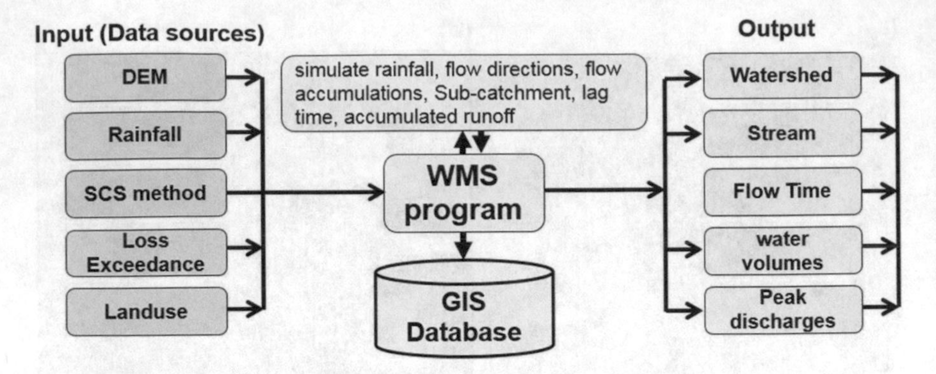

Fig. 17.4 Methodology of the study showing all conducted steps

which include the model HEC-1. Where a case study of these models was presented on the basins that are in the storage of the Ethiopian GERD.

The study collected data on the region and built a digital database using GIS, hydrologic, and climate station data in the Upper Blue Nile Basin. The rain-simulation was carried out with multiple potentials (15, 30, 50, 100 mm), where the Blue Nile basin above the Ethiopian GERD was found to be extremely dangerous (extreme) in the event of rain of more than 50 mm. This is due to the long water flow in the basin, where water flow in the basin continues for more than 39 h.

The outputs are digital floodplain maps showing both magnitude and depth of flooding in major rivers. Data was standardized and transformed into the same projection geodatabase. The data used by ArcGIS to construct the main parametric models were: DEM and drainage, Land cover/use, Soil data, Geology and meteorological data.

17.2.5 WMS Model: Capabilities and Limitations

WMS is pre/post handling software based on GIS that supports numerous hydrological/hydraulic and water quality models commonly used by water resource technicians around the globe. It offers a user-friendly interface to develop the required input files for these models, as well as some graphics and animation skills to view the resulting output from these models, if appropriate (Nelson et al. 2005). WMS is an extensive graphical modeling environment for all stages of hydrology and hydraulic hydrology (Fig. 17.5). WMS involves strong instruments for automating modeling procedures such as automated basin delineation, geometric parameter calculations, GIS overlay computations (CN, depth of rainfall, sections of HSPF, coefficients of roughness, etc.), cross-section extraction from field information. WMS, version 7.1, supports hydrologic modeling with HEC-1, TR-20, TR-55, Rational Method, NFF, MODRAT, and HSPF. Hydraulic models supported include HEC-RAS, SMPDBK,

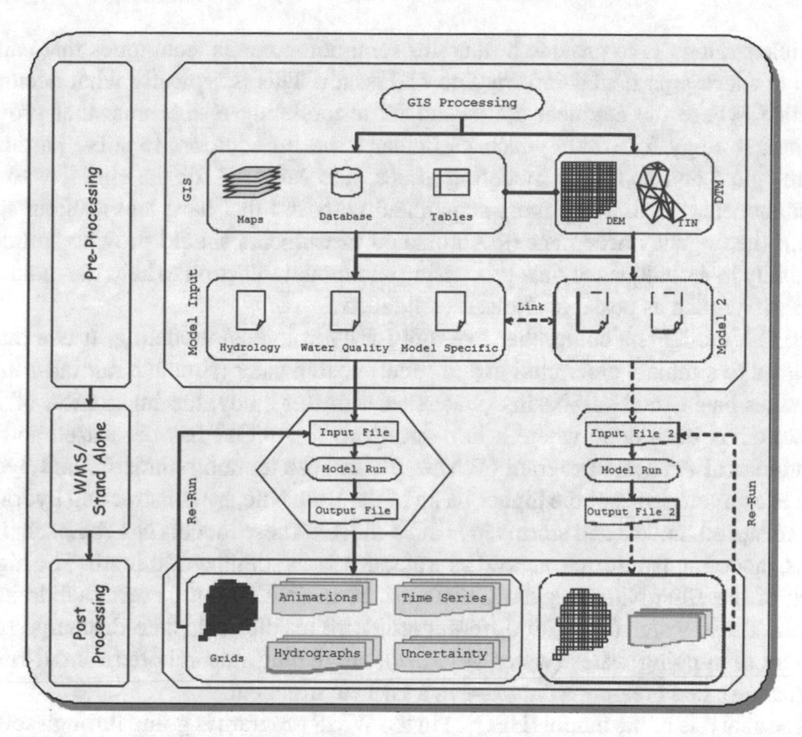

Fig. 17.5 Conceptual representations of watershed modeling system (WMS) and water quality models interface (Nelson 2004)

and CE-QUAL-W2. It is now possible to model two-dimensional integrated hydrology (including hydraulic channels and groundwater interaction) with GSSHA (EMS-I 2004).

The WMS tutorials (Nelson 2004) give users a clear step-by-step instruction on building a User Control Input (UCI) file. However, since all catchments are distinct, for some particular watersheds there are some technical problems that may develop. In this chapter, these technical problems will be discussed in an attempt to address some of those concerns, if not all.

The conceptual representation of WMS is shown in Fig. 17.5. As we can see, the modeling method begins with pre-processing where WMS is usually used to produce input files of particular models using accessible information sources. Those input files can then be used either in WMS or in a stand-alone version to operate the model. Later, for some models, WMS offers some post-processing instruments that allow technicians to view outputs.

While instruments such as WMS provide a mechanism for facilitating parts of the hydrological and hydraulic modeling process, no fully independent techniques are accessible to automate it. Too many differences exist between information, issue information and localized circumstances in order to produce a single algorithm that can adapt to all circumstances and fully automate this complex process. A more sensible strategy is to provide a range of semi-autonomous techniques that can be used to address particular information and issues. This is typically what occurs in practice, where the engineer can select those modeling instruments that provide the best strategy for solving each particular issue. In addition, to solve particular hydrological and hydraulic modeling issues, it is essential for the engineer to use those models that they are most acquainted with and that have an excellent track record (i.e., agency acceptance). Automated instruments should provide sufficient flexibility in techniques of data processing and model selection to help the modeling process as much as possible (Nelson et al. 2005).

HEC-1 model is a comprehensive tool for hydrological modeling. It is a model designed to simulate precipitation and runoff of rainwater (simulate rainfall-runoff) for water basins and sub-basins (watershed). In this study, the integration of GIS software has been done where a tool Archydro in ArcGIS has been used and the Hydrological Analysis Program (WMS). To analyze the morphometric and hydrological characteristics of the higher basin of the Blue Nile, by constructing hydraulic and collapsed models and storms in HEC-1 model. These models can represent flow paths, modeling landforms, as well as water depth and drainage strength. The higher basin of the Blue Nile over the GERD has been divided into several subdrainage basins. These watershed with different areas are used to calculate discharge time, amount of running water (water volumes), simulating rain runoff (rainfall-runoff simulation), and creating a database in a GIS environment.

The analysis of the model (HEC-1) in the WMS program is going through several steps, the most important of which is to provide the DEM model with a coordinate system of Universal Transverse Mercator. Then the TOPAZ tool in WMS was used to compute flow directions and flow accumulations; followed by the extraction of a Streams and then the division of the upper basin of the Blue Nile over the Renaissance dam into subbasins (Sub-catchments). This is to take into account the variation in land use, soil type and topography of each sub-basin, to calculate the amount of water flowing from each basin accurately, to estimate the amount of flooding over the entire basin and to determine the amount of final stream in front of the Ethiopian GERD Dam.

17.2.6 Pre-processing

The WDM file should involve all the attributes that define the data in each data set to be stored. It should also hold data from the time series for the relevant data sets. WMS can be used for model pre-processing once the WDM file is in location and ready for manipulation. In using WMS to produce the User Control Input (UCI) file, few significant steps are discussed as follows.

17.2.6.1 Watershed Delineation

Defining a watershed is no longer a cumbersome method; WMS very simply and straightforwardly delineates a watershed. Watershed delineation can usually take a very short time in the order of a minute or a few minutes, depending on the size of the watershed and the resolution of the underlying topographical data. WMS can delineate a Triangulated Irregular Network (TIN) or a Digital Elevation Model (DEM) watershed, as well as a map-based delineation (Nelson 2004).

17.2.6.2 Incorporating Land Use Data

HSPF effectively divides the watershed into separate sections of soil and reaches/reservoirs of water. These sections can be based on many characteristics of the hydro-topo-geostructure. One of the vital feature-based maps in the segmentation of the watershed is the map of land use.

There are currently three distinct forms in which land-use information can be imported into WMS, either as land-use coverage, land-use grid, or land-use shapefile. The import of land use information into a WMS project can be summarized as follows:

1. Land use coverage: A new coverage of type "Land Use" must be developed in the map module. Then the shapefile for soil use should be open while the application for land use is the active coverage. A key step to be taken when importing land use into the project is to map the land-use code (LU Code) to land use before the import process is completed.
2. Land use grid: While the drainage coverage is the active coverage in the map module, it is important to import the land use grid into the project.
3. Land use shapefile: Add the land use shapefile to the project in the GIS module.

17.2.6.3 Segmenting the Watershed

Based on the imported land use, WMS is prepared to begin segmenting the watershed at this point. It only requires a table of characteristics used to map information on land use to the characteristics needed for modeling. Table 17.1 demonstrates a table of attributes that WMS uses land use information to segment the watershed. However, for this table, WMS needs a particular format (* .tbl) (Nelson 2004).

The WMS function "Apply Parameters to Segments" allows users to enter parameters for one segment and copy those parameters entered into other sections, thereby reducing the quantity of data entry time. For most parameters, there are some suggested values (Nelson 2004). As we will see later, these values could alter as we calibrate the model. The current WMS version (Nelson 2004) displays a context-sensitive description of each parameter in the status bar when mouse activity occurs in the parameter edit field. This description can be entered with a range of plausible

Table 17.1 Example of an LU code attribute table

LU Code	Description
14	Rainfed croplands
20	Mosaic cropland (50–70%)/vegetation (grassland/shrubland/forest) (20–50%)
30	Mosaic vegetation (grassland/shrubland/forest) (50–70%)/cropland (20–50%)
40	Closed to open (>15%) broadleaved evergreen or semi-deciduous forest (>5 m)
60	Open (15–40%) broadleaved deciduous forest/woodland (>5 m)
110	Mosaic forest or shrubland (50–70%)/grassland (20–50%)
130	Closed to open (>15%) (broadleaved or needle leaved, evergreen or deciduous) shrubland (<5 m)
210	Water bodies

values. In addition, some instruments, either database-related (USEPA 1999) or GIS-related (Al-Abed and Whiteley 2002), can be used to obtain most of the necessary HSPF parameters using land use and soil information. Currently, WMS does not support any of these tools.

17.2.6.4 Fine Tuning

Modelers are urged to begin fine-tuning their model once segment operations are defined. The definition of operations for the reach/reservoir parts of the watershed is one of the most significant steps. As required, modelers can also aggregate and disaggregate segments and identify external sources and objectives (Nelson 2004). External sources, objectives and mass connections are used to move between operating pairs in the same INGRP or between individual activities and internal sources/targets (Bicknell et al. 2001).

17.2.7 Input Data of Model

In this study, data on DEM models were collected at a 90-m resolution called SRTM and land-use data. SCS Method, composite Curve Numbers (CN) for infiltration capacity, and runoff were chosen as a methodology. The amount of rain falling on each basin has been entered and the values of the composite curve (CN) calculated based on soil quality (Soil), land use/cover. Also, calculate the time of concentration, calculate the lag time, and the slope. WMS calculates the values of the composite curve (CN) for each of the sub-basins in a mechanical manner based on soil quality and land-use data. CN values in different sub-basins were 79, so accurate results could be produced about the period during which the stream continues (Time of Peak), and the amount of discharge per second (Peak discharges) and the total amount of running water per basin for each basin.

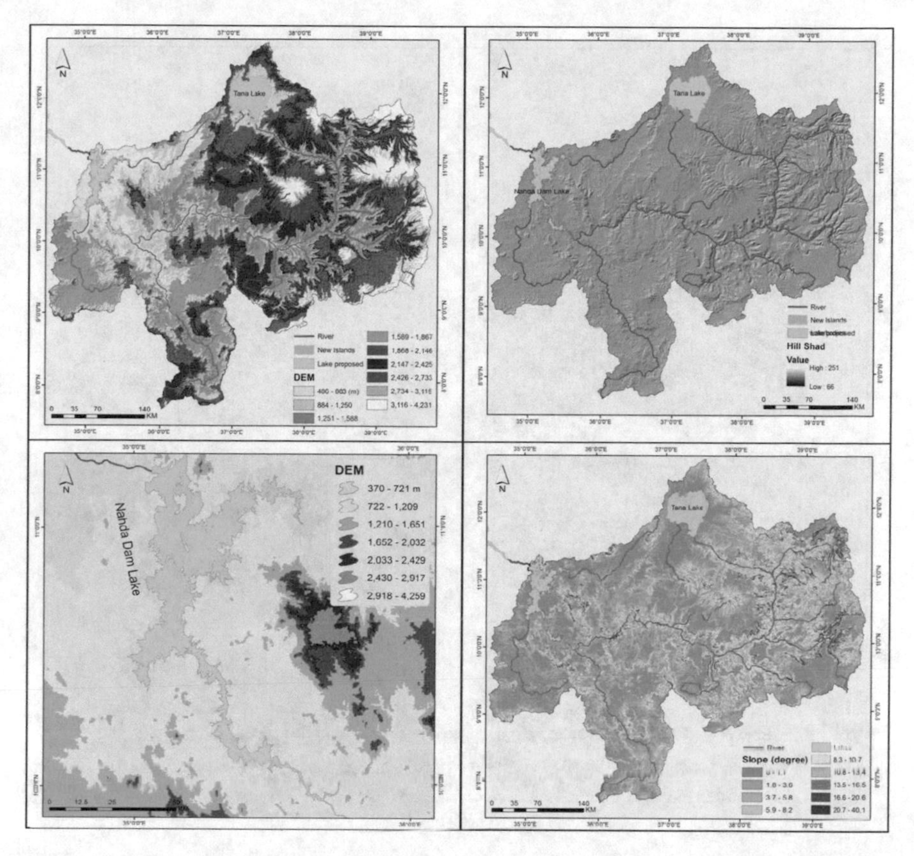

Fig. 17.6 Digital elevation model (DEM) and Hill Shad and slope map of the study area

17.2.7.1 Digital Elevation Model

Before conducting any operation of a simulation file preparation HMS, the DEM of the study region, where its function is fundamental in the physical characterization and calculation of the parameters, is crucial (Fig. 17.6).

17.2.7.2 Land Use

Considering the specific demands of the chosen modular mix, specifically the NRCS CN method as a production function, it was an inevitable way to produce a land-use map for the entire research region. The data intended to be found in this map should, however, be genuine to the classification recognized by NRCS. Therefore, we had to connect NRCS courses with information collected from all recognized bibliographic data dealing with this portion (Fig. 17.7).

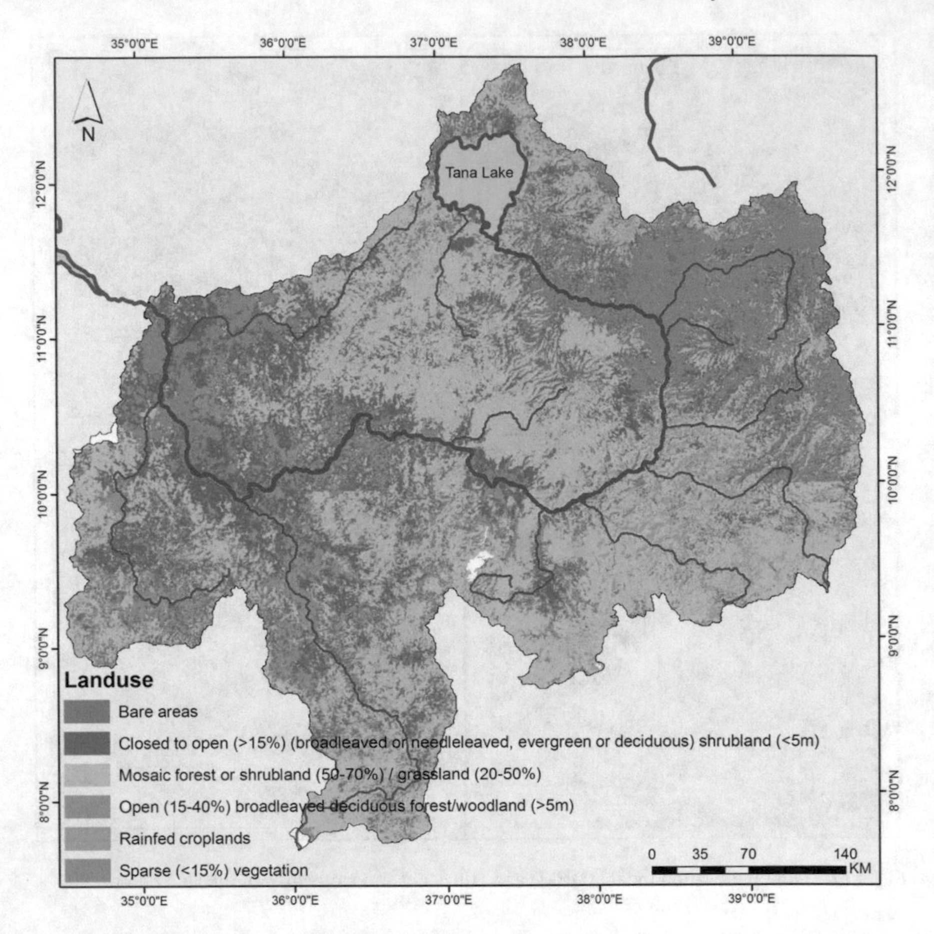

Fig. 17.7 Land use map of the study area

17.2.7.3 Rainfall Distribution and Processing

Rainfall is the main factor affecting the hydrological model. Rainfall (precipitation) data has been collected and analyzed from the climatic stations such as Bahar Dar, Combolcha, and Gondar in the study area. It was found that the depths of the rain during the days of August range from 0, 15, 30 to 50 mm. Thus, more than one scenario of the state of the basin and the amount of water running when rain falls in the different depths (15, 30, 50 and 100 mm) has been developed. The amount of runoff and discharge time for each category of rainfall was recognized as such. Based on the Type II-24 h model to represent rainfall, which is designed to design rain distribution (accumulated rainfall) within the WMS program on the basin within 24 h and the accumulated runoff. It is a model in which rain cries are assumed to occur every 6 min, 24 h a day, although most rain occurs during the 6-h average, the 6-min intervals have been used to enter precipitation data, as they provide the highest possible level of detail.

For each case, rain should be considered to have fallen on the watershed during the day this event happened in the form of rainfall height. The primary factor influencing the hydrological model is precipitation. Together with the bibliographic evaluation, precipitation information gathered from three distinct stations permitted the study of rainfall conduct (because there are no intensity values available). This information also made it possible to delineate the region's Intensity Duration Frequency Curves, IDF.

17.2.7.4 Soil Groups (CN)

One of the basic parameters to calculate the CN is the definition of the hydrological sets of soil and land uses, since the value of CN depends on both of them. The SCS method determined four hydrological sets of soil due to the speed of water transition inflow rate through. Those sets are (A-B-C-D), and each has its characteristics of runoff.

The watershed of Upper Blue Nile is presented in those Hydrological soil groups since the Hydrological soil group (A) represents an area of almost 49,860 km^2 and it presented in the high permeable soil, while the Hydrological soil group (C) covers an area of 76,184 km^2, and the Hydrological Group of Soil (D) covers an area of 44,233 km^2.

Soil properties affect the relationship between rainfall and runoff by affecting the rate of leakage. The soil is divided into four main groups based on leakage rates (A, B, C, D) and the group (D) is our area of concern in the upper basin of the Blue Nile above the Renaissance dam. The soil group (D) has a high potential running due to very slow leakage rates of 1.3 mm/h (0.052 per hour). This soil is primarily made of mud, and is most important because it is a clay soil that has a high water level and is swelling up in its moist parts. The most important features of the group soil (D) are that they consist of clay loam, sandy loam, sandy clay loam, clay loam, or clay (Fig. 17.8).

17.2.8 Delineation of Drainage System

Identifying the drainage system where a flood can occur is vital (Fig. 17.9). This will assist in identifying surface water basin features and their impact on surface run-off behavior. In addition, the drainage system's morphometric features play an important part in regulating the water flow regime between these streams and their effect on the flood cycle. GIS digital applications, such as Global Mapper software and WMS watershed modeling systems, have been implemented directly as they are capable of automated stream and water basin tracing. Production of drainage network and

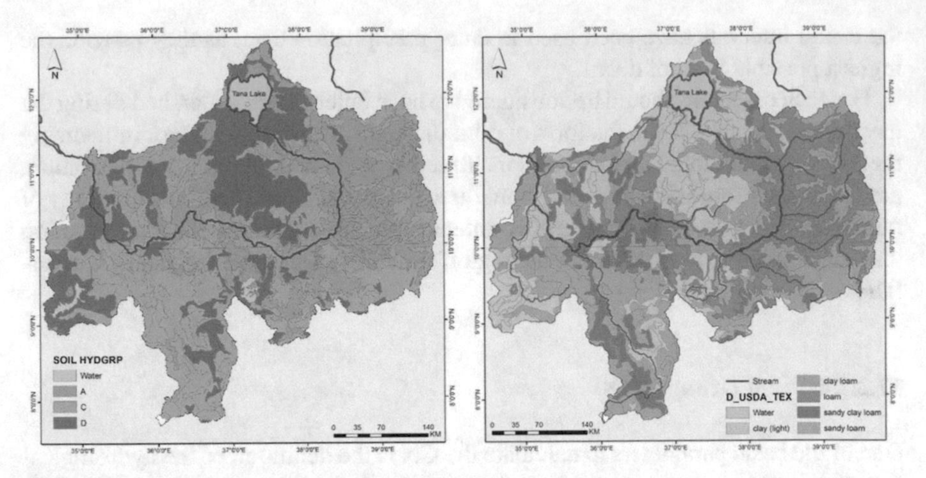

Fig. 17.8 Soil groups (A, C, D) of the study area

catchment area in GIS form helps to apply various measurement processes and data integration as well as can assist modify geospatial information whenever necessary.

17.2.8.1 3D-Ground Surface Creation

With the help of Global Mapper Software, 3D-Ground surfaces for the research region were prepared using DEM (Digital Elevation Model). The DEM file for the research region was downloaded from ASTER GDEM Worldwide Elevation Data (1 arc second resolution). The surface features such as slopes, depressions, mountains, and Wadi's have been closely examined using sophisticated methods such as the "waterdrop" to determine the places of outlets in the reservoirs. DEM Model and contour lines maps (Fig. 17.10) demonstrate the mountain nature of the study area as the contour lines are very close (Fig. 17.10), and the contributing sub-basins are connected by nearly right angles to the mainstream.

17.2.8.2 Basins Delineation

Basin delineation is conducted in conjunction with outlets and streams by evaluating flow instructions. A flow path from each cell's middle is traced until a stream (or outlet straight) or DEM's edge is hit. Those flowing into a stream and thus outlet are assigned to a basin and those hitting the edge of the DEM are not allocated to a basin. Cells that are not in a basin are called the NULL basin.

With the help of the Watershed Modeling System (WMS) program, the watershed and its sub-basins delineation method were carried out. Imported into WMS was the DEM surface produced in Global Mapper and tiny gaps in depression were filled. Then the network of the Flow path was created using the integrated TOPAZ Flow

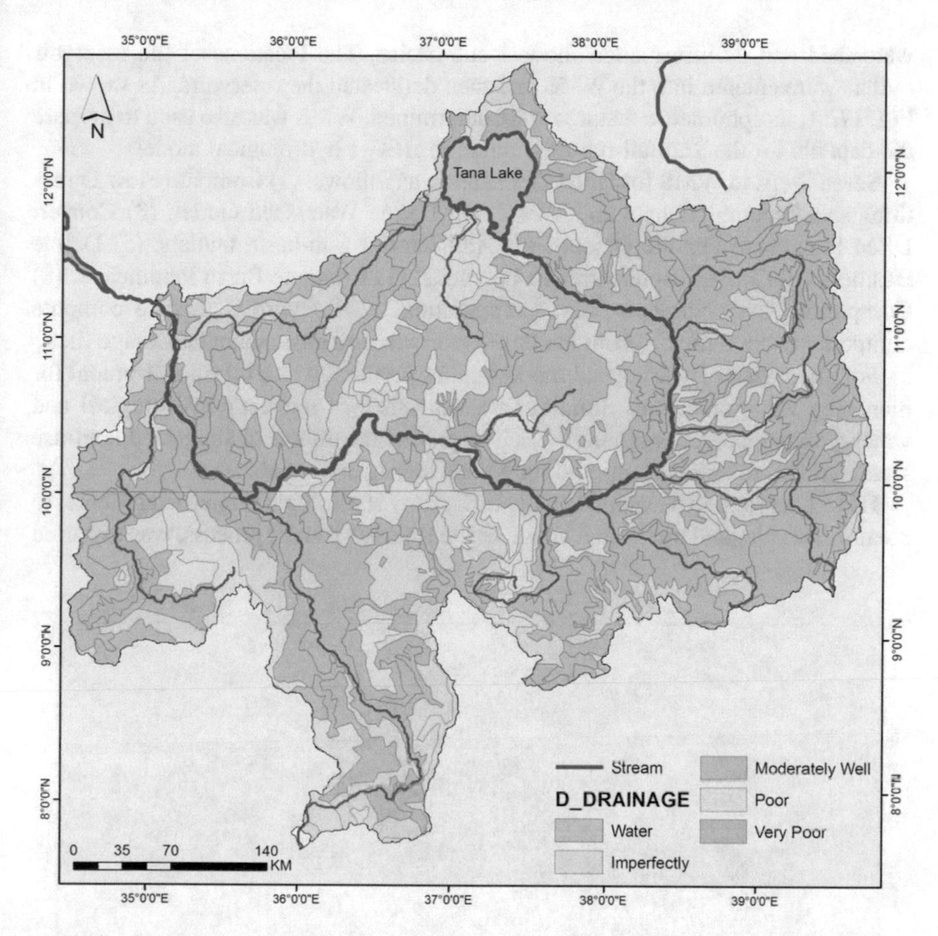

Fig. 17.9 The drainage system for the study area

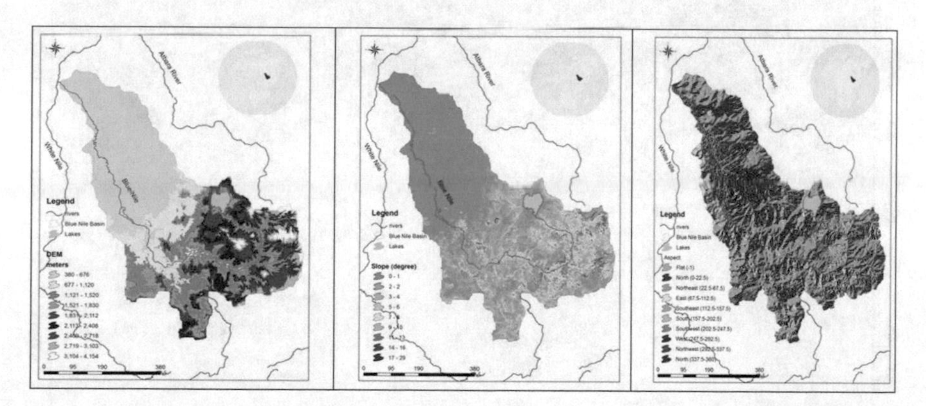

Fig. 17.10 DEM, slope, and aspect model for the study area

watershed and its information module sub-basins. The locations of the reservoir outlets were entered into the WMS and then delineated the reservoirs. As shown in Fig. 17.11, morphometric features were determined. WMS was also used to prepare the data file for the Rainfall-runoff simulation HEC-1 hydrological model.

Seven Steps in WMS for DEM Delineation as follows: (1) Compute Flow Directions and Accumulations with TOPAZ. (2) Define Watershed Outlet. (3) Convert DEM Streams to Feature Objects. (4) Add Interior Sub-basin Outlets. (5) Define Basin(s). (6) Convert Boundaries to Polygons. (7) Compute Basin Parameters. (8) Composite Curve Numbers (CN) Computation WMS-Hydro is used to compute composite curve numbers from land use, soil type, and basin boundary shape files.

For many years, hydrological modeling has performed as a useful instrument for managing water resources. Simulating a watershed of interest's hydrological and water quality conduct is generally used to predict the effects of suggested land-use situations and to assess short- and long-term management policies.

HEC-1, a program developed by the U.S. Army Hydrologic Engineering Center to create a hydrological model and simulate the Rainfall-Runoff process, was exported

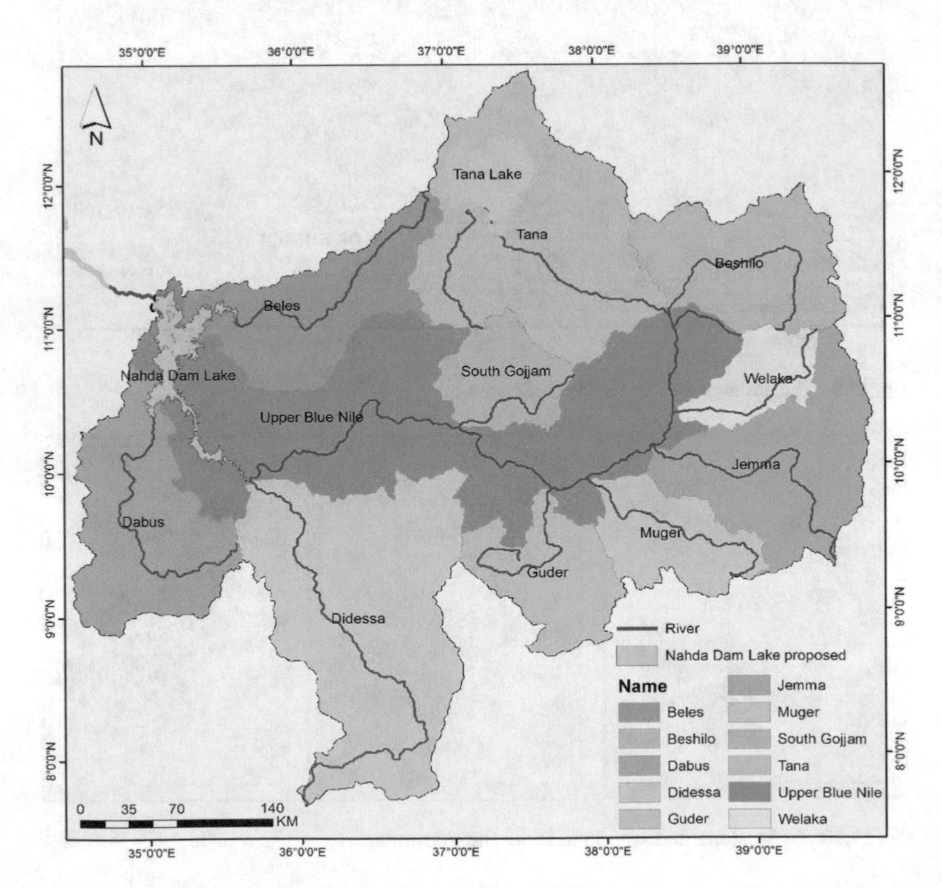

Fig. 17.11 Delineated basins, which are computed by WMS

to the project area sub-basins. Soil type, land use, and other characteristics of the basin were gathered and used in the program to determine the amount of basin curve, CN, and loss of rainfall. The curve number is regarded to be a function of multiple variables including hydrological soil groups, cover type, therapy (i.e. management practice), hydrological condition, background runoff condition, and catchment impervious region. The curve number therefore ranges (theoretically) from 0 to 100, with a practical range from 40 to 98 (Seybert 2006). Since the curve number (CN) model neglects precipitation intensities, CN values can be anticipated to decline with higher storm depth, but eventually approach a steady value (Hawkins 1993). Taking into account that within group (A), natural desert landscaping, the soil in the study area was chosen to be 63. Using rainfall depths predicted in the HEC-1 model for 50 and 100 years, results were acquired in the form of outlet hydrographs and peak flows were defined at the catchment point shown in Fig. 17.11 for both return periods.

A mapping table that relates land use and soil ID to curve numbers for soil kinds A, B, C and D is described in addition to the three shapefiles used to calculate CN. The steps for calculating composite CN in WMS are: (1) Delineate a watershed. (2) Create/Read soil type and land use data (grid or polygon coverages). (3) Read in a table relating land use and soil type to curve number. Figure 17.11 illustrates the delineated sub-basins Name for each sub-basin as prepared by WMS.

17.2.9 Morphometric Analysis of the Basins

Any basin's effectiveness can be assessed through its drainage system morphometric parameters. Table 17.2 shows the primary morphometric parameters considered. The morphometric parameters show a complete area of 307.45 km^2 in the catchment region. There are nine sub-basins in the region, each of which has an elongated shape that adds to enormous transmission losses. The average overland slope implies the natural mountain characteristics of the study area that accelerates overland flow over small distances (1055–1080 m). The basin elevation from the most upstream edge to the outlet varies from 1409 to 1451 m which indicates an overall slope of 0.0416–0.0627 m/m for Dabus basin and Beles basin respectively. The average slope of the main channel varies from 0.013 to 0.0177 m/m. This slope is relatively high compared to slopes of main channels in sedimentary rocks regions (0.01 m/m).

17.2.10 Hydrologic Model Development—HEC-1

Using "Data Export to HMS" functions, the model was exported to an HEC-1 project file after the basin model was completed and delineated with WMS. The designs consist of 11 basins covering the entire basin, reach and model parameters earlier identified. HMS needs the definition of a meteorological model as well as the control requirements in relation to the basin model. The design hyetographs for various

Table 17.2 Shows the primary morphometric parameters

Name	Area	Basin slop	MFDIST	MFDSLOPE	MSTDIST	MSTSLOPE	LAGTIME	TC	ADJSLOPE
Dabus	14,708.5	0.0416	295,380	0.0046	291,466	0.0045	38.1	63.6	24.3
Upper Blue Nile	41,555.8	0.0936	796,322	0.0040	790,494	0.0033	56.3	93.8	21.3
Beles	12,404.0	0.0628	289,858	0.0057	285,236	0.0048	30.6	51.0	30.0
Didessa	28,244.3	0.0671	442,140	0.0044	436,397	0.0041	41.5	69.2	23.3
Tana	24,652.3	0.0641	333,409	0.0047	326,323	0.0038	33.9	56.5	25.0
Guder	8148.5	0.0924	183,270	0.0059	179,649	0.0057	17.5	29.1	31.2
Muger	7415.3	0.1012	204,874	0.012	201,253	0.0097	18.3	30.4	59.3
Beshilo	12,169.3	0.1566	212,941	0.0122	205,819	0.0094	15.1	25.2	64.3
Jemma	14,730.0	0.1126	248,663	0.0082	241,920	0.0070	20.2	33.7	43.1
Welaka	4736.8	0.1148	164,988	0.0153	156,745	0.0128	14.4	24.0	81.0
South Gojjam	6065.0	0.0771	125,816	0.0226	120,695	0.0167	14.2	23.6	119.3

statistical return phases were obtained from the earlier prepared IDF curves using the frequency storm technique in the meteorological model. For a return period of 10 and 25 years, six hours (360 min) storms with an intensity location of 33% were developed and an SCS storm of 24 h was used for 50 and 100 years return. All distribution of storms is as per Type IA SCS.

17.2.11 Storage Capacity of a GERD Dam

Calculating the storage capacity of a dam reservoir needs numerous datasets that roughly define the lake's place, height, size and depth (i.e. bathymetry). The surface area and spatially variable depths for the advanced lake will, of course, differ significantly at distinct stages of the dam operation. Realistically, to estimate the storage capacity and its effect on downstream hydrology, scenarios of distinct heights for the dam and resulting lake level should be modeled. But, owing to its expected impact on Egyptian water supplies, the assessment of peak storage capability for the reconnaissance dam gets a specific concern from both the public and hydrologists. The first important information on the dam place (Fig. 17.12) was given by the appearance of the building site for the dam on the Rapideye satellite image of 20 July 2019. The SRTM DEM tiles were pre-processed individually and then mosaiced to analyze the parameters of the reservoir. For each pixel whose value is −32,768, the pre-processing has begun to assign an elevation value. In order to modify the values of the artifacts to their neighboring pixels, the DEM was then filled in, hence increasing the accuracy of generating hydrologically correct parameters from these DEMs. The drainage networks obtained from the DEM needed the "flow direction" calculation from the "filled" DEM and the calculated "flow accumulation" from the

Fig. 17.12 Image acquired on 20 July 2019 shows locations of the Renaissance Dam and its spillway dam. https://www.planet.com/explorer/

"flow direction" grid. On the satellite images, the resulting drainage networks and related sub-catchments were overlaid to verify that they correspond to visible wadis and streams. This comparison is essential in order to guarantee DEM's precision in simulating the topographical elevation and the resulting forms and geometry of the lake.

17.3 Results and Discussions

The results of this paper can be discussed as follows:

17.3.1 Hydrologic Model Evaluation

Depending on the annual hydrograph form at the outlet of the reservoir (Fig. 17.13), the flow begins 3 h lag after the storm and then reaches its peak after 7 h. These two findings fully match the data gathered with individuals in the village during site visits (flood incident May 2014). The precipitation began around 6:00 a.m., according to eyewitnesses, while the stream came at 10:00 a.m. and continued for 6–7 h at intensive flow.

17.3.1.1 Hydrological Properties

The concentration time of the watershed Tc = 7.0 h is close to the assumed storm event duration of 6 h while the calibration of Paris is required according to the storm distribution of Type IA. As per hydrograph, the Lag Time "TL" is 7 h in the bridge section, whereas according to eyewitness it is 4 h, which means that additional calibration is required for the distribution of storm intensity (more data should be collected for storm events). Peak rate factor (K') is a critical parameter for hydrograph shape determination. It is used to depict the hydrograph shape impact of watershed storage. Watersheds with little or no storage impacts are allocated high K' values, and low K' values are allocated to watersheds with important ponding impacts. Comparing the shape of the hydrograph to the shape factor of the SCS-Method, the maximum rate factor is between $300 < K' < 484$; indicating that the storage and preservation of the watershed are important.

Hydrological properties study has shown that the hydrograph has steep climbing limbs, sharp peaks, and mild recession limbs. In reaction to intense rainfall and overland flow dominance in runoff generation during dry weather, steep rising limb occurs, while mild recession limb reflects the long-lived nature of runoff and tiny losses in transmission. Further research is needed to create a more extensive knowledge of the study area's hydrological features by calibrating the model based on more flood occurrences.

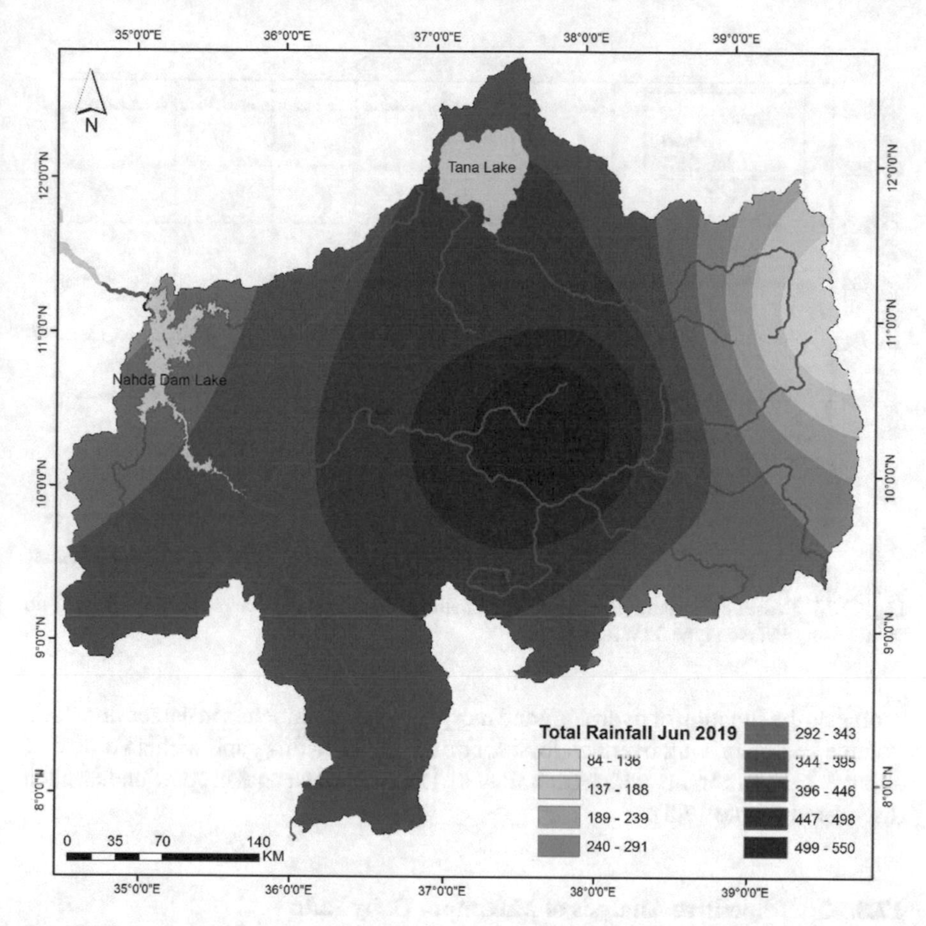

Fig. 17.13 Total rainfall at June 2019. https://livingatlas.arcgis.com/waterbalance/

17.3.1.2 Rain Simulation for Current and Future Reality

In this study, a rain simulation has been carried out with multiple possibilities that represent the current and future reality in the upper basin of the Blue Nile above the Renaissance dam and these are 15, 30, 50 mm, which represent the present reality. As for future reality, rain may fall with potential 100 mm, and the higher basin of the Blue Nile over the GERD dam is considered very dangerous in case of rain of more than 50 mm. This is because of the long flow of water in the basin, where the flow of water in the valleys continues for more than (39 h) (Table 17.3).

Table 17.3 illustrates the results of the 50-year hydrograph return period produced by HEC-1. Results show that the quantity of the hydrograph is about 4 million m^3 for sub-basin 1 with a maximum flow of about 87.5 m^3/s. Converting the flow over the sub-basin region into runoff depth outcomes in a rainfall storm of 23.5 mm out of 102 mm. In other words, losses constitute approximately 79 mm (i.e. 77%), which

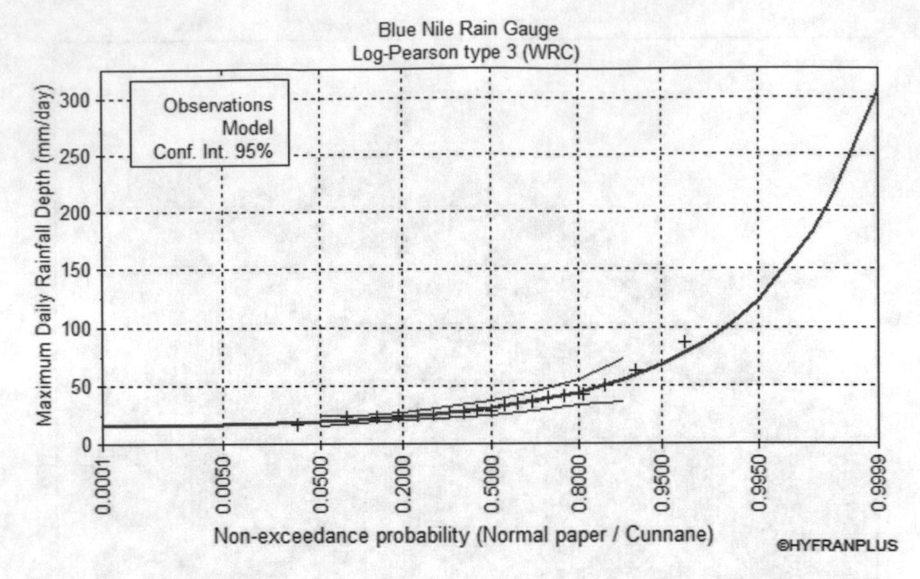

Fig. 17.14 Possible distribution curve of Debremarcos station data to determine the depth of rain values using Pearson type 3 (WRC)

promotes the function of hydrology and morphometric parameters in decreasing flood volume and increasing overland losses. For the outlet hydrograph with a volume of about 7.233 million m^3 with a peak flow of 157 m^3/s, the same proportion has been discovered (Table 17.3).

17.3.1.3 Repetitive Analysis of Maximum Daily Rain

The precise quantification of rainfall on the water basin is one of the most important factors that accurately aid the calculations of torrents collected from those rains. It is the correct basis for water statistics and the probability of the occurrence of torrents. The proper distribution of rainfall and torrents stations contributes to reliable information covering the whole region and leads to the avoidance of readings or recording between stations. Rainfall intensity is the main and influential factor in the formation of torrents that must be taken into account when planning construction and development projects.

Statistical analysis of the maximum daily rainfall values and the use and testing of different probability distributions to obtain rain value at different frequency times was performed by reference to station records from NASA's space information site (https://power.larc.nasa.gov/data-access-viewer/).

Knowledge of the density of rainfalls and the determination of the depth of rain for different iterative periods (2–3–5–10–20–50–100 years). This is done using the statistical analysis program (HydranPlus) and the application of different statistical distributions.

Table 17.3 The parameters of the selected basins in the study area

Upper Blue Nile sub-basins	Rain 15 (mm)			Rain 30 (mm)			Rain 50 (mm)			Rain 100 (mm)		
	Peak (m^3/s)	Time (min)	Volume (Flow (m^3)	Peak (m^3/s)	Time (min)	Volume (flow (m^3))	Peak (m^3/s)	Time (min)	Volume (flow (m^3))	Peak (m^3/s)	Time (min)	Volume (flow (m^3)
Dabus	3615	3	477,084	292	3315	47,639,144	1157	3300	188,344,115	4390	3255	714,470,033
Beles	3135	2.8	402,345	278	2820	40,176,500	1101	2790	158,840,099	4182	2775	602,548,706
Didessa	3825	5.5	916,145	545	3540	91,482,191	2155	3525	361,680,135	8178	3450	1,372,007,922
Guder	2370	3	264,302	281	2025	26,392,955	1118	1950	104,346,094	4286	1905	395,829,459
Muger	2475	2.5	240,519	235	2145	24,017,807	935	2070	94,955,768	3574	2040	360,208,025
Jemma	2595	4.5	477,792	430	2265	47,710,406	1709	2190	188,625,848	6523	2175	715,538,745
Welaka	2220	2	153,648	181	1905	15,342,152	723	1815	60,656,089	2781	1755	230,094,567
Beshilo	2265	4.8	394,723	448	1950	39,415,338	1786	1860	155,830,791	6861	1815	591,133,044
South Gojjam	2205	2.6	196,728	235	1890	19,644,494	939	1800	77,665,681	3614	1740	294,619,248
Tana	3480	5.1	799,627	503	3165	79,847,724	1992	3150	315,682,571	7560	3120	1,197,519,456
Upper Blue Nile	2745	30	5,670,832	2952	2490	566,265,203	11670	2370	2,238,762,078	44330	2340	8,492,585,048

Table 17.4 Maximum precipitation per day in Debremarcos station

Year	Month	Day	Max precipitation per day (mm)
2000	7	24	35.18
2001	6	30	22.89
2002	6	28	41.61
2003	8	8	23.79
2004	4	5	17.16
2005	7	29	26.83
2006	6	29	35.01
2007	7	9	27.68
2008	8	3	25.01
2009	10	11	26.85
2010	8	3	28.05
2011	7	19	23.29
2012	7	17	24.64
2013	7	13	49.13
2014	7	30	33.19
2015	6	23	39.18
2016	5	10	62.27
2017	7	20	86.11
2018	8	14	40.95

https://power.larc.nasa.gov/data-access-viewer/

17.3.1.4 Analysis of Debremarcos Stabilization Station for Rain Depth

Data on rainfall for Debremarcos were provided from 2000 to 2018, about 19 years as shown in Table 17.4. By applying different probability distributions of rainfall information to the station, the distribution was found that the log-Pearson type 3 (WRC) is the most appropriate distribution of this information. Therefore, this is used to perform a probability analysis to determine the rainfall depth values of different recursive times (Tables 17.5 and 17.6). From the analysis, the rainfall depth value for a 100-year iterative period was found to be 103 mm. The probability of rain depth over 50 years is 86 mm, and during two years is 30.2 mm. Figure 17.14 shows the potential distribution curve of station data using Pearson type 3 (WRC).

17.3.1.5 Outlet Hydrograph Curve

Simulate hydrographic analyses when 15, 30, 50, and 100 mm of rain falls within 24 h are presented in Fig. 17.15.

Table 17.5 Possible analysis to determine the depth of rain values by using Pearson type 3 (WRC)

Basic statistics	
Minimum	17.2
Maximum	86.1
Average	35.2
Standard deviation	16.4
Median	28.1
Coefficient of variation (Cv)	0.465
Skewness coefficient (Cs)	1.98
Kurtosis coefficient (Ck)	5.49

Table 17.6 The results of the Pearson type 3 (WRC) patch for a Debremarcos station

Time	Q	Possible amount of rain, XT	Stander deviation	Confidence interval (95%)
100	0.99	103	44.9	N/D
50	0.98	86.1	29.8	N/D
20	0.95	67	16	N/D
10	0.9	54.7	9.47	36.1–73.3
5	0.8	43.8	5.64	32.7–54.8
3	0.6667	36.6	4.1	28.5–44.6
2	0.5	30.7	3.17	24.4–36.9

The analysis of hydrographic curve is shown at the output of the upper basin of the Blue Nile above the GERD dam in the event of a 15 mm rainfall in one day. The results indicate that the water Volumes was about 5,670,832 m^3 with a peak discharge of about 2745 m^3/s in the upper basin of the Blue Nile above the Renaissance dam (Fig. 17.15).

This figure illustrates a simulation of the water flow in the upper basin of the Blue Nile above the Renaissance dam in various rain lumps between 15 to 150 mm to represent reality and the future. The upper basin of the Blue Nile is divided over the Renaissance dam into 10 subbasins, the main source of water flows in the Upper Blue Nile basin above the GERD Dam, where the water of these sub-basins will be controlled by the Ethiopian Renaissance Dam. Analysis of the hydrographic curve results of the higher basin of the Blue Nile above the GERD dam in the event of rainfall of 15 mm shows the presence of water flows of 5,670,832 m^3 of running water. This will be controlled by the Ethiopian GERD Dam.

In the event of a simulation with 30 mm rainfall within 24 h, the water Volumes discharge in the upper tub of the Blue Nile above the Renaissance dam will be approximately 566,265,203 m^3 with a peak discharge or flow of 2952 m^3/s, the upper basin of the Blue Nile above the Renaissance dam.

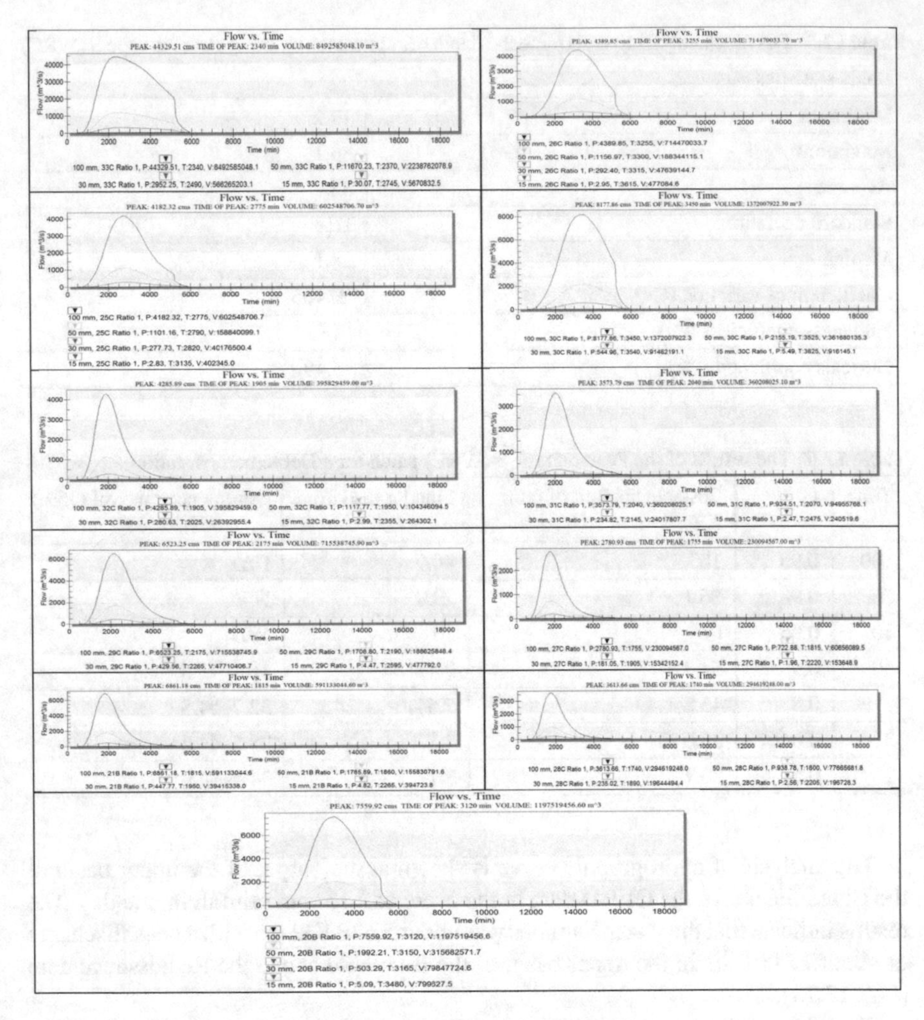

Fig. 17.15 Outlet hydrograph created by HEC-1 Upper of Blue Nile

The simulation of 50 mm rainfall in 24 h shows that the water volume in the upper tub of the Blue Nile over the GERD dam will be 2,238,762,078 m^3 with a peak flow of 11670 m^3/s in the upper basin of the Blue Nile above the Renaissance dam.

In some years, over the GERD dam, the upper basin of the Blue Nile falls more than 50 mm of rain in 24 h and in case the volume of water flow is too large, causing floods and surface runoff that is not uncommon for what falls annually. Thus, a 100 mm rainfall simulation was made within 24 h (Fig. 17.15). If 100 mm of rain occurs within 24 h, then an 8,492,585,048 m^3 of water occurs, with a peak flow of approximately 14329 m^3/s. The flow of water into the basin begins after 30 min from the rainstorm (rainfall storm) and then reaches its peak after 45 h from the rainstorm. The runoff of the pelvis lasts for 5000 min, i.e. 83 h.

17.3.2 Dam's Storage Ability

The topography and elevation of the Renaissance Dam area vary from 506 to 1777 m asl. A narrow floodplain area is the construction site of the main dam, bounded by two high shoulders (1222 and 945 m asl) of the huge basement rocks. The development of the dam on this site (508 m asl) can only support a lake with a maximum depth of 100 m (i.e. a maximum lake level of 606 m asl). This is because, apart from the main course of the Blue Nile (Fig. 17.16), the lake at 606 m asl stage will spill over its shoreline into the Rosaries downstream through a southeastern tributary canal. For this reason, the spillway dam is being built south of the Renaissance Dam to regulate the overflow and increase the primary dam's storage ability. The Renaissance Dam alone could not support the creation of a reservoir below 100 m, but the development of an auxiliary dam in the saddle area between the hillslopes on the western side of the Blue Nile will allow the construction of the main dam above 100 m above the stage of the floodplain. Again, the spillway dam can reach a maximum height of 80 m as another overflow occurs at the height of 686 m amsl and therefore requires a second spillway dam further upstream at that site.

The Renaissance Dam's GIS-simulated reservoir at 606 m asl (i.e. 100 m maximum water depth) showed that the lake would cover roughly 745 km^2 and that the quantity of water stored would reach km^3. In fact, the abstraction of 17.5 km^3 of water to fill the reservoir will influence the downstream net annual flow that will remain until the dam is built. The effect will be immediately linked to the duration of the building period and the strength of the basin's summer flood seasons. In addition, spillway dam building will greatly improve storage ability and result in net annual evaporation loss owing to the increased surface area. For example, the development of a 30 m high spillway dam will enable the Renaissance Dam to reach a height of

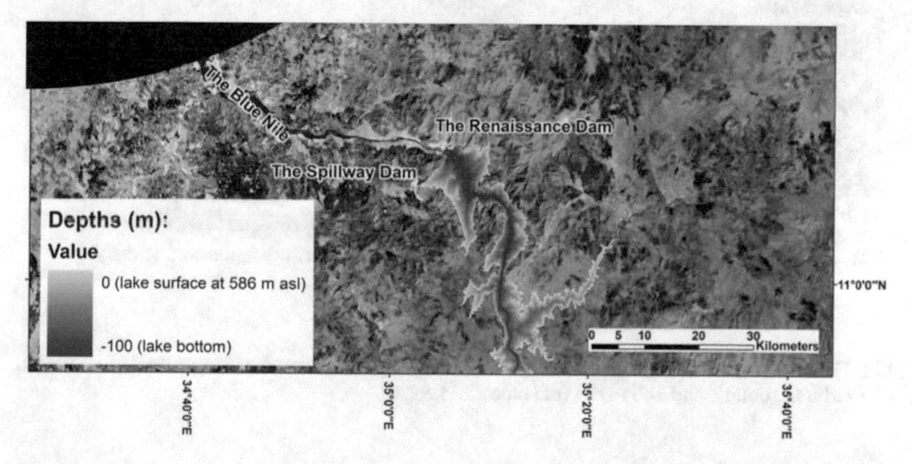

Fig. 17.16 Simulation of the subsequent lake when approximately 100 m high is the Renaissance Dam. The lake will store water of up to 17.5 km^3, covering 745 km^3

130 m, thus storing 56 km^3 of water in the advanced lake, which will cover about 1560 km^2.

The dam will stand at 147 m height and the spillway dam will be 45 m high, and the storage capacity will reach 74 km^3, as shown by published data on the Renaissance Dam (Fig. 17.17). The approximately storage capability is shown in Table 17.7.

Remote sensing and DEM information were important to monitor continuing dynamic changes in big surface water bodies; to extract hydrological parameters and to model the water balance (El Bastawesy et al. 2008). The use of this strategy was motivated by the enhanced accessibility of satellite information with elevated and moderate multi-temporal resolution. The precision of simulation and hydrological modeling depends on the quality of the DEM used, the accessibility of long flow and rainfall information for calibration and ground truthing.

It should be observed, however, that certain constraints have been experienced in the growth of the GIS model conceived herein. The SRTM-DEM (Fig. 17.18) is the accessible elevation information for this region, which was obtained in 2001 and is yet to be developed. Consequently, for precision, simulated dam lake at different concentrations must be analyzed somehow.

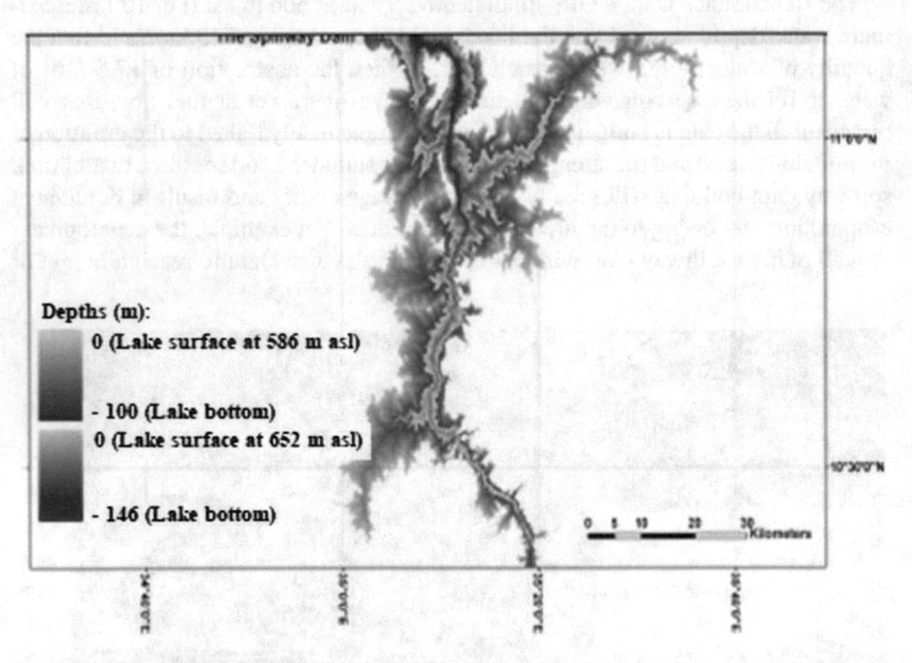

Fig. 17.17 Comparison of two scenarios for the resulting lake if the dam is constructed 100 m tall (the red scale color), and 146 m tall (the blue scale color)

Table 17.7 The estimated rating curves for some scenarios of the Renaissance Dam and its developed lake

Maximum lake depth (m)	Surface area (km^2)	Maximum storage capacity (km^3)	Average annual loss (km^3) due to evaporation (924 mm/year)
80	442	9.6	0.4
100	745	17.5	0.7
130	1560	56	1.4
146	1954	80.5	1.8
180	3130	173	2.9

17.3.3 Analysis of the Storage Capacity of the Renaissance Dam in Case of Rainfall of 100 mm Within 24 h

The Ethiopian GERD Dam is designed to be 150 m high above a 500-m high altitude area, which is illustrated by the analysis of the characteristics of the dam's height, position, and the hydrological characteristics of the Blue Nile basin and the drainage slope (hydrograph Curve) when rainfall of 100 mm within one day. The storage characteristics of the Ethiopian Renaissance dam vary from one level to another. Table 17.8 shows that if the water is full in front of the dam up to 555 m asl (50 m from the height of the dam), the dam will have a capacity of 890,386,955.6 m^3. If the water is full in front of the dam up to 630 m asl (130 m above the height of the dam), the storage capacity will be 35,053,628,668 m^3.

Figure 17.19 shows the hydrographic curve of storage capacity in front of the Ethiopian GERD Dam, where the storage capacity increases in front of the dam and decreases the trend toward the upstream, where the water's rise increases and the depth of the water decreases and thus the storage capacity decreases.

17.3.4 The River Islands Sank and New Island Appear

The formation of the Ethiopian GERD Dam Lake is similar to the High Dam Lake in Egypt, both of which are water storage in front of the dams and are industrial lakes that have led to geological and morphological changes in the region. The Ethiopian GERD dam is built on the Blue Nile stream near the Ethiopian-Sudanese border, which is designed to rise 150 m above the surface of the earth. The land on which it was decided to build is about 500 m from the sea level. The dam will consist of a giant freshwater lake extending from the 500 to 650 m asl, which will be about 200 km from the dam's position to the end of the 650-m contour line along the Blue Nile.

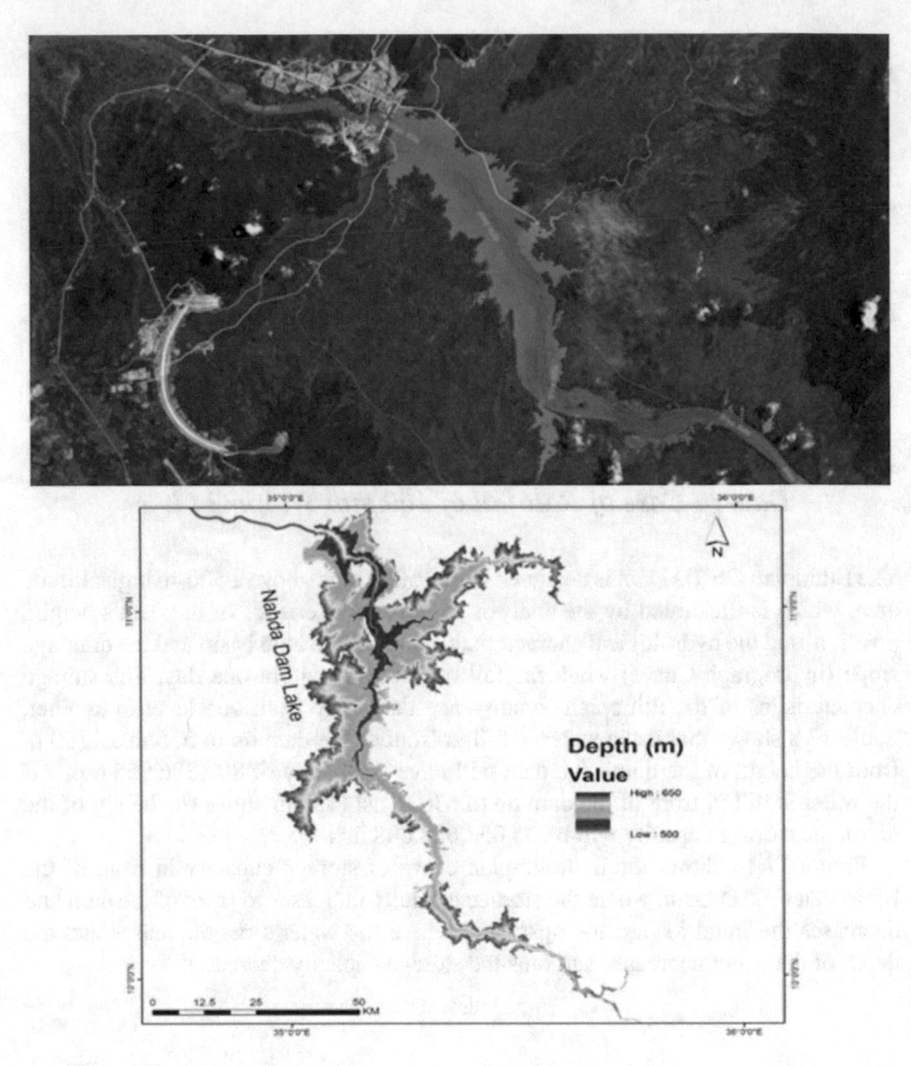

Fig. 17.18 The simulated lake of Dam. The high coincidence between the lake as appear on the Rapideye image (top) and the SRTM-DEM simulation indicate the reliability of estimation https://www.eorc.jaxa.jp/ALOS/en/aw3d30/data/index.htm

The water storage in front of the Ethiopian GERD Dam will consist of a 200 km water lake with an average width of about 10,748 km and a maximum lake width of about 37,941 km and a minimum lake width of about 902 m. The construction of the dam will lead to geomorphological changes in the area where many river islands as well as the 200-km-long river stream of the Blue Nile will sink (Fig. 17.20). The average width of the current river stream in the position of the GERD Dam Lake is about 336 m, with the maximum width of the river reaching 744 m and the lowest width of the river course at about 65 m.

Table 17.8 Storage capacity of the renaissance dam in case of rainfall of 100 mm within 24 h

Elevation (m)	Storage (m^3)
521.6	1,424,987.75
528.36	21,990,630.05
535.12	87,664,752.73
541.87	231,925,914.7
548.63	480,460,217
555.39	890,386,955.6
562.15	1,492,919,969
568.91	2,336,196,970
575.66	3,485,486,840
582.42	5,065,121,711
589.18	7,178,167,682
595.94	9,850,903,542
602.69	13,173,507,356
609.45	17,262,869,354
616.21	22,186,304,913
622.97	28,065,319,162
629.73	35,053,628,668
636.48	43,193,391,712
643.24	52,514,410,926
650	63,033,051,467

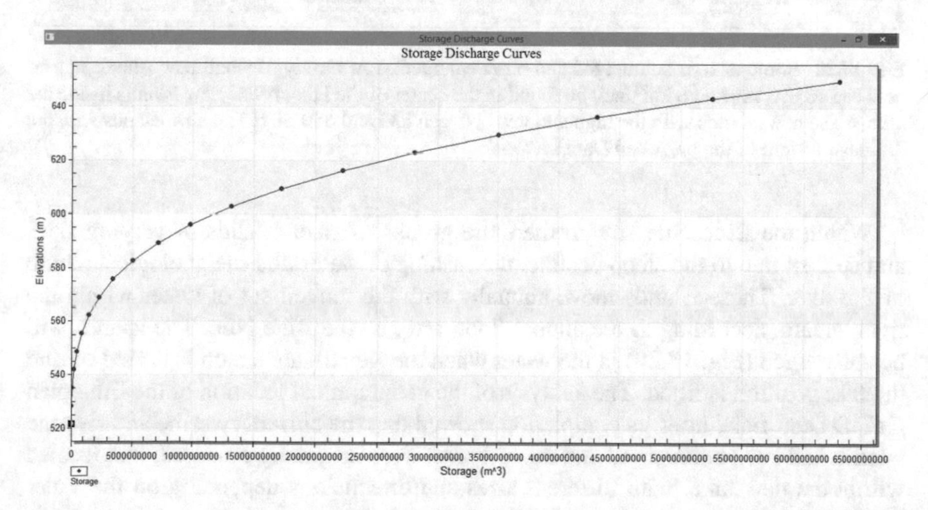

Fig. 17.19 The hydrographic curve of storage capacity in front of the Ethiopian Renaissance dam

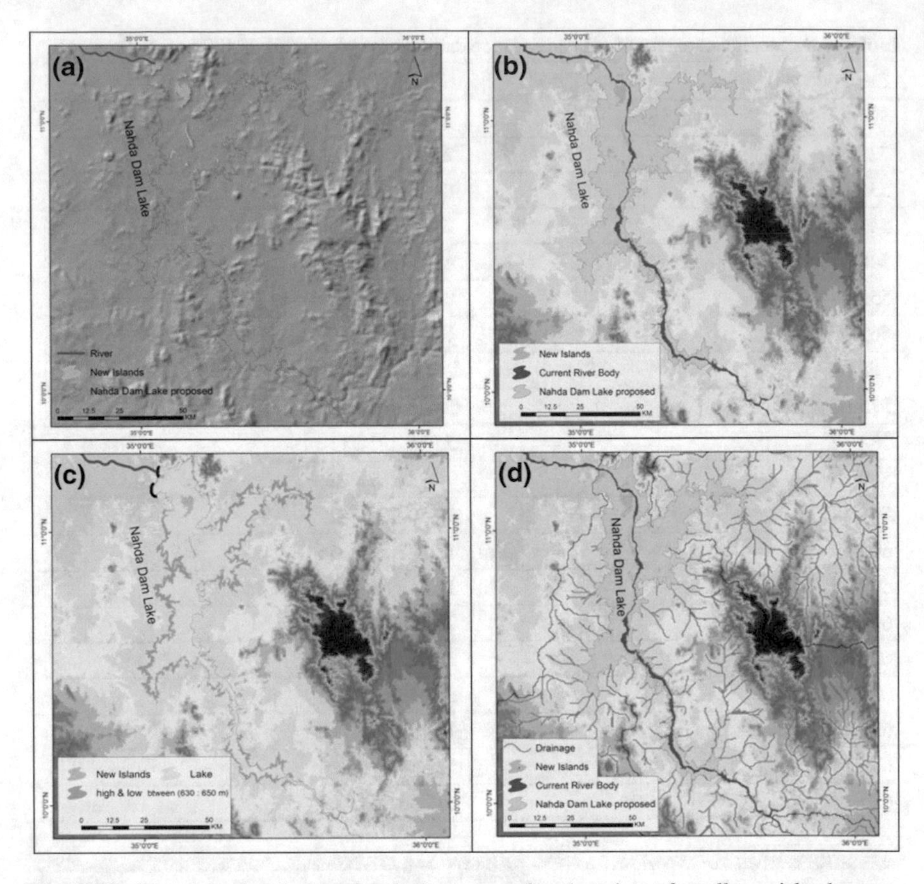

Fig. 17.20 Some islands submerged and other emerged: **a** A variety of small new islands appear between 18,393 m^2 and 6 km^2 will be found in the center of the lake. **b** The new islands inside the lake. **c** The new islands with the high and low between 630 and 650 m. **d** The new islands with the drainage system of the proposed Dam Lake

Within the Blue Nile stream there are groups of river islands of varying sizes and spaces; due to the steep decline, the rushing of water and the geological nature of the river. These islands move annually with the movement of water within the river stream, according to the annual flood force in the Blue Nile. The islands will be submerged (Fig. 17.20) in the water when the Renaissance dam is turned on and the lake position is filled. The analysis of the geographical location of the Ethiopian GERD Dam Lake after its completion showed that the current riverine and riverine islands had been submerged and new islands, were originally emerged, that located within the new lake, with different sizes and dimensions depending on the water level in the lake. In the center of the GERD Dam Lake, a variety of small new islands between 18,393 m^2 and 6 km^2 will be found in the center of the lake, about 77 islands except for one which will be about 27 km^2 and will appear in the middle of the lake in front of the Renaissance dam.

17.3.5 Assessing Seismic Risk in Ethiopia

Assessing 'hazard'—as a forward-looking idea—implies evaluating prospective occurrences, quantifying the probability that they will happen, and assessing the future implications of this (UNISDR 2013). It also includes assessment of past events and consequences and is contingent on availability and quality of data.

90% of seismicity and volcanic activity in Ethiopia is linked to the rift scheme in East Africa. The East African Rift System is a 50–60 km wide area of volcanoes and faults stretching north to south in Eastern Africa for over 3000 km, from northern Ethiopia to the southern Zambezi. It cuts in a NE-SW direction through Ethiopia (Ayele and Kulhánek 2007). The seismic energy is released as micro-earthquakes (Ayele and Kulhánek 2007).

Figure 17.21 shows the number of earthquake threat in Ethiopia based on information from the 2013 Global Assessment Report on Disaster Risk Reduction. The yellow to red color in this map reflects the degree of earthquake hazard, with red being the most important and yellow less important. The green is the degree of vulnerability to earthquake loss with more important loss of darker green significance.

17.3.5.1 Seismic and Earthquake Hazard in Ethiopia

The active Great Rift Valley allows Ethiopia vulnerable to two seismic risk kinds: earthquakes and volcanic eruptions. It's not at danger from tsunamis as a landlocked nation. Using information from the EM-DAT database of one of the best-known catastrophe databases. A total of ten earthquakes and eruptions occurred between 1900 and 2013, resulting in a total of 93 fatalities, 165 wounded, 420 homeless and

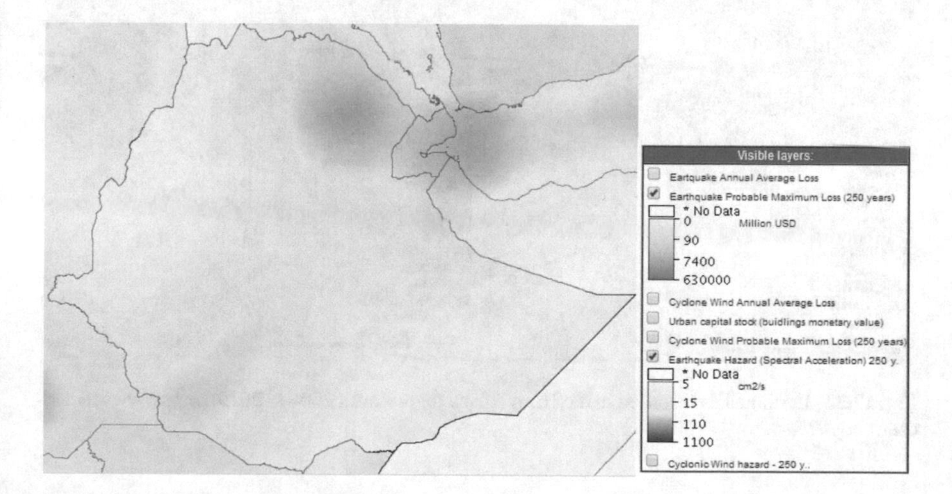

Fig. 17.21 Earthquake risk for Ethiopia (Herbert 2014)

11,000 affected. These are estimated to have an economic cost of more than US$7 million. https://www.emdat.be/result-country-profile?disgroup=natural&country=eth&period=1900$2013.

Earthquakes happen along the rift scheme in Ethiopia. Figure 17.22 shows Earthquakes recorded in the Horn Africa region from 2004 to 2019 (https://www.emsc-csem.org/#2). The red dot size describes earthquake magnitude—varying from 3.5 to 7.2.

According to the EM-DAT Database, a total of seven earthquakes occurred between 1900 and 2013, killing a total of 24 individuals, affecting 585 individuals and causing financial harm of more than US$7 million (Herbert 2014). Since Addis Ababa is 75–100 km from the western edge of the Ethiopian Rift Valley, earthquakes often affect it.

Figure 17.23 shows Ethiopia's major cities in relation to seismic hazard. Notably the three most populous cities—Addis Ababa (3 million), Dire Dawa (273,600), and Mek'ele (271,600) 16—are found in the most seismically hazardous areas—marked in yellow in the centre of the country and categorized as having a "medium" risk of seismic hazard. The cities Addis Ababa, Nazret, Dire Dawa and Awassa are very near main fault lines (e.g. the Wonji fault, Nazret fault, Addis-Ambo-Ghedo fault, and Fil Woha fault) where many earthquakes have previously occurred (Kinde 2002).

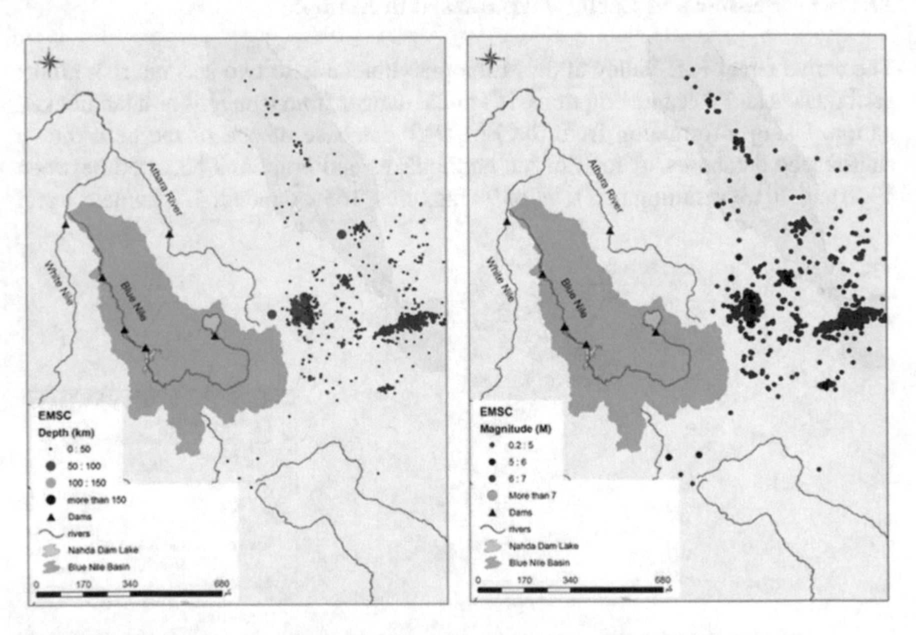

Fig. 17.22 Earthquakes recorded in the Horn Africa region from 2004 to 2019. https://www.emsc-csem.org/#2

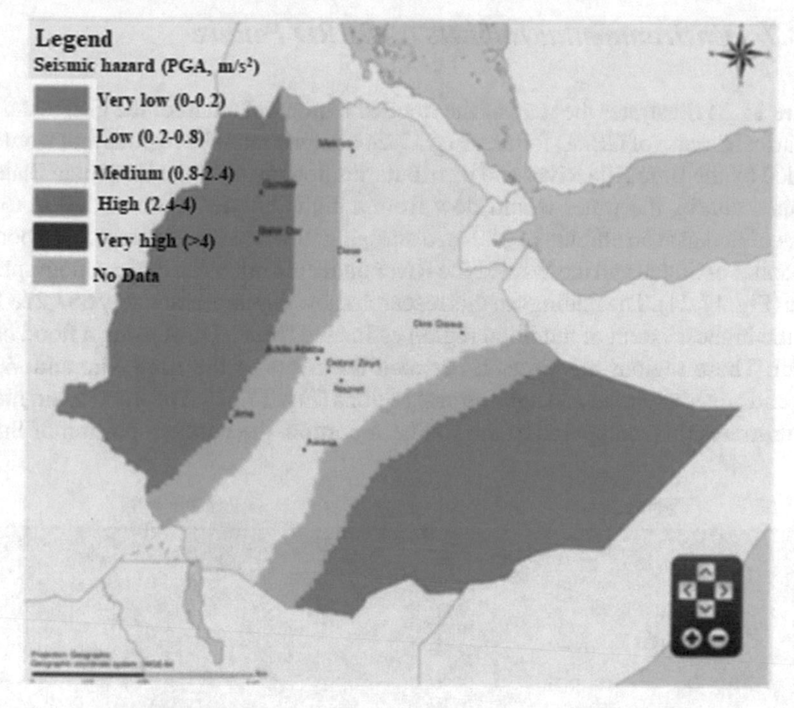

Fig. 17.23 Seismic distribution map, Ethiopia (https://www.whoeatlas.org/africa/countries/ethiopia/ethiopia-seismic-map.html)

17.3.6 Expected Risks from the Dam's Construction

Safety surveys on dams have shown that the Ethiopian GERD Dam's safety factor is only 1.5° from 9°. Studies on dam safety verify that there are 340 dams in the globe, resulting in the collapse of a total of 300 dams. The big percentage of these dams were big dams of the size of the Ethiopian dam and the High Dam, but it is unlikely that the High Dam could collapse because it is a rocky dam made up of rocks stacked on top of each other in layers.

Some are coated with cement, but GERD Dam is made of concrete and is at danger of collapse, as well as the rocky cumulus dam that can be repaired in a few months. Thus, if the dam collapses, Ethiopia will be safe. Ethiopia has already constructed several river dams, three of which have failed due to the complexity of the area's nature, the force of rushing water and the steep slope, and the weakness of the soil on which the dams are constructed. It is more probable that the GERD Dam will collapse. Experts said the dam was created to collapse. The safety of the dam is very low. This implies that owing to the weight of the water behind it and the weakness of the region in particular, any earthquake will damage this dam and it is a seismic zone adjacent to the African groove.

17.3.7 Environmental Impacts of GERD Failure

Figure 17.24 illustrates the map of the flooded region in Sudan for the GERD failure scenario. In a case of GERD failure, Fig. 17.24 demonstrates the regions that would be flooded by the Blue Nile River and its tributaries flowing from the Ethiopian Plateau. In this scenario, the water would flow from a high altitude region of 520 m to the plains of Sudan at an altitude of 480 m, destroying the Renaissance Dam and flooding the capital of Sudan and the White Nile River under the influence of the topographical slope (Fig. 17.24). The findings of this research show that approximately 667,228 km^2 was the highest extent of the flood region estimated from a DEM using a flood basin model. These regions are situated between the rivers of the Blue Nile and White Nile and cover all urban and agricultural regions (Fig. 17.24). The flood basin model demonstrates that, compared to the southern portion, the northern portion of Sudan

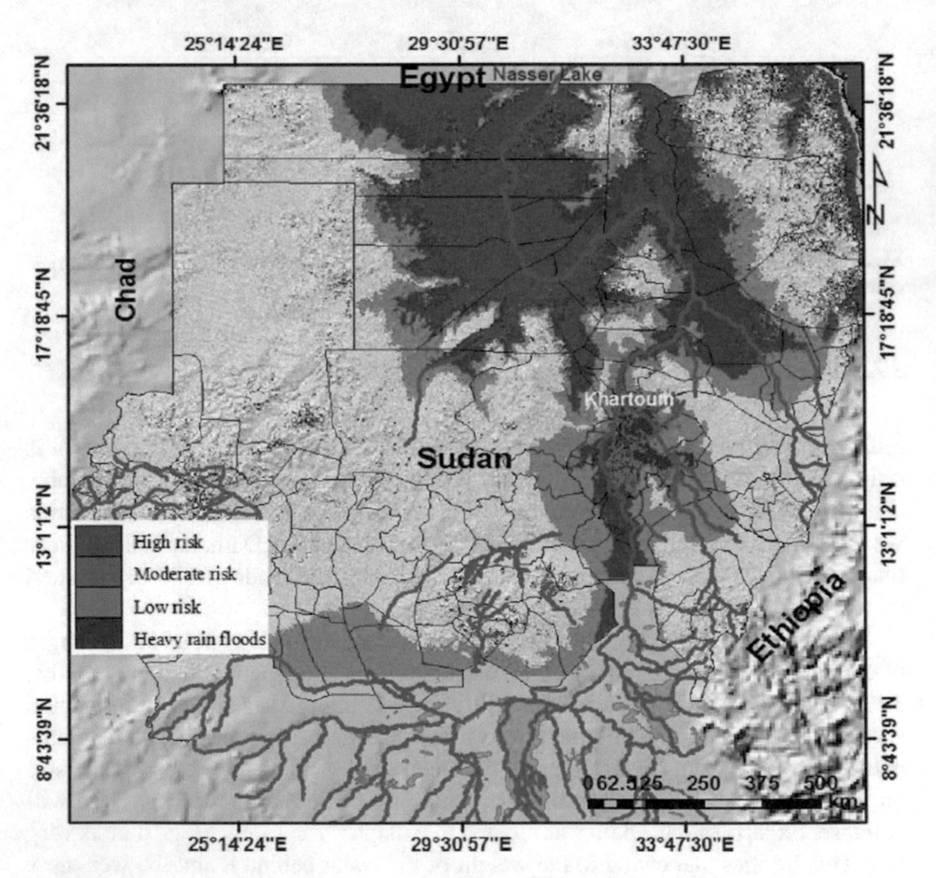

Fig. 17.24 Simulated from DEM using the flood basin model, the Blue Nile River flooded regions of Sudan show the regions that will be impacted by flooding in the event of GERD failure. During the wet seasons, the blue polygons show the flooded regions with rainfall water (Mohamed and Elmahdy 2017)

is a lowland region. Thus, in the southern portion of Sudan, the river-flooded region is much bigger than in the south. In the southern portion of Sudan, however, harm to houses and farms would be much more serious than that in the north owing to the elevated water flow rate. This situation shows the need to build a saddle dam on the GERD to avoid water from pouring out of the dam's south-west side (left wall).

17.4 Conclusions and Recommendations

In this study, using remote sensing and GIS with the help of the Watershed Modeling System (WMS) Program and Global Mapper software, the watershed and its sub-basins delineation process were carried out. Analysis of morphometric parameters stated that elevated infiltration levels contribute to the morphometric features of the watershed in high-velocity floods. The HEC-HMS program conducted hydrological modeling which simulated the rainfall-runoff process using the curve number model. In the outlet of the catchment area sub-basins as well as at the catchment outlet, flood hydrographs were developed. HEC-RAS program researched the method of delineation of the floodplain.

Four objectives are identified in this study. The following are the conclusions related to these specific objectives. First objective was to analyze the dam's storage ability. The Renaissance Dam's GIS-simulated reservoir at 606 m asl (i.e. 100 m maximum water depth) showed that the lake would cover roughly 745 km^2 and that the quantity of water stored would reach 74 billion cubic meters. The specific conclusions are the following:

1. The expected 50- and 100-year rainfall depths are 102 and 117 mm respectively.
2. High-velocity floods with high infiltration rates lead to the morphometric features of the watershed.
3. Flood volumes of 7.233 and 9.88 million m^3 for 50- and 100-year return periods.
4. The peak floods are 157 and 223.6 m^3/s for both return periods with a runoff coefficient of 23% and 31.5%.
5. Hydrologic losses represent 77% and 68.5% of the 50 and 100 return period's floods.

Second objective was to delineate the River Islands Sank and New Island Appear. The islands will be submerged in the water when the Renaissance dam is turned on and the lake position is filled. The analysis of the geographical location of the Ethiopian GERD Dam Lake after its completion showed that the current riverine and riverine islands had been submerged and new islands, were originally emerged, that located within the new lake, with different sizes and dimensions depending on the water level in the lake. In the center of the GERD Dam Lake, a variety of small new islands between 18393 m^2 and 6 km^2 will be found in the center of the lake, about 77 islands except for one which will be about 27 km^2 and will appear in the middle of the lake in front of the Renaissance dam. Third objective was to assess the Seismic and Earthquake Hazard in Ethiopia. According to the EM-DAT Database, a total of seven earthquakes occurred between 1900 and 2013, killing a total of 24 individuals, affecting 585 individuals. Final objective was to map and simulate the flooded area in Sudan and Egypt in the case of GERD failure. Safety surveys on dams have shown that the Ethiopian GERD Dam's safety factor is only 1.5 degrees from 9 degrees. It is more probable that the GERD Dam will collapse. Experts said the dam was created to collapse. The safety of the dam is very low. This implies that owing to the weight of the water behind it and the weakness of the region in particular, any earthquake will damage this dam and it is a seismic zone adjacent to the African groove. In a case of GERD failure, it demonstrates the regions that would be flooded by the Blue Nile River and its tributaries flowing from the Ethiopian Plateau. In this scenario, the water would flow from a high altitude region of 520 m to the plains of Sudan at an altitude of 480 m, destroying the Renaissance Dam and flooding the capital of Sudan and the White Nile River under the influence of the topographical slope. The findings of this research show that approximately 667,228 km^2 was the highest extent of the flood region estimated from a DEM using a flood basin model. These regions are situated between the rivers of the Blue Nile and White Nile and cover all urban and agricultural regions.

Sensitivity analyzes for both the hydrological and hydraulic models have shown that the results are representative and appropriate for design purposes. The hydraulic modeling findings were integrated into a representative flood hazard map for catchment together with a thorough assessment of the study area's hydrological features.

The research advises the following to keep sustainable development in future:

1. Since flood volumes range from 7.233 million to 9.88 million m^3, the research does not suggest the construction of floodwater storage dams.
2. Floodwaters promote downstream socio-economic operations and recharge groundwater aquifers.
3. The flood plain should be declared risk areas.
4. It is suggested that detailed surveys and field investigations quantify assets that would be detrimental to the 100 floods.
5. Daily precipitation meteorological network, air pressure, air temperature, wind velocity and direction and ripple measurement stations as well as flood warning devices are suggested.

References

Abtew W, Melesse AM, Dessalegne T (2009) Spatial, inter and intra-annual variability of the Upper Blue Nile basin rainfall. Hydrol Process 23:3075–3082

Al-Abed NA, Whiteley HR (2002) Calibration of the hydrological simulation program fortran (HSPF) model using automatic calibration and geographical information systems. Hydrol Process 16:3169–3188

Andrew ME, Wulder MA, Nelson TA (2014) Potential contributions of remote sensing to ecosystem service assessments. Prog Phys Geography 38:328–353

Ayele A, Kulhánek O (2007) Spatial and temporal variations of seismicity in the Horn of Africa from 1960 to 1993. Geophys J Int 130(3):805–810

Bicknell BR, Imhoff JC, Kittle JL, Jobes TH, Donigian AS (2001) Hydrological simulation program-Fortran (HSPF), version 12, user's manual. AQUA TERRA Consultants, Mountain View, California, USA

Conway D (2000) The climate and hydrology of the Upper Blue Nile River. Geogr J 166:49–62

Dow D, Beyth M, Hailu T (1971) Palaeozoic glacial rocks recently discovered in Northern Ethiopia. Geol Mag 108:53–60

Echogdali FZ, Boutaleb S, Jauregui J, Elmouden A (2018) Cartography of flooding hazard in semi-arid climate: the case of Tata Valley (South-East of Morocco). J Geogr Nat Disasters 8:1–11

El Bastawesy M, Khalaf FI, Arafat SM (2008) The use of remote sensing and GIS for the estimation of water loss from Tushka lakes, southwestern desert, Egypt. J Afr Earth Sci 52(3):73–80

Elmahdy S, Marghany M, Mohamed M (2016) Application of a weighted spatial probability model in GIS to analyse landslides in Penang Island, Malaysia. Geomatics Nat Hazards Risk 7:345–359

EMS-I, EMS, Inc. (2004) Watershed modeling system, version 7.1. URL: https://www.ems-i.com/WMS/WMS_Overview/wms_overview.html

Fenicia F, Kavetski D, Savenije HHG, Clark MP, Schoups G, Pfister L et al (2013) Catchment properties, function, and conceptual model representation: is there a correspondence? Hydrol Process 28:2451–2467

Hall J, Grey D, Garrick D, Fung F, Brown C, Dadson S et al (2014) Coping with the curse of freshwater variability. Science 346:429–430

Hawkins RH (1993) Asymptotic determination of runoff curve numbers from data. J Irrig Drain Eng ASCE 119(2):334–345

Herbert S (2014) Assessing seismic risk in Ethiopia. GSDRC Helpdesk Research Report 1087, GSDRC, University of Birmingham, Birmingham, UK

IPoE (2013) International panel of experts (IPoE) on Grand Ethiopian Renaissance Dam Project (GERDP). Final Report, Addis Ababa, Ethiopia, 31st May 2013. Available: https://www.internationalrivers.org

Johnston RM, McCartney M (2010) Inventory of water storage types in the Blue Nile and Volta river basins, vol 140. IWMI

Kazmin V (1972) Geological map of Ethiopia, 1:2,000,000 scale. Hunting Surveys, Addis Ababa

Kinde S (2002) Earthquake risks in Addis Ababa and other major Ethiopian cities—will the country be caught off-guarded? MediaETHIOPIA

Matkan A, Shakiba A, Pourali H, Azari H (2009) Flood early warning with integration of hydrologic and hydraulic models, RS and GIS (case study: Madarsoo Basin, Iran). World Appl Sci J 6:1698–1704

McConnell RB (1972) Geological development of the rift system of eastern Africa. Bull Geol Soc Am Boulder 83:2549–2572

Mohamed M (2016) Mapping of tecto-lineaments and investigate their association with earthquakes in Egypt: a hybrid approach using remote sensing data. Geomatics Nat Hazards Risk 7:600–619

Mohamed M, Elmahdy S (2017) Remote sensing of the Grand Ethiopian Renaissance Dam: a hazard and environmental impacts assessment. Geomatics Nat Hazards Risk 8(2):1225–1240

Nelson EJ (2004) Watershed modeling system, version 7.1, tutorial. Department of Civil and Environmental Engineering, Environmental Modeling Research Laboratory, Brigham Young University, Provo, Utah, USA

Nelson EJ, Wallace RM, Smemoe C, Salah A (2005) Automated hydrologic and hydraulic modeling using watershed modeling system (WMS). In: International conference of UNESCO FLANDERS FIT FRIEND-NILE project, towards a better cooperation, 12–15 Nov 2005, Sharm El-Sheikh, Egypt (in press)

Omran E-SE, Negm A (2019) Environmental impacts of the GERD project on Egypt's Aswan high dam lake and mitigation and adaptation options. In: Negm AM, Abdel-Fattah S (eds) Grand Ethiopian Renaissance Dam versus Aswan high dam: a view from Egypt. Springer International Publishing, Cham, pp 175–196

Qafari G (2004) Flood hazard zoning using GIS (case study: Babolrood River, Mazandaran Province, Iran). M.Sc thesis of watershed management, 126 p, Faculty of Natural Resources of University of Mazandaran

Rientjes T, Haile A, Kebede E, Mannaerts C, Habib E, Steenhuis T (2011) Changes in land cover, rainfall and stream flow in Upper Gilgel Abbay catchment, Blue Nile basin—Ethiopia. Hydrol Earth Syst Sci 15:1979–1989

Seybert TA (2006) Storm water management for land development. Wiley, Hodoken, NJ

Sheffield J, Wood EF, Roderick ML (2012) Little change in global drought over the past 60 years. Nature 491:435–438

Taye MT, Willems P (2011) Influence of climate variability on representative QDF predictions of the upper Blue Nile basin. J Hydrol Hydromech 411:355–365

Teferi E, Uhlenbrook S, Bewket W, Wenninger J, Simane B (2010) The use of remote sensing to quantify wetland loss in the Choke Mountain range, Upper Blue Nile basin, Ethiopia. Hydrol Earth Syst Sci 14(12)

Tekleab S, Uhlenbrook S, Mohamed Y, Savenije H, Temesgen M, Wenninger J (2011) Water balance modeling of Upper Blue Nile catchments using a top-down approach. Hydrol Earth Syst Sci 15(7)

Uhlenbrook S, Mohamed Y, Gragne A (2010) Analyzing catchment behavior through catchment modeling in the Gilgel Abay, Upper Blue Nile River basin, Ethiopia. Hydrol Earth Syst Sci 14:2153–2165

UNISDR (2013) Gar global risk assessment: data, sources and usage. UNISDR

USEPA (1999) HSPFParm: an interactive database of HSPF model parameters, version 1.0. Office of Water, EPA-823-R-99004

Veilleux J (2013) The human security dimensions of dam development: the Grand Ethiopian Renaissance Dam. Glob Dialogue (Online) 15:42

Wang H, Guan H, Gutiérrez-Jurado HA, Simmons CT (2014) Examination of water budget using satellite products over Australia. J Hydrol 511:546–554

Wolela A (2012) Diagenetic evolution and reservoir potential of the Barremian-Cenomanian Debre Libanose Sandstone, Blue Nile (Abay) Basin, Ethiopia. Cretac Res 36(4):83–95

Chapter 18
Quantitative Analysis of Shoreline Dynamics Along the Mediterranean Coastal Strip of Egypt. Case Study: Marina El-Alamein Resort

Wiame W. M. Emam and Kareem M. Soliman

Abstract This chapter aimed at monitoring, analyzing, and quantifying shoreline dynamics along the Egyptian Mediterranean coastal strip from Sidi Abd El-Rahman to El-Arish over 3 decades (1987–2017) using remote sensing (RS) and geographical information system (GIS) technology, with special reference to Marina El-Alamein resort's shoreline. Besides developing predictive shoreline scenarios along Marina El-Alamein shoreline (for years 2023, 2057, and 2107). On-screen digitization technique was applied upon Landsat images (1987–2017) to detect areas of great erosion and accretion using ArcGIS 10.1. Incorporating DSAS 4.3 within ArcGIS facilitated quantification of shoreline changes and calculating erosion/accretion rates. Results generated shoreline change maps for the entire strip for 30 years. Quantitatively, the entire coastal strip from Sidi Abd El-Rahman to El-Arish had dynamically changed over time. Constructing hard structures along Marina El-Alamein shoreline during the last 3 decades interrupted greatly the shoreline stability where seaward shifting was predominant. The overall trend of shoreline change rate indicated very high accretion with an average of 2.6 $\pm$ 4.64 m/year (EPR) and 2.54 $\pm$ 4.22 m/year (LRR). Shoreline evolution model predicted that the western inlet might be closed in the future. Consequently, careful planning coupled with long-term considerations is required when developing coastal engineering structures to ensure the sustainability of available natural resources.

Keywords Egyptian Mediterranean shoreline · Marina El-Alamein coast · Landsat, ArcGIS 10.1, DSAS 4.3 · Accretion/erosion · End-point rate (EPR) · Linear regression rate (LRR) · Predictive shoreline scenarios

---

W. W. M. Emam (✉) · K. M. Soliman
Zoology Department, Faculty of Science, Ain Shams University, Cairo, Egypt
e-mail: dr_wiame2006@yahoo.com

K. M. Soliman
e-mail: dr_kariemsoliman@yahoo.com

S. F. Elbeih et al. (eds.), *Environmental Remote Sensing in Egypt*,
Springer Geophysics, https://doi.org/10.1007/978-3-030-39593-3_18

18.1 Introduction

US Army Corps of Engineers (2008) defined the term "shoreline" as the physical interface between land and sea. The shoreline is a highly dynamic feature, and the magnitude of change fluctuates from point to another. Both anthropogenic (as, coastal development) and natural pressures (as, coastal erosion and sediment transport) disturb greatly the sensitive nature of coastal zones, particularly shoreline morphodynamics, which in turn influence the three pillars of sustainable development (socio-economic environment) in that area (Fig. 18.1). Shores are subjected frequently to erosion and deposition. Coastal erosion (landward retreat) is an incrementing threat facing more than 70% of the worlds' shores (Appeaning-Addo et al. 2011). In terms of time, it can be either permanent long-term or short-term erosion (Sheeja and Ajay-Gokul 2016).

Remote sensing (RS) technology, as a source of information, offered the availability of coherent time series of satellite image required for shoreline monitoring, in which the same area can be observed repeatedly at scales impractical to cover with traditional field techniques (Ahmed and Barale 2014). In coastal waters, shoreline development was successfully analyzed from Landsat images (Yu et al. 2011; Ahmad and Lakhan 2012; Huang et al. 2012; Mukhopadhyay 2012; Emam 2016 and Fu et al. 2017).

The availability of huge remote sensing image repositories demands appropriate resources to explore this data (Manjula et al. 2010). Geographic information system (GIS), as a tool, offers a suitable platform that creates, integrates, stores, visualizes, and analyzes geographically referenced data to solve problems, make decisions, and manage ecosystems properly. Adding Digital Shoreline Analysis

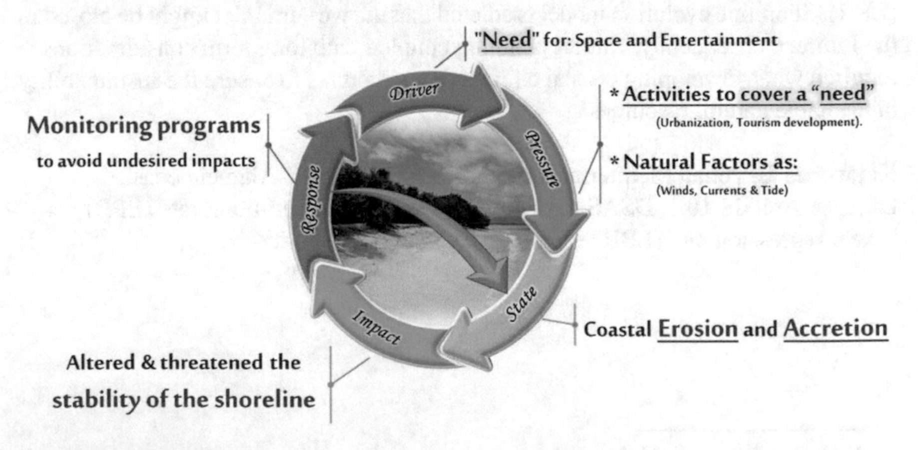

Fig. 18.1 DPSIR framework with feedback loop between responses. In coastal ecosystems, both anthropogenic and natural pressures can ultimately change the ecosystem's state threating the stability of the shoreline. Consequently, continuous monitoring programs are required to avoid undesired impacts and to ensure the sustainability of the destination

System (DSAS) within ArcGIS facilitates calculating rate-of-change statistics (erosion/accretion rates) from multiple historic shoreline positions through vector data (Thieler et al. 2009). Integrating satellite images and GIS with spatial models as DSAS proved to be of great importance in studying shoreline dynamics in coastal areas of USA, Turkey, Italy, Cameroon, Ghana, India, Bangladesh, and Vietnam (Moussaid et al. 2015; Hegde and Akshaya 2015).

In Egypt, when human intervention has begun to establish villages and resorts on the coast, shoreline morphodynamics have changed greatly threatening their investments and the sustainability state of the destination. Marina, also known as Marina El Alamein, is the first Egyptian tourist resort in the northern coast area overlooking the Mediterranean Sea, 94 km west of Alexandria, owned by the New Urban Communities Authority. It covers an area of about 15.9 km^2 with eight hotels, most notably Porto Marina Hotel and Ocean Blue Resort. Besides, it has the first international yacht berth in the eastern part of North Africa.

Despite the great importance of Marina El-Alamein resort, the area suffered from a lack of studies in auditing, analyzing and quantifying shoreline change rates using high precision techniques. Consequently, this chapter aimed in generating a reliable, up-to-date digital database of shoreline dynamics along the Egyptian Mediterranean coastal strip from Sidi Abd El-Rahman to El-Arish with special reference to Marina El-Alamein resort's shoreline. In conserving coastal ecosystems from possible future losses and developing sustainable managerial policies, periodic monitoring of regional shoreline dynamics besides a precise quantification and modelling of chronological shoreline shifting rates are a prerequisite. This study is the first in analyzing and quantifying shoreline changes of El-Alamein coast for short-term (years) and long-term trends (decades), besides developing predictive shoreline scenarios using Landsat satellite images, ArcGIS 10.1 software and digital shoreline analysis system (DSAS) to help in mitigating the impacts of coastal tourism and ensure the sustainability state of the destination.

18.2 Study Area

The study area, viz. the Mediterranean northern coast of Egypt (Fig. 18.2) is uniquely referenced on latitude 30° 58′ 12.953″–31° 9′ 5.404″ N and longitude 28° 45′ 57.793″–33° 49′ 8.144″ E. It extends for about 550 km along the Mediterranean Sea from Sidi Abd El-Rahman (west) to El-Arish (east) at the Egypt-Gaza border.

The coastal strip embraces deltaic sediments, sand dunes, lakes and lagoons, salt marshes, mud flats, and rocky beaches (Hereher 2015). Geomorphologically this coastline is split into three different regions as follows:

- The western North coast (100 km long from Sidi Abd El-Rahman to Alexandria), which is a rocky beach consisting of calcareous rocks with vertical cliffs.

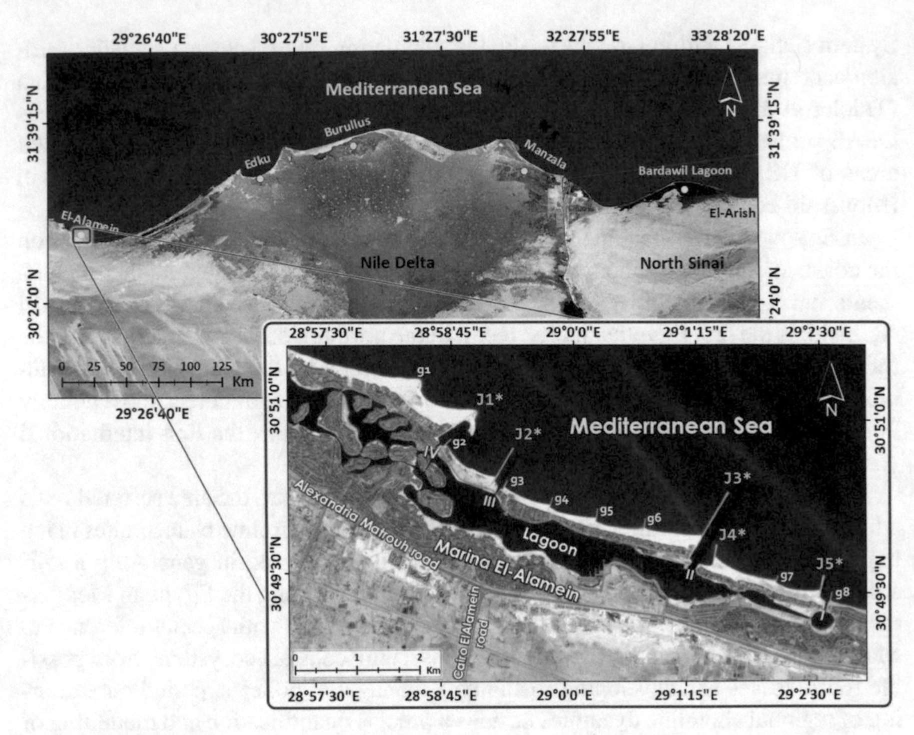

Fig. 18.2 Location of study area, (**J**): jetties; (**g**): groins; (**I, II, III, IV**): inlets

- The middle region represents the Nile Delta (250 km long from Alexandria to Port Said), hosts the majority of the country's population. It consists of flat deltaic sandy beaches, coastal flats, sabkhas, sand sheets, and three coastal lagoons (Edku, Burullus, and Manzala).
- The eastern North Sinai coastal plain (200 km long), which is a coastal sand dune field interrupted by the hypersaline shallow El-Bardawil Lagoon.

Marina El-Alamein (Fig. 18.2) is a touristic village located near the town of El-Alamein in Matrouh Governorate along the northwestern Egyptian Mediterranean coast, extending for about 20.37 km in length and about 300 km away from Cairo. The village is uniquely referenced on latitude 30° 51′ 22.31″–30° 49′ 17.91″ N and longitude 28° 57′ 11″–29° 3′ 45.69″ E. It embraces El-Alamein Lagoon (an artificial lagoon) which is separated from the Mediterranean Sea by an elongated sand barrier bisected with four inlets (I, II, III, and IV) to ensure inflow of seawater.

18.3 Data and Method Analysis

18.3.1 Landsat Dataset

The study used two sets of Landsat satellite images (TM 1987 and OLI 2017), each composed of five images, to examine long-term spatiotemporal shoreline dynamics signifying the Egyptian Mediterranean coastal strip from Sidi Abd El-Rahman to El-Arish (Table 18.1). Further attention was given to investigate in details the shoreline dynamics of Marina El-Alamein resort over 3 decades from 1987 to 2017. Consequently, nine more Landsat images (TM 1999, ETM$^+$ 2001, ETM$^+$ 2003, ETM$^+$ 2005, ETM$^+$ 2007, ETM$^+$ 2009, ETM$^+$ 2011, OLI 2013, and OLI 2015) signifying the resort were downloaded (Table 18.1).

All images were chosen to be in spring season at near acquisition dates to eliminate the effects of seasonal differences. The study downloaded Landsat images as a zipped file at no cost from the United States Geological Survey (USGS). They were chosen to be in good quality (IQ = 9), cloud-free (less than 10%) and level-1 precision terrain corrected (L1TP) to the Universal Transverse Mercator (UTM) map projection system; zone 36 and 35 north on the World Geodetic Datum of 1984 (WGS 84) (Table

Table 18.1 Landsat satellite data used in the present study as obtained from USGS

Satellite	Sensor	Path/row	Date	UTM zone
Landsat 5	TM	175/38	18–June–1987	36
		176/38	22–April–1987	36
		177/38	16–June–1987	36
		177/39	15–May–1987	36
		178/39	07–June–1987	35
			05–April–1999	
Landsat 7	ETM$^+$	178/39	20–May–2001	35
			26–May–2003	
			31–May–2005	
			03–April–2007	
			10–May–2009	
			14–April–2011	
Landsat 8	OLI_TIRS	178/39	27–April–2013	35
			03–May–2015	
		175/39	04–June–2017	36
		176/38	26–May–2017	36
		177/38	02–June–2017	36
		177/39	18–June–2017	36
		178/39	24–May–2017	35

18.1). USGS rectified all images with total root mean square error (RMSE) less than 0.44 m. A false color composite was used.

18.3.2 Shoreline Digitization

Careful on-screen digitization for the entire coastal strip under study was applied upon two Landsat images (1987 and 2017), and upon nine more Landsat images (1999, 2001, 2003, 2005, 2007, 2009, 2011, 2013, and 2015) for Marina El-Alamein resort (Fig. 18.3) using ArcGIS 10.1 software and saved separately as shapefiles. The scale of shoreline depicted in this study ranged from 1:5,000 to 1:20,000. The digitized line features were overlaid to generate shoreline change map and highlight the dynamic movement of the shoreline in terms of advancing by sedimentation or retreating by erosion.

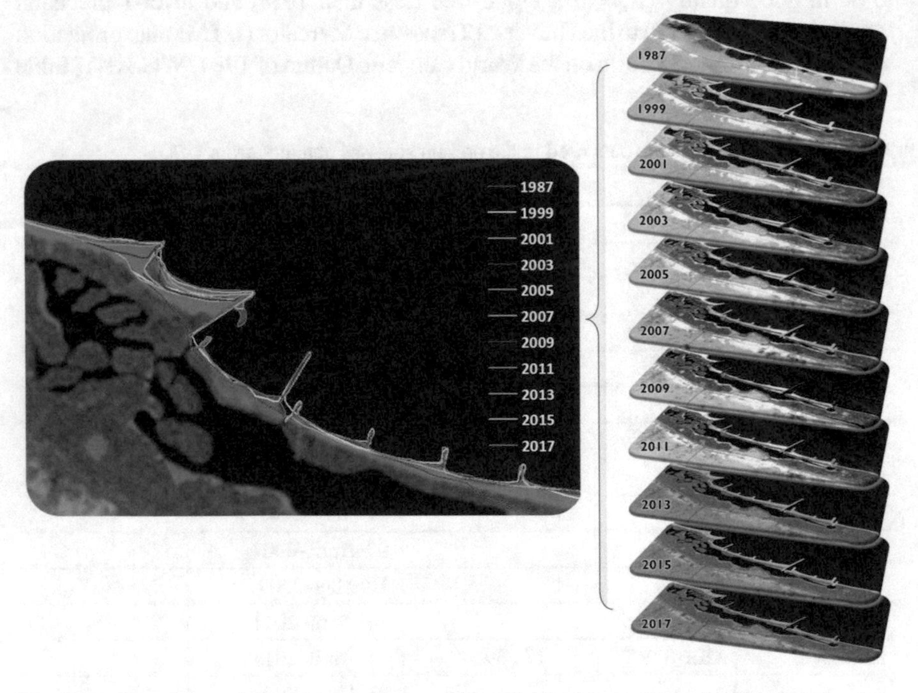

Fig. 18.3 Superimposed historic shoreline positions digitized from eleven Landsat images along Marina El-Alamein shoreline

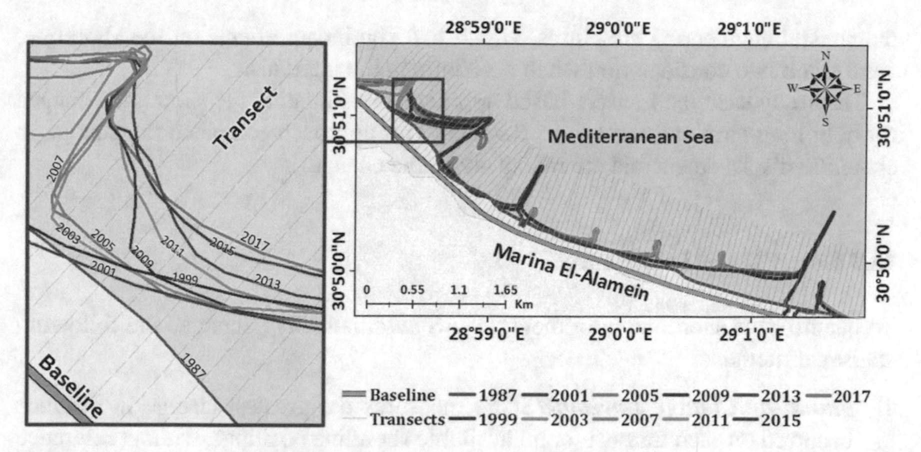

Fig. 18.4 DSAS casted transects orthogonal to the onshore baseline at 70 m intervals, intersecting the eleven digitized shoreline vectors along Marina El-Alamein coast

18.3.3 Digital Shoreline Analysis System (DSAS)

Digital Shoreline Analysis System (DSAS) version 4.3 was utilized to perform the shoreline analysis. DSAS is a freely available toolbox extension to ESRI ArcGIS v.10 that calculates rate-of-change statistics from multiple historic shoreline positions through vector data (Thieler et al. 2009). The process of analyzing data using DSAS involves three steps: first, constructing baseline parallel to the shoreline from which transects will originate perpendicularly, second, setting transect parameters that splits shoreline into intervals and finally setting and calculating the rate of change statistics at each transect. Prior to shoreline analysis, the study created geodatabase including the digitized shoreline vectors besides the hypothetical onshore baseline buffered 150 m from the shoreline position for the year 1987. DSAS casted 7911 transects orthogonal to the onshore baseline at 70 m intervals to intersect all the digitized shoreline positions (Fig. 18.4). Transects extended from Sidi Abd El-Rahman (west) to El-Arish (east) at the Egypt-Gaza border.

18.3.4 Shoreline Change Analysis

Long-term shoreline changes were calculated at each transect for 30 years (1987–2017) along the entire coastal strip. Moreover, ten different time frames of short-term changes were analyzed separately (1987–1999, 1999–2001, 2001–2003, 2003–2005, 2005–2007, 2007–2009, 2009–2011, 2011–2013, 2013–2015, and 2015–2017) along Marina El-Alamein coast in order to determine whether the erosion rates have accelerated or decelerated between the individual spans. The construction dates of the coastal engineering structures, which had significant effects on the shoreline, were taken into consideration when determining the intervals.

The study used the transect-based approach in estimating the shoreline changes for both long-term and short-term changes. This approach computed the change in shoreline displacement and the rate of shoreline change.

18.3.4.1 Change in Distance

To quantify the shoreline movement, DSAS automatically generated the following statistical methods.

i. ***Shoreline Change Envelope(SCE)*** measures the greatest change in distance occurred on each transect for all available shoreline positions, without reference to their specific dates. It provides the envelope of variability.
ii. ***Net Shoreline Movement (NSM)*** to obtain the distance between the oldest and the youngest shorelines.

18.3.4.2 Change in the Rate Calculation

DSAS automatically generated the following statistical methods.

i. ***End Point Rate (EPR)*** converts the net shoreline movement into an annual rate of shoreline change by dividing the distance of shoreline movement by the time taken between the oldest and the youngest shoreline positions.
ii. ***Linear Regression Rate (LRR)*** determines a rate-of-change statistic by fitting the most suitable regression line crossing all shorelines at each transect (Eq. 18.1), and it does not take erroneous values in calculating the shoreline change rate (Kankara et al. 2014).

$$\mathbf{Y} = \mathbf{m} * \mathbf{X} + \mathbf{b} \tag{18.1}$$

where; **m**: is the slope of the line = LRR (m/y), **b**: is a constant value, **X**: is the independent variable = time (in years), and **Y**: is the dependent variable = shoreline shift (m) (Thieler et al. 2009).

Moreover, the values of correlation coefficient of determination (LR^2) were computed for each transect. LR^2 values range from 0.0 to 1.0 and measure the strength of the estimated regression rate, LRR. According to the degree of their strength, the significance of shoreline change rate is identified.

In case of erosion, a negative (−) value is assigned, whereas positive (+) values indicate accretion. According to Natesan et al. (2015), the study categorized the current results of erosion/accretion rates for Marina El-Alamein coast into seven classes (Table 18.2).

Table 18.2 Shoreline classification based on EPR & LRR

Shoreline classification	Rate of change (m/year)
Very High Accretion (VHA)	$> +2$
High Accretion (HA)	$> +1 \rightarrow \leq +2$
Medium Accretion (MA)	$>0 \rightarrow \leq +1$
No change	0
Medium Erosion (ME)	$< 0 \rightarrow \geq -1$
High Erosion (HE)	$<-1 \rightarrow \geq -2$
Very High Erosion (VHE)	<-2

EPR, End-Point Rate; **LRR**, Linear Regression Rate

18.3.5 Modelling Shoreline Rate of Change

Shoreline-change rate analysis was a prerequisite in predicting an accurate future shoreline position. Consequently, Linear Regression (LR) models were implemented to predict future shoreline. The position of the future shoreline (Y) for a given data is estimated using the rate of shoreline movement (LRR), the time interval between observed and predicted shoreline (or date of the year = X), and model intercept that can be expressed as in Eqs. (18.2) and (18.3):

$$\textbf{Future shoreline position} = (\textbf{LRR} * \textbf{Year}) + \textbf{Intercept} \tag{18.2}$$

LRR intercept can be calculated as:

$$\textbf{Intercept} = \textbf{Y} - (\textbf{LRR} * \textbf{X}) \tag{18.3}$$

where; "**Y**": denotes the shoreline shift (m), and "**X**": represent time (in years).

Prior to the prediction of future shoreline, the positional shift in the model (calculated shoreline of 2017) was validated with respect to the actual image (extracted shoreline of 2017). The validation (location error in model estimated shoreline) was carried out in terms of RMSE. After model calibration, the model was applied to predict Marina Al-Alamein shoreline for 2023, 2057, and 2107.

18.4 Spatiotemporal Shoreline Dynamics

Mapping and analyzing quantitatively the changes in the position of shorelines at different timescales aid in detecting areas of great erosion and deposition (Al-Bakri 1996), modelling shoreline morphodynamics (Maiti and Bhattacharya 2009), and spatial planning and management.

18.4.1 The Egyptian Mediterranean Coastal Strip from Sidi Abd El-Rahman to El-Arish

The study used two sets of Landsat satellite images (1987 and 2017) to extract the shoreline, analyze, and quantify the spatiotemporal dynamics signifying the Egyptian Mediterranean coastal strip from Sidi Abd El-Rahman to El-Arish. Moreover, shoreline rate-of-change map along the entire strip was generated.

Shoreline Change Envelope (SCE)

SCE measures the distance between the shoreline farthest from and closest to the baseline at each transect. Table (18.3) quantified the amount of change in shoreline position (SCE) along the Egyptian Mediterranean coastal strip from Sidi Abd El-Rahman to El-Arish during the period 1987–2017 at each transect. SCE results indicated that the greatest displacement in shoreline position was about 1717.66 m with an average of 102.17 ± 175.78 m in 30 years. About 14% of transects displayed stability in shoreline positions, whereas, 76% of transects experienced shoreline displacement up to 250 m.

End Point Rate (EPR)

Unlike SCE, EPR results designate the direction of change per year rather than the total amount of shoreline change in position during the period from 1987 to 2017. EPR considers only two positions; the youngest (2017) and the oldest (1987). In terms of stability and erosion/accretion pattern, EPR revealed three classes of coasts

Table 18.3 Descriptive summary statistics of the change in shoreline position (SCE) for the Egyptian Mediterranean coastal strip from Sidi Abd El-Rahman to El-Arish during the period 1987–2017

SCE (m)					
	Transects		Min.	Max.	Av.±SD
Interval	No.	%			
No change	1079	13.6	–	–	–
0 < 250	5977	75.5	0.02	249.12	66.11 ± 59.44
250 < 500	664	8.39	250.2	497.03	347.48 ± 70.16
500 < 750	67	0.84	505.2	742.6	599 ± 74.26
750 < 1000	44	0.55	759.3	988.2	874.1 ± 70.2
1000 < 1250	40	0.5	1002	1246	1101 ± 91.08
1250 < 1500	21	0.26	1273	1486	1367 ± 68.98
1500 < 1750	19	0.24	1511	1717.66	1620 ± 70.62
Total	7911	100	0	1717.66	102.17 ± 175.78

Fig. 18.5 Shoreline rate of change map generated along the Egyptian Mediterranean coastal strip from Sidi Abd El-Rahman to El-Arish during the period 1987–2017

(erosional, accretional, and stable) along the Egyptian Mediterranean coastal strip from Sidi Abd El-Rahman to El-Arish. Erosional coasts constituted about 51% of transects, accretional coasts about 35% transects, and 14% of transects remained with no appreciable change in shoreline position (Fig. 18.5). The overall trend of shoreline change rate indicated medium erosion with an average of -0.484 ± 0.28 m/year (EPR).

According to Cataudella et al. (2015), the Mediterranean accommodates about 400 coastal lagoons covering a surface area more than 6400 km^2 in fifteen countries. Counting on number of coastal lagoons, Italy comes in the first rank with 198 lagoons of about 1675.75 km^2. On the other hand, Egypt hosts only five Mediterranean coastal lagoons with a total surface area more than 1990 km^2. About 40% of the coastal strip under study hosts those lagoons, namely; El-Alamein (artificial lagoon on the western coast), Edku, Burullus, Manzala (deltaic coast), and Bardawil Lagoon (Sinai coast) (Fig. 18.2).

Edku Lagoon

Referenced on latitude 31° 16′ 31.316″–31° 18′ 41.669″ N and longitude 30° 6′ 58.066″ 30° 16′ 10.284″ E. The lagoon extends for about 15 km in length (Fig. 18.6). EPR results indicated that along Edku shoreline, about 78% of transects displayed accretion, 16% of transects remained unchanged, and 6% transects experienced evidence of erosion (Fig. 18.6). The overall trend of shoreline change rate indicated very high accretion with an average of 2.919 ± 0.55 m/year (EPR) during the period 1987–2017.

Burullus Lagoon

Referenced on latitude 31° 27′ 50.147″–31° 36′ 13.174″ N and longitude 30° 33′ 19.018″–31° 6′ 1.513″ E. The lagoon extends for about 52.2 km in length (Fig. 18.7). EPR results indicated that along Burullus shoreline, about 69% of tran-

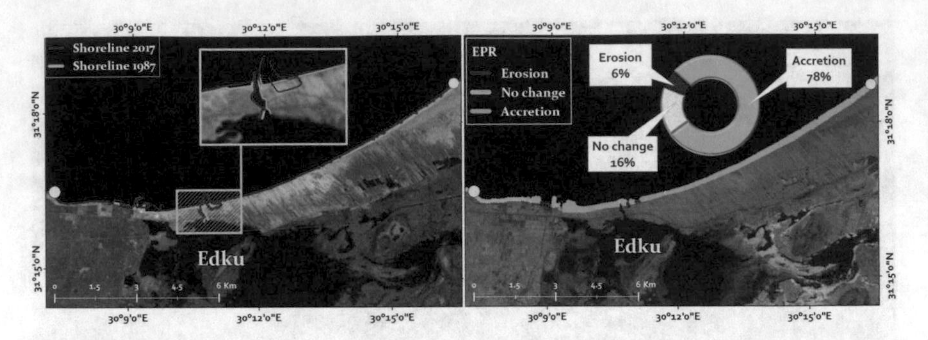

Fig. 18.6 Shoreline rate of change map generated along Edku Lagoon during the period 1987–2017

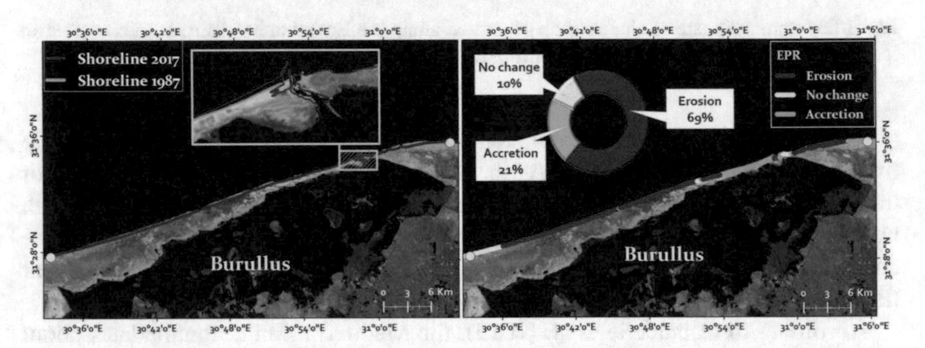

Fig. 18.7 Shoreline rate of change map generated along Burullus Lagoon during the period 1987–2017

sects displayed erosion, 21% of transects experienced evidence of accretion, and 10% transects remained unchanged (Fig. 18.7). The overall trend of shoreline change rate indicated very high erosion with an average of -2.95 ± 0.69 m/year (EPR) during the period 1987–2017.

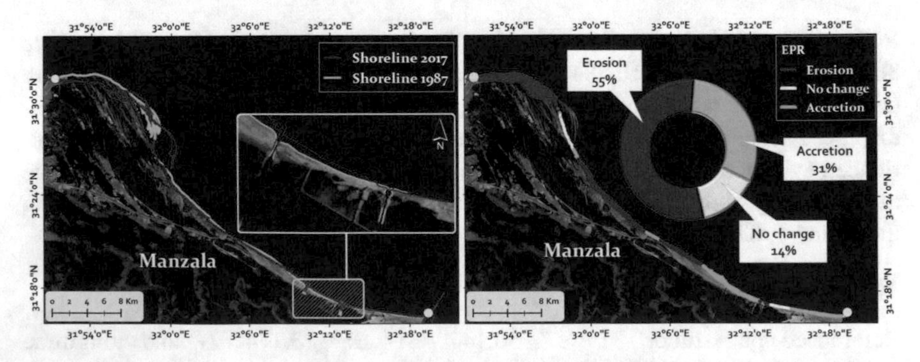

Fig. 18.8 Shoreline rate of change map generated along Manzala Lagoon during the period 1987–2017

Manzala Lagoon

Referenced on latitude 31° 31′ 23.868″–31° 16′ 35.157″ N and longitude 31° 51′ 15.853″–32° 19′ 15.351″ E. Manzala lagoon extends for about 44.1 km in length (Fig. 18.8). EPR results indicated that along Manzala shoreline, about 55% of transects displayed erosion, 14% of transects remained unchanged, and 31% transects experienced evidence of deposition (Fig. 18.8). The overall trend of shoreline change rate indicated very high erosion with an average of −27.4 ± 13.8 m/year (EPR) during the period 1987–2017.

Bardawil Lagoon

Referenced on latitude 31° 3′ 37.121″–31° 6′ 54.395″ N and longitude 32° 39′ 59.282″–33° 35′ 20.54″ E. The lagoon extends for about 88.5 m in length (Fig. 18.9). EPR results indicated that about 71% of transects displayed erosion, 7% of transects remained unchanged, and 22% transects experienced evidence of deposition along Bardawil shoreline (Fig. 18.9). The overall trend of shoreline change

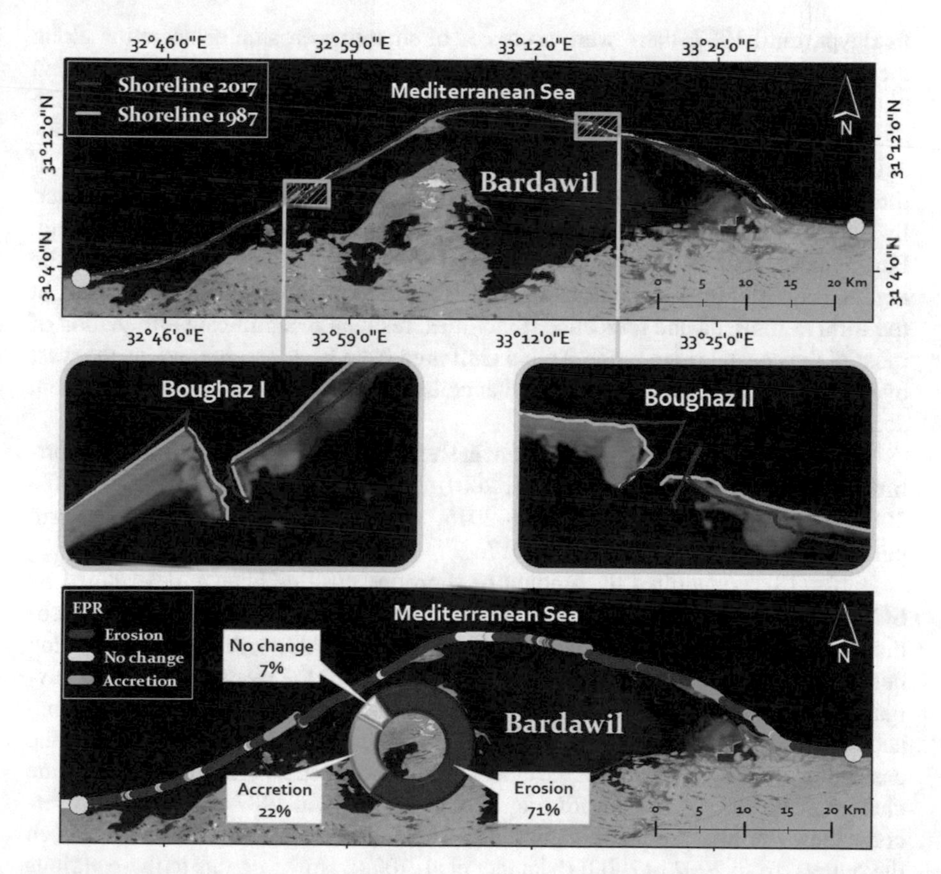

Fig. 18.9 Shoreline rate of change map generated along Bardawil Lagoon during the period 1987–2017

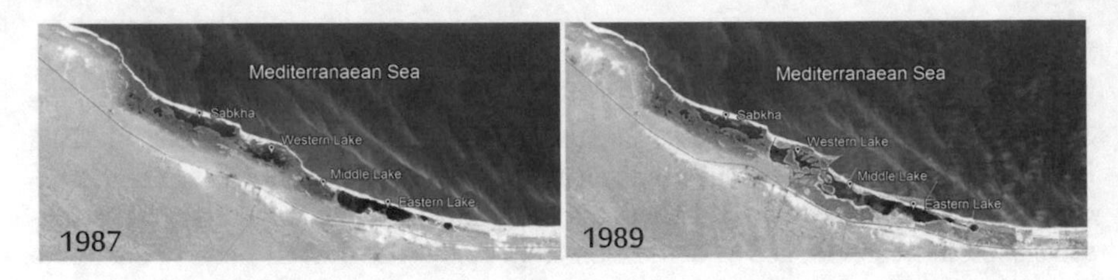

Fig. 18.10 Google Earth Pro photos for years 1987 and 1989 along Marina El-Alamein coast

rate indicated high erosion with an average of -1.46 ± 0.26 m/year (EPR) during the period 1987–2017.

18.4.2 Marina El-Alamein resort's shoreline

In Egypt, until 1987, there were no traces of shoreline erosion or accretion along the northwestern coast of Alexandria and even Matrouh. Only in late 1980s, when human intervention has begun to establish villages and resorts on the coast, shoreline morphodynamics had changed greatly threatening their investments and the sustainability state of the destination. In 1987, Marina El-Alamein lagoon was formed of three closed hypersaline sabkhas separated from the Mediterranean Sea by a barrier. In 1989, these sabkhas were connected into one lagoon for recreation and yachting. Furthermore, four inlets have been dredged to ensure inflow of seawater into the lagoon (Fig. 18.10). El-Asmar et al. (2012) found that anthropogenic activities, in the form of touristic and recreational facilities, resulted in significant elimination of most of the coastal ridge in the Arab's Gulf area from El-Alamein towards the west of Alexandria. Moreover, they added that constructing Marina Resort in El-Alamein accelerated shoreline erosion.

The study analyzed Marina Al-Alamein shoreline dynamics in ten different short-time frames (1987–1999, 1999–2001, 2001–2003, 2003–2005, 2005–2007, 2007–2009, 2009–2011, 2011–2013, 2013–2015, and 2015–2017), besides, long-term period for the period from 1987 to 2017.

Table (18.4) quantified the amount of shoreline position change determined by SCE. Iskander et al. 2008 stated that constructing five offshore rubble jetties perpendicular to the shoreline during the period from 1989 to 1993, with lengths fluctuating between 350 and 1250 m (Fanos 2004), was to stabilize the inlets and improve navigation. These jetties interrupted the prevailing eastward-longshore current causing accretion on the updrift side (west of the inlets) and erosion on the downdrift side (the east side). That is why results of the current study indicated that the greatest shoreline change SCE (463.35 m; seaward migration) was between 1987 and 1999. To overcome these resultant problems, eight short groynes have been built in the area between the outlets (from 2002 to 2003) (Iskander et al. 2008). However, due to these various coastal engineering constructions, a remarkable shoreline retreat between 1984 and

Table 18.4 Amount of shoreline position change according to the SCE method for Marina El-Alamein coast

SCE (m)								
Period	Min.	Max.	Av. ±SD	Number of Transects				
				<50	50 < 100	100 < 200	200 < 300	300 < 650
1987–1999	0	463.35	65.55 ± 94.28	109	34	10	15	6
1999–2001	0	73.68	8.98 ± 16.43	168	5	–	–	–
2001–2003	0	106.97	14.68 ± 20.18	156	11	1	–	–
2003–2005	0	173.65	8.93 ± 22.77	167	3	4	–	–
2005–2007	0	75.65	3.63 ± 13.53	167	7	–	–	–
2007–2009	0	174.04	2.96 ± 19.33	171	–	3	–	–
2009–2011	0	67.57	2.50 ± 9.94	172	2	–	–	–
2011–2013	0	120.14	8.92 ± 14.05	169	2	1	–	–
2013–2015	0	79.63	6.21 ± 14.63	169	5	–	–	–
2015–2017	0	69.60	4.98 ± 8.89	170	1	–	–	–
1987–2017	0	626.76	99.47 ± 126.08	92	38	11	16	17

2007 occurred with ranges between –48 and –60 m (El-Asmar et al. 2012). Consequently, precise up-to-date quantitative scientific information on rates and spatial distribution of Marina El-Alamein shoreline modifications was required to guide coastal decision-makers in understanding the shoreline dynamics of the study area.

Table (18.5) analyzed the rate of shoreline change (EPR) along Marina El-Alamein coast in ten periods. The overall trend of EPR in Marina El-Alamein coast indicated that during the intervals 1987–1999, 2001–2003, and 2013–2015, the coast suffered very high accretion (VHA) with average EPR values of 3.96, 6.31, and 2.49 m/year, respectively. On the other hand, the greatest amount of erosion occurred only between 2011–2013 followed by 1999–2001, with an average rate of −1.11 m/year and −0.82 m/year, respectively.

The study calculated the rate of change statistics for the period from 1987 to 2017, along Marina El-Alamein coast, using both EPR and LRR method to compare the results (Table 18.6). EPR considers only two positions, the youngest and the oldest, whereas LRR considers all the 11 years shoreline position. About 72% of

Table 18.5 Analysis of EPR short-term changes in Marina El-Alamein coast

Period	Accretion (m/year)			Erosion (m/year)			Av. (m/year)	Overall EPR trend
	Min.	Max.	Av. ± SD	Min.	Max.	Av. ± SD		
1987–1999	0.03	39.18	9.50 ± 9.58	−10.4	−0.02	−2.65 ± 2.15	3.96	VHA
1999–2001	1.97	29.42	12.82 ± 7.7	−34.7	−1.02	−13.64 ± 8.6	−0.82	ME
2001–2003	0.25	53.05	14.6 ± 10.3	−15.6	−0.16	−4.80 ± 4.43	6.31	VHA
2003–2005	0.02	86.23	18.1 ± 22.6	−33.9	−0.04	−8.90 ± 8.06	0.75	MA
2005–2007	1.25	41.09	20.8 ± 13.6	−30.5	−0.05	−7.77 ± 15.2	1.62	HA
2007–2009	0.21	82.82	27.3 ± 32.2	ND	ND	ND	1.41	HA
2009–2011	0.91	35.03	14.6 ± 10.8	−5.40	−0.24	−2.82 ± 3.65	1.23	HA
2011–2013	0.57	31.00	10.06 ± 8.6	−59.1	−0.19	−6.22 ± 7.03	−1.11	HE
2013–2015	0.51	39.49	13.1 ± 9.09	−36.6	−0.02	−7.26 ± 13.3	2.49	VHA
2015–2017	0.26	14.38	5.33 ± 3.98	−33.8	−0.38	−5.73 ± 6.07	0.01	MA

Table 18.6 Analysis of EPR and LRR long-term shoreline changes (1987–2017) for Marina El-Alamein coast

Long-term shoreline changes (1987 – 2017)		EPR (m/yr)				LRR (m/yr)			
		Transect		Min – Max (m/yr)	Av. ±SD	Transect		Min – Max (m/yr)	Av. ±SD
		No.	%			No.	%		
Accretion (m/yr)	Very high	59	33.9	2.13 – 20.92	7.7 ± 4.59	61	35.1	2.2 – 19.55	7.14 ± 3.96
	High	27	15.5	1.11 – 2.04	1.54 ± 0.29	24	13.7	1.11 – 2.08	1.54 ± 0.32
	Medium	39	22.4	0.02 – 1	0.57 ± 0.28	40	22.9	0.02 – 1.04	0.47 ± 0.31
		125			4.18 ± 4.65	125			3.93 ± 4.20
No Change		8	4.59	0	0	9	5.17	0	0
Erosion (m/yr)	Medium	17	9.7	-0.99 – 0.05	-0.4 ± 0.3	16	9.19	-0.90 – -0.07	-0.34 ± 0.25
	High	14	8.04	-1.87 – -1.22	-1.5 ± 0.2	17	9.7	-1.94 – -1.04	-1.53 ± 0.28
	Very high	10	5.7	-3.10 – -2.06	-2.5 ± 0.36	7	4.02	-2.69 – -2.13	-2.39 ± 0.21
		41			-1.3 ± 0.9	40			-1.21 ± 0.81
Total		174	100	-3.10 – 20.92	2.6 ± 4.64	174	100	-2.69 – 19.5	2.54 ± 4.22
Overall trend		Very High Accretion							

transects displayed accretion (Fig. 18.11). The overall trend of shoreline change rate indicated very high accretion with an average of 2.6 ± 4.64 m/year (EPR) and 2.54 ± 4.22 m/year (LRR).

In terms of accuracy, the performance of LRR was superior to EPR. EPR and LRR show an average difference of 0.13 m/year. Moreover, the coefficient of determination (LR^2) used in estimating LRR change rate ranged from 0.5 to 0.98 in about 59.17% of transects. Consequently, LRR has been chosen in predicting future shoreline of Marina El-Alamein in the short term 2023 and long term 2057 and 2107 (Fig. 18.12). The predicted shorelines in 2023, 2057, and 2107 indicated that the overall shoreline trend is shifting towards the sea. Transects from 24 to 49 and from 96 to 117 will be

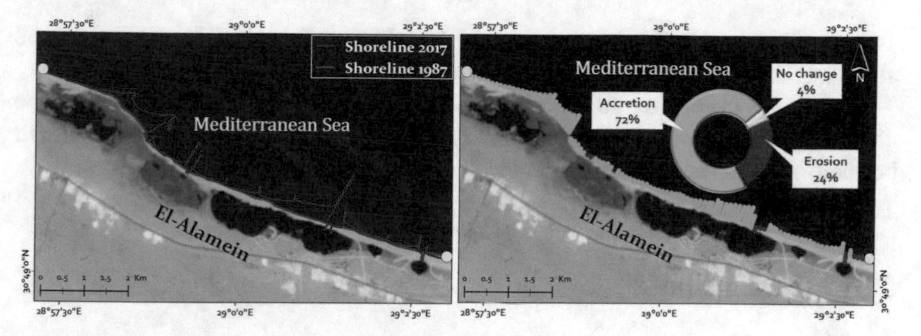

Fig. 18.11 Shoreline rate of change map generated along Marina El-Alamein coast during the period 1987–2017

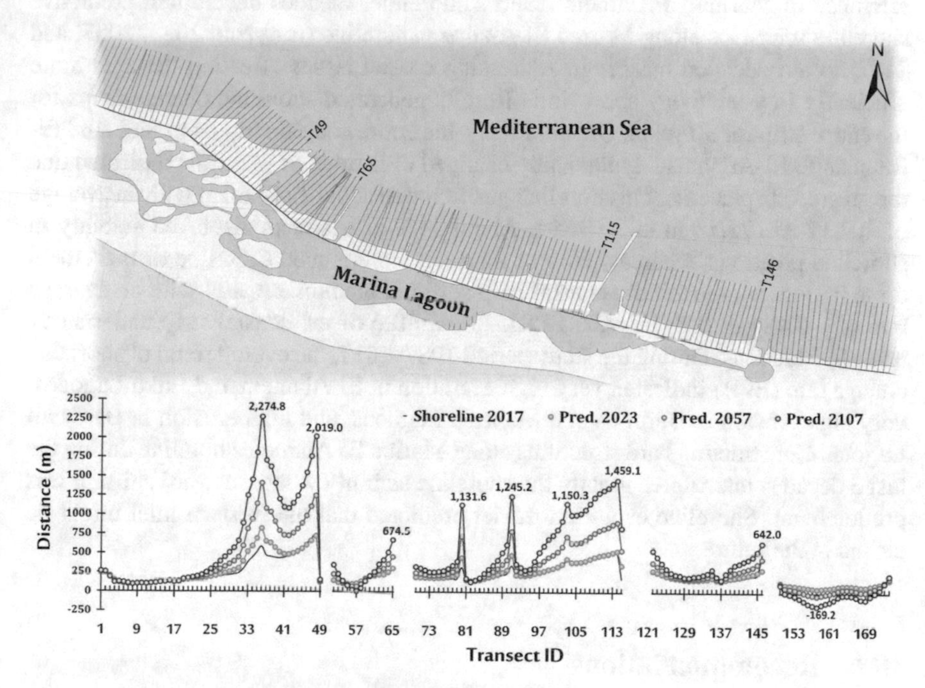

Fig. 18.12 Actual Shoreline for year 2017 and predicted shorelines for years 2023, 2057, and 2107 along Marina El-Alamein coast

exposed to high accretion, and the western inlet will be closed. Moreover, transects from 55 to 59 and from 151 to 171 will be eroded.

In six years during the period from 2017–2023, the average (±SD) accretion and erosion will be 34.81 ± 5.42 m and −14.35 ± 4.8 m, respectively. The greatest accretion will be about 820 m at transect 36 in the western side.

After 40 years, in 2057, the average (±SD) accretion and erosion will be 154.24 ± 15.7 m and −42.83 ± 4.48 m, respectively. The greatest accretion will be about

1409.3 m at transect 36 in the western side; whereas, the greatest erosion will be about −95.5 m at transect 158 at the eastern part of Marina El-Alamein.

During 2107 (i.e. after 90 years), it is predicted that the greatest accretion will be about 2274 m at transect 36 east to the first groin; whereas, the greatest erosion will be about −169 m at transect 158 at the eastern part of Marina El-Alamein coast.

18.5 Conclusion

The study provided a comprehensive historical shoreline analysis for the Egyptian Mediterranean coastal strip from Sidi Abd El-Rahman to El-Arish, with special reference to Marina El-Alamein resort's shoreline. Besides developing predictive shoreline scenarios along Marina El-Alamein shoreline (for years 2023, 2057, and 2107) to aid decision makers in addressing coastal issues affecting the area more efficiently in a relatively short time. Results generated shoreline change-maps for the entire strip for 30 years. Quantitatively, the entire coastal strip from Sidi Abd El-Rahman to El-Arish had dynamically changed over time. SCE results indicated that the greatest displacement in shoreline position was about 1717.66 m with an average of 102.17 ± 175.78 m in 30 years. About 14% of transects displayed stability in shoreline positions, whereas, 76% of transects experienced shoreline displacement up to 250 m. Moreover, the coastal strip suffered medium erosion with an average rate of −0.484 ± 0.28 m/year (EPR). About 40% of the coastal strip under study hosts five lagoons. During the study period 1987–2017, the overall trend of shoreline change rate (EPR) indicated very high accretion in El-Alamein and Edku Lagoons, very high erosion in Burullus and Manzala Lagoons, and high erosion in Bardawil Lagoon. Constructing hard structures along Marina El-Alamein shoreline during the last 3 decades interrupted greatly the shoreline stability where seaward shifting was predominant. Shoreline evolution model predicted that the western inlet might be closed in the future.

18.6 Recommendations

Strategic environmental impact assessment coupled with long-term considerations are required when developing coastal engineering structures to ensure the sustainability of available natural resources. Furthermore, continuous monitoring programs, including permanent satellite monitoring of the coastal zone, for evaluating possible shoreline changes caused by the sea level rise and changes in wind-wave regime should be adopted.

Acknowledgements Authors would like to thank the United States Geological Survey (USGS) (http://earthexplorer.usgs.gov/) for offering satellite data at no cost. Special appreciation and thanks are given to Professor Dr. Waheed Emam for sharing his pearls of wisdom during the course of this research.

References

Ahmed MH, Barale V (2014) Satellite surveys of lagoon and coastal waters in the southeastern Mediterranean area. In: Barale V, Gade M (eds) Remote sensing of the African seas. Springer, New York, Heidelberg, Dordrecht, London, 436 pp

Ahmad SR, Lakhan VC (2012) GIS-based analysis and modeling of coastline advance and retreat along the coast of Guyana. Mar Geodesy 35:1–15

Al Bakri D (1996) A geomorphological approach to sustainable planning and management of the coastal zone of Kuwait. Geomorphology 17:323–337

Appeaning-Addo K, Jayson-Quashigah PN, Kufogbe KS (2011) Quantitative analysis of shoreline change using medium resolution satellite imagery in Keta, Ghana. Mar Sci 1(1):1–9

Cataudella S, Crosetti D, Massa F (2015) Mediterranean coastal lagoons: sustainable management and interactions among aquaculture, capture fisheries and the environment. Studies and reviews. General Fisheries Commission for the Mediterranean No. 95. FAO, Rome, 278 pp

El-Asmar HM, Ahmed MH, Taha MMN, Assal EM (2012) Human impacts on geological and cultural heritage in the coastal zone west of Alexandria to Al-Alamein, Egypt. Geoheritage 4:263–274

Emam WWM (2016) Management plan for enhancing Bardawil Lagoon productivity using remote sensing and geographic information system. Ph.D. Thesis, Zoology Department, Faculty of Science, Ain Shams University, 335 pp

Fanos AM (2004) Problems facing El-Alamein Marina tourist center. Technical report (in Arabic), CORI, Egypt

Fu Y, Guo Q, Wu X, Fang H, Pan Y (2017) Analysis and prediction of changes in coastline morphology in the Bohai Sea, China, using remote sensing. Sustainability 9:900. https://doi.org/doi:10.3390/su9060900

Hegde AV, Akshaya BJ (2015) Shoreline transformation study of Karnataka coast: geospatial approach. Aquat Proc 4:151–156

Hereher ME (2015) Coastal vulnerability assessment for Egypt's Mediterranean coast. Geomat Nat Hazards Risk 6(4):342–355

Huang H, Liu Y, Qiu Z (2012) Morphodynamic evolution of the Xiaoqing river mouth: a Huanghe river-derived mixed energy estuary. Chin J Oceanol Limnol 30:889–904

Iskander MM, Abo Zed AI, El Sayed WR, Fanos AM (2008) Existing marine coastal problems, western Mediterranean coast, Egypt. Emir J Eng Res 13(3):27–35

Kankara RS, Selvan SC, Rajan B, Arockiaraj S (2014) An adaptive approach to monitor the shoreline changes in ICZM framework: a case study of Chennai coast. Indian J Geo-Mar Sci 43(7):1266–1271

Maiti S, Bhattacharya AK (2009) Shoreline change analysis and its application to prediction: a remote sensing and statistics based approach. Mar Geol 257:11–23

Manjula KR, Jyothi S, Varma SAK (2010) Digitizing the forest resource map using ArcGIS. Int J Comput Sci Issues 7(6):300–306

Moussaid J, Fora AA, Zourarah B, Maanan M (2015) Using automatic computation to analyze the rate of shoreline change on the Kenitra coast, Morocco. Ocean Eng 102(1):71–77

Mukhopadhyay A (2012) Automatic shoreline detection and future prediction: a case study on Puri coast, Bay of Bengal, India. Eur J Remote sens 45:201–213

Natesan U, Parthasarathy A, Vishnunath R, Kumar GEJ, Ferrer VA (2015) Monitoring long-term shoreline changes along Tamil Nadu, India using geospatial techniques. Aquat Proc 4:325–332

Sheeja PS, Ajay-Gokul AJ (2016) Application of digital shoreline analysis system in coastal erosion assessment. Int J Eng Sci Comput 6(6):7876–7883

Thieler ER, Himmelstoss EA, Zichichi JL, Ergul A (2009) Digital shoreline analysis system (DSAS) version 4.0—an ArcGIS extension for calculating shoreline change. Open-File report. U.S. Geological Survey Report No. 2008-1278

US Army Corps of Engineers (2008) Coastal Engineering Manual, Part IV, EM 1110-2-1100

Yu K, Hu C, Muller-Karger F (2011) Shoreline changes in west-central Florida between 1987 and 2008 from landsat observations. Int J Remote Sens 32(23):8299–8313

Part VI
Conclusions

Chapter 19
Update, Conclusions, and Recommendations of "Environmental Remote Sensing in Egypt"

Salwa F. Elbeih, Ahmed M. El-Zeiny, Abdelazim M. Negm and Andrey Kostianoy

Abstract The current chapter highlights the main conclusions and recommendations of the chapters presented in the book. Also, some findings from recently published research works related to the Environmental Remote Sensing in Egypt are discussed. This chapter contains information on remote sensing applications in the field of environmental applications. The topics covered in the book include: environmental applications of remote sensing, radar remote sensing applications, monitoring changes in natural ecosystems, groundwater exploration, monitoring and protection of Egyptian Northern Lakes, environmental hazards threatening Lake Nasser, oil pollution in the Mediterranean and Red Seas, modeling of climate changes, land degradation and desertification, landscapes of Egypt, bathymetry modeling, sediment capacity, shoreline dynamics, monitoring of the Nile River. In addition, a set of recommendations for future research work is pointed out to direct the future research towards the importance of using advanced remote sensing techniques in the field of environment for the sake of its sustainability and protection.

S. F. Elbeih (✉)
Engineering Applications Department, National Authority for Remote Sensing and Space Sciences (NARSS), Cairo 1564 Alf Maskan, Egypt
e-mail: saelbeih@gmail.com; saelbeih@narss.sci.eg

A. M. El-Zeiny
Environmental Studies Department, National Authority for Remote Sensing and Space Sciences (NARSS), Cairo 1564 Alf Maskan, Egypt
e-mail: aelzeny@narss.sci.eg

A. M. Negm
Water and Water Structures Engineering Department, Faculty of Engineering, Zagazig University, Zagazig 44519, Egypt
e-mail: amnegm85@yahoo.com; amnegm@zu.edu.eg

A. Kostianoy
P. P. Shirshov Institute of Oceanology, Russian Academy of Sciences, 36, Nakhimovsky Pr., Moscow 117997, Russia
e-mail: kostianoy@gmail.com

S.Yu. Witte Moscow University, 12, Build. 1, 2nd Kozhukhovsky Pr., Moscow 115432, Russia

S. F. Elbeih et al. (eds.), *Environmental Remote Sensing in Egypt*,
Springer Geophysics, https://doi.org/10.1007/978-3-030-39593-3_19

Keywords Egypt · Environment · Remote sensing · Monitoring · Water resources · Climate change · Oil pollution · Ecosystem · Water quality · Desertification · Red Sea · Nile River · Mediterranean Sea · Lakes · Radar · Landscape

19.1 Introduction

Environmental applications of remote sensing depend basically on using the electromagnetic spectrum for detection, measurement and monitoring at a distance. Most of remote sensing applications undergo these types of analysis. Monitoring environmental resources is one of the essential applications that support the sustainable development for a region through assessing the spatiotemporal changes in the resource. Achieving the sustainable development goals of Egypt for the year 2030 and later cannot be achieved without sustaining the environment with all its surrounding elements. The next section will present a brief discussion of the important findings of some recently published studies on the Environmental Remote Sensing in Egypt. Then the main conclusions and recommendations of the book chapters in addition to few recommendations for researchers and decision makers are presented.

19.2 An Update

Recently in Egypt, there are numerous environmental applications utilizing remotely sensed data. These studies basically depend on using electromagnetic spectrum for detection, measurement and monitoring at a distance. To maintain and perform effective policies to conserve resources and to sustain territorial development of Egypt, there is an urgent need to assess the environmental changes as a result of natural and/or anthropogenic factors. A recent study of El-Zeiny and Effat (2017) was conducted to monitor and map spatiotemporal changes in Land Use/Land Cover (LULC) and Land Surface Temperature (LST) in El-Fayoum governorate and its districts using Landsat data and GIS. Finding of this study showed that the annual rate of land reclamation in the governorate exceeds the annual land loss, as a result of urban sprawl, which is explained by the continuous increase in the agricultural lands during the whole period of study. In this study, authors concluded that remote sensing and GIS techniques could successfully be used to assess spatiotemporal environmental impacts of planned developmental projects and uncontrolled human activities on LULC and relationship with LST.

An interesting up-to-date study of Ibrahim et al. (2009) mapped Urban Heat Islands (UHIs) and assessed the associated environmental characteristics in Qalyubia Governorate using remote sensing and GIS. They found that the low mean values of NDVI and MNDWI were observed in association with UHIs regions which confirmed the positive impact of green cover and water bodies in eliminating UHIs phenomena.

They concluded that the decrease in green spaces and water bodies as well as the increase in urban density lead to increasing the intensity and widespread of UHIs phenomenon. Finally, the study recommended the necessity to consider results of the present study for urban designers, planners, and architects in designing and planning urban communities.

Concerning air quality, a recent paper of Li et al. (2019) has been just published. This research addresses the aerosol characteristics and variability over Cairo and the Greater Nile Delta region over the last 20 years using an integrative multi-sensor approach of remotely sensed and PM10 ground data. The results of this paper show the validity of using Multi-angle Imaging Spectroradiometer (MISR) and Moderate Resolution Imaging Spectroradiometer (MODIS) sensors on the Terra and Aqua platforms for quantitative aerosol optical depth (AOD) assessment as compared to Ozone Monitoring Instrument (OMI), Sea-viewing Wide Field-of-view Sensor (SeaWiFS), and POLarization and Directionality of the Earth's Reflectances (POLDER). In addition, extracted MISR-based aerosol products have been proven to be quite effective in investigating the characteristics of mixed aerosols. In this paper, daily AErosol RObotic NETwork (AERONET) AOD observations were collected and classified using K-means unsupervised machine learning algorithms, showing five typical patterns of aerosols in the region under investigation.

One of the latest papers on environmental monitoring of the coastal resources was El-Zeiny et al. (2016). In this paper, remote sensing and GIS techniques were employed to monitor and quantify annual changes of the Damietta shoreline along the Mediterranean Sea, Egypt. It was found that the coastal landforms are highly dynamic; however urbanization and population rapidly increase on these areas due to the abundant natural resources. The study recorded that the construction of breakwaters in the study area has decreased the annual rate of erosion. This decrease was synchronized with an increase in the annual accreted areas from 0.234 km^2 to 1.152 km^2 during 1997–2011 and 2011–2014, respectively.

A newly published paper on mangrove ecosystem detection using hyperspectral data sets is Basheer et al. (2019). The presence of hyperspectral remote sensing techniques can potentially improve the ability to measure the spectral signature of mangrove to differentiate mangrove from the other vegetation and to get detailed information about this ecosystem. This study has been carried out for mapping, monitoring and managing the Red Sea mangrove ecosystems through measuring their spectral properties using advanced hyperspectral remote sensing techniques. The hyperspectral signatures of *A. marina mangrove* at the different sites showed that mangroves recorded a high reflectance at the visible and NIR region of the spectrum than the other regions and there are similarities at certain wavelengths and some differences at other wavelengths used for differentiation between mangroves in various environments.

Another paper conducted in the Red Sea coast investigated the coral reefs using remote sensing by Khaled et al. (2019). Results of this study shows that during the last 42 years, the coral reef cover decreased by 6.21 km^2 while the built coastal area increased by 13.4 km^2. These observations were used to compute total economic value (TEV) of coral reef habitats and the cost of degradation in terms of physical losses of coral reef area which is equal to about18.63$ Billion.

The environmental risk of mosquito was discussed in a recent study of El-Zeiny et al. (2017) at Suez Canal zone on basis of Landsat images and GIS modeling. This study was an attempt to develop a GIS-based model to detect mosquito breeding habitats at Suez Canal Zone. Results of this study revealed that *Culex pipiens* and *Ochlerotatus detritus* are the most abundant species in Suez Canal Zone. The developed prediction model achieved an accuracy of 80.95% and increased to 100% at some sites. This study showed the maximum predicted area located in Port Said governorate. Such kind of studies provide the baseline information for decision makers to take necessary optimal control strategies to mitigate mosquito nuisance, proliferation and potential diseases transmission.

Another up-to-date paper of Sowilem et al. (2019) investigated the spatial distribution of mosquito breeding sites within the Dakhla Oasis of the Western Desert of Egypt. Results of this paper showed that the main vector of Malaria disease (*Anopheles pharoensis* and *Anopheles sergentii*), as well as the *Culex pipiens*, which is the main vector of filarial disease are abundant. Further, the geo-environmental setting and the discharge of increasing cultivated areas develop considerable waterlogging and pond areas, which are favorable breeding sites of mosquito. They finally concluded that mosquito larval populations fluctuated with the dynamics of vegetation cover in Dakhla.

Exploitation of land and water resources has increased rapidly in Alexandria Governorate, paralleling regional population and industrial growth. Encroachment of urban areas and increase pollution sources are typically considered to have detrimental effects on Lake Maryut ecosystems. Lake Maryut suffering from different effects including nutrient loading, chemical substances, contaminated sediment, invasive aquatic plants, and wholesale hydrologic alterations, eutrophication is the most widespread pressure impacting on the lake. Therefore, Selim and El Raey (2018) proposed and conducted a study that aims to propose three restoration alternative strategies for Lake Maryut based on the information derived from Remote Sensing, GIS, and field survey. The proposed alternatives are mechanical, biological, and biomanipulation alternatives. The result obtained by RAPID IMPACT ASSESSMENT MATRIX (RIAM Software) indicates that the most suitable alternatives for Lake Maryut restoration plan are the mechanical alternative, and biological alternative. On the other hand the "bio-manipulation alternatives" has some significant positive impacts on the surrounding environment, and with some negative impacts especially on biological and ecological component.

El-Zeiny and El-Kafrawy (2017) conducted a recent study to explore the use of Landsat data and GIS for assessing water pollution at Burullus Lake, Egypt. The methodology of this paper included the usage of three previously developed water quality empirical models for biological oxygen demand (BOD), total nitrogen (TN) and total phosphorus (TP) on the calibrated image. Further, a GIS model was generated to identify map pollution degree of the lake. Findings of this study confirmed that the Lake water is subjected to pollution from multiple sources; particularly domestic and agricultural drains. They concluded that Burullus Lake is extensively subjected to interrupting human activities which have a great negative impact on water quality. They added that data observation techniques and water quality empirical models were successful in assessing and mapping water pollution.

More interestingly, a recently published paper of El-Zeiny et al. (2019) was carried out to assess pollution, its sources, impacts and proposed solutions in Qaroun Lake using field survey, laboratory analyses and geospatial techniques. They used the Environmental Protection Agency (EPA) threshold limits, contamination factor, degree of contamination and ecological risk index to assess water and sediment pollution. Landsat OLI integrated with Sentinel-2 imagery were processed to assess the land uses surrounding Qaroun Lake. The processed images showed that Qaroun Lake is surrounded by variable interrupting land uses. A remarkable fluctuation was recorded in most of the studied characteristics. They found that metal levels are impacted and positively correlated with the Lake salinity. Finally the study concluded that the wastewater discharge into Qaroun Lake increased levels of various pollutants to serious grades. To reduce pollution and accelerate Qaroun Lake recovery, they proposed two integrated wastewater treatment plants must be established on El-Bats and El-Wadi drains.

An up-to-date study of El-Alfy et al. (2019) was conducted to assess the levels of toxic heavy metals and organochlorine pesticides (OCPs) in water, sediments and *Phragmites australis* of freshwater and marine environments at Rosetta area. Potential sources of pollution were spatially assessed using remote sensing and geographic information system techniques. Remote sensing results showed that drainage canals, cultivated and urbanized zones are the major sources of contamination in the studied area. This study showed that using integrated remote sensing and chemical analyses could provide a regional and cost-effective assessment tool of environmental toxicity in fresh-saline water interface.

For soil analyses in regard of environment, an up-to-date paper of El-Zeiny and Effat (2019) has been just published. The paper highlighted the impact of irrigation using wastewater on soil characteristics in El-Fayoum Governorate using remote sensing and GIS techniques. Soil samples were collected and analyzed for pH, Electric conductivity (EC), Organic Carbon (OC), Total Nitrogen (TN), Total Phosphorus (TP), Ca, K, Na and Mg. Results of this study referred to a fluctuation in soil characteristics among different districts, as a result of variation and severity of activities. The study referred that the variation of soil characteristics is greatly impacted by irrigation using wastewater particularly from El-Bats and El-Wadi drains. The conclusion went to the point that the agricultural land problems have greatly been impacted by lack of sufficient water supply and to the role of using the geospatial techniques along with soil analyses to facilitate studying the environmental problems of soil.

Recently, Abowaly et al. (2018) assessed the effect of the urban area stretch on agricultural lands of the Egyptian Northern Nile Delta using Landsat TM, Landsat ETM + and Landsat OLI satellite images. This study showed that the urban area growth throughout the 1984–2016 was on the expense of the soils that have good capability with amount of 18 km^2. The urban area growth over the non-capable soils (barren land) was very large. The study summarized that the urban sprawl represents one of the main soil loss and degradation processes in the Nile Delta.

For soil spectral investigation, an article is released by Mohamed et al. (2018) to explore the potentiality of predicting soil properties based on spectroscopic measurements. Authors found that the VIS–NIR reflection spectroscopy reduces the cost and time, therefore has a wonderful ability and potential use as a rapid soil analysis for both precision soil management and assessing soil quality.

In 2019, Sayed et al., employed a multi-criteria analysis model supported by GIS to select optimum locations for extracting groundwater using solar energy. The GIS-based model was used to generate an informative map for the solar energy distribution in Moghra Oasis, Western Desert of Egypt. The output of the model revealed that the optimum site for the installation of the photovoltaic panels would be near to the Nile Delta and outside the oil fields, Qattara Depression, and Moghra Lake.

19.3 Conclusions

The following conclusions are mainly extracted from the chapters presented in this volume:

1. An increase in the number of publications about “Remote sensing in Egypt” from 181 documents in the period 2000–2009 to 517 documents in the period 2010–2019 (according to the Scopus database).
2. The academic disciplines and fields of RS studies cover areas in geology, agriculture, engineering, environment, water, marine sciences, mineral resources, and space archaeology.
3. Using MODIS Aqua EVI instead of NDVI data for the inventory and mapping of agricultural lands in Egypt. Time series analysis of satellite data revealed the strong pattern of the Egyptian cultivated lands which is estimated to be 10.15 million feddan on the17th of January 2019 (4.09% of Egypt area).
4. From the geomorphological features, islands of the River Nile are considered distincting more in the River Nile in Egypt. It amounts 507 islands, which occupies areas of 144.05 km^2.
5. Egypt paid a great attention for the establishment of new settlements and land reclamation projects, to overcome the over population crisis.
6. Integration of geomorphology, remote sensing and GIS has been used for exploration of surface and groundwater.
7. Trends of environmental changes for water and aquatic vegetation in major lakes and lagoons along the northern coast were identified.
8. Compiling the first database of spectral reflectance from agriculture crops and soil using ground-based spectroradiometer and later used for crop classification.
9. Tourist economy in coastal areas has been linked to environmental sensitivity using indices obtained from remote sensing data.
10. Qualitative interpretation of SAR images reveals subsurface features that do not appear in optical and thermal infrared images where the subsurface penetration of radar signal has served in locating potential groundwater sites.

11. Using the L-band data available from the Japanese ALOS-PALSAR system to verify archaeological information in selected sites.
12. Considerable changes in LULC due to grazing, agricultural activities, environmental pollution and reclamation activities from aquaculture, agricultural wastes rich in fertilizers. In addition, sand dunes have negatively impacted the water bodies at the northern part of Egypt.
13. Expansion of the salt marsh areas along the northwestern coast of Egypt during 1984–1990 and then declined between 1990 and 2014, which could be due to anthropogenic activities and climatic changes.
14. Dynamic changes in the spatial distribution of the major plant communities in Moghra Oasis between 1984 and 2011.
15. Radar microwave, thermal infrared data and SRTM DEMs are ideal for detection and mapping of water-bearing structures.
16. SRTM DEMs reveal buried ancient river courses and lake basins that may have acted as preferential flow paths and sites for subsurface water in arid and hyper-arid sandy deserts.
17. The Northern Lakes in Egypt had been increasingly subjected to intensive and diverse development activities including fishing, aquaculture industry, dumping wastes, land reclamation, lakes drying, urbanization, saltpan, and recreational uses.
18. Remote sensing proves to be very successful in monitoring the water quality, vegetation types, and ecological changes along the Manzala Lake.
19. Lake Nasser, as Egypt national water reservoir, is exposed to several environmental and anthropogenic hazards. Of these hazards, the construction of dams on the upper reaches of the Nile comes on the top of them where the Ethiopian Renaissance Dam is the most dangerous.
20. The most polluted area in the Mediterranean waters of Egypt is a region located 18 nm northeastward of Port Said which is explained by ship traffic from Port Said and shipping activities related to offshore installations.
21. Increased ship traffic and offshore gas/oil exploration and production in the region of the Nile Delta represent a serious potential threat to its environment.
22. A series of satellite images acquired in 2017–2019 from different SAR, optical and infrared sensors were used to detect a number of oil spills in the Northern Red Sea where the most polluted area is the Port of Suez.
23. The water around Hurghada and Sharm el-Sheikh resorts looks quite clean, but even though small-size oil spills occur as well.
24. Medium resolution optical and infrared imagery revealed mesoscale water dynamics in the form of cyclonic and anticyclonic eddies, dipoles and intrusions, which can redistribute oil pollution across and along the Red Sea, and contribute to coastal pollution.
25. NDVI monthly data obtained from MODIS-Terra from February 2000 to December 2018 and provided by the NASA Giovanni online data system v.4.30, with a spatial resolution of about 5 km, showed that during two decades there is a steady rise in the NDVI values from 2% for the whole country to 25% for the important agricultural areas located eastward of Faris and Nagaa Al Hajar.

26. A remote sensing based framework has been proposed for predicting water quality of different water sources by integrating satellite data and ground-based measurements.
27. Multi-spectral satellite imaging allows to follow different measurable parameters related to state of soil, vegetation and climate.
28. Determining the environmental sensitivity to desertification, defines the total magnitude of expected desertification impacts linked with different dominating processes. The provisional FAO/UNEP methodology insinuate on each process status, rate, inherent risk and hazard.
29. Under the effect of both climate change and the effect of human impact, some landscapes in Egypt were inherited from past wet environmental conditions, and new features were developed. Therefore, several types of landscapes were recognized, and were classified into two groups; physical and anthropogenic types.
30. The entire coastal strip from Sidi Abd El-Rahman to El-Arish had dynamically changed over time. About 14% of transects displayed stability in shoreline positions, whereas, 76% of transects experienced shoreline displacement up to 250 m.
31. During the period 1987–2017, the overall trend of shoreline change rate (EPR) indicated very high accretion in El-Alamein and Edku Lagoons, very high erosion in Burullus and Manzala Lagoons, and high erosion in Bardawil Lagoon.

19.4 Recommendations

The following recommendations are mainly extracted from the chapters presented in this volume:

1. Encouraging governmental and non-governmental sponsors to fund the grant projects that cover the application of remote sensing in various sectors of Egypt's economy.
2. Exerting additional efforts by the Egyptian government to ensure the widespread application of remote sensing in various research and industrial fields.
3. Developing more institutions, laboratories, and research centers to incorporate scientists and specialists that can apply remote sensing and related technologies.
4. Expand the application of satellite remote sensing that can monitor and investigate the Mediterranean and Red Seas surrounding Egypt.
5. Promote international collaborations in the field of satellite remote sensing with different countries and international organizations.
6. Reinforcing the collaboration between the Egyptian remote sensing community and the MENA counterparts to study regional phenomena such as sand storms, pollutant transport, water resources management and impacts of regional climate changes.

7. Improve partnership with international organizations for capacity building in terms of adjusting the scheme of the research and the output to the common needs of the partners.
8. Exploring the utilization of the coarse-resolution data from passive microwave and scatterometer systems. So, far, none of these systems have been widely used in Egypt because of their coarse resolution.
9. Exploring applications in earth systems sciences that contribute to the international efforts in understanding interactions between the earth's components and track/model impacts of the current global warming.
10. Using remote sensing data for supporting short-term forecast models such as weather, pollutant transport and land surface processes (e.g. energy balance, carbon emission, and nutrient fluxes) on a local to a regional scale.
11. In Egypt, more applications using SAR data are expected to emerge as open access data, processing software and image interpretation capabilities.
12. Conducting field measurements of surface parameters coinciding with the SAR overpasses or shortly after the overpasses if the surface conditions remain unchanged.
13. Using coarse resolution data from scatterometer and passive microwave data to retrieve information at regional scale.
14. Integrating remote sensing data with field observations to map the distribution of different mosquito species.
15. Employing remote sensing to detect the illegal application of agricultural areas in the Nile Delta and Nile Valley.
16. Establishing necessary policies and regulations for sustainable natural resources management.
17. Using remote sensing techniques by land and water managers and policymakers to perform soil erosion risk assessment.
18. Assessing the carbon storage capacity of salt marshes in the coastal lands of Egypt.
19. Using radar microwave, thermal infrared data and SRTM DEMs data for hydrological investigations.
20. Delineation of former rivers and their basins might aid to future exploration for groundwater, oil and gas resources, and help to inform neotectonic, archaeological, as well as paleoclimate research efforts across the Egyptian deserts as well as the rest of the Great Sahara in North Africa.
21. Implementation of an integrated management program for development of the lakes and adjacent areas taking into account the capability of recent remote sensing techniques in identifying and assessing changes and law enforcement.
22. Periodic monitoring and assessment are necessary for controlling deterioration of the coastal zones especially in cases of absence of main tools for management; legal frame, corrective actions, technical and socioeconomic factors.
23. Developing a decentralized remote sensing and geographic information system (GIS) capability to collect and upgrade available data and to help decision makers on both local and national scales.

24. Carrying out a sensitivity analysis for all coastal lakes in Egypt to assess vulnerabilities to various problems including pollution, lake reclamation, erosion of lakes, illegal catching of fish fry, etc. Developing a contingency plan for emergency measures to protect the lakes.
25. Further analysis of detailed topographic and bathymetric data, e.g. openings and interaction between drains and basins for elaboration of an enhanced model that would permit detailed analysis of the lakes hydrodynamics.
26. Using remote sensing as a valuable tool for studying surface water quality by integrating all available data into an easily accessible data system to improve remote sensing images potential as tool for resource managers.
27. Setting up regulations and guidelines for coastal development incorporating the adaptation of the impacts of the sea level rise.
28. Designing engineering structures to protect high risk vulnerable areas to inundation by introducing appropriate measures to mitigate penetration of seawater towards the cultivated soil of the Nile Delta.
29. Encouraging the application of nonstructural erosion protection techniques as part of a Living Shorelines approach applied by the UNDP's "Living with the Sea" Project in the Nile Delta of Egypt.
30. Preserving the natural protection environments by stabilizing dunes, creating or restoring wetlands, prohibiting land infill of wetlands and preventing quarrying and destruction of the shore-parallel carbonate ridges.
31. An operational monitoring model to be developed to follow at field level elements contributing to desertification sensitivity.
32. Applying different models for bathymetry estimation from high-resolution multispectral image such as World-View 3; this would increase the accuracy of the mapping bathymetry from satellite images.
33. Studying the possibility of removing sediment deposits from AHDL to be used in agricultural land reclamation and to identify how these sediments affect crop yields.
34. Strategic environmental impact assessment coupled with long-term considerations are required when developing coastal engineering structures to ensure the sustainability of available natural resources.
35. Continuous monitoring programs for evaluating possible shoreline changes should be adopted.
36. Establishing an authority to supervise all physical and anthropogenic aspects of Lake Nasser and regularly monitor the Lake region from space for protection purposes. It can be called Lake Nasser Authority for Protection and Development and it should be an independent authority, have the right to take measures to protect the Lake, and suggest new projects to develop new viable rural and urban centres in the Lake region.
37. Establishment of the International Satellite Monitoring Center to perform permanent satellite monitoring of oil pollution and water dynamics in the Northern Red Sea. As for the Mediterranean coastal zone of Egypt, this Center should combine integrated satellite monitoring of oil pollution in both seas, as well as

provide the authorities with other useful information for coastal zone management, tourism and port development, urban planning. It can also be used for monitoring of water resources, vegetation, desertification for the whole territory of Egypt subjected to climate change. Real-time satellite monitoring will help to avoid potential catastrophes resulting from large oil spills, and for sustainable development of the coastal tourism business in Egypt

38. An institutional capability in the form of a multidisciplinary National Center for Climate Change. It should have components of Water Resources Needs for Adaptation, Agricultural Resources Needs for Adaptation, Coastal Resources Needs for Adaptation

Acknowledgements The authors of this chapter would like to acknowledge the participation of all authors who contributed in this volume and make it a reality. It is worth mention that A. Kostianoy was partially supported in the framework of the P. P. Shirshov Institute of Oceanology RAS budgetary financing (Project N 149-2019-0004).

Also, A. M. Negm would like to acknowledge the partial financial support from the Academy of Scientific Research and Technology (ASRT) of Egypt via the bilateral collaboration Italian (CNR)–Egyptian (ASRT) project titled “Experimentation of the new Sentinel missions for the observation of inland water bodies on the course of the Nile River”.

References

Abowaly ME, Moghanm FS, El Nahry AH, Shalaby A, Khedr HS (2018) Assessment of land use changes and its impact on agricultural soils in the North Nile delta region of Egypt using GIS and remote sensing. Egypti J Soil Sci 58(3):359–372

Basheer MA, El Kafrawy SB, Mekawy AA (2019) Identification of mangrove plant using hyperspectral remote sensing data along the Red Sea, Egypt. Egypt J Aquat Biol Fish 23(1):27–36

El-Alfy MA, Hasballah AF, El-Hamid HTA, El-Zeiny AM (2019) Toxicity assessment of heavy metals and organochlorine pesticides in freshwater and marine environments, Rosetta area, Egypt using multiple approaches. Sustain Environ Res 29:19. https://doi.org/10.1186/s42834-019-0020-9

El-Zeiny AM, Effat HA (2017) Environmental monitoring of spatiotemporal change in land use/land cover and its impact on land surface temperature in El-Fayoum governorate, Egypt. Remote Sens Appl Soc Environ 8:266–277

El-Zeiny AM, Effat HA (2019) Environmental analysis of soil characteristics in El-Fayoum governorate using geomatics approach. Environ Monit Assess 191:463. https://doi.org/10.1007/s10661-019-7587-9

El-Zeiny A, El-Kafrawy S (2017) Assessment of water pollution induced by human activities in Burullus Lake using Landsat 8 operational land imager and GIS. Egypt J Remote Sens Space Sci 20:S49–S56

El-Zeiny A, Gad A, El-Gammal M, Ibrahim M (2016) Space-borne technology for monitoring temporal changes along Damietta shoreline, Northern Egypt. Int J Adv Res 4(1):459–468. ISSN 2320-5407

El-Zeiny A, El-Hefni A, Sowilem M (2017) Geospatial techniques for environmental modeling of mosquito breeding habitats at Suez Canal Zone, Egypt. Egypt J Remote Sens Space Sci 20:283–293

El-Zeiny AM, El Kafrawy SB, Ahmed MH (2019) Geomatics based approach for assessing Qaroun Lake pollution. Egypt J Remote Sens Space Sci. https://doi.org/10.1016/j.ejrs.2019.07.003
Ibrahim MS, El-Gammal MI, Shalaby AA, El-Zeiny AM, Rostom NG (2009) Environmental and spatial assessment of urban heat Islands in Qalyubia Governorate, Egypt. Egypt J Soil Sci 9(2):1–15
Khaled M, Muller-Karger F, Obuid-Allah A, Ahmed M, El-Kafrawy S (2019) Using landsat data to assess the status of coral reefs cover along the red sea coast, Egypt. Int J Ecotoxicol Ecobiol 4(1):17–31
Li W, Ali E, Abou El-Magd I, Mourad M, El-Askary H (2019) Studying the impact on urban health over the greater delta region in Egypt due to aerosol variability using optical characteristics from satellite observations and ground-based aeronet measurements. Remote Sens 11(17):1998. https://doi.org/10.3390/rs11171998
Mohamed ES, Saleh AM, Belal AB, Gad A (2018) Application of near-infrared reflectance for quantitative assessment of soil properties. Egypt J Remote Sens Space Sci 21(1):1–14
Sayed E, Riad P, Elbeih S, Hagras M, Hassan A (2019) Multi criteria analysis for groundwater management using solar energy in Moghra Oasis, Egypt. Egypt J Remote Sens Space Sci 22(3):227–235
Selim N, El Raey M (2018) EIA For Lake Maryut sustainable development alternatives using GIS, RS, and RIAM software. Assiut Univ Bull Environ Res 21(1)
Sowilem MM, El-Zeiny AM, Mohamed ES (2019) Mosquito larval species and geographical information system (GIS) mapping of environmental vulnerable areas, Dakhla Oasis, Egypt. Int J Environ Clim Change 9(1):17–28